POWER SYSTEM ANALYSIS AND DESIGN

FOURTH EDITION

J. DUNCAN GLOVER
FAILURE ELECTRICAL, LLC

MULUKUTLA S. SARMA
NORTHEASTERN UNIVERSITY

THOMAS J. OVERBYE
UNIVERSITY OF ILLINOIS

CENGAGE
Learning™

Australia · Brazil · Canada · Mexico · Singapore · Spain · United Kingdom · United States

CENGAGE
Learning™

**Power System Analysis and Design,
Fourth Edition**
J. Duncan Glover
Mulukutla S. Sarma
Thomas J. Overbye

Director, Global Engineering Program:
    Chris Carson

Senior Developmental Editor: Hilda Gowans

Production Manager: Renate McCloy

Production Service: RPK Editorial
    Services, Inc.

Copyeditor: Harlan James

Proofreader: Erin Wagner

Indexer: Shelly Gerger-Knechtl

Compositor: Asco Typesetters

Creative Director: Angela Cluer

Internal Designer: Carmela Pereira

Cover Designer: Andrew Adams

Cover Image: Mark Atkins/Shutterstock

Permissions Coordinator: Melody Tolson

Library of Congress Control Number: 2007920218

U.S. Student Edition:

ISBN-13: 978-0-534-54884-1
ISBN-10: 0-534-54884-9

Cengage Learning
200 First Stamford Place, Suite 400
Stamford, CT 06902
USA

Cengage Learning is a leading provider of customized learning solutions
with office locations around the globe, including Singapore, the United
Kingdom, Australia, Mexico, Brazil, and Japan. Locate your local office at:
international.cengage.com/region.

Cengage Learning products are represented in Canada
by Nelson Education Ltd.

For your course and learning solutions, visit
academic.cengage.com/engineering.

Purchase any of our products at your local college store or at our preferred
online store **www.ichapters.com**.

Printed in the United States of America
3 4 5 6 7   11 10 09 08

TO ANNA ELIZABETH, OWEN JOSEPH, AND THOSE TO FOLLOW

i thank You God for most this amazing
day: for the leaping greenly spirits of trees
and a blue true dream of sky; and for everything
which is natural which is infinite which is yes

(i who have died am alive again today,
and this is the sun's birthday; this is the birth
day of life and of love and wings: and of the gay
great happening illimitably earth)

how should tasting touching hearing seeing
breathing any—lifted from the no
of all nothing—human merely being
doubt unimaginable You?

(now the ears of my ears awake and
now the eyes of my eyes are opened)

E. E. Cummings

CONTENTS

PREFACE

The objective of this book is to present methods of power system analysis and design, particularly with the aid of a personal computer, in sufficient depth to give the student the basic theory at the undergraduate level. The approach is designed to develop students' thinking process, enabling them to reach a sound understanding of a broad range of topics related to power system engineering, while motivating their interest in the electrical power industry. Because we believe that fundamental physical concepts underlie creative engineering and form the most valuable and permanent part of an engineering education, we highlight physical concepts while giving due attention to mathematical techniques. Both theory and modeling are developed from simple beginnings so that they can be readily extended to new and complex situations.

Thomas J. Overbye, *University of Illinois at Urbana-Champaign*, is welcomed as a new co-author of this edition of the text. He is one of the creators of PowerWorld Simulator, a user-friendly extremely visual power system analysis and simulation software package that is integrated into this text. Professor Overbye's expertise in power system operations, control, stability, computational algorithms, and education help to insure the text remains up to date and student oriented.

This edition of the text also features the following: (1) updated case studies for 10 chapters along with 3 case studies from the previous edition describing present-day, practical applications and new technologies along with ample coverage of the ongoing restructuring of the electric utility industry; (2) an updated PowerWorld Simulator package; and (3) updated problems at the end of chapters 2–9.

One of the most challenging aspects of engineering education is giving students an intuitive feel for the systems they are studying. Engineering systems are, for the most part, complex. While paper-and-pencil exercises can be quite useful for highlighting the fundamentals, they often fall short in imparting the desired intuitive insight. To help provide this insight, a CD enclosed with the book contains PowerWorld Simulator, which is used to integrate computer-based examples, problems, and design projects throughout the text.

PowerWorld Simulator was originally developed at the University of Illinois at Urbana–Champaign to teach the basics of power systems to nontechnical people involved in the electricity industry, with version 1.0 introduced in June 1994. The program's interactive and graphical design made it an immediate hit as an educational tool, but a funny thing happened—its interactive and graphical design also appealed to engineers doing analysis of real power systems. To meet the needs of a growing group of users, Power-World Simulator was commercialized in 1996 by the formation of Power-World Corporation. Thus while retaining its appeal for education, over the years PowerWorld Simulator has evolved into a top-notch analysis package, able to handle power systems of any size. PowerWorld Simulator is now used throughout the power industry, with a range of users encompassing universities, utilities of all sizes, government regulators, power marketers, and consulting firms.

In integrating PowerWorld Simulator with the text, our design philosophy has been to use the software to extend, rather than replace, the fully worked examples provided in previous editions. Therefore, except when the problem size makes it impractical, each PowerWorld Simulator example includes a fully worked hand solution of the problem along with a PowerWorld Simulator case. This format allows students to simultaneously see the details of how a problem is solved and a computer implementation of the solution. The added benefit from PowerWorld Simulator is its ability to easily extend the example. Through its interactive design, students can quickly vary example parameters and immediately see the impact such changes have on the solution. By reworking the examples with the new parameters, students get immediate feedback on whether they understand the solution process. The interactive and visual design of PowerWorld Simulator also makes it an excellent tool for instructors to use for in-class demonstrations. With the many examples and problem cases contained on the CD, instructors can easily demonstrate many of the text topics. Additional PowerWorld Simulator functionality is introduced in the text examples, problems, and design projects.

The text is intended to be fully covered in a two-semester or three-quarter course offered to seniors and first-year graduate students. The organization of chapters and individual sections is flexible enough to give the instructor sufficient latitude in choosing topics to cover, especially in a one-semester course. The text is supported by an ample number of worked examples covering most of the theoretical points raised. The many problems to be worked with a calculator as well as problems to be worked using a personal computer have been expanded in this edition.

As background for this course, it is assumed that students have had courses in electric network theory (including transient analysis) and ordinary differential equations and have been exposed to linear systems, matrix algebra, and computer programming. In addition, it would be helpful, but not necessary, to have had an electric machines course.

After an introduction to the history of electric power systems along with present and future trends, Chapter 2 on fundamentals orients the students to the terminology and serves as a brief review. The chapter reviews phasor concepts, power, and single-phase as well as three-phase circuits.

Chapters 3 through 6 examine power transformers, transmission-line parameters, steady-state operation of transmission lines, and power flows including the Newton–Raphson method. These chapters provide a basic understanding of power systems under balanced three-phase, steady-state, normal operating conditions.

Chapters 7 through 10, which cover symmetrical faults, symmetrical components, unsymmetrical faults, and system protection, come under the general heading of power system short-circuit protection. Chapter 11 is a self-contained chapter on power system controls, including turbine-generator controls, load-frequency control, economic dispatch, and optimal power flow.

The last two chapters examine transient operation of transmission lines, including surge protection; and transient stability, which includes the swing equation, the equal-area criterion, and multimachine stability. These self-contained chapters come under the general heading of power system transients.

ACKNOWLEDGMENTS

The material in this text was gradually developed to meet the needs of classes taught at universities in the United States and abroad over the past 30 years. The 13 chapters were written by the first author, J. Duncan Glover, *Failure Electrical LLC*, who is indebted to many people who helped during the planning and writing of this book. The profound influence of earlier texts written on power systems, particularly by W. D. Stevenson, Jr., and the developments made by various outstanding engineers are gratefully acknowledged. Details of sources can only be made through references at the end of each chapter, as they are otherwise too numerous to mention.

Co-author Thomas Overbye updated Chapter 6 (*Power Flows*), Chapter 11 (*Power System Controls*), and Chapter 13 (*Transient Stability*) of this edition of the text. He also provided the examples and problems using Power-World Simulator as well as three design projects. Co-author Mulukutla Sarma, *Northeastern University*, contributed to end-of-chapter problems.

We commend Christopher Carson and Hilda Gowans of Cengage Learning and Rose Kernan of RPK Editorial Services Inc., for their broad knowledge, skills, and ingenuity in publishing this edition. The reviewers of the fourth edition are as follows: Robert C. Degeneff, *Rensselaer Polytechnic Institute*; Venkata Dinavahi, *University of Alberta*; Richard G. Farmer, *Arizona State University*; Steven M. Hietpas, *South Dakota State University*; M. Hashem Nehrir, *Montana State University*; Anil Pahwa, *Kansas State University*; and Ghadir Radman; *Tennessee Technical University*.

The following reviewers made substantial contributions to the third edition: Sohrab Asgarpoor, *University of Nebraska–Lincoln*; Mariesa L. Crow, *University of Missouri–Rolla*; Ilya Y. Grinberg, *State University of New York, College at Buffalo*; Iqbal Husain, *The University of Akron*; W. H. Kersting, *New Mexico State University*; John A. Palmer, *Colorado School of Mines*; Satish J. Ranada, *New Mexico State University*; and Shyama C. Tandon, *California Polytechnic State University*.

The following reviewers made substantial contributions to the second edition: Max D. Anderson, *University of Missouri–Rolla*; Sohrab Asgarpoor, *University of Nebraska–Lincoln*; Kaveh Ashenayi, *University of Tulsa*; Richard D. Christie, Jr., *University of Washington*; Mariesa L. Crow, *University of Missouri–Rolla*; Richard G. Farmer, *Arizona State University*; Saul Goldberg, *California Polytechnic University*; Clifford H. Grigg, *Rose-Hulman Institute of Technology*; Howard B. Hamilton, *University of Pittsburgh*; Leo Holzenthal, Jr., *University of New Orleans*; Walid Hubbi, *New Jersey Institute of Technology*; Charles W. Isherwood, *University of Massachusetts–Dartmouth*; W. H. Kersting, *New Mexico State University*; Wayne E. Knabach, *South Dakota State University*; Pierre-Jean Lagace, *IREQ Institut de Reserche d'Hydro–Quebec*; James T. Lancaster, *Alfred University*; Kwang Y. Lee, *Pennsylvania State University*; Mohsen Lotfalian, *University of Evansville*; Rene B. Marxheimer, *San Francisco State University*, Lamine Mili, *Virginia Polytechnic Institute and State University*; Osama A. Mohammed, *Florida International University*; Clifford C. Mosher, *Washington State University*, Anil Pahwa, *Kansas State University*; M. A. Pai, *University of Illinois at Urbana–Champaign*; R. Ramakumar, *Oklahoma State University*; Teodoro C. Robles, *Milwaukee School of Engineering*, Ronald G. Schultz, *Cleveland State University*; Stephen A. Sebo, *Ohio State University*; Raymond Shoults, *University of Texas at Arlington*, Richard D. Shultz, *University of Wisconsin at Platteville*; Charles Slivinsky, *University of Missouri–Columbia*; John P. Stahl, *Ohio Northern University*; E. K. Stanek, *University of Missouri–Rolla*; Robert D. Strattan, *University of Tulsa*; Tian-Shen Tang, *Texas A&M University–Kingsville*; S. S. Venkata, *University of Washington*; Francis M. Wells, *Vanderbilt University*; Bill Wieserman, *University of Pennsylvania–Johnstown*; Stephen Williams, *U.S. Naval Postgraduate School*; and Salah M. Yousif, *California State University–Sacramento*.

In addition, the following reviewers made many contributions to the first edition: Frederick C. Brockhurst, *Rose-Hulman Institute of Technology*; Bell A. Cogbill. *Northeastern University*; Saul Goldberg, *California Polytechnic State University*; Mack Grady, *University of Texas at Austin*; Leonard F. Grigsby, *Auburn University*; Howard Hamilton, *University of Pittsburgh*; William F. Horton, *California Polytechnic State University*; W. H. Kersting, *New Mexico State University*; John Pavlat, *Iowa State University*; R. Ramakumar, *Oklahoma State University*; B. Don Russell, *Texas A&M*; Sheppard Salon, *Rensselaer Polytechnic Institute*; Stephen A. Sebo, *Ohio State University*; and Dennis O. Wiitanen, *Michigan Technological University*.

In conclusion, the objective in writing this text and the accompanying software package will have been fulfilled if the book is considered to be student-oriented, comprehensive, and up to date, with consistent notation and necessary detailed explanation at the level for which it is intended.

LIST OF SYMBOLS, UNITS, AND NOTATION

Symbol	Description
a	operator $1/120°$
a_t	transformer turns ratio
A	area
A	transmission line parameter
A	symmetrical components transformation matrix
B	loss coefficient
B	frequency bias constant
B	phasor magnetic flux density
B	transmission line parameter
C	capacitance
C	transmission line parameter
D	distance
D	transmission line parameter
E	phasor source voltage
E	phasor electric field strength
f	frequency
G	conductance
G	conductance matrix
H	normalized inertia constant
H	phasor magnetic field intensity
$i(t)$	instantaneous current
I	current magnitude (rms unless otherwise indicated)
I	phasor current
I	vector of phasor currents
j	operator $1/90°$
J	moment of inertia
l	length
l	length
L	inductance
L	inductance matrix
N	number (of buses, lines, turns, etc.)
p.f.	power factor
$p(t)$	instantaneous power

Symbol	Description
P	real power
q	charge
Q	reactive power
r	radius
R	resistance
R	turbine-governor regulation constant
R	resistance matrix
s	Laplace operator
S	apparent power
S	complex power
t	time
T	period
T	temperature
T	torque
$v(t)$	instantaneous voltage
V	voltage magnitude (rms unless otherwise indicated)
V	phasor voltage
V	vector of phasor voltages
X	reactance
X	reactance matrix
Y	phasor admittance
Y	admittance matrix
Z	phasor impedance
Z	impedance matrix
α	angular acceleration
α	transformer phase shift angle
β	current angle
β	area frequency response characteristic
δ	voltage angle
δ	torque angle
ε	permittivity
Γ	reflection or refraction coefficient

Symbol	Description	Symbol	Description
λ	magnetic flux linkage	θ	impedance angle
λ	penalty factor	θ	angular position
Φ	magnetic flux	μ	permeability
ρ	resistivity	v	velocity of propagation
τ	time in cycles	ω	radian frequency
τ	transmission line transit time		

SI Units

A	ampere	
C	coulomb	
F	farad	
H	henry	
Hz	hertz	
J	joule	
kg	kilogram	
m	meter	
N	newton	
rad	radian	
s	second	
S	siemen	
VA	voltampere	
var	voltampere reactive	
W	watt	
Wb	weber	
Ω	ohm	

English Units

BTU	British thermal unit
cmil	circular mil
ft	foot
hp	horsepower
in	inch
mi	mile

Notation

Lowercase letters such as v(t) and i(t) indicate instantaneous values.

Uppercase letters such as V and I indicate rms values.

Uppercase letters in italic such as V and I indicate rms phasors.

Matrices and vectors with real components such as **R** and **I** are indicated by boldface type.

Matrices and vectors with complex components such as **Z** and **I** are indicated by boldface italic type.

Superscript T denotes vector or matrix transpose.

Asterisk (*) denotes complex conjugate.

■ indicates the end of an example and continuation of text.

PW highlights problems that utilize PowerWorld Simulator.

1300 MW coal-fired power plant (Courtesy of American Electric Power Company)

INTRODUCTION

Electrical engineers are concerned with every step in the process of generation, transmission, distribution, and utilization of electrical energy. The electric utility industry is probably the largest and most complex industry in the world. The electrical engineer who works in that industry will encounter challenging problems in designing future power systems to deliver increasing amounts of electrical energy in a safe, clean, and economical manner.

The objectives of this chapter are to review briefly the history of the electric utility industry, to discuss present and future trends in electric power systems, to describe the restructuring of the electric utility industry, and to introduce PowerWorld Simulator—a power system analysis and simulation software package.

CASE STUDY   The following article describes the restructuring of the electric utility industry that has been taking place in the United States and the impacts on an aging transmission infrastructure. Independent power producers, increased competition in the generation sector, and open access for generators to the U.S. transmission system have changed the way the transmission system is utilized. The need for investment in new transmission and transmission technologies, for further refinements in restructuring, and for training and education systems to replenish the workforce are discussed [7].

The Future Beckons: Will the Electric Power Industry Heed the Call?

CHRISTOPHER E. ROOT

Over the last four decades, the U.S. electric power industry has undergone unprecedented change. In the 1960s, regulated utilities generated and delivered power within a localized service area. The decade was marked by high load growth and modest price stability. This stood in sharp contrast to the wild increases in the price of fuel oil, focus on energy conservation, and slow growth of the 1970s. Utilities quickly put the brakes on generation expansion projects, switched to coal or other nonoil fuel sources, and significantly cut back on the expansion of their networks as load growth slowed to a crawl. During the 1980s, the economy in many regions of the country began to rebound. The 1980s also brought the emergence of independent power producers and the deregulation of the natural gas wholesale markets and pipelines. These developments resulted in a significant increase in natural gas transmission into the northeastern United States and in the use of natural gas as the preferred fuel for new generating plants.

During the last ten years, the industry in many areas of the United States has seen increased competition in the generation sector and a fundamental shift in the role of the nation's electric transmission system, with the 1996 enactment of the Federal Energy Regulatory Commission (FERC) Order No. 888, which mandated open access for generators to

the nation's transmission system. And while prices for distribution and transmission of electricity remained regulated, unregulated energy commodity markets have developed in several regions. FERC has supported these changes with rulings leading to the formation of independent system operators (ISOs) and regional transmission organizations (RTOs) to administer the electricity markets in several regions of the United States, including New England, New York, the Mid-Atlantic, the Midwest, and California.

The transmission system originally was built to deliver power from a utility's generator across town to its distribution company. Today, the transmission system is being used to deliver power across states or entire regions. As market forces increasingly determine the location of generation sources, the transmission grid is being asked to play an even more important role in markets and the reliability of the system. In areas where markets have been restructured, customers have begun to see significant benefits. But full delivery of restructuring's benefits is being impeded by an inadequate, underinvested transmission system.

If the last 30 years are any indication, the structure of the industry and the increasing demands placed on the nation's transmission infrastructure and the people who operate and manage it are likely to continue unabated. In order to meet the challenges of the future, to continue to maintain the stable, reliable, and efficient system we have known for more than a century and to support the con-

("The Future Beckons", Christopher E. Root, Supplement to IEEE Power & Energy (May/June 2006) pg. 58–65.)

tinued development of efficient competitive markets, U.S. industry leaders must address three significant issues:

- an aging transmission system suffering from substantial underinvestment, which is exacerbated by an out-of-date industry structure
- the need for a regulatory framework that will spur independent investment, ownership, and management of the nation's grid
- an aging workforce and the need for a succession plan to ensure the existence of the next generation of technical expertise in the industry.

ARE WE SPENDING ENOUGH?

In areas that have restructured power markets, substantial benefits have been delivered to customers in the form of lower prices, greater supplier choice, and environmental benefits, largely due to the development and operation of new, cleaner generation. There is, however, a growing recognition that the delivery of the full value of restructuring to customers has been stalled by an inadequate transmission system that was not designed for the new demands being placed on it. In fact, investment in the nation's electricity infrastructure has been declining for decades. Transmission investment has been falling for a quarter century at an average rate of almost US$50 million a year (in constant 2003 U.S. dollars), though there has been a small upturn in the last few years. Transmission investment has not kept up with load growth or generation investment in recent years, nor has it been sufficiently expanded to accommodate the advent of regional power markets (see Figure 1).

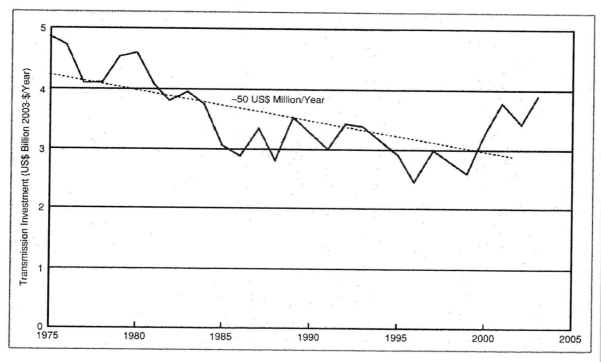

Figure 1

Annual transmission investments by investor-owned utilities, 1975–2003. (Source: Eric Hirst, "U.S. Transmission Capacity: Present Status and Future Prospects," 2004. Graph used with permission from the Edison Electric Institute, 2004. All rights reserved.)

TABLE I Transmission investment in the United
States and in international competitive markets.

Country	Investment in High Voltage Transmission (>230 kV) Normalized by Load for 2004–2008 (in US$M/GW/year)	Number of Transmission-Owning Entities
New Zealand	22.0	1
England & Wales (NGT)	16.5	1
Denmark	12.5	2
Spain	12.3	1
The Netherlands	12.0	1
Norway	9.2	1
Poland	8.6	1
Finland	7.2	1
United States	4.6 (based on representative data from EEI)	450 (69 in EEI)

Outlooks for future transmission development vary, with Edison Electric Institute (EEI) data suggesting a modest increase in expected transmission investment and other sources forecasting a continued decline. Even assuming EEI's projections are realized, this level of transmission investment in the United States is dwarfed by that of other international competitive electricity markets, as shown in Table I, and is expected to lag behind what is needed.

The lack of transmission investment has led to a high (and increasing in some areas) level of congestion-related costs in many regions. For instance, total uplift for New England is in the range of US$169 million per year, while locational installed capacity prices and reliability must-run charges are on the rise. In New York, congestion costs have increased substantially, from US$310 million in 2001 to US$525 million in 2002, US$688 million in 2003, and US$629 million in 2004. In PJM Interconnection (PJM), an RTO that administers electricity markets for all or parts of 14 states in the Northeast, Midwest, and Mid-Atlantic, congestion costs have con-

tinued to increase, even when adjusted to reflect PJM's expanding footprint into western and southern regions.

Because regions do not currently quantify the costs of constraints in the same way, it is difficult to make direct comparisons from congestion data between regions. However, the magnitude and upward trend of available congestion cost data indicates a significant and growing problem that is increasing costs to customers.

THE SYSTEM IS AGING

While we are pushing the transmission system harder, it is not getting any younger. In the northeastern United States, the bulk transmission system operates primarily at 345 kV. The majority of this system originally was constructed during the 1960s and into the early 1970s, and its substations, wires, towers, and poles are, on average, more than 40 years old. (Figure 2 shows the age of National Grid's U.S. transmission structures.) While all utilities have maintenance plans in place for these systems, ever-increasing congestion levels in many areas are making it increasingly difficult to schedule circuit outages for routine upgrades.

The combination of aging infrastructure, increased congestion, and the lack of significant expansion in transmission capacity has led to the need to carefully prioritize maintenance and construction, which in turn led to the evolution of the science of asset management, which many utilities have adopted. Asset management entails quantifying the risks of not doing work as a means to ensure that the highest priority work is performed. It has significantly helped the industry in maintaining reliability. As the assets continue to age, this combination of engineering, experience, and business risk will grow in importance to the industry. If this is not done well, the impact on utilities in terms of reliability and asset replacement will be significant.

And while asset management techniques will help in managing investment, the age issue undoubtedly will require substantial reinvestment at some point to replace the installed equipment at the end of its lifetime.

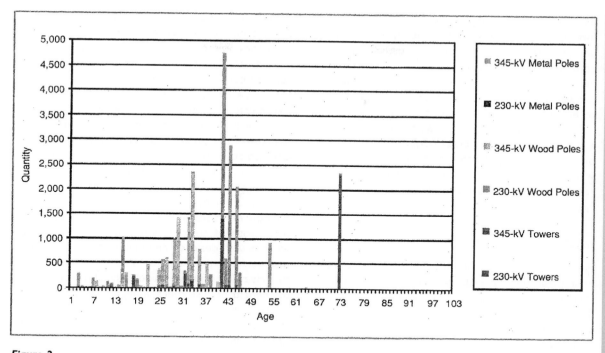

Figure 2
Age of National Grid towers and poles.

TECHNOLOGY WILL HAVE A ROLE

The expansion of the transmission network in the United States will be very difficult, if not impossible, if the traditional approach of adding new overhead lines continues. Issues of land availability, concerns about property values, aesthetics, and other licensing concerns make siting new lines a difficult proposition in many areas of the United States. New approaches to expansion will be required to improve the transmission networks of the future.

Where new lines are the only answer, more underground solutions will be chosen. In some circumstances, superconducting cable will become a viable option. There are several companies, including National Grid, installing short superconducting lines to gain experience with this newly available technology and solve real problems. While it is reasonable to expect this solution to become more prevalent, it is important to recognize that it is not inexpensive.

Technology has an important role to play in utilizing existing lines and transmission corridors to increase capacity. Lightweight, high-temperature overhead conductors are now becoming available for line upgrades without significant tower modifications. Monitoring systems for real-time ratings and better computer control schemes are providing improved information to control room operators to run the system at higher load levels. The development and common use of static var compensators for voltage and reactive control, and the general use of new solid-state equipment to solve real problems are just around the corner and should add a new dimension to the traditional wires and transformers approach to addressing stability and short-term energy storage issues.

These are just a few examples of some of the

exciting new technologies that will be tools for the future. It is encouraging that the development of new and innovative solutions to existing problems continues. In the future, innovation must take a leading role in developing solutions to transmission problems, and it will be important for the regulators to encourage the use of new techniques and technologies. Most of these new technologies have a higher cost than traditional solutions, which will place increasing pressure on capital investment. It will be important to ensure that appropriate cost recovery mechanisms are developed to address this issue.

INDUSTRY STRUCTURE

Another factor contributing to underinvestment in the transmission system is the tremendous fragmentation that exists in the U.S. electricity industry. There are literally hundreds of entities that own and operate transmission. The United States has more than 100 separate control areas and more than 50 regulators that oversee the nation's grid. The patchwork of ownership and operation lies in stark contrast to the interregional delivery demands that are being placed on the nation's transmission infrastructure.

Federal policymakers continue to encourage transmission owners across the nation to join RTOs. Indeed, RTO/ISO formation was intended to occupy a central role in carrying forward FERC's vision of restructuring, and an extraordinary amount of effort has been expended in making this model work. While RTOs/ISOs take a step toward an independent, coordinated transmission system, it remains unclear whether they are the best long-term solution to deliver efficient transmission system operation while ensuring reliability and delivering value to customers.

Broad regional markets require policies that facilitate and encourage active grid planning, management, and the construction of transmission upgrades both for reliability and economic needs. A strong transmission infrastructure or network platform would allow greater fuel diversity, more stable and competitive energy prices, and the relaxation (and

perhaps ultimate removal) of administrative mechanisms to mitigate market power. This would also allow for common asset management approaches to the transmission system. The creation of independent transmission companies (ITCs), i.e., companies that focus on the investment in and operation of transmission independent of generation interests, would be a key institutional step toward an industry structure that appropriately views transmission as a facilitator of robust competitive electricity markets. ITCs recognize transmission as an enabler of competitive electricity markets. Policies that provide a more prominent role for such companies would align the interests of transmission owners/operators with those of customers, permitting the development of well-designed and enduring power markets that perform the function of any market, namely, to drive the efficient allocation of resources for the benefit of customers. In its policy statement released in June 2005, FERC reiterated its commitment to ITC formation to support improving the performance and efficiency of the grid.

Having no interest in financial outcomes within a power market, the ITC's goal is to deliver maximum value to customers through transmission operation and investment. With appropriate incentives, ITCs will pursue opportunities to leverage relatively small expenditures on transmission construction and management to create a healthy market and provide larger savings in the supply portion of customer's bills. They also offer benefits over nonprofit RTO/ISO models, where the incentives for efficient operation and investment may be less focused.

An ideal industry structure would permit ITCs to own, operate, and manage transmission assets over a wide area. This would allow ITCs to access economies of scale in asset investment, planning, and operations to increase throughout and enhance reliability in the most cost-effective manner. This structure would also avoid ownership fragmentation within a single market, which is a key obstacle to the introduction of performance-based rates that benefit customers by aligning the interests of transmission companies and customers in reducing congestion. This approach to "horizontal integration" of the transmission sector under a single regulated for-profit entity is key to establishing an industry structure that recognizes the transmission system as a market enabler and provider of infrastructure to support effective competitive markets. Market administration would be contracted out to another (potentially nonprofit) entity while generators, other suppliers, demand response providers, and load serving entities (LSEs) would all compete and innovate in fully functioning markets, delivering still-increased efficiency and more choices for customers.

REGULATORY ISSUES

The industry clearly shoulders much of the responsibility for determining its own future and for taking the steps necessary to ensure the robustness of the nation's transmission system. However, the industry also operates within an environment governed by substantial regulatory controls. Therefore, policy-makers also will have a significant role in helping to remove the obstacles to the delivery of the full benefits of industry restructuring to customers. In order to ensure adequate transmission investment and the expansion of the system as appropriate, the following policy issues must be addressed:

- *Regional planning:* Because the transmission system is an integrated network, planning for system needs should occur on a regional basis. Regional planning recognizes that transmission investment and the benefits transmission can deliver to customers are regional in nature rather than bounded by state or service area lines. Meaningful regional planning processes also take into account the fact that transmission provides both reliability and economic benefits. Comprehensive planning processes provide for mechanisms to pursue regulated transmission solutions for reliability and economic needs in the event that the market fails to respond or is identified as unlikely to respond to these needs in a timely manner. In areas where regional system planning processes have been implemented, such as New England and PJM, progress is being made towards identifying and building transmission projects that will address regional needs and do so in a way that is cost effective for customers.

- *Cost recovery and allocation:* Comprehensive regional planning processes that identify needed transmission projects must be accompanied by cost recovery and allocation mechanisms that recognize the broad benefits of transmission and its role in supporting and enabling regional electricity markets. Mechanisms that allocate the costs of transmission investment broadly view transmission as the regional market enabler it is and should be, provide greater certainty and reduce delays in cost recovery, and, thus, remove obstacles to provide further incentives for the owners and operators of transmission to make such investment.

- *Certainty of rate recovery and state cooperation:* It is critical that transmission owners are assured certain and adequate rate recovery under a regional planning process. Independent administration of the planning processes will assure that transmission enhancements required for reliability and market efficiency do not unduly burden retail customers with additional costs. FERC and the states must work together to provide for certainty in rate recovery from ultimate customers through federal and state jurisdictional rates.

- *Incentives to encourage transmission investment, independence, and consolidation:* At a time when a significant increase in transmission investment is needed to ensure reliability, produce an adequate platform for competitive power markets and regional electricity commerce, and to promote fuel diversity and renewable sources of supply, incentives not only for investment but also for independence and consolidation of transmission are needed and warranted. Incentives should be designed to promote transmission organizations that acknowledge the benefits to customers of varying degrees of transmission independence and reward that independence accordingly. These incentives may take the form of enhanced rates of return or other financial incentives for assets managed, operated, and/or owned by an ITC.

The debate about transmission regulation will continue. Ultimately, having the correct mixture of incentives and reliability standards will be a critical factor that will determine whether or not the nation's grid can successfully tie markets together and improve the overall reliability of the bulk transmission system in the United States. The future transmission system must be able to meet the needs of customers reliably and support competitive markets that provide them with electricity efficiently. Failure to invest in the transmission system now will mean an increased likelihood of reduced reliability and higher costs to customers in the future.

WORKFORCE OF THE FUTURE

Clearly, the nation's transmission system will need considerable investment and physical work due to age, growth of the use of electricity, changing markets, and how the networks are used. As previously noted, this will lead to a required significant increase in capital spending. But another critical resource is beginning to become a concern to many in the industry, specifically the continued availability of qualified power system engineers.

Utility executives polled by the Electric Power Research Institute in 2003 estimated that 50% of the technical workforce will reach retirement in the next 5–10 years. This puts the average age near 50, with many utilities still hiring just a few college graduates each year. Looking a few years ahead, at the same time when a significant number of power engineers will be considering retirement, the need for them will be significantly increasing. The supply of power engineers will have to be great enough to replace the large numbers of those retiring in addition to the number required to respond to the anticipated increase in transmission capital spending.

Today, the number of universities offering power engineering programs has decreased. Some universities, such as Rensselaer Polytechnic Institute, no longer have separate power system engineering departments. According to the IEEE, the number of power system engineering graduates has dropped from approximately 2,000 per year in the 1980s to 500 today. Overall, the number of engineering graduates has dropped 50% in the last 15 years. Turning this situation around will require a long-term effort by many groups working together, including utilities, consultants, manufacturers, universities, and groups such as the IEEE Power Engineering Society (PES).

Part of the challenge is that utilities are competing for engineering students against other industries, such as telecommunications or computer software development, that are perceived as being more glamorous or more hip than the power industry and have no problem attracting large numbers of new engineers.

For the most part, the power industry has not done a great job of selling itself. Too often, headlines focus on negatives such as rate increases, power outages, and community relations issues related to a proposed new generation plant or transmission line. To a large extent, the industry also has become a victim of its own success by delivering electricity so reliably that the public generally takes it for granted, which makes the good news more difficult to tell. It is incumbent upon the industry to take a much more proactive role in helping its public—including talented engineering students—understand the dedication, commitment, ingenuity, and innovation that is required to keep the nation's electricity system humming. PES can play an important role in this.

On a related note, as the industry continues to develop new, innovative technologies, they should be documented and showcased to help generate excitement about the industry among college-age engineers and help attract them to power system engineering.

The utilities, consultants, and manufacturers must strengthen their relationships with strong technical institutions to continue increasing support for electrical engineering departments to offer power systems classes at the undergraduate level. In some cases, this may even require underwriting a class. Experience at National Grid has shown that when support for a class is guaranteed, the number of students who sign up typically is greater than expected. The industry needs to further support these efforts by offering presentations to students on the complexity of the power system, real problems that need to be solved, and the impact that a reliable, cost-efficient power system has on society. Sponsoring more student internships and research projects will introduce additional students and faculty to the unique challenges of the industry. In the future, the industry will have to hire more nonpower engineers and train them in the specifics of power system engineering or rely on hiring from overseas.

Finally, the industry needs to cultivate relationships with universities to assist in developing professors who are knowledgeable about the industry. This can take the form of research work, consulting, and teaching custom programs for the industry. National Grid has developed relationships with several northeastern U.S. institutions that are offering courses for graduate engineers who may not have power backgrounds. The courses can be offered online, at he university, or on site at the utility.

This problem will only get worse if industry leaders do not work together to resolve it. The industry's future depends on its ability to anticipate what lies ahead and the development of the necessary human resources to meet the challenges.

CONCLUSIONS

The electric transmission system plays a critical role in the lives of the people of the United States. It is an ever-changing system both in physical terms and how it is operated and regulated. These changes must be recognized and actions developed accordingly. Since the industry is made up of many organizations that share the system, it can be difficult to agree on action plans.

There are a few points on which all can agree. The first is that the transmission assets continue to get older and investment is not keeping up with needs when looking over a future horizon. The issue will only get worse as more lines and substations exceed the 50-year age mark. Technology development and application undoubtedly will increase as engineers look for new and creative ways to combat the congestion issues and increased electrical demand—and new overhead transmission lines will be only one of the solutions considered.

The second is that it will be important for further refinement in the restructuring of the industry to occur. The changes made since the late 1990s have delivered benefits to customers in the Northeast in the form of lower energy costs and access to greater competitive electric markets. Regulators and policymakers should recognize that independently owned, operated, managed, and widely planned networks are important to solving future problems most efficiently. Having a reliable, regional, uncongested transmission system will enable a healthy competitive marketplace.

The last, but certainly not least, concern is with the industry's future workforce. Over the last year, there has been significant discussion of the issue, but it will take a considerable effort by many to guide the future workforce into a position of appreciating the electricity industry and desiring to enter it and to ensure that the training and education systems are in place to develop the new engineers who will be required to upgrade and maintain the electric power system.

The industry has many challenges, but it also has great resources and a good reputation. Through the efforts of many and by working together through organizations such as PES, the industry can move forward to the benefit of the public and the United States as a whole.

ACKNOWLEDGMENTS

The following National Grid staff members contributed to this article: Jackie Barry, manager, transmission communications; Janet Gail Besser, vice president, regulatory affairs, U.S. Transmission; Mary Ellen Paravalos, director, regulatory policy, U.S. Transmission; Joseph Rossignoli, principal analyst, regulatory policy, U.S. Transmission.

FOR FURTHER READING

National Grid, "Transmission: The critical link. Delivering the promise of industry restructuring to customers," June 2005 [Online]. Available: http://www.nationalgridus.com/transmission_the_critical_link/

E. Hirst, "U.S. transmission capacity: Present status and future prospects," Edison Electric Inst. and U.S. Dept. Energy, Aug. 2004.

Consumer Energy Council of America, "Keeping the power flowing: Ensuring a strong transmission system to support consumer needs for cost-effectiveness, security and reliability," Jan. 2005 [Online]. Available: http://www.cecarf.org

"Electricity sector framework for the future," Electric Power Res. Inst., Aug. 2003.

J. R. Borland, "A shortage of talent," *Transmission Distribution World*, Sep. 1, 2002.

BIOGRAPHY

Christopher E. Root is senior vice president of Transmission and Distribution (T&D) Technical Services of National Grid's U.S. business. He oversees the T&D technical services organization in New England and New York. He received a B.S. in electrical engineering from Northeastern University, Massachusetts, and a master's in engineering from Rensselaer Polytechnic Institute, New York. In 1997, he completed the Program for Management Development from the Harvard University Graduate School of Business. He is a registered Professional Engineer in the states of Massachusetts and Rhode Island and is a Senior Member of the IEEE.

1.1

HISTORY OF ELECTRIC POWER SYSTEMS

In 1878, Thomas A. Edison began work on the electric light and formulated the concept of a centrally located power station with distributed lighting serving a surrounding area. He perfected his light by October 1879, and the opening of his historic Pearl Street Station in New York City on September 4, 1882, marked the beginning of the electric utility industry (see Figure 1.1). At Pearl Street, dc generators, then called dynamos, were driven by steam engines to supply an initial load of 30 kW for 110-V incandescent lighting to 59 customers in a 1-square-mile area. From this beginning in 1882 through 1972, the electric utility industry grew at a remarkable pace—a growth based on continuous reductions in the price of electricity due primarily to technological acomplishment and creative engineering.

The introduction of the practical dc motor by Sprague Electric, as well as the growth of incandescent lighting, promoted the expansion of Edison's dc systems. The development of three-wire 220-V dc systems allowed load to increase somewhat, but as transmission distances and loads continued to increase, voltage problems were encountered. These limitations of maximum distance and load were overcome in 1885 by William Stanley's development of a commercially practical transformer. Stanley installed an ac distribution system in Great Barrington, Massachusetts, to supply 150 lamps. With the transformer, the ability to transmit power at high voltage with corresponding lower current and lower line-voltage drops made ac more attractive than dc. The first single-phase ac line in the United States operated in 1889 in Oregon, between Oregon City and Portland—21 km at 4 kV.

The growth of ac systems was further encouraged in 1888 when Nikola

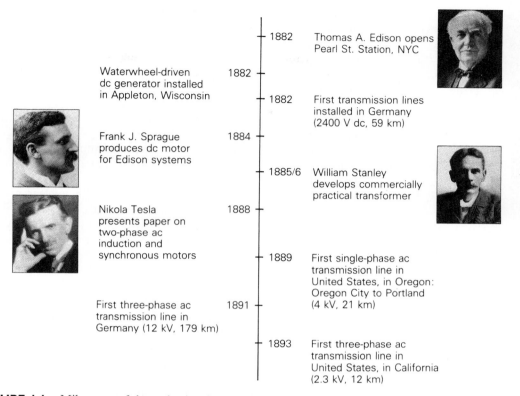

FIGURE 1.1   Milestones of the early electric utility industry [1] (H. M. Rustebakke et al., Electric Utility Systems Practice, 4th Ed. (New York: Wiley, 1983). Photos courtesy of Westinghouse Historical Collection.)

Tesla presented a paper at a meeting of the American Institute of Electrical Engineers describing two-phase induction and synchronous motors, which made evident the advantages of polyphase versus single-phase systems. The first three-phase line in Germany became operational in 1891, transmitting power 179 km at 12 kV. The first three-phase line in the United States (in California) became operational in 1893, transmitting power 12 km at 2.3 kV. The three-phase induction motor conceived by Tesla went on to become the workhorse of the industry.

In the same year that Edison's steam-driven generators were inaugurated, a waterwheel-driven generator was installed in Appleton, Wisconsin. Since then, most electric energy has been generated in steam-powered and in water-powered (called hydro) turbine plants. Today, steam turbines account for more than 85% of U.S. electric energy generation, whereas hydro turbines account for about 7%. Gas turbines are used in some cases to meet peak loads.

Steam plants are fueled primarily by coal, gas, oil, and uranium. Of these, coal is the most widely used fuel in the United States due to its abundance in the country. Although many of these coal-fueled power plants were converted to oil during the early 1970s, that trend has been reversed back to coal since the 1973/74 oil embargo, which caused an oil shortage and created a national desire to reduce dependency on foreign oil. In 2004, approximately 50% of electricity in the United States was generated from coal [2].

In 1957, nuclear units with 90-MW steam-turbine capacity, fueled by uranium, were installed, and today nuclear units with 1312-MW steam-turbine capacity are in service. In 2004, approximately 19% of electricity in the United States was generated from uranium from 103 nuclear power plants. However, the growth of nuclear capacity in the United States has been halted by rising construction costs, licensing delays, and public opinion. Although there are no emissions associated with nuclear power generation, there are safety issues and environmental issues, such as the disposal of used nuclear fuel and the impact of heated cooling-tower water on aquatic habitats. Future technologies for nuclear power are concentrated on safety and environmental issues [2].

Starting in the 1990s, the choice of fuel for new power plants in the United States has been natural gas due to its availability and low cost as well as the higher efficiency, lower emissions, shorter construction-lead times, safety, and lack of controversy associated with power plants that use natural gas. Natural gas is used to generate electricity by the following processes: (1) gas combustion turbines use natural gas directly to fire the turbine; (2) steam turbines burn natural gas to create steam in a boiler, which is then run through the steam turbine; (3) combined cycle units use a gas combustion turbine by burning natural gas, and the hot exhaust gases from the combustion turbine are used to boil water that operates a steam turbine; and (4) fuel cells powered by natural gas generate electricity using electrochemical reactions by passing streams of natural gas and oxidants over electrodes that are separated by an electrolyte. In 2004, approximately 19% of electricity in the United States was generated from natural gas [2].

In 2004 in the United States approximately 7% of electricity was generated by water (hydro power) and 3% by oil [2]. Other types of electric power generation are also being used, including wind-turbine generators; geothermal power plants, wherein energy in the form of steam or hot water is extracted from the earth's upper crust; solar cell arrays; and tidal power plants. These sources of energy cannot be ignored, but they are not expected to supply a large percentage of the world's future energy needs. On the other hand, nuclear fusion energy just may. Substantial research efforts have shown nuclear fusion energy to be a promising technology for producing safe, pollution-free, and economical electric energy later in the 21st century and beyond. The fuel consumed in a nuclear fusion reaction is deuterium, of which a virtually inexhaustible supply is present in seawater.

The early ac systems operated at various frequencies including 25, 50, 60, and 133 Hz. In 1891, it was proposed that 60 Hz be the standard frequency in the United States. In 1893, 25-Hz systems were introduced with the

FIGURE 1.2

Growth of U.S. electric
energy consumption
[1, 2, 3, 5] (H. M.
Rustebakke et al.,
Electric Utility Systems
Practice, 4th Ed. (New
York: Wiley, 1983).)

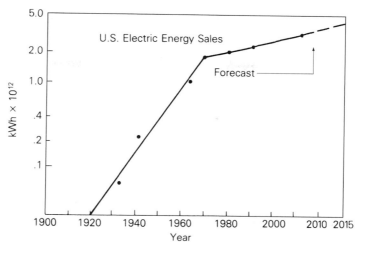

FIGURE 1.2

Growth of U.S. electric
energy consumption
[1, 2, 3, 5] (H. M.
Rustebakke et al.,
Electric Utility Systems
Practice, 4th Ed. (New
York: Wiley, 1983).)

synchronous converter. However, these systems were used primarily for railroad electrification (and many are now retired) because they had the disadvantage of causing incandescent lights to flicker. In California, the Los Angeles Department of Power and Water operated at 50 Hz, but converted to 60 Hz when power from the Hoover Dam became operational in 1937. In 1949, Southern California Edison also converted from 50 to 60 Hz. Today, the two standard frequencies for generation, transmission, and distribution of electric power in the world are 60 Hz (in the United States, Canada, Japan, Brazil) and 50 Hz (in Europe, the former Soviet republics, South America except Brazil, India, also Japan). The advantage of 60-Hz systems is that generators, motors, and transformers in these systems are generally smaller than 50-Hz equipment with the same ratings. The advantage of 50-Hz systems is that transmission lines and transformers have smaller reactances at 50 Hz than at 60 Hz.

As shown in Figure 1.2, the rate of growth of electric energy in the United States was approximately 7% per year from 1902 to 1972. This corresponds to a doubling of electric energy consumption every 10 years over the 70-year period. In other words, every 10 years the industry installed a new electric system equal in energy-producing capacity to the total of what it had built since the industry began. The annual growth rate slowed after the oil embargo of 1973/74. Kilowatt-hour consumption in the United States increased by 3.4% per year from 1972 to 1980, and by 2.1% per year from 1980 to 2000.

Along with increases in load growth, there have been continuing increases in the size of generating units (Table 1.1). The principal incentive to build larger units has been economy of scale—that is, a reduction in installed cost per kilowatt of capacity for larger units. However, there have also been steady improvements in generation efficiency. For example, in 1934 the average heat rate for steam generation in the U.S. electric industry was 17,950

	Hydroelectric Generators		Generators Driven by Single-Shaft, 3600 r/min Fossil-Fueled Steam Turbines	
TABLE 1.1 Growth of generator sizes in the United States [1]	Size (MVA)	Year of Installation	Size (MVA)	Year of Installation
	4	1895	5	1914
	108	1941	50	1937
	158	1966	216	1953
	232	1973	506	1963
	615	1975	907	1969
	718	1978	1120	1974

BTU/kWh, which corresponds to 19% efficiency. By 1991, the average heat rate was 10,367 BTU/kWh, which corresponds to 33% efficiency. These improvements in thermal efficiency due to increases in unit size and in steam temperature and pressure, as well as to the use of steam reheat, have resulted in savings in fuel costs and overall operating costs.

There have been continuing increases, too, in transmission voltages (Table 1.2). From Edison's 220-V three-wire dc grid to 4-kV single-phase and 2.3-kV three-phase transmission, ac transmission voltages in the United States have risen progressively to 150, 230, 345, 500, and now 765 kV. And ultra-high voltages (UHV) above 1000 kV are now being studied. The incentives for increasing transmission voltages have been: (1) increases in transmission distance and transmission capacity, (2) smaller line-voltage drops, (3) reduced line losses, (4) reduced right-of-way requirements per MW transfer, and (5) lower capital and operating costs of transmission. Today, one 765-kV three-phase line can transmit thousands of megawatts over hundreds of kilometers.

The technological developments that have occurred in conjunction with ac transmission, including developments in insulation, protection, and control, are in themselves important. The following examples are noteworthy:

1. The suspension insulator
2. The high-speed relay system, currently capable of detecting short-circuit currents within one cycle (0.017 s)
3. High-speed, extra-high-voltage (EHV) circuit breakers, capable of interrupting up to 63-kA three-phase short-circuit currents within two cycles (0.033 s)
4. High-speed reclosure of EHV lines, which enables automatic return to service within a fraction of a second after a fault has been cleared
5. The EHV surge arrester, which provides protection againsttransient overvoltages due to lightning strikes and line-switching operations

TABLE 1.2

History of increases in three-phase transmission voltages in the United States [1]

Voltage (kV)	Year of Installation
2.3	1893
44	1897
150	1913
165	1922
230	1923
287	1935
345	1953
500	1965
765	1969

6. Power-line carrier, microwave, and fiber optics as communication mechanisms for protecting, controlling, and metering transmission lines

7. The principle of insulation coordination applied to the design of an entire transmission system

8. Energy control centers with supervisory control and data acquisition (SCADA) and with automatic generation control (AGC) for centralized computer monitoring and control of generation, transmission, and distribution

9. Automated distribution features, including reclosers and remotely controlled sectionalizing switches with fault-indicating capability, along with automated mapping/facilities management (AM/FM) and geographic information systems (GIS) for quick isolation and identification of outages and for rapid restoration of customer services

10. Digital relays capable of circuit breaker control, data logging, fault locating, self-checking, fault analysis, remote query, and relay event monitoring/recording.

In 1954, the first modern high-voltage dc (HVDC) transmission line was put into operation in Sweden between Vastervik and the island of Gotland in the Baltic sea; it operated at 100 kV for a distance of 100 km. The first HVDC line in the United States was the $\pm$400-kV, 1360-km Pacific Intertie line installed between Oregon and California in 1970. As of 2000, four other HVDC lines up to 400 kV and five back-to-back ac-dc links had been installed in the United States, and a total of 30 HVDC lines up to 533 kV had been installed worldwide.

For an HVDC line embedded in an ac system, solid-state converters at both ends of the dc line operate as rectifiers and inverters. Since the cost of an HVDC transmission line is less than that of an ac line with the same capacity, the additional cost of converters for dc transmission is offset when the line is long enough. Studies have shown that overhead HVDC transmission is economical in the United States for transmission distances longer than about 600 km.

In the United States, electric utilities grew first as isolated systems, with new ones continuously starting up throughout the country. Gradually, however, neighboring electric utilities began to interconnect, to operate in parallel. This improved both reliability and economy. Figure 1.3 shows major 230-kV and higher-voltage, interconnected transmission in the United States in 2000. An interconnected system has many advantages. An interconnected utility can draw upon another's rotating generator reserves during a time of need (such as a sudden generator outage or load increase), thereby maintaining continuity of service, increasing reliability, and reducing the total number of generators that need to be kept running under no-load conditions. Also,

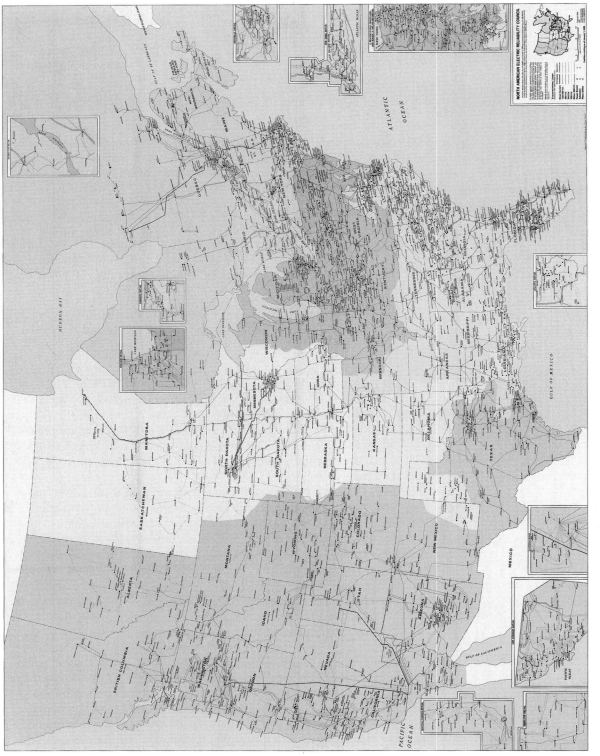

FIGURE 1.3 Major transmission in the United States—2000 [8]. © North American Electric Reliability Council. Reprinted with permission.

interconnected utilities can schedule power transfers during normal periods to take advantage of energy-cost differences in respective areas, load diversity, time zone differences, and seasonal conditions. For example, utilities whose generation is primarily hydro can supply low-cost power during high-water periods in spring/summer, and can receive power from the interconnection during low-water periods in fall/winter. Interconnections also allow shared ownership of larger, more efficient generating units.

While sharing the benefits of interconnected operation, each utility is obligated to help neighbors who are in trouble, to maintain scheduled intertie transfers during normal periods, and to participate in system frequency regulation.

In addition to the benefits/obligations of interconnected operation, there are disadvantages. Interconnections, for example, have increased fault currents that occur during short circuits, thus requiring the use of circuit breakers with higher interrupting capability. Furthermore, although overall system reliability and economy have improved dramatically through interconnection, there is a remote possibility that an initial disturbance may lead to a regional blackout, such as the one that occurred in August 2003 in the northeastern United States and Canada.

1.2

PRESENT AND FUTURE TRENDS

Present trends indicate that the United States is becoming more electrified as it shifts away from a dependence on the direct use of fossil fuels. The electric power industry advances economic growth, promotes business development and expansion, provides solid employment opportunities, enhances the quality of life for its users, and powers the world. Increasing electrification in the United States is evidenced in part by the ongoing digital revolution. According to the Edison Electric Institute, analysts use a term called "electricity intensity" to relate electricity use to the gross domestic product (GDP). Since 1960, the intensity of electricity use in the United States, measured by electricity consumption per dollar of real GDP, has increased by more than 25%. By comparison, the overall intensity of energy use (including electricity and the direct use of fossil fuels) has decreased by more than 40% over the same time period [4].

As shown in Figure 1.2, the growth rate in the use of electricity in the United States is projected to increase by about 2% per year from 2004 to 2015 [3, 5]. Although electricity forecasts for the next ten years are based on economic and social factors that are subject to change, 2% annual growth rate is considered necessary to generate the GDP anticipated over that period. Variations in longer-term forecasts of 1.5-to-2.5% annual growth from 2004 to 2030 are based on low-to-high ranges in economic growth. Average

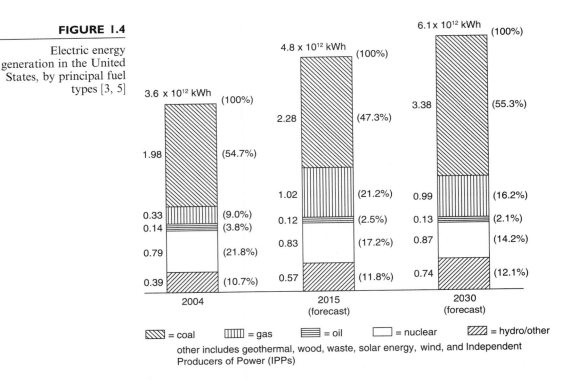

FIGURE 1.4

Electric energy generation in the United States, by principal fuel types [3, 5]

= coal   = gas   = oil   = nuclear   = hydro/other

other includes geothermal, wood, waste, solar energy, wind, and Independent Producers of Power (IPPs)

delivered electricity prices are projected to initially decline from 7.6 cents per kilowatt-hour (2004 dollars) in 2004 to a low of 7.1 cents per kilowatt-hour in 2015, as a result of declines in natural gas and coal prices. After 2015, average delivered electricity prices are projected to increase to 7.5 cents per kilowatt-hour in 2030 [5].

Figure 1.4 shows the percentages of various fuels used to meet U.S. electric energy requirements for 2004 and those projected for 2015 and 2030. Several trends are apparent in the chart. One is the increasing use of coal. This trend is due primarily to the large amount of U.S. coal reserves, which, according to some estimates, is sufficient to meet U.S. energy needs for the next 500 years. Implementation of public policies that have been proposed to reduce carbon dioxide emissions and air pollution could reverse this trend. Another trend is the increase in consumption of natural gas with gas-fired turbines that are safe, clean, and more efficient than competing technologies. Regulatory policies to lower greenhouse gas emissions could accelerate a switchover from coal to gas, but that would require an increasing supply of deliverable natural gas [10]. A percentage decrease in nuclear fuel consumption is also evident. No new nuclear plant has been ordered in the United States for more than 25 years. The projected growth from 0.79×10^{12} kwh in 2004 to 0.87×10^{12} kWh in 2030 in nuclear generation is based on uprates at existing plants and some new nuclear capacity that is cost competitive. Safety

FIGURE 1.5

Installed generating
capability in the United
States by principal fuel
types [3, 5]

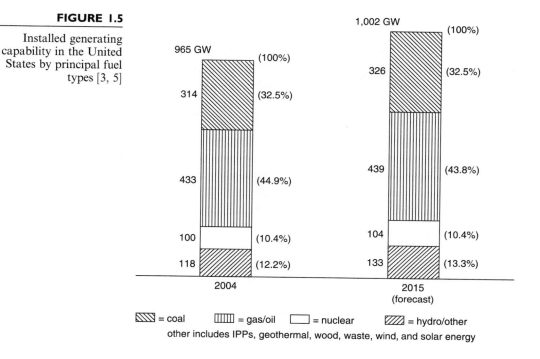

= coal   = gas/oil   = nuclear   = hydro/other
other includes IPPs, geothermal, wood, waste, wind, and solar energy

concerns will require passive or inherently safe reactor designs with standardized, modular construction of nuclear units. Also shown in Figure 1.4 is
a small percentage increase in electricity generation from hydro and renewable energy sources including electricity generated from geothermal, wood,
waste, and solar energy.

Figure 1.5 shows the 2004 and projected 2015 U.S. generating capability by principal fuel type. As shown, total U.S. generating capacity is projected to reach 1,002 GW (1 GW = 1000 MW) by the year 2015. This represents a 0.4% annual projected growth in generating capacity, which is less
than the 2% annual projected growth in electric energy production and consumption. As a result, generating capacity reserve margins are currently
shrinking. Generating resources in the near term (2005–2009) with sufficient
generating capacity reserve margins will be adequate to meet customer demand throughout North America, provided new generating facilities are constructed as anticipated. Generating resource adequacy in the longer term
(2010–2014) is more uncertain and depends on the following factors: timely
completion of planned capacity additions; ability to obtain necessary siting
and environmental permits; ability to obtain financial backing; price and
supply of fuel; and political and regulatory actions [6].

As of 2004, there were 162,979 circuit miles of existing transmission
(230 kV and above) in the United States. An additional 9,953 circuit miles of
new transmission that is projected for the 2005–2014 time frame, will in-

crease the total to 172,932 U.S. circuit miles (230 kV and above). The North American Reliability Council (NERC) generally expects the transmission system in North America to perform reliably during 2005–2014. While the system has come under increasing strain over the past several years because of lack of transmission investment, an aging transmission infrastructure, increased load demand, and tighter transmission operating margins, it is NERC's opinion that reliability can be maintained and should not be threatened if the electric utility industry adheres to NERC reliability standards [6].

However, specific areas and problems have been identified in each of NERC's seasonal and long-term reliability assessments. Transmission congestion and constraints impede the transfer of electric power within and between geographical regions. As the electric utility industry continues to undergo restructuring, the transmission system is being adapted as a delivery system for commercial energy sales. New transmission limitations could appear in unexpected locations as new generation is installed and market-driven energy transactions shift generation patterns. Restructuring policies that encourage greater investment in the transmission grid would enhance transmission reliability. Having and maintaining reliable, regional, uncongested transmission systems will also enable a competitive marketplace [6, 7].

Growth in distribution construction roughly correlates with growth in electric energy construction. During the last two decades, many U.S. utilities converted older 2.4-, 4.1-, and 5-kV primary distribution systems to 12 or 15 kV. The 15-kV voltage class is widely preferred by U.S. utilities for new installations; 25 kV, 34.5 kV, and higher primary distribution voltages are also utilized. Secondary distribution reduces the voltage for utilization by commercial and residential customers. Common secondary distribution voltages in the United States are 240/120 V, single-phase, three-wire; 208Y/120 V, three-phase, four-wire; and 480Y/277 V, three-phase, four-wire.

Utility executives polled by the Electric Power Research Institute in 2003 estimated that 50% of the electric-utility technical workforce in the United States will reach retirement in the next five to ten years. And according to the IEEE, the number of U.S. power system engineering graduates has dropped from approximately 2,000 per year in the 1980s to 500 in 2006. The continuing availability of qualified power system engineers is a critical resource to ensure that transmission and distribution systems are maintained and operated efficiently and reliably [7].

1.3

ELECTRIC UTILITY INDUSTRY STRUCTURE

The case study at the beginning of this chapter describes the restructuring of the electric utility industry that has been ongoing in the United States. The previous structure of large, vertically integrated monopolies that existed until the last decade of the twentieth century is being replaced by a horizontal structure with generating companies, transmission companies, and distribution companies as separate business facilities.

In 1992, the United States Congress passed the Energy Policy Act, which has shifted and continues to further shift regulatory power from the state level to the federal level. The 1992 Energy Policy Act mandates the Federal Energy Regulatory Commission (FERC) to ensure that adequate transmission and distribution access is available to Exempt Wholesale Generators (EWGs) and nonutility generation (NUG). In 1996, FERC issued the "MegaRule," which regulates Transmission Open Access (TOA).

TOA was mandated in order to facilitate competition in wholesale generation. As a result, a broad range of Independent Power Producers (IPPs) and cogenerators now submit bids and compete in energy markets to match electric energy supply and demand. In the future, the retail structure of power distribution may resemble the existing structure of the telephone industry; that is, consumers would choose which generator to buy power from. Also, with demand-side metering, consumers would know the retail price of electric energy at any given time and choose when to purchase it.

Overall system reliability has become a major concern as the electric utility industry adapts to the new horizontal structure. The North American Electric Reliability Council (NERC), which was created after the 1965 Northeast blackout, is responsible for maintaining system standards and reliability. NERC coordinates its efforts with FERC and other organizations such as the Edison Electric Institute (EEI) [8].

As shown in Figure 1.3, the transmission system in North America is interconnected in a large power grid known as the North American Power Systems Interconnection. NERC divides this grid into ten geographic regions known as coordinating councils (such as WSCC, the Western Systems Coordinating Council) or power pools (such as MAPP, the Mid-Continent Area Power Pool). The councils or pools consist of several neighboring utility companies that jointly perform regional planning studies and operate jointly to schedule generation.

The basic premise of TOA is that transmission owners treat all transmission users on a nondiscriminatory and comparable basis. In December 1999, FERC issued Order 2000, which calls for companies owning transmission systems to put transmission systems under the control of Regional Transmission Organizations (RTOs). Several of the NERC regions have either established Independent System Operators (ISOs) or planned for ISOs to operate the transmission system and facilitate transmission services. Maintenance of the transmission system remains the responsibility of the transmission owners.

At the time of the August 14, 2003 blackout in the northeastern United States and Canada, NERC reliability standards were voluntary. In August 2005, the U.S. Federal government passed the *Energy Policy Act of 2005*, which authorizes the creation of an electric reliability organization (ERO) with the statutory authority to enforce compliance with reliability standards among all market participants. On July 20, 2006 FERC issued an order certifying NERC as the ERO for the United States. NERC is also seeking recognition as the ERO from governmental authorities in Canada [8].

The objectives of electric utility restructuring are to increase competition, decrease regulation, and in the long run lower consumer prices. There is a concern that the benefits from breaking up the old vertically integrated utilities will be unrealized if the new unbundled generation and transmission companies are able to exert market power. Market power refers to the ability of one seller or group of sellers to maintain prices above competitive levels for a significant period of time, which could be done via collusion or by taking advantage of operational anomalies that create and exploit transmission congestion. Market power can be eliminated by independent supervision of generation and transmission companies, by ensuring that there are an ample number of generation companies, by eliminating transmission congestion, and by creating a truly competitive market, where the spot price at each node (bus) in the transmission system equals the marginal cost of providing energy at that node, where the energy provider is any generator bidding into the system [9].

1.4

COMPUTERS IN POWER SYSTEM ENGINEERING

As electric utilities have grown in size and the number of interconnections has increased, planning for future expansion has become increasingly complex. The increasing cost of additions and modifications has made it imperative that utilities consider a range of design options, and perform detailed studies of the effects on the system of each option, based on a number of assumptions: normal and abnormal operating conditions, peak and off-peak loadings, and present and future years of operation. A large volume of network data must also be collected and accurately handled. To assist the engineer in this power system planning, digital computers and highly developed computer programs are used. Such programs include power-flow, stability, short-circuit, and transients programs.

Power-flow programs compute the voltage magnitudes, phase angles, and transmission-line power flows for a network under steady-state operating conditions. Other results, including transformer tap settings and generator reactive power outputs, are also computed. Today's computers have sufficient storage and speed to efficiently compute power-flow solutions for networks with 100,000 buses and 150,000 transmission lines. High-speed printers then print out the complete solution in tabular form for analysis by the planning engineer. Also available are interactive power-flow programs, whereby power-flow results are displayed on computer screens in the form of single-line diagrams; the engineer uses these to modify the network with a mouse or from a keyboard and can readily visualize the results. The computer's large storage and high-speed capabilities allow the engineer to run the many different cases necessary to analyze and design transmission and generation-expansion options.

Stability programs are used to study power systems under disturbance conditions to determine whether synchronous generators and motors remain in synchronism. System disturbances can be caused by the sudden loss of a generator or transmission line, by sudden load increases or decreases, and by short circuits and switching operations. The stability program combines power-flow equations and machine-dynamic equations to compute the angular swings of machines during disturbances. The program also computes critical clearing times for network faults, and allows the engineer to investigate the effects of various machine parameters, network modifications, disturbance types, and control schemes.

Short-circuits programs are used to compute three-phase and line-to-ground faults in power system networks in order to select circuit breakers for fault interruption, select relays that detect faults and control circuit breakers, and determine relay settings. Short-circuit currents are computed for each relay and circuit-breaker location, and for various system-operating conditions such as lines or generating units out of service, in order to determine minimum and maximum fault currents.

Transients programs compute the magnitudes and shapes of transient overvoltages and currents that result from lightning strikes and line-switching operations. The planning engineer uses the results of a transients program to determine insulation requirements for lines, transformers, and other equipment, and to select surge arresters that protect equipment against transient overvoltages.

Other computer programs for power system planning include relay-coordination programs and distribution-circuits programs. Computer programs for generation-expansion planning include reliability analysis and loss-of-load probability (LOLP) programs, production cost programs, and investment cost programs.

1.5

POWERWORLD SIMULATOR

PowerWorld Simulator (PowerWorld) version 12_GSO is a commercial-grade power system analysis and simulation package that accompanies this text. The purposes of integrating PowerWorld with the text are to provide computer solutions to examples in the text, to extend the examples, to demonstrate topics covered in the text, to provide a software tool for more realistic design projects, and to provide the readers with experience using a commercial grade power system analysis package. To use this software package, first install PowerWorld on your computer by following the instructions on the CD that accompanies the text. Note, while the book comes with the most recent version at the time of publication, in the future updated versions of this software can be downloaded at www.powerworld.com/GloverSarmaOverbye. The remainder of this section provides the necessary details to get up and running with PowerWorld.

EXAMPLE 1.1   Introduction to PowerWorld Simulator

After installing PowerWorld, double-click on the PW icon to start the program. Power system analysis requires, of course, that the user provide the program with a model of the power system. With PowerWorld you can either build a new case (model) from scratch or start from an existing case. Initially we'll start from an existing case. Select **File, Open Case**. This displays the Open Dialog. Select the Example 1_1 case in the Chapter 1 directory, and then click Open. The display should look similar to Figure 1.6.

For users familiar with electric circuit schematics it is readily apparent that Figure 1.6 does NOT look like a traditional schematic. This is because

FIGURE 1.6

Example power system

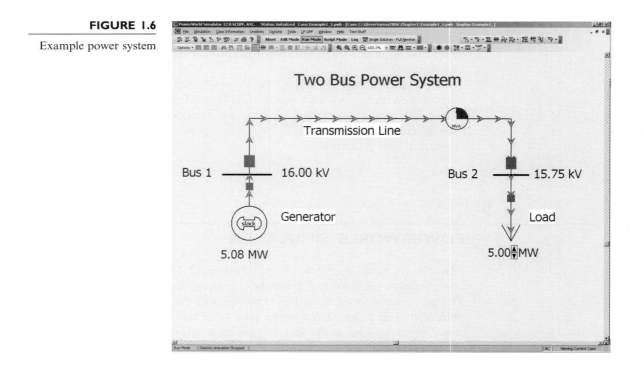

the system is drawn in what is called one-line diagram form. A brief explanation is in order. Electric power systems range in size from small dc systems with peak power demands of perhaps a few milliwatts (mW) to large continent-spanning interconnected ac systems with peak demands of hundreds of Gigawatts (GW) of demand (1 GW = 1×10^9 Watt). The subject of this book and also PowerWorld are the high-voltage, high-power, interconnected ac systems. Almost without exception these systems operate using three-phase ac power at either 50 or 60 Hz. As discussed in Section 2.6, full analysis of an arbitrary three-phase system requires consideration of each of the three phases. Drawing such systems in full schematic form quickly gets excessively complicated. Thankfully, during normal operation three-phase systems are usually balanced. This permits the system to be accurately modeled as an equivalent single-phase system (the details are discussed in Chapter 8, *Symmetrical Components*). Most power system analysis packages, including PowerWorld, use this approach. Then connections between devices are then drawn with a single line joining the system devices, hence the term "one-line" diagram. However, do keep in mind that the actual systems are three phase.

Figure 1.6 illustrates how the major power system components are represented in PowerWorld. Generators are shown as a circle with a "dog-bone" rotor, large arrows represent loads, and transmission lines are simply drawn as lines. In power system terminology, the nodes at which two or more devices join are called buses. In PowerWorld think lines usually represent buses; the bus voltages are shown in kilovolts (kV) in the fields immediately to the right of the buses. In addition to voltages, power engineers are also concerned with how power flows through the system (the solution of the power flow problem is covered in Chapter 6, *Power Flows*). In PowerWorld, power flows can be visualized with arrows superimposed on the generators, loads, and transmission lines. The size and speed of the arrows indicates the direction of flow.

One of the unique aspects of PowerWorld is its ability to animate power systems. To start the animation, select **Simulation** from the main menu, the select **Solve and Animate** (i.e., **Simulation, Solve and Animate**). The one-line should spring to life! While the one-line is being animated you can interact with the system. Figure 1.6 represents a simple power system in which a generator is supplying power to a load through a 16 kV distribution system feeder. The solid red blocks on the line and load represent circuit breakers. To open a circuit breaker simply click on it. Since the load is series connected to the generator, clicking on any of the circuit breakers isolates the load from the generator resulting in a blackout. To restore the system click again on the circuit breaker to close it and then again select **Simulation, Solve and Animate**. To vary the load click on the up or down arrows between the load value and the "MW" field. Note that because of the impedance of the line, the load's voltage drops as the load is increased.

You can view additional information about most of the elements on the one-line by right-clicking on them. For example right-clicking on the generator symbol brings up a local menu of additional information about the generator, while right-clicking on the transmission line brings up local menu of information about the line. The meaning of many of these fields will become clearer as you progress through the book. To modify the display itself simply right-click on a blank area of the one-line. This displays the one-line local menu. Select **Oneline Display Options** to display the Oneline Display Options Dialog. From this dialog you can customize many of the display features. For example, to change the animated flow arrow color select the "Animated Flows" from the options shown on the left side of the dialog. Then click on the green colored box next to the "Actual MW" field (towards the bottom of the dialog) on the to change its color. There are several techniques for panning and/or zooming on the one-line. One method to pan is to press the keyboard arrow keys in the direction you would like to move. To zoom just hold down the Ctrl key while pressing the up arrow to zoom in, or the down arrow to zoom out. Alternatively, you can drag the one-line by clicking and holding the left mouse button down and then moving the mouse—the one-line should follow. To go to a favorite view, from the one-line local menu select the **Go To View** to view a list of saved views.

If you would like to retain your changes after you exit PowerWorld you need to save the results. To do this select **File, Save Case As** and then enter a different file name so as to not overwrite the initial case. One important note: PowerWorld actually saves the information associated with the power system model itself in a different file from the information associated with the one-line. The power system model is stored in *.pwb files (PowerWorld binary file) while the one-line display information is stored in *.pwd files (PowerWorld display file). For all the cases discussed in this book, the names of both files should be the same (excepting the different extensions). The reason for the dual file system is to provide flexibility. With large system models, it is quite common for a system to be displayed using multiple one-line diagrams. Furthermore, a single one-line diagram might be used at different times to display information about different cases.

EXAMPLE 1.2   PowerWorld Simulator—Edit Mode

PowerWorld has two major modes of operations. The Run Mode, which was just introduced, is used for running simulations and performing analysis. The Edit Mode, which is used for modifying existing cases and building new cases, is introduced in this example. To switch to the Edit Mode click on the **Edit Mode** button in the top center of the display. We'll use the edit mode to add an additional bus and load as well as two new lines to the system given in Example 1.1.

When switching to the Edit Mode, notice that the main menu changes slightly, several new toolbars appear, and that the one-line now has a superimposed grid to help with alignment (the grid can be customized using the

FIGURE 1.7

Example 1.2—Edit
Mode view with new bus

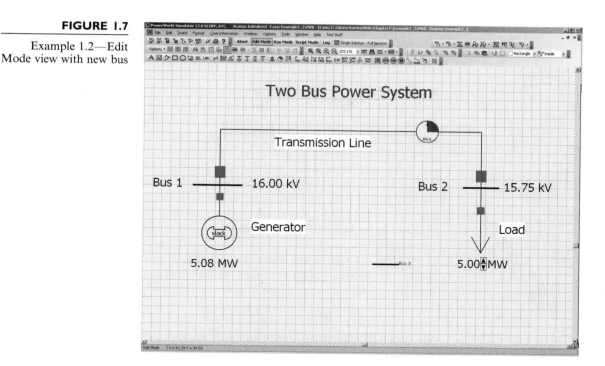

Grid/Highlight Unlinked options category on the Oneline Display Options Dialog). In the Edit Mode, we will first add a new bus to the system. This can be done graphically by first selecting **Insert, Bus** and then moving the mouse to the desired one line location and clicking (note the **Insert** menu is only available in the Edit Mode). The Bus Options dialog then appears. This dialog is used to set the bus parameters. For now leave all the bus fields at their default values, except set Bus Name to "Bus 3" and set the nominal voltage to "16.0; note that the number for this new bus was automatically set to the one greater than the highest bus number in the case. The one-line should look similar to Figure 1.7. You may wish to save you case now to avoid losing your changes.

By default, when a new bus is inserted a "bus field" is also inserted. Bus fields are used to show information about buses on the one-lines. In this case the new field shows the bus name, although initially in rather small fonts. To change the field's font size click on the field to select it, and then select **Format, Font** to display the Font dialog. To make it easier to see, change the font's size to a larger value. Since we would also like to see the bus voltage magnitude, we need to add an additional bus field. Select **Insert, Field, Bus Field** and then click near bus 3. This displays the Bus Field Options dialog. Make sure the bus number is set to 3, and that the "Type of Field" is Bus Voltage. Again resize with the **Format, Font** dialog.

Next, we'll insert some load at bus 3. This can be done graphically by selecting **Insert, Load**, and then clicking on bus 3. The Load Options dialog then appears, allowing you to set the load parameters. Note that the load was automatically assigned to bus 3. Leave all the fields at their default values, except set the orientation to "Down", and enter 5.0 in the Constant Power column MW Value field. As the name implies, a constant power load treats the load power as being independent of bus voltage; constant power loads models are commonly used in power system analysis. By default PowerWorld "anchors" each load symbol to its bus. This is a handy feature when changing a drawing since when you drag the bus the load and all associated fields move as well. Note that two fields showing the load's real (MW) and reactive (Mvar) power were also auto-inserted with the load. Since we won't be needing the reactive right now, select this field and then select **Edit, Delete** to remove it. You should also resize the MW field using the **Format, Font** command.

Now we need to join the bus 3 load to the rest of the system. We'll do this by adding a line from bus 2 to bus 3. Select **Insert, Transmission Line** and then click on bus 2. This begins the line drawing. During line drawing PowerWorld adds a new line segment for each mouse click. After adding several segments place the cursor on bus 3 and double-click. The Transmission Line/Transformer Options dialog appears allowing you to set the line parameters. Note that PowerWorld should have auto set the "from" and "to" bus numbers based upon the starting and ending buses (buses 2 and 3). If these values have not been set automatically then you probably did not click exactly on bus 2 or bus 3; manually enter the values. Next, set the line's Series Resistance (R) field to 0.3, the Series Reactance (X) field to 0.6, and the MVA Limits Limit A field to 20 (the details of transformer and transmission line modeling is covered in Chapters 3 through 5). Select OK to close the dialog. Note that Simulator also auto-inserted two circuit breakers and a round "pie chart" symbol. The pie charts are used to show the percentage loading of the line. You can change the display size for these objects by right-clicking on them to display their option dialogs.

EXAMPLE 1.3   PowerWorld Simulator—Run Mode

Next, we need to switch back to Run Mode to animate the new system developed in Example 1.2. Click on the **Run Mode** button and then select **Simulation, Solve and Animate** to start the simulation. You should see the arrows flow from bus 1 to bus 2 to bus 3. Note that the total generation is now about 10.4 MW, with 10 MW flowing to the two loads and 0.4 MW lost to the wire resistance. To add the load variation arrows to the bus 3 load click on the load MW field (not the load arrow itself) to display the field's local menu. Select **Load Field Information Dialog** to view the Load Field Options dialog. Set the "Delta per Mouse Click" field to "1.0", which will change the load

FIGURE 1.8

Example 1.3—new
three-bus system

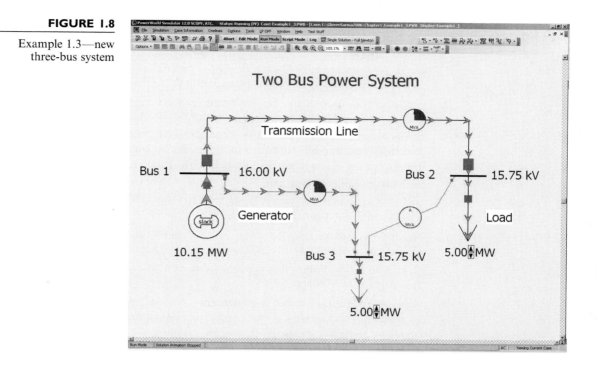

by one MW per click on the up/down arrows. You may also like to set the
"Digits to Right of Decimal" to 2 to see more digits in the load field. Be sure
to save your case. The new system now has one generator and two loads. The
system is still radial, meaning that a break anywhere on the wire joining bus
1 to bus 2 would result in a blackout of all the loads. Radial power systems
are quite common in the lower voltage distribution systems. At higher voltage levels, networked systems are typically used. In a networked system, each
load has at least two possible sources of power. We can convert our system to
a networked system simply by adding a new line from bus 1 to bus 3. To do
this switch back to Edit Mode and then repeat the previous line insertion
process except you should start at bus 1 and end at bus 3; use the same line
parameters as for the bus 2 to 3 line. Your final system should look similar to
the system shown in Figure 1.8. Note that now you can open any single line
and still supply both loads—a nice increase in reliability!

With this introduction you now have the skills necessary to begin using
PowerWorld to interactively learn about power systems. If you'd like to take
a look at some of the larger systems you'll be studying, open PowerWorld
case Example 6.13. This case models a power system with 37 buses. Notice
that when you open any line in the system the flow of power immediately re-
distributes to continue to meet the total load demand.

REFERENCES

1. H. M. Rustebakke et al., *Electric Utility Systems Practice*, 4th ed. (New York: Wiley, 1983). Photos courtesy of Westinghouse Historical Collection.

2. *Statistical Yearbook of the Electric Utility Industry—2004 Data.* (Washington, DC: Edison Electric Institute, www.eei.org, September 2005).

3. North American Electric Reliability Council (NERC), *Electricity Supply and Demand for 2005* (Princeton, NJ: NERC, www.nerc.com, September 2005).

4. *Key Facts about the Electric Power Industry* (Washington, DC: Edison Electric Institute, www.eei.org, August 2006).

5. *Annual Energy Outlook 2006* (Washington, DC: Energy Information Administration, www.eia.doe.gov, February 2006).

6. North American Electric Reliability Council (NERC), *2005 Long-Term Reliability Assessment* (Princeton, NJ: NERC, www.nerc.com, September 2005).

7. C. E. Root, "The Future Beckons," *Supplement to IEEE Power & Energy Magazine*, *4*, 3 (May/June 2006), pp. 58–65.

8. *About NERC* (Princeton, NJ: www.nerc.com, August 2006).

9. T. J. Overbye and J. Weber, "Visualizing the Electric Grid," *IEEE Spectrum*, *38*, 2 (February 2001), pp. 52–58.

2

FUNDAMENTALS

The objective of this chapter is to review basic concepts and establish terminology and notation. In particular, we review phasors, instantaneous power, complex power, network equations, and elementary aspects of balanced three-phase circuits. Students who have already had courses in electric network theory and basic electric machines should find this chapter to be primarily refresher material.

CASE STUDY   The following article describes the recent penetration of distributed generation into electric utility generation markets and forecasts of distributed generation market growth. Throughout most of the twentieth century, electric utility companies built increasingly larger generating plants, primarily hydro or thermal (using coal, gas, oil, or nuclear fuel) that connected to transmission systems. At the end of the twentieth century, following the ongoing deregulation of the electric utility industry with increased competition in the United States and in other countries, smaller generation sources that connect directly to distribution systems have emerged. Distributed generation sources include renewable technologies (including geothermal, ocean tides, solar, and wind) and nonrenewable technologies (including internal combustion engines, combustion turbines, combined cycle, microturbines, and fuel cells) [5].

Distributed Generation: Semantic Hype or the Dawn of a New Era?

HANS B. PÜTTGEN, PAUL R. MACGREGOR, AND FRANK C. LAMBERT

As the electric utility industry continues to restructure, driven both by rapidly evolving regulatory environments and by market forces, the emergence of a number of new generation technologies also profoundly influences the industry's outlook. While it is certainly true that government public policies and regulations have played a major role in the rapidly growing rate at which distributed generation is penetrating the market, it is also the case that a number of technologies have reached a development stage allowing for large-scale implementation within existing electric utility systems.

At the onset of any discussion related to distributed generation, one question begs to be answered: Is the fact that electric power producing facilities are distributed actually a new and revolutionary concept? Have power plants not always been located across broad expanses of land? The answer to these questions clearly is that electric power plants have always been sited all across the service territories of the utilities owning them. Hence, the opening question: As with many so-called innovations that have been put forward during the recent past, is the entire concept of dis-

tributed generation a simple semantic marketing hype or are we actually at the dawn of a new electric power generation era? We believe that a new electric power production industry is emerging, and that it will rely on a broad array of new technologies. This article sets the stage for further coverage of distributed generation to appear in future issues of *IEEE Power & Energy Magazine*.

PRESENT POWER PRODUCTION SITUATION

Since the beginning of the twentieth century, the backbone of the electric power industry structure has been large utilities operating within well-defined geographical territories and within local market monopolies under the scrutiny of various regulatory bodies. Traditionally, these utilities own the generation, transmission, and distribution facilities within their assigned service territories; they finance the construction of these facilities and then incorporate the related capital costs in their rate structure which is subsequently approved by the relevant regulatory bodies. The technologies deployed and the siting of the new facilities are generally also subject to regulatory approval.

Three major types of power plants have been constructed primarily:

("Distributed Generation: Semantic Hype or ..." by Hans B. Püttgen, Paul R. MacGregor, and Frank C. Lambert, IEEE Power & Energy (Jan/Feb 2003) pg. 22–29.)

TABLE I   Worldwide installed capacity (GW) by 1 January 2000. (Source: Energy Information Administration.)

Region	Thermal	Hydro	Nuclear	Other/Renew	Total
North America	642	176	109	18	945
Central and South America	64	112	2	3	181
Western Europe	353	142	128	10	633
Eastern Europe and former USSR	298	80	48	0	426
Middle East	94	4	0	0	98
Africa	73	20	2	0	95
Asia and Oceania	651	160	69	4	884
Total	2,175	694	358	35	3,262
Percentage	66.6	21.3	11.0	1.1	100

- hydro, either run-of-the-river facilities or various types of dams
- thermal, using either coal, oil, or gas
- nuclear.

Until the end of the twentieth century, other generation technologies only had an incidental impact.

Table 1 shows the installed capacities on a worldwide basis at the end of the twentieth century.

As we look into the future, all three technologies mentioned above have their own sets of problems associated with them:

- Given their friendly environmental impact, hydro power plants are most often the preferred generation technology wherever and whenever feasible. However, the identification of feasible new sites in highly industrialized countries is becoming increasingly difficult. In highly developed countries, where the cost-attractive traditional hydro facility sites have been almost entirely built, some power plants could be, and are, reconfigured to become pumped-storage facilities. On the other hand, while hydro electric power production is saturating within industrialized countries, it represents very significant development opportunities in several developing regions of the world. While hydro power plants do not create any pollution related to their daily operation, they do bring significant environmental and often societal upheaval when they are constructed. Recently completed facilities or on-going construction projects in South America and Asia have been, and remain, at the center of controversies that go far beyond the national boundaries of their home nations.

- Even though several pollution-abatement technologies are being successfully implemented, often at significant capital and operational costs, fossil fuel thermal power plants bring operating pollution problems that are becoming increasingly difficult to ignore. The emergence of a broad array of "green power" marketing initiatives provides yet another indication of the growing concern regarding air pollution. While some parts of the world have significant coal reserves, a growing concern is the depletion of the world's increasingly scarce oil and gas reserves for the purpose of electricity production. Future generations will most probably need our remaining carbon resources to fulfill materials production requirements as opposed to as a raw energy source.

- Except for a few economically emerging regions of the world, it is safe to observe that nuclear power production, using existing technologies, will decrease during the coming decades as old plants are retired and are not being replaced. Several European countries, such as Germany and Sweden, have enacted laws to accelerate the decommissioning of existing nuclear power plants. However, emerging technologies, such as the pebble bed technology, which allow for a highly standardized manufacturing of the power plants with modular installed capacities, may revive the nuclear power industry as will most probably be required within any generation mix that is free of fossil fuels.

As the technologies evolved, ever larger power production units were constructed allowing their operators to take full advantage of construction-cost economies of scale to provide a more cost-attractive generation mix to their customers. However, siting these ever larger facilities has become increasingly difficult. Hydro facilities must be sited as dictated by geography, even if this means displacing very large population centers and/or permanently and seriously affecting the local ecology. Since it is more convenient to transport energy in its electric form, fossil thermal plants are generally sited either close to raw fuel sources or to fuel conversion/ treatment facilities. The pollution concerns mentioned earlier dictate their siting far away from population centers. A broad range of environmental concerns mandate that nuclear power plants be located far away from population centers.

These siting issues, as well as the need to share these large power production facilities within a formalized market structure, have required the construction of large, complex, and capital-intensive electric power transmission networks. These transmission networks have become an increasing source of concern as their sustained development becomes a problem from a right-of-way point of view and as their economic operation comes in limbo under a reregulated electric utility industry. Ecological and environmental protection concerns, as well as political pressure, also often mandate that new transmission facilities be constructed underground, which even further compounds the issue by imposing often unbearable construction cost impediments.

As the industry enters the competitive arena, fewer and fewer corporations are capable of taking on the financing of the construction of large electric power plants at costs far exceeding a billion dollars. Under the present economic and investment cli-

mate, with its almost exclusive focus on short-term results, the justification of a multibillion dollar investment with a pay-back period measured in decades has become virtually impossible. In several industrialized countries, aggressive public policies backed by strict regulatory mandates are such that electric power production within the confines of vertically integrated utilities has most probably been relegated to the past, while a true highly diversified electric power production industry is the future.

WHAT IS DISTRIBUTED GENERATION?

Before launching into an overview of distributed generation, it is appropriate to put forward a definition or at least an operational confine related to distributed generation. It is generally agreed upon that any electric power production technology that is such that it is integrated within distribution systems fits under the distributed generation umbrella. The designations "distributed" and "dispersed" are used interchangeably.

One can further categorize distributed generation technologies as renewable and nonrenewable. Renewable technologies include:

- solar, photovoltaic or thermal
- wind
- geothermal
- ocean.

Nonrenewable technologies include:

- internal combustion engine, ice
- combined cycle
- combustion turbine
- microturbines
- fuel cell.

Distributed generation should not to be confused with renewable generation. Distributed generation technologies may be renewable or not; in fact, some distributed generation technologies could, if fully deployed, significantly contribute to present air pollution problems.

The increased market penetration of distributed generation has also been the advent of an electric power production industry. Many, if not most, of the players in this industry are not the traditional electric utilities; in fact, several of these new players actually are spin-offs of the traditional utilities. Electric power production facilities that do not belong to electric utilities are referred to as nonutility generators (NUGs). The rapid emergence of NUGs is illustrated by the fact that, starting during the early 1990s, more generation capacity is added each year in the United States by NUGs than by traditional electric utilities. NUGs represented 5% of the installed generation capability in the United States at the beginning of the 1990s; by the end of the decade, the proportion had grown to 20% as it grew from less that 40 GW to more than 150 GW. These statistics also take into account the fact that several large electric utilities have actually spun off their generation capabilities within separate corporate entities, while they have remained as what has now been referred to as "wire companies."

CAPABILITY RATINGS AND SYSTEM INTERFACES

While future issues of *IEEE Power & Energy Magazine* will focus on specific distributed generation technologies, it is useful to broadly mention the range of capabilities for the various technologies generally falling under the distributed generation category (Table 2). The electric power network interface, which plays a major role when considering the net-work operation aspects related to dispersed generation, is also listed in Table 2.

MARKET PENETRATION

While reliable and representative historic data are difficult to produce, distributed generation market penetration is expected to increase dramatically during the next few years (Figure 1).

Leading manufacturers, market research organizations, and consulting entities of the industry project that the distributed generation market will be between US$10 and 30 billion by the year 2010.

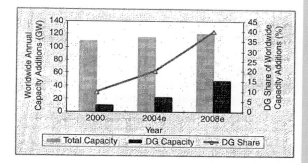

Figure 1
Distributed generation market growth. (Source: Merrill Lynch and the U.S. Energy Information Agency (EIA), January 2001.)

TABLE 2   Distributed generation capabilities and system interfaces.

Technology	Typical Capability Ranges	Utility Interface
Solar, photovoltaic	A few W to several hundred kW	dc to ac converter
Wind	A few hundred W to a few MW	asynchronous generator
Geothermal	A few hundred kW to a few MW	synchronous generator
Ocean	A few hundred kW to a few MW	four-quadr. synchronous machine
ICE	A few hundred kW to tens of MW	synchr. generator or ac to ac converter
Combined cycle	A few tens of MW to several hundred MW	synchronous generator
Combustion turbine	A few MW to hundreds of MW	synchronous generator
Microturbines	A few tens of kW to a few MW	ac to ac converter
Fuel cells	A few tens of kW to a few tens of MW	dc to ac converter

Since another article is this issue of *IEEE Power & Energy Magazine* is on green power, the following discussion focuses on three nonrenewable technologies that have significant immediate or short-term potential.

Internal combustion engine generators

At the present time, the predominant distributed generation technology is represented by internal combustion engines (ICE) driving standard electric generators. By 1996, over 600,000 units were installed in the United States with a combined installed capacity exceeding 100,000 MW. While the unit sizes ranged from a few kW to well over several MW, almost 70% of all units had installed capacities ranging from 10 to 200 kW. The vast majority of these units were installed to serve as backup generators for sensitive loads (such as special manufacturing facilities, large information processing centers, hospitals, airports, military installations, large office towers, hotels, etc.) for which long-duration energy supply failures would have catastrophic consequences.

These units represent a significant potential energy production resource in view of their very low load cycles. Several corporations, including existing utilities, are starting to offer remote management services for these units such that they can become revenue sources for their owners. While such increased energy production using backup ICE generators would enable delaying construction of new generation capacity, such utilization of engine generators creates location-specific environmental issues associated with the equipment's operational characteristics as well as potential utility interconnection issues.

Specifically, engine generators feature high levels of NO_x emissions and represent a potential noise nuisance to their immediate surroundings. While noise abatement materials and enclosures may be applied at fairly low costs to address the latter issue, the remedies for NO_x emissions, such as selective catalytic reduction (SCR), are quite expensive. The capital cost of adding an SCR system to an engine generator can double its installed cost. The sheer size of the SCR system can make the installation in-

550 kW internal combustion engine generator installed in the basement of an office building. Cooling is on the far right; the engine is in the middle with the generator on the left with the control module on the far left. The vertical exhaust is in the middle.

feasible for many existing engine generators located in restrictive building enclosures.

From an electric utility perspective, as potential incremental generation capacity, distributed engine generators, with their low installed costs and fairly high operational costs, represent super-peaking capacity that could be economically dispatched, albeit only a few hundred hours per year. Under such peaking operational scenarios, the total contribution of NO_x emissions by engine generators, as a percentage of the total from all generation, would be fairly low. The deployment of this large and untapped peak generation capacity presently is largely prohibited in most industrialized countries by existing environmental protection regulations. Since this significant generation capacity is already installed, its release could be almost immediate after regulatory relief is promulgated.

With regards to potential utility interconnection issues, most existing engine generators are sized to provide power to critical and emergency loads only, i.e., only to a fraction of the total on-site load. These engine generators, when operated during non-emergencies, would only be reducing on-site peak demand, and the supply requirements from the utility system and generally would not be injecting power back into the utility network. With proper switchgear and breaker configurations, these engine generators may be operated in parallel with the local electric utility system without significant implications to the utility system operation or personnel safety.

Fuel cells

Fuel cells probably represent the power production technology receiving the most development attention. Each individual fuel cell consists of an electrolyte that is "sandwiched" between fuel and oxidant electrodes. The fuel typically is hydrogen and the oxidant typically is oxygen. The fuel cell produces electricity directly by way of various chemical reactions without an intermediate conversion into mechanical energy. While some particular application fuel cells directly use hydrogen as the raw source of energy, the hydrogen fuel is typically extracted from some form of fossil fuel.

Individual fuel cells are combined in various series and parallel configurations to constitute a fuel cell system. The fuel cell system, which produces dc electricity, is connected to the local utility system by way of a power electronic dc to ac converter.

Five 200 kW fuel cells. (Courtesy UTC Fuel Cells.)

Fuel cells are developed for both mobile and stationary applications. The mobile applications are already being deployed for buses and are in various experimental stages for automobiles. Stationary systems are being installed in residential and commercial applications in the United States and Europe.

The major types of fuel cells are designated by the type of electrolytes used.

- Alkaline fuel cell (AFC): This is one of the earliest fuel cell technologies that has been successfully deployed during many NASA shuttle missions. AFCs use a liquid solution of potassium hydroxide as the electrolyte with an operating temperature of 70–90 °C. The lower operating temperature facilitates rapid startup of the unit. One of the major disadvantages of this technology is its intolerance of CO_2 and the requirement to install expensive CO_2 scrubbers.
- Polymer electrolyte membrane or proton exchange membrane (PEM): This fuel cell technology utilizes a solid polymer as the electrolyte. The polymer is an excellent conductor of protons and an insulator of electrons; it does not require liquid management. This unit features a low operating temperature of 70–90 °C, which facilities rapid startup. The PEM fuel cell has a high power density and is a leading candidate for portable power, mobile, and residential sector applications.
- Solid oxide fuel cell (SOFC): A solid ceramic material is used for the electrolyte at operating temperatures of 600–1,000 °C. This high operating temperature, while hampering rapid startup as required for most mobile applications, helps to increase the efficiency and frees up the SOFC to use a variety of fuels without a separate reformer. This technology is primarily targeted at medium and large-scale stationary power generation applications.
- Molten carbonate fuel cell (PAFC): A molten carbonate salt mixture is used for the electrolyte and requires operating temperatures of 600–1,000 °C. This technology is targeted at medium- and large-scale stationary power generation applications.
- Phosphoric acid fuel cell (PAFC): A liquid phosphoric acid contained in a Teflon matrix is used as the electrolyte for these fuel cells. The operating temperature is 175–200 °C to facilitate the removal of water from the electrolyte.

This technology is very tolerant to impurities in the fuel stream and is the most mature in terms of system development and commercialization. Over 200 stationary units with a typical capacity of 200 kW have been installed in the United States.

Microturbines

Microturbines are essentially very small combustion turbines, individually of the size of a refrigerator, that are often packaged in multiunit systems. In most configurations, the microturbine is a single-shaft machine with the compressor and turbine mounted on the same shaft as the electric generator. With a single rotating shaft, gearboxes and associated parts are eliminated, helping to improve manufacturing costs and operational reliability. The rather high rotational speeds vary in the range from 50,000 to 120,000 rpm, depending on the output capacity of the microturbine. This high-frequency output is first rectified and then converted to 50 or 60 Hz.

Despite lower operational temperatures than those of combustion turbines, microturbines produce energy with efficiencies in the 25 to 30% range. These efficiencies are made possible by deploying, for example, a heat recuperation system that transfers waste heat energy from the exhaust stream back into the incoming air stream. The generator is cooled by airflow into the turbine, thereby eliminating the need for liquid cooling equipment and associated auxiliary power requirements. Some microturbines use air bearings, thereby eliminating the need for oil systems and their associated power requirements.

Microturbines are capable of burning a number of fuels at high- and low-pressure levels, including natural gas, waste (sour) gas, landfill gas, or propane. Regardless of the fuel, microturbines have demonstrated that they feature very low air pollution emissions, particularly NO_x emissions, at about 1/100th of the level of diesel-fired ICEs. Microturbines emit significantly lower noise levels and generate far less vibration than ICEs.

Two primary concerns are associated with microturbines that could impact their rate of market adoption: capital cost and equipment lifetime. Spe-

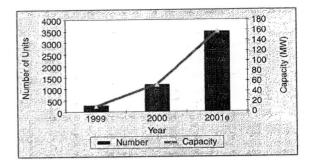

Figure 2
Microturbine sales growth. (Source: Primen, January 2001.)

cifically, the capital cost of a microturbine, on a per installed kW basis, can be several times that of an ICE, and the projected equipment life time, measured in operational hours before replacement, is several times shorter than for an ICE. The combined result of these impediments is a significantly higher life cycle cost compared to other distributed generation technologies, such as ICEs.

Nevertheless, microturbine sales, driven by environmental concerns and niche applications such as landfills, have been increasing dramatically since the commercial introduction of a 30 kW model in 1999 (Figure 2).

POTENTIAL GENERATION MIX ISSUES

When considering a significant market penetration of distributed generation technologies, it is important to keep in mind that several of them, such as solar and wind, are not dispatchable by man. The generation mix issue is likely to first come to a head in conjunction with wind power, which is rapidly being developed in several countries around the world. Among the leading countries are: Germany, United States, Spain, Denmark, and India. Some 1,600 MW of wind power were installed in the United States in 2001, with the state of Texas leading the way; Texas is forecasting that over 10% of its electricity demand will be supplied by wind power by the year 2010. Germany, which presently has over 11,000 MW of installed wind power, is putting increased emphasis on the technology as its socialist-green party coalition seeks to curtail and eventually eliminate its nuclear power production facilities.

30 kW microturbine. (Capstone Turbine Corp.)

The Danish wind power situation is particularly interesting. While Germany, the United States, and Spain each have more installed wind power capacity, Denmark leads the pack in terms of relative market penetration. In 1996, the Danish government set forth a national energy long-range plan that called for 1,500 MW of installed wind power by the year 2005; this goal was already reached by the end of the year 1999. By the year 2030, wind power is expected to reach 5,500 MW, such that 50% of the electricity demands will be satisfied using wind power. Since Denmark does not have a lot of uninhabited land and since the available land is precious for agricultural endeavors, most of the new wind power capacity is added by way of wind farms installed at sea. Such installations at sea, while bringing significant challenges in terms of underwater cable systems, partially alleviate the environmental concern of noise.

Two major problems have to be addressed when such aggressive goals are set for nondispatchable distributed generation, such wind power production:

- Wind power is very cyclical and also unpredictable. Stand-by energy must be available when wind energy is not available. Such stand-by energy can either be provided by near-by systems, such as Germany or Sweden in the case of Denmark, or by other energy storage technologies. Germany is also rapidly developing its wind power capabilities while curtailing its thermal production facilities; the same is true for Sweden. Denmark's geography is such that the construction of pumped-storage hydro facilities is unrealistic. As a result, Denmark may have to revise downward its ambitious wind power plans due to the lack of available backup energy resources.
- Wind generators overwhelming feature induction-asynchronous generators. While these machines are particularly well suited to the variable speed nature of wind machines, they can not operate without reactive power support from the network to which they are connected. As a result, the Danish utilities are faced with significant reactive power support issues which will only become worse as wind power penetration increases.

NETWORK CONSIDERATIONS

Distributed generation technologies are overwhelmingly connected to existing electric power delivery systems at the distribution level. One of their significant benefits is that they are modular enough to be conveniently integrated within electric distribution systems, thereby relieving some of the necessity to invest in transmission system expansion.

However, significant penetration within existing electric distribution systems is not without a new set of problems. The following are among the key issues that must be addressed.

Power quality
Several of the distributed generation technologies rely on some form of power electronic device in conjunction with the distribution network interface, be it ac-to-ac or dc-to-ac converters. All of these devices inject currents that are not perfectly sinusoidal. The resulting harmonic distortion, if not properly contained and filtered, can bring serious operational difficulties to the loads connected on the same distribution system. Existing standards have been enacted to limit the harmonic content acceptable in conjunction with various power elec-

tronic loads; similar standards are required for distributed generation systems and are under various stages of preparedness.

Reactive power coordination

Distributed generation, implemented at the distribution level, i.e., close to the load, can bring significant relief to the reactive coordination by providing close proximity reactive power support at the distribution level, provided the proper network interface technology is used and that proper system configuration has taken place. However, wind generation actually contributes to worsen the reactive coordination problem. Most wind generators feature asynchronous induction generators that are ideally suited to the variable speed characteristics of wind machines but that must rely on the network to which they are connected for reactive power support.

Reliability and reserve margin

Several distributed generation technologies are such that their production levels depend on mother nature (wind and solar) or are such that their availability is subject to the operational priorities of their owners. The requirement to use sophisticated power electronic network interfaces may affect the plant's availability. As a result, the issue of reliability comes to the forefront along with the necessity to maintain sufficient generation reserve margins. Traditionally, the vertically integrated utility was also responsible for the availability of sufficient reserve margins to ensure adequate system reliability. Under a highly distributed generation ownership scenario, assignment of reserve margin maintenance increasingly will become a problem unless a market-driven solution is put forward.

Reliability and network redundancy

Most electric distribution systems feature a radial network configuration as opposed to the meshed structure adopted at transmission levels. As a result, network redundancy becomes an issue when significant distributed generation is connected directly to distribution systems since single line outages could completely curtail the availability of generation facilities.

Safety

Distribution system protection schemes typically are designed to rapidly isolate faults occurring either at load locations or on the line itself. The assumption is that, if the distribution line is disconnected somewhere between the fault and the feeding substation, then repair work can safely proceed. Clearly, if distributed generation is connected on the same distribution feeder, then significantly more sophisticated protective relaying schemes must be designed and implemented to properly protect not only the personnel working on the lines but also the loads connected to them.

Accountability

A daunting problem is looming over the "brave new electric utility industry" in its restructured configuration: Who will the customer call when the lights go out? The local "wire company" might arguably answer, "my wires are just fine, thank you." The existence of local transmission company may not even be known by the end-user. The power producer might arguably respond, "please refer your inquiry to your local wire company, with which we have a service contract." The resolution of this all-important question is still very much open for debate.

PUBLIC POLICY AND REGULATORY IMPACT

While it is not our intent to revisit the various regulatory environments that are driving the electric utility industry, it is worth summarizing their overall framework. The actual implementation of public policy varies from one country to another; however, the overall philosophies of these public policies share several common goals and outcomes. Overall, the impetus of these policies are such that they:

- Aim to create a competitive environment where the customer will eventually have a choice between several electric energy providers. The rate at which competitive markets are opened up varies significantly from one country to the next and even from one region to the next within a particular country.

- Aim to encourage the broad access to the electric power production arena by a wide range of players, particularly nonutility entities. These policies have often resulted in the launch of corporations that own electric power production facilities all round the world as if they were any other type of production facility.
- Often result in the creation of electric utilities that only own and operate electric power delivery systems. These wire companies sell their services to the power producing corporations to enable the delivery of electric energy between them and their customers using the wire company infrastructure.
- Have, in effect, resulted in a situation in which the transmission systems are increasingly in limbo between production companies and distribution companies.

Public policies and regulations are often different from one region of a country to another. Such lack of uniformity does not facilitate the penetration of new generation technologies. As some of the early and more aggressive deregulation experiments, which should be more appropriately referred to as "reregulation" experiments, have failed, they are being reconsidered by the legislatures having enacted them and by the regulators tasked with enforcing them. Such "time-of-the-day" public policies and regulations also represent a manor hindrance to accelerated distributed generation market penetration.

It is important not to overlook the tax incentive impact on the development of emerging dispersed generation technologies. In some instances, such as photovoltaics and, to some extent, wind power, the construction and subsequent operation of distributed generation facilities are almost entirely driven by tax incentives, which often vary significantly from one region to the next and from one year to the next. They generally are provided at two levels:

- Construction tax incentives, either in the form of an upfront grant or in the form of accelerated depreciation schedules. In some extreme, although rare, occasions, new gener-

ation facilities have been constructed to harvest the tax incentives and then almost never operated.
- Operational tax incentives, generally in the form of revenue tax abatements. In some occasions, distributed generation facilities have ceased operation only a few short years after construction as the tax incentive structure has changed.

STANDARDS

Important and mission-critical work is on-going in the standards arena under the umbrella of the *IEEE Standard for interconnecting Distributed Resources with Electric Power Systems* (IEEE Standard P1547). The tenth draft of the standard was balloted recently, with an overwhelmingly positive affirmative outcome. It is expected that the IEEE Standards Board will approve the document in early 2003, which would represent a major milestone in the development of distributed generation.

Three working groups are developing companion standards and guides as follows:

- *IEEE Standard for Conformance Test Procedures for Equipment Interconnecting Distributed Resources with Electric Power Systems* (IEEE Standard P1547.1) will provide manufacturers and users with a common set of test procedures to verify that the equipment to be deployed will meet the requirements of IEEE Standard 1547. This includes type, production, and commissioning tests to provide repeatable results, independent of test location, and flexibility to accommodate a variety of distributed resources technologies.
- *IEEE Applications Guide for Interconnecting Distributed Resources with Electric Power Systems* (IEEE Standard P1547.2) will provide the technical background and application details to support the understanding of IEEE Standard 1547. The background and rationale of the technical requirements will be discussed in terms of the operation of the distributed resource interconnection with the electric power system. This document will include technical descrip-

tions as well as schematics, applications guidance, and interconnection examples to enhance the use of IEEE Standard 1547.

- *IEEE Guide for Monitoring, Information Exchange, and Control of Distributed Resources with Electric Power Systems* (IEEE Standard P1547.3) will facilitate the interoperability of one or more distributed resources interconnected with the electric power system. It will describe functionality, parameters, and methodology for monitoring, information exchange, and control for the interconnected distributed resources with, or associated with, electric power systems.

FURTHER READING

IEEE Power Eng. Rev., vol. 22, pp. 5–23, Mar. 2002. Series of articles on harnessing the power of hydro.

IEEE Power Eng. Rev., vol. 22, pp. 4–18, Sep. 2002, and pp. 21–28, Oct. 2002. Series of articles on wind power.

Proc. IEEE, special issue "2001: An energy odyssey," vol. 89, Dec. 2001.

H. B. Püttgen, D. R. Volzka, M. I. Olken, "Restructuring and reregulation of the U.S. electric utility industry," *IEEE Power Eng. Rev.*, vol. 21, pp. 8–10, Feb. 2001.

R. Mandelbaum, "Reap the wild wind," pp. 34–39, *IEEE Spectr.*, vol. 39, Oct. 2002.

BIOGRAPHIES

Hans B. Püttgen is Georgia Power professor and vice chair within the School of Electrical and Computer Engineering at the Georgia Institute of Technology. He has been at Georgia Tech since 1981. He is also the director and Management Board chair of Georgia Tech's National Electric Energy Test, Research, and Application Center (NEETRAC). He serves as president of Georgia Tech Lorraine. He received his Ingénieur Diplômé degree from the Swiss Federal Institute of Technology, completed his graduate business administration and management education at the University of Lausanne, and received his Ph.D. in electric power engineering from the University of Florida. He is active in PES and serves as president-elect.

Paul R. MacGregor is CEO of Delfin Energy. Previously, he has served as executive vice president for Altra Energy Technologies, vice president for Energy Imperium, business development manager for EDS Utilities, and product manager for both Power Technologies and General Electric. He has authored over 35 technical papers, was named as one of three finalists of the Eta Kappa Nu Outstanding Young Electrical Engineer Award, and was elected to the Georgia Tech Council of Outstanding Young Engineering Alumni. He graduated from the Georgia Institute of Technology with B.S., M.S., and Ph.D. degrees in electrical engineering as well as a M.S. in technology and science policy. He has served as chair, vice chair, and treasurer for the Schenectady Chapter of IEEE and is the past chair of an IEEE PES technical committee working group on Investment Strategies.

Frank C. Lambert is the Electrical Systems program manager at NEETRAC at the Georgia Institute of Technology. He received B.S. and M.S. degrees in electric power engineering from Georgia Tech. He worked for Georgia Power Company from 1973 until 1995, gaining experience in distribution and transmission engineering, operations, and management. He joined NEETRAC in 1996 to manage the Electrical Systems Research Program and is active in PES, where he serves on several working groups in the Distribution Subcommittee.

2.1

PHASORS

A sinusoidal voltage or current at constant frequency is characterized by two parameters: a maximum value and a phase angle. A voltage

$$v(t) = V_{max} \cos(\omega t + \delta) \qquad (2.1.1)$$

has a maximum value V_{max} and a phase angle δ when referenced to $\cos(\omega t)$. The root-mean-square (rms) value, also called *effective value*, of the sinusoidal voltage is

$$V = \frac{V_{max}}{\sqrt{2}} \qquad (2.1.2)$$

Euler's identity, $e^{j\phi} = \cos\phi + j\sin\phi$, can be used to express a sinusoid in terms of a phasor. For the above voltage,

$$v(t) = \text{Re}[V_{max}e^{j(\omega t + \delta)}]$$
$$= \text{Re}[\sqrt{2}(Ve^{j\delta})e^{j\omega t}] \qquad (2.1.3)$$

where $j = \sqrt{-1}$ and Re denotes "real part of." The rms phasor representation of the voltage is given in three forms—exponential, polar, and rectangular:

$$V = \underbrace{Ve^{j\delta}}_{\text{exponential}} = \underbrace{V\underline{/\delta}}_{\text{polar}} = \underbrace{V\cos\delta + jV\sin\delta}_{\text{rectangular}} \qquad (2.1.4)$$

A phasor can be easily converted from one form to another. Conversion from polar to rectangular is shown in the phasor diagram of Figure 2.1. Euler's identity can be used to convert from exponential to rectangular form. As an example, the voltage

$$v(t) = 169.7 \cos(\omega t + 60°) \quad \text{volts} \qquad (2.1.5)$$

has a maximum value $V_{max} = 169.7$ volts, a phase angle $\delta = 60°$ when referenced to $\cos(\omega t)$, and an rms phasor representation in polar form of

$$V = 120\underline{/60°} \quad \text{volts} \qquad (2.1.6)$$

Also, the current

$$i(t) = 100 \cos(\omega t + 45°) \quad \text{A} \qquad (2.1.7)$$

has a maximum value $I_{max} = 100$ A, an rms value $I = 100/\sqrt{2} = 70.7$ A, a phase angle of $45°$, and a phasor representation

$$I = 70.7\underline{/45°} = 70.7e^{j45} = 50 + j50 \quad \text{A} \qquad (2.1.8)$$

The relationships between the voltage and current phasors for the three passive elements—resistor, inductor, and capacitor—are summarized in Fig-

FIGURE 2.1

Phasor diagram for converting from polar to rectangular form

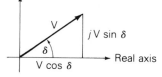

Imaginary axis

V

$jV \sin \delta$

δ

$V \cos \delta$

Real axis

FIGURE 2.2

Summary of
relationships between
phasors V and I for
constant R, L, and C
elements with sinusoidal-
steady-state excitation

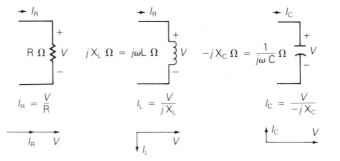

ure 2.2, where sinusoidal-steady-state excitation and constant values of R, L, and C are assumed.

When voltages and currents are discussed in this text, lowercase letters such as $v(t)$ and $i(t)$ indicate instantaneous values, uppercase letters such as V and I indicate rms values, and uppercase letters in italics such as V and I indicate rms phasors. When voltage or current values are specified, they shall be rms values unless otherwise indicated.

2.2

INSTANTANEOUS POWER IN SINGLE-PHASE AC CIRCUITS

Power is the rate of change of energy with respect to time. The unit of power is a watt, which is a joule per second. Instead of saying that a load absorbs energy at a rate given by the power, it is common practice to say that a load absorbs power. The instantaneous power in watts absorbed by an electrical load is the product of the instantaneous voltage across the load in volts and the instantaneous current into the load in amperes. Assume that the load voltage is

$$v(t) = \text{V}_{\max} \cos(\omega t + \delta) \quad \text{volts} \tag{2.2.1}$$

We now investigate the instantaneous power absorbed by purely resistive, purely inductive, purely capacitive, and general RLC loads. We also introduce the concepts of real power, power factor, and reactive power. The physical significance of real and reactive power is also discussed.

PURELY RESISTIVE LOAD

For a purely resistive load, the current into the load is in phase with the load voltage, $I = V/\text{R}$, and the current into the resistive load is

$$i_R(t) = I_{Rmax} \cos(\omega t + \delta) \quad A \tag{2.2.2}$$

where $I_{Rmax} = V_{max}/R$. The instantaneous power absorbed by the resistor is

$$
\begin{aligned}
p_R(t) &= v(t)i_R(t) = V_{max}I_{Rmax} \cos^2(\omega t + \delta) \\
&= \tfrac{1}{2}V_{max}I_{Rmax}\{1 + \cos[2(\omega t + \delta)]\} \\
&= VI_R\{1 + \cos[2(\omega t + \delta)]\} \quad W
\end{aligned}
\tag{2.2.3}
$$

As indicated by (2.2.3), the instantaneous power absorbed by the resistor has an average value

$$P_R = VI_R = \frac{V^2}{R} = I_R^2 R \quad W \tag{2.2.4}$$

plus a double-frequency term $VI_R \cos[2(\omega t + \delta)]$.

PURELY INDUCTIVE LOAD

For a purely inductive load, the current lags the voltage by 90°, $I_L = V/(jX_L)$, and

$$i_L(t) = I_{Lmax} \cos(\omega t + \delta - 90°) \quad A \tag{2.2.5}$$

where $I_{Lmax} = V_{max}/X_L$, and $X_L = \omega L$ is the inductive reactance. The instantaneous power absorbed by the inductor is*

$$
\begin{aligned}
p_L(t) &= v(t)i_L(t) = V_{max}I_{Lmax} \cos(\omega t + \delta) \cos(\omega t + \delta - 90°) \\
&= \tfrac{1}{2}V_{max}I_{Lmax} \cos[2(\omega t + \delta) - 90°] \\
&= VI_L \sin[2(\omega t + \delta)] \quad W
\end{aligned}
\tag{2.2.6}
$$

As indicated by (2.2.6), the instantaneous power absorbed by the inductor is a double-frequency sinusoid with *zero* average value.

PURELY CAPACITIVE LOAD

For a purely capacitive load, the current leads the voltage by 90°, $I_C = V/(-jX_C)$, and

$$i_C(t) = I_{Cmax} \cos(\omega t + \delta + 90°) \quad A \tag{2.2.7}$$

where $I_{Cmax} = V_{max}/X_C$, and $X_C = 1/(\omega C)$ is the capacitive reactance. The instantaneous power absorbed by the capacitor is

$$
\begin{aligned}
p_C(t) &= v(t)i_C(t) = V_{max}I_{Cmax} \cos(\omega t + \delta) \cos(\omega t + \delta + 90°) \\
&= \tfrac{1}{2}V_{max}I_{Cmax} \cos[2(\omega t + \delta) + 90°)] \\
&= -VI_C \sin[2(\omega t + \delta)] \quad W
\end{aligned}
\tag{2.2.8}
$$

The instantaneous power absorbed by a capacitor is also a double-frequency sinusoid with *zero* average value.

GENERAL RLC LOAD

For a general load composed of RLC elements under sinusoidal-steady-state excitation, the load current is of the form

$$i(t) = I_{max} \cos(\omega t + \beta) \quad A \tag{2.2.9}$$

The instantaneous power absorbed by the load is then*

$$
\begin{aligned}
p(t) = v(t)i(t) &= V_{max}I_{max} \cos(\omega t + \delta) \cos(\omega t + \beta) \\
&= \tfrac{1}{2} V_{max}I_{max}\{\cos(\delta - \beta) + \cos[2(\omega t + \delta) - (\delta - \beta)]\} \\
&= VI \cos(\delta - \beta) + VI \cos(\delta - \beta) \cos[2(\omega t + \delta)] \\
&\quad + VI \sin(\delta - \beta) \sin[2(\omega t + \delta)]
\end{aligned}
$$

$$p(t) = VI \cos(\delta - \beta)\{1 + \cos[2(\omega t + \delta)]\} + VI \sin(\delta - \beta) \sin[2(\omega t + \delta)]$$

Letting $I \cos(\delta - \beta) = I_R$ and $I \sin(\delta - \beta) = I_X$ gives

$$p(t) = \underbrace{VI_R\{1 + \cos[2(\omega t + \delta)]\}}_{p_R(t)} + \underbrace{VI_X \sin[2(\omega t + \delta)]}_{p_X(t)} \tag{2.2.10}$$

As indicated by (2.2.10), the instantaneous power absorbed by the load has two components: One can be associated with the power $p_R(t)$ absorbed by the resistive component of the load, and the other can be associated with the power $p_X(t)$ absorbed by the reactive (inductive or capacitive) component of the load. The first component $p_R(t)$ in (2.2.10) is identical to (2.2.3), where $I_R = I \cos(\delta - \beta)$ is the component of the load current in phase with the load voltage. The phase angle $(\delta - \beta)$ represents the angle between the voltage and current. The second component $p_X(t)$ in (2.2.10) is identical to (2.2.6) or (2.2.8), where $I_X = I \sin(\delta - \beta)$ is the component of load current 90° out of phase with the voltage.

REAL POWER

Equation (2.2.10) shows that the instantaneous power $p_R(t)$ absorbed by the resistive component of the load is a double-frequency sinusoid with average value P given by

$$P = VI_R = VI \cos(\delta - \beta) \quad W \tag{2.2.11}$$

---

*Use the identity: $\cos A \cos B = \tfrac{1}{2}[\cos(A - B) + \cos(A + B)]$.

The *average power* P is also called *real power* or *active power*. All three terms indicate the same quantity P given by (2.2.11).

POWER FACTOR

The term $\cos(\delta - \beta)$ in (2.2.11) is called the *power factor*. The phase angle $(\delta - \beta)$, which is the angle between the voltage and current, is called the *power factor angle*. For dc circuits, the power absorbed by a load is the product of the dc load voltage and the dc load current; for ac circuits, the average power absorbed by a load is the product of the rms load voltage V, rms load current I, and the power factor $\cos(\delta - \beta)$, as shown by (2.2.11). For inductive loads, the current lags the voltage, which means β is less than δ, and the power factor is said to be *lagging*. For capacitive loads, the current leads the voltage, which means β is greater than δ, and the power factor is said to be *leading*. By convention, the power factor $\cos(\delta - \beta)$ is positive. If $|\delta - \beta|$ is greater than 90°, then the reference direction for current may be reversed, resulting in a positive value of $\cos(\delta - \beta)$.

REACTIVE POWER

The instantaneous power absorbed by the reactive part of the load, given by the component $p_X(t)$ in (2.2.10), is a double-frequency sinusoid with zero average value and with amplitude Q given by

$$Q = VI_X = VI \sin(\delta - \beta) \quad \text{var} \tag{2.2.12}$$

The term Q is given the name *reactive power*. Although it has the same units as real power, the usual practice is to define units of reactive power as volt-amperes reactive, or var.

EXAMPLE 2.1   Instantaneous, real, and reactive power; power factor

The voltage $v(t) = 141.4 \cos(\omega t)$ is applied to a load consisting of a 10-Ω resistor in parallel with an inductive reactance $X_L = \omega L = 3.77\ \Omega$. Calculate the instantaneous power absorbed by the resistor and by the inductor. Also calculate the real and reactive power absorbed by the load, and the power factor.

SOLUTION   The circuit and phasor diagram are shown in Figure 2.3(a). The load voltage is

$$V = \frac{141.4}{\sqrt{2}} \underline{/0^\circ} = 100\underline{/0^\circ} \quad \text{volts}$$

The resistor current is

FIGURE 2.3

Circuit and phasor diagram for Example 2.1

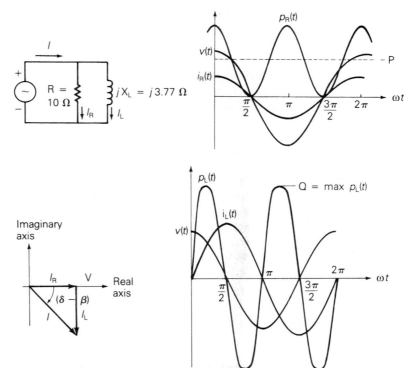

(a) Circuit and phasor diagram   (b) Waveforms

$$I_R = \frac{V}{R} = \frac{100}{10}\underline{/0°} = 10\underline{/0°} \quad \text{A}$$

The inductor current is

$$I_L = \frac{V}{jX_L} = \frac{100}{(j3.77)}\underline{/0°} = 26.53\underline{/-90°} \quad \text{A}$$

The total load current is

$$I = I_R + I_L = 10 - j26.53 = 28.35\underline{/-69.34°} \quad \text{A}$$

The instantaneous power absorbed by the resistor is, from (2.2.3),

$$p_R(t) = (100)(10)[1 + \cos(2\omega t)]$$
$$= 1000[1 + \cos(2\omega t)] \quad \text{W}$$

The instantaneous power absorbed by the inductor is, from (2.2.6),

$$p_L(t) = (100)(26.53)\sin(2\omega t)$$
$$= 2653\sin(2\omega t) \quad \text{W}$$

The real power absorbed by the load is, from (2.2.11),

$$P = VI \cos(\delta - \beta) = (100)(28.53) \cos(0° + 69.34°)$$
$$= 1000 \quad W$$

(*Note*: P is also equal to $VI_R = V^2/R$.)
The reactive power absorbed by the load is, from (2.2.12),

$$Q = VI \sin(\delta - \beta) = (100)(28.53) \sin(0° + 69.34°)$$
$$= 2653 \quad var$$

(*Note*: Q is also equal to $VI_L = V^2/X_L$.)
The power factor is

$$p.f. = \cos(\delta - \beta) = \cos(69.34°) = 0.3528 \quad lagging$$

Voltage, current, and power waveforms are shown in Figure 2.3(b).

Note that $p_R(t)$ and $p_X(t)$, given by (2.2.10), are strictly valid only for a parallel R-X load. For a general RLC load, the voltages across the resistive and reactive components may not be in phase with the source voltage $v(t)$, resulting in additional phase shifts in $p_R(t)$ and $p_X(t)$ (see Problem 2.13). However, (2.2.11) and (2.2.12) for P and Q are valid for a general RLC load.

■

PHYSICAL SIGNIFICANCE OF REAL AND REACTIVE POWER

The physical significance of real power P is easily understood. The total energy absorbed by a load during a time interval T, consisting of one cycle of the sinusoidal voltage, is PT watt-seconds (Ws). During a time interval of n cycles, the energy absorbed is $P(nT)$ watt-seconds, all of which is absorbed by the resistive component of the load. A kilowatt-hour meter is designed to measure the energy absorbed by a load during a time interval $(t_2 - t_1)$, consisting of an integral number of cycles, by integrating the real power P over the time interval $(t_2 - t_1)$.

The physical significance of reactive power Q is not as easily understood. Q refers to the maximum value of the instantaneous power absorbed by the reactive component of the load. The instantaneous reactive power, given by the second term $p_X(t)$ in (2.2.10), is alternately positive and negative, and it expresses the reversible flow of energy to and from the reactive component of the load. Q may be positive or negative, depending on the sign of $(\delta - \beta)$ in (2.2.12). Reactive power Q is a useful quantity when describing the operation of power systems (this will become evident in later chapters). As one example, shunt capacitors can be used in transmission systems to deliver reactive power and thereby increase voltage magnitudes during heavy load periods (see Chapter 5).

2.3

COMPLEX POWER

For circuits operating in sinusoidal-steady-state, real and reactive power are conveniently calculated from complex power, defined below. Let the voltage across a circuit element be $V = \mathrm{V}\underline{/\delta}$, and the current into the element be $I = \mathrm{I}\underline{/\beta}$. Then the complex power S is the product of the voltage and the conjugate of the current:

$$S = VI^* = [\mathrm{V}\underline{/\delta}][\mathrm{I}\underline{/\beta}]^* = \mathrm{VI}\underline{/\delta - \beta}$$
$$= \mathrm{VI}\cos(\delta - \beta) + j\mathrm{VI}\sin(\delta - \beta) \qquad (2.3.1)$$

where $(\delta - \beta)$ is the angle between the voltage and current. Comparing (2.3.1) with (2.2.11) and (2.2.12), S is recognized as

$$S = P + jQ \qquad (2.3.2)$$

The magnitude $S = \mathrm{VI}$ of the complex power S is called the *apparent power*. Although it has the same units as P and Q, it is common practice to define the units of apparent power S as voltamperes or VA. The real power P is obtained by multiplying the apparent power $S = \mathrm{VI}$ by the power factor p.f. $= \cos(\delta - \beta)$.

The procedure for determining whether a circuit element absorbs or delivers power is summarized in Figure 2.4. Figure 2.4(a) shows the *load convention*, where the current *enters* the positive terminal of the circuit element, and the complex power *absorbed* by the circuit element is calculated from (2.3.1). This equation shows that, depending on the value of $(\delta - \beta)$, P may have either a positive or negative value. If P is positive, then the circuit element absorbs positive real power. However, if P is negative, the circuit element absorbs negative real power, or alternatively, it delivers positive real power. Similarly, if Q is positive, the circuit element in Figure 2.4(a) absorbs positive reactive power. However, if Q is negative, the circuit element absorbs negative reactive power, or it delivers positive reactive power.

FIGURE 2.4

Load and generator conventions

(a) *Load convention*. Current *enters* positive terminal of circuit element. If P is positive, then positive real power is *absorbed*. If Q is positive, then positive reactive power is *absorbed*. If P (Q) is negative, then positive real (reactive) power is *delivered*.

(b) *Generator convention*. Current *leaves* positive terminal of the circuit element. If P is positive, then positive real power is *delivered*. If Q is positive, then positive reactive power is *delivered*. If P (Q) is negative, then positive real (reactive) power is *absorbed*.

Figure 2.4(b) shows the *generator convention*, where the current *leaves* the positive terminal of the circuit element, and the complex power *delivered* is calculated from (2.3.1). When P is positive (negative) the circuit element *delivers* positive (negative) real power. Similarly, when Q is positive (negative), the circuit element *delivers* positive (negative) reactive power.

EXAMPLE 2.2   Real and reactive power, delivered or absorbed

A single-phase voltage source with $V = 100\underline{/130°}$ volts delivers a current $I = 10\underline{/10°}$ A, which leaves the positive terminal of the source. Calculate the source real and reactive power, and state whether the source delivers or absorbs each of these.

SOLUTION   Since I leaves the positive terminal of the source, the generator convention is assumed, and the complex power delivered is, from (2.3.1),

$$S = VI^* = [100\underline{/130°}][10\underline{/10°}]^*$$

$$S = 1000\underline{/120°} = -500 + j866$$

$$P = \text{Re}[S] = -500 \quad \text{W}$$

$$Q = \text{Im}[S] = +866 \quad \text{var}$$

where Im denotes "imaginary part of." The source absorbs 500 W and delivers 866 var. Readers familiar with electric machines will recognize that one example of this source is a synchronous motor. When a synchronous motor operates at a leading power factor, it absorbs real power and delivers reactive power.   ∎

The *load convention* is used for the RLC elements shown in Figure 2.2. Therefore, the complex power *absorbed* by any of these three elements can be calculated as follows. Assume a load voltage $V = \text{V}\underline{/\delta}$. Then, from (2.3.1),

$$\text{resistor: } S_R = VI_R^* = [\text{V}\underline{/\delta}]\left[\frac{\text{V}}{\text{R}}\underline{/-\delta}\right] = \frac{\text{V}^2}{\text{R}} \qquad (2.3.3)$$

$$\text{inductor: } S_L = VI_L^*[\text{V}\underline{/\delta}]\left[\frac{\text{V}}{-j\text{X}_L}\underline{/-\delta}\right] = +j\frac{\text{V}^2}{\text{X}_L} \qquad (2.3.4)$$

$$\text{capacitor: } S_C = VI_C^* = [\text{V}\underline{/\delta}]\left[\frac{\text{V}}{j\text{X}_C}\underline{/-\delta}\right] = -j\frac{\text{V}^2}{\text{X}_C} \qquad (2.3.5)$$

From these complex power expressions, the following can be stated:

A (positive-valued) resistor absorbs (positive) real power, $P_R = \text{V}^2/\text{R}$ W, and zero reactive power, $Q_R = 0$ var.

An inductor absorbs zero real power, $P_L = 0$ W, and positive reactive power, $Q_L = \text{V}^2/\text{X}_L$ var.

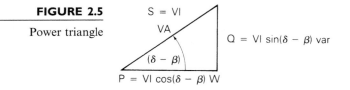

FIGURE 2.5

Power triangle

A capacitor *absorbs* zero real power, $P_C = 0$ W, and *negative* reactive power, $Q_C = -V^2/X_C$ var. Alternatively, a capacitor *delivers positive* reactive power, $+V^2/X_C$.

For a general load composed of RLC elements, complex power S is also calculated from (2.3.1). The real power $P = \text{Re}(S)$ absorbed by a passive load is always positive. The reactive power $Q = \text{Im}(S)$ absorbed by a load may be either positive or negative. When the load is inductive, the current lags the voltage, which means β is less than δ in (2.3.1), and the reactive power absorbed is positive. When the load is capacitive, the current leads the voltage, which means β is greater than δ, and the reactive power absorbed is negative; or, alternatively, the capacitive load delivers positive reactive power.

Complex power can be summarized graphically by use of the power triangle shown in Figure 2.5. As shown, the apparent power S, real power P, and reactive power Q form the three sides of the power triangle. The power factor angle $(\delta - \beta)$ is also shown, and the following expressions can be obtained:

$$S = \sqrt{P^2 + Q^2} \tag{2.3.6}$$

$$(\delta - \beta) = \tan^{-1}(Q/P) \tag{2.3.7}$$

$$Q = P \tan(\delta - \beta) \tag{2.3.8}$$

$$\text{p.f.} = \cos(\delta - \beta) = \frac{P}{S} = \frac{P}{\sqrt{P^2 + Q^2}} \tag{2.3.9}$$

EXAMPLE 2.3   Power triangle and power factor correction

A single-phase source delivers 100 kW to a load operating at a power factor of 0.8 lagging. Calculate the reactive power to be delivered by a capacitor connected in parallel with the load in order to raise the source power factor to 0.95 lagging. Also draw the power triangle for the source and load. Assume that the source voltage is constant, and neglect the line impedance between the source and load.

FIGURE 2.6

Circuit and power
triangle for Example 2.3

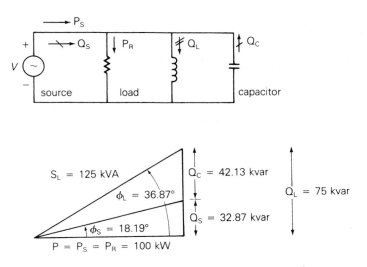

SOLUTION The circuit and power triangle are shown in Figure 2.6. The real power $P = P_S = P_R$ delivered by the source and absorbed by the load is not changed when the capacitor is connected in parallel with the load, since the capacitor delivers only reactive power Q_C. For the load, the power factor angle, reactive power absorbed, and apparent power are

$$\theta_L = (\delta - \beta_L) = \cos^{-1}(0.8) = 36.87°$$

$$Q_L = P \tan \theta_L = 100 \tan(36.87°) = 75 \quad \text{kvar}$$

$$S_L = \frac{P}{\cos \theta_L} = 125 \quad \text{kVA}$$

After the capacitor is connected, the power factor angle, reactive power delivered, and apparent power of the source are

$$\theta_S = (\delta - \beta_S) = \cos^{-1}(0.95) = 18.19°$$

$$Q_S = P \tan \theta_S = 100 \tan(18.19°) = 32.87 \quad \text{kvar}$$

$$S_S = \frac{P}{\cos \theta_S} = \frac{100}{0.95} = 105.3 \quad \text{kVA}$$

The capacitor delivers

$$Q_C = Q_L - Q_S = 75 - 32.87 = 42.13 \quad \text{kvar}$$

The method of connecting a capacitor in parallel with an inductive load is known as *power factor correction*. The effect of the capacitor is to increase the power factor of the source that delivers power to the load. Also, the source apparent power S_S decreases. As shown in Figure 2.6, the source apparent power for this example decreases from 125 kVA without the ca-

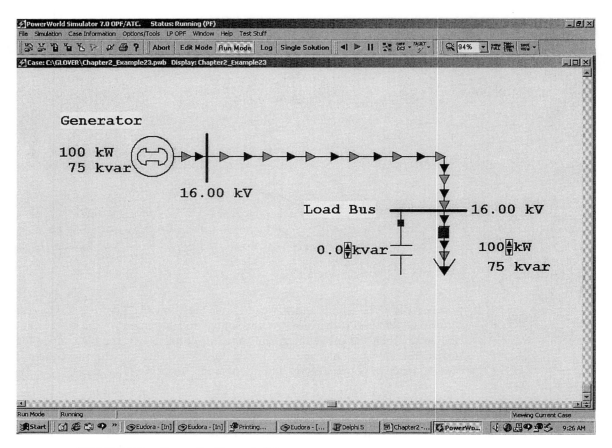

FIGURE 2.7   Screen for Example 2.3

pacitor to 105.3 kVA with the capacitor. The source current $I_S = S_S/V$ also decreases. When line impedance between the source and load is included, the decrease in source current results in lower line losses and lower line-voltage drops. The end result of power factor correction is improved efficiency and improved voltage regulation.

To see an animated view of this example, open PowerWorld Simulator case Example 2_3. On the main menu, select **Simulation, Solve, and Animate** to begin the simulation. The speed and size of the green arrows are proportional to the real power supplied to the load bus, and the blue arrows are proportional to the reactive power. Here reactive compensation can be supplied in discrete 20-kVar steps by clicking on the arrows in the capacitor's kvar field, and the load can be varied by clicking on the arrows in the load field. Notice that increasing the reactive compensation decreases both the reactive power flow on the supply line and the kVA power supplied by the generator; the real power flow is unchanged. ∎

2.4

NETWORK EQUATIONS

For circuits operating in sinusoidal-steady-state, Kirchhoff's current law (KCL) and voltage law (KVL) apply to phasor currents and voltages. Thus the sum of all phasor currents entering any node is zero and the sum of the phasor-voltage drops around any closed path is zero. Network analysis techniques based on Kirchhoff's laws, including nodal analysis, mesh or loop analysis, superposition, source transformations, and Thévenin's theorem or Norton's theorem, are useful for analyzing such circuits.

Various computer solutions of power system problems are formulated from nodal equations, which can be systematically applied to circuits. The circuit shown in Figure 2.8, which is used here to review nodal analysis, is assumed to be operating in sinusoidal-steady-state; source voltages are represented by phasors E_{S1}, E_{S2}, and E_{S3}; circuit impedances are specified in ohms. Nodal equations are written in the following three steps:

STEP 1   For a circuit with $(N + 1)$ nodes (also called buses), select one bus as the reference bus and define the voltages at the remaining buses with respect to the reference bus.

The circuit in Figure 2.8 has four buses—that is, $N + 1 = 4$ or $N = 3$. Bus 0 is selected as the reference bus, and bus voltages V_{10}, V_{20}, and V_{30} are then defined with respect to bus 0.

STEP 2   Transform each voltage source in series with an impedance to an equivalent current source in parallel with that impedance. Also, show admittance values instead of impedance values on the circuit diagram. Each current source is equal to the voltage source divided by the source impedance.

FIGURE 2.8

Circuit diagram for reviewing nodal analysis

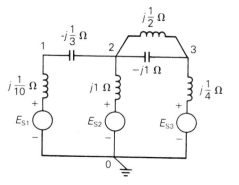

FIGURE 2.9

Circuit of Figure 2.8 with equivalent current sources replacing voltage sources. Admittance values are also shown.

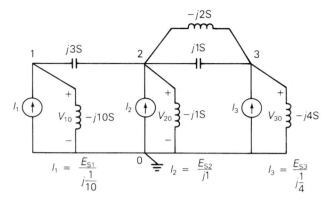

In Figure 2.9 equivalent current sources I_1, I_2, and I_3 are shown, and all impedances are converted to corresponding admittances.

STEP 3   Write nodal equations in matrix format as follows:

$$
\begin{bmatrix}
Y_{11} & Y_{12} & Y_{13} & \cdots & Y_{1N} \\
Y_{21} & Y_{22} & Y_{23} & \cdots & Y_{2N} \\
Y_{31} & Y_{32} & Y_{33} & \cdots & Y_{3N} \\
\vdots & \vdots & \vdots & & \vdots \\
Y_{N1} & Y_{N2} & Y_{N3} & \cdots & Y_{NN}
\end{bmatrix}
\begin{bmatrix}
V_{10} \\
V_{20} \\
V_{30} \\
\vdots \\
V_{N0}
\end{bmatrix}
=
\begin{bmatrix}
I_1 \\
I_2 \\
I_3 \\
\vdots \\
I_N
\end{bmatrix}
\quad (2.4.1)
$$

Using matrix notation, (2.4.1) becomes

$$YV = I \qquad (2.4.2)$$

where Y is the $N \times N$ bus admittance matrix, V is the column vector of N bus voltages, and I is the column vector of N current sources. The elements Y_{kn} of the bus admittance matrix Y are formed as follows:

diagonal elements: $Y_{kk} =$ sum of admittances connected to bus k
$(k = 1, 2, \ldots, N)$ (2.4.3)

off-diagonal elements: $Y_{kn} = -($sum of admittances connected between buses k and $n)$ $(k \neq n)$ (2.4.4)

The diagonal element Y_{kk} is called the *self-admittance* or the *driving-point admittance* of bus k, and the off-diagonal element Y_{kn} for $k \neq n$ is called the *mutual admittance* or the *transfer admittance* between buses k and n. Since $Y_{kn} = Y_{nk}$, the matrix Y is symmetric.

For the circuit of Figure 2.9, (2.4.1) becomes

$$
\begin{bmatrix}
(j3 - j10) & -(j3) & 0 \\
-(j3) & (j3 - j1 + j1 - j2) & -(j1 - j2) \\
0 & -(j1 - j2) & (j1 - j2 - j4)
\end{bmatrix}
\begin{bmatrix}
V_{10} \\
V_{20} \\
V_{30}
\end{bmatrix}
$$

$$
=
\begin{bmatrix}
I_1 \\
I_2 \\
I_3
\end{bmatrix}
$$

$$
j
\begin{bmatrix}
-7 & -3 & 0 \\
-3 & 1 & 1 \\
0 & 1 & -5
\end{bmatrix}
\begin{bmatrix}
V_{10} \\
V_{20} \\
V_{30}
\end{bmatrix}
=
\begin{bmatrix}
I_1 \\
I_2 \\
I_3
\end{bmatrix}
\tag{2.4.5}
$$

The advantage of this method of writing nodal equations is that a digital computer can be used both to generate the admittance matrix Y and to solve (2.4.2) for the unknown bus voltage vector V. Once a circuit is specified with the reference bus and other buses identified, the circuit admittances and their bus connections become computer input data for calculating the elements Y_{kn} via (2.4.3) and (2.4.4). After Y is calculated and the current source vector I is given as input, standard computer programs for solving simultaneous linear equations can then be used to determine the bus voltage vector V.

When double subscripts are used to denote a voltage in this text, the voltage shall be that at the node identified by the first subscript with respect to the node identified by the second subscript. For example, the voltage V_{10} in Figure 2.9 is the voltage at node 1 with respect to node 0. Also, a current I_{ab} shall indicate the current from node a to node b. Voltage polarity marks $(+/-)$ and current reference arrows $(\rightarrow)$ are not required when double subscript notation is employed. The polarity marks in Figure 2.9 for V_{10}, V_{20}, and V_{30}, although not required, are shown for clarity. The reference arrows for sources I_1, I_2, and I_3 in Figure 2.9 are required, however, since single subscripts are used for these currents. Matrices and vectors shall be indicated in this text by boldface type (for example, Y or V).

2.5

BALANCED THREE-PHASE CIRCUITS

In this section we introduce the following topics for balanced three-phase circuits: Y connections, line-to-neutral voltages, line-to-line voltages, line currents, Δ loads, Δ–Y conversions, and equivalent line-to-neutral diagrams.

FIGURE 2.10

Circuit diagram of a three-phase Y-connected source feeding a balanced-Y load

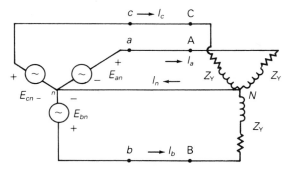

BALANCED-Y CONNECTIONS

Figure 2.10 shows a three-phase Y-connected (or "wye-connected") voltage source feeding a balanced-Y-connected load. For a Y connection, the neutrals of each phase are connected. In Figure 2.10 the source neutral connection is labeled bus n and the load neutral connection is labeled bus N. The three-phase source is assumed to be ideal since source impedances are neglected. Also neglected are the line impedances between the source and load terminals, and the neutral impedance between buses n and N. The three-phase load is *balanced*, which means the load impedances in all three phases are identical.

BALANCED LINE-TO-NEUTRAL VOLTAGES

In Figure 2.10, the terminal buses of the three-phase source are labeled a, b, and c, and the source line-to-neutral voltages are labeled E_{an}, E_{bn}, and E_{cn}. The source is *balanced* when these voltages have equal magnitudes and an equal 120°-phase difference between any two phases. An example of balanced three-phase line-to-neutral voltages is

FIGURE 2.11

Phasor diagram of balanced positive-sequence line-to-neutral voltages with E_{an} as the reference

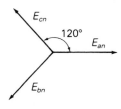

$$E_{an} = 10\underline{/0°}$$
$$E_{bn} = 10\underline{/-120°} = 10\underline{/+240°} \qquad (2.5.1)$$
$$E_{cn} = 10\underline{/+120°} = 10\underline{/-240°} \quad \text{volts}$$

where the line-to-neutral voltage magnitude is 10 volts and E_{an} is the reference phasor. The phase sequence is called *positive sequence* or *abc* sequence when E_{an} leads E_{bn} by 120° and E_{bn} leads E_{cn} by 120°. The phase sequence is called *negative sequence* or *acb* sequence when E_{an} leads E_{cn} by 120° and E_{cn} leads E_{bn} by 120°. The voltages in (2.5.1) are positive-sequence voltages, since E_{an} leads E_{bn} by 120°. The corresponding phasor diagram is shown in Figure 2.11.

BALANCED LINE-TO-LINE VOLTAGES

The voltages E_{ab}, E_{bc}, and E_{ca} between phases are called line-to-line voltages. Writing a KVL equation for a closed path around buses a, b, and n in Figure 2.10,

$$E_{ab} = E_{an} - E_{bn} \qquad (2.5.2)$$

For the line-to-neutral voltages of (2.5.1),

$$E_{ab} = 10\underline{/0°} - 10\underline{/-120°} = 10 - 10\left[\frac{-1 - j\sqrt{3}}{2}\right]$$

$$E_{ab} = \sqrt{3}(10)\left(\frac{\sqrt{3} + j1}{2}\right) = \sqrt{3}(10\underline{/30°}) \quad \text{volts} \qquad (2.5.3)$$

Similarly, the line-to-line voltages E_{bc} and E_{ca} are

$$E_{bc} = E_{bn} - E_{cn} = 10\underline{/-120°} - 10\underline{/+120°}$$
$$= \sqrt{3}(10\underline{/-90°}) \quad \text{volts} \qquad (2.5.4)$$

$$E_{ca} = E_{cn} - E_{an} = 10\underline{/+120°} - 10\underline{/0°}$$
$$= \sqrt{3}(10\underline{/150°}) \quad \text{volts} \qquad (2.5.5)$$

The line-to-line voltages of (2.5.3)–(2.5.5) are also balanced, since they have equal magnitudes of $\sqrt{3}(10)$ volts and 120° displacement between any two phases. Comparison of these line-to-line voltages with the line-to-neutral voltages of (2.5.1) leads to the following conclusion:

In a balanced three-phase Y-connected system with positive-sequence sources, the line-to-line voltages are $\sqrt{3}$ times the line-to-neutral voltages and lead by 30°. That is,

$$E_{ab} = \sqrt{3}E_{an}\underline{/+30°}$$
$$E_{bc} = \sqrt{3}E_{bn}\underline{/+30°} \qquad (2.5.6)$$
$$E_{ca} = \sqrt{3}E_{cn}\underline{/+30°}$$

This very important result is summarized in Figure 2.12. In Figure 2.12(a) each phasor begins at the origin of the phasor diagram. In Figure 2.12(b) the line-to-line voltages form an equilateral triangle with vertices labeled a, b, c corresponding to buses a, b, and c of the system; the line-to-neutral voltages begin at the vertices and end at the center of the triangle, which is labeled n for neutral bus n. Also, the clockwise sequence of the vertices abc in Figure 2.12(b) indicates positive-sequence voltages. In both diagrams, E_{an} is the reference. However, the diagrams could be rotated to align with any other reference.

Since the balanced line-to-line voltages form a closed triangle in Figure 2.12, their sum is zero. In fact, the sum of line-to-line voltages $(E_{ab} + E_{bc} + E_{ca})$

FIGURE 2.12

Positive-sequence
line-to-neutral and
line-to-line voltages in a
balanced three-phase
Y-connected system

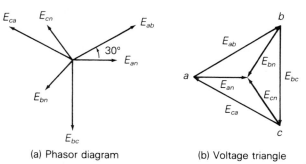

(a) Phasor diagram   (b) Voltage triangle

is *always* zero, even if the system is unbalanced, since these voltages form a closed path around buses a, b, and c. Also, in a balanced system the sum of the line-to-neutral voltages $(E_{an} + E_{bn} + E_{cn})$ equals zero.

BALANCED LINE CURRENTS

Since the impedance between the source and load neutrals in Figure 2.10 is neglected, buses n and N are at the same potential, $E_{nN} = 0$. Accordingly, a separate KVL equation can be written for each phase, and the line currents can be written by inspection:

$$I_a = E_{an}/Z_Y$$
$$I_b = E_{bn}/Z_Y \qquad (2.5.7)$$
$$I_c = E_{cn}/Z_Y$$

For example, if each phase of the Y-connected load has an impedance $Z_Y = 2\underline{/30°} \ \Omega$, then

$$I_a = \frac{10\underline{/0°}}{2\underline{/30°}} = 5\underline{/-30°} \quad \text{A}$$

$$I_b = \frac{10\underline{/-120°}}{2\underline{/30°}} = 5\underline{/-150°} \quad \text{A} \qquad (2.5.8)$$

$$I_c = \frac{10\underline{/+120°}}{2\underline{/30°}} = 5\underline{/90°} \quad \text{A}$$

The line currents are also balanced, since they have equal magnitudes of 5 A and 120° displacement between any two phases. The neutral current I_n is determined by writing a KCL equation at bus N in Figure 2.10.

$$I_n = I_a + I_b + I_c \qquad (2.5.9)$$

Using the line currents of (2.5.8),

FIGURE 2.13

Phasor diagram of line currents in a balanced three-phase system

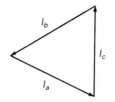

$$I_n = 5\underline{/-30°} + 5\underline{/-150°} + 5\underline{/90°}$$

$$I_n = 5\left(\frac{\sqrt{3} - j1}{2}\right) + 5\left(\frac{-\sqrt{3} - j1}{2}\right) + j5 = 0 \qquad (2.5.10)$$

The phasor diagram of the line currents is shown in Figure 2.13. Since these line currents form a closed triangle, their sum, which is the neutral current I_n, is zero. In general, the sum of any balanced three-phase set of phasors is zero, since balanced phasors form a closed triangle. Thus, although the impedance between neutrals n and N in Figure 2.10 is assumed to be zero, the neutral current will be zero for *any* neutral impedance ranging from short circuit (0 Ω) to open circuit (∞ Ω), as long as the system is balanced. If the system is not balanced—which could occur if the source voltages, load impedances, or line impedances were unbalanced—then the line currents will not be balanced and a neutral current I_n may flow between buses n and N.

BALANCED Δ LOADS

Figure 2.14 shows a three-phase Y-connected source feeding a balanced-Δ-connected (or "delta-connected") load. For a balanced-Δ connection, equal load impedances Z_Δ are connected in a triangle whose vertices form the buses, labeled A, B, and C in Figure 2.14. The Δ connection does not have a neutral bus.

Since the line impedances are neglected in Figure 2.14, the source line-to-line voltages are equal to the load line-to-line voltages, and the Δ-load currents I_{AB}, I_{BC}, and I_{CA} are

$$I_{AB} = E_{ab}/Z_\Delta$$

$$I_{BC} = E_{bc}/Z_\Delta \qquad (2.5.11)$$

$$I_{CA} = E_{ca}/Z_\Delta$$

For example, if the line-to-line voltages are given by (2.5.3)–(2.5.5) and if $Z_\Delta = 5\underline{/30°}$ Ω, then the Δ-load currents are

FIGURE 2.14

Circuit diagram of a Y-connected source feeding a balanced-Δ load

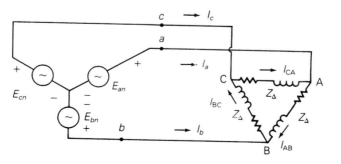

$$I_{AB} = \sqrt{3}\left(\frac{10/30°}{5/30°}\right) = 3.464/0° \quad A$$

$$I_{BC} = \sqrt{3}\left(\frac{10/-90°}{5/30°}\right) = 3.464/-120° \quad A \tag{2.5.12}$$

$$I_{CA} = \sqrt{3}\left(\frac{10/150°}{5/30°}\right) = 3.464/+120° \quad A$$

Also, the line currents can be determined by writing a KCL equation at each bus of the Δ load, as follows:

$$I_a = I_{AB} - I_{CA} = 3.464/0° - 3.464/120° = \sqrt{3}(3.464/-30°)$$

$$I_b = I_{BC} - I_{AB} = 3.464/-120° - 3.464/0° = \sqrt{3}(3.464/-150°) \tag{2.5.13}$$

$$I_c = I_{CA} - I_{BC} = 3.464/120° - 3.464/-120° = \sqrt{3}(3.464/+90°)$$

Both the Δ-load currents given by (2.5.12) and the line currents given by (2.5.13) are balanced. Thus the sum of balanced Δ-load currents $(I_{AB} + I_{BC} + I_{CA})$ equals zero. The sum of line currents $(I_a + I_b + I_c)$ is always zero for a Δ-connected load even if the system is unbalanced, since there is no neutral wire. Comparison of (2.5.12) and (2.5.13) leads to the following conclusion:

For a balanced-Δ load supplied by a balanced positive-sequence source, the line currents into the load are $\sqrt{3}$ times the Δ-load currents and lag by 30°. That is,

$$I_a = \sqrt{3}I_{AB}/-30°$$

$$I_b = \sqrt{3}I_{BC}/-30° \tag{2.5.14}$$

$$I_c = \sqrt{3}I_{CA}/-30°$$

This result is summarized in Figure 2.15.

FIGURE 2.15

Phasor diagram of line currents and load currents for a balanced-Δ load

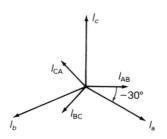

FIGURE 2.16

Δ–Y conversion for
balanced loads

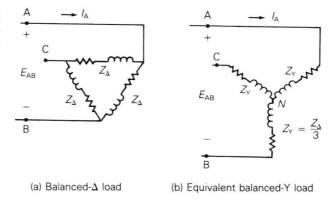

(a) Balanced-Δ load   (b) Equivalent balanced-Y load

Δ–Y CONVERSION FOR BALANCED LOADS

Figure 2.16 shows the conversion of a balanced-Δ load to a balanced-Y load. If balanced voltages are applied, then these loads will be equivalent as viewed from their terminal buses A, B, and C when the line currents into the Δ load are the same as the line currents into the Y load. For the Δ load,

$$I_A = \sqrt{3} I_{AB} \underline{/-30^\circ} = \frac{\sqrt{3} E_{AB} \underline{/-30^\circ}}{Z_\Delta} \qquad (2.5.15)$$

and for the Y load,

$$I_A = \frac{E_{AN}}{Z_Y} = \frac{E_{AB} \underline{/-30^\circ}}{\sqrt{3} Z_Y} \qquad (2.5.16)$$

Comparison of (2.5.15) and (2.5.16) indicates that I_A will be the same for both the Δ and Y loads when

$$Z_Y = \frac{Z_\Delta}{3} \qquad (2.5.17)$$

Also, the other line currents I_B and I_C into the Y load will equal those into the Δ load when $Z_Y = Z_\Delta/3$, since these loads are balanced. Thus a balanced-Δ load can be converted to an equivalent balanced-Y load by dividing the Δ-load impedance by 3. The angles of these Δ- and equivalent Y-load impedances are the same. Similarly, a balanced-Y load can be converted to an equivalent balanced-Δ load using $Z_\Delta = 3Z_Y$.

EXAMPLE 2.4   Balanced Δ and Y loads

A balanced, positive-sequence, Y-connected voltage source with $E_{ab} = 480\underline{/0^\circ}$ volts is applied to a balanced-Δ load with $Z_\Delta = 30\underline{/40^\circ}$ Ω. The line impedance between the source and load is $Z_L = 1\underline{/85^\circ}$ Ω for each phase. Calculate the line currents, the Δ-load currents, and the voltages at the load terminals.

FIGURE 2.17

Circuit diagram for
Example 2.4

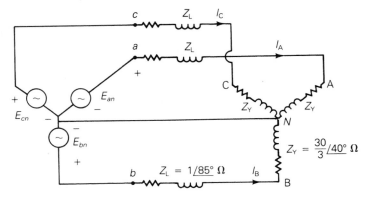

SOLUTION   The solution is most easily obtained as follows. First, convert
the Δ load to an equivalent Y. Then connect the source and Y-load neutrals
with a zero-ohm neutral wire. The connection of the neutral wire has no
effect on the circuit, since the neutral current $I_n = 0$ in a balanced system.
The resulting circuit is shown in Figure 2.17. The line currents are

$$I_A = \frac{E_{an}}{Z_L + Z_Y} = \frac{\dfrac{480}{\sqrt{3}} \underline{/-30^\circ}}{1\underline{/85^\circ} + \dfrac{30}{3}\underline{/40^\circ}}$$

$$= \frac{277.1\underline{/-30^\circ}}{(0.0872 + j0.9962) + (7.660 + j6.428)} \qquad (2.5.18)$$

$$= \frac{277.1\underline{/-30^\circ}}{(7.748 + j7.424)} = \frac{277.1\underline{/-30^\circ}}{10.73\underline{/43.78^\circ}} = 25.83\underline{/-73.78^\circ} \quad A$$

$$I_B = 25.83\underline{/166.22^\circ} \quad A$$

$$I_C = 25.83\underline{/46.22^\circ} \quad A$$

The Δ-load currents are, from (2.5.14),

$$I_{AB} = \frac{I_a}{\sqrt{3}}\underline{/+30^\circ} = \frac{25.83}{\sqrt{3}}\underline{/-73.78^\circ + 30^\circ} = 14.91\underline{/-43.78^\circ} \quad A$$

$$I_{BC} = 14.91\underline{/-163.78^\circ} \quad A \qquad (2.5.19)$$

$$I_{CA} = 14.91\underline{/+76.22^\circ} \quad A$$

The voltages at the load terminals are

$$E_{AB} = Z_\Delta I_{AB} = (30\underline{/40^\circ})(14.91\underline{/-43.78^\circ}) = 447.3\underline{/-3.78^\circ}$$

$$E_{BC} = 447.3\underline{/-123.78^\circ} \qquad (2.5.20)$$

$$E_{CA} = 447.3\underline{/116.22^\circ} \quad \text{volts}$$

∎

FIGURE 2.18

Equivalent line-to-neutral diagram for the circuit of Example 2.4

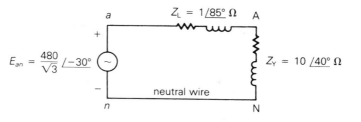

EQUIVALENT LINE-TO-NEUTRAL DIAGRAMS

When working with balanced three-phase circuits, only one phase need be analyzed. Δ loads can be converted to Y loads, and all source and load neutrals can be connected with a zero-ohm neutral wire without changing the solution. Then one phase of the circuit can be solved. The voltages and currents in the other two phases are equal in magnitude to and $\pm 120°$ out of phase with those of the solved phase. Figure 2.18 shows an equivalent line-to-neutral diagram for one phase of the circuit in Example 2.4.

When discussing three-phase systems in this text, voltages shall be rms line-to-line voltages unless otherwise indicated. This is standard industry practice.

2.6

POWER IN BALANCED THREE-PHASE CIRCUITS

In this section, we discuss instantaneous power and complex power for balanced three-phase generators and motors and for balanced-Y and Δ-impedance loads.

INSTANTANEOUS POWER: BALANCED THREE-PHASE GENERATORS

Figure 2.19 shows a Y-connected generator represented by three voltage sources with their neutrals connected at bus n and by three identical generator impedances Z_g. Assume that the generator is operating under balanced steady-state conditions with the instantaneous generator terminal voltage given by

$$v_{an}(t) = \sqrt{2}V_{LN} \cos(\omega t + \delta) \quad \text{volts} \tag{2.6.1}$$

and with the instantaneous current leaving the positive terminal of phase a given by

$$i_a(t) = \sqrt{2}I_L \cos(\omega t + \beta) \quad \text{A} \tag{2.6.2}$$

FIGURE 2.19

Y-connected generator

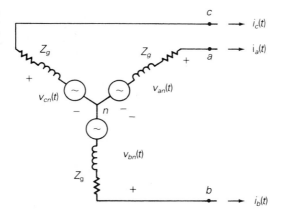

where V_{LN} is the rms line-to-neutral voltage and I_L is the rms line current. The instantaneous power $p_a(t)$ delivered by phase a of the generator is

$$p_a(t) = v_{an}(t)i_a(t)$$

$$= 2V_{LN}I_L \cos(\omega t + \delta) \cos(\omega t + \beta)$$

$$= V_{LN}I_L \cos(\delta - \beta) + V_{LN}I_L \cos(2\omega t + \delta + \beta) \quad \text{W} \qquad (2.6.3)$$

Assuming balanced operating conditions, the voltages and currents of phases b and c have the same magnitudes as those of phase a and are $\pm 120°$ out of phase with phase a. Therefore the instantaneous power delivered by phase b is

$$p_b(t) = 2V_{LN}I_L \cos(\omega t + \delta - 120°) \cos(\omega t + \beta - 120°)$$

$$= V_{LN}I_L \cos(\delta - \beta) + V_{LN}I_L \cos(2\omega t + \delta + \beta - 240°) \quad \text{W} \quad (2.6.4)$$

and by phase c,

$$p_c(t) = 2V_{LN}I_L \cos(\omega t + \delta + 120°) \cos(\omega t + \beta + 120°)$$

$$= V_{LN}I_L \cos(\delta - \beta) + V_{LN}I_L \cos(2\omega t + \delta + \beta + 240°) \quad \text{W} \quad (2.6.5)$$

The total instantaneous power $p_{3\phi}(t)$ delivered by the three-phase generator is the sum of the instantaneous powers delivered by each phase. Using (2.6.3)–(2.6.5):

$$p_{3\phi}(t) = p_a(t) + p_b(t) + p_c(t)$$

$$= 3V_{LN}I_L \cos(\delta - \beta) + V_{LN}I_L[\cos(2\omega t + \delta + \beta)$$

$$+ \cos(2\omega t + \delta + \beta - 240°)$$

$$+ \cos(2\omega t + \delta + \beta + 240°)] \quad \text{W} \qquad (2.6.6)$$

The three cosine terms within the brackets of (2.6.6) can be represented by a *balanced* set of three phasors. Therefore, the sum of these three terms is zero for any value of δ, for any value of β, and for all values of t. Equation (2.6.6) then reduces to

$$p_{3\phi}(t) = P_{3\phi} = 3V_{LN}I_L \cos(\delta - \beta) \quad W \tag{2.6.7}$$

Equation (2.6.7) can be written in terms of the line-to-line voltage V_{LL} instead of the line-to-neutral voltage V_{LN}. Under balanced operating conditions,

$$V_{LN} = V_{LL}/\sqrt{3} \quad \text{and} \quad P_{3\phi} = \sqrt{3}V_{LL}I_L \cos(\delta - \beta) \quad W \tag{2.6.8}$$

Inspection of (2.6.8) leads to the following conclusion:

> The total instantaneous power delivered by a three-phase generator under balanced operating conditions is not a function of time, but a constant, $p_{3\phi}(t) = P_{3\phi}$.

INSTANTANEOUS POWER: BALANCED THREE-PHASE MOTORS AND IMPEDANCE LOADS

The total instantaneous power absorbed by a three-phase motor under balanced steady-state conditions is also a constant. Figure 2.19 can be used to represent a three-phase motor by reversing the line currents to enter rather than leave the positive terminals. Then (2.6.1)–(2.6.8), valid for power *delivered* by a generator, are also valid for power *absorbed* by a motor. These equations are also valid for the instantaneous power absorbed by a balanced three-phase impedance load.

COMPLEX POWER: BALANCED THREE-PHASE GENERATORS

The phasor representations of the voltage and current in (2.6.1) and (2.6.2) are

$$V_{an} = V_{LN}\underline{/\delta} \quad \text{volts} \tag{2.6.9}$$
$$I_a = I_L\underline{/\beta} \quad A \tag{2.6.10}$$

where I_a leaves positive terminal "a" of the generator. The complex power S_a delivered by phase a of the generator is

$$S_a = V_{an}I_a^* = V_{LN}I_L\underline{/(\delta - \beta)}$$
$$= V_{LN}I_L \cos(\delta - \beta) + jV_{LN}I_L \sin(\delta - \beta) \tag{2.6.11}$$

Under balanced operating conditions, the complex powers delivered by phases b and c are identical to S_a, and the total complex power $S_{3\phi}$ delivered by the generator is

$$S_{3\phi} = S_a + S_b + S_c = 3S_a$$

$$= 3V_{LN}I_L \underline{/(\delta - \beta)}$$

$$= 3V_{LN}I_L \cos(\delta - \beta) + j3V_{LN}I_L \sin(\delta - \beta) \tag{2.6.12}$$

In terms of the total real and reactive powers,

$$S_{3\phi} = P_{3\phi} + jQ_{3\phi} \tag{2.6.13}$$

where

$$P_{3\phi} = \text{Re}(S_{3\phi}) = 3V_{LN}I_L \cos(\delta - \beta)$$

$$= \sqrt{3}V_{LL}I_L \cos(\delta - \beta) \quad \text{W} \tag{2.6.14}$$

and

$$Q_{3\phi} = \text{Im}(S_{3\phi}) = 3V_{LN}I_L \sin(\delta - \beta)$$

$$= \sqrt{3}V_{LL}I_L \sin(\delta - \beta) \quad \text{var} \tag{2.6.15}$$

Also, the total apparent power is

$$S_{3\phi} = |S_{3\phi}| = 3V_{LN}I_L = \sqrt{3}V_{LL}I_L \quad \text{VA} \tag{2.6.16}$$

COMPLEX POWER: BALANCED THREE-PHASE MOTORS

The preceding expressions for complex, real, reactive, and apparent power *delivered* by a three-phase generator are also valid for the complex, real, reactive, and apparent power *absorbed* by a three-phase motor.

COMPLEX POWER: BALANCED-Y AND BALANCED-Δ IMPEDANCE LOADS

Equations (2.6.13)–(2.6.16) are also valid for balanced-Y and -Δ impedance loads. For a balanced-Y load, the line-to-neutral voltage across the phase *a* load impedance and the current entering the positive terminal of that load impedance can be represented by (2.6.9) and (2.6.10). Then (2.6.11)–(2.6.16) are valid for the power absorbed by the balanced-Y load.

For a balanced-Δ load, the line-to-line voltage across the phase *a–b* load impedance and the current into the positive terminal of that load impedance can be represented by

$$V_{ab} = V_{LL}\underline{/\delta} \quad \text{volts} \tag{2.6.17}$$

$$I_{ab} = I_\Delta\underline{/\beta} \quad \text{A} \tag{2.6.18}$$

where V_{LL} is the rms line-to-line voltage and I_Δ is the rms Δ-load current. The complex power S_{ab} absorbed by the phase a–b load impedance is then

$$S_{ab} = V_{ab}I_{ab}^* = V_{LL}I_\Delta\underline{/(\delta - \beta)} \tag{2.6.19}$$

The total complex power absorbed by the Δ load is

$$S_{3\phi} = S_{ab} + S_{bc} + S_{ca} = 3S_{ab}$$
$$= 3V_{LL}I_\Delta\underline{/(\delta - \beta)}$$
$$= 3V_{LL}I_\Delta\cos(\delta - \beta) + j3V_{LL}I_\Delta\sin(\delta - \beta) \tag{2.6.20}$$

Rewriting (2.6.19) in terms of the total real and reactive power,

$$S_{3\phi} = P_{3\phi} + jQ_{3\phi} \tag{2.6.21}$$
$$P_{3\phi} = \text{Re}(S_{3\phi}) = 3V_{LL}I_\Delta\cos(\delta - \beta)$$
$$= \sqrt{3}V_{LL}I_L\cos(\delta - \beta) \quad \text{W} \tag{2.6.22}$$
$$Q_{3\phi} = \text{Im}(S_{3\phi}) = 3V_{LL}I_\Delta\sin(\delta - \beta)$$
$$= \sqrt{3}V_{LL}I_L\sin(\delta - \beta) \quad \text{var} \tag{2.6.23}$$

where the Δ-load current I_Δ is expressed in terms of the line current $I_L = \sqrt{3}I_\Delta$ in (2.6.22) and (2.6.23). Also, the total apparent power is

$$S_{3\phi} = |S_{3\phi}| = 3V_{LL}I_\Delta = \sqrt{3}V_{LL}I_L \quad \text{VA} \tag{2.6.24}$$

Equations (2.6.21)–(2.6.24) developed for the balanced-Δ load are identical to (2.6.13)–(2.6.16).

2.7

ADVANTAGES OF BALANCED THREE-PHASE VERSUS SINGLE-PHASE SYSTEMS

Figure 2.20 shows three separate single-phase systems. Each single-phase system consists of the following identical components: (1) a generator represented by a voltage source and a generator impedance Z_g; (2) a forward and return conductor represented by two series line impedances Z_L; (3) a load represented by an impedance Z_Y. The three single-phase systems, although completely separated, are drawn in a Y configuration in the figure to illustrate two advantages of three-phase systems.

Each separate single-phase system requires that *both* the forward and return conductors have a current capacity (or *ampacity*) equal to or greater than the load current. However, if the source and load neutrals in Figure 2.20 are connected to form a three-phase system, and if the source voltages are balanced with equal magnitudes and with 120° displacement between phases,

FIGURE 2.20

Three single-phase
systems

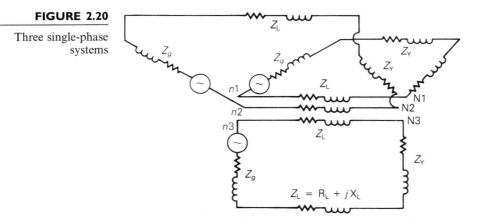

$$Z_L = R_L + jX_L$$

then the neutral current will be zero [see (2.5.10)] and the three neutral conductors can be removed. Thus, the balanced three-phase system, while delivering the same power to the three load impedances Z_Y, requires only half the number of conductors needed for the three separate single-phase systems. Also, the total I^2R line losses in the three-phase system are only half those of the three separate single-phase systems, and the line-voltage drop between the source and load in the three-phase system is half that of each single-phase system. Therefore, one advantage of balanced three-phase systems over separate single-phase systems is reduced capital and operating costs of transmission and distribution, as well as better voltage regulation.

Some three-phase systems such as Δ-connected systems and three-wire Y-connected systems do not have any neutral conductor. However, the majority of three-phase systems are four-wire Y-connected systems, where a grounded neutral conductor is used. Neutral conductors are used to reduce transient overvoltages, which can be caused by lightning strikes and by line-switching operations, and to carry unbalanced currents, which can occur during unsymmetrical short-circuit conditions. Neutral conductors for transmission lines are typically smaller in size and ampacity than the phase conductors because the neutral current is nearly zero under normal operating conditions. Thus, the cost of a neutral conductor is substantially less than that of a phase conductor. The capital and operating costs of three-phase transmission and distribution systems with or without neutral conductors are substantially less than those of separate single-phase systems.

A second advantage of three-phase systems is that the total instantaneous electric power delivered by a three-phase generator under balanced steady-state conditions is (nearly) constant, as shown in Section 2.6. A three-phase generator (constructed with its field winding on one shaft and with its three-phase windings equally displaced by 120° on the stator core) will also have a nearly constant mechanical input power under balanced steady-state conditions, since the mechanical input power equals the electrical output power plus the small generator losses. Furthermore, the mechanical shaft torque, which equals mechanical input power divided by mechanical radian frequency ($T_{mech} = P_{mech}/\omega_m$) is nearly constant.

On the other hand, the equation for the instantaneous electric power delivered by a single-phase generator under balanced steady-state conditions is the same as the instantaneous power delivered by one phase of a three-phase generator, given by $p_a(t)$ in (2.6.3). As shown in that equation, $p_a(t)$ has two components: a constant and a double-frequency sinusoid. Both the mechanical input power and the mechanical shaft torque of the single-phase generator will have corresponding double-frequency components that create shaft vibration and noise, which could cause shaft failure in large machines. Accordingly, most electric generators and motors rated 5 kVA and higher are constructed as three-phase machines in order to produce nearly constant torque and thereby minimize shaft vibration and noise.

PROBLEMS

SECTION 2.1

2.1   Given the complex numbers $A_1 = 5\underline{/60°}$ and $A_2 = -3 - j4$, (a) convert A_1 to rectangular form; (b) convert A_2 to polar and exponential form; (c) calculate $A_3 = (A_1 + A_2)$, giving your answer in polar form; (d) calculate $A_4 = A_1 A_2$, giving your answer in rectangular form; (e) calculate $A_5 = A_1/(A_2^*)$, giving your answer in exponential form.

2.2   Convert the following instantaneous currents to phasors, using $\cos(\omega t)$ as the reference. Give your answers in both rectangular and polar form.
(a) $i(t) = 400\sqrt{2}\cos(\omega t - 30°)$;
(b) $i(t) = 5\sin(\omega t + 15°)$;
(c) $i(t) = 4\cos(\omega t - 30°) + 5\sqrt{2}\sin(\omega t + 15°)$.

2.3   The instantaneous voltage across a circuit element is $v(t) = 678.8\sin(\omega t - 15°)$ volts, and the instantaneous current entering the positive terminal of the circuit element is $i(t) = 200\cos(\omega t - 5°)$ A. For both the current and voltage, determine (a) the maximum value, (b) the rms value, (c) the phasor expression, using $\cos(\omega t)$ as the reference.

2.4   For the single-phase circuit shown in Figure 2.21, $I = 10\underline{/0°}$ A. (a) Compute the phasors I_1, I_2, and V. (b) Draw a phasor diagram showing I, I_1, I_2, and V.

FIGURE 2.21

Circuit for Problem 2.4

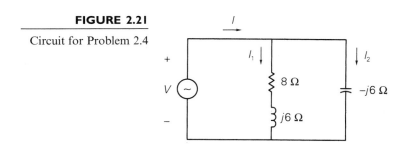

2.5 A 60-Hz, single-phase source with $V = 277\underline{/30°}$ volts is applied to a circuit element. (a) Determine the instantaneous source voltage. Also determine the phasor and instantaneous currents entering the positive terminal if the circuit element is (b) a 20-Ω resistor, (c) a 10-mH inductor, (d) a capacitor with 25-Ω reactance.

2.6 (a) Transform $v(t) = 100 \cos(377t - 30°)$ to phasor form. Comment on whether $\omega = 377$ appears in your answer. (b) Transform $V = 100\underline{/20°}$ to instantaneous form. Assume that $\omega = 377$. (c) Add the two sinusoidal functions $a(t)$ and $b(t)$ of the same frequency given as follows: $a(t) = A\sqrt{2} \cos(\omega t + \alpha)$ and $b(t) = B\sqrt{2} \cos(\omega t + \beta)$. Use phasor methods and obtain the resultant $c(t)$. Does the resultant have the same frequency?

2.7 Let a 100-V sinusoidal source be connected to a series combination of a 3-Ω resistor, an 8-Ω inductor, and a 4-Ω capacitor. (a) Draw the circuit diagram. (b) Compute the series impedance. (c) Determine the current I delivered by the source. Is the current lagging or leading the source voltage? What is the power factor of this circuit?

2.8 Consider the circuit shown in Figure 2.22 in time domain. Convert the entire circuit into phasor domain.

FIGURE 2.22

Circuit for Problem 2.8

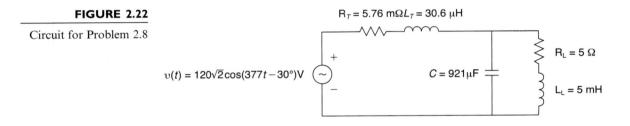

2.9 For the circuit shown in Figure 2.23, compute the voltage across the load terminals.

FIGURE 2.23

Circuit for Problem 2.9

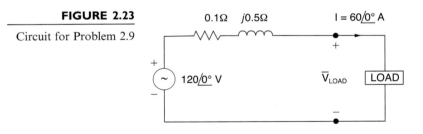

SECTION 2.2

2.10 For the circuit element of Problem 2.3, calculate (a) the instantaneous power absorbed, (b) the real power (state whether it is delivered or absorbed), (c) the reactive power (state whether delivered or absorbed), (d) the power factor (state whether lagging or leading).

[*Note*: By convention the power factor $\cos(\delta - \beta)$ is positive. If $|\delta - \beta|$ is greater than 90°, then the reference direction for current may be reversed, resulting in a positive value of $\cos(\delta - \beta)$].

2.11 Referring to Problem 2.5, determine the instantaneous power, real power, and reactive power absorbed by: (a) the 20-Ω resistor, (b) the 10-mH inductor, (c) the capacitor with 25-Ω reactance. Also determine the source power factor and state whether lagging or leading.

2.12 The voltage $v(t) = 678.8 \cos(\omega t + 45°)$ volts is applied to a load consisting of a 10-Ω resistor in parallel with a capacitive reactance $X_C = 25$ Ω. Calculate (a) the instantaneous power absorbed by the resistor, (b) the instantaneous power absorbed by the capacitor, (c) the real power absorbed by the resistor, (d) the reactive power delivered by the capacitor, (e) the load power factor.

2.13 Repeat Problem 2.12 if the resistor and capacitor are connected in series.

2.14 A single-phase source is applied to a two-terminal, passive circuit with equivalent impedance $Z = 2.0\underline{/-45°}$ Ω measured from the terminals. The source current is $i(t) = 4\sqrt{2} \cos(\omega t)$ kA. Determine the (a) instantaneous power, (b) real power, and (c) reactive power delivered by the source. (d) Also determine the source power factor.

2.15 Let a voltage source $v(t) = 4 \cos(\omega t + 60°)$ be connected to an impedance $Z = 2\underline{/30°}$ Ω. (a) Given the operating frequency to be 60 Hz, determine the expressions for the current and instantaneous power delivered by the source as functions of time. (b) Plot these functions along with $v(t)$ on a single graph for comparison. (c) Find the frequency and average value of the instantaneous power.

2.16 A single-phase, 120-V (rms), 60-Hz source supplies power to a series R-L circuit consisting of R = 10 Ω and L = 40 mH. (a) Determine the power factor of the circuit and state whether it is lagging or leading. (b) Determine the real and reactive power absorbed by the load. (c) Calculate the peak magnetic energy W_{int} stored in the inductor by using the expression $W_{int} = L(I_{rms})^2$ and check whether the reactive power $Q = \omega W$ is satisfied. (*Note*: The instantaneous magnetic energy storage fluctuates between zero and the peak energy. This energy must be sent twice each cycle to the load from the source by means of reactive power flows.)

SECTION 2.3

2.17 Consider a load impedance of $Z = j\omega L$ connected to a voltage V let the current drawn be I.
(a) Develop an expression for the reactive power Q in terms of ω, L, and I, from complex power considerations.
(b) Let the instantaneous current be $i(t) = \sqrt{2}I \cos(\omega t + \theta)$. Obtain an expression for the instantaneous power $p(t)$ into L, and then express it in terms of Q.
(c) Comment on the average real power P supplied to the inductor and the instantaneous power supplied.

2.18 Let a series R-L-C network be connected to a source voltage V, drawing a current I.
(a) In terms of the load impedance $Z = Z < Z$, find expressions for P and Q, from complex power considerations.
(b) Express $p(t)$ in terms of P and Q, by choosing $i(t) = \sqrt{2}I \cos \omega t$.
(c) For the case of $Z = R + j\omega L + 1/j\omega c$, interpret the result of part (b) in terms of P, Q_L, and Q_C. In particular, if $\omega^2 LC = 1$, when the inductive and capacitive reactances cancel, comment on what happens.

2.19 Consider a single-phase load with an applied voltage $v(t) = 150\sqrt{2}\cos(\omega t + 10°)$ volts and load current $i(t) = 5\sqrt{2}\cos(\omega t - 50°)$ A. (a) Determine the power triangle. (b) Find the power factor and specify whether it is lagging or leading. (c) Calculate the reactive power supplied by capacitors in parallel with the load that correct the power factor to 0.9 lagging.

2.20 A circuit consists of two impedances, $Z_1 = 20\underline{/30°}$ Ω and $Z_2 = 14.14\underline{/-45°}$ Ω, in parallel, supplied by a source voltage $V = 100\underline{/60°}$ volts. Determine the power triangle for each of the impedances and for the source.

2.21 An industrial plant consisting primarily of induction motor loads absorbs 1000 kW at 0.7 power factor lagging. (a) Compute the required kVA rating of a shunt capacitor to improve the power factor to 0.9 lagging. (b) Calculate the resulting power factor if a synchronous motor rated 1000 hp with 90% efficiency operating at rated load and at unity power factor is added to the plant instead of the capacitor. Assume constant voltage. (1 hp = 0.746 kW)

2.22 The real power delivered by a source to two impedances, $Z_1 = 3 + j5$ Ω and $Z_2 = 10$ Ω, connected in parallel, is 2000 W. Determine (a) the real power absorbed by each of the impedances and (b) the source current.

2.23 A single-phase source has a terminal voltage $V = 120\underline{/0°}$ volts and a current $I = 25\underline{/30°}$ A, which leaves the positive terminal of the source. Determine the real and reactive power, and state whether the source is delivering or absorbing each.

2.24 A source supplies power to the following three loads connected in parallel: (1) a lighting load drawing 10 kW, (2) an induction motor drawing 10 kVA at 0.90 power factor lagging, and (3) a synchronous motor operating at 10 hp, 85% efficiency and 0.95 power factor leading (1 hp = 0.746 kW). Determine the real, reactive, and apparent power delivered by the source. Also, draw the source power triangle.

2.25 Consider the series R-L-C circuit of Problem 2.7 and calculate the complex power absorbed by each of the elements R, L, and C, as well as the complex power absorbed by the total load. Draw the resultant power triangle. Check whether the complex power delivered by the source equals the total complex power absorbed by the load.

2.26 A small manufacturing plant is located 2 km down a transmission line, which has a series reactance of 0.5 Ω/km. The line resistance is negligible. The line voltage at the plant is $480\underline{/0°}$ V (rms), and the plant consumes 120 kW at 0.85 power factor lagging. Determine the voltage and power factor at the sending end of the transmission line by using (a) a complex power approach and (b) a circuit analysis approach.

2.27 An industrial load consisting of a bank of induction motors consumes 50 kW at a power factor of 0.8 lagging from a 220-V, 60-Hz, single-phase source. By placing a bank of capacitors in parallel with the load, the resultant power factor is to be raised to 0.95 lagging. Find the net capacitance of the capacitor bank in μF that is required.

2.28 Three loads are connected in parallel across a single-phase source voltage of 240 V (RMS).
Load 1 absorbs 12 kW and 6.667 kVAR;
Load 2 absorbs 4 kVA at 0.96PF leading;
Load 3 absorbs 15 kW at unity power factor.
Calculate the equivalent impedance, Z, for the three parallel loads, for two cases:
(i) Series combination of R and X, and (ii) parallel combination of R and X.

2.29 Modeling the transmission lines as inductors, with $Sij = S_{ji}^*$,
Compute S_{13}, S_{31}, S_{23}, S_{32}, and S_{G3}, in Figure 2.24. (*Hint*: complex power balance
holds good at each bus, statisfying KCL.)

FIGURE 2.24

System diagram for
Problem 2.29

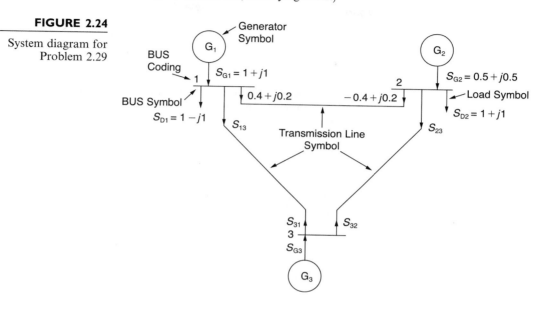

2.30 Figure 2.25 shows three loads connected in parallel across a 1000-V (RMS), 60-Hz
single-phase source.
Load 1: Inductive load, 125 kVA, 0.28PF lagging
Load 2: Capacitive load, 10 kW, 40 kVAR
Load 3: Resistive load, 15 kW
(a) Determine the total kW, kvar, kva, and supply power factor.
(b) In order to improve the power factor to 0.8 lagging, a capacitor of negligible resistance is connected in parallel with the above loads. Find the KVAR rating of that capacitor and the capacitance in μf.
Comment on the magnitude of the supply current after adding the capacitor.

FIGURE 2.25

Circuit for Problem 2.30

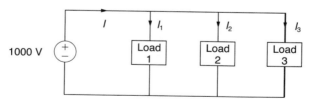

2.31 Consider two interconnected voltage sources connected by a line of impedance
$Z = jx \, \Omega$, as shown in Figure 2.26.
(a) Obtain expressions for P_{12} and Q_{12}.
(b) Determine the maximum power transfer and the condition for it to occur.

FIGURE 2.26

Circuit for Problem 2.31

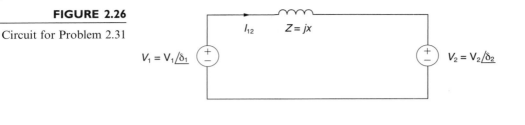

$V_1 = V_1\underline{/\delta_1}$ I_{12} $Z = jx$ $V_2 = V_2\underline{/\delta_2}$

PW   **2.32**   In PowerWorld Simulator Problem 2.32 (see Figure 2.27) a 10 MW/5 Mvar load is supplied at 20 kV through a feeder with an impedance of $1.5 + j3\ \Omega$. The load is compensated with a capacitor whose output, Q_{cap}, can be varied in 0.5 Mvar steps between 0 and 10.0 Mvars. What value of Q_{cap} minimizes the real power line losses? What value of Q_{cap} minimizes the MVA power flow into the feeder?

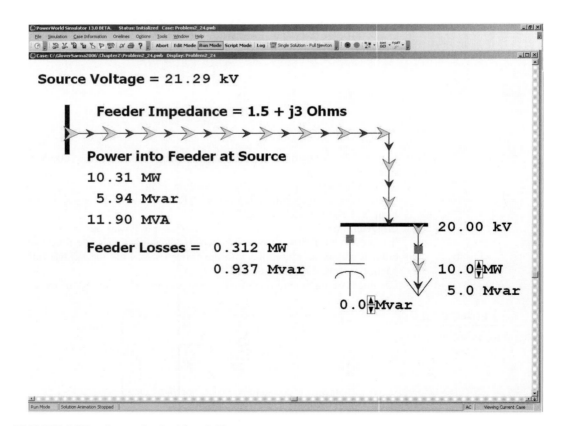

FIGURE 2.27   Screen for Problem 2.32

PW   **2.33**   For the system from Problem 2.32, plot the real and reactive line losses as Q_{cap} is varied between 0 and 10.0 Mvars.

PW   **2.34**   For the system from Problem 2.32, assume that half the time the load is 10 MW/5 Mvar, and for the other half it is 20 MW/10 Mvar. What single value of Q_{cap} would minimize the average losses? Assume that Q_{cap} can only be varied in 0.5 Mvar steps.

SECTION 2.4

2.35   For the circuit shown in Figure 2.28, convert the voltage sources to equivalent current sources and write nodal equations in matrix format using bus 0 as the reference bus. Do not solve the equations.

FIGURE 2.28
_____

Circuit diagram for
Problems 2.35 and 2.36

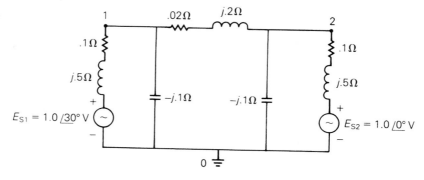

2.36   For the circuit shown in Figure 2.28, write a computer program that uses the sources, impedances, and bus connections as input data to (a) compute the 2×2 bus admittance matrix $\mathbf{Y}$, (b) convert the voltage sources to current sources and compute the vector of source currents into buses 1 and 2.

2.37   Determine the 4×4 bus admittance matrix and write nodal equations in matrix format for the circuit shown in Figure 2.29. Do not solve the equations.

FIGURE 2.29
_____

Circuit for Problem 2.37

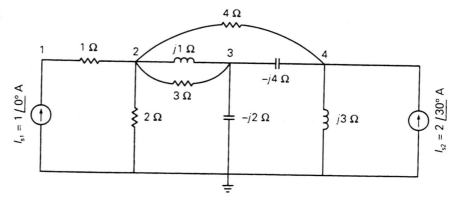

2.38   Given the impedance diagram of a simple system as shown in Figure 2.30, draw the admittance diagram for the system and develop the 4×4 bus admittance matrix $\mathbf{Y}_{bus}$ by inspection.

FIGURE 2.30

System diagram for
Problem 2.38

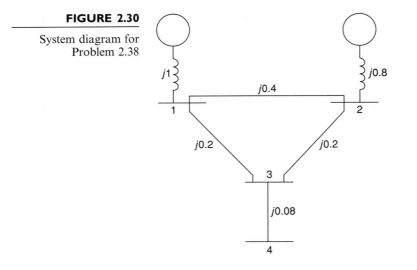

2.39 (a) Given the circuit diagram in Figure 2.31 showing admittances and current sources at nodes 3 and 4, set up the nodal equations in matrix format. (b) If the parameters are given by: $Y_a = -j0.8$ S, $Y_b = -j4.0$ S, $Y_c = -j4.0$ S, $Y_d = -j8.0$ S, $Y_e = -j5.0$ S, $Y_f = -j2.5$ S, $Y_g = -j0.8$ S, $I_3 = 1.0\underline{/-90°}$ A, and $I_4 = 0.62\underline{/-135°}$ A, set up the nodal equations and suggest how you would go about solving for the voltages at the nodes.

FIGURE 2.31

Circuit diagram for
Problem 2.39

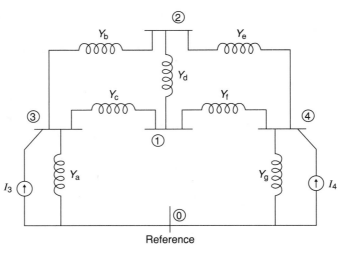

SECTIONS 2.5 AND 2.6

2.40 A balanced three-phase 208-V source supplies a balanced three-phase load. If the line current I_A is measured to be 10 A and is in phase with the line-to-line voltage V_{BC}, find the per-phase load impedance if the load is (a) Y-connected, (b) Δ-connected.

2.41 A three-phase 25-kVA, 208-V, 60-Hz alternator, operating under balanced steady-state conditions, supplies a line current of 20 A per phase at a 0.8 lagging power factor and at rated voltage. Determine the power triangle for this operating condition.

2.42 A balanced Δ-connected impedance load with $(12 + j9)$ Ω per phase is supplied by a balanced three-phase 60-Hz, 208-V source. (a) Calculate the line current, the total real and reactive power absorbed by the load, the load power factor, and the apparent load power. (b) Sketch a phasor diagram showing the line currents, the line-to-line source voltages, and the Δ-load currents. Assume positive sequence and use V_{ab} as the reference.

2.43 A three-phase line, which has an impedance of $(2 + j4)$ Ω per phase, feeds two balanced three-phase loads that are connected in parallel. One of the loads is Y-connected with an impedance of $(30 + j40)$ Ω per phase, and the other is Δ-connected with an impedance of $(60 - j45)$ Ω per phase. The line is energized at the sending end from a 60-Hz, three-phase, balanced voltage source of $120\sqrt{3}$ V (rms, line-to-line). Determine (a) the current, real power, and reactive power delivered by the sending-end source; (b) the line-to-line voltage at the load; (c) the current per phase in each load; and (d) the total three-phase real and reactive powers absorbed by each load and by the line. Check that the total three-phase complex power delivered by the source equals the total three-phase power absorbed by the line and loads.

2.44 Two balanced three-phase loads that are connected in parallel are fed by a three-phase line having a series impedance of $(0.4 + j2.7)$ Ω per phase. One of the loads absorbs 560 kVA at 0.707 power factor lagging, and the other 132 kW at unity power factor. The line-to-line voltage at the load end of the line is $2200\sqrt{3}$ V. Compute (a) the line-to-line voltage at the source end of the line, (b) the total real and reactive power losses in the three-phase line, and (c) the total three-phase real and reactive power supplied at the sending end of the line. Check that the total three-phase complex power delivered by the source equals the total three-phase complex power absorbed by the line and loads.

2.45 Two balanced Y-connected loads, one drawing 10 kW at 0.8 power factor lagging and the other 15 kW at 0.9 power factor leading, are connected in parallel and supplied by a balanced three-phase Y-connected, 480-V source. (a) Determine the source current. (b) If the load neutrals are connected to the source neutral by a zero-ohm neutral wire through an ammeter, what will the ammeter read?

2.46 Three identical impedances $Z_\Delta = 20\underline{/60°}$ Ω are connected in Δ to a balanced three-phase 480-V source by three identical line conductors with impedance $Z_L = (0.8 + j0.6)$ Ω per line. (a) Calculate the line-to-line voltage at the load terminals. (b) Repeat part (a) when a Δ-connected capacitor bank with reactance $(-j20)$ Ω per phase is connected in parallel with the load.

2.47 Two three-phase generators supply a three-phase load through separate three-phase lines. The load absorbs 30 kW at 0.8 power factor lagging. The line impedance is $(1.4 + j1.6)$ Ω per phase between generator G1 and the load, and $(0.8 + j1)$ Ω per phase between generator G2 and the load. If generator G1 supplies 15 kW at 0.8 power factor lagging, with a terminal voltage of 460 V line-to-line, determine (a) the voltage at the load terminals, (b) the voltage at the terminals of generator G2, and (c) the real and reactive power supplied by generator G2. Assume balanced operation.

2.48 Two balanced Y-connected loads in parallel, one drawing 15 kW at 0.6 power factor lagging and the other drawing 10 kVA at 0.8 power factor leading, are supplied by a balanced, three-phase, 480-volt source. (a) Draw the power triangle for each load and for the combined load. (b) Determine the power factor of the combined load and state whether lagging or leading. (c) Determine the magnitude of the line current from the source. (d) Δ-connected capacitors are now installed in parallel with the combined load. What value of capacitive reactance is needed in each leg of the Δ to make the source power factor unity? Give your answer in Ω. (e) Compute the magnitude of the current in each capacitor and the line current from the source.

2.49 Figure 2.32 gives the general Δ–Y transformation. (a) Show that the general transformation reduces to that given in Figure 2.16 for a balanced three-phase load. (b) Determine the impedances of the equivalent Y for the following Δ impedances: $Z_{AB} = j10$, $Z_{BC} = j20$, and $Z_{CA} = -j25\ \Omega$.

FIGURE 2.32

General Δ–Y
transformation

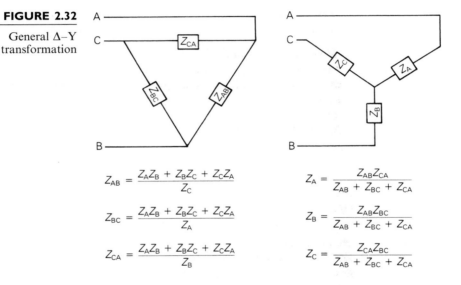

$$Z_{AB} = \frac{Z_A Z_B + Z_B Z_C + Z_C Z_A}{Z_C}$$

$$Z_{BC} = \frac{Z_A Z_B + Z_B Z_C + Z_C Z_A}{Z_A}$$

$$Z_{CA} = \frac{Z_A Z_B + Z_B Z_C + Z_C Z_A}{Z_B}$$

$$Z_A = \frac{Z_{AB} Z_{CA}}{Z_{AB} + Z_{BC} + Z_{CA}}$$

$$Z_B = \frac{Z_{AB} Z_{BC}}{Z_{AB} + Z_{BC} + Z_{CA}}$$

$$Z_C = \frac{Z_{CA} Z_{BC}}{Z_{AB} + Z_{BC} + Z_{CA}}$$

2.50 Consider the balanced three-phase system shown in Figure 2.33. Determine $v_1(t)$ and $i_2(t)$. Assume positive phase sequence.

FIGURE 2.33

Circuit for Problem 2.50

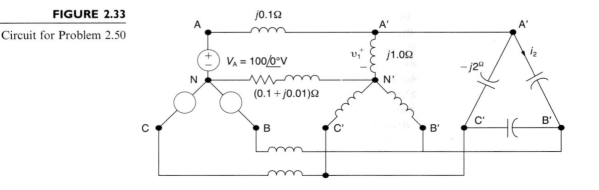

2.51   A three-phase line with an impedance of $(0.2 + j1.0)\Omega$/phase feeds three balanced three-phase loads connected in parallel.
Load 1: Absorbs a total of 150 kW and 120 kVAR; Load 2: Delta connected with an impedance of $(150 - j48)\Omega$/phase; Load 3: 120 kVA at 0.6 PF leading. If the line-to-neutral voltage at the load end of the line is 2000 v (RMS), determine the magnitude of the line-to-line voltage at the source end of the line.

2.52   A balanced three-phase load is connected to a 4.16-kV, three-phase, four-wire, grounded-wye dedicated distribution feeder. The load can be modeled by an impedance of $Z_L = (4.7 + j9)\Omega$/phase, wye-connected. The empedance of the phase conductors is $(0.3 + j1)\Omega$. Determine the following by using the phase A to neutral voltage as a reference and assume positive phase sequence:
(a) Line currents for phases A, B, and C.
(b) Line-to-neutral voltages for all three phases at the load.
(c) Apparent, active, and reactive power dissipated per phase, and for all three phases in the load.
(d) Active power losses per phase and for all three phases in the phase conductors.

REFERENCES

1.   W. H. Hayt, Jr., and J. E. Kemmerly, *Engineering Circuit Analysis*, 7ᵗʰ ed. (New York: McGraw-Hill, 2006).

2.   W. A. Blackwell and L. L. Grigsby, *Introductory Network Theory* (Boston: PWS, 1985).

3.   A. E. Fitzgerald, D. E. Higginbotham, and A. Grabel, *Basic Electrical Engineering* (New York: McGraw-Hill, 1981).

4.   W. D. Stevenson, Jr., *Elements of Power System Analysis*, 4th ed. (New York: McGraw-Hill, 1982).

5.   H. Püttgen, P. MacGregor, and F. Lambert, "Distributed Generation: Semantic Hype or the Dawn of a New Era," *IEEE Power & Energy Magazine*, *1*, 1 (January/February 2003), pp. 22–29.

Core and coil assemblies of a three-phase 20.3 $kV\Delta/345$ kVY step-up transformer. This oil-immersed transformer is rated 325 MVA self-cooled (OA)/542 MVA forced oil, forced air-cooled (FOA)/607 MVA forced oil, forced air-cooled (FOA) (Courtesy of General Electric)

3

POWER TRANSFORMERS

The power transformer is a major power system component that permits economical power transmission with high efficiency and low series-voltage drops. Since electric power is proportional to the product of voltage and current, low current levels (and therefore low I^2R losses and low IZ voltage drops) can be maintained for given power levels via high voltages. Power transformers transform ac voltage and current to optimum levels for generation, transmission, distribution, and utilization of electric power.

The development in 1885 by William Stanley of a commercially practical transformer was what made ac power systems more attractive than dc power systems. The ac system with a transformer overcame voltage problems encountered in dc systems as load levels and transmission distances increased. Today's modern power transformers have nearly 100% efficiency, with ratings up to and beyond 1300 MVA.

In this chapter, we review basic transformer theory and develop equivalent circuits for practical transformers operating under sinusoidal-steady-state conditions. We look at models of single-phase two-winding, three-phase two-winding, and three-phase three-winding transformers, as well as autotransformers and regulating transformers. Also, the per-unit system, which simplifies power system analysis by eliminating the ideal transformer winding in transformer equivalent circuits, is introduced in this chapter and used throughout the remainder of the text.

CASE STUDY   The following article describes how electric utilities are focusing on extending the life of power transformers and paper-insulated-lead-covered cables. The life of power transformers can be extended by continuously removing aging and degradation products from the transformer oil as they are formed. Transformer health can be established through online monitoring, tests, inspections, and knowledge of the transformer's age, known problems and operating history [8].

Life Extension and Condition Assessment

NICK DOMINELLI, AVARAL RAO, AND PRABHA KUNDUR

Utilities rely heavily on their infrastructure to provide reliable power supply to customers. Some of the infrastructure in North America and Europe is getting old, and hence its reliability is seriously compromised. A large proportion of the assets are nearing their designed end of life or have already exceeded it. In some utilities, more than 40% of the substation transformers are more than 40 years old. Some of the underground distribution paper-insulated lead-covered (PILC) medium-voltage cables remain in service even though they have surpassed their design life. The replacement of components of older infrastructure will be prolonged as utilities are faced with budget and staff constraints.

The current trend is to optimize asset utilization by operating existing equipment at ever-higher capacity levels, in many cases, exceeding nameplate ratings in order to defer capital investment in new facilities or in the refurbishment of existing facilities. Thus, many utilities have focused their efforts on developing methods to assess the condition of equipment in order to extend the life of existing infrastructure. In many cases, utilities are focusing their efforts on extending the life of power transformers and PILC cables because these components offer the greatest payback.

Extending the life of transformers and PILC cables will be possible only if utilities can assess their current condition or implement continuous monitoring measures. Over the last ten years, we have carried out many projects to address these issues. We report here some of these activities.

LIFE EXTENSION OF TRANSFORMERS

Continuous online oil purification
The insulation system of high-voltage power transformers consists of oil, paper, and other cellulose-

("Life Extension and Condition Assessment" by Nick Dominelli, Avaral Rao, and Prabha Kundur, IEEE Power & Energy (May/June 2006) pg. 24–35.)

based solids. Throughout its service life, the insulation is subjected to numerous stresses, including high temperatures, vibration, electric fields, and exposure to moisture, oxygen, acids, and other chemical contaminants.

As a result, over time a gradual loss in mechanical and dielectric properties will eventually compromise the unit's reliability. The degradation is primarily a chemical process that is substantially accelerated by heat and the presence of oxygen and moisture. Moisture is particularly detrimental to paper, as it will initiate hydrolysis and scission of the cellulose chain. Oxygen attacks both paper and oil, producing acids and other polar compounds that promote further degradation. Ideally, these materials should be removed continuously so that they do not accumulate to the point where they can cause irreversible damage to the paper. In addition, continuous removal promotes the migration of adsorbed products from the paper to the oil, making the process more effective than traditional periodic reclamation.

It is generally accepted that the useful life of transformers is determined by the residual strength of the cellulose insulation and not by its oil properties. This is because the oil can be reconditioned by conventional treatment or simply replaced, but the paper degrades through an irreversible process of depolymerization. Therefore, by maintaining the oil at near-new conditions at all times, utilities can significantly prolong transformer life and considerably reduce the risk of dielectric breakdown, which is blamed for many of the extra-high-voltage power transformer failures.

Laboratory studies

Following extensive laboratory work, we found that a combination of selective adsorbents and hollow fiber modules was effective in removing a multitude of harmful products, including moisture, acids, polar oxidation products, oil and paper degradation products, oxygen, particles, and dissolved metals. Based on these results, a multipurpose online oil purification unit was built and tested. This consisted of particulate filters and adsorbent bed cartridges to remove acids, contaminants, and degradation products. Oxygen was removed by using a hollow fiber

membrane module that allowed dissolved gases to permeate through the walls of the membrane and into the housing. If the gases and other products are continuously removed from the shell side by applying a vacuum or a stripping gas, equilibrium will never be achieved and the oil can be completely degassed.

Two approaches were investigated to remove moisture from transformer oil. The first used a desiccant and was aimed at making treated oil virtually moisture free. The second used a hollow fiber membrane module, which was already being used for degassing. In this case, the moisture removal from the transformer oil per pass would be only partial (up to 50%).

Figure 1 shows the multipurpose online continuous oil conditioning system prototype. The system was instrumented with gauges, thermocouples, and temperature and pressure switches, and plumbed

Figure 1
A multipurpose continuous online oil purification unit prototype.

TABLE 1   Test results of the multipurpose online continuous oil conditioning system prototype.

Volume Treated (Liters)	H_2O ppm	IFT dyn/cm	NN mgKOH/g	PF %	Breakdown kV	Col	DBPC %	Cu ppm	T_{in} °C
Incoming Oil	3	36.8	0.014	0.935	32	1	0.22	0.07	51.9
110.7	<1	43.2	<0.1	0.042	57	1	0.23	0.001	51.9
575	<1	41.5	<0.1	0.079	62	1	0.23	0.01	41.5
799	<1	42.1	<0.1	0.051	60.9	1	—	0.009	55.5
1018	<1	41.6	<0.01	0.06	—	1	—	—	47.2

with bypass and diverting valves to allow the system to operate in the degassing/adsorption mode or only in the adsorption mode.

We evaluated this prototype unit in our high-voltage lab using a 500-kV power transformer. The properties of the oil at the outlet of the conditioning unit, as a function of treated oil volume, are shown in Table 1. The results show that the properties of the treated oil are equal to or better than those of new oil. It is significant to notice that the oxidation inhibitor (DBPC) content is not depleted.

Development and evaluation of dedicated field units

Simpler or single-purpose units were developed and built for field application. Several of these prototype units have been developed and undergone extensive lab evaluation. These are stand-alone units designed for online unattended operation. They are equipped with fail-safe features and can be installed on an energized transformer in a few hours. The units are mounted on light two- or four-wheeled carts for portability in the substation. These are briefly described below.

- *Decontamination/dehydration:* This unit is capable of continuously removing particles, moisture, oil, and paper degradation products and oxidation precursors from transformer oil.
- *Dehydration:* This unit is capable of continuously removing moisture and particulates from transformer oil and is well suited for transformer dry-outs and to keep them dry at all times.
- *Degassing/dehydration:* This unit is capable of continuously removing dissolved gases, moisture, and particulates from transformer oil.

The first two units have been installed in the field (Figure 2) and have undergone testing. The results are shown in Table 2.

We are currently conducting accelerated aging tests with custom-built equipment to simulate power transformer conditions. These tests will quantify the technical and economic benefits of online continuous transformer oil purification. We are also extending the technique to remove coking precursors from arced load tap changer (LTC) oils. This will

Figure 2
A dehydration unit (left) and decontamination system (right) installed on 60-kV field transformers.

TABLE 2 Field results for the decontamination system installed on the 60-kV transformer.

Oil Volume Treated	IFT dyn/cm	PF %	NN mg/KOH	Breakdown kV	Polar Compounds
30	20.5	0.89	0.07	37	2,050
12,917*	29.9	0.28	0.01	56	1,197
31,620	34.2	0.07	0.01	51	323
219,806*	35.5	0.07	—	61	201
487,802*	38.7	0.05	—	62	102
506,897	41.8	0.06	—	49	59
New Oil	39.5	<0.1	<0.01	45–55	90–100

*adsorbent cartridge replaced.

help prevent premature failures due to coking and increase the maintenance interval for LTCs.

EQUIPMENT HEALTH RATING OF POWER TRANSFORMERS

Utility operation and planning personnel need tools to evaluate the state of their equipment in order to decide whether to reuse, refurbish, or replace it. We have developed a computer program to assist transformer specialists to assess transformers' health consistently and reproducibly. The approach and logic of the program are designed to emulate that of the expert.

The program automatically calculates and updates a transformer's health rating (EHRt) based on information from tests, inspections, operating history, known problems, age, etc. A specialist can review and edit all the ratings at any time and then provide diagnostics and prescriptions to the program. The software then generates technical prescriptions that include an EHRt rating, required intervention, work duration, cost, technical benefits, and consequences. The program can also provide an overview of transformer health ratings for a selected station or substation or for the entire company. A brief description of its functions follows.

Database
The program uses data from a database composed of the following categories: nameplate, known problems, age and operating/maintenance history, laboratory tests (DGA, moisture in oil, furans and phenols, dielectric breakdown, IFT, DP of paper),

field tests (Doble, Core megger, winding insulation, winding resistance), internal inspections (leads, winding, insulation, core, yoke, etc.), and external inspection of all accessories (bushings, OLTC, DETC, tank, etc.).

Condition indicators
The program divides the transformer into two categories: its main components (or indispensable parts), consisting of core and coil, tank, de-energized tap changer, on-load tap changer, and accessories (replaceable parts), consisting of bushings, oil, radiator/heat exchanger, oil preservation system, pumps, fans, and gauges. These are collectively known as *maintainable items*.

Each maintainable item has a series of tests (ti) and inspections associated with it for which the program calculates a condition indicator C_i (ti). These are combined to obtain separate indicators for the main components C_i (main) and the accessories C_i (acc.), which are then combined to give the production unit condition indicator C_i (pu). A condition indicator is also derived from each of the other data categories, including age and operating history C_i (age) and known problems C_i (kp). The condition indicator for known problems is derived from a list of known, documented conditions that are expected to shorten the transformer's lifespan. The C_i (age) is further refined by taking into account the transformer's loading and maintenance history. These condition indicators are then further combined to arrive at a final EHRt rating. The process is shown in Figure 3.

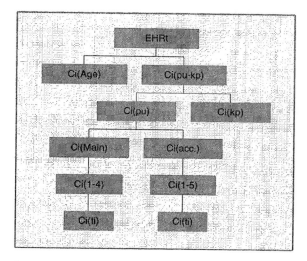

Figure 3
A diagram showing the hierarchy for combining condition indicators.

Figure 4
A top-level screen for EHR showing diagnostics and prescriptions.

The program assigns four levels of EHRt values as illustrated in Table 3.

Display

The results are displayed at four levels. The top menu bar has buttons that allow the user to view details about specific equipment, to see summaries of test results or histograms of EHRt for a station or company, to view the technical prescriptions, and to enter data or edit tables and lists of Ci values, thresholds, etc. The other windows show: equipment information; EHRt ratings, along with color bars for the values; diagnostics and prescription; the evaluator and date; and comments. By selecting the condition indicator, the user can drill down to the second level screen, while the equipment selection is used to select specific equipment or all equipment at a location. A top-level display screen is shown in Figure 4.

The second level shows the condition indicators for all main components and accessories from tests, inspections, and age and known problems, as shown in Figure 5. All Cis are displayed in color-coded bars and as numerical values. Each component, along with known problems and age, is linked to the third-level menu. If there are no results to display, the bar charts will be blank. Double-clicking any of the underlined maintainable items will bring up the third-level screen for the selected equipment.

TABLE 3   Transformer EHRt rating levels and definitions.

Code	Name	Rating	Color	Definition
G	Good	100–75	Green	No noticeable deterioration or defects
F	Fair	74–50	Yellow	Some deterioration or defects function not affected
P	Poor	49–20	Orange	Serious deterioration or defects in at least some portions of the asset
U	Unsatisfactory	<20	Red	Extensive deterioration or defects no longer functions as designed

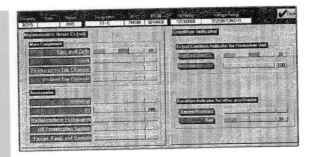

Figure 5
A second-level screen for EHR showing condition indicators.

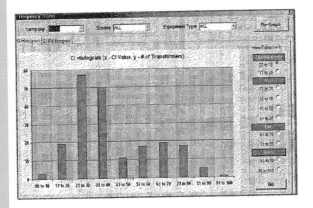

Figure 6
A histogram of condition indicators for a substation.

The third level shows the condition indicators for each test Ci (ti) associated with each component as well as diagnosis and prescriptions for that component. The fourth-level screen shows the actual test results and the associated condition indicator values, along with a selection of diagnostics and prescriptions to choose from. An important feature of the program is to provide global views of equipment health ratings. Figure 6 is a histogram of the EHRt rating distribution for an entire station.

Review process
The review process normally starts at the lowest-level screen and works its way up to the top screen.

During the review process, the specialist can also assign diagnostic conditions and prescriptions for each test or inspection result. These are combined and carried forward to the higher-level screen all the way up to the top-level screen. For each prescription, the expert may also complete a technical prescription sheet to provide managers with estimates of repair costs and work duration and information on the possible consequences of not doing or delaying the work.

THE USE OF CHEMICAL TRACERS TO DETERMINE THE EXTENT OF WEAR IN LTCS

Most power transformers use LTCs to regulate the voltage output. These are mechanical devices and prone to wear. It has been estimated that a third of all transformer failures can be traced to faulty LTCs.

Failure due to excessive contact wear may be averted if the extent of wear can be monitored.

Principle and approach
Two approaches were considered for detecting excessive contact wear. Both approaches involve embedding a tracer material at a predetermined depth into the contact. When the contact wears down sufficiently, the tracer material becomes exposed. In the first approach, the tracer material is released into the fluid, where it can be detected by periodic off-line analysis. The second approach, better suited for online monitoring of arcing contacts, is to embed a dissimilar tracer metal in the contacts. The spectrum of the light emitted from the arc depends on the contact material. When the tracer metal is exposed to the arc, it will emit light at discrete wavelengths that can be detected by a specific sensor. This is the principle for online monitoring and is illustrated in Figures 7 and 8. The light emitted during arcing is filtered before being detected by a photodiode or photo multiplier. The optical filter is selected to allow only discrete wavelengths emitted by the tracer metal. In Figure 7, no signal is detected since the tracer material is not exposed, whereas in Figure 8, a signal is detected, indicating that the tracer material has been exposed.

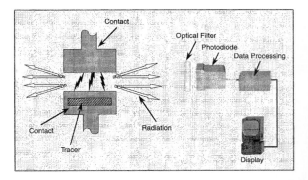

Figure 7
Online contact wear monitoring device (tracer metal not exposed).

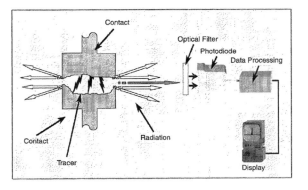

Figure 8
Online contact wear monitoring device (tracer metal exposed).

Selection of suitable tracer materials

Based on preliminary laboratory work, we selected several candidate tracer materials for further evaluation for off-line monitoring. These included magnesium metal, perfluorocarbon liquids, and nanocrystal materials that fluoresce at a discreet wavelength. For online monitoring, lithium metal was selected since it has a strong emission bands in the red region of the spectrum (510 nm and 680 nm).

Laboratory evaluation with an LTC

A three-phase load tap changer (Federal Pioneer 25 kV) was set up to mimic field conditions as they relate to arcing contacts.

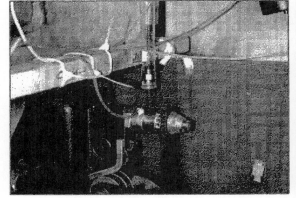

Figure 9
Fiber-optic setup in the LTC for online monitoring.

TABLE 4 Custom contacts designed to hold tracer compounds.

Name	Tracer Material	Rating
Mg	Magnesium metal	Stationary contact
Fluoro	Perfluoro	Stationary contact
Litihm	Lithium metal	Stationary contact
Nano	Nanocrystals	Moving contact
Bromo	Bromo compound	Moving contact
Teflon	PTFE-Teflon	Moving contact

The switching rate was set at five times per minute. A fiber-optic cable was installed adjacent to the contact with the lithium implant (Figure 9). This was connected to a UV/VIS (ultraviolet/visible) spectrometer and a laptop computer for online monitoring.

As part of our study, custom contacts were designed with cavities to hold the tracer compounds. These contacts have a relatively thin wear layer to reduce the time that it takes for the tracer compound to show up in the oil. They were designated as shown in Table 4. The contacts were installed in the LTC (three moving and three stationary).

Testing and monitoring program

Following the collection of initial oil samples, the LTC was energized, and the accelerated switching

Figure 10
Arcing contacts after 69,000 switching operations.

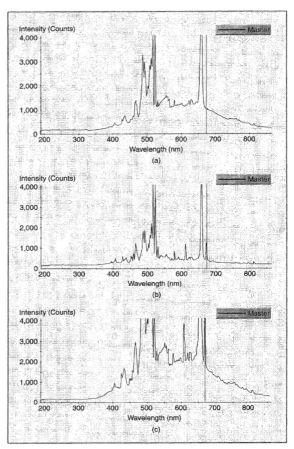

Figure 11
Emission spectrum of arcing contact (a) at start (b) after approximately 80,000 switches, and (c) as in (b) with a more intense arc.

program commenced. The switching and sampling program continued for 69,000 switch operations. At that point, the LTC was de-energized, the oil drained, and the contacts examined. A picture of the worn-out contacts is shown in Figure 10, which shows that three of the six contacts had been breached. These included the two containing organic liquid tracers and the one containing nanocrystals. The three breached contacts were replaced with two conventional contacts and one with nanocrystals, the oil was returned to the LTC, and the switching program resumed.

Analysis of the oil for the organic tracer compounds clearly revealed their presence, indicating that the contacts had been breached after about 60,000 switching operations. The same result was obtained for the nanocrystals.

Online monitoring of the arcing contact with the lithium tracer indicated that it had been breached after the switching program was resumed. Figure 11 shows typical emission spectra of the contact at discreet intervals of the switching program. The characteristic emission line monitored at 671 nm is marked with a green line. We can see that no emission is present before the contact was breached [Figure 11(a)]. However, it is clearly present in Figure 11(b) and (c), presumably after breaching.

Field trials and beta testing program
We are currently carrying out field tests at a substation in our system using a voltage regulator that has been equipped with custom-designed contacts suitable for both off-line and online monitoring. The unit is set up for accelerated switching in order to breach the contacts in a short interval.

In addition, we are participating in an EPRI-sponsored program to test the Informer Series contacts in order to validate product reliability under field service conditions. This program is still in progress, and the results will be reported at a later date.

CONDITION ASSESSMENT OF PILC CABLES

The integrity of the PILC cables used in the underground distribution system depends on the condition of its paper insulation and lead sheath. During the operation of PILC cables, heating occurs as current is transmitted through the conductor. Consequently, the temperature of the insulation and lead sheathing slowly increases. Since the loading is cyclical, the PILC cables are exposed to heating and cooling cycles. The rise in temperature and thermal cycling can introduce cracks on the lead sheath of PILC cables (Figure 12). From 2000–2003, Powertech developed many diagnostic tests to assess the condition of both the oil-paper insulation system and the lead sheath. We explored electrical and chemical diagnostic techniques to assess the condition of oil-paper insulation and developed nondestructive testing (NDT) methods to evaluate the condition of the lead sheath.

The lead sheath of medium-voltage PILC cables is perhaps the most critical factor in determining the integrity of the cable, since most cable failures are directly related to lead sheath damage. This damage usually occurs in the form of cracks due to the synergistic effects of creep and fatigue or of surface damage from mechanical abrasion and corrosion. Once a crack occurs on the lead sheath, moisture from the surrounding underground environment invariably penetrates into the oil-paper insulation,

degrading its dielectric strength and subsequently leading to cable failure. Nearly half of the damage on the lead sheath of PILC cables develops in the cable chambers (manholes). If the condition of the lead sheath can be detected before cracks occur, then measures can be taken to avoid such failures. Since PILC cables are easily accessible in manholes, inspections of the lead sheath can be performed there using NDT methods.

The lead sheath of PILC cables is commonly visually inspected for damage. This technique is quick and inexpensive, but it cannot detect defects that are invisible to the naked eye. Hence, there is a need to identify better NDT methods to inspect the lead sheath of PILC cables. NDT methods such as dye penetrant inspection, eddy current inspection, and in situ metallographic examination are used to examine the surface of the nonferrous metallic components. We have successfully applied these NDT methods to examine the surface of the lead sheath of PILC cables. These three techniques are explained next.

Liquid dye penetrant inspection

Liquid dye penetrant inspection is a nondestructive testing method used to detect surface breaking defects in any nonporous material. Liquid dye penetrant is applied to the surface and drawn into surface defects by capillary action. Once a preset dwell time has passed, excess liquid is removed, and a developer is applied to draw out the penetrant from the defects. A visual inspection is then performed. The liquid dye may be either visible or fluorescent. If a visible dye is used, cracks or surface defects show up as red visible indications; if a fluorescent dye is used, surface defects are visible under ultraviolet (UV) light. Figure 13 illustrates the liquid dye penetrant inspection procedure.

The fluorescent dye penetrant inspection method has worked well in detecting surface defects on the lead sheath. A surface crack 1-mm deep on the lead sheath of a PILC cable sample was intentionally created, and the surface of the lead sheath was examined using the fluorescent dye penetrant inspection method. The defect is clearly visible under ultraviolet (UV) light (Figure 14).

Figure 12
Cracked lead sheath of a PILC cable.

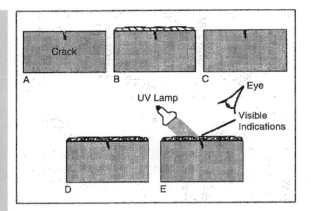

Figure 13
A = surface crack; B = liquid dye penetrant is applied on the surface; C = excess dye is wiped off, leaving dye in the crack; D = developer is applied on the surface; and E = cracks show up as colored indications (red under ambient light or fluorescent indications under ultraviolet light).

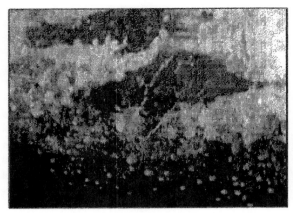

Figure 15
The detection of cracks on the lead sheath of PILC cable through florescent dye penetrant inspection.

Figure 14
An artificial crack (1-mm deep) on the lead sheath of a PILC cable sample.

Since the lead sheath of the PILC cable is accessible in manholes, the fluorescent dye inspection method can be easily adopted for field inspections. The British Columbia Hydro distribution system introduced this technique to assess the condition of the lead sheath of old distribution PILC cables. One

of the findings of that examination is shown in Figure 15. The bright lines correspond to cracks on the surface of the lead sheath. These are very fine cracks that are not visible to the naked eye. Thus, the fluorescent dye penetrant examination method is an excellent live-line inspection tool to assess the condition of the lead sheath of PILC cables.

Eddy current inspection
Eddy current inspection is a nondestructive testing method based on the principle of electromagnetic induction of metallic materials to detect surface and near-surface discontinuities. When alternating current (ac) current flows in a coil (probe coil) close to the surface of a conducting material, an eddy current is induced on its surface. These currents flow in closed loops. The magnitude and phase of the eddy current affect the impedance of the coil. A surface crack immediately underneath the coil interrupts the eddy current, which in turn increases the impedance of the coil. This phenomenon makes it possible to monitor the voltage across the coil and, thus, to detect surface cracks on the conducting material. Changes in the eddy current are displayed on an impedance phase diagram in a cathode ray tube (CRT). Changes caused by the instrument operator, such as the distance between the probe coil and test piece, will cause a horizontal shift (lift-off)

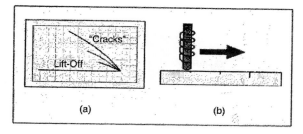

Figure 16
(a) Display on CRT when (b) moving the eddy-current probe over a series of simulated cracks of varying depths.

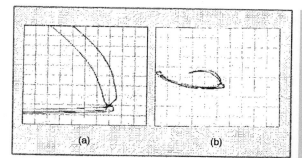

Figure 17
(a) Impedance phase diagram of 3-mm-deep crack on the lead sheath. (b) Impedance phase diagram of a 3-mm-deep crack on the lead sheath with a plastic cover.

in the spot forming the trace, while the presence of any cracks or flaws will cause the spot to move vertically. The output of the trace is shown in Figure 16.

The most important part of the eddy current inspection is the selection of its probe coil. The shape of the probe coil largely depends on the purpose of the inspection and the shape of the part being inspected. A "pencil probe" is a single-coil surface probe, which is commonly used for the inspection of surface defects.

Powertech used an eddy current instrument with a pencil probe to develop a technique to inspect the lead sheaths. Initial experiments were carried out on sheaths of small samples in the laboratory to optimize the probe. After many iterations, we found that a simple transformer-type differential probe was best suited to inspect the surface condition of the lead sheaths of PILC cables. A good signal-to-noise response was obtained between 50–500 kHz frequencies. A calibration sample of the PILC cable was made to optimize the eddy current inspection technique. Using a precision knife used for woodcarving and model-building crafts, three cracklike incisions (1-, 2-, and 3-mm deep) were made on the surface of the lead sheath. The impedance phase diagram of one of the cracks on the lead sheath is shown in Figure 17(a). To find the range of the eddy current inspection technique, the same cable sample was inspected after covering the cracked area with a plastic jacket. Figure 17(b) shows the impedance phase diagram of the crack on the lead sheath with

the jacket. This observation clearly indicates that the eddy current inspection can detect cracks on the lead sheath of PILC cables with or without the plastic jacket.

Currently, efforts are underway to assess the use of this technique for live-line inspections of the lead sheath of PILC cables in the field.

In situ metallographic examination
Metallurgical microstructure, such as grain size, plays an important role in the life of the lead sheath of PILC cable. The grain size of the lead sheath can be assessed through metallographic examination. In situ metallographic examination, in accordance with the ASTM (American Society for Testing and Materials) E 1351 standard, has been extensively used to evaluate the metallographic features of metallic components. This is a nondestructive sampling procedure that records and preserves the microstructure as a negative relief on a plastic film.

We have adopted this technique, originally proposed by CIGRE Working Group 21.05, to assess the condition of the lead sheath of aging PILC cables through grain size measurements. The main purpose of the examination is to determine whether the lead sheath has undergone substantial grain growth during service. The grain size of the lead sheath is known to play an important role in resisting the synergistic effects of creep and fatigue damage.

Some experiments of in situ metallographic

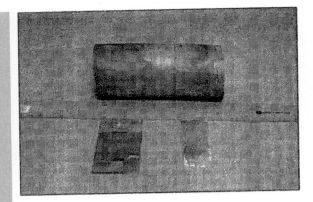

Figure 18
Lead sheath sample: replicating film and replica mounted on the glass plate.

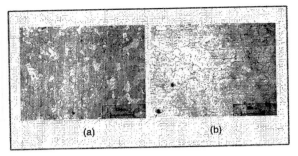

(a)   (b)

Figure 19
(a) The microstructure of the lead sheath from the conventional method. (b) The microstructure Of the lead sheath from the replica.

examination were carried out on cable samples removed from the field. The surface of the lead sheath, approximately 12 mm × 18 mm, was cleaned using an industrial solvent such as acetone. Subsequently, the cleaned surface was hand-ground and polished to obtain a mirror-finish surface. The polished surface was then etched using cotton wool soaked with a freshly prepared solution containing 75 mm glacial acetic acid and 25 ml hydrogen peroxide. The etching solution was allowed to remain on the surface for 20–30 s, after which the surface was cleaned thoroughly with distilled water to remove any residue left by the etching solution and then dried using isopropyl alcohol. Subsequently, a replica of the etched surface was taken using cellulose acetate replicating film and acetone. The replica was permanently mounted on a glass slide using adhesive tape, and the replicating surface was coated with a thin layer of gold palladium in a vacuum coating unit. Figure 18 shows the prepared cable sample, along with the replicating film and replica.

The microstructure obtained from this replica was compared with the actual microstructure obtained through the conventional metallographic technique. Figure 19(a) and (b) shows these two microstructures and demonstrates that the in situ metallographic technique can replicate the microstructure of the lead sheath.

Figure 20
Microstructure of the lead sheath of a cable using the in situ metallographic technique.

Recently, this technique has been used to assess the condition of the lead sheath of a cable in the field. The clarity of the microstructure obtained from the replica, shown in Figure 20, appears to be very good. Hence, it is feasible to use the in situ metallographic technique to assess the condition of the lead sheath of PILC cables. Hence, this technique can be successfully used in the field to assess the grain size and condition of the lead sheath of PILC cables.

CONCLUSIONS

Extending the life of aging infrastructure requires the use of noninvasive condition assessment and online preventive techniques that can be easily implemented in the field. The condition of in-service transformers and the lead sheath of aging underground PILC cables can be assessed using innovative inspection and monitoring techniques.

The useful life of power transformers can be extended by continuously removing aging and degradation products from the oil as they are formed. The health of transformers can be established with the help of custom software that uses a combination of oil and field tests, inspections, age, operating conditions, and known problems. The erosion of the LTC arc tip of the transformer can be easily monitored off-line through the use of chemical tracers or online by embedding lithium in the contacts and examining the emitted arc light.

The lead sheath of aging PILC cables can be assessed using conventional nondestructive inspection techniques. Surface cracks or defects on the lead sheath of PILC cables can be detected using the fluorescent dye penetrant inspection method. The surface of the lead sheath of PILC cables can also be examined using an eddy current inspection method. The microstructure of the lead sheath can be assessed using the in situ metallographic (replica) inspection technique.

Such condition monitoring, assessment, and life extension techniques will help utilities decide whether an aging transformer or underground PILC distribution cable can be operated reliably beyond its design life.

FOR FURTHER READING

S. Kovacevic, N. Dominelli, and B. Ward, "Transformer life extension by continuous on-line oil conditioning," presented at the IEEE International Symposium on Electrical Insulation (ISEI), EIC/EMCW Conference, Indianapolis, IN, Sep. 2003.

N. Dominelli, B. Nichols, and B. Ward, "Development and evaluation of chemical tracers for LTC wear indicators," presented at the EPRI Substation Equipment Diagnostics Conference, New Orleans, LA, Feb. 2004.

N. Dominelli, M. Lau, D. Olan, and J. Newell, "Equipment health rating of power transformers," presented at the 2004 IEEE International Symposium on Electrical Insulation, Indianapolis, IN, Sep. 2004.

"Assessment of paper-insulated lead-covered cable condition—electrical, chemical and metallurgical condition," EPRI, Palo Alto, CA, EPRI Rep. 1001724, Mar. 2003.

CIGRE WG21.05, "Diagnostic methods for HV paper cables and accessories," *Electra*, no. 176, pp. 25–51, Feb. 1998.

Standard Practice for Production and Evaluation of Filed Metallographic Replicas, ASTM Standard E1351-01.

V. Buchholz, M. Colwell, J. P. Crine, A. Rao, S. Cherukupalli, and B. S. Bernstein, "Condition assessment of distribution PILC cables," presented at the 2001 IEEE/PES Transmission and Distribution Conference, Atlanta, GA, Oct.–Nov. 2001.

BIOGRAPHIES

Nick Dominelli is the director of applied chemistry with Powertech Labs Inc., a research and technology company located in Surrey, British Columbia, Canada. He is currently involved in research and development in the areas of transformer diagnostics, online transformer oil monitoring and conditioning, insulating paper degradation, aging properties of insulating oils, sulfur hexafluoride gas, and PCB destruction and decontamination. He is a member of the ASTM (American Society for Testing and Materials) D23 Subcommittee D27-03 on Electrical Insulating Fluids.

Avaral Rao is the director of the materials engineering business unit at Powertech Labs Inc. He holds a Ph.D. in metallurgical engineering from the University of British Columbia, Canada, and his primary field of expertise is nondestructive testing and evaluation, failure analysis, welding technology, fracture mechanics, and cavitation erosion. He has authored 33 publications on these subjects. He is a registered professional engineer in the province of British Columbia, Canada, and is a member of the American Society for Non-destructive Testing, the American Welding Society, and the American Society for Testing and Materials.

Prabha Kundur holds a Ph.D. in electrical engineering from the University of Toronto, Canada, and has over 33 years of experience in the electric power industry. He is currently the president and chief executive officer of Powertech Labs Inc. He has also served as adjunct professor at the University of Toronto since 1979 and at the University of British Columbia since 1994.

Kundur has a long record of service and leadership in the IEEE Power Engineering Society (PES). He has chaired numerous committee and working groups of the PES and was elected a Fellow of the IEEE in 1985. He is the immediate past-chair of the IEEE Power System Dynamic Performance Committee and is currently the PES vice president for Education/Industry Relation Activities. He is also active in CIGRE and is currently the chair of the CIGRE Study Committee C4 on System Technical Performance. He is the recipient of the 1997 IEEE Nikola Tesla Award, the 1999 CIGRE Technical Committee Award, and the 2005 IEEE PES Charles Concordia Power System Engineering Award. He has been awarded two honorary degrees: Doctor Honoris Causa by the University Politechnica of Bucharest, Romania, in 2003, and Doctor of Engineering Honoris Causa by the University of Waterloo, Canada, in 2004.

3.1

THE IDEAL TRANSFORMER

Figure 3.1 shows a basic single-phase two-winding transformer, where the two windings are wrapped around a magnetic core [1, 2, 3]. It is assumed here that the transformer is operating under sinusoidal-steady-state excitation. Shown in the figure are the phasor voltages E_1 and E_2 across the windings, and the phasor currents I_1 entering winding 1, which has N_1 turns, and I_2 leaving winding 2, which has N_2 turns. A phasor flux Φ_c set up in the core and a magnetic field intensity phasor H_c are also shown. The core has a

FIGURE 3.1

Basic single-phase two-winding transformer

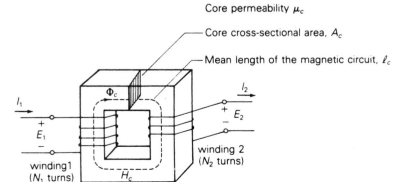

FIGURE 3.2

Schematic representation
of a single-phase two-
winding transformer

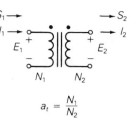

$$a_t = \frac{N_1}{N_2}$$

cross-sectional area denoted A_c, a mean length of the magnetic circuit l_c, and a magnetic permeability μ_c, assumed constant.

For an ideal transformer, the following are assumed:

1. The windings have zero resistance; therefore, the I^2R losses in the windings are zero.

2. The core permeability μ_c is infinite, which corresponds to zero core reluctance.

3. There is no leakage flux; that is, the entire flux Φ_c is confined to the core and links both windings.

4. There are no core losses.

A schematic representation of a two-winding transformer is shown in Figure 3.2. Ampere's and Faraday's laws can be used along with the preceding assumptions to derive the ideal transformer relationships. Ampere's law states that the tangential component of the magnetic field intensity vector integrated along a closed path equals the net current enclosed by that path; that is,

$$\oint H_{\tan} \, dl = I_{\text{enclosed}} \tag{3.1.1}$$

If the core center line shown in Figure 3.1 is selected as the closed path, and if H_c is constant along the path as well as tangent to the path, then (3.1.1) becomes

$$H_c l_c = N_1 I_1 - N_2 I_2 \tag{3.1.2}$$

Note that the current I_1 is enclosed N_1 times and I_2 is enclosed N_2 times, one time for each turn of the coils. Also, using the right-hand rule*, current I_1 contributes to clockwise flux but current I_2 contributes to counterclockwise flux. Thus, in (3.1.2) the net current enclosed is $N_1 I_1 - N_2 I_2$. For constant core permeability μ_c, the magnetic flux density B_c within the core, also constant, is

---

*The right-hand rule for a coil is as follows: Wrap the fingers of your right hand around the coil in the direction of the current. Your right thumb then points in the direction of the flux.

$$B_c = \mu_c H_c \quad \text{Wb/m}^2 \tag{3.1.3}$$

and the core flux Φ_c is

$$\Phi_c = B_c A_c \quad \text{Wb} \tag{3.1.4}$$

Using (3.1.3) and (3.1.4) in (3.1.2) yields

$$N_1 I_1 - N_2 I_2 = l_c B_c / \mu_c = \left(\frac{l_c}{\mu_c A_c}\right) \Phi_c \tag{3.1.5}$$

We define core reluctance R_c as

$$R_c = \frac{l_c}{\mu_c A_c} \tag{3.1.6}$$

Then (3.1.5) becomes

$$N_1 I_1 - N_2 I_2 = R_c \Phi_c \tag{3.1.7}$$

Equation (3.1.7) can be called "Ohm's law" for the magnetic circuit, wherein the net magnetomotive force mmf $= N_1 I_1 - N_2 I_2$ equals the product of the core reluctance R_c and the core flux Φ_c. Reluctance R_c, which impedes the establishment of flux in a magnetic circuit, is analogous to resistance in an electric circuit. For an ideal transformer, μ_c is assumed infinite, which, from (3.1.6), means that R_c is zero, and (3.1.7) becomes

$$N_1 I_1 = N_2 I_2 \tag{3.1.8}$$

In practice, power transformer windings and cores are contained within enclosures, and the winding directions are not visible. One way of conveying winding information is to place a dot at one end of each winding such that when current enters a winding at the dot, it produces an mmf acting in the *same* direction. This dot convention is shown in the schematic of Figure 3.2. The dots are conventionally called *polarity marks*.

Equation (3.1.8) is written for current I_1 *entering* its dotted terminal and current I_2 *leaving* its dotted terminal. As such, I_1 and I_2 are *in phase*, since $I_1 = (N_2/N_1)I_2$. If the direction chosen for I_2 were reversed, such that both currents entered their dotted terminals, then I_1 would be 180° *out of phase* with I_2.

Faraday's law states that the voltage $e(t)$ induced across an N-turn winding by a time-varying flux $\phi(t)$ linking the winding is

$$e(t) = N \frac{d\phi(t)}{dt} \tag{3.1.9}$$

Assuming a sinusoidal-steady-state flux with constant frequency ω, and representing $e(t)$ and $\phi(t)$ by their phasors E and Φ, (3.1.9) becomes

$$E = N(j\omega)\Phi \tag{3.1.10}$$

For an ideal transformer, the entire flux is assumed to be confined to the core, linking both windings. From Faraday's law, the induced voltages across the windings of Figure 3.1 are

$$E_1 = N_1(j\omega)\Phi_c \tag{3.1.11}$$

$$E_2 = N_2(j\omega)\Phi_c \tag{3.1.12}$$

Dividing (3.1.11) by (3.1.12) yields

$$\frac{E_1}{E_2} = \frac{N_1}{N_2} \tag{3.1.13}$$

or

$$\frac{E_1}{N_1} = \frac{E_2}{N_2} \tag{3.1.14}$$

The dots shown in Figure 3.2 indicate that the voltages E_1 and E_2, both of which have their $+$ polarities at the dotted terminals, are in phase. If the polarity chosen for one of the voltages in Figure 3.1 were reversed, then E_1 would be $180°$ out of phase with E_2.

The turns ratio a_t is defined as follows:

$$a_t = \frac{N_1}{N_2} \tag{3.1.15}$$

Using a_t in (3.1.8) and (3.1.14), the basic relations for an ideal single-phase two-winding transformer are

$$E_1 = \left(\frac{N_1}{N_2}\right)E_2 = a_t E_2 \tag{3.1.16}$$

$$I_1 = \left(\frac{N_2}{N_1}\right)I_2 = \frac{I_2}{a_t} \tag{3.1.17}$$

Two additional relations concerning complex power and impedance can be derived from (3.1.16) and (3.1.17) as follows. The complex power entering winding 1 in Figure 3.2 is

$$S_1 = E_1 I_1^* \tag{3.1.18}$$

Using (3.1.16) and (3.1.17),

$$S_1 = E_1 I_1^* = (a_t E_2)\left(\frac{I_2}{a_t}\right)^* = E_2 I_2^* = S_2 \tag{3.1.19}$$

As shown by (3.1.19), the complex power S_1 entering winding 1 equals the complex power S_2 leaving winding 2. That is, an ideal transformer has no real or reactive power loss.

If an impedance Z_2 is connected across winding 2 of the ideal transformer in Figure 3.2, then

$$Z_2 = \frac{E_2}{I_2} \tag{3.1.20}$$

This impedance, when measured from winding 1, is

$$Z_2' = \frac{E_1}{I_1} = \frac{a_t E_2}{I_2/a_t} = a_t^2 Z_2 = \left(\frac{N_1}{N_2}\right)^2 Z_2 \qquad (3.1.21)$$

Thus, the impedance Z_2 connected to winding 2 is referred to winding 1 by multiplying Z_2 by a_t^2, the square of the turns ratio.

EXAMPLE 3.1   Ideal, single-phase two-winding transformer

A single-phase two-winding transformer is rated 20 kVA, 480/120 V, 60 Hz. A source connected to the 480-V winding supplies an impedance load connected to the 120-V winding. The load absorbs 15 kVA at 0.8 p.f. lagging when the load voltage is 118 V. Assume that the transformer is ideal and calculate the following:

   a. The voltage across the 480-V winding.

   b. The load impedance.

   c. The load impedance referred to the 480-V winding.

   d. The real and reactive power supplied to the 480-V winding.

SOLUTION

   a. The circuit is shown in Figure 3.3, where winding 1 denotes the 480-V winding and winding 2 denotes the 120-V winding. Selecting the load voltage E_2 as the reference,

$$E_2 = 118\underline{/0^\circ} \text{ V}$$

The turns ratio is, from (3.1.13),

$$a_t = \frac{N_1}{N_2} = \frac{E_{1\text{rated}}}{E_{2\text{rated}}} = \frac{480}{120} = 4$$

and the voltage across winding 1 is

$$E_1 = a_t E_2 = 4(118\underline{/0^\circ}) = 472\underline{/0^\circ} \text{ V}$$

FIGURE 3.3

$S_1 \rightarrow$

$I_1 \rightarrow$

$\rightarrow S_2 = 15{,}000 \underline{/36.87^\circ}$ VA

$\rightarrow I_2 = 127.12 \underline{/{-}36.87^\circ}$ A

E_1

Z_2 $E_2 = 118 \underline{/0^\circ}$ V

$N_1 \quad N_2$

$a_t = \dfrac{N_1}{N_2} = 4$

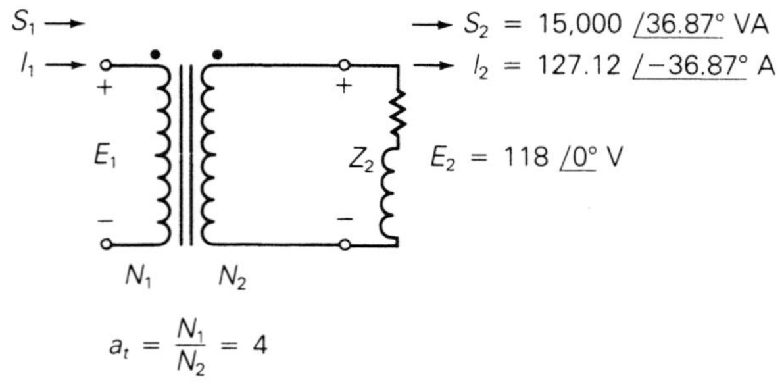

b. The complex power S_2 absorbed by the load is

$$S_2 = E_2 I_2^* = 118 I_2^* = 15,000\underline{/\cos^{-1}(0.8)} = 15,000\underline{/36.87°} \quad \text{VA}$$

Solving, the load current I_2 is

$$I_2 = 127.12\underline{/-36.87°} \quad \text{A}$$

The load impedance Z_2 is

$$Z_2 = \frac{E_2}{I_2} = \frac{118\underline{/0°}}{127.12\underline{/-36.87°}} = 0.9283\underline{/36.87°} \quad \Omega$$

c. From (3.1.21), the load impedance referred to the 480-V winding is

$$Z_2' = a_t^2 Z_2 = (4)^2 (0.9283\underline{/36.87°}) = 14.85\underline{/36.87°} \quad \Omega$$

d. From (3.1.19)

$$S_1 = S_2 = 15,000\underline{/36.87°} = 12,000 + j9000$$

Thus, the real and reactive powers supplied to the 480-V winding are

$$P_1 = \text{Re } S_1 = 12,000 \text{ W} = 12 \text{ kW}$$

$$Q_1 = \text{Im } S_1 = 9000 \text{ var} = 9 \text{ kvar} \qquad \blacksquare$$

Figure 3.4 shows a schematic of a conceptual single-phase, phase-shifting transformer. This transformer is not an idealization of an actual transformer since it is physically impossible to obtain a complex turns ratio. It will be used later in this chapter as a mathematical model for representing phase shift of three-phase transformers. As shown in Figure 3.4, the complex turns ratio a_t is defined for the phase-shifting transformer as

$$a_t = \frac{e^{j\phi}}{1} = e^{j\phi} \tag{3.1.22}$$

FIGURE 3.4

Schematic representation of a conceptual single-phase, phase-shifting transformer

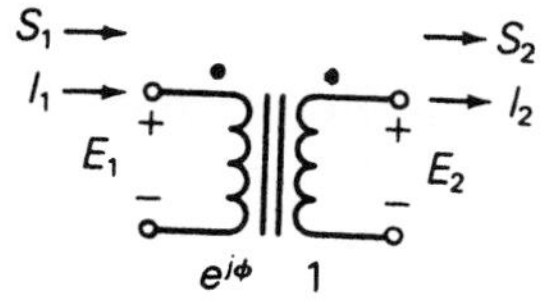

$$a_t = e^{j\phi}$$

$$E_1 = a_t E_2 = e^{j\phi} E_2$$

$$I_1 = \frac{I_2}{a_t^*} = e^{j\phi} I_2$$

$$S_1 = S_2$$

$$Z_2' = Z_2$$

where ϕ is the phase-shift angle. The transformer relations are then

$$E_1 = a_t E_2 = e^{j\phi} E_2 \tag{3.1.23}$$

$$I_1 = \frac{I_2}{a_t^*} = e^{j\phi} I_2 \tag{3.1.24}$$

Note that the phase angle of E_1 leads the phase angle of E_2 by ϕ. Similarly, I_1 leads I_2 by the angle ϕ. However, the magnitudes are unchanged; that is, $|E_1| = |E_2|$ and $|I_1| = |I_2|$.

From these two relations, the following two additional relations are derived:

$$S_1 = E_1 I_1^* = (a_t E_2)\left(\frac{I_2}{a_t^*}\right)^* = E_2 I_2^* = S_2 \tag{3.1.25}$$

$$Z_2' = \frac{E_1}{I_1} = \frac{a_t E_2}{\dfrac{1}{a_t^*} I_2} = |a_t|^2 Z_2 = Z_2 \tag{3.1.26}$$

Thus, impedance is unchanged when it is referred from one side of an ideal phase-shifting transformer to the other. Also, the ideal phase-shifting transformer has no real or reactive power losses since $S_1 = S_2$.

Note that (3.1.23) and (3.1.24) for the phase-shifting transformer are the same as (3.1.16) and (3.1.17) for the ideal physical transformer except for the complex conjugate (*) in (3.1.24). The complex conjugate for the phase-shifting transformer is required to make $S_1 = S_2$ (complex power into winding 1 equals complex power out of winding 2), as shown in (3.1.25).

3.2

EQUIVALENT CIRCUITS FOR PRACTICAL TRANSFORMERS

Figure 3.5 shows an equivalent circuit for a practical single-phase two-winding transformer, which differs from the ideal transformer as follows:

FIGURE 3.5

Equivalent circuit of a practical single-phase two-winding transformer

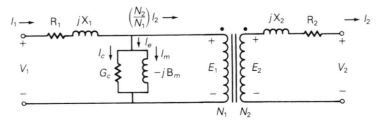

1. The windings have resistance.

2. The core permeability μ_c is finite.

3. The magnetic flux is not entirely confined to the core.

4. There are real and reactive power losses in the core.

The resistance R_1 is included in series with winding 1 of the figure to account for I^2R losses in this winding. A reactance X_1, called the leakage reactance of winding 1, is also included in series with winding 1 to account for the leakage flux of winding 1. This leakage flux is the component of the flux that links winding 1 but does not link winding 2; it causes a voltage drop $I_1(jX_1)$, which is proportional to I_1 and leads I_1 by 90°. There is also a reactive power loss $I_1^2 X_1$ associated with this leakage reactance. Similarly, there is a resistance R_2 and a leakage reactance X_2 in series with winding 2.

Equation (3.1.7) shows that for finite core permeability μ_c, the total mmf is not zero. Dividing (3.1.7) by N_1 and using (3.1.11), we get

$$I_1 - \left(\frac{N_2}{N_1}\right)I_2 = \frac{R_c}{N_1}\Phi_c = \frac{R_c}{N_1}\left(\frac{E_1}{j\omega N_1}\right) = -j\left(\frac{R_c}{\omega N_1^2}\right)E_1 \qquad (3.2.1)$$

Defining the term on the right-hand side of (3.2.1) to be I_m, called *magnetizing* current, it is evident that I_m lags E_1 by 90°, and can be represented by a shunt inductor with susceptance $B_m = \left(\dfrac{R_c}{\omega N_1^2}\right)$ mhos.* However, in reality there is an additional shunt branch, represented by a resistor with conductance G_c mhos, which carries a current I_c, called the *core loss* current. I_c is in phase with E_1. When the core loss current I_c is included, (3.2.1) becomes

$$I_1 - \left(\frac{N_2}{N_1}\right)I_2 = I_c + I_m = (G_c - jB_m)E_1 \qquad (3.2.2)$$

The equivalent circuit of Figure 3.5, which includes the shunt branch with admittance $(G_c - jB_m)$ mhos, satisfies the KCL equation (3.2.2). Note that when winding 2 is open ($I_2 = 0$) and when a sinusoidal voltage V_1 is applied to winding 1, then (3.2.2) indicates that the current I_1 will have two components: the core loss current I_c and the magnetizing current I_m. Associated with I_c is a real power loss $I_c^2/G_c = E_1^2 G_c$ W. This real power loss accounts for both hysteresis and eddy current losses within the core. Hysteresis loss occurs because a cyclic variation of flux within the core requires energy dissipated as heat. As such, hysteresis loss can be reduced by the use of special high grades of alloy steel as core material. Eddy current loss occurs because induced currents called eddy currents flow within the magnetic core perpendicular to the flux. As such, eddy current loss can be reduced by con-

---

* The units of admittance, conductance, and susceptance, which in the SI system are siemens (with symbol S), are also called mhos (with symbol ℧) or ohms^{-1} (with symbol Ω^{-1}).

FIGURE 3.6

Equivalent circuits for a
practical single-phase
two-winding transformer

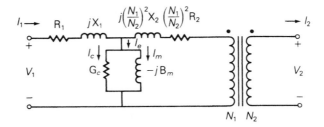

(a) R_2 and X_2 are referred to winding 1

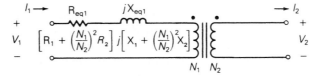

(b) Neglecting exciting current

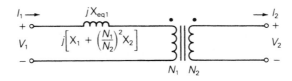

(c) Neglecting exciting current and I^2R winding loss

structing the core with laminated sheets of alloy steel. Associated with I_m is a reactive power loss $I_m^2/B_m = E_1^2 B_m$ var. This reactive power is required to magnetize the core. The phasor sum $(I_c + I_m)$ is called the *exciting* current I_e.

Figure 3.6 shows three alternative equivalent circuits for a practical single-phase two-winding transformer. In Figure 3.6(a), the resistance R_2 and leakage reactance X_2 of winding 2 are referred to winding 1 via (3.1.21). In Figure 3.6(b), the shunt branch is omitted, which corresponds to neglecting the exciting current. Since the exciting current is usually less than 5% of rated current, neglecting it in power system studies is often valid unless transformer efficiency or exciting current phenomena are of particular concern. For large power transformers rated more than 500 kVA, the winding resistances, which are small compared to the leakage reactances, can often be neglected, as shown in Figure 3.6(c).

Thus, a practical transformer operating in sinusoidal steady state is equivalent to an ideal transformer with external impedance and admittance branches, as shown in Figure 3.6. The external branches can be evaluated from short-circuit and open-circuit tests, as illustrated by the following example.

EXAMPLE 3.2   Transformer short-circuit and open-circuit tests

A single-phase two-winding transformer is rated 20 kVA, 480/120 volts, 60 Hz. During a short-circuit test, where rated current at rated frequency is applied to the 480-volt winding (denoted winding 1), with the 120-volt winding (winding 2) shorted, the following readings are obtained: $V_1 = 35$ volts, $P_1 = 300$ W. During an open-circuit test, where rated voltage is applied to winding 2, with winding 1 open, the following readings are obtained: $I_2 = 12$ A, $P_2 = 200$ W.

 a. From the short-circuit test, determine the equivalent series impedance $Z_{eq1} = R_{eq1} + jX_{eq1}$ referred to winding 1. Neglect the shunt admittance.

 b. From the open-circuit test, determine the shunt admittance $Y_m = G_c - jB_m$ referred to winding 1. Neglect the series impedance.

SOLUTION

 a. The equivalent circuit for the short-circuit test is shown in Figure 3.7(a), where the shunt admittance branch is neglected. Rated current for winding 1 is

$$I_{1rated} = \frac{S_{rated}}{V_{1rated}} = \frac{20 \times 10^3}{480} = 41.667 \quad A$$

R_{eq1}, Z_{eq1}, and X_{eq1} are then determined as follows:

$$R_{eq1} = \frac{P_1}{I_{1rated}^2} = \frac{300}{(41.667)^2} = 0.1728 \quad \Omega$$

FIGURE 3.7

Circuits for Example 3.2

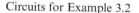

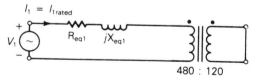

480 : 120

(a) Short-circuit test (neglecting shunt admittance)

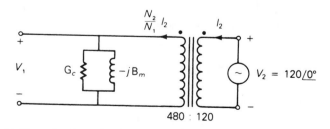

480 : 120

(b) Open-circuit test (neglecting series impedance)

$$|Z_{eq1}| = \frac{V_1}{I_{1rated}} = \frac{35}{41.667} = 0.8400 \quad \Omega$$

$$X_{eq1} = \sqrt{Z_{eq1}^2 - R_{eq1}^2} = 0.8220 \quad \Omega$$

$$Z_{eq1} = R_{eq1} + jX_{eq1} = 0.1728 + j0.8220 = 0.8400\underline{/78.13°} \quad \Omega$$

b. The equivalent circuit for the open-circuit test is shown in Figure 3.7(b), where the series impedance is neglected. From (3.1.16),

$$V_1 = E_1 = a_t E_2 = \frac{N_1}{N_2} V_{2rated} = \frac{480}{120}(120) = 480 \text{ volts}$$

G_c, Y_m, and B_m are then determined as follows:

$$G_c = \frac{P_2}{V_1^2} = \frac{200}{(480)^2} = 0.000868 \quad S$$

$$|Y_m| = \frac{\left(\dfrac{N_2}{N_1}\right)I_2}{V_1} = \frac{\left(\dfrac{120}{480}\right)(12)}{480} = 0.00625 \quad S$$

$$B_m = \sqrt{Y_m^2 - G_c^2} = \sqrt{(0.00625)^2 - (0.000868)^2} = 0.00619 \quad S$$

$$Y_m = G_c - jB_m = 0.000868 - j0.00619 = 0.00625\underline{/-82.02°} \quad S$$

Note that the equivalent series impedance is usually evaluated at rated current from a short-circuit test, and the shunt admittance is evaluated at rated voltage from an open-circuit test. For small variations in transformer operation near rated conditions, the impedance and admittance values are often assumed constant. ∎

The following are not represented by the equivalent circuit of Figure 3.5:

1. Saturation

2. Inrush current

3. Nonsinusoidal exciting current

4. Surge phenomena

They are briefly discussed in the following sections.

SATURATION

In deriving the equivalent circuit of the ideal and practical transformers, we have assumed constant core permeability μ_c and the linear relationship $B_c = \mu_c H_c$ of (3.1.3). However, the relationship between B and H for ferro-

FIGURE 3.8

B–H curves for M-5
grain-oriented electrical
steel 0.012 in. thick
(Reprinted with
permission of AK
Steel Corporation)

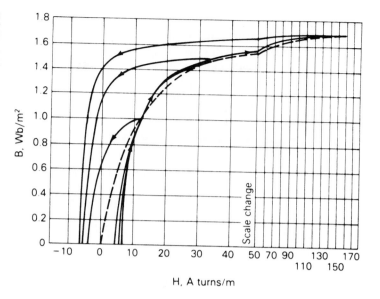

magnetic materials used for transformer cores is nonlinear and multivalued.
Figure 3.8 shows a set of B–H curves for a grain-oriented electrical steel typically used in transformers. As shown, each curve is multivalued, which is
caused by hysteresis. For many engineering applications, the B–H curves can
be adequately described by the dashed line drawn through the curves in
Figure 3.8. Note that as H increases, the core becomes saturated; that is, the
curves flatten out as B increases above 1 Wb/m². If the magnitude of the
voltage applied to a transformer is too large, the core will saturate and a high
magnetizing current will flow. In a well-designed transformer, the applied
peak voltage causes the peak flux density in steady state to occur at the knee
of the B–H curve, with a corresponding low value of magnetizing current.

INRUSH CURRENT

When a transformer is first energized, a transient current much larger than
rated transformer current can flow for several cycles. This current, called *inrush current*, is nonsinusoidal and has a large dc component. To understand
the cause of inrush, assume that before energization, the transformer core is
magnetized with a residual flux density $B(0) = 1.5$ Wb/m² (near the knee of
the dotted curve in Figure 3.8). If the transformer is then energized when the
source voltage is positive and increasing, Faraday's law, (3.1.9), will cause
the flux density $B(t)$ to increase further, since

$$B(t) = \frac{\phi(t)}{A} = \frac{1}{NA} \int_0^t e(t) \, dt + B(0)$$

As B(t) moves into the saturation region of the B–H curve, large values of H(t) will occur, and, from Ampere's law, (3.1.1), corresponding large values of current $i(t)$ will flow for several cycles until it has dissipated. Since normal inrush currents can be as large as abnormal short-circuit currents in transformers, transformer protection schemes must be able to distinguish between these two types of currents.

NONSINUSOIDAL EXCITING CURRENT

When a sinusoidal voltage is applied to one winding of a transformer with the other winding open, the flux $\phi(t)$ and flux density B(t) will, from Faraday's law, (3.1.9), be very nearly sinusoidal in steady state. However, the magnetic field intensity H(t) and the resulting exciting current will not be sinusoidal in steady state, due to the nonlinear B–H curve. If the exciting current is measured and analyzed by Fourier analysis techniques, one finds that it has a fundamental component and a set of odd harmonics. The principal harmonic is the third, whose rms value is typically about 40% of the total rms exciting current. However, the nonsinusoidal nature of exciting current is usually neglected unless harmonic effects are of direct concern, because the exciting current itself is usually less than 5% of rated current for power transformers.

SURGE PHENOMENA

When power transformers are subjected to transient overvoltages caused by lightning or switching surges, the capacitances of the transformer windings have important effects on transient response. Transformer winding capacitances and response to surges are discussed in Chapter 12.

3.3

THE PER-UNIT SYSTEM

Power-system quantities such as voltage, current, power, and impedance are often expressed in per-unit or percent of specified base values. For example, if a base voltage of 20 kV is specified, then the voltage 18 kV is $(18/20) = 0.9$ per unit or 90%. Calculations can then be made with per-unit quantities rather than with the actual quantities.

One advantage of the per-unit system is that by properly specifying base quantities, the transformer equivalent circuit can be simplified. The ideal transformer winding can be eliminated, such that voltages, currents, and external impedances and admittances expressed in per-unit do not change when they are referred from one side of a transformer to the other. This can be a significant advantage even in a power system of moderate size, where hundreds of transformers may be encountered. The per-unit system allows us to avoid the possibility of making serious calculation errors when referring quantities from one side of a transformer to the other. Another advantage of the per-unit system is that the per-unit impedances of electrical equipment of similar type usually lie within a narrow numerical range when the equipment ratings are used as base values. Because of this, per-unit impedance data can be checked rapidly for gross errors by someone familiar with per-unit quantities. In addition, manufacturers usually specify the impedances of machines and transformers in per-unit or percent of nameplate rating.

Per-unit quantities are calculated as follows:

$$\text{per-unit quantity} = \frac{\text{actual quantity}}{\text{base value of quantity}} \tag{3.3.1}$$

where *actual quantity* is the value of the quantity in the actual units. The base value has the same units as the actual quantity, thus making the per-unit quantity dimensionless. Also, the base value is always a real number. Therefore, the angle of the per-unit quantity is the same as the angle of the actual quantity.

Two independent base values can be arbitrarily selected at one point in a power system. Usually the base voltage V_{baseLN} and base complex power $S_{base1\phi}$ are selected for either a single-phase circuit or for one phase of a three-phase circuit. Then, in order for electrical laws to be valid in the per-unit system, the following relations must be used for other base values:

$$P_{base1\phi} = Q_{base1\phi} = S_{base1\phi} \tag{3.3.2}$$

$$I_{base} = \frac{S_{base1\phi}}{V_{baseLN}} \tag{3.3.3}$$

$$Z_{base} = R_{base} = X_{base} = \frac{V_{baseLN}}{I_{base}} = \frac{V_{baseLN}^2}{S_{base1\phi}} \tag{3.3.4}$$

$$Y_{base} = G_{base} = B_{base} = \frac{1}{Z_{base}} \tag{3.3.5}$$

In (3.3.2)–(3.3.5) the subscripts LN and 1ϕ denote "line-to-neutral" and "per-phase," respectively, for three-phase circuits. These equations are also valid for single-phase circuits, where subscripts can be omitted.

By convention, we adopt the following two rules for base quantities:

1. The value of $S_{base1\phi}$ is the same for the entire power system of concern.

2. The ratio of the voltage bases on either side of a transformer is selected to be the same as the ratio of the transformer voltage ratings.

With these two rules, a per-unit impedance remains unchanged when referred from one side of a transformer to the other.

EXAMPLE 3.3   Per-unit impedance: single-phase transformer

A single-phase two-winding transformer is rated 20 kVA, 480/120 volts, 60 Hz. The equivalent leakage impedance of the transformer referred to the 120-volt winding, denoted winding 2, is $Z_{eq2} = 0.0525\underline{/78.13°}$ Ω. Using the transformer ratings as base values, determine the per-unit leakage impedance referred to winding 2 and referred to winding 1.

SOLUTION   The values of S_{base}, V_{base1}, and V_{base2} are, from the transformer ratings,

$$S_{base} = 20 \text{ kVA}, \qquad V_{base1} = 480 \text{ volts}, \qquad V_{base2} = 120 \text{ volts}$$

Using (3.3.4), the base impedance on the 120-volt side of the transformer is

$$Z_{base2} = \frac{V_{base2}^2}{S_{base}} = \frac{(120)^2}{20,000} = 0.72 \quad \Omega$$

Then, using (3.3.1), the per-unit leakage impedance referred to winding 2 is

$$Z_{eq2p.u.} = \frac{Z_{eq2}}{Z_{base2}} = \frac{0.0525\underline{/78.13°}}{0.72} = 0.0729\underline{/78.13°} \quad \text{per unit}$$

If Z_{eq2} is referred to winding 1,

$$Z_{eq1} = a_t^2 Z_{eq2} = \left(\frac{N_1}{N_2}\right)^2 Z_{eq2} = \left(\frac{480}{120}\right)^2 (0.0525\underline{/78.13°})$$

$$= 0.84\underline{/78.13°} \quad \Omega$$

The base impedance on the 480-volt side of the transformer is

$$Z_{base1} = \frac{V_{base1}^2}{S_{base}} = \frac{(480)^2}{20,000} = 11.52 \quad \Omega$$

and the per-unit leakage reactance referred to winding 1 is

$$Z_{eq1p.u.} = \frac{Z_{eq1}}{Z_{base1}} = \frac{0.84\underline{/78.13°}}{11.52} = 0.0729\underline{/78.13°} \text{ per unit} = Z_{eq2p.u.}$$

Thus, the *per-unit* leakage impedance remains unchanged when referred from winding 2 to winding 1. This has been achieved by specifying

$$\frac{V_{base1}}{V_{base2}} = \frac{V_{rated1}}{V_{rated2}} = \left(\frac{480}{120}\right)$$ ■

Figure 3.9 shows three per-unit circuits of a single-phase two-winding transformer. The ideal transformer, shown in Figure 3.9(a), satisfies the per-unit relations $E_{1p.u.} = E_{2p.u.}$, and $I_{1p.u.} = I_{2p.u.}$, which can be derived as follows. First divide (3.1.16) by V_{base1}:

$$E_{1p.u.} = \frac{E_1}{V_{base1}} = \frac{N_1}{N_2} \times \frac{E_2}{V_{base1}} \tag{3.3.6}$$

FIGURE 3.9

Per-unit equivalent
circuits of a single-phase
two-winding transformer

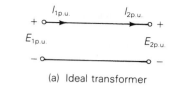

(a) Ideal transformer

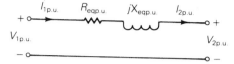

(b) Neglecting exciting current

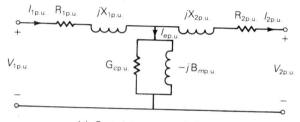

(c) Complete representation

Then, using $V_{\text{base1}}/V_{\text{base2}} = V_{\text{rated1}}/V_{\text{rated2}} = N_1/N_2$,

$$E_{1\text{p.u.}} = \frac{N_1}{N_2}\frac{E_2}{\left(\dfrac{N_1}{N_2}\right)V_{\text{base2}}} = \frac{E_2}{V_{\text{base2}}} = E_{2\text{p.u.}} \tag{3.3.7}$$

Similarly, divide (3.1.17) by I_{base1}:

$$I_{1\text{p.u.}} = \frac{I_1}{I_{\text{base1}}} = \frac{N_2}{N_1}\frac{I_2}{I_{\text{base1}}} \tag{3.3.8}$$

Then, using $I_{\text{base1}} = S_{\text{base}}/V_{\text{base1}} = S_{\text{base}}/[(N_1/N_2)V_{\text{base2}}] = (N_2/N_1)I_{\text{base2}}$,

$$I_{1\text{p.u.}} = \frac{N_2}{N_1}\frac{I_2}{\left(\dfrac{N_2}{N_1}\right)I_{\text{base2}}} = \frac{I_2}{I_{\text{base2}}} = I_{2\text{p.u.}} \tag{3.3.9}$$

Thus, the ideal transformer winding in Figure 3.2 is eliminated from the per-unit circuit in Figure 3.9(a). The per-unit leakage impedance is included in Figure 3.9(b), and the per-unit shunt admittance branch is added in Figure 3.9(c) to obtain the complete representation.

When only one component, such as a transformer, is considered, the nameplate ratings of that component are usually selected as base values. When several components are involved, however, the system base values may be different from the nameplate ratings of any particular device. It is then necessary to convert the per-unit impedance of a device from its nameplate ratings to the system base values. To convert a per-unit impedance from "old" to "new" base values, use

$$Z_{\text{p.u.new}} = \frac{Z_{\text{actual}}}{Z_{\text{basenew}}} = \frac{Z_{\text{p.u.old}} Z_{\text{baseold}}}{Z_{\text{basenew}}} \tag{3.3.10}$$

or, from (3.3.4),

$$Z_{\text{p.u.new}} = Z_{\text{p.u.old}} \left(\frac{V_{\text{baseold}}}{V_{\text{basenew}}} \right)^2 \left(\frac{S_{\text{basenew}}}{S_{\text{baseold}}} \right) \tag{3.3.11}$$

EXAMPLE 3.4   Per-unit circuit: three-zone single-phase network

Three zones of a single-phase circuit are identified in Figure 3.10(a). The zones are connected by transformers T_1 and T_2, whose ratings are also shown. Using

FIGURE 3.10

Circuits for Example 3.4

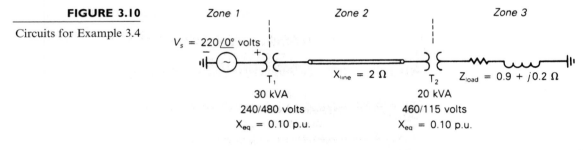

(a) Single-phase circuit

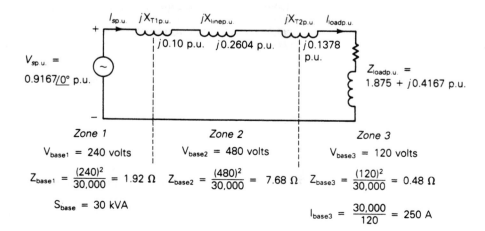

(b) Per-unit circuit

base values of 30 kVA and 240 volts in zone 1, draw the per-unit circuit and determine the per-unit impedances and the per-unit source voltage. Then calculate the load current both in per-unit and in amperes. Transformer winding resistances and shunt admittance branches are neglected.

SOLUTION   First the base values in each zone are determined. $S_{base} = 30$ kVA is the same for the entire network. Also, $V_{base1} = 240$ volts, as specified for zone 1. When moving across a transformer, the voltage base is changed in proportion to the transformer voltage ratings. Thus,

$$V_{base2} = \left(\frac{480}{240}\right)(240) = 480 \quad \text{volts}$$

and

$$V_{base3} = \left(\frac{115}{460}\right)(480) = 120 \quad \text{volts}$$

The base impedances in zones 2 and 3 are

$$Z_{base2} = \frac{V_{base2}^2}{S_{base}} = \frac{480^2}{30,000} = 7.68 \quad \Omega$$

and

$$Z_{base3} = \frac{V_{base3}^2}{S_{base}} = \frac{120^2}{30,000} = 0.48 \quad \Omega$$

and the base current in zone 3 is

$$I_{base3} = \frac{S_{base}}{V_{base3}} = \frac{30,000}{120} = 250 \quad \text{A}$$

Next, the per-unit circuit impedances are calculated using the system base values. Since $S_{base} = 30$ kVA is the same as the kVA rating of transformer T_1, and $V_{base1} = 240$ volts is the same as the voltage rating of the zone 1 side of transformer T_1, the per-unit leakage reactance of T_1 is the same as its nameplate value, $X_{T1p.u.} = 0.1$ per unit. However, the per-unit leakage reactance of transformer T_2 must be converted from its nameplate rating to the system base. Using (3.3.11) and $V_{base2} = 480$ volts,

$$X_{T2p.u.} = (0.10)\left(\frac{460}{480}\right)^2 \left(\frac{30,000}{20,000}\right) = 0.1378 \quad \text{per unit}$$

Alternatively, using $V_{base3} = 120$ volts,

$$X_{T2p.u.} = (0.10)\left(\frac{115}{120}\right)^2 \left(\frac{30,000}{20,000}\right) = 0.1378 \quad \text{per unit}$$

which gives the same result. The line, which is located in zone 2, has a per-unit reactance

$$X_{linep.u.} = \frac{X_{line}}{Z_{base2}} = \frac{2}{7.68} = 0.2604 \quad \text{per unit}$$

and the load, which is located in zone 3, has a per-unit impedance

$$Z_{\text{loadp.u.}} = \frac{Z_{\text{load}}}{Z_{\text{base3}}} = \frac{0.9 + j0.2}{0.48} = 1.875 + j0.4167 \quad \text{per unit}$$

The per-unit circuit is shown in Figure 3.10(b), where the base values for each zone, per-unit impedances, and the per-unit source voltage are shown. The per-unit load current is then easily calculated from Figure 3.10(b) as follows:

$$
\begin{aligned}
I_{\text{loadp.u.}} = I_{\text{sp.u.}} &= \frac{V_{\text{sp.u.}}}{j(X_{\text{T1p.u.}} + X_{\text{linep.u.}} + X_{\text{T2p.u.}}) + Z_{\text{loadp.u.}}} \\
&= \frac{0.9167\underline{/0°}}{j(0.10 + 0.2604 + 0.1378) + (1.875 + j0.4167)} \\
&= \frac{0.9167\underline{/0°}}{1.875 + j0.9149} = \frac{0.9167\underline{/0°}}{2.086\underline{/26.01°}} \\
&= 0.4395\underline{/-26.01°} \quad \text{per unit}
\end{aligned}
$$

The actual load current is

$$I_{\text{load}} = (I_{\text{loadp.u.}})I_{\text{base3}} = (0.4395\underline{/-26.01°})(250) = 109.9\underline{/-26.01°} \quad \text{A}$$

Note that the per-unit equivalent circuit of Figure 3.10(b) is relatively easy to analyze, since ideal transformer windings have been eliminated by proper selection of base values.   ∎

Balanced three-phase circuits can be solved in per-unit on a per-phase basis after converting Δ-load impedances to equivalent Y impedances. Base values can be selected either on a per-phase basis or on a three-phase basis. Equations (3.3.1)–(3.3.5) remain valid for three-phase circuits on a per-phase basis. Usually $S_{\text{base3}\phi}$ and V_{baseLL} are selected, where the subscripts 3ϕ and LL denote "three-phase" and "line-to-line," respectively. Then the following relations must be used for other base values:

$$S_{\text{base1}\phi} = \frac{S_{\text{base3}\phi}}{3} \tag{3.3.12}$$

$$V_{\text{baseLN}} = \frac{V_{\text{baseLL}}}{\sqrt{3}} \tag{3.3.13}$$

$$S_{\text{base3}\phi} = P_{\text{base3}\phi} = Q_{\text{base3}\phi} \tag{3.3.14}$$

$$I_{\text{base}} = \frac{S_{\text{base1}\phi}}{V_{\text{baseLN}}} = \frac{S_{\text{base3}\phi}}{\sqrt{3}V_{\text{baseLL}}} \tag{3.3.15}$$

$$Z_{\text{base}} = \frac{V_{\text{baseLN}}}{I_{\text{base}}} = \frac{V_{\text{baseLN}}^2}{S_{\text{base1}\phi}} = \frac{V_{\text{baseLL}}^2}{S_{\text{base3}\phi}} \tag{3.3.16}$$

$$R_{\text{base}} = X_{\text{base}} = Z_{\text{base}} = \frac{1}{Y_{\text{base}}} \tag{3.3.17}$$

EXAMPLE 3.5   **Per-unit and actual currents in balanced three-phase networks**

As in Example 2.5, a balanced-Y-connected voltage source with $E_{ab} = 480\underline{/0^\circ}$ volts is applied to a balanced-Δ load with $Z_\Delta = 30\underline{/40^\circ}$ Ω. The line impedance between the source and load is $Z_L = 1\underline{/85^\circ}$ Ω for each phase. Calculate the per-unit and actual current in phase a of the line using $S_{base3\phi} = 10$ kVA and $V_{baseLL} = 480$ volts.

SOLUTION   First, convert Z_Δ to an equivalent Z_Y; the equivalent line-to-neutral diagram is shown in Figure 2.17. The base impedance is, from (3.3.16),

$$Z_{base} = \frac{V_{baseLL}^2}{S_{base3\phi}} = \frac{(480)^2}{10,000} = 23.04 \quad \Omega$$

The per-unit line and load impedances are

$$Z_{Lp.u.} = \frac{Z_L}{Z_{base}} = \frac{1\underline{/85^\circ}}{23.04} = 0.04340\underline{/85^\circ} \quad \text{per unit}$$

and

$$Z_{Yp.u.} = \frac{Z_Y}{Z_{base}} = \frac{10\underline{/40^\circ}}{23.04} = 0.4340\underline{/40^\circ} \quad \text{per unit}$$

Also,

$$V_{baseLN} = \frac{V_{baseLL}}{\sqrt{3}} = \frac{480}{\sqrt{3}} = 277 \quad \text{volts}$$

and

$$E_{anp.u.} = \frac{E_{an}}{V_{baseLN}} = \frac{277\underline{/-30^\circ}}{277} = 1.0\underline{/-30^\circ} \quad \text{per unit}$$

The per-unit equivalent circuit is shown in Figure 3.11. The per-unit line current in phase a is then

FIGURE 3.11

Circuit for Example 3.5

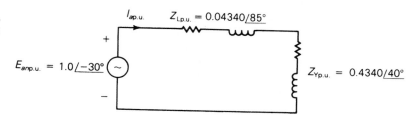

$$I_{ap.u.} = \frac{E_{anp.u.}}{Z_{Lp.u.} + Z_{Yp.u.}} = \frac{1.0\underline{/-30°}}{0.04340\underline{/85°} + 0.4340\underline{/40°}}$$

$$= \frac{1.0\underline{/-30°}}{(0.00378 + j0.04323) + (0.3325 + j0.2790)}$$

$$= \frac{1.0\underline{/-30°}}{0.3362 + j0.3222} = \frac{1.0\underline{/-30°}}{0.4657\underline{/43.78°}}$$

$$= 2.147\underline{/-73.78°} \quad \text{per unit}$$

The base current is

$$I_{base} = \frac{S_{base3\phi}}{\sqrt{3}V_{baseLL}} = \frac{10,000}{\sqrt{3}(480)} = 12.03 \quad A$$

and the actual phase a line current is

$$I_a = (2.147\underline{/-73.78°})(12.03) = 25.83\underline{/-73.78°} \quad A \qquad \blacksquare$$

3.4

THREE-PHASE TRANSFORMER CONNECTIONS AND PHASE SHIFT

Three identical single-phase two-winding transformers may be connected to form a three-phase bank. Four ways to connect the windings are Y–Y, Y–Δ, Δ–Y, and Δ–Δ. For example, Figure 3.12 shows a three-phase Y–Y bank. Figure 3.12(a) shows the core and coil arrangements. The American standard for marking three-phase transformers substitutes H1, H2, and H3 on the high-voltage terminals and X1, X2, and X3 on the low-voltage terminals in place of the polarity dots. Also, in this text, we will use uppercase letters ABC to identify phases on the high-voltage side of the transformer and lowercase letters abc to identify phases on the low-voltage side of the transformer. In Figure 3.12(a) the transformer high-voltage terminals H1, H2, and H3 are connected to phases A, B, and C, and the low-voltage terminals X1, X2, and X3 are connected to phases a, b, and c, respectively.

Figure 3.12(b) shows a schematic representation of the three-phase Y–Y transformer. Windings on the same core are drawn in parallel, and the phasor relationship for balanced positive-sequence operation is shown. For example, high-voltage winding H1–N is on the same magnetic core as low-voltage winding X1–n in Figure 3.12(b). Also, V_{AN} is in phase with V_{an}. Figure 3.12(c) shows a single-line diagram of a Y–Y transformer. A single-line diagram shows one phase of a three-phase network with the neutral wire omitted and with components represented by symbols rather than equivalent circuits.

The phases of a Y–Y or a Δ–Δ transformer can be labeled so there is

FIGURE 3.12

Three-phase two-
winding Y–Y
transformer bank

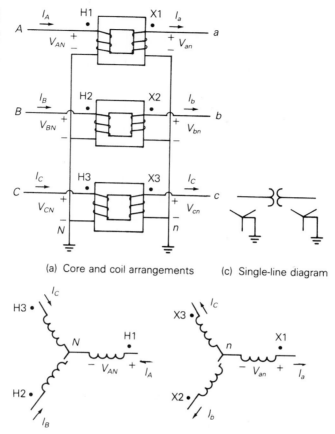

(a) Core and coil arrangements   (c) Single-line diagram

(b) Schematic representation showing phasor
relationship for positive sequence operation

no phase shift between corresponding quantities on the low- and high-voltage windings. However, for Y–Δ and Δ–Y transformers, there is always a phase shift. Figure 3.13 shows a Y–Δ transformer. The labeling of the windings and the schematic representation are in accordance with the American standard, which is as follows:

> In either a Y–Δ or Δ–Y transformer, positive-sequence quantities on the high-voltage side shall lead their corresponding quantities on the low-voltage side by 30°.

As shown in Figure 3.13(b), V_{AN} leads V_{an} by 30°.

The positive-sequence phasor diagram shown in Figure 3.13(b) can be constructed via the following five steps, which are also indicated in Figure 3.13:

FIGURE 3.13

Three-phase two-winding Y–Δ transformer bank

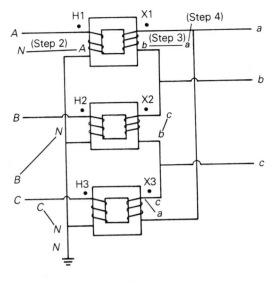

(a) Core and coil arrangement

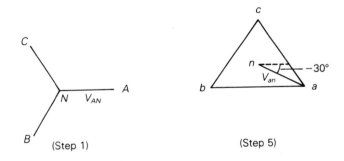

(Step 1)   (Step 5)

(b) Positive-sequence phasor diagram

STEP 1 Assume that balanced positive-sequence voltages are applied to the Y winding. Draw the positive-sequence phasor diagram for these voltages.

STEP 2 Move phasor *A–N* next to terminals *A–N* in Figure 3.13(a). Identify the ends of this line in the same manner as in the phasor diagram. Similarly, move phasors *B–N* and *C–N* next to terminals *B–N* and *C–N* in Figure 3.13(a).

STEP 3 For each single-phase transformer, the voltage across the low-voltage winding must be in phase with the voltage across the high-voltage winding, assuming an ideal transformer. Therefore, draw a line next to each low-voltage winding parallel to the corresponding line already drawn next to the high-voltage winding.

STEP 4   Label the ends of the lines drawn in Step 3 by inspecting the polarity marks. For example, phase A is connected to dotted terminal H1, and A appears on the *right* side of line A–N. Therefore, phase a, which is connected to dotted terminal X1, must be on the *right* side, and b on the left side of line a–b. Similarly, phase B is connected to dotted terminal H2, and B is *down* on line B–N. Therefore, phase b, connected to dotted terminal X2, must be *down* on line b–c. Similarly, c is *up* on line c–a.

STEP 5   Bring the three lines labeled in Step 4 together to complete the phasor diagram for the low-voltage Δ winding. Note that V_{AN} leads V_{an} by 30° in accordance with the American standard.

EXAMPLE 3.6   Phase shift in Δ–Y transformers

Assume that balanced negative-sequence voltages are applied to the high-voltage windings of the Y–Δ transformer shown in Figure 3.13. Determine the negative-sequence phase shift of this transformer.

SOLUTION   The negative-sequence diagram, shown in Figure 3.14, is constructed from the following five steps, as outlined above:

STEP 1   Draw the phasor diagram of balanced negative-sequence voltages, which are applied to the Y winding.

STEP 2   Move the phasors A–N, B–N, and C–N next to the high-voltage Y windings.

STEP 3   For each single-phase transformer, draw a line next to the low-voltage winding that is parallel to the line drawn in Step 2 next to the high-voltage winding.

STEP 4   Label the lines drawn in Step 3. For example, phase B, which is connected to dotted terminal H2, is shown *up* on line B–N; therefore phase b, which is connected to dotted terminal X2, must be *up* on line b–c.

STEP 5   Bring the lines drawn in Step 4 together to form the negative-sequence phasor diagram for the low-voltage Δ winding.

As shown in Figure 3.14, the high-voltage phasors *lag* the low-voltage phasors by 30°. Thus the negative-sequence phase shift is the reverse of the positive-sequence phase shift.   ∎

The Δ–Y transformer is commonly used as a generator step-up transformer, where the Δ winding is connected to the generator terminals and the Y

FIGURE 3.14

Example 3.6—
Construction of
negative-sequence
phasor diagram for Y–Δ
transformer bank

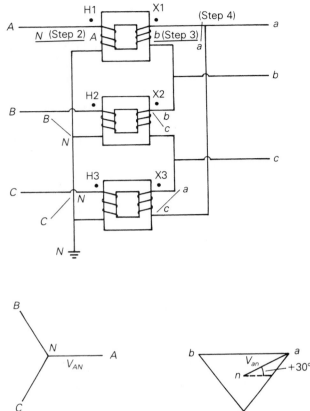

winding is connected to a transmission line. One advantage of a high-voltage Y winding is that a neutral point N is provided for grounding on the high-voltage side. With a permanently grounded neutral, the insulation requirements for the high-voltage transformer windings are reduced. The high-voltage insulation can be graded or tapered from maximum insulation at terminals ABC to minimum insulation at grounded terminal N. One advantage of the Δ winding is that the undesirable third harmonic magnetizing current, caused by the nonlinear core B–H characteristic, remains trapped inside the Δ winding. Third harmonic currents are (triple-frequency) zero-sequence currents, which cannot enter or leave a Δ connection, but can flow within the Δ. The Y–Y transformer is seldom used because of difficulties with third harmonic exciting current.

   The Δ–Δ transformer has the advantage that one phase can be removed for repair or maintenance while the remaining phases continue to operate as

FIGURE 3.15

Transformer core
configurations

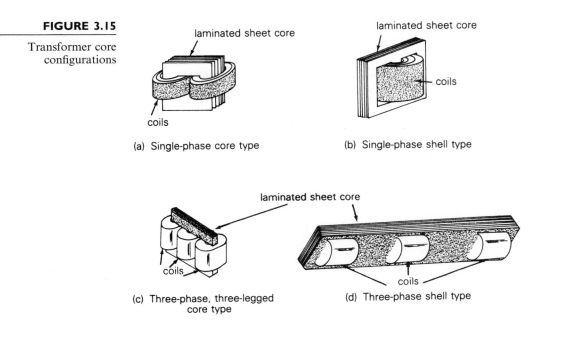

(a) Single-phase core type

(b) Single-phase shell type

(c) Three-phase, three-legged core type

(d) Three-phase shell type

a three-phase bank. This *open-Δ* connection permits balanced three-phase operation with the kVA rating reduced to 58% of the original bank (see Problem 3.36).

Instead of a bank of three single-phase transformers, all six windings may be placed on a common three-phase core to form a three-phase transformer, as shown in Figure 3.15. The three-phase core contains less iron than the three single-phase units; therefore it costs less, weighs less, requires less floor space, and has a slightly higher efficiency. However, a winding failure would require replacement of an entire three-phase transformer, compared to replacement of only one phase of a three-phase bank.

3.5

PER-UNIT EQUIVALENT CIRCUITS OF BALANCED THREE-PHASE TWO-WINDING TRANSFORMERS

Figure 3.16(a) is a schematic representation of an ideal Y–Y transformer grounded through neutral impedances Z_N and Z_n. Figure 3.16(b) shows the per-unit equivalent circuit of this ideal transformer for balanced three-phase operation. Throughout the remainder of this text, per-unit quantities will be used unless otherwise indicated. Also, the subscript "p.u.," used to indicate a per-unit quantity, will be omitted in most cases.

FIGURE 3.16

Ideal Y–Y transformer

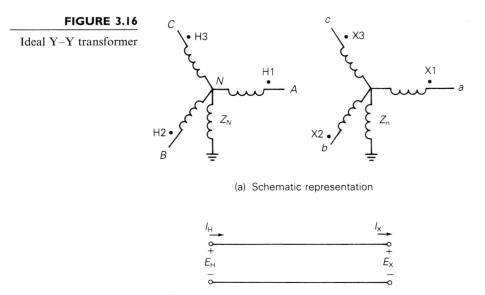

(a) Schematic representation

(b) Per-unit equivalent circuit for balanced three-phase operation

By convention, we adopt the following two rules for selecting base quantities:

1. A common S_{base} is selected for both the H and X terminals.

2. The ratio of the voltage bases V_{baseH}/V_{baseX} is selected to be equal to the ratio of the rated line-to-line voltages $V_{ratedHLL}/V_{ratedXLL}$.

When balanced three-phase currents are applied to the transformer, the neutral currents are zero and there are no voltage drops across the neutral impedances. Therefore, the per-unit equivalent circuit of the ideal Y–Y transformer, Figure 3.16(b), is the same as the per-unit single-phase ideal transformer, Figure 3.9(a).

The per-unit equivalent circuit of a practical Y–Y transformer is shown in Figure 3.17(a). This network is obtained by adding external impedances to the equivalent circuit of the ideal transformer, as in Figure 3.9(c).

The per-unit equivalent circuit of the Y–Δ transformer, shown in Figure 3.17(b), includes a phase shift. For the American standard, the positive-sequence voltages and currents on the high-voltage side of the Y–Δ transformer lead the corresponding quantities on the low-voltage side by 30°. The phase shift in the equivalent circuit of Figure 3.17(b) is represented by the phase-shifting transformer of Figure 3.4.

The per-unit equivalent circuit of the Δ–Δ transformer, shown in Figure 3.17(c), is the same as that of the Y–Y transformer. It is assumed that the windings are labeled so there is no phase shift. Also, the per-unit impedances do not depend on the winding connections, but the base voltages do.

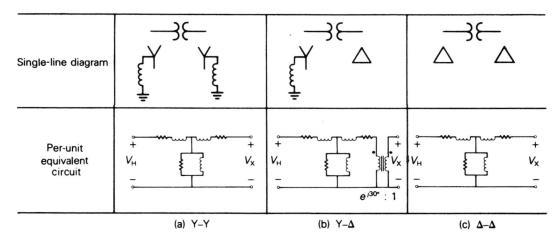

Single-line diagram		
Per-unit equivalent circuit		
(a) Y–Y	(b) Y–Δ	(c) Δ–Δ

FIGURE 3.17   Per-unit equivalent circuits of practical Y–Y, Y–Δ, and Δ–Δ transformers for balanced three-phase operation

EXAMPLE 3.7   Voltage calculations: balanced Y–Y and Δ–Y transformers

Three single-phase two-winding transformers, each rated 400 MVA, 13.8/199.2 kV, with leakage reactance $X_{eq} = 0.10$ per unit, are connected to form a three-phase bank. Winding resistances and exciting current are neglected. The high-voltage windings are connected in Y. A three-phase load operating under balanced positive-sequence conditions on the high-voltage side absorbs 1000 MVA at 0.90 p.f. lagging, with $V_{AN} = 199.2\underline{/0°}$ kV. Determine the voltage V_{an} at the low-voltage bus if the low-voltage windings are connected (a) in Y, (b) in Δ.

SOLUTION   The per-unit network is shown in Figure 3.18. Using the transformer bank ratings as base quantities, $S_{base3\phi} = 1200$ MVA, $V_{baseHLL} = 345$ kV, and $I_{baseH} = 1200/(345\sqrt{3}) = 2.008$ kA. The per-unit load voltage and load current are then

$$V_{AN} = 1.0\underline{/0°} \quad \text{per unit}$$

$$I_A = \frac{1000/(345\sqrt{3})}{2.008}\underline{/-\cos^{-1}0.9} = 0.8333\underline{/-25.84°} \quad \text{per unit}$$

a. For the Y–Y transformer, Figure 3.18(a),

$$I_a = I_A = 0.8333\underline{/-25.84°} \quad \text{per unit}$$

$$V_{an} = V_{AN} + (jX_{eq})I_A$$

$$= 1.0\underline{/0°} + (j0.10)(0.8333\underline{/-25.84°})$$

$$= 1.0 + 0.08333\underline{/64.16°} = 1.0363 + j0.0750 = 1.039\underline{/4.139°}$$

$$= 1.039\underline{/4.139°} \quad \text{per unit}$$

FIGURE 3.18

Per-unit network for
Example 3.7

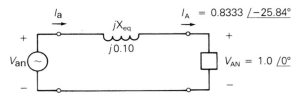

(a) Y-connected low-voltage windings

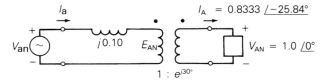

(b) Δ-connected low-voltage windings

Further, since $V_{baseXLN} = 13.8$ kV for the low-voltage Y windings, $V_{an} = 1.039(13.8) = 14.34$ kV, and

$$V_{an} = 14.34\underline{/4.139°} \quad \text{kV}$$

b. For the Δ–Y transformer, Figure 3.18(b),

$$E_{an} = e^{-j30°} V_{AN} = 1.0\underline{/-30°} \quad \text{per unit}$$

$$I_a = e^{-j30°} I_A = 0.8333\underline{/-25.84° - 30°} = 0.8333\underline{/-55.84°} \quad \text{per unit}$$

$$V_{an} = E_{AN} + (jX_{eq})I_A = 1.0\underline{/-30°} + (j0.10)(0.8333\underline{/-55.84°})$$

$$V_{an} = 1.039\underline{/-25.861°} \quad \text{per unit}$$

Further, since $V_{baseXLN} = 13.8/\sqrt{3} = 7.967$ kV for the low-voltage Δ windings, $V_{an} = (1.039)(7.967) = 8.278$ kV, and

$$V_{an} = 8.278\underline{/-25.861°} \quad \text{kV} \qquad ■$$

EXAMPLE 3.8 **Per-unit voltage drop and per-unit fault current: balanced three-phase transformer**

A 200-MVA, 345-kVΔ/34.5-kV Y substation transformer has an 8% leakage reactance. The transformer acts as a connecting link between 345-kV transmission and 34.5-kV distribution. Transformer winding resistances and exciting current are neglected. The high-voltage bus connected to the transformer is assumed to be an ideal 345-kV positive-sequence source with negligible source impedance. Using the transformer ratings as base values, determine:

a. The per-unit magnitudes of transformer voltage drop and voltage at the low-voltage terminals when rated transformer current at 0.8 p.f. lagging enters the high-voltage terminals

b. The per-unit magnitude of the fault current when a three-phase-to-ground bolted short circuit occurs at the low-voltage terminals

SOLUTION In both parts (a) and (b), only balanced positive-sequence current will flow, since there are no imbalances. Also, because we are interested only in voltage and current magnitudes, the Δ–Y transformer phase shift can be omitted.

a. As shown in Figure 3.19(a),

$$V_{drop} = I_{rated}X_{eq} = (1.0)(0.08) = 0.08 \quad \text{per unit}$$

and

$$V_{an} = V_{AN} - (jX_{eq})I_{rated}$$
$$= 1.0\underline{/0^\circ} - (j0.08)(1.0\underline{/-36.87^\circ})$$
$$= 1.0 - (j0.08)(0.8 - j0.6) = 0.952 - j0.064$$
$$= 0.954\underline{/-3.85^\circ} \quad \text{per unit}$$

b. As shown in Figure 3.19(b),

$$I_{SC} = \frac{V_{AN}}{X_{eq}} = \frac{1.0}{0.08} = 12.5 \quad \text{per unit}$$

Under rated current conditions [part (a)], the 0.08 per-unit voltage drop across the transformer leakage reactance causes the voltage at the low-voltage terminals to be 0.954 per unit. Also, under three-phase short-circuit conditions [part

FIGURE 3.19

Circuits for Example 3.8

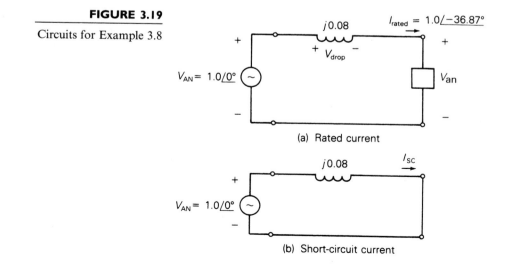

(a) Rated current

(b) Short-circuit current

(b)], the fault current is 12.5 times the rated transformer current. This example illustrates a compromise in the design or specification of transformer leakage reactance. A low value is desired to minimize voltage drops, but a high value is desired to limit fault currents. Typical transformer leakage reactances are given in Table A.2 in the Appendix. ■

3.6

THREE-WINDING TRANSFORMERS

Figure 3.20(a) shows a basic single-phase three-winding transformer. The ideal transformer relations for a two-winding transformer, (3.1.8) and (3.1.14), can easily be extended to obtain corresponding relations for an ideal three-winding transformer. In actual units, these relations are

$$N_1 I_1 = N_2 I_2 + N_3 I_3 \tag{3.6.1}$$

$$\frac{E_1}{N_1} = \frac{E_2}{N_2} = \frac{E_3}{N_3} \tag{3.6.2}$$

where I_1 enters the dotted terminal, I_2 and I_3 leave dotted terminals, and E_1, E_2, and E_3 have their $+$ polarities at dotted terminals. In per-unit, (3.6.1) and (3.6.2) are

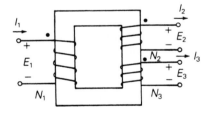

(a) Basic core and coil configuration

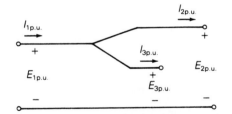

(b) Per-unit equivalent circuit—ideal transformer

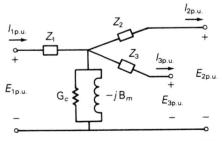

(c) Per-unit equivalent circuit—practical transformer

FIGURE 3.20 Single-phase three-winding transformer

$$I_{1\text{p.u.}} = I_{2\text{p.u.}} + I_{3\text{p.u.}} \tag{3.6.3}$$

$$E_{1\text{p.u.}} = E_{2\text{p.u.}} = E_{3\text{p.u.}} \tag{3.6.4}$$

where a common S_{base} is selected for all three windings, and voltage bases are selected in proportion to the rated voltages of the windings. These two per-unit relations are satisfied by the per-unit equivalent circuit shown in Figure 3.20(b). Also, external series impedance and shunt admittance branches are included in the practical three-winding transformer circuit shown in Figure 3.20(c). The shunt admittance branch, a core loss resistor in parallel with a magnetizing inductor, can be evaluated from an open-circuit test. Also, when one winding is left open, the three-winding transformer behaves as a two-winding transformer, and standard short-circuit tests can be used to evaluate per-unit leakage impedances, which are defined as follows:

Z_{12} = per-unit leakage impedance measured from winding 1, with winding 2 shorted and winding 3 open

Z_{13} = per-unit leakage impedance measured from winding 1, with winding 3 shorted and winding 2 open

Z_{23} = per-unit leakage impedance measured from winding 2, with winding 3 shorted and winding 1 open

From Figure 3.20(c), with winding 2 shorted and winding 3 open, the leakage impedance measured from winding 1 is, neglecting the shunt admittance branch,

$$Z_{12} = Z_1 + Z_2 \tag{3.6.5}$$

Similarly,

$$Z_{13} = Z_1 + Z_3 \tag{3.6.6}$$

and

$$Z_{23} = Z_2 + Z_3 \tag{3.6.7}$$

Solving (3.6.5)–(3.6.7),

$$Z_1 = \tfrac{1}{2}(Z_{12} + Z_{13} - Z_{23}) \tag{3.6.8}$$

$$Z_2 = \tfrac{1}{2}(Z_{12} + Z_{23} - Z_{13}) \tag{3.6.9}$$

$$Z_3 = \tfrac{1}{2}(Z_{13} + Z_{23} - Z_{12}) \tag{3.6.10}$$

Equations (3.6.8)–(3.6.10) can be used to evaluate the per-unit series impedances Z_1, Z_2, and Z_3 of the three-winding transformer equivalent circuit from the per-unit leakage impedances Z_{12}, Z_{13}, and Z_{23}, which, in turn, are determined from short-circuit tests.

Note that each of the windings on a three-winding transformer may have a *different* kVA rating. If the leakage impedances from short-circuit tests are expressed in per-unit based on winding ratings, they must first be converted to per-unit on a common S_{base} *before* they are used in (3.6.8)–(3.6.10).

EXAMPLE 3.9   Three-winding single-phase transformer: per-unit impedances

The ratings of a single-phase three-winding transformer are

> winding 1: 300 MVA, 13.8 kV
>
> winding 2: 300 MVA, 199.2 kV
>
> winding 3: 50 MVA, 19.92 kV

The leakage reactances, from short-circuit tests, are

> $X_{12} = 0.10$ per unit on a 300-MVA, 13.8-kV base
>
> $X_{13} = 0.16$ per unit on a 50-MVA, 13.8-kV base
>
> $X_{23} = 0.14$ per unit on a 50-MVA, 199.2-kV base

Winding resistances and exciting current are neglected. Calculate the impedances of the per-unit equivalent circuit using a base of 300 MVA and 13.8 kV for terminal 1.

SOLUTION $S_{base} = 300$ MVA is the same for all three terminals. Also, the specified voltage base for terminal 1 is $V_{base1} = 13.8$ kV. The base voltages for terminals 2 and 3 are then $V_{base2} = 199.2$ kV and $V_{base3} = 19.92$ kV, which are the rated voltages of these windings. From the data given, $X_{12} = 0.10$ per unit was measured from terminal 1 using the same base values as those specified for the circuit. However, $X_{13} = 0.16$ and $X_{23} = 0.14$ per unit on a 50-MVA base are first converted to the 300-MVA circuit base.

$$X_{13} = (0.16)\left(\frac{300}{50}\right) = 0.96 \quad \text{per unit}$$

$$X_{23} = (0.14)\left(\frac{300}{50}\right) = 0.84 \quad \text{per unit}$$

Then, from (3.6.8)–(3.6.10),

> $X_1 = \frac{1}{2}(0.10 + 0.96 - 0.84) = 0.11 \quad$ per unit
>
> $X_2 = \frac{1}{2}(0.10 + 0.84 - 0.96) = -0.01 \quad$ per unit
>
> $X_3 = \frac{1}{2}(0.84 + 0.96 - 0.10) = 0.85 \quad$ per unit

FIGURE 3.21

Circuit for Example 3.9

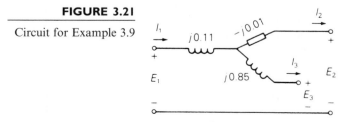

The per-unit equivalent circuit of this three-winding transformer is shown in Figure 3.21. Note that X_2 is negative. This illustrates the fact that X_1, X_2, and X_3 are *not* leakage reactances, but instead are equivalent reactances derived from the leakage reactances. Leakage reactances are always positive.

Note also that the node where the three equivalent circuit reactances are connected does not correspond to any physical location within the transformer. Rather, it is simply part of the equivalent circuit representation. ■

EXAMPLE 3.10   **Three-winding three-phase transformer: balanced operation**

Three transformers, each identical to that described in Example 3.9, are connected as a three-phase bank in order to feed power from a 900-MVA, 13.8-kV generator to a 345-kV transmission line and to a 34.5-kV distribution line. The transformer windings are connected as follows:

> 13.8-kV   windings (X): Δ, to generator
> 199.2-kV   windings (H): solidly grounded Y, to 345-kV line
> 19.92-kV   windings (M): grounded Y through $Z_n = j0.10$ Ω,
>   to 34.5-kV line

The positive-sequence voltages and currents of the high- and medium-voltage Y windings lead the corresponding quantities of the low-voltage Δ winding by 30°. Draw the per-unit network, using a three-phase base of 900 MVA and 13.8 kV for terminal X. Assume balanced positive-sequence operation.

SOLUTION   The per-unit network is shown in Figure 3.22. $V_{baseX} = 13.8$ kV, which is the rated line-to-line voltage of terminal X. Since the M and H windings are Y-connected, $V_{baseM} = \sqrt{3}(19.92) = 34.5$ kV, and $V_{baseH} = \sqrt{3}(199.2) = 345$ kV, which are the rated line-to-line voltages of the M and H windings. Also, a phase-shifting transformer is included in the network. The neutral impedance is not included in the network, since there is no neutral current under balanced operation. ■

FIGURE 3.22

Per-unit network for
Example 3.10

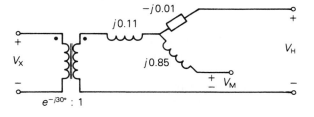

FIGURE 3.23

Ideal single-phase
transformers

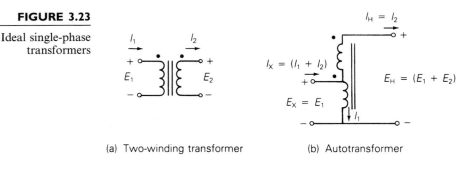

(a) Two-winding transformer   (b) Autotransformer

3.7

AUTOTRANSFORMERS

A single-phase two-winding transformer is shown in Figure 3.23(a) with two separate windings, which is the usual two-winding transformer; the same transformer is shown in Figure 3.23(b) with the two windings connected in series, which is called an *autotransformer*. For the usual transformer [Figure 3.23(a)] the two windings are coupled magnetically via the mutual core flux. For the autotransformer [Figure 3.23(b)] the windings are both electrically and magnetically coupled. The autotransformer has smaller per-unit leakage impedances than the usual transformer; this results in both smaller series-voltage drops (an advantage) and higher short-circuit currents (a disadvantage). The autotransformer also has lower per-unit losses (higher efficiency), lower exciting current, and lower cost if the turns ratio is not too large. The electrical connection of the windings, however, allows transient overvoltages to pass through the autotransformer more easily.

EXAMPLE 3.11   Autotransformer: single-phase

The single-phase two-winding 20-kVA, 480/120-volt transformer of Example 3.3 is connected as an autotransformer, as in Figure 3.23(b), where winding 1 is the 120-volt winding. For this autotransformer, determine (a) the voltage ratings E_X and E_H of the low- and high-voltage terminals, (b) the kVA rating, and (c) the per-unit leakage impedance.

SOLUTION

a. Since the 120-volt winding is connected to the low-voltage terminal, $E_X = 120$ volts. When $E_X = E_1 = 120$ volts is applied to the low-voltage terminal, $E_2 = 480$ volts is induced across the 480-volt winding, neglecting the voltage drop across the leakage impedance. Therefore, $E_H = E_1 + E_2 = 120 + 480 = 600$ volts.

b. As a normal two-winding transformer rated 20 kVA, the rated current of the 480-volt winding is $I_2 = I_H = 20{,}000/480 = 41.667$ A. As an autotransformer, the 480-volt winding can carry the same current. Therefore, the kVA rating $S_H = E_H I_H = (600)(41.667) = 25$ kVA. Note also that when $I_H = I_2 = 41.667$ A, a current $I_1 = \dfrac{480}{120}(41.667) = 166.7$ A is induced in the 120-volt winding. Therefore, $I_X = I_1 + I_2 = 208.3$ A (neglecting exciting current) and $S_X = E_X I_X = (120)(208.3) = 25$ kVA, which is the same rating as calculated for the high-voltage terminal.

c. From Example 3.3, the leakage impedance is $0.0729\underline{/78.13°}$ per unit as a normal, two-winding transformer. As an autotransformer, the leakage impedance *in ohms* is the same as for the normal transformer, since the core and windings are the same for both (only the external winding connections are different). However, the base impedances are different. For the high-voltage terminal, using (3.3.4),

$$Z_{\text{baseHold}} = \frac{(480)^2}{20{,}000} = 11.52 \quad \Omega \quad \text{as a normal transformer}$$

$$Z_{\text{baseHnew}} = \frac{(600)^2}{25{,}000} = 14.4 \quad \Omega \quad \text{as an autotransformer}$$

Therefore, using (3.3.10),

$$Z_{\text{p.u.new}} = (0.0729\underline{/78.13°})\left(\frac{11.52}{14.4}\right) = 0.05832\underline{/78.13°} \quad \text{per unit}$$

For this example, the rating is 25 kVA, 120/600 volts as an autotransformer versus 20 kVA, 120/480 volts as a normal transformer. The autotransformer has both a larger kVA rating and a larger voltage ratio for the same cost. Also, the per-unit leakage impedance of the autotransformer is smaller. However, the increased high-voltage rating as well as the electrical connection of the windings may require more insulation for both windings. ∎

3.8

TRANSFORMERS WITH OFF-NOMINAL TURNS RATIOS

It has been shown that models of transformers that use per-unit quantities are simpler than those that use actual quantities. The ideal transformer winding is eliminated when the ratio of the selected voltage bases equals the ratio of the voltage ratings of the windings. In some cases, however, it is impossible to select voltage bases in this manner. For example, consider the two transformers connected in parallel in Figure 3.24. Transformer T_1 is rated 13.8/345 kV and T_2 is rated 13.2/345 kV. If we select $V_{\text{baseH}} = 345$ kV, then

FIGURE 3.24

Two transformers
connected in parallel

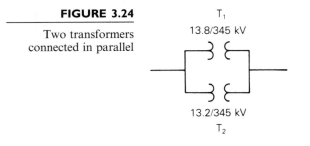

T_1
13.8/345 kV

13.2/345 kV
T_2

transformer T_1 requires $V_{baseX} = 13.8$ kV and T_2 requires $V_{baseX} = 13.2$ kV. It is clearly impossible to select the appropriate voltage bases for both transformers.

To accommodate this situation, we will develop a per-unit model of a transformer whose voltage ratings are not in proportion to the selected base voltages. Such a transformer is said to have an "off-nominal turns ratio." Figure 3.25(a) shows a transformer with rated voltages V_{1rated} and V_{2rated}, which satisfy

$$V_{1rated} = a_t V_{2rated} \qquad (3.8.1)$$

where a_t is assumed, in general, to be either real or complex. Suppose the selected voltage bases satisfy

$$V_{base1} = b V_{base2} \qquad (3.8.2)$$

Defining $c = a_t/b$, (3.8.1) can be rewritten as

$$V_{1rated} = b \left(\frac{a_t}{b} \right) V_{2rated} = bc\, V_{2rated} \qquad (3.8.3)$$

Equation (3.8.3) can be represented by two transformers in series, as shown in Figure 3.25(b). The first transformer has the same ratio of rated winding voltages as the ratio of the selected base voltages, b. Therefore, this transformer has a standard per-unit model, as shown in Figure 3.9 or 3.17. We will assume that the second transformer is ideal, and all real and reactive losses are associated with the first transformer. The resulting per-unit model is shown in Figure 3.25(c), where, for simplicity, the shunt-exciting branch is neglected. Note that if $a_t = b$, then the ideal transformer winding shown in this figure can be eliminated, since its turns ratio $c = (a_t/b) = 1$.

The per-unit model shown in Figure 3.25(c) is perfectly valid, but it is not suitable for some of the computer programs presented in later chapters because these programs do not accommodate ideal transformer windings. An alternative representation can be developed, however, by writing nodal equations for this figure as follows:

FIGURE 3.25

Transformer with
off-nominal turns ratio

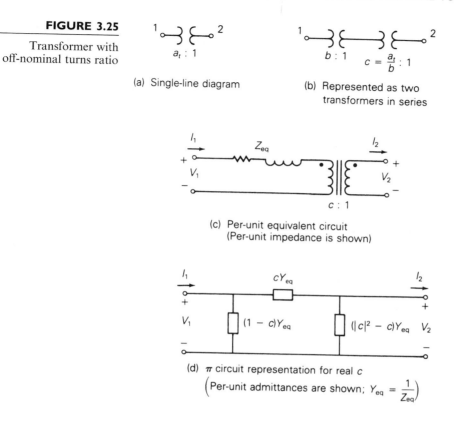

(a) Single-line diagram

(b) Represented as two
transformers in series

(c) Per-unit equivalent circuit
(Per-unit impedance is shown)

(d) π circuit representation for real c
$\left(\text{Per-unit admittances are shown; } Y_{eq} = \dfrac{1}{Z_{eq}}\right)$

$$\begin{bmatrix} I_1 \\ -I_2 \end{bmatrix} = \begin{bmatrix} Y_{11} & Y_{12} \\ Y_{21} & Y_{22} \end{bmatrix} \begin{bmatrix} V_1 \\ V_2 \end{bmatrix} \tag{3.8.4}$$

where both I_1 and $-I_2$ are referenced *into* their nodes in accordance with the nodal equation method (Section 2.4). Recalling two-port network theory, the admittance parameters of (3.8.4) are, from Figure 3.23(c)

$$Y_{11} = \left.\frac{I_1}{V_1}\right|_{V_2=0} = \frac{1}{Z_{eq}} = Y_{eq} \tag{3.8.5}$$

$$Y_{22} = \left.\frac{-I_2}{V_2}\right|_{V_1=0} = \frac{1}{Z_{eq}/|c|^2} = |c|^2 Y_{eq} \tag{3.8.6}$$

$$Y_{12} = \left.\frac{I_1}{V_2}\right|_{V_1=0} = \frac{-cV_2/Z_{eq}}{V_2} = -c Y_{eq} \tag{3.8.7}$$

$$Y_{21} = \left.\frac{-I_2}{V_1}\right|_{V_2=0} = \frac{-c^*I_1}{V_1} = -c^* Y_{eq} \tag{3.8.8}$$

Equations (3.8.4)–(3.8.8) with real or complex c are convenient for representing transformers with off-nominal turns ratios in the computer programs presented later. Note that when c is complex, Y_{12} is not equal to Y_{21}, and the preceding admittance parameters cannot be synthesized with a passive RLC circuit. However, the π network shown in Figure 3.25(d), which has the same admittance parameters as (3.8.4)–(3.8.8), can be synthesized for real c. Note also that when $c = 1$, the shunt branches in this figure become open circuits (zero per unit mhos), and the series branch becomes Y_{eq} per unit mhos (or Z_{eq} per unit ohms).

EXAMPLE 3.12   **Tap-changing three-phase transformer: per-unit positive-sequence network**

A three-phase generator step-up transformer is rated 1000 MVA, 13.8 kV Δ/345 kV Y with $Z_{eq} = j0.10$ per unit. The transformer high-voltage winding has $\pm 10\%$ taps. The system base quantities are

$$S_{\text{base}3\phi} = 500 \quad \text{MVA}$$

$$V_{\text{baseXLL}} = 13.8 \quad \text{kV}$$

$$V_{\text{baseHLL}} = 345 \quad \text{kV}$$

Determine the per-unit equivalent circuit for the following tap settings:

a. Rated tap

b. -10% tap (providing a 10% voltage decrease for the high-voltage winding)

Assume balanced positive-sequence operation. Neglect transformer winding resistance, exciting current, and phase shift.

SOLUTION

a. Using (3.8.1) and (3.8.2) with the low-voltage winding denoted winding 1,

$$a_t = \frac{13.8}{345} = 0.04 \qquad b = \frac{V_{\text{baseXLL}}}{V_{\text{baseHLL}}} = \frac{13.8}{345} = a_t \qquad c = 1$$

From (3.3.11)

$$Z_{\text{p.u.new}} = (j0.10)\left(\frac{500}{1000}\right) = j0.05 \quad \text{per unit}$$

The per-unit equivalent circuit, not including winding resistance, exciting current, and phase shift is:

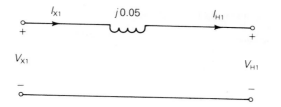

(Per-unit impedance is shown)

b. Using (3.8.1) and (3.8.2),

$$a_t = \frac{13.8}{345(0.9)} = 0.04444 \qquad b = \frac{13.8}{345} = 0.04$$

$$c = \frac{a_t}{b} = \frac{0.04444}{0.04} = 1.1111$$

From Figure 3.23(d),

$$c\,Y_{eq} = 1.1111\left(\frac{1}{j0.05}\right) = -j22.22 \quad \text{per unit}$$

$$(1-c)\,Y_{eq} = (-0.11111)(-j20) = +j2.222 \quad \text{per unit}$$

$$(|c|^2 - c)\,Y_{eq} = (1.2346 - 1.1)(-j20) = -j2.469 \quad \text{per unit}$$

The per-unit positive-sequence network is:

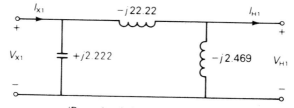

(Per-unit admittances are shown)

 Open PowerWorld Simulator case Example 3_12 and select **Simulation, Solve and Animate** to see an animated view of this LTC transformer example. Initially the generator/step-up transformer feeds a 500 MW/100 Mvar load. As is typical in practice, the transformer's taps are adjusted in discrete steps, with each step changing the tap ratio by 0.625% (hence a 10% change requires 16 steps). Click on arrows next to the transformer's tap to manually adjust the tap by one step. Note that changing the tap directly changes the load voltage.

 Because of the varying voltage drops caused by changing loads, LTCs are often operated to automatically regulate a bus voltage. This is particularly true when they are used as step-down transformers. To place the example transformer on automatic control, click on the "Manual" field. This toggles the transformer control mode to automatic. Now the transformer manually

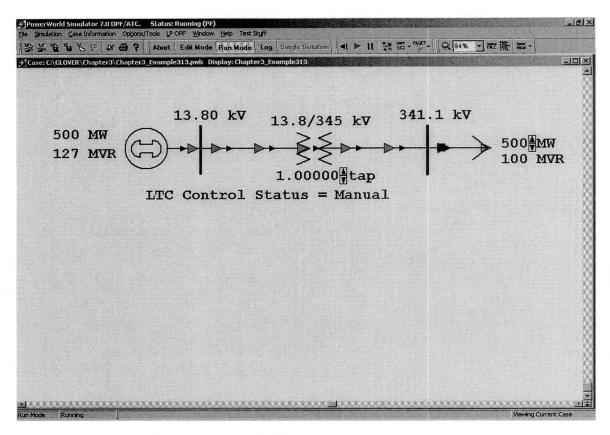

Screen for Example 3.12

changes its tap ratio to maintain the load voltage within a specified voltage range, between 0.995 and 1.005 per unit in this case. To see the LTC in automatic operation use the load arrows to vary the load, particularly the Mvar field, noting that the LTC changes to keep the load's voltage within the specified deadband. ∎

The three-phase regulating transformers shown in Figures 3.26 and 3.27 can be modeled as transformers with off-nominal turns ratios. For the voltage-magnitude-regulating transformer shown in Figure 3.26, adjustable voltages ΔV_{an}, ΔV_{bn}, and ΔV_{cn}, which have equal magnitudes ΔV and which are in phase with the phase voltages V_{an}, V_{bn}, and V_{cn}, are placed in the series link between buses a–a', b–b', and c–c'. Modeled as a transformer with an off-nominal turns ratio (see Figure 3.25), $c = (1 + \Delta V)$ for a voltage-magnitude increase toward bus abc, or $c = (1 + \Delta V)^{-1}$ for an increase toward bus $a'b'c'$.

For the phase-angle-regulating transformer in Figure 3.27, the series voltages ΔV_{an}, ΔV_{bn}, and ΔV_{cn} are ±90° out of phase with the phase voltages

FIGURE 3.26

An example of a
voltage-magnitude-
regulating transformer

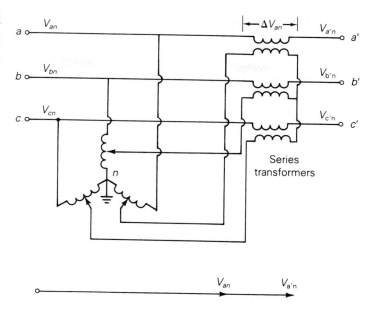

FIGURE 3.26

V_{an}, V_{bn}, and V_{cn}. The phasor diagram in Figure 3.27 indicates that each of
the bus voltages $V_{a'n}$, $V_{b'n}$, and $V_{c'n}$ has a phase shift that is approximately
proportional to the magnitude of the added series voltage. Modeled as a
transformer with an off-nominal turns ratio (see Figure 3.25), $c \approx 1\underline{/\alpha}$ for a
phase increase toward bus abc or $c \approx 1\underline{/-\alpha}$ for a phase increase toward bus
$a'b'c'$.

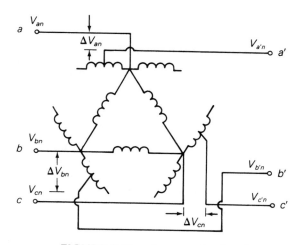

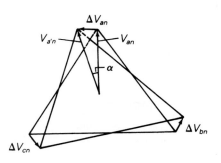

FIGURE 3.27   An example of a phase-angle-regulating transformer. Windings drawn in parallel are
on the same core

EXAMPLE 3.13 **Voltage-regulating and phase-shifting three-phase transformers**

Two buses abc and $a'b'c'$ are connected by two parallel lines L1 and L2 with positive-sequence series reactances $X_{L1} = 0.25$ and $X_{L2} = 0.20$ per unit. A regulating transformer is placed in series with line L1 at bus $a'b'c'$. Determine the 2×2 bus admittance matrix when the regulating transformer (a) provides a 0.05 per-unit increase in voltage magnitude toward bus $a'b'c'$ and (b) advances the phase $3°$ toward bus $a'b'c'$. Assume that the regulating transformer is ideal. Also, the series resistance and shunt admittance of the lines are neglected.

SOLUTION The circuit is shown in Figure 3.28.

a. For the voltage-magnitude-regulating transformer, $c = (1 + \Delta V)^{-1} = (1.05)^{-1} = 0.9524$ per unit. From (3.7.5)–(3.7.8), the admittance parameters of the regulating transformer in series with line L1 are

$$Y_{11L1} = \frac{1}{j0.25} = -j4.0$$

$$Y_{22L1} = (0.9524)^2(-j4.0) = -j3.628$$

$$Y_{12L1} = Y_{21L1} = (-0.9524)(-j4.0) = j3.810$$

For line L2 alone,

$$Y_{11L2} = Y_{22L2} = \frac{1}{j0.20} = -j5.0$$

$$Y_{12L2} = Y_{21L2} = -(-j5.0) = j5.0$$

Combining the above admittances in parallel,

$$Y_{11} = Y_{11L1} + Y_{11L2} = -j4.0 - j5.0 = -j9.0$$

$$Y_{22} = Y_{22L1} + Y_{22L2} = -j3.628 - j5.0 = -j8.628$$

$$Y_{12} = Y_{21} = Y_{12L1} + Y_{12L2} = j3.810 + j5.0 = j8.810 \quad \text{per unit}$$

FIGURE 3.28

Positive-sequence circuit for Example 3.13

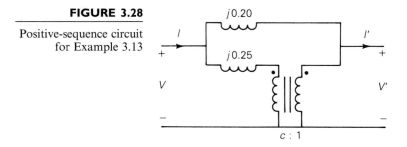

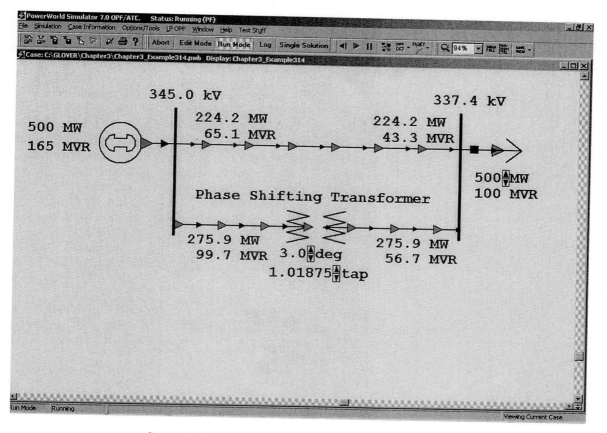

Screen for Example 3.13

b. For the phase-angle-regulating transformer, $c = 1\underline{/-\alpha} = 1\underline{/-3^\circ}$. Then, for this regulating transformer in series with line L1,

$$Y_{11\text{L}1} = \frac{1}{j0.25} = -j4.0$$

$$Y_{22\text{L}1} = |1.0\underline{/-3^\circ}|^2 (-j4.0) = -j4.0$$

$$Y_{12\text{L}1} = -(1.0\underline{/-3^\circ})(-j4.0) = 4.0\underline{/87^\circ} = 0.2093 + j3.9945$$

$$Y_{21\text{L}1} = -(1.0\underline{/-3^\circ})^*(-j4.0) = 4.0\underline{/93^\circ} = -0.2093 + j3.9945$$

The admittance parameters for line L2 alone are given in part (a) above. Combining the admittances in parallel,

$$Y_{11} = Y_{22} = -j4.0 - j5.0 = -j9.0$$

$$Y_{12} = 0.2093 + j3.9945 + j5.0 = 0.2093 + j8.9945$$

$$Y_{21} = -0.2093 + j3.9945 + j5.0 = -0.2093 + j8.9945 \quad \text{per unit}$$

An animated view of this example is provided in PowerWorld Simulator case Example 3_13. In this case, the transformer and a parallel transmission line are assumed to be supplying power from a 345-kV generator to a 345-kV load, with an initial phase angle of 3 degrees. First select **Simulation, Solve and Animate**, then click on the arrows next to the phase angle to change the angle in one-degree steps, or by the tap field to change the LTC tap in 0.625% steps. Notice that changing the phase angle primarily changes the real power flow, whereas changing the LTC tap changes the reactive power flow. In this example, the line flow fields are showing the absolute value of the real or reactive power flow; the direction of the flow is indicated with arrows. Traditional power flow programs usually indicate power flow direction using a convention that flow into the line or other device is assumed to be positive. Using this convention with the figure, the flows on the left side of the line would be positive and those on the right would be negative. You can display results in PowerWorld Simulator using this convention by unchecking the "Use Absolute Values for MW/Mvar Line Flows" field on the Display Options page of the Oneline Display Options dialog. ∎

Note that a voltage-magnitude-regulating transformer controls the *reactive* power flow in the series link in which it is installed, whereas a phase-angle-regulating transformer controls the *real* power flow (see Problem 3.59).

PROBLEMS

SECTION 3.1

3.1 (a) An ideal single-phase two-winding transformer with turns ratio $a_t = N_1/N_2$ is connected with a series impedance Z_2 across winding 2. If one wants to replace Z_2, with a series impedance Z_1 across winding 1 and keep the terminal behavior of the two circuits to be identical, find Z_1 in terms of Z_2.
(b) Would the above result be true if instead of a series impedance there is a shunt impedance?
(c) Can one refer a ladder network on the secondary (2) side to the primary (1) side simply by multiplying every impendance by a_t^2?

3.2 An ideal transformer with $N_1 = 2000$ and $N_2 = 500$ is connected with an impedance Z_{2_2} across winding 2, called secondary. If $V_1 = 1000 \underline{/0°}$ V and $I_1 = 5 \underline{/-30°}$ A, determine V_2, I_2, Z_2, and the impedance Z_2', which is the value of Z_2 referred to the primary side of the transformer.

3.3 Consider an ideal transformer with $N_1 = 3000$ and $N_2 = 500$ turns. Let winding 1 be connected to a source whose voltage is $e_1(t) = 100(1 - |t|)$ volts for $-1 \leq t \leq 1$ and $e_1(t) = 0$ for $|t| > 1$ second. A 3-farad capacitor is connected across winding 2. Sketch $e_1(t)$, $e_2(t)$, $i_1(t)$, and $i_2(t)$ versus time t.

3.4 A single-phase 100-kVA, 2400/240-volt, 60-Hz distribution transformer is used as a step-down transformer. The load, which is connected to the 240-volt secondary winding, absorbs 80 kVA at 0.8 power factor lagging and is at 230 volts. Assuming an ideal transformer, calculate the following: (a) primary voltage, (b) load impedance, (c) load impedance referred to the primary, and (d) the real and reactive power supplied to the primary winding.

3.5 Rework Problem 3.4 if the load connected to the 240-V secondary winding absorbs 110 kVA under short-term overload conditions at 0.85 power factor leading and at 230 volts.

3.6 For a conceptual single-phase, phase-shifting transformer, the primary voltage leads the secondary voltage by 30°. A load connected to the secondary winding absorbs 50 kVA at 0.9 power factor leading and at a voltage $E_2 = 277\underline{/0°}$ volts. Determine (a) the primary voltage, (b) primary and secondary currents, (c) load impedance referred to the primary winding, and (d) complex power supplied to the primary winding.

3.7 Consider a source of voltage $v(t) = 10\sqrt{2}\,\sin(2t)$V, with an internal resistance of 1800 Ω. A transformer that can be considered as ideal is used to couple a 50-Ω resistive load to the source. (a) Determine the transformer primary-to-secondary turns ratio required to ensure maximum power transfer by matching the load and source resistances. (b) Find the average power delivered to the load, assuming maximum power transfer.

3.8 For the circuit shown in Figure 3.29, determine $v_{out}(t)$.

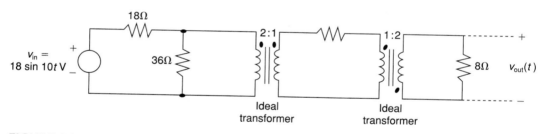

FIGURE 3.29   Problem 3.8

SECTION 3.2

3.9 A single-phase transformer has 2000 turns on the primary winding and 500 turns on the secondary. Winding resistances are $R_1 = 2\ \Omega$ and $R_2 = 0.125\ \Omega$; leakage reactances are $X_1 = 8\ \Omega$ and $X_2 = 0.5\ \Omega$. The resistance load on the secondary is 12 Ω.
(a) If the applied voltage at the terminals of the primary is 1000 V, determine V_2 at the load terminals of the transformer, neglecting magnetizing current.
(b) If the voltage regulation is defined as the difference between the voltage magnitude at the load terminals of the transformer at full load and at no load in percent of full-load voltage with input voltage held constant, compute the percent voltage regulation.

3.10 A single-phase step-down transformer is rated 15 MVA, 66 kV/11.5 kV. With the 11.5 kV winding short-circuited, rated current flows when the voltage applied to the primary is 5.5 kV. The power input is read as 100 kW. Determine R_{eq1} and X_{eq1} in ohms referred to the high-voltage winding.

3.11 For the transformer in Problem 3.10, the open-circuit test with 11.5 kV applied results in a power input of 65 kW and a current of 30 A. Compute the values for G_c and B_m in siemens referred to the high-voltage winding. Compute the efficiency of the transformer for a load of 10 MW at 0.8 PF lagging at rated voltage.

3.12 The following data are obtained when open-circuit and short-circuit tests are performed on a single-phase, 50-kVA, 2400/240-volt, 60-Hz distribution transformer.

	VOLTAGE (volts)	CURRENT (amperes)	POWER (watts)
Measurements on low-voltage side with high-voltage winding open	240	5.97	213
Measurements on high-voltage side with low-voltage winding shorted	60.0	20.8	750

(a) Neglecting the series impedance, determine the exciting admittance referred to the high-voltage side. (b) Neglecting the exciting admittance, determine the equivalent series impedance referred to the high-voltage side. (c) Assuming equal series impedances for the primary and referred secondary, obtain an equivalent T-circuit referred to the high-voltage side.

3.13 A single-phase 100-kVA, 2400/240-volt, 60-Hz distribution transformer has a 2-ohm equivalent leakage reactance and a 6000-ohm magnetizing reactance referred to the high-voltage side. If rated voltage is applied to the high-voltage winding, calculate the open-circuit secondary voltage. Neglect I^2R and G_c^2V losses. Assume equal series leakage reactances for the primary and referred secondary.

3.14 A single-phase 50-kVA, 2400/240-volt, 60-Hz distribution transformer is used as a step-down transformer at the load end of a 2400-volt feeder whose series impedance is $(1.0 + j2.0)$ ohms. The equivalent series impedance of the transformer is $(1.0 + j2.5)$ ohms referred to the high-voltage (primary) side. The transformer is delivering rated load at 0.8 power factor lagging and at rated secondary voltage. Neglecting the transformer exciting current, determine (a) the voltage at the transformer primary terminals, (b) the voltage at the sending end of the feeder, and (c) the real and reactive power delivered to the sending end of the feeder.

3.15 Rework Problem 3.14 if the transformer is delivering rated load at rated secondary voltage and at (a) unity power factor, (b) 0.8 power factor leading. Compare the results with those of Problem 3.14.

3.16 A single-phase, 50-kVA, 2400/240-V, 60-Hz distribution transformer has the following parameters:

Resistance of the 2400-V winding: $R_1 = 0.75 \ \Omega$

Resistance of the 240-V winding: $R_2 = 0.0075 \ \Omega$

Leakage reactance of the 2400-V winding: $X_1 = 1.0 \ \Omega$

Leakage reactance of the 240-V winding: $X_2 = 0.01 \ \Omega$

Exciting admittance on the 240-V side $= 0.003 - j0.02 \ S$

(a) Draw the equivalent circuit referred to the high-voltage side of the transformer. (b) Draw the equivalent circuit referred to the low-voltage side of the transformer. Show the numerical values of impedances on the equivalent circuits.

3.17   The transformer of Problem 3.16 is supplying a rated load of 50 kVA at a rated secondary voltage of 240 V and at 0.8 power factor lagging. Neglecting the transformer exciting current, (a) Determine the input terminal voltage of the transformer on the high-voltage side. (b) Sketch the corresponding phasor diagram. (c) If the transformer is used as a step-down transformer at the load end of a feeder whose impedance is $0.5 + j2.0\ \Omega$, find the voltage V_S and the power factor at the sending end of the feeder.

SECTION 3.3

3.18   Using the transformer ratings as base quantities, work Problem 3.13 in per-unit.

3.19   Using the transformer ratings as base quantities, work Problem 3.14 in per-unit.

3.20   Using base values of 20 kVA and 115 volts in zone 3, rework Example 3.4.

3.21   Rework Example 3.5, using $S_{base3\phi} = 100$ kVA and $V_{baseLL} = 600$ volts.

3.22   A balanced Y-connected voltage source with $E_{ag} = 277\underline{/0^\circ}$ volts is applied to a balanced-Y load in parallel with a balanced-Δ load, where $Z_Y = 30 + j10$ and $Z_\Delta = 45 - j25$ ohms. The Y load is solidly grounded. Using base values of $S_{base1\phi} = 5$ kVA and $V_{baseLN} = 277$ volts, calculate the source current I_a in per-unit and in amperes.

3.23   Figure 3.30 shows the one-line diagram of a three-phase power system. By selecting a common base of 100 MVA and 22 kV on the generator side, draw an impedance diagram showing all impedances including the load impedance in per-unit. The data are given as follows:

$$
\begin{array}{llll}
G: & 90\ \text{MVA} & 22\ \text{kV} & x = 0.18\ \text{per unit} \\
T1: & 50\ \text{MVA} & 22/220\ \text{kV} & x = 0.10\ \text{per unit} \\
T2: & 40\ \text{MVA} & 220/11\ \text{kV} & x = 0.06\ \text{per unit} \\
T3: & 40\ \text{MVA} & 22/110\ \text{kV} & x = 0.064\ \text{per unit} \\
T4: & 40\ \text{MVA} & 110/11\ \text{kV} & x = 0.08\ \text{per unit} \\
M: & 66.5\ \text{MVA} & 10.45\ \text{kV} & x = 0.185\ \text{per unit}
\end{array}
$$

Lines 1 and 2 have series reactances of 48.4 and 65.43 Ω, respectively. At bus 4, the three-phase load absorbs 57 MVA at 10.45 kV and 0.6 power factor lagging.

FIGURE 3.30

Problem 3.23

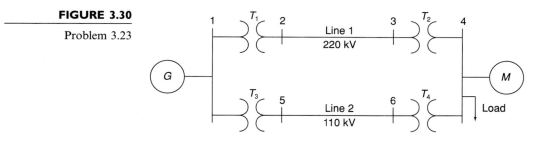

3.24 For Problem 3.18, the motor operates at full load, at 0.8 power factor leading, and at a terminal voltage of 10.45 kV. Determine (a) the voltage at bus 1, the generator bus, and (b) the generator and motor internal EMFs.

3.25 Consider a single-phase electric system shown in Figure 3.31.
Transformers are rated as follows:
X–Y 15 MVA, 13.8/138 kV, leakage reactance 10%
Y–Z 15 MVA, 138/69 kV, leakage reactance 8%
With the base in circuit Y chosen as 15 MVA, 138 kV, determine the per-unit impedance of the 500 Ω resistive load in circuit Z, referred to circuits Z, Y, and X. Neglecting magnetizing currents, transformer resistances, and line impedances, draw the impedance diagram in per unit.

FIGURE 3.31

Single-phase electric
system for Problem 3.25

3.26 A bank of three single-phase transformers, each rated 30 MVA, 38.1/3.81 kV, are connected in Y–Δ with a balanced load of three 1 – Ω, wye-connected resistors. Choosing a base of 90 MVA, 66 kV for the high-voltage side of the three-phase transformer, specify the base for the low-voltage side. Compute the per-unit resistance of the load on the base for the low-voltage side. Also, determine the load resistance in ohms referred to the high-voltage side and the per-unit value on the chosen base.

3.27 A three-phase transformer is rated 500 MVA, 220 Y/22 Δ kV. The wye-equivalent short-circuit impedance, considered equal to the leakage reactance, measured on the low-voltage side is 0.1 Ω. Compute the per-unit reactance of the transformer. In a system in which the base on the high-voltage side of the transformer is 100 MVA, 230 kV, what value of the per-unit reactance should be used to represent this transformer?

3.28 For the system shown in Figure 3.32, draw an impedance diagram in per unit, by choosing 100 kVA to be the base kVA and 2400 V as the base voltage for the generators.

FIGURE 3.32

System for Problem 3.28

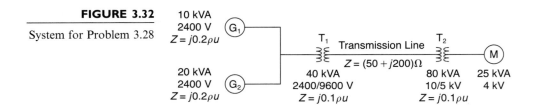

3.29   Consider three ideal single-phase transformers (with a voltage gain of η) put together as a delta-wye three-phase bank as shown in Figure 3.33. Assuming positive-sequence voltages for V_{an}, V_{bn}, and V_{cn}, find $V_{a'n'}$, $V_{b'n'}$, and $V_{c'n'}$ in terms of V_{an}, V_{bn}, and V_{cn}, respectively.

(a) Would such relationships hold for the line voltages as well?

(b) Looking into the current relationships, express I_a', I_b', and I_c' in terms of I_a, I_b, and I_c, respectively.

(c) Let S' and S be the per-phase complex power output and input, respectively. Find S' in terms of S.

FIGURE 3.33

Δ–Y connection for Problem 3.29

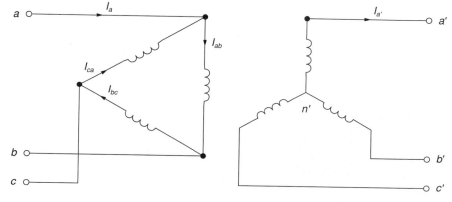

3.30   Reconsider Problem 3.29. If V_{an}, V_{bn}, and V_{cn} are a negative-sequence set, how would the voltage and current relationships change?

(a) If C_1 is the complex positive-sequence voltage gain in Problem 3.29, and C_2 is the negative sequence complex voltage gain, express the relationship between C_1 and C_2.

3.31   If positive-sequence voltages are assumed and the wye-delta connection is considered, again with ideal transformers as in Problem 3.29, find the complex voltage gain C_3.

(a) What would the gain be for a negative-sequence set?

(b) Comment on the complex power gain.

(c) When terminated in a symmetric wye-connected load, find the referred impedance Z_L', the secondary impedance Z_L referred to primary (i.e., the per-phase driving-point impedance on the primary side), in terms of Z_L and the complex voltage gain C.

SECTION 3.4

3.32   Determine the positive- and negative-sequence phase shifts for the three-phase transformers shown in Figure 3.34.

3.33   Consider the three single-phase two-winding transformers shown in Figure 3.35. The high-voltage windings are connected in Y. (a) For the low-voltage side, connect the windings in Δ, place the polarity marks, and label the terminals a, b, and c in accordance with the American standard. (b) Relabel the terminals a', b', and c' such that V_{AN} is 90° out of phase with $V_{a'n}$ for positive sequence.

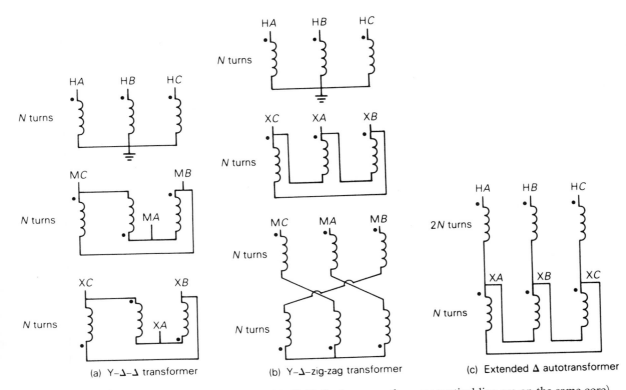

(a) Y–Δ–Δ transformer   (b) Y–Δ–zig-zag transformer   (c) Extended Δ autotransformer

FIGURE 3.34   Problems 3.32 and 3.52 (Coils drawn on the same vertical line are on the same core)

FIGURE 3.35

Problem 3.33

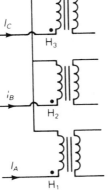

3.34   Three single-phase, two-winding transformers, each rated 450 MVA, 20 kV/288.7 kV, with leakage reactance $X_{eq} = 0.10$ per unit, are connected to form a three-phase bank. The high-voltage windings are connected in Y with a solidly grounded neutral. Draw the per-unit equivalent circuit if the low-voltage windings are connected (a) in Δ with American standard phase shift, (b) in Y with an open neutral. Use the transformer ratings as base quantities. Winding resistances and exciting current are neglected.

3.35   Consider a bank of three single-phase two-winding transformers whose high-voltage terminals are connected to a three-phase, 13.8-kV feeder. The low-voltage terminals are connected to a three-phase substation load rated 2.4 MVA and 2.3 kV. Determine the required voltage, current, and MVA ratings of both windings of each transformer, when the high-voltage/low-voltage windings are connected (a) Y–Δ, (b) Δ–Y, (c) Y–Y, and (d) Δ–Δ.

3.36   Three single-phase two-winding transformers, each rated 25 MVA, 34.5/13.8 kV, are connected to form a three-phase Δ–Δ bank. Balanced positive-sequence voltages are applied to the high-voltage terminals, and a balanced, resistive Y load connected to the low-voltage terminals absorbs 75 MW at 13.8 kV. If one of the single-phase transformers is removed (resulting in an open-Δ connection) and the balanced load is simultaneously reduced to 43.3 MW (57.7% of the original value), determine (a) the load voltages V_{an}, V_{bn}, and V_{cn}; (b) load currents I_a, I_b, and I_c; and (c) the MVA supplied by each of the remaining two transformers. Are balanced voltages still applied to the load? Is the open-Δ transformer overloaded?

3.37   Three single-phase two-winding transformers, each rated 25 MVA, 38.1/3.81 kV, are connected to form a three-phase Y–Δ bank with a balanced Y-connected resistive load of 0.6 Ω per phase on the low-voltage side. By choosing a base of 75 MVA (three phase) and 66 kV (line-to-line) for the high voltage side of the transformer bank, specify the base quantities for the low-voltage side. Determine the per-unit resistance of the load on the base for the low-voltage side. Then determine the load resistance R_L in ohms referred to the high-voltage side and the per-unit value of this load resistance on the chosen base.

3.38   Consider a three-phase generator rated 300 MVA, 23 kV, supplying a system load of 240 MVA and 0.9 power factor lagging at 230 kV through a 330 MVA, 23 Δ/230 Y-kV step-up transformer with a leakage reactance of 0.11 per unit. (a) Neglecting the exciting current and choosing base values at the load of 100 MVA and 230 kV, find the phasor currents I_A, I_B, and I_C supplied to the load in per unit. (b) By choosing the load terminal voltage V_A as reference, specify the proper base for the generator circuit and determine the generator voltage V as well as the phasor currents I_a, I_b, and I_c, from the generator. (*Note:* Take into account the phase shift of the transformer.) (c) Find the generator terminal voltage in kV and the real power supplied by the generator in MW. (d) By omitting the transformer phase shift altogether, check to see whether you get the same magnitude of generator terminal voltage and real power delivered by the generator.

SECTION 3.5

3.39   The leakage reactance of a three-phase, 500-MVA, 345 Y/23 Δ-kV transformer is 0.09 per unit based on its own ratings. The Y winding has a solidly grounded neutral. Draw the per-unit equivalent circuit. Neglect the exciting admittance and assume American standard phase shift.

3.40   Choosing system bases to be 360/24 kV and 100 MVA, redraw the per-unit equivalent circuit for Problem 3.39.

3.41   Consider the single-line diagram of the power system shown in Figure 3.36. Equipment ratings are:

generator 1:	750 MVA, 18 kV, $X'' = 0.2$ per unit
generator 2:	750 MVA, 18 kV, $X'' = 0.2$
synchronous motor 3:	1500 MVA, 20 kV, $X'' = 0.2$
3-phase Δ–Y transformers T_1, T_2, T_3, T_4:	750 MVA, 500 kV Y/20 kV Δ, X = 0.1
3-phase Y–Y transformer T_5:	1500 MVA, 500 kV Y/20 kV Y, X = 0.1

Neglecting resistance, transformer phase shift, and magnetizing reactance, draw the equivalent reactance diagram. Use a base of 100 MVA and 500 kV for the 40-ohm line. Determine the per-unit reactances.

FIGURE 3.36

Problems 3.41 and 3.42

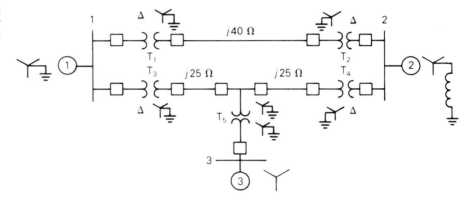

3.42   For the power system in Problem 3.41, the synchronous motor absorbs 1200 MW at 0.8 power factor leading with the bus 3 voltage at 18 kV. Determine the bus 1 and bus 2 voltages in kV. Assume that generators 1 and 2 deliver equal real powers and equal reactive powers. Also assume a balanced three-phase system with positive-sequence sources.

3.43   Three single-phase transformers, each rated 10 MVA, 66.4/12.5 kV, 60 Hz, with an equivalent series reactance of 0.12 per unit divided equally between primary and secondary, are connected in a three-phase bank. The high-voltage windings are Y connected and their terminals are directly connected to a 115-kV three-phase bus. The secondary terminals are all shorted together. Find the currents entering the high-voltage terminals and leaving the low-voltage terminals if the low-voltage windings are (a) Y connected, (b) Δ connected.

3.44   A 100-MVA, 13.2-kV three-phase generator, which has a positive-sequence reactance of 1.2 per unit on the generator base, is connected to a 110-MVA, 13.2 Δ/115 Y-kV step-up transformer with a series impedance of $(0.005 + j0.1)$ per unit on its own base. (a) Calculate the per-unit generator reactance on the transformer base. (b) The load at the transformer terminals is 80 MW at unity power factor and at 115 kV. Choosing the transformer high-side voltage as the reference phasor, draw a phasor diagram for this condition. (c) For the condition of part (b), find the transformer low-side voltage and the generator internal voltage behind its reactance. Also compute the generator output power and power factor.

3.45   Figure 3.37 shows a one-line diagram of a system in which the three-phase generator is rated 300 MVA, 20 kV with a subtransient reactance of 0.2 per unit and with its neutral grounded through a 0.4-Ω reactor. The transmission line is 64 km long with a series reactance of 0.5 Ω/km. The three-phase transformer T_1 is rated 350 MVA, 230/ 20 kV with a leakage reactance of 0.1 per unit. Transformer T_2 is composed of three single-phase transformers, each rated 100 MVA, 127/13.2 kV with a leakage reactance of 0.1 per unit. Two 13.2-kV motors M_1 and M_2 with a subtransient reactance of 0.2 per unit for each motor represent the load. M_1 has a rated input of 200 MVA with its neutral grounded through a 0.4-Ω current-limiting reactor. M_2 has a rated input of 100 MVA with its neutral not connected to ground. Neglect phase shifts associated with the transformers. Choose the generator rating as base in the generator circuit and draw the positive-sequence reactance diagram showing all reactances in per unit.

FIGURE 3.37

Problems 3.45 and 3.46

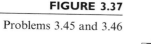
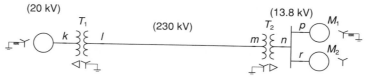

3.46   The motors M_1 and M_2 of Problem 3.45 have inputs of 120 and 60 MW, respectively, at 13.2 kV, and both operate at unity power factor. Determine the generator terminal voltage and voltage regulation of the line. Neglect transformer phase shifts.

3.47   Consider the one-line diagram shown in Figure 3.38. The three-phase transformer bank is made up of three identical single-phase transformers, each specified by $X_l = 0.24$ Ω (on the low-voltage side), negligible resistance and magnetizing current, and turns ratio $\eta = N_2/N_1 = 10$. The transformer bank is delivering 100 MW at 0.8 PF lagging to a substation bus whose voltage is 230 kV.
(a) Determine the primary current magnitude, primary voltage (line-to-line) magnitude, and the three-phase complex power supplied by the generator. Choose the line-to-neutral voltage at the bus, $V_{a'n'}$, as the reference. Account for the phase shift, and assume positive-sequence operation.
(b) Find the phase shift between the primary and secondary voltages.

FIGURE 3.38

One-line diagram for
Problem 3.47

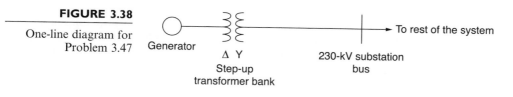

3.48   With the same transformer banks as in Problem 3.47, Figure 3.39 shows the one-line diagram of a generator, a step-up transformer bank, a transmission line, a step-down

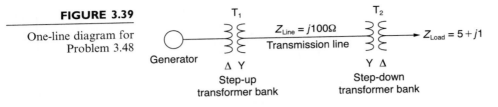

transformer bank, and an impedance load. The generator terminal voltage is 15 kV
(line-to-line).
(a) Draw the per-phase equivalent circuit, accounting for phase shifts for positive-
sequence operation.
(b) By choosing the line-to-neutral generator terminal voltage as the reference, determine the magnitudes of the generator current, transmission-line current, load current,
and line-to-line load voltage. Also, find the three-phase complex power delivered to
the load.

3.49 Consider the single-line diagram of a power system shown in Figure 3.40 with equipment ratings given below:

Generator G_1:	50 MVA, 13.2 kV, $x = 0.15$ pu
Generator G_2:	20 MVA, 13.8 kV, $x = 0.15$ pu
3-phase Δ–Y transformer T_1:	80 MVA, 13.2 Δ/165 Y kV, $X = 0.1$ pu
3-phase Y–Δ transformer T_2:	40 MVA, 165 Y/13.8 Δ kV, $X = 0.1$ pu
Load:	40 MVA, 0.8 PF lagging, operating at 150 kV

Choose a base of 100 MVA for the system and 132-kV base in the transmission-line
circuit. Let the load be modeled as a parallel combination of resistance and inductance. Neglect transformer phase shifts.
Draw a per-phase equivalent circuit of the system showing all impedances in per unit.

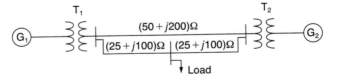

SECTION 3.6

3.50 A single-phase three-winding transformer has the following parameters: $Z_1 = Z_2 =$
$Z_3 = 0 + j0.06$, $G_c = 0$, and $B_m = 0.2$ per unit. Three identical transformers, as
described, are connected with their primaries in Y (solidly grounded neutral) and
with their secondaries and tertiaries in Δ. Draw the per-unit sequence networks of this
transformer bank.

3.51 The ratings of a three-phase three-winding transformer are:

Primary (1):	Y connected, 66 kV, 20 MVA
Secondary (2):	Y connected, 13.2 kV, 15 MVA
Tertiary (3):	Δ connected, 2.3 kV, 5 MVA

Neglecting winding resistances and exciting current, the per-unit leakage reactances are:

$$X_{12} = 0.08 \text{ on a } 20\text{-MVA}, 66\text{-kV base}$$

$$X_{13} = 0.10 \text{ on a } 20\text{-MVA}, 66\text{-kV base}$$

$$X_{23} = 0.09 \text{ on a } 15\text{-MVA}, 13.2\text{-kV base}$$

(a) Determine the per-unit reactances X_1, X_2, X_3 of the equivalent circuit on a 20-MVA, 66-kV base at the primary terminals. (b) Purely resistive loads of 12 MW at 13.2 kV and 5 MW at 2.3 kV are connected to the secondary and tertiary sides of the transformer, respectively. Draw the per-unit impedance diagram, showing the per-unit impedances on a 20-MVA, 66-kV base at the primary terminals.

3.52 Draw the per-unit equivalent circuit for the transformers shown in Figure 3.34. Include ideal phase-shifting transformers showing phase shifts determined in Problem 3.32. Assume that all windings have the same kVA rating and that the equivalent leakage reactance of any two windings with the third winding open is 0.10 per unit. Neglect the exciting admittance.

3.53 The ratings of a three-phase, three-winding transformer are:

Primary: Y connected, 66 kV, 15 MVA

Secondary: Y connected, 13.2 kV, 10 MVA

Tertiary: Δ connected, 2.3 kV, 5 MVA

Neglecting resistances and exciting current, the leakage reactances are:

$$X_{PS} = 0.07 \text{ per unit on a } 15\text{-MVA}, 66\text{-kV base}$$

$$X_{PT} = 0.09 \text{ per unit on a } 15\text{-MVA}, 66\text{-kV base}$$

$$X_{ST} = 0.08 \text{ per unit on a } 10\text{-MVA}, 13.2\text{-kV base}$$

Determine the per-unit reactances of the per-phase equivalent circuit using a base of 15 MVA and 66 kV for the primary.

3.54 An infinite bus, which is a constant voltage source, is connected to the primary of the three-winding transformer of Problem 3.53. A 7.5-MVA, 13.2-kV synchronous motor with a subtransient reactance of 0.2 per unit is connected to the transformer secondary. A 5-MW, 2.3-kV three-phase resistive load is connected to the tertiary. Choosing a base of 66 kV and 15 MVA in the primary, draw the impedance diagram of the system showing per-unit impedances. Neglect transformer exciting current, phase shifts, and all resistances except the resistive load.

SECTION 3.7

3.55 A single-phase 15-kVA, 2400/240-volt, 60-Hz two-winding distribution transformer is connected as an autotransformer to step up the voltage from 2400 to 2640 volts. (a) Draw a schematic diagram of this arrangement, showing all voltages and currents when delivering full load at rated voltage. (b) Find the permissible kVA rating of the autotransformer if the winding currents and voltages are not to exceed the rated values as a two-winding transformer. How much of this kVA rating is transformed by magnetic induction? (c) The following data are obtained from tests carried out on the transformer when it is connected as a two-winding transformer:

Open-circuit test with the low-voltage terminals excited:
Applied voltage = 240 V, Input current = 0.68 A, Input power = 105 W.

Short-circuit test with the high-voltage terminals excited:
Applied voltage = 120 V, Input current = 6.25 A, Input power = 330 W.

Based on the data, compute the efficiency of the autotransformer corresponding to full load, rated voltage, and 0.8 power factor lagging. Comment on why the efficiency is higher as an autotransformer than as a two-winding transformer.

3.56   Three single-phase two-winding transformers, each rated 3 kVA, 220/110 volts, 60 Hz, with a 0.10 per-unit leakage reactance, are connected as a three-phase extended Δ autotransformer bank, as shown in Figure 3.31(c). The low-voltage Δ winding has a 110 volt rating. (a) Draw the positive-sequence phasor diagram and show that the high-voltage winding has a 479.5 volt rating. (b) A three-phase load connected to the low-voltage terminals absorbs 6 kW at 110 volts and at 0.8 power factor lagging. Draw the per-unit impedance diagram and calculate the voltage and current at the high-voltage terminals. Assume positive-sequence operation.

3.57   A two-winding single-phase transformer rated 60 kVA, 240/1200 V, 60 Hz, has an efficiency of 0.96 when operated at rated load, 0.8 power factor lagging. This transformer is to be utilized as a 1440/1200-V step-down autotransformer in a power distribution system. (a) Find the permissible kVA rating of the autotransformer if the winding currents and voltages are not to exceed the ratings as a two-winding transformer. Assume an ideal transformer. (b) Determine the efficiency of the autotransformer with the kVA loading of part (a) and 0.8 power factor leading.

3.58   A single-phase two-winding transformer rated 90 MVA, 80/120 kV is to be connected as an autotransformer rated 80/200 kV. Assume that the transformer is ideal. (a) Draw a schematic diagram of the ideal transformer connected as an autotransformer, showing the voltages, currents, and dot notation for polarity. (b) Determine the permissible kVA rating of the autotransformer if the winding currents and voltages are not to exceed the rated values as a two-winding transformer. How much of the kVA rating is transferred by magnetic induction?

SECTION 3.8

3.59   The two parallel lines in Example 3.13 supply a balanced load with a load current of $1.0/\underline{-30°}$ per unit. Determine the real and reactive power supplied to the load bus from each parallel line with (a) no regulating transformer, (b) the voltage-magnitude-regulating transformer in Example 3.13(a), and (c) the phase-angle-regulating transformer in Example 3.13(b). Assume that the voltage at bus *abc* is adjusted so that the voltage at bus $a'b'c'$ remains constant at $1.0/\underline{0°}$ per unit. Also assume positive sequence. Comment on the effects of the regulating transformers.

PW   **3.60**   PowerWorld Simulator case Problem 3.60 duplicates Example 3.13 except that a resistance term of 0.05 per unit has been added to the first line and 0.04 per unit to the second. Since the system is no longer lossless, a field showing the real power losses has also been added to the one-line. With the LTC tap fixed at 1.05, plot the real power losses as the phase shift angle is varied from −10 to +10 degrees. What value of phase shift minimizes the system losses?

PW   **3.61**   Repeat Problem 3.60, except keep the phase-shift angle fixed at 3.0 degrees, while varying the LTC tap between 0.9 and 1.1. What tap value minimizes the real power losses?

3.62 Rework Example 3.12 for a +10% tap, providing a 10% increase for the high-voltage winding.

3.63 A 23/230-kV step-up transformer feeds a three-phase transmission line, which in turn supplies a 150-MVA, 0.8 lagging power factor load through a step-down 230/23-kV transformer. The impedance of the line and transformers at 230 kV is $18 + j60$ Ω. Determine the tap setting for each transformer to maintain the voltage at the load at 23 kV.

3.64 The per-unit equivalent circuit of two transformers T_a and T_b connected in parallel, with the same nominal voltage ratio and the same reactance of 0.1 per unit on the same base, is shown in Figure 3.41. Transformer T_b has a voltage-magnitude step-up toward the load of 1.05 times that of T_b (that is, the tap on the secondary winding of T_a is set to 1.05). The load is represented by $0.8 + j0.6$ per unit at a voltage $V_2 = 1.0/0°$ per unit. Determine the complex power in per unit transmitted to the load through each transformer. Comment on how the transformers share the real and reactive powers.

FIGURE 3.41

Problem 3.64

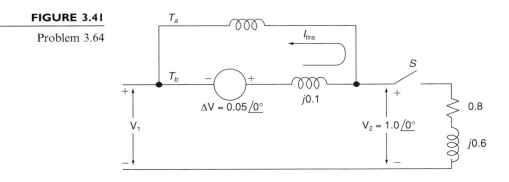

3.65 Reconsider Problem 3.64 with the change that now T_b includes both a transformer of the same turns ratio as T_a and a regulating transformer with a 3° phase shift. On the base of T_a, the impedance of the two components of T_b is $j0.1$ per unit. Determine the complex power in per unit transmitted to the load through each transformer. Comment on how the transformers share the real and reactive powers.

CASE STUDY QUESTIONS

A. The case study for this chapter describes ways to extend the life of aging power transformers and paper-insulated-lead-covered (PILC) cables. What is the impact of an aging power-system equipment on power system outages—their frequency and duration—and power system reliability?

B. What are the advantages of extending the useful service life of power transformers and PILC cables?

REFERENCES

1. R. Feinberg, *Modern Power Transformer Practice* (New York: Wiley, 1979).

2. A. C. Franklin and D. P. Franklin, *The J & P Transformer Book*, 11th ed. (London: Butterworths, 1983).

3. W. D. Stevenson, Jr., *Elements of Power System Analysis*, 4th ed. (New York: McGraw-Hill, 1982).

4. J. R. Neuenswander, *Modern Power Systems* (Scranton, PA: International Textbook Company, 1971).

5. M. S. Sarma, *Electric Machines* (Dubuque, IA: Brown, 1985).

6. A. E. Fitzgerald, C. Kingsley, and S. Umans, *Electric Machinery*, 4th ed. (New York: McGraw-Hill, 1983).

7. O. I. Elgerd, *Electric Energy Systems: An Introduction* (New York: McGraw-Hill, 1982).

8. N. Dominelli and P. Kundur, "Life Extension and Condition Assessment," *IEEE Power & Energy Magazine*, *4*, 3 (May/June 2006), pp. 24–35.

4

TRANSMISSION LINE PARAMETERS

In this chapter, we discuss the four basic transmission-line parameters: series resistance, series inductance. shunt capacitance, and shunt conductance. We also investigate transmission-line electric and magnetic fields.

Series resistance accounts for ohmic (I^2R) line losses. Series impedance, including resistance and inductive reactance, gives rise to series-voltage drops along the line. Shunt capacitance gives rise to line-charging currents. Shunt conductance accounts for V^2G line losses due to leakage currents between conductors or between conductors and ground. Shunt conductance of overhead lines is usually neglected.

Although the ideas developed in this chapter can be applied to underground transmission and distribution, the primary focus here is on overhead lines. Underground transmission in the United States presently accounts for less than 1% of total transmission, and is found mostly in large cities or under waterways. There is, however, a large application for underground cable in distribution systems.

CASE STUDY   Two transmission articles are presented here. The first article covers transmission conductor technologies including conventional conductors, high-temperature conductors, and emerging conductor technologies [10]. Conventional conductors include the aluminum conductor steel reinforced (ACSR), the homogeneous all aluminum alloy conductor (AAAC), the aluminum conductor alloy reinforced (ACAR), and others. High-temperature conductors are based on aluminum-zirconium alloys that resist the annealing effects of high temperatures. Emerging conductor designs make use of composite material technology. The second article describes American Electric Power's (AEP's) new Wyoming-Jacksons Ferry transmission line, which became operational in June 2006 [12]. This 90-mile, 765-kV line extends from AEP's Wyoming Station near Oceana WV to Jacksons Ferry Station near Pulaski VA. The new line uses a number of innovative technologies including the first 765-kV six-bundle conductor configuration in North America, which reduces the audible noise level of the project to approximately half that of earlier lines.

Transmission Line Conductor Design Comes of Age

ART J. PETERSON JR. AND SVEN HOFFMANN

Deregulation and competition have changed power flows across transmission networks significantly. Meanwhile, demand for electricity continues to grow, as do the increasing challenges of building new transmission circuits. As a result, utilities need innovative ways to increase circuit capacities to reduce congestion and maintain reliability.

National Grid is monitoring transmission conductor technologies with the intent of testing and deploying innovative conductor technologies within the United States over the next few years. In the UK, National Grid has been using conductor replacement as a means of increasing circuit capacity since the mid 1980s, most recently involving the high-temperature, low-sag "Gap-type" conductor. As a first step in developing a global conductor deployment strategy, National Grid embarked on an overall assessment of overhead transmission line

conductor technologies, examining innovative, and emerging technologies.

About National Grid

National Grid USA is a subsidiary of National Grid Transco, an international energy-delivery business with principal activities in the regulated electric and gas industries. National Grid is the largest transmission business in the northeast United States, as well as one of the 10 largest electric utilities in the United States National Grid achieved this by combining New England Electric System, Eastern Utilities Associates and Niagara Mohawk between March 2000 and January 2002. Its electricity-delivery network includes 9000 miles (14,484 km) of transmission lines and 72,000 miles (115,872 km) of distribution lines.

National Grid UK is the owner, operator and developer of the high-voltage electricity transmission network in England and Wales, comprising approximately 9000 circuit-miles of overhead line and 600 circuit-miles of underground cable at 275 and 400 kV, connecting more than 300 substations.

("Transmission Line Conductor Design Comes of Age" by Art J. Peterson Jr. and Sven Hoffmann, Transmission & Distribution World Magazine, (Aug/2006).)

De-stranding the Gap conductor for field installation.

Re-stranding of conductor.

CONVENTIONAL CONDUCTORS

The reality is that there is no single "wonder material." As such, the vast majority of overhead line conductors are nonhomogeneous (made up of more than one material). Typically, this involves a high-strength core material surrounded by a high-conductivity material. The most common conductor type is the aluminum conductor steel reinforced (ACSR), which has been in use for more than 80 years. By varying the relative cross-sectional areas of steel and aluminum, the conductor can be made stronger at the expense of conductivity (for areas with high ice loads, for example), or it can be made more conductive at the expense of strength where it's not required.

More recently, in the last 15 to 20 years, the homogeneous all-aluminum alloy conductor (AAAC) has become quite popular, especially for National Grid in the UK where it is now the standard conductor type employed for new and refurbished lines. Conductors made up of this alloy (a heat treatable aluminum-magnesium-silicon alloy) are, for the same diameter as an ACSR, stronger, lighter, and more conductive although they are a little more expensive and have a higher expansion coefficient. However, their high strength-to-weight ratio allows them to be strung to much lower initial sags, which allows higher operating temperatures. The resulting tension levels are relatively high, which could result in increased vibration and early fatigue of the conductors. In the UK, with favorable terrain, wind conditions and dampers, these tensions are accept-

able and have allowed National Grid to increase the capacities of some lines by up to 50%.

For the purpose of this article, the three materials mentioned so far—steel, aluminum and aluminum alloy—are considered to be the materials from which conventional conductors are made. The ACSR and AAAC are two examples of such conductors. Other combinations available include aluminum conductor alloy reinforced (ACAR), aluminum alloy conductor steel reinforced (AACSR) and the less common all-aluminum conductor (AAC).

Conductors of these materials also are available in other forms, such as compacted conductors, where the strands are shaped so as not to leave any voids within the conductor's cross section (a standard conductor uses round strands), increasing the amount of conducting material without increasing the diameter. These conductors are designated trapezoidal-wire (TW) or, for example, ACSR/TW and AACSR/TW. Other shaped conductors are available that have noncircular cross sections designed to minimize the effects of wind-induced motions and vibrations.

HIGH-TEMPERATURE CONDUCTORS

Research in Japan in the 1960s produced a series of aluminum-zirconium alloys that resisted the annealing effects of high temperatures. These alloys can retain their strength at temperatures up to 230 °C (446 °F). The most common of these alloys—TA1, ZTA1 and XTA1—are the basis of a variety of high-temperature conductors.

Clamp used for Gap conductor.

The thermal expansion coefficients of all the conventional steel-cored conductors are governed by both materials together, resulting in a value between that of the steel and that of the aluminum. This behavior relies on the fact that both components are carrying mechanical stress.

However, because the expansion coefficient of aluminum is twice that of steel, stress will be increasingly transferred to the steel core as the conductor's temperature rises. Eventually the core bears all the stress in the conductor. From this point on, the conductor as a whole essentially takes on the expansion coefficient of the core. For a typical 54/7 ACSR (54 aluminum strands, 7 steel) this transition point (also known as the "knee-point") occurs around 100 °C (212 °F).

For lines built to accommodate relatively large sags, the T-aluminum conductor, steel reinforced (TACSR) conductor was developed. (This is essentially identical to ACSR but uses the heat-resistant aluminum alloy designated TA1). Because this conductor can be used at high temperatures with no strength loss, advantage can be taken of the low-sag behavior above the knee-point.

If a conductor could be designed with a core that exhibited a lower expansion coefficient than steel, or that exhibited a lower knee-point temperature, more advantage could be taken of the high-temperature alloys. A conductor that exhibits both of these properties uses Invar, an alloy of iron and nickel. Invar has an expansion coefficient about one-third of steel (2.8 microstrain per Kelvin up to 100 °C, and 3.6 over 100 °C, as opposed to 11.5 for steel). T-aluminum conductor Invar reinforced (TACIR) is capable of operation up to 150 °C (302 °F), with ZTACIR and XTACIR capable of 210 °C (410 °F) and 230 °C (446 °F), respectively.

Further, the transition temperature, although

dependent on many factors, is typically lower than that for an ACSR, allowing use of the high temperatures within lower sag limits than required for the TACSR conductors. One disadvantage of this conductor is that Invar is considerably weaker than steel. Therefore, for high-strength applications (to resist ice loading, for example), the core needs to make up a greater proportion of the conductor's area, reducing or even negating the high-temperature benefits. As a result, the ACIR-type conductors are used in favorable areas in Japan and Asia, but are not commonly used in the United States or Europe.

There will still be instances, however, where insufficient clearance is available to take full advantage of the transitional behavior of the ACIR conductors. A conductor more suitable for uprating purposes would exhibit a knee-point at much lower temperatures. Two conductors are available that exhibit this behavior: the Gap-type conductor and a variant of the ACSR that uses fully annealed aluminum.

Developed in Japan during the 1970s, Gap-type ZT-aluminum conductor steel reinforced (GZTACSR) uses heat-resistant aluminum over a steel core. It has been used in Japan, Saudi Arabia, and Malaysia, and is being extensively implemented by National Grid in the UK. The principle of the Gap-type conductor is that it can be tensioned on the steel core alone during erection. A small annular Gap exists between a high-strength steel core and the first layer of trapezoidal-shaped aluminum strands, which allows this to be achieved. The result is a conductor with a knee-point at the erection temperature. Above this, thermal expansion is that of steel (11.5 microstrain per Kelvin), while below it is that of a comparable ACSR (approximately 18). This construction allows for low-sag properties above the erection temperature and good strength below it as the aluminum alloy can take up significant load.

For example, the application of GZTACSR by National Grid in the UK allowed a 90 °C (194 °F) rated 570 mm^2 AAAC to be replaced with a 620 mm^2 GZTACSR (Matthew). The Gap-type conductor, being of compacted construction, actually had a smaller diameter than the AAAC, despite having a larger nominal area. The low-sag properties allowed

Semi-strain assembly installed on line in a rural area of the UK.

a rated temperature of 170 °C (338 °F) and gave a 30% increase in rating for the same sag.

The principal drawback of the Gap-type conductor is its complex installation procedure, which requires destranding the aluminum alloy to properly install on the joints. There is also the need for "semi-strain" assemblies for long line sections (typically every five spans). Experience in the UK has shown that a Gap-type conductor requires about 25% more time to install than an ACSR.

A semi-strain assembly is, in essence, a pair of back-to-back compression anchors at the bottom of a suspension insulator set. It is needed to avoid potential problems caused by the friction that developes between the steel core and the aluminum layers when using running blocks. This helps to prevent the steel core from hanging up within the conductor.

During 1999 and 2000, in the UK, National Grid installed 8 km (single circuit) of Matthew GZTACSR. Later this year and continuing through to next year, National Grid will be refurbishing a 60 km (37-mile) double-circuit (120 circuit-km) route in the UK with Matthew.

A different conductor of a more standard construction is aluminum conductor steel supported (ACSS), formerly known as SSAC. Introduced in the 1980s, this conductor uses fully annealed aluminum around a steel core. The steel core provides the entire conductor support. The aluminum strands are "dead soft," thus the conductor may be operated at temperatures in excess of 200 °C without loss of strength. The maximum operating temperature of the conductor is limited by the coating used on the steel core. Conventional galvanized coatings deteriorate rapidly at temperatures above 245 °C (473 °F). If a zinc-5% aluminum mischmetal alloy coated steel core is used, temperatures of 250 °C are possible.

Since the fully annealed aluminum cannot support significant stress, the conductor has a thermal expansion similar to that of steel. Tension in the aluminum strands is normally low. This helps to improve the conductor's self-damping characteristics and helps to reduce the need for dampers.

For some applications there will be concern over the lack of strength in the aluminum, as well as the possibility of damage to the relatively soft outer layers. However, ACSS is available as ACSS/TW, improving, its strength. ACSS requires special care when installing. The soft annealed aluminum wires can be easily damaged and "bird-caging" can occur. As with the other high-temperature conductors, the heat requires the use of special suspension clamps, high-temperature deadends, and high-temperature splices to avoid hardware damage.

EMERGING CONDUCTOR TECHNOLOGIES

Presently, all the emerging designs have one thing in common—the use of composite material technology.

Aluminum conductor carbon fiber reinforced (ACFR) from Japan makes use of the very-low-expansion coefficient of carbon fiber, resulting in a conductor with a lower knee-point of around 70 °C (158 °F). The core is a resin-matrix composite containing carbon fiber. This composite is capable of withstanding temperatures up to 150 °C. The ACFR is about 30% lighter and has an expansion coefficient (above the knee-point) that is 8% that of an ACSR of the same stranding, giving a rating increase of around 50% with no structural work required.

Meanwhile, in the United States, 3M has developed the Aluminum Conductor Composite Re-

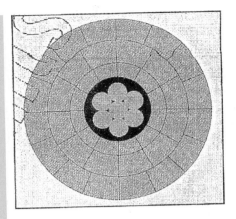

A cross section of the Gap conductor.

inforced (ACCR). The core is an aluminum-matrix composite containing alumina fibers, with the outer layers made from a heat-resistant aluminum alloy. As with the ACFR, the low-expansion coefficient of the core contributes to a fairly low knee-point, allowing the conductor to make full use of the heat resistant alloy within existing sag constraints. Depending on the application, rating increases between 50% and 200% are possible as the conductor can be rated up to 230 °C.

Also in the United States, two more designs based on glass-fiber composites are emerging. Composite Technology Corp. (CTC; Irvine, California, U.S.) calls it the aluminum conductor composite core (ACCC), and W. Brandt Goldsworthy and Associates (Torrance, California) are developing composite reinforced aluminum conductor (CRAC). These conductors are expected to offer between 40% and 100% increases in ratings.

Over the next few years, National Grid plans to install ACSS and the Gap conductor techology within its U.S. transmission system. Even a test span of one or more of the new composite conductors is being considered.

Art J. Peterson Jr. is a senior engineer in National Grid's transmission line engineering and project management department in Syracuse, New York. Peterson received a BS degree in physics from Le Moyne College in Syracuse; a MS degree in physics from Clarkson University in Potsdam, New York; a M. Eng. degree in nuclear engineering from Pennsylvania State University in State College, Pennsylvania; and a Ph.D. in organization and management from Capella University in Minneapolis, Minnesota. He has 20 years of experience in electric generation and transmission.
Art.Peterson@us.ngrid.com

Sven Hoffmann is the circuits forward policy team leader in National Grid's asset strategy group in Coventry, United Kingdom. Hoffmann has a bachelor's in engineering degree from the University of Birmingham in England. He is a chartered engineer with the Institution of Electrical Engineers, and the UK Regular Member for CIGRE Study Committee B2. Hoffmann has been working at National Grid, specializing in thermal and mechanical aspects of overhead lines for eight years.
sven.hoffmann@uk.ngrid.com

Mammoth 765-kV Project Winds Through Appalachian Mountains

JIM HAUNTY

AEP is constructing the Wyoming-Jacksons Ferry line having addressed community, environmental, regulatory, and design concerns.

("Mammoth 765-kV Project" by Jim Haunty, Transmission & Distribution World Magazine, (Feb/2006).)

American electric power (AEP) operates more than 2000 miles (3219 km) of 765-kV transmission lines in the United States, from the southeastern shore of Lake Michigan to the rolling red clay Piedmont country of central Virginia. Yet, for a critical 90-mile (145-km) 765-kV transmission addition

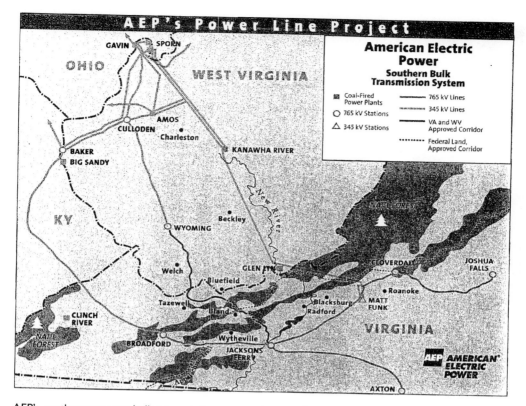

AEP's southeastern area bulk transmission system showing the new 765-kV line from Wyoming Station to Jacksons Ferry Station.

nearing completion in just a few months in AEP's southern West Virginia and southwest Virginia service areas, it's as though AEP (Columbus, Ohio, U.S.) started from scratch, challenging assumptions of the past and taking a fresh look at all aspects of this project: engineering, right-of-way management, procurement, scheduling and construction practices.

AEP's Wyoming-Jacksons Ferry line—the first major high-voltage supply line since 1973—stretches from its Wyoming station near Oceana, West Virginia, south to Jacksons Ferry Station near Wytheville and Pulaski, Virginia. Two major radial extensions of AEP's 765-kV system were built in Virginia, but the new line is the first transmission line in more than 30 years from AEP's major generating plants on the Kanawha and Ohio rivers to southern West Virginia, western Virginia and eastern Kentucky. But not for want of trying.

A SLOW START

AEP announced plans in March 1990 for a 110-mile 765-kV line to Roanoke, Virginia. This proposal evolved into the Wyoming-Jacksons Ferry line during regulatory proceedings in Virginia. An exhaustive, up-and-down permit review process before two state utility commissions and the U.S. Forest Service (USFS) consumed almost 13 years. As early as December 1995, the State Corporation Commission (SCC), Virginia's utility regulatory body, issued an interim order in which it cited a "compelling need for additional capacity" to serve southwest

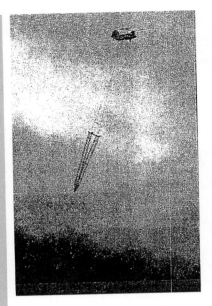

Columbia Helicopter's Chinook transporting guyed-V tower masts to tower erection sites.

Guyed-V tower mast base mounting detail.

Virginia and said that the proposed transmission line was the best alternative to meet the need. Finally, in December 2002 the USFS issued an environmental impact statement permitting 11 miles (18 km) of the line to pass through the Jefferson National Forest in western Virginia. Following appeals and litigation by opponents of the line, as well as the SCC's approval of AEP's final line design and right-of-way clearing plans and early right-of-way acquisition, right-of-way clearing began in December 2003. Electricity is scheduled to begin flowing by late June 2006. The expression "started from scratch" was used to make a point about AEP's innovations in connection with this largest transmission project currently underway in the United States. In reality, AEP was building upon almost 40 years of experience in the operation of 765-kV lines and almost 90 years of designing, building and operating transmission lines. AEP's first major transmission line was the Windsor-to-Canton 138-kV line from a mine-mouth generating plant in the northern panhandle of West Virginia to the vital steel industries of northeast Ohio during World War I.

IT STARTS WITH THE WIRE

With the Wyoming-Jacksons Ferry line, it all starts with the wire (795 kcm ACSR 45/7-Tern), and more specifically, with the number and arrangement of conductors. AEP's existing 765-kV system consists of two types of conductor bundles per phase. Each phase has four conductors. When AEP started its

Crews on the tower lift the six-bundle wire off the stringing block and attach it to the yoke plate.

765-kV system in the late 1960s, each of the three phases consisted of four 954 kcm-Rail wires at a diameter of 1.165 inches (3 cm) per wire. When additions were made to this system in the mid-1970s, a larger (1352 kcm-Dipper) wire size (1.385 inches [3.5 cm]) per wire was selected and used for the remainder of the system. The result was 1098 miles (1767 km) of the original size and 924 miles (1487 km) of the larger wire, until Wyoming-Jacksons Ferry.

To improve efficiency by reducing corona and to reduce audible noise associated with the existing four-wire configuration at higher elevations in foul weather, AEP analyzed and tested various wire configurations and decided to install a new six-wire 795-kcm bundle on the new line. In fact, in 1995 the company installed a 1-mile (1.6-km) section of six-wire bundles on its Jacksons Ferry-Axton line in the Blue Ridge Mountains near Floyd, Virginia, south of Roanoke. This trial section of line proved invaluable during the siting and hearing process as local residents and state regulators were able to observe first hand how the new design cut the audible noise nearly in half from older generations of 765-kV lines.

Actual operation of this section of line and lab tests of the new design confirmed its improved operation; this important threshold decision begat a whole series of snowball actions.

Among the initial challenges associated with the new six-wire design was how to arrange the six conductors. What would be the most efficient shape or geometry of the six wires? With four wires, it was fairly simple, a square. Only the spacer-dampers changed with the four-wire design. Initially, a coiled wire spacer in the shape of a square was employed. Later, a stronger, solid X-shaped spacer was used to replace damaged square spacers.

Where the existing four-wire arrangement measures 18 inches (46 cm) per side, the new bundle—in the shape of a symmetrical hexagon—is held in place by a 30-inch (67-cm)-diameter spacer-damper. Of course, the new design was chosen only after rigorous lab and field testing for strength and corona.

New conductor bundle geometry and spacer hardware drove the next challenge—electrical clearance from the tower structure itself. This required new hardware and new connections. And these required design and testing.

STRUCTURES ARE NEXT

AEP's existing guyed-V aluminum and self-supporting four-legged steel towers have proven to be a very efficient and acceptable design. The high cost of aluminum, and the fact that the market for aluminum tower construction had dried up, drove

Specially designed "six-wire conductor motorized buggies" applying spacer-dampers after final sag/clip-end.

Crew prepares for pulling wire by installing grips that will attach the wire to the running board lying on the ground. The running board is pulled through the blocks laying the six conductors in the proper wheel of the stringing block.

the company to galvanized steel for its guyed-V towers on the Wyoming-Jacksons Ferry line. While the same basic structure geometry was retained, a new family of structures was designed to incorporate new connections and details.

Prototypes of six new structure designs were assembled at the Electric Power Research Institute (EPRI) Solutions test facility in Haslet, Texas, U.S. These six structures represented a sample of the 12 types of towers used for the actual line. These prototypes were tested under varying loading conditions at the EPRI facility. Incremental improvements were made as a result. This type of structural testing is a balance between economics and engineering. The objective is to reach a level of comfort and confidence that a certain design, at a reasonable cost, will work.

AEP's existing 765-kV system uses some 2.5 million ceramic insulators. The company invited bids for both ceramic and polymer insulators. AEP has used polymer insulators selectively in some applications and extensively on its 345-kV transmission system serving its western service area in Texas, Oklahoma, Arkansas and Louisiana.

After an exhaustive analysis of the risks of using the new material and the costs and benefits of polymer insulators, AEP opted for polymer on the 223 guyed-V towers and 70 dead-end towers because it is less expensive, lighter and easier to handle. Ceramic insulators were installed on the 40 self-supporting suspension towers.

AEP's guyed-V towers use four stranded steel guys that are anchored by grouted strands (up to a 1.625-inch [4.128-cm] diameter) at a depth of 55 ft to 90 ft (17 m to 27 m) into soil and bedrock in the mountainous terrain of West Virginia and Virginia. This is a unique design developed by AEP for earlier generations of 765-kV transmission construction.

AESTHETICS AND ACCESS

Because of the protracted 13-year permitting and siting approval process involving regulatory commissions in Virginia and West Virginia and the USFS and local opposition to this project, building the line itself would require great care and sensitivity. AEP

The completed line is shown here traversing the rolling and mountainous landscape of Tazewell County, Virginia. The longest conductor span on the project is 3948 ft (1203 m).

engaged a team of land-use experts from Virginia Tech and West Virginia universities who went through a significant process of analysis to recommend a 1000-ft (305-m)-wide corridor that minimized environmental and visual impacts to a landscape with cherished scenic and cultural values. They hiked and drove thousands of miles observing, taking notes and snapping photos, and they created computer images of towers imposed on the landscape to simulate impacts. Working with property owners and regulatory agencies, AEP engineers selected an optimum 200-ft (61-m)-wide right of way within the 1000-ft corridor.

AEP selected PAR Electrical Contractors Inc. (Kansas City, Missouri, U.S.), part of Quanta Services Inc., as its line construction contractor and brought PAR in early as a partner in construction preplanning. Due in part to the sensitivities cited previously regarding land-disturbing activities, AEP decided to award separate specialty contracts, one for right-of-way and tower site clearing (Phillips & Jordan Inc.; Wilmington, North Carolina, U.S.) and the second for access road construction and maintenance and land reclamation (Orders Construction Co.; St. Albans, West Virginia).

AEP had committed during the permitting process to minimize clearing, that is, to remove only tall-growing incompatible species and leave lower-growing species and to selectively apply herbicides exclusively by hand from the ground in the future. For example, the right of way was not cleared where conductor-to-ground clearance exceeded 100 ft (30.5 m).

The road subcontractor used, and in some cases improved, existing roads for access to tower sites. In all, 160 miles (257 km) of new roads were constructed in the mountainous terrain. The road contractor was also responsible for maintaining countless public and private roads during construction.

While cranes were used primarily to erect towers, helicopters were used to transport workers and materials to these sites, as well as for many other uses. In fact, helicopters were used wherever possible, particularly for completing nine guyed-V towers in inaccessible and restricted access areas, including one tower site in a roadless area of the Jefferson National Forest. Helicopters also were invaluable in transporting and setting guyed-V tower masts. In addition, they were used for hanging insulators and other hardware, replenishing wire buggies with spacer-dampers, flying and threading lead lines, and transporting steel bundles and conductor reels.

INNOVATION FOR THE FUTURE

The Wyoming-Jacksons Ferry line represents a solution to the need for additional electrical capacity in a part of AEP's service area. But innovations embodied in the physical aspects of the line—new wire bundle design and new tower series—confirm a commitment on AEP's part to push beyond the status quo and provide technical leadership for the electric utility industry. As AEP electrical engineers, high-voltage system planners and transmission engineers undertook the task of designing this new line in 1991 by looking to improve on the past, future transmission planners will look to advances made with Wyoming-Jacksons Ferry as their launching pad. **TDW**

EDITOR'S NOTE

This is the first in a series of articles on AEP's Wyoming-Jacksons Ferry 765-kV project. Future articles will highlight construction, engineering and the six-wire bundle configuration used on the project. More information about the Wyoming-Jacksons Ferry project can be found online at www.apcocustomer.com/news/765kv/default.asp.

Jim Haunty served as vice president of Transmission Capital Improvements for AEP until he retired on Dec. 31, 2005. He was responsible for engineering, design, construction and project management for the company's 11-state transmission line and T&D substation system capital project. After service in the U.S. Navy, Haunty entered this industry in 1967 with (at that time) AB Chance Steel Poles Constructors Division (Houston, Texas). His AEP career first began in late 1969 when he started with Columbus Southern Ohio Electric (CSOE) as supervisor civil engineering design. He left CSOE and returned in 1974, and has since held a variety of management positions with AEP. Haunty holds a BSCE degree from Purdue University in West Lafayette, Indiana.

Experts validate need for transmission line

The winter of 2005–2006 should be the last one that customers of American Electric Power (AEP) operating companies in southern West Virginia and southwest Virginia have to live without a much-needed, long-proposed and long-delayed major transmission line now under construction. AEP's Wyoming-Jacksons Ferry 765-kV line is scheduled for completion in a few months following a two-and-a-half-year construction project. Electricity should start to flow on the 90-mile (145-km) line in late June 2006.

Announced in 1990, the line—originally a 110-mile line (177 km) to Cloverdale Station at Roanoke, Virginia—had been proposed by AEP system planners many years beforehand to meet utility industry reliability standards. Despite its need having been recognized by two state regulatory agencies and all their outside consultants, routing, and permitting issues delayed approval of the line for many years.

In its order approving the line and routing in May 2001, the Virginia State Corporation Commission (SCC) acknowledged that AEP's filing had engendered a protracted and highly contested proceeding, and that the original and rerouted corridors had attracted substantial and understandable opposition. At the same time, the SCC acknowledged that the existing transmission system in southwest Virginia was seriously overloaded. The SCC accepted the hearing examiner's finding that 32 different system operating contingencies violated single- or double-contingency criteria, and provide clear and compelling evidence that the situation was critical.

AEP's Virginia service area is transmission dependent. The last major transmission reinforcement for the area came in 1973 with the completion of the Jacksons Ferry-Cloverdale 765-kV line. The region is a large importer of power from northerly AEP plants. And the gap between regional generation and customer load and peak demands more than doubled between 1973 and 1996.

In order to eliminate the gap and provide reliable transmission capability for future growth that meets industry standards, AEP proposed extending its 765-kV system from Wyoming Station in West Virginia to Virginia. By the time the new transmission line is in service in mid-2006, transmission planners estimate that the peak demands will be more than 180% greater than the 2512-W peak of 1973. Many computer models and contingency scenarios indicated the loss of major transmission facilities at peak load conditions could result in unscheduled outages and cascading blackouts.

Key events in the history of AEP's Wyoming-Jacksons Ferry 765-kV transmission line

March 1990
AEP announces plans to build 130-mile (209-km), 765-kV transmission line from its Wyoming station near Oceana, West Virginia, to its Cloverdale Station near Roanoke, Virginia. The proposed line would be the first major reinforcement of the bulk transmission system in the area since the original system was completed in 1973.

March 1991–February 1993
Applications for permits and certificates filed with utility regulatory agencies in Virginia and West Virginia and the U.S. Forest Service (USFS).

December 1995
Virginia State Corporation Commission (SCC) issues interim order citing compelling need for additional electric capacity in the region and directs AEP to file additional information.

June 1996
USFS issues draft environmental impact statement denying AEP permission to cross federal lands. Earlier, in May 1996, the U.S. Park Service recommended denial of AEP's proposed crossing of the New River because the area was under study as a federally protected scenic river.

May 1998
Public Service Commission (PSC) of West Virginia approves construction of Wyoming-Cloverdale line.

September 1998
SCC directs AEP to study and report on an alternate routing, from Wyoming to AEP's Jacksons Ferry Station.

May 2001
SCC approves construction of the 90-mile (145-km) Wyoming-Jacksons Ferry line.

March 2002
PSC amends its 1998 order to approve new route.

December 2002
USFS issues environmental impact statement and recommends the line be allowed to cross federal lands.

December 2003
Right-of-way clearing begins, following right-of-way acquisition and design.

April 2004
Tower foundation construction begins, with first tower erected in August.

October 2005
Final tower completed. Electricity scheduled to flow by mid-2006.

4.1

TRANSMISSION LINE DESIGN CONSIDERATIONS

An overhead transmission line consists of conductors, insulators, support structures, and, in most cases, shield wires.

CONDUCTORS

Aluminum has replaced copper as the most common conductor metal for overhead transmission. Although a larger aluminum cross-sectional area is required to obtain the same loss as in a copper conductor, aluminum has a lower cost and lighter weight. Also, the supply of aluminum is abundant, whereas that of copper is limited.

One of the most common conductor types is aluminum conductor, steel-reinforced (ACSR), which consists of layers of aluminum strands surrounding a central core of steel strands (Figure 4.1). Stranded conductors are easier to manufacture, since larger conductor sizes can be obtained by simply adding successive layers of strands. Stranded conductors are also easier to handle and more flexible than solid conductors, especially in larger sizes. The use of steel strands gives ACSR conductors a high strength-to-weight ratio. For purposes of heat dissipation, overhead transmission-line conductors are bare (no insulating cover).

Other conductor types include the all-aluminum conductor (AAC), all-aluminum-alloy conductor (AAAC), aluminum conductor alloy-reinforced (ACAR), and aluminum-clad steel conductor (Alumoweld). Higher-temperature conductors capable of operation in excess of 150 °C include the aluminum conductor steel supported (ACSS), which uses fully annealed aluminum around a steel core, and the gap-type ZT-aluminum conductor (GTZACSR) which uses heat-resistant aluminum over a steel core with a small annular gap between the steel and first layer of aluminum strands. Emerging technologies use composite materials, including the aluminum conductor carbon reinforced (ACFR), whose core is a resinmatrix composite containing carbon fiber, and the aluminum conductor composite reinforced (ACCR), whose core is an aluminum-matrix containing aluminum fibers [10].

FIGURE 4.1

Typical ACSR conductor

54/7 Cardinal

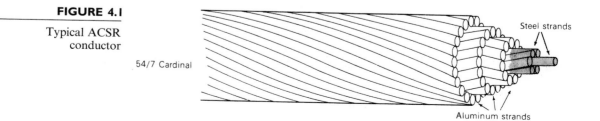

Steel strands

Aluminum strands

FIGURE 4.2

A 765-kV transmission line with self-supporting lattice steel towers (Courtesy of the American Electric Power Company)

FIGURE 4.3

A 345-kV double-circuit transmission line with self-supporting lattice steel towers (Courtesy of NSTAR, formerly Boston Edison Company)

EHV lines often have more than one conductor per phase; these conductors are called a *bundle*. The 765-kV line in Figure 4.2 has four conductors per phase, and the 345-kV double-circuit line in Figure 4.3 has two conductors per phase. Bundle conductors have a lower electric field strength at the conductor surfaces, thereby controlling corona. They also have a smaller series reactance.

INSULATORS

Insulators for transmission lines above 69 kV are suspension-type insulators, which consist of a string of discs, typically porcelain. The standard disc (Figure 4.4) has a 10-in. (0.254-m) diameter, $5\frac{3}{4}$-in. (0.146-m) spacing between centers of adjacent discs, and a mechanical strength of 7500 kg. The 765-kV line in Figure 4.2 has two strings per phase in a V-shaped arrangement,

FIGURE 4.4

Cut-away view of a
standard insulator disc
for suspension insulator
strings (Courtesy of
Ohio Brass)

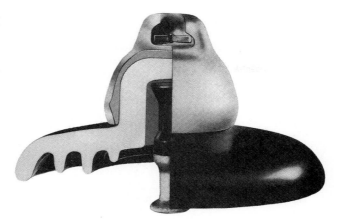

FIGURE 4.4

Cut-away view of a
standard insulator disc
for suspension insulator
strings (Courtesy of
Ohio Brass)

FIGURE 4.5

Wood frame structure
for a 345-kV line
(Courtesy of NSTAR,
formerly Boston Edison
Company)

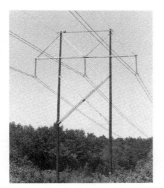

which helps to restrain conductor swings. The 345-kV line in Figure 4.5 has
one vertical string per phase. The number of insulator discs in a string increases with line voltage (Table 4.1). Other types of discs include larger units
with higher mechanical strength and fog insulators for use in contaminated
areas.

SUPPORT STRUCTURES

Transmission lines employ a variety of support structures. Figure 4.2 shows a
self-supporting, lattice steel tower typically used for 500- and 765-kV lines.
Double-circuit 345-kV lines usually have self-supporting steel towers with the
phases arranged either in a triangular configuration to reduce tower height or in
a vertical configuration to reduce tower width (Figure 4.3). Wood frame configurations are commonly used for voltages of 345 kV and below (Figure 4.5).

Nominal Voltage	Phase Conductors				
		Aluminum Cross-Section Area per Conductor (ACSR) (kcmil)	Bundle Spacing (cm)	Minimum Clearances	
(kV)	Number of Conductors per Bundle			Phase-to-Phase (m)	Phase-to-Ground (m)
69	1	—	—	—	—
138	1	300–700	—	4 to 5	—
230	1	400–1000	—	6 to 9	—
345	1	2000–2500	—	6 to 9	7.6 to 11
345	2	800–2200	45.7	6 to 9	7.6 to 11
500	2	2000–2500	45.7	9 to 11	9 to 14
500	3	900–1500	45.7	9 to 11	9 to 14
765	4	900–1300	45.7	13.7	12.2

Nominal Voltage	Suspension Insulator String		Shield Wires		
	Number of Strings per Phase	Number of Standard Insulator Discs per Suspension String	Type	Number	Diameter (cm)
(kV)					
69	1	4 to 6	Steel	0, 1 or 2	—
138	1	8 to 11	Steel	0, 1 or 2	—
230	1	12 to 21	Steel or ACSR	1 or 2	1.1 to 1.5
345	1	18 to 21	Alumoweld	2	0.87 to 1.5
345	1 and 2	18 to 21	Alumoweld	2	0.87 to 1.5
500	2 and 4	24 to 27	Alumoweld	2	0.98 to 1.5
500	2 and 4	24 to 27	Alumoweld	2	0.98 to 1.5
765	2 and 4	30 to 35	Alumoweld	2	0.98

SHIELD WIRES

Shield wires located above the phase conductors protect the phase conductors against lightning. They are usually high- or extra-high-strength steel, Alumoweld, or ACSR with much smaller cross section than the phase conductors. The number and location of the shield wires are selected so that almost all lightning strokes terminate on the shield wires rather than on the phase conductors. Figures 4.2, 4.3, and 4.5 have two shield wires. Shield wires are grounded to the tower. As such, when lightning strikes a shield wire, it flows harmlessly to ground, provided the tower impedance and tower footing resistance are small.

The decision to build new transmission is based on power-system planning studies to meet future system requirements of load growth and new generation. The points of interconnection of each new line to the system, as well as the power and voltage ratings of each, are selected based on these studies. Thereafter, transmission-line design is based on optimization of electrical, mechanical, environmental, and economic factors.

ELECTRICAL FACTORS

Electrical design dictates the type, size, and number of bundle conductors per phase. Phase conductors are selected to have sufficient thermal capacity to meet continuous, emergency overload, and short-circuit current ratings. For EHV lines, the number of bundle conductors per phase is selected to control the voltage gradient at conductor surfaces, thereby reducing or eliminating corona.

Electrical design also dictates the number of insulator discs, vertical or V-shaped string arrangement, phase-to-phase clearance, and phase-to-tower clearance, all selected to provide adequate line insulation. Line insulation must withstand transient overvoltages due to lightning and switching surges, even when insulators are contaminated by fog, salt, or industrial pollution. Reduced clearances due to conductor swings during winds must also be accounted for.

The number, type, and location of shield wires are selected to intercept lightning strokes that would otherwise hit the phase conductors. Also, tower footing resistance can be reduced by using driven ground rods or a buried conductor (called *counterpoise*) running parallel to the line. Line height is selected to satisfy prescribed conductor-to-ground clearances and to control ground-level electric field and its potential shock hazard.

Conductor spacings, types, and sizes also determine the series impedance and shunt admittance. Series impedance affects line-voltage drops, I^2R losses, and stability limits (Chapters 5, 13). Shunt admittance, primarily capacitive, affects line-charging currents, which inject reactive power into the power system. Shunt reactors (inductors) are often installed on lightly loaded EHV lines to absorb part of this reactive power, thereby reducing overvoltages.

MECHANICAL FACTORS

Mechanical design focuses on the strength of the conductors, insulator strings, and support structures. Conductors must be strong enough to support a specified thickness of ice and a specified wind in addition to their own weight. Suspension insulator strings must be strong enough to support the phase conductors with ice and wind loadings from tower to tower (span length). Towers that satisfy minimum strength requirements, called suspension towers, are designed to support the phase conductors and shield wires with ice and wind loadings, and, in some cases, the unbalanced pull due to breakage of one or two conductors. Dead-end towers located every mile or so satisfy the maximum strength requirement of breakage of all conductors on one side of the tower. Angles in the line employ angle towers with intermediate strength. Conductor vibrations, which can cause conductor fatigue failure and damage to towers, are also of concern. Vibrations are controlled by adjustment of conductor tensions, use of vibration dampers, and—for bundle conductors—large bundle spacing and frequent use of bundle spacers.

ENVIRONMENTAL FACTORS

Environmental factors include land usage and visual impact. When a line route is selected, the effect on local communities and population centers, land values, access to property, wildlife, and use of public parks and facilities must all be considered. Reduction in visual impact is obtained by aesthetic tower design and by blending the line with the countryside. Also, the biological effects of prolonged exposure to electric and magnetic fields near transmission lines is of concern. Extensive research has been and continues to be done in this area.

ECONOMIC FACTORS

The optimum line design meets all the technical design criteria at lowest overall cost, which includes the total installed cost of the line as well as the cost of line losses over the operating life of the line. Many design factors affect cost. Utilities and consulting organizations use digital computer programs combined with specialized knowledge and physical experience to achieve optimum line design.

4.2

RESISTANCE

The dc resistance of a conductor at a specified temperature T is

$$R_{dc,T} = \frac{\rho_T l}{A} \quad \Omega \tag{4.2.1}$$

where ρ_T = conductor resistivity at temperature T

l = conductor length

A = conductor cross-sectional area

Two sets of units commonly used for calculating resistance, SI and English units, are summarized in Table 4.2. In English units, conductor cross-sectional area is expressed in circular mils (cmil). One inch equals 1000 mils

Quantity	Symbol	SI Units	English Units
Resistivity	ρ	Ωm	Ω-cmil/ft
Length	ℓ	m	ft
Cross-sectional area	A	m^2	cmil
dc resistance	$R_{dc} = \dfrac{\rho\ell}{A}$	Ω	Ω

and 1 cmil equals $\pi/4$ sq mil. A circle with diameter D in., or (D in.) (1000 mil/in.) = 1000 D mil = d mil, has an area

$$A = \left(\frac{\pi}{4}\,D^2\,\text{in.}^2\right)\left(1000\,\frac{\text{mil}}{\text{in.}}\right)^2 = \frac{\pi}{4}(1000\,D)^2 = \frac{\pi}{4}d^2 \quad \text{sq mil}$$

or

$$A = \left(\frac{\pi}{4}d^2\,\text{sq mil}\right)\left(\frac{1\,\text{cmil}}{\pi/4\,\text{sq mil}}\right) = d^2 \quad \text{cmil} \tag{4.2.2}$$

Resistivity depends on the conductor metal. Annealed copper is the international standard for measuring resistivity ρ (or conductivity σ, where $\sigma = 1/\rho$). Resistivity of conductor metals is listed in Table 4.3. As shown, hard-drawn aluminum, which has 61% of the conductivity of the international standard, has a resistivity at 20 °C of 17.00 Ω-cmil/ft or 2.83 × 10^{-8} Ωm.

Conductor resistance depends on the following factors:

1. Spiraling

2. Temperature

3. Frequency ("skin effect")

4. Current magnitude—magnetic conductors

These are described in the following paragraphs.

Material	% Conductivity	$\rho_{20\,°C}$ Resistivity at 20 °C Ωm × 10^{-8}	Ω-cmil/ft	T Temperature Constant °C
Copper:				
Annealed	100%	1.72	10.37	234.5
Hard-drawn	97.3%	1.77	10.66	241.5
Aluminum				
Hard-drawn	61%	2.83	17.00	228.1
Brass	20–27%	6.4–8.4	38–51	480
Iron	17.2%	10	60	180
Silver	108%	1.59	9.6	243
Sodium	40%	4.3	26	207
Steel	2–14%	12–88	72–530	180–980

For stranded conductors, alternate layers of strands are spiraled in opposite directions to hold the strands together. Spiraling makes the strands 1 or 2% longer than the actual conductor length. As a result, the dc resistance of a stranded conductor is 1 or 2% larger than that calculated from (4.2.1) for a specified conductor length.

Resistivity of conductor metals varies linearly over normal operating temperatures according to

$$\rho_{T2} = \rho_{T1}\left(\frac{T_2 + T}{T_1 + T}\right) \tag{4.2.3}$$

where ρ_{T2} and ρ_{T1} are resistivities at temperatures T_2 and T_1 °C, respectively. T is a temperature constant that depends on the conductor material, and is listed in Table 4.3.

The ac resistance or *effective* resistance of a conductor is

$$R_{ac} = \frac{P_{loss}}{|I|^2} \quad \Omega \tag{4.2.4}$$

where P_{loss} is the conductor real power loss in watts and I is the rms conductor current. For dc, the current distribution is uniform throughout the conductor cross section, and (4.2.1) is valid. However, for ac, the current distribution is nonuniform. As frequency increases, the current in a solid cylindrical conductor tends to crowd toward the conductor surface, with smaller current density at the conductor center. This phenomenon is called *skin effect*. A conductor with a large radius can even have an oscillatory current density versus the radial distance from the conductor center.

With increasing frequency, conductor loss increases, which, from (4.2.4), causes the ac resistance to increase. At power frequencies (60 Hz), the ac resistance is at most a few percent higher than the dc resistance. Conductor manufacturers normally provide dc, 50-Hz, and 60-Hz conductor resistance based on test data (see Appendix Tables A.3 and A.4).

For magnetic conductors, such as steel conductors used for shield wires, resistance depends on current magnitude. The internal flux linkages, and therefore the iron or magnetic losses, depend on the current magnitude. For ACSR conductors, the steel core has a relatively high resistivity compared to the aluminum strands, and therefore the effect of current magnitude on ACSR conductor resistance is small. Tables on magnetic conductors list resistance at two current levels (see Table A.4).

EXAMPLE 4.1   **Stranded conductor: dc and ac resistance**

Table A.3 lists a 4/0 copper conductor with 12 strands. Strand diameter is 0.1328 in. For this conductor:

    a. Verify the total copper cross-sectional area of 211,600 cmil.

    b. Verify the dc resistance at 50 °C of 0.302 Ω/mi. Assume a 2% increase in resistance due to spiraling.

    c. From Table A.3, determine the percent increase in resistance at 60 Hz versus dc.

SOLUTION

a. The strand diameter is $d = (0.1328$ in.$)$ $(1000$ mil/in.$) = 132.8$ mil, and, from (4.2.2), the strand area is d^2 cmil. Using four significant figures, the cross-sectional area of the 12-strand conductor is

$$A = 12d^2 = 12(132.8)^2 = 211,600 \quad \text{cmil}$$

which agrees with the value given in Table A.3.

b. Using (4.2.3) and hard-drawn copper data from Table 4.3,

$$\rho_{50\,°C} = 10.66 \left(\frac{50 + 241.5}{20 + 241.5} \right) = 11.88 \quad \Omega\text{-cmil/ft}$$

From (4.2.1), the dc resistance at $50\,°C$ for a conductor length of 1 mile (5280 ft) is

$$R_{dc,\,50\,°C} = \frac{(11.88)(5280 \times 1.02)}{211,600} = 0.302 \quad \Omega/\text{mi}$$

which agrees with the value listed in Table A.3.

c. From Table A.3,

$$\frac{R_{60\text{ Hz},\,50\,°C}}{R_{dc,\,50\,°C}} = \frac{0.303}{0.302} = 1.003 \qquad \frac{R_{60\text{ Hz},\,25\,°C}}{R_{dc,\,25\,°C}} = \frac{0.278}{0.276} = 1.007$$

Thus, the 60-Hz resistance of this conductor is about 0.3–0.7% higher than the dc resistance. The variation of these two ratios is due to the fact that resistance in Table A.3 is given to only three significant figures. ■

4.3

CONDUCTANCE

Conductance accounts for real power loss between conductors or between conductors and ground. For overhead lines, this power loss is due to leakage currents at insulators and to corona. Insulator leakage current depends on the amount of dirt, salt, and other contaminants that have accumulated on insulators, as well as on meteorological factors, particularly the presence of moisture. Corona occurs when a high value of electric field strength at a conductor surface causes the air to become electrically ionized and to conduct. The real power loss due to corona, called *corona loss*, depends on meteorological conditions, particularly rain, and on conductor surface irregularities. Losses due to insulator leakage and corona are usually small compared to conductor I^2R loss. Conductance is usually neglected in power system studies because it is a very small component of the shunt admittance.

4.4

INDUCTANCE: SOLID CYLINDRICAL CONDUCTOR

The inductance of a magnetic circuit that has a constant permeability μ can be obtained by determining the following:

1. Magnetic field intensity H, from Ampere's law

2. Magnetic flux density B $(B = \mu H)$

3. Flux linkages λ

4. Inductance from flux linkages per ampere $(L = \lambda/I)$

As a step toward computing the inductances of more general conductors and conductor configurations, we first compute the internal, external, and total inductance of a solid cylindrical conductor. We also compute the flux linking one conductor in an array of current-carrying conductors.

Figure 4.6 shows a 1-meter section of a solid cylindrical conductor with radius r, carrying current I. For simplicity, assume that the conductor (1) is sufficiently long that end effects are neglected, (2) is nonmagnetic $(\mu = \mu_0 = 4\pi \times 10^{-7}$ H/m), and (3) has a uniform current density (skin effect is neglected). From (3.1.1), Ampere's law states that

$$\oint H_{\tan} \, dl = I_{\text{enclosed}} \tag{4.4.1}$$

To determine the magnetic field inside the conductor, select the dashed circle of radius $x < r$ shown in Figure 4.6 as the closed contour for Ampere's law. Due to symmetry, H_x is constant along the contour. Also, there is no radial component of H_x, so H_x is tangent to the contour. That is, the conductor has a concentric magnetic field. From (4.4.1), the integral of H_x around the selected contour is

$$H_x(2\pi x) = I_x \qquad \text{for } x < r \tag{4.4.2}$$

FIGURE 4.6

Internal magnetic field
of a solid cylindrical
conductor

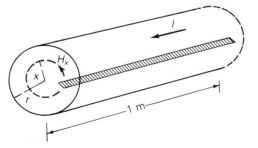

where I_x is the portion of the total current enclosed by the contour. Solving (4.4.2)

$$H_x = \frac{I_x}{2\pi x} \quad \text{A/m} \tag{4.4.3}$$

Now assume a uniform current distribution within the conductor, that is

$$I_x = \left(\frac{x}{r}\right)^2 I \quad \text{for } x < r \tag{4.4.4}$$

Using (4.4.4) in (4.4.3)

$$H_x = \frac{xI}{2\pi r^2} \quad \text{A/m} \tag{4.4.5}$$

For a nonmagnetic conductor, the magnetic flux density B_x is

$$B_x = \mu_0 H_x = \frac{\mu_0 xI}{2\pi r^2} \quad \text{Wb/m}^2 \tag{4.4.6}$$

The differential flux $d\Phi$ per-unit length of conductor in the cross-hatched rectangle of width dx shown in Figure 4.6 is

$$d\Phi = B_x \, dx \quad \text{Wb/m} \tag{4.4.7}$$

Computation of the differential flux linkage $d\lambda$ in the rectangle is tricky since only the fraction $(x/r)^2$ of the total current I is linked by the flux. That is,

$$d\lambda = \left(\frac{x}{r}\right)^2 d\Phi = \frac{\mu_0 I}{2\pi r^4} x^3 \, dx \quad \text{Wb-t/m} \tag{4.4.8}$$

Integrating (4.4.8) from $x = 0$ to $x = r$ determines the total flux linkages λ_{int} inside the conductor

$$\lambda_{\text{int}} = \int_0^r d\lambda = \frac{\mu_0 I}{2\pi r^4} \int_0^r x^3 \, dx = \frac{\mu_0 I}{8\pi} = \frac{1}{2} \times 10^{-7} I \quad \text{Wb-t/m} \tag{4.4.9}$$

The internal inductance L_{int} per-unit length of conductor due to this flux linkage is then

$$\text{L}_{\text{int}} = \frac{\lambda_{\text{int}}}{I} = \frac{\mu_0}{8\pi} = \frac{1}{2} \times 10^{-7} \quad \text{H/m} \tag{4.4.10}$$

Next, in order to determine the magnetic field outside the conductor, select the dashed circle of radius $x > r$ shown in Figure 4.7 as the closed contour for Ampere's law. Noting that this contour encloses the entire current I, integration of (4.4.1) yields

$$H_x(2\pi x) = I \tag{4.4.11}$$

which gives

$$H_x = \frac{I}{2\pi x} \quad \text{A/m} \quad x > r \tag{4.4.12}$$

FIGURE 4.7

External magnetic field of a solid cylindrical conductor

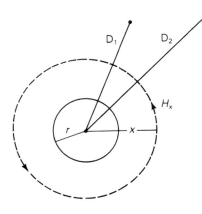

Outside the conductor, $\mu = \mu_0$ and

$$B_x = \mu_0 H_x = (4\pi \times 10^{-7})\frac{I}{2\pi x} = 2 \times 10^{-7}\frac{I}{x} \quad \text{Wb/m}^2 \tag{4.4.13}$$

$$d\Phi = B_x \, dx = 2 \times 10^{-7}\frac{I}{x} \, dx \quad \text{Wb/m} \tag{4.4.14}$$

Since the entire current I is linked by the flux outside the conductor,

$$d\lambda = d\Phi = 2 \times 10^{-7}\frac{I}{x} \, dx \quad \text{Wb-t/m} \tag{4.4.15}$$

Integrating (4.4.15) between two external points at distances D_1 and D_2 from the conductor center gives the external flux linkage λ_{12} between D_1 and D_2:

$$\lambda_{12} = \int_{D_1}^{D_2} d\lambda = 2 \times 10^{-7} I \int_{D_1}^{D_2} \frac{dx}{x}$$

$$= 2 \times 10^{-7} I \ln\left(\frac{D_2}{D_1}\right) \quad \text{Wb-t/m} \tag{4.4.16}$$

The external inductance L_{12} per-unit length due to the flux linkages between D_1 and D_2 is then

$$L_{12} = \frac{\lambda_{12}}{I} = 2 \times 10^{-7} \ln\left(\frac{D_2}{D_1}\right) \quad \text{H/m} \tag{4.4.17}$$

The total flux λ_P linking the conductor out to external point P at distance D is the sum of the internal flux linkage, (4.4.9), and the external flux linkage, (4.4.16) from $D_1 = r$ to $D_2 = D$. That is

$$\lambda_P = \frac{1}{2} \times 10^{-7} I + 2 \times 10^{-7} I \ln\frac{D}{r} \tag{4.4.18}$$

Using the identity $\frac{1}{2} = 2 \ln e^{1/4}$ in (4.4.18), a more convenient expression for λ_P is obtained:

$$\lambda_P = 2 \times 10^{-7} I \left(\ln e^{1/4} + \ln \frac{D}{r} \right)$$

$$= 2 \times 10^{-7} I \ln \frac{D}{e^{-1/4} r}$$

$$= 2 \times 10^{-7} I \ln \frac{D}{r'} \quad \text{Wb-t/m} \tag{4.4.19}$$

where

$$r' = e^{-1/4} r = 0.7788 r \tag{4.4.20}$$

Also, the total inductance L_P due to both internal and external flux linkages out to distance D is

$$L_P = \frac{\lambda_P}{I} = 2 \times 10^{-7} \ln \left(\frac{D}{r'} \right) \quad \text{H/m} \tag{4.4.21}$$

Finally, consider the array of M solid cylindrical conductors shown in Figure 4.8. Assume that each conductor m carries current I_m referenced out of the page. Also assume that the sum of the conductor currents is zero— that is,

$$I_1 + I_2 + \cdots + I_M = \sum_{m=1}^{M} I_m = 0 \tag{4.4.22}$$

The flux linkage λ_{kPk}, which links conductor k out to point P due to current I_k, is, from (4.4.19),

$$\lambda_{kPk} = 2 \times 10^{-7} I_k \ln \frac{D_{Pk}}{r'_k} \tag{4.4.23}$$

FIGURE 4.8

Array of M solid cylindrical conductors

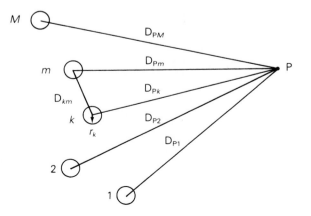

Note that λ_{kPk} includes both internal and external flux linkages due to I_k. The flux linkage λ_{kPm}, which links conductor k out to P due to I_m, is, from (4.4.16),

$$\lambda_{kPm} = 2 \times 10^{-7} I_m \ln \frac{D_{Pm}}{D_{km}} \tag{4.4.24}$$

In (4.4.24) we use D_{km} instead of $(D_{km} - r_k)$ or $(D_{km} + r_k)$, which is a valid approximation when D_{km} is much greater than r_k. It can also be shown that this is a good approximation even when D_{km} is small. Using superposition, the total flux linkage λ_{kP}, which links conductor k out to P due to all the currents, is

$$\lambda_{kP} = \lambda_{kP1} + \lambda_{kP2} + \cdots + \lambda_{kPM}$$

$$= 2 \times 10^{-7} \sum_{m=1}^{M} I_m \ln \frac{D_{Pm}}{D_{km}} \tag{4.4.25}$$

where we define $D_{kk} = r_k' = e^{-1/4} r_k$ when $m = k$ in the above summation. Equation (4.4.25) is separated into two summations:

$$\lambda_{kP} = 2 \times 10^{-7} \sum_{m=1}^{M} I_m \ln \frac{1}{D_{km}} + 2 \times 10^{-7} \sum_{m=1}^{M} I_m \ln D_{Pm} \tag{4.4.26}$$

Removing the last term from the second summation we get:

$$\lambda_{kP} = 2 \times 10^{-7} \left[\sum_{m=1}^{M} I_m \ln \frac{1}{D_{km}} + \sum_{m=1}^{M-1} I_m \ln D_{Pm} + I_M \ln D_{PM} \right] \tag{4.4.27}$$

From (4.4.22),

$$I_M = -(I_1 + I_2 + \cdots + I_{M-1}) = -\sum_{m=1}^{M-1} I_m \tag{4.4.28}$$

Using (4.4.28) in (4.4.27)

$$\lambda_{kP} = 2 \times 10^{-7} \left[\sum_{m=1}^{M} I_m \ln \frac{1}{D_{km}} + \sum_{m=1}^{M-1} I_m \ln D_{Pm} - \sum_{m=1}^{M-1} I_m \ln D_{PM} \right]$$

$$= 2 \times 10^{-7} \left[\sum_{m=1}^{M} I_m \ln \frac{1}{D_{km}} + \sum_{m=1}^{M-1} I_m \ln \frac{D_{Pm}}{D_{PM}} \right] \tag{4.4.29}$$

Now, let λ_k equal the total flux linking conductor k out to infinity. That is, $\lambda_k = \lim_{P \to \infty} \lambda_{kP}$. As P $\to \infty$, all the distances D_{Pm} become equal, the ratios D_{Pm}/D_{PM} become unity, and $\ln(D_{Pm}/D_{PM}) \to 0$. Therefore, the second summation in (4.4.29) becomes zero as P $\to \infty$, and

$$\lambda_k = 2 \times 10^{-7} \sum_{m=1}^{M} I_m \ln \frac{1}{D_{km}} \quad \text{Wb-t/m} \tag{4.4.30}$$

Equation (4.4.30) gives the total flux linking conductor k in an array of M conductors carrying currents $I_1, I_2, \ldots, I_M$, whose sum is zero. This equation is valid for either dc or ac currents. λ_k is a dc flux linkage when the currents are dc, and λ_k is a phasor flux linkage when the currents are phasor representations of sinusoids.

4.5

INDUCTANCE: SINGLE-PHASE TWO-WIRE LINE AND THREE-PHASE THREE-WIRE LINE WITH EQUAL PHASE SPACING

The results of the previous section are used here to determine the inductances of two relatively simple transmission lines: a single-phase two-wire line and a three-phase three-wire line with equal phase spacing.

Figure 4.9(a) shows a single-phase two-wire line consisting of two solid cylindrical conductors x and y. Conductor x with radius r_x carries phasor current $I_x = I$ referenced out of the page. Conductor y with radius r_y carries return current $I_y = -I$. Since the sum of the two currents is zero, (4.4.30) is valid, from which the total flux linking conductor x is

$$
\begin{aligned}
\lambda_x &= 2 \times 10^{-7}\left(I_x \ln \frac{1}{D_{xx}} + I_y \ln \frac{1}{D_{xy}}\right) \\
&= 2 \times 10^{-7}\left(I \ln \frac{1}{r_x'} - I \ln \frac{1}{D}\right) \\
&= 2 \times 10^{-7} I \ln \frac{D}{r_x'} \quad \text{Wb-t/m}
\end{aligned}
\tag{4.5.1}
$$

where $r_x' = e^{-1/4} r_x = 0.7788 r_x$.

The inductance of conductor x is then

$$
L_x = \frac{\lambda_x}{I_x} = \frac{\lambda_x}{I} = 2 \times 10^{-7} \ln \frac{D}{r_x'} \quad \text{H/m per conductor}
\tag{4.5.2}
$$

FIGURE 4.9

Single-phase two-wire line

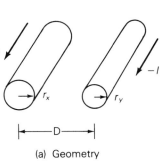

(a) Geometry

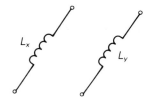

(b) Inductances

Similarly, the total flux linking conductor y is

$$\lambda_y = 2 \times 10^{-7} \left(I_x \ln \frac{1}{D_{yx}} + I_y \ln \frac{1}{D_{yy}} \right)$$

$$= 2 \times 10^{-7} \left(I \ln \frac{1}{D} - I \ln \frac{1}{r'_y} \right)$$

$$= -2 \times 10^{-7} I \ln \frac{D}{r'_y} \qquad (4.5.3)$$

and

$$L_y = \frac{\lambda_y}{I_y} = \frac{\lambda_y}{-I} = 2 \times 10^{-7} \ln \frac{D}{r'_y} \quad \text{H/m per conductor} \qquad (4.5.4)$$

The total inductance of the single-phase circuit, also called *loop inductance*, is

$$L = L_x + L_y = 2 \times 10^{-7} \left(\ln \frac{D}{r'_x} + \ln \frac{D}{r'_y} \right)$$

$$= 2 \times 10^{-7} \ln \frac{D^2}{r'_x r'_y}$$

$$= 4 \times 10^{-7} \ln \frac{D}{\sqrt{r'_x r'_y}} \quad \text{H/m per circuit} \qquad (4.5.5)$$

Also, if $r'_x = r'_y = r'$, the total circuit inductance is

$$L = 4 \times 10^{-7} \ln \frac{D}{r'} \quad \text{H/m per circuit} \qquad (4.5.6)$$

The inductances of the single-phase two-wire line are shown in Figure 4.9(b).

Figure 4.10(a) shows a three-phase three-wire line consisting of three solid cylindrical conductors a, b, c, each with radius r, and with equal phase spacing D between any two conductors. To determine inductance, assume balanced positive-sequence currents I_a, I_b, I_c that satisfy $I_a + I_b + I_c = 0$. Then (4.4.30) is valid and the total flux linking the phase a conductor is

FIGURE 4.10

Three-phase three-wire line with equal phase spacing

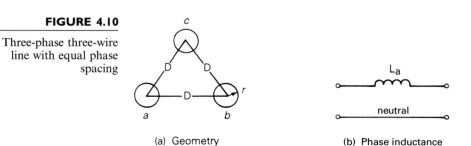

(a) Geometry

(b) Phase inductance

$$\lambda_a = 2 \times 10^{-7} \left(I_a \ln \frac{1}{r'} + I_b \ln \frac{1}{D} + I_c \ln \frac{1}{D} \right)$$

$$= 2 \times 10^{-7} \left[I_a \ln \frac{1}{r'} + (I_b + I_c) \ln \frac{1}{D} \right] \qquad (4.5.7)$$

Using $(I_b + I_c) = -I_a$,

$$\lambda_a = 2 \times 10^{-7} \left(I_a \ln \frac{1}{r'} - I_a \ln \frac{1}{D} \right)$$

$$= 2 \times 10^{-7} I_a \ln \frac{D}{r'} \quad \text{Wb-t/m} \qquad (4.5.8)$$

The inductance of phase a is then

$$L_a = \frac{\lambda_a}{I_a} = 2 \times 10^{-7} \ln \frac{D}{r'} \quad \text{H/m per phase} \qquad (4.5.9)$$

Due to symmetry, the same result is obtained for $L_b = \lambda_b/I_b$ and for $L_c = \lambda_c/I_c$. However, only one phase need be considered for balanced three-phase operation of this line, since the flux linkages of each phase have equal magnitudes and 120° displacement. The phase inductance is shown in Figure 4.10(b).

4.6

INDUCTANCE: COMPOSITE CONDUCTORS, UNEQUAL PHASE SPACING, BUNDLED CONDUCTORS

The results of Section 4.5 are extended here to include composite conductors, which consist of two or more solid cylindrical subconductors in parallel. A stranded conductor is one example of a composite conductor. For simplicity we assume that for each conductor, the subconductors are identical and share the conductor current equally.

Figure 4.11 shows a single-phase two-conductor line consisting of two composite conductors x and y. Conductor x has N identical subconductors, each with radius r_x and with current (I/N) referenced out of the page. Similarly, conductor y consists of M identical subconductors, each with radius r_y and with return current $(-I/M)$. Since the sum of all the currents is zero, (4.4.30) is valid and the total flux Φ_k linking subconductor k of conductor x is

$$\Phi_k = 2 \times 10^{-7} \left[\frac{I}{N} \sum_{m=1}^{N} \ln \frac{1}{D_{km}} - \frac{I}{M} \sum_{m=1'}^{M} \ln \frac{1}{D_{km}} \right] \qquad (4.6.1)$$

Since only the fraction $(1/N)$ of the total conductor current I is linked by this flux, the flux linkage λ_k of (the current in) subconductor k is

FIGURE 4.11

Single-phase two-
conductor line with
composite conductors

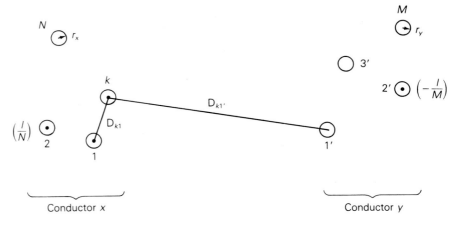

$$\lambda_k = \frac{\Phi_k}{N} = 2 \times 10^{-7} I \left[\frac{1}{N^2} \sum_{m=1}^{N} \ln \frac{1}{D_{km}} - \frac{1}{NM} \sum_{m=1'}^{M} \ln \frac{1}{D_{km}} \right] \qquad (4.6.2)$$

The total flux linkage of conductor x is

$$\lambda_x = \sum_{k=1}^{N} \lambda_k = 2 \times 10^{-7} I \sum_{k=1}^{N} \left[\frac{1}{N^2} \sum_{m=1}^{N} \ln \frac{1}{D_{km}} - \frac{1}{NM} \sum_{m=1'}^{M} \ln \frac{1}{D_{km}} \right]$$

$$(4.6.3)$$

Using $\ln A^\alpha = \alpha \ln A$ and $\sum \ln A_k = \ln \prod A_k$ (sum of $\ln s = \ln$ of products), (4.6.3) can be rewritten in the following form:

$$\lambda_x = 2 \times 10^{-7} I \ln \prod_{k=1}^{N} \frac{\left(\prod\limits_{m=1'}^{M} D_{km} \right)^{1/NM}}{\left(\prod\limits_{m=1}^{N} D_{km} \right)^{1/N^2}} \qquad (4.6.4)$$

and the inductance of conductor x, $L_x = \dfrac{\lambda_x}{I}$, can be written as

$$L_x = 2 \times 10^{-7} \ln \frac{D_{xy}}{D_{xx}} \quad \text{H/m per conductor} \qquad (4.6.5)$$

where

$$D_{xy} = \sqrt[MN]{\prod_{k=1}^{N} \prod_{m=1'}^{M} D_{km}} \qquad (4.6.6)$$

$$D_{xx} = \sqrt[N^2]{\prod_{k=1}^{N} \prod_{m=1}^{N} D_{km}} \qquad (4.6.7)$$

D_{xy}, given by (4.6.6), is the MNth root of the product of the MN distances from the subconductors of conductor x to the subconductors of conductor y. Associated with each subconductor k of conductor x are the M distances $D_{k1}{}', D_{k2}{}', \ldots, D_{kM}$ to the subconductors of conductor y. For N subconductors in conductor x, there are therefore MN of these distances. D_{xy} is called the *geometric mean distance* or GMD between conductors x and y.

Also, D_{xx}, given by (4.6.7), is the N^2 root of the product of the N^2 distances between the subconductors of conductor x. Associated with each subconductor k are the N distances $D_{k1}, D_{k2}, \ldots, D_{kk} = r', \ldots, D_{kN}$. For N subconductors in conductor x, there are therefore N^2 of these distances. D_{xx} is called the *geometric mean radius* or GMR of conductor x.

Similarly, for conductor y,

$$L_y = 2 \times 10^{-7} \ln \frac{D_{xy}}{D_{yy}} \quad \text{H/m per conductor} \tag{4.6.8}$$

where

$$D_{yy} = \sqrt[M^2]{\prod_{k=1'}^{M} \prod_{m=1'}^{M} D_{km}} \tag{4.6.9}$$

D_{yy}, the GMR of conductor y, is the M^2 root of the product of the M^2 distances between the subconductors of conductor y. The total inductance L of the single-phase circuit is

$$L = L_x + L_y \quad \text{H/m per circuit} \tag{4.6.10}$$

EXAMPLE 4.2   GMR, GMD, and inductance: single-phase two-conductor line

Expand (4.6.6), (4.6.7), and (4.6.9) for $N = 3$ and $M = 2'$. Then evaluate L_x, L_y, and L in H/m for the single-phase two-conductor line shown in Figure 4.12.

FIGURE 4.12

Single-phase
two-conductor
line for Example 4.2

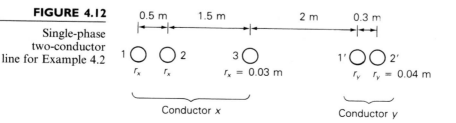

SOLUTION   For $N = 3$ and $M = 2'$, (4.6.6) becomes

$$\mathbf{D}_{xy} = \sqrt[6]{\prod_{k=1}^{3} \prod_{m=1'}^{2'} \mathbf{D}_{km}}$$

$$= \sqrt[6]{\prod_{k=1}^{3} \mathbf{D}_{k1'}\mathbf{D}_{k2'}}$$

$$= \sqrt[6]{(\mathbf{D}_{11'}\mathbf{D}_{12'})(\mathbf{D}_{21'}\mathbf{D}_{22'})(\mathbf{D}_{31'}\mathbf{D}_{32'})}$$

Similarly, (4.6.7) becomes

$$\mathbf{D}_{xx} = \sqrt[9]{\prod_{k=1}^{3} \prod_{m=1}^{3} \mathbf{D}_{km}}$$

$$= \sqrt[9]{\prod_{k=1}^{3} \mathbf{D}_{k1}\mathbf{D}_{k2}\mathbf{D}_{k3}}$$

$$= \sqrt[9]{(\mathbf{D}_{11}\mathbf{D}_{12}\mathbf{D}_{13})(\mathbf{D}_{21}\mathbf{D}_{22}\mathbf{D}_{23})(\mathbf{D}_{31}\mathbf{D}_{32}\mathbf{D}_{33})}$$

and (4.6.9) becomes

$$\mathbf{D}_{yy} = \sqrt[4]{\prod_{k=1'}^{2'} \prod_{m=1'}^{2'} \mathbf{D}_{km}}$$

$$= \sqrt[4]{\prod_{k=1'}^{2'} \mathbf{D}_{k1'}\mathbf{D}_{k2'}}$$

$$= \sqrt[4]{(\mathbf{D}_{1'1'}\mathbf{D}_{1'2'})(\mathbf{D}_{2'1'}\mathbf{D}_{2'2'})}$$

Evaluating $\mathbf{D}_{xy}$, $\mathbf{D}_{xx}$, and $\mathbf{D}_{yy}$ for the single-phase two-conductor line shown in Figure 4.12,

$$D_{11'} = 4 \text{ m} \qquad D_{12'} = 4.3 \text{ m} \qquad D_{21'} = 3.5 \text{ m}$$

$$D_{22'} = 3.8 \text{ m} \qquad D_{31'} = 2 \text{ m} \qquad D_{32'} = 2.3 \text{ m}$$

$$D_{xy} = \sqrt[6]{(4)(4.3)(3.5)(3.8)(2)(2.3)} = 3.189 \text{ m}$$

$$D_{11} = D_{22} = D_{33} = r'_x = e^{-1/4}r_x = (0.7788)(0.03) = 0.02336 \text{ m}$$

$$D_{21} = D_{12} = 0.5 \text{ m}$$

$$D_{23} = D_{32} = 1.5 \text{ m}$$

$$D_{31} = D_{13} = 2.0 \text{ m}$$

$$D_{xx} = \sqrt[9]{(0.02336)^3(0.5)^2(1.5)^2(2.0)^2} = 0.3128 \text{ m}$$

$$D_{1'1'} = D_{2'2'} = r'_y = e^{-1/4}r_y = (0.7788)(0.04) = 0.03115 \text{ m}$$

$$D_{1'2'} = D_{2'1'} = 0.3 \text{ m}$$

$$D_{yy} = \sqrt[4]{(0.03115)^2(0.3)^2} = 0.09667 \text{ m}$$

Then, from (4.6.5), (4.6.8), and (4.6.10):

$$L_x = 2 \times 10^{-7} \ln\left(\frac{3.189}{0.3128}\right) = 4.644 \times 10^{-7} \quad \text{H/m per conductor}$$

$$L_y = 2 \times 10^{-7} \ln\left(\frac{3.189}{0.09667}\right) = 6.992 \times 10^{-7} \quad \text{H/m per conductor}$$

$$L = L_x + L_y = 1.164 \times 10^{-6} \quad \text{H/m per circuit} \qquad \blacksquare$$

It is seldom necessary to calculate GMR or GMD for standard lines. The GMR of standard conductors is provided by conductor manufacturers and can be found in various handbooks (see Appendix Tables A.3 and A.4). Also, if the distances between conductors are large compared to the distances between subconductors of each conductor, then the GMD between conductors is approximately equal to the distance between conductor centers.

EXAMPLE 4.3   **Inductance and inductive reactance: single-phase line**

A single-phase line operating at 60 Hz consists of two 4/0 12-strand copper conductors with 5 ft spacing between conductor centers. The line length is 20 miles. Determine the total inductance in H and the total inductive reactance in Ω.

SOLUTION   The GMD between conductor centers is $D_{xy} = 5$ ft. Also, from Table A.3, the GMR of a 4/0 12-strand copper conductor is $D_{xx} = D_{yy} = 0.01750$ ft. From (4.6.5) and (4.6.8),

$$L_x = L_y = 2 \times 10^{-7} \ln\left(\frac{5}{0.01750}\right)\frac{\text{H}}{\text{m}} \times 1609\frac{\text{m}}{\text{mi}} \times 20 \text{ mi}$$

$$= 0.03639 \quad \text{H per conductor}$$

The total inductance is

$$L = L_x + L_y = 2 \times 0.03639 = 0.07279 \quad \text{H per circuit}$$

and the total inductive reactance is

$$X_L = 2\pi f L = (2\pi)(60)(0.07279) = 27.44 \quad \Omega \text{ per circuit} \qquad \blacksquare$$

FIGURE 4.13

Completely transposed three-phase line

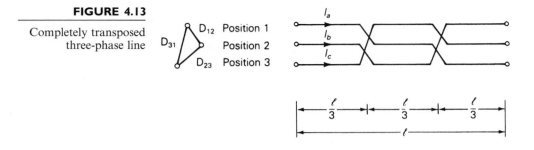

To calculate inductance for three-phase lines with stranded conductors and equal phase spacing, r' is replaced by the conductor GMR in (4.5.9). If the spacings between phases are unequal, then balanced positive-sequence flux linkages are not obtained from balanced positive-sequence currents. Instead, unbalanced flux linkages occur, and the phase inductances are unequal. However, balance can be restored by exchanging the conductor positions along the line, a technique called *transposition*.

Figure 4.13 shows a completely transposed three-phase line. The line is transposed at two locations such that each phase occupies each position for one-third of the line length. Conductor positions are denoted 1, 2, 3 with distances D_{12}, D_{23}, D_{31} between positions. The conductors are identical, each with GMR denoted D_S. To calculate inductance of this line, assume balanced positive-sequence currents I_a, I_b, I_c, for which $I_a + I_b + I_c = 0$. Again, (4.4.30) is valid, and the total flux linking the phase a conductor while it is in position 1 is

$$\lambda_{a1} = 2 \times 10^{-7} \left[I_a \ln \frac{1}{D_S} + I_b \ln \frac{1}{D_{12}} + I_c \ln \frac{1}{D_{31}} \right] \quad \text{Wb-t/m} \quad (4.6.11)$$

Similarly, the total flux linkage of this conductor while it is in positions 2 and 3 is

$$\lambda_{a2} = 2 \times 10^{-7} \left[I_a \ln \frac{1}{D_S} + I_b \ln \frac{1}{D_{23}} + I_c \ln \frac{1}{D_{12}} \right] \quad \text{Wb-t/m} \quad (4.6.12)$$

$$\lambda_{a3} = 2 \times 10^{-7} \left[I_a \ln \frac{1}{D_S} + I_b \ln \frac{1}{D_{31}} + I_c \ln \frac{1}{D_{23}} \right] \quad \text{Wb-t/m} \quad (4.6.13)$$

The average of the above flux linkages is

$$\lambda_a = \frac{\lambda_{a1}\left(\frac{l}{3}\right) + \lambda_{a2}\left(\frac{l}{3}\right) + \lambda_{a3}\left(\frac{l}{3}\right)}{l} = \frac{\lambda_{a1} + \lambda_{a2} + \lambda_{a3}}{3}$$

$$= \frac{2 \times 10^{-7}}{3} \left[3I_a \ln \frac{1}{D_S} + I_b \ln \frac{1}{D_{12}D_{23}D_{31}} + I_c \ln \frac{1}{D_{12}D_{23}D_{31}} \right] \quad (4.6.14)$$

Using $(I_b + I_c) = -I_a$ in (4.6.14),

$$\lambda_a = \frac{2 \times 10^{-7}}{3} \left[3I_a \ln \frac{1}{D_S} - I_a \ln \frac{1}{D_{12}D_{23}D_{31}} \right]$$

$$= 2 \times 10^{-7} I_a \ln \frac{\sqrt[3]{D_{12}D_{23}D_{31}}}{D_S} \quad \text{Wb-t/m} \tag{4.6.15}$$

and the average inductance of phase a is

$$L_a = \frac{\lambda_a}{I_a} = 2 \times 10^{-7} \ln \frac{\sqrt[3]{D_{12}D_{23}D_{31}}}{D_S} \quad \text{H/m per phase} \tag{4.6.16}$$

The same result is obtained for $L_b = \lambda_b/I_b$ and for $L_c = \lambda_c/I_c$. However, only one phase need be considered for balanced three-phase operation of a completely transposed three-phase line. Defining

$$D_{eq} = \sqrt[3]{D_{12}D_{23}D_{31}} \tag{4.6.17}$$

we have

$$L_a = 2 \times 10^{-7} \ln \frac{D_{eq}}{D_S} \quad \text{H/m} \tag{4.6.18}$$

D_{eq}, the cube root of the product of the three-phase spacings, is the geometric mean distance between phases. Also, D_S is the conductor GMR for stranded conductors, or r' for solid cylindrical conductors.

EXAMPLE 4.4   **Inductance and inductive reactance: three-phase line**

A completely transposed 60-Hz three-phase line has flat horizontal phase spacing with 10 m between adjacent conductors. The conductors are 1,590,000 cmil ACSR with 54/3 stranding. Line length is 200 km. Determine the inductance in H and the inductive reactance in Ω.

SOLUTION   From Table A.4, the GMR of a 1,590,000 cmil 54/3 ACSR conductor is

$$D_S = 0.0520 \text{ ft} \frac{1 \text{ m}}{3.28 \text{ ft}} = 0.0159 \text{ m}$$

Also, from (4.6.17) and (4.6.18),

$$D_{eq} = \sqrt[3]{(10)(10)(20)} = 12.6 \text{ m}$$

$$L_a = 2 \times 10^{-7} \ln \left(\frac{12.6}{0.0159} \right) \frac{\text{H}}{\text{m}} \times \frac{1000 \text{ m}}{\text{km}} \times 200 \text{ km}$$

$$= 0.267 \quad \text{H}$$

The inductive reactance of phase a is

$$X_a = 2\pi f L_a = 2\pi(60)(0.267) = 101 \quad \Omega$$

■

FIGURE 4.14

Bundle conductor configurations

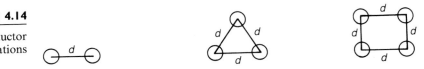

It is common practice for EHV lines to use more than one conductor per phase, a practice called *bundling*. Bundling reduces the electric field strength at the conductor surfaces, which in turn reduces or eliminates corona and its results: undesirable power loss, communications interference, and audible noise. Bundling also reduces the series reactance of the line by increasing the GMR of the bundle.

Figure 4.14 shows common EHV bundles consisting of two, three, or four conductors. The three-conductor bundle has its conductors on the vertices of an equilateral triangle, and the four-conductor bundle has its conductors on the corners of a square. To calculate inductance, D_S in (4.6.18) is replaced by the GMR of the bundle. Since the bundle constitutes a composite conductor, calculation of bundle GMR is, in general, given by (4.6.7). If the conductors are stranded and the bundle spacing d is large compared to the conductor outside radius, each stranded conductor is first replaced by an equivalent solid cylindrical conductor with GMR $= D_S$. Then the bundle is replaced by one equivalent conductor with GMR $= D_{SL}$, given by (4.6.7) with $n = 2, 3,$ or 4 as follows:

Two-conductor bundle:

$$D_{SL} = \sqrt[4]{(D_S \times d)^2} = \sqrt{D_S d} \qquad (4.6.19)$$

Three-conductor bundle:

$$D_{SL} = \sqrt[9]{(D_S \times d \times d)^3} = \sqrt[3]{D_S d^2} \qquad (4.6.20)$$

Four-conductor bundle:

$$D_{SL} = \sqrt[16]{(D_S \times d \times d \times d\sqrt{2})^4} = 1.091 \sqrt[4]{D_S d^3} \qquad (4.6.21)$$

The inductance is then

$$L_a = 2 \times 10^{-7} \ln \frac{D_{eq}}{D_{SL}} \quad \text{H/m} \qquad (4.6.22)$$

If the phase spacings are large compared to the bundle spacing, then sufficient accuracy for D_{eq} is obtained by using the distances between bundle centers.

EXAMPLE 4.5   Inductive reactance: three-phase line with bundled conductors

Each of the 1,590,000 cmil conductors in Example 4.4 is replaced by two 795,000 cmil ACSR 26/2 conductors, as shown in Figure 4.15. Bundle spac-

FIGURE 4.15

Three-phase bundled conductor line for Example 4.5

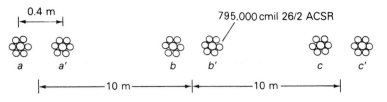

ing is 0.40 m. Flat horizontal spacing is retained, with 10 m between adjacent bundle centers. Calculate the inductive reactance of the line and compare it with that of Example 4.4.

SOLUTION   From Table A.4, the GMR of a 795,000 cmil 26/2 ACSR conductor is

$$D_S = 0.0375 \text{ ft} \times \frac{1 \text{ m}}{3.28 \text{ ft}} = 0.0114 \text{ m}$$

From (4.6.19), the two-conductor bundle GMR is

$$D_{SL} = \sqrt{(0.0114)(0.40)} = 0.0676 \quad \text{m}$$

Since $D_{eq} = 12.6$ m is the same as in Example 4.4,

$$L_a = 2 \times 10^{-7} \ln\left(\frac{12.6}{0.0676}\right)(1000)(200) = 0.209 \quad \text{H}$$

$$X_a = 2\pi f L_1 = (2\pi)(60)(0.209) = 78.8 \quad \Omega$$

The reactance of the bundled line, 78.8 Ω, is 22% less than that of Example 4.4, even though the two-conductor bundle has the same amount of conductor material (that is, the same cmil per phase). One advantage of reduced series line reactance is smaller line-voltage drops. Also, the loadability of medium and long EHV lines is increased (see Chapter 5).   ∎

4.7

SERIES IMPEDANCES: THREE-PHASE LINE WITH NEUTRAL CONDUCTORS AND EARTH RETURN

In this section, we develop equations suitable for computer calculation of the series impedances, including resistances and inductive reactances, for the three-phase overhead line shown in Figure 4.16. This line has three phase conductors a, b, and c, where bundled conductors, if any, have already been replaced by equivalent conductors, as described in Section 4.6. The line also has N neutral conductors denoted $n1, n2, \ldots, nN$.* All the neutral conductors

---

*Instead of *shield wire* we use the term *neutral conductor*, which applies to distribution as well as transmission lines.

FIGURE 4.16

Three-phase
transmission line with
earth replaced by earth
return conductors

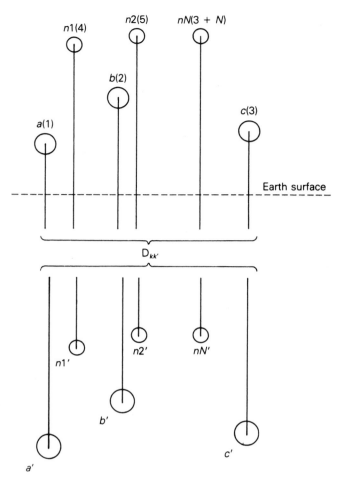

are connected in parallel and are grounded to the earth at regular intervals
along the line. Any isolated neutral conductors that carry no current are
omitted. The phase conductors are insulated from each other and from earth.

If the phase currents are not balanced, there may be a return current in
the grounded neutral wires and in the earth. The earth return current will
spread out under the line, seeking the lowest impedance return path. A classic
paper by Carson [4], later modified by others [5, 6], shows that the earth can
be replaced by a set of "earth return" conductors located directly under the
overhead conductors, as shown in Figure 4.16. Each earth return conductor
carries the negative of its overhead conductor current, has a GMR denoted
$D_{k'k'}$, distance $D_{kk'}$ from its overhead conductor, and resistance $R_{k'}$ given
by:

TABLE 4.4	Type of Earth	Resistivity (Ωm)	$D_{kk'}$ (m)
Earth resistivities and 60-Hz equivalent conductor distances	Sea water	0.01–1.0	8.50–85.0
	Swampy ground	10–100	269–850
	Average damp earth	100	850
	Dry earth	1000	2690
	Pure slate	10^7	269,000
	Sandstone	10^9	2,690,000

$$D_{k'k'} = D_{kk} \quad \text{m} \tag{4.7.1}$$

$$D_{kk'} = 658.5\sqrt{\rho/f} \quad \text{m} \tag{4.7.2}$$

$$R_{k'} = 9.869 \times 10^{-7} f \quad \Omega/\text{m} \tag{4.7.3}$$

where ρ is the earth resistivity in ohm-meters and f is frequency in hertz. Table 4.4 lists earth resistivities and 60-Hz equivalent conductor distances for various types of earth. It is common practice to select $\rho = 100 \ \Omega$m when actual data are unavailable.

Note that the GMR of each earth return conductor, $D_{k'k'}$, is the same as the GMR of its corresponding overhead conductor, D_{kk}. Also, all the earth return conductors have the same distance $D_{kk'}$ from their overhead conductors and the same resistance $R_{k'}$.

For simplicity, we renumber the overhead conductors from 1 to $(3 + N)$, beginning with the phase conductors, then overhead neutral conductors, as shown in Figure 4.16. Operating as a transmission line, the sum of the currents in all the conductors is zero. That is,

$$\sum_{k=1}^{(6+2N)} I_k = 0 \tag{4.7.4}$$

Equation (4.4.30) is therefore valid, and the flux linking overhead conductor k is

$$\lambda_k = 2 \times 10^{-7} \sum_{m=1}^{(3+N)} I_m \ln \frac{D_{km'}}{D_{km}} \quad \text{Wb-t/m} \tag{4.7.5}$$

In matrix format, (4.7.5) becomes

$$\lambda = \mathbf{L}I \tag{4.7.6}$$

where

λ is a $(3 + N)$ vector

I is a $(3 + N)$ vector

$\mathbf{L}$ is a $(3 + N) \times (3 + N)$ matrix whose elements are:

$$L_{km} = 2 \times 10^{-7} \ln \frac{D_{km'}}{D_{km}} \tag{4.7.7}$$

FIGURE 4.17

Circuit representation of
series-phase impedances

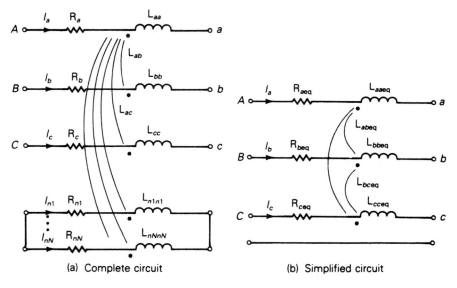

(a) Complete circuit   (b) Simplified circuit

When $k = m$, $\mathbf{D}_{kk}$ in (4.7.7) is the GMR of (bundled) conductor k. When $k \neq m$, $\mathbf{D}_{km}$ is the distance between conductors k and m.

A circuit representation of a 1-meter section of the line is shown in Figure 4.17(a). Using this circuit, the vector of voltage drops across the conductors is:

$$
\begin{bmatrix}
E_{Aa} \\
E_{Bb} \\
E_{Cc} \\
0 \\
0 \\
\vdots \\
0
\end{bmatrix}
= (\mathbf{R} + j\omega \mathbf{L})
\begin{bmatrix}
I_a \\
I_b \\
I_c \\
I_{n1} \\
\vdots \\
I_{nN}
\end{bmatrix}
\tag{4.7.8}
$$

where $\mathbf{L}$ is given by (4.7.7) and $\mathbf{R}$ is a $(3 + N) \times (3 + N)$ matrix of conductor resistances.

$$
\mathbf{R} =
\begin{bmatrix}
(\mathbf{R}_a + \mathbf{R}_{k'})\mathbf{R}_{k'} \cdots & & \mathbf{R}_{k'} \\
\mathbf{R}_{k'}(\mathbf{R}_b + \mathbf{R}_{k'})\mathbf{R}_{k'} \cdots & & \vdots \\
& (\mathbf{R}_c + \mathbf{R}_{k'})\mathbf{R}_{k'} \cdots & \\
& (\mathbf{R}_{n1} + \mathbf{R}_{k'})\mathbf{R}_{k'} \cdots & \\
& & \ddots \\
\mathbf{R}_{k'} & & (\mathbf{R}_{nN} + \mathbf{R}_{k'})
\end{bmatrix}
\ \Omega/\text{m}
\tag{4.7.9}
$$

The resistance matrix of (4.7.9) includes the resistance R_k of each overhead conductor and a mutual resistance $R_{k'}$ due to the image conductors. R_k of each overhead conductor is obtained from conductor tables such as Appendix Table A.3 or A.4, for a specified frequency, temperature, and current. $R_{k'}$ of all the image conductors is the same, as given by (4.7.3).

Our objective now is to reduce the $(3 + N)$ equations in (4.7.8) to three equations, thereby obtaining the simplified circuit representations shown in Figure 4.17(b). We partition (4.7.8) as follows:

$$
\begin{bmatrix} E_{Aa} \\ E_{Bb} \\ E_{Cc} \\ \hline 0 \\ \cdots \\ 0 \end{bmatrix}
=
\begin{bmatrix}
\overbrace{\begin{matrix} Z_{11} & Z_{12} & Z_{13} \\ Z_{21} & Z_{22} & Z_{23} \\ Z_{31} & Z_{32} & Z_{33} \end{matrix}}^{Z_A} & \vdots & \overbrace{\begin{matrix} Z_{14} & \cdots & Z_{1(3+N)} \\ Z_{24} & \cdots & Z_{2(3+N)} \\ Z_{34} & \cdots & Z_{3(3+N)} \end{matrix}}^{Z_B} \\
\hline
\underbrace{\begin{matrix} Z_{41} & Z_{42} & Z_{43} \\ & & \\ Z_{(3+N)1} & Z_{(3+N)2} & Z_{(3+N)3} \end{matrix}}_{Z_C} & \vdots & \underbrace{\begin{matrix} Z_{44} & \cdots & Z_{4(3+N)} \\ & & \\ Z_{(3+N)4} & \cdots & Z_{(3+N)(3+N)} \end{matrix}}_{Z_D}
\end{bmatrix}
\begin{bmatrix} I_a \\ I_b \\ I_c \\ \hline I_{n1} \\ I_{nN} \\ \vdots \end{bmatrix}
$$

(4.7.10)

The diagonal elements of this matrix are

$$
Z_{kk} = R_k + R_{k'} + j\omega 2 \times 10 \ln \frac{D_{kk'}}{D_{kk}} \quad \Omega/\text{m}
$$

(4.7.11)

And the off-diagonal elements, for $k \neq m$, are

$$
Z_{km} = R_{k'} + j\omega 2 \times 10 \ln \frac{D_{km'}}{D_{km}} \quad \Omega/\text{m}
$$

(4.7.12)

Next, (4.7.10) is partitioned as shown above to obtain

$$
\begin{bmatrix} E_P \\ 0 \end{bmatrix} = \begin{bmatrix} Z_A & Z_B \\ \hline Z_C & Z_D \end{bmatrix} \begin{bmatrix} I_P \\ I_n \end{bmatrix}
$$

(4.7.13)

where

$$
E_P = \begin{bmatrix} E_{Aa} \\ E_{Bb} \\ E_{Cc} \end{bmatrix}; \quad I_P = \begin{bmatrix} I_a \\ I_b \\ I_c \end{bmatrix}; \quad I_n = \begin{bmatrix} I_{n1} \\ \vdots \\ I_{nN} \end{bmatrix}
$$

E_P is the three-dimensional vector of voltage drops across the phase conductors (including the neutral voltage drop). I_P is the three-dimensional vector of phase currents and I_n is the N vector of neutral currents. Also, the $(3 + N) \times (3 + N)$ matrix in (4.7.10) is partitioned to obtain the following matrices:

Z_A with dimension 3×3

Z_B with dimension $3 \times N$

Z_C with dimension $N \times 3$

Z_D with dimension $N \times N$

Equation (4.7.13) is rewritten as two separate matrix equations:

$$E_P = Z_A I_P + Z_B I_n \tag{4.7.14}$$

$$0 = Z_C I_P + Z_D I_n \tag{4.7.15}$$

Solving (4.7.15) for I_n,

$$I_n = -Z_D^{-1} Z_C I_P \tag{4.7.16}$$

Using (4.7.16) in (4.7.14):

$$E_P = [Z_A - Z_B Z_D^{-1} Z_C] I_P \tag{4.7.17}$$

or

$$E_P = Z_P I_P \tag{4.7.18}$$

where

$$Z_P = Z_A - Z_B Z_D^{-1} Z_C \tag{4.7.19}$$

Equation (4.7.17), the desired result, relates the phase-conductor voltage drops (including neutral voltage drop) to the phase currents. Z_P given by (4.7.19) is the 3×3 series-phase impedance matrix, whose elements are denoted

$$Z_P = \begin{bmatrix} Z_{aaeq} & Z_{abeq} & Z_{aceq} \\ Z_{abeq} & Z_{bbeq} & Z_{bceq} \\ Z_{aceq} & Z_{bceq} & Z_{cceq} \end{bmatrix} \quad \Omega/m \tag{4.7.20}$$

If the line is completely transposed, the diagonal and off-diagonal elements are averaged to obtain

$$\hat{Z}_P = \begin{bmatrix} \hat{Z}_{aaeq} & \hat{Z}_{abeq} & \hat{Z}_{abeq} \\ \hat{Z}_{abeq} & \hat{Z}_{aaeq} & \hat{Z}_{abeq} \\ \hat{Z}_{abeq} & \hat{Z}_{abeq} & \hat{Z}_{aaeq} \end{bmatrix} \quad \Omega/m \tag{4.7.21}$$

where

$$\hat{Z}_{aaeq} = \tfrac{1}{3}(Z_{aaeq} + Z_{bbeq} + Z_{cceq}) \tag{4.7.22}$$

$$\hat{Z}_{abeq} = \tfrac{1}{3}(Z_{abeq} + Z_{aceq} + Z_{bceq}) \tag{4.7.23}$$

4.8

ELECTRIC FIELD AND VOLTAGE: SOLID CYLINDRICAL CONDUCTOR

The capacitance between conductors in a medium with constant permittivity ε can be obtained by determining the following:

1. Electric field strength E, from Gauss's law

2. Voltage between conductors

3. Capacitance from charge per unit volt ($C = q/V$)

As a step toward computing capacitances of general conductor configurations, we first compute the electric field of a uniformly charged, solid cylindrical conductor and the voltage between two points outside the conductor. We also compute the voltage between two conductors in an array of charged conductors.

Gauss's law states that the total electric flux leaving a closed surface equals the total charge within the volume enclosed by the surface. That is, the normal component of electric flux density integrated over a closed surface equals the charge enclosed:

$$\oiint D_\perp \, ds = \oiint \varepsilon E_\perp \, ds = Q_{\text{enclosed}} \tag{4.8.1}$$

where $D_\perp$ denotes the normal component of electric flux density, $E_\perp$ denotes the normal component of electric field strength, and ds denotes the differential surface area. From Gauss's law, electric charge is a source of electric fields. Electric field lines originate from positive charges and terminate at negative charges.

Figure 4.18 shows a solid cylindrical conductor with radius r and with charge q coulombs per meter (assumed positive in the figure), uniformly distributed on the conductor surface. For simplicity, assume that the conductor is (1) sufficiently long that end effects are negligible, and (2) a perfect conductor (that is, zero resistivity, $\rho = 0$).

Inside the perfect conductor, Ohm's law gives $E_{\text{int}} = \rho J = 0$. That is, the internal electric field E_{int} is zero. To determine the electric field outside the conductor, select the cylinder with radius $x > r$ and with 1-meter length, shown in Figure 4.18, as the closed surface for Gauss's law. Due to the uniform charge distribution, the electric field strength E_x is constant on the cylinder. Also, there is no tangential component of E_x, so the electric field is radial to the conductor. Then, integration of (4.8.1) yields

$$\varepsilon E_x (2\pi x)(1) = q(1)$$

$$E_x = \frac{q}{2\pi\varepsilon x} \quad \text{V/m} \tag{4.8.2}$$

where, for a conductor in free space, $\varepsilon = \varepsilon_0 = 8.854 \times 10^{-12}$ F/m.

FIGURE 4.18

Perfectly conducting
solid cylindrical
conductor with uniform
charge distribution

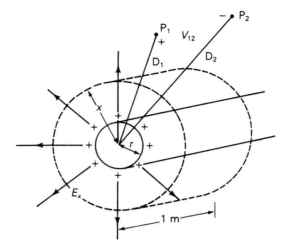

A plot of the electric field lines is also shown in Figure 4.18. The direction of the field lines, denoted by the arrows, is from the positive charges where the field originates, to the negative charges, which in this case are at infinity. If the charge on the conductor surface were negative, then the direction of the field lines would be reversed.

Concentric cylinders surrounding the conductor are constant potential surfaces. The potential difference between two concentric cylinders at distances D_1 and D_2 from the conductor center is

$$V_{12} = \int_{D_1}^{D_2} E_x \, dx \tag{4.8.3}$$

Using (4.8.2) in (4.8.1),

$$V_{12} = \int_{D_1}^{D_2} \frac{q}{2\pi\varepsilon x} \, dx = \frac{q}{2\pi\varepsilon} \ln \frac{D_2}{D_1} \quad \text{volts} \tag{4.8.4}$$

Equation (4.8.4) gives the voltage V_{12} between two points, P_1 and P_2, at distances D_1 and D_2 from the conductor center, as shown in Figure 4.18. Also, in accordance with our notation, V_{12} is the voltage at P_1 with respect to P_2. If q is positive and D_2 is greater than D_1, as shown in the figure, then V_{12} is positive; that is, P_1 is at a higher potential than P_2. Equation (4.8.4) is also valid for either dc or ac. For ac, V_{12} is a phasor voltage and q is a phasor representation of a sinusoidal charge.

Now apply (4.8.4) to the array of M solid cylindrical conductors shown in Figure 4.19. Assume that each conductor m has an ac charge q_m C/m uniformly distributed along the conductor. The voltage V_{kim} between conductors k and i due to the charge q_m acting alone is

$$V_{kim} = \frac{q_m}{2\pi\varepsilon} \ln \frac{D_{im}}{D_{km}} \quad \text{volts} \tag{4.8.5}$$

FIGURE 4.19

Array of M solid
cylindrical conductors

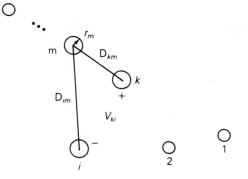

where $D_{mm} = r_m$ when $k = m$ or $i = m$. In (4.8.5) we have neglected the distortion of the electric field in the vicinity of the other conductors, caused by the fact that the other conductors themselves are constant potential surfaces. V_{kim} can be thought of as the voltage between cylinders with radii D_{km} and D_{im} concentric to conductor m at points on the cylinders remote from conductors, where there is no distortion.

Using superposition, the voltage V_{ki} between conductors k and i due to all the changes is

$$V_{ki} = \frac{1}{2\pi\varepsilon} \sum_{m=1}^{M} q_m \ln \frac{D_{im}}{D_{km}} \quad \text{volts} \tag{4.8.6}$$

4.9

CAPACITANCE: SINGLE-PHASE TWO-WIRE LINE AND THREE-PHASE THREE-WIRE LINE WITH EQUAL PHASE SPACING

The results of the previous section are used here to determine the capacitances of the two relatively simple transmission lines considered in Section 4.5, a single-phase two-wire line and a three-phase three-wire line with equal phase spacing.

First we consider the single-phase two-wire line shown in Figure 4.9. Assume that the conductors are energized by a voltage source such that conductor x has a uniform charge q C/m and, assuming conservation of charge, conductor y has an equal quantity of negative charge $-q$. Using (4.8.6) with $k = x$, $i = y$, and $m = x, y$,

$$V_{xy} = \frac{1}{2\pi\varepsilon} \left[q \ln \frac{D_{yx}}{D_{xx}} - q \ln \frac{D_{yy}}{D_{xy}} \right]$$

$$= \frac{q}{2\pi\varepsilon} \ln \frac{D_{yx}D_{xy}}{D_{xx}D_{yy}} \tag{4.9.1}$$

Using $D_{xy} = D_{yx} = D$, $D_{xx} = r_x$, and $D_{yy} = r_y$, (4.9.1) becomes

$$V_{xy} = \frac{q}{\pi \varepsilon} \ln \frac{D}{\sqrt{r_x r_y}} \quad \text{volts} \qquad (4.9.2)$$

For a 1-meter line length, the capacitance between conductors is

$$C_{xy} = \frac{q}{V_{xy}} = \frac{\pi \varepsilon}{\ln \left(\dfrac{D}{\sqrt{r_x r_y}} \right)} \quad \text{F/m line-to-line} \qquad (4.9.3)$$

and if $r_x = r_y = r$,

$$C_{xy} = \frac{\pi \varepsilon}{\ln(D/r)} \quad \text{F/m line-to-line} \qquad (4.9.4)$$

If the two-wire line is supplied by a transformer with a grounded center tap, then the voltage between each conductor and ground is one-half that given by (4.9.2). That is,

$$V_{xn} = V_{yn} = \frac{V_{xy}}{2} \qquad (4.9.5)$$

and the capacitance from either line to the grounded neutral is

$$C_n = C_{xn} = C_{yn} = \frac{q}{V_{xn}} = 2C_{xy}$$

$$= \frac{2\pi \varepsilon}{\ln(D/r)} \quad \text{F/m line-to-neutral} \qquad (4.9.6)$$

Circuit representations of the line-to-line and line-to-neutral capacitances are shown in Figure 4.20. Note that if the neutral is open in Figure 4.20(b), the two line-to-neutral capacitances combine in series to give the line-to-line capacitance.

Next consider the three-phase line with equal phase spacing shown in Figure 4.10. We shall neglect the effect of earth and neutral conductors here. To determine the positive-sequence capacitance, assume positive-sequence charges q_a, q_b, q_c such that $q_a + q_b + q_c = 0$. Using (4.8.6) with $k = a$, $i = b$, and $m = a, b, c$, the voltage V_{ab} between conductors a and b is

$$V_{ab} = \frac{1}{2\pi \varepsilon} \left[q_a \ln \frac{D_{ba}}{D_{aa}} + q_b \ln \frac{D_{bb}}{D_{ab}} + q_c \ln \frac{D_{bc}}{D_{ac}} \right] \qquad (4.9.7)$$

Using $D_{aa} = D_{bb} = r$, and $D_{ab} = D_{ba} = D_{ca} = D_{cb} = D$, (4.9.7) becomes

FIGURE 4.20

Circuit representation of capacitances for a single-phase two-wire line

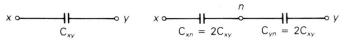

(a) Line-to-line capacitance   (b) Line-to-neutral capacitances

$$V_{ab} = \frac{1}{2\pi\varepsilon}\left[q_a \ln \frac{D}{r} + q_b \ln \frac{r}{D} + q_c \ln \frac{D}{D}\right]$$

$$= \frac{1}{2\pi\varepsilon}\left[q_a \ln \frac{D}{r} + q_b \ln \frac{r}{D}\right] \quad \text{volts} \qquad (4.9.8)$$

Note that the third term in (4.9.8) is zero because conductors a and b are equidistant from conductor c. Thus, conductors a and b lie on a constant potential cylinder for the electric field due to q_c.

Similarly, using (4.8.6) with $k = a$, $i = c$, and $m = a, b, c$, the voltage V_{ac} is

$$V_{ac} = \frac{1}{2\pi\varepsilon}\left[q_a \ln \frac{D_{ca}}{D_{aa}} + q_b \ln \frac{D_{cb}}{D_{ab}} + q_c \ln \frac{D_{cc}}{D_{ac}}\right]$$

$$= \frac{1}{2\pi\varepsilon}\left[q_a \ln \frac{D}{r} + q_b \ln \frac{D}{D} + q_c \ln \frac{r}{D}\right]$$

$$= \frac{1}{2\pi\varepsilon}\left[q_a \ln \frac{D}{r} + q_c \ln \frac{r}{D}\right] \quad \text{volts} \qquad (4.9.9)$$

Recall that for balanced positive-sequence voltages,

$$V_{ab} = \sqrt{3}\,V_{an}\underline{/+30^\circ} = \sqrt{3}\,V_{an}\left[\frac{\sqrt{3}}{2} + j\frac{1}{2}\right] \qquad (4.9.10)$$

$$V_{ac} = -V_{ca} = \sqrt{3}\,V_{an}\underline{/-30^\circ} = \sqrt{3}\,V_{an}\left[\frac{\sqrt{3}}{2} - j\frac{1}{2}\right] \qquad (4.9.11)$$

Adding (4.9.10) and (4.9.11) yields

$$V_{ab} + V_{ac} = 3V_{an} \qquad (4.9.12)$$

Using (4.9.8) and (4.9.9) in (4.9.12),

$$V_{an} = \frac{1}{3}\left(\frac{1}{2\pi\varepsilon}\right)\left[2q_a \ln \frac{D}{r} + (q_b + q_c) \ln \frac{r}{D}\right] \qquad (4.9.13)$$

and with $q_b + q_c = -q_a$,

$$V_{an} = \frac{1}{2\pi\varepsilon}q_a \ln \frac{D}{r} \quad \text{volts} \qquad (4.9.14)$$

The capacitance-to-neutral per line length is

$$C_{an} = \frac{q_a}{V_{an}} = \frac{2\pi\varepsilon}{\ln(D/r)} \quad \text{F/m line-to-neutral} \qquad (4.9.15)$$

Due to symmetry, the same result is obtained for $C_{bn} = q_b/V_{bn}$ and $C_{cn} = q_c/V_{cn}$. For balanced three-phase operation, however, only one phase need be considered. A circuit representation of the capacitance-to-neutral is shown in Figure 4.21.

FIGURE 4.21

Circuit representation of the capacitance-to-neutral of a three-phase line with equal phase spacing

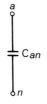

4.10

CAPACITANCE: STRANDED CONDUCTORS, UNEQUAL PHASE SPACING, BUNDLED CONDUCTORS

Equations (4.9.6) and (4.9.15) are based on the assumption that the conductors are solid cylindrical conductors with zero resistivity. The electric field inside these conductors is zero, and the external electric field is perpendicular to the conductor surfaces. Practical conductors with resistivities similar to those listed in Table 4.3 have a small internal electric field. As a result, the external electric field is slightly altered near the conductor surfaces. Also, the electric field near the surface of a stranded conductor is not the same as that of a solid cylindrical conductor. However, it is normal practice when calculating line capacitance to replace a stranded conductor by a perfectly conducting solid cylindrical conductor whose radius equals the outside radius of the stranded conductor. The resulting error in capacitance is small since only the electric field near the conductor surfaces is affected.

Also, (4.8.2) is based on the assumption that there is uniform charge distribution. But conductor charge distribution is nonuniform in the presence of other charged conductors. Therefore (4.9.6) and (4.9.15), which are derived from (4.8.2), are not exact. However, the nonuniformity of conductor charge distribution can be shown to have a negligible effect on line capacitance.

For three-phase lines with unequal phase spacing, balanced positive-sequence voltages are not obtained with balanced positive-sequence charges. Instead, unbalanced line-to-neutral voltages occur, and the phase-to-neutral capacitances are unequal. Balance can be restored by transposing the line such that each phase occupies each position for one-third of the line length. If equations similar to (4.9.7) for V_{ab} as well as for V_{ac} are written for each position in the transposition cycle, and are then averaged and used in (4.9.12)–(4.9.14), the resulting capacitance becomes

$$C_{an} = \frac{2\pi\varepsilon}{\ln(D_{eq}/r)} \quad \text{F/m} \tag{4.10.1}$$

where

$$D_{eq} = \sqrt[3]{D_{ab}D_{bc}D_{ac}} \tag{4.10.2}$$

Figure 4.22 shows a bundled conductor line with two conductors per bundle. To determine the capacitance of this line, assume balanced positive-sequence charges q_a, q_b, q_c for each phase such that $q_a + q_b + q_c = 0$. Assume that the conductors in each bundle, which are in parallel, share the charges equally. Thus conductors a and a' each have the charge $q_a/2$. Also assume that the phase spacings are much larger than the bundle spacings so that D_{ab} may be used instead of $(D_{ab} - d)$ or $(D_{ab} + d)$. Then, using (4.8.6) with $k = a$, $i = b$, $m = a, a', b, b', c, c'$,

FIGURE 4.22

Three-phase line with
two conductors per
bundle

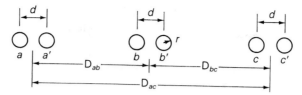

$$V_{ab} = \frac{1}{2\pi\varepsilon}\left[\frac{q_a}{2}\ln\frac{D_{ba}}{D_{aa}} + \frac{q_a}{2}\ln\frac{D_{ba'}}{D_{aa'}} + \frac{q_b}{2}\ln\frac{D_{bb}}{D_{ab}}\right.$$

$$\left. + \frac{q_b}{2}\ln\frac{D_{bb'}}{D_{ab'}} + \frac{q_c}{2}\ln\frac{D_{bc}}{D_{ac}} + \frac{q_c}{2}\ln\frac{D_{bc'}}{D_{ac'}}\right]$$

$$= \frac{1}{2\pi\varepsilon}\left[\frac{q_a}{2}\left(\ln\frac{D_{ab}}{r} + \ln\frac{D_{ab}}{d}\right) + \frac{q_b}{2}\left(\ln\frac{r}{D_{ab}} + \ln\frac{d}{D_{ab}}\right)\right.$$

$$\left. + \frac{q_c}{2}\left(\ln\frac{D_{bc}}{D_{ac}} + \ln\frac{D_{bc}}{D_{ac}}\right)\right]$$

$$= \frac{1}{2\pi\varepsilon}\left[q_a\ln\frac{D_{ab}}{\sqrt{rd}} + q_b\ln\frac{\sqrt{rd}}{D_{ab}} + q_c\ln\frac{D_{bc}}{D_{ac}}\right] \quad (4.10.3)$$

Equation (4.10.3) is the same as (4.9.7), except that D_{aa} and D_{bb} in (4.9.7) are replaced by $\sqrt{rd}$ in this equation. Therefore, for a transposed line, derivation of the capacitance would yield

$$C_{an} = \frac{2\pi\varepsilon}{\ln(D_{eq}/D_{SC})} \quad \text{F/m} \quad (4.10.4)$$

where

$$D_{SC} = \sqrt{rd} \quad \text{for a two-conductor bundle} \quad (4.10.5)$$

Similarly,

$$D_{SC} = \sqrt[3]{rd^2} \quad \text{for a three-conductor bundle} \quad (4.10.6)$$

$$D_{SC} = 1.091\sqrt[4]{rd^3} \quad \text{for a four-conductor bundle} \quad (4.10.7)$$

Equation (4.10.4) for capacitance is analogous to (4.6.22) for inductance. In both cases D_{eq}, given by (4.6.17) or (4.10.2), is the geometric mean of the distances between phases. Also, (4.10.5)–(4.10.7) for D_{SC} are analogous to (4.6.19)–(4.6.21) for D_{SL}, except that the conductor outside radius r replaces the conductor GMR D_S.

The current supplied to the transmission-line capacitance is called *charging current*. For a single-phase circuit operating at line-to-line voltage $V_{xy} = V_{xy}\underline{/0°}$, the charging current is

$$I_{chg} = Y_{xy}V_{xy} = j\omega C_{xy}V_{xy} \quad \text{A} \quad (4.10.8)$$

As shown in Chapter 2, a capacitor delivers reactive power. From (2.3.5), the reactive power delivered by the line-to-line capacitance is

$$Q_C = \frac{V_{xy}^2}{X_c} = Y_{xy}V_{xy}^2 = \omega C_{xy}V_{xy}^2 \quad \text{var} \tag{4.10.9}$$

For a completely transposed three-phase line that has balanced positive-sequence voltages with $V_{an} = V_{LN}\underline{/0°}$, the phase a charging current is

$$I_{chg} = YV_{an} = j\omega C_{an}V_{LN} \quad \text{A} \tag{4.10.10}$$

and the reactive power delivered by phase a is

$$Q_{C1\phi} = YV_{an}^2 = \omega C_{an}V_{LN}^2 \quad \text{var} \tag{4.10.11}$$

The total reactive power supplied by the three-phase line is

$$Q_{C3\phi} = 3Q_{C1\phi} = 3\omega C_{an}V_{LN}^2 = \omega C_{an}V_{LL}^2 \quad \text{var} \tag{4.10.12}$$

EXAMPLE 4.6   **Capacitance, admittance, and reactive power supplied: single-phase line**

For the single-phase line in Example 4.3, determine the line-to-line capacitance in F and the line-to-line admittance in S. If the line voltage is 20 kV, determine the reactive power in kvar supplied by this capacitance.

SOLUTION   From Table A.3, the outside radius of a 4/0 12-strand copper conductor is

$$r = \frac{0.552}{2} \text{ in.} \times \frac{1 \text{ ft}}{12 \text{ in.}} = 0.023 \text{ ft}$$

and from (4.9.4),

$$C_{xy} = \frac{\pi(8.854 \times 10^{-12})}{\ln\left(\dfrac{5}{0.023}\right)} = 5.169 \times 10^{-12} \quad \text{F/m}$$

or

$$C_{xy} = 5.169 \times 10^{-12} \frac{\text{F}}{\text{m}} \times 1609 \frac{\text{m}}{\text{mi}} \times 20 \text{ mi} = 1.66 \times 10^{-7} \quad \text{F}$$

and the shunt admittance is

$$Y_{xy} = j\omega C_{xy} = j(2\pi 60)(1.66 \times 10^{-7})$$
$$= j6.27 \times 10^{-5} \quad \text{S line-to-line}$$

From (4.10.9),

$$Q_C = (6.27 \times 10^{-5})(20 \times 10^3)^2 = 25.1 \quad \text{kvar} \qquad \blacksquare$$

EXAMPLE 4.7   **Capacitance and shunt admittance; charging current and reactive power supplied: three-phase line**

For the three-phase line in Example 4.5, determine the capacitance-to-neutral in F and the shunt admittance-to-neutral in S. If the line voltage is 345 kV, determine the charging current in kA per phase and the total reactive power in Mvar supplied by the line capacitance. Assume balanced positive-sequence voltages.

SOLUTION   From Table A.4, the outside radius of a 795,000 cmil 26/2 ACSR conductor is

$$r = \frac{1.108}{2} \text{ in.} \times 0.0254 \frac{\text{m}}{\text{in.}} = 0.0141 \quad \text{m}$$

From (4.10.5), the equivalent radius of the two-conductor bundle is

$$D_{SC} = \sqrt{(0.0141)(0.40)} = 0.0750 \quad \text{m}$$

$D_{eq} = 12.6$ m is the same as in Example 4.5. Therefore, from (4.10.4),

$$C_{an} = \frac{(2\pi)(8.854 \times 10^{-12})}{\ln\left(\frac{12.6}{0.0750}\right)} \frac{\text{F}}{\text{m}} \times 1000 \frac{\text{m}}{\text{km}} \times 200 \text{ km}$$

$$= 2.17 \times 10^{-6} \quad \text{F}$$

The shunt admittance-to-neutral is

$$Y_{an} = j\omega C_{an} = j(2\pi 60)(2.17 \times 10^{-6})$$

$$= j8.19 \times 10^{-4} \quad \text{S}$$

From (4.10.10),

$$I_{chg} = |I_{chg}| = (8.19 \times 10^{-4})\left(\frac{345}{\sqrt{3}}\right) = 0.163 \quad \text{kA/phase}$$

and from (4.10.12),

$$Q_{C3\phi} = (8.19 \times 10^{-4})(345)^2 = 97.5 \quad \text{Mvar} \qquad \blacksquare$$

4.11

SHUNT ADMITTANCES: LINES WITH NEUTRAL CONDUCTORS AND EARTH RETURN

In this section, we develop equations suitable for computer calculation of the shunt admittances for the three-phase overhead line shown in Figure 4.16. We approximate the earth surface as a perfectly conducting horizontal plane,

FIGURE 4.23

Method of images

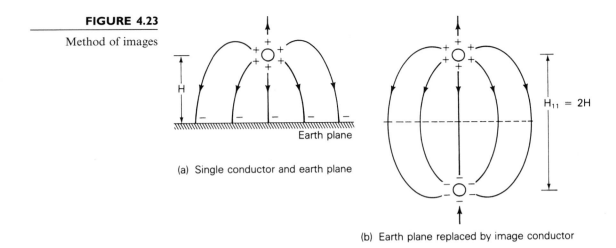

(a) Single conductor and earth plane

(b) Earth plane replaced by image conductor

even though the earth under the line may have irregular terrain and re-sistivities as shown in Table 4.4.

The effect of the earth plane is accounted for by the *method of images,* described as follows. Consider a single conductor with uniform charge distri-bution and with height H above a perfectly conducting earth plane, as shown in Figure 4.23(a). When the conductor has a positive charge, an equal quan-tity of negative charge is induced on the earth. The electric field lines will originate from the positive charges on the conductor and terminate at the negative charges on the earth. Also, the electric field lines are perpendicular to the surfaces of the conductor and earth.

Now replace the earth by the image conductor shown in Figure 4.23(b), which has the same radius as the original conductor, lies directly below the original conductor with conductor separation $H_{11} = 2H$, and has an equal quantity of negative charge. The electric field above the dashed line repre-senting the location of the removed earth plane in Figure 4.23(b) is identical to the electric field above the earth plane in Figure 4.23(a). Therefore, the voltage between any two points above the earth is the same in both figures.

EXAMPLE 4.8   Effect of earth on capacitance: single-phase line

If the single-phase line in Example 4.6 has flat horizontal spacing with 18-ft average line height, determine the effect of the earth on capacitance. Assume a perfectly conducting earth plane.

SOLUTION   The earth plane is replaced by a separate image conductor for each overhead conductor, and the conductors are charged as shown in Figure 4.24. From (4.8.6), the voltage between conductors x and y is

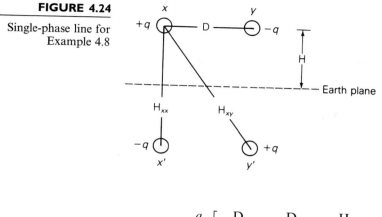

FIGURE 4.24

Single-phase line for Example 4.8

$$V_{xy} = \frac{q}{2\pi\varepsilon}\left[\ln\frac{D_{yx}}{D_{xx}} - \ln\frac{D_{yy}}{D_{xy}} - \ln\frac{H_{yx}}{H_{xx}} + \ln\frac{H_{yy}}{H_{xy}}\right]$$

$$= \frac{q}{2\pi\varepsilon}\left[\ln\frac{D_{yx}D_{xy}}{D_{xx}D_{yy}} - \ln\frac{H_{yx}H_{xy}}{H_{xx}H_{yy}}\right]$$

$$= \frac{q}{\pi\varepsilon}\left[\ln\frac{D}{r} - \ln\frac{H_{xy}}{H_{xx}}\right]$$

The line-to-line capacitance is

$$C_{xy} = \frac{q}{V_{xy}} = \frac{\pi\varepsilon}{\ln\dfrac{D}{r} - \ln\dfrac{H_{xy}}{H_{xx}}}\quad \text{F/m}$$

Using $D = 5$ ft, $r = 0.023$ ft, $H_{xx} = 2H = 36$ ft, and $H_{xy} = \sqrt{(36)^2 + (5)^2} = 36.346$ ft,

$$C_{xy} = \frac{\pi(8.854 \times 10^{-12})}{\ln\dfrac{5}{0.023} - \ln\dfrac{36.346}{36}} = 5.178 \times 10^{-12}\quad \text{F/m}$$

compared with 5.169×10^{-12} F/m in Example 4.6. The effect of the earth plane is to slightly increase the capacitance. Note that as the line height H increases, the ratio H_{xy}/H_{xx} approaches 1, $\ln(H_{xy}/H_{xx}) \to 0$, and the effect of the earth becomes negligible. ∎

For the three-phase line with N neutral conductors shown in Figure 4.25, the perfectly conducting earth plane is replaced by a separate image conductor for each overhead conductor. The overhead conductors a, b, c, $n1$, $n2, \ldots, nN$ carry charges $q_a, q_b, q_c, q_{n1}, \ldots, q_{nN}$, and the image conductors a', b', c', $n1', \ldots, nN'$ carry charges $-q_a, -q_b, -q_c, -q_{n1}, \ldots, -q_{nN}$. Applying (4.8.6) to determine the voltage $V_{kk'}$ between any conductor k and its image conductor k',

FIGURE 4.25

Three-phase line with
neutral conductors and
with earth plane
replaced by image
conductors

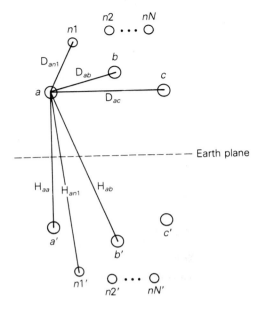

$$V_{kk'} = \frac{1}{2\pi\varepsilon} \left[\sum_{m=a}^{nN} q_m \ln \frac{H_{km}}{D_{km}} - \sum_{m=a}^{nN} q_m \ln \frac{D_{km}}{H_{km}} \right]$$

$$= \frac{2}{2\pi\varepsilon} \sum_{m=a}^{nN} q_m \ln \frac{H_{km}}{D_{km}} \qquad (4.11.1)$$

where $D_{kk} = r_k$ and D_{km} is the distance between overhead conductors k and m. H_{km} is the distance between overhead conductor k and image conductor m. By symmetry, the voltage V_{kn} between conductor k and the earth is one-half of $V_{kk'}$.

$$V_{kn} = \frac{1}{2} V_{kk'} = \frac{1}{2\pi\varepsilon} \sum_{m=a}^{nN} q_m \ln \frac{H_{km}}{D_{km}} \qquad (4.11.2)$$

where

$$k = a, b, c, n1, n2, \ldots, nN$$

$$m = a, b, c, n1, n2, \ldots, nN$$

Since all the neutral conductors are grounded to the earth,

$$V_{kn} = 0 \qquad \text{for } k = n1, n2, \ldots, nN \qquad (4.11.3)$$

In matrix format, (4.11.2) and (4.11.3) are

$$
\begin{bmatrix}
V_{an} \\
V_{bn} \\
V_{cn} \\
0 \\
\vdots \\
0
\end{bmatrix}
=
\left[
\begin{array}{ccc|ccc}
P_{aa} & P_{ab} & P_{ac} & P_{an1} & \cdots & P_{anN} \\
P_{ba} & P_{bb} & P_{bc} & P_{bn1} & \cdots & P_{bnN} \\
P_{ca} & P_{cb} & P_{cc} & P_{cn1} & \cdots & P_{cnN} \\
\hline
P_{n1a} & P_{n1b} & P_{n1c} & P_{n1n1} & \cdots & P_{n1nN} \\
\vdots & & & & & \vdots \\
P_{nNa} & P_{nNb} & P_{nNc} & P_{nNn1} & \cdots & P_{nNnN}
\end{array}
\right]
\begin{bmatrix}
q_a \\
q_b \\
q_c \\
q_{n1} \\
\vdots \\
q_{nN}
\end{bmatrix}
\tag{4.11.4}
$$

where the top braces label P_A (columns a,b,c) and P_B (columns $n1,\dots,nN$), and the bottom braces label P_C and P_D.

The elements of the $(3 + N) \times (3 + N)$ matrix $\mathbf{P}$ are

$$
P_{km} = \frac{1}{2\pi\varepsilon} \ln \frac{H_{km}}{D_{km}} \quad \text{m/F}
\tag{4.11.5}
$$

where

$$
k = a, b, c, n1, \dots, nN
$$
$$
m = a, b, c, n1, \dots, nN
$$

Equation (4.11.4) is now partitioned as shown above to obtain

$$
\begin{bmatrix} V_P \\ 0 \end{bmatrix} =
\left[\begin{array}{c|c} \mathbf{P}_A & \mathbf{P}_B \\ \hline \mathbf{P}_C & \mathbf{P}_D \end{array} \right]
\begin{bmatrix} q_P \\ q_n \end{bmatrix}
\tag{4.11.6}
$$

V_P is the three-dimensional vector of phase-to-neutral voltages. q_P is the three-dimensional vector of phase-conductor charges and q_n is the N vector of neutral conductor charges. The $(3 + N) \times (3 + N)\mathbf{P}$ matrix is partitioned as shown in (4.11.4) to obtain:

$\mathbf{P}_A$ with dimension 3×3

$\mathbf{P}_B$ with dimension $3 \times N$

$\mathbf{P}_C$ with dimension $N \times 3$

$\mathbf{P}_D$ with dimension $N \times N$

Equation (4.11.6) is rewritten as two separate equations:

$$
V_P = \mathbf{P}_A q_P + \mathbf{P}_B q_n
\tag{4.11.7}
$$
$$
0 = \mathbf{P}_C q_P + \mathbf{P}_D q_n
\tag{4.11.8}
$$

Then (4.11.8) is solved for q_n, which is used in (4.11.7) to obtain

$$
V_P = (\mathbf{P}_A - \mathbf{P}_B \mathbf{P}_D^{-1} \mathbf{P}_C) q_P
\tag{4.11.9}
$$

or

$$
q_P = \mathbf{C}_P V_P
\tag{4.11.10}
$$

where

$$\mathbf{C_P} = (\mathbf{P}_A - \mathbf{P}_B\mathbf{P}_D^{-1}\mathbf{P}_C)^{-1} \quad \text{F/m} \tag{4.11.11}$$

Equation (4.11.10), the desired result, relates the phase-conductor charges to the phase-to-neutral voltages. $\mathbf{C_P}$ is the 3×3 matrix of phase capacitances whose elements are denoted

$$\mathbf{C_P} = \begin{bmatrix} C_{aa} & C_{ab} & C_{ac} \\ C_{ab} & C_{bb} & C_{bc} \\ C_{ac} & C_{bc} & C_{cc} \end{bmatrix} \quad \text{F/m} \tag{4.11.12}$$

It can be shown that $\mathbf{C_P}$ is a symmetric matrix whose diagonal terms C_{aa}, C_{bb}, C_{cc} are positive, and whose off-diagonal terms C_{ab}, C_{bc}, C_{ac} are negative. This indicates that when a positive line-to-neutral voltage is applied to one phase, a positive charge is induced on that phase and negative charges are induced on the other phases, which is physically correct.

If the line is completely transposed, the diagonal and off-diagonal elements of $\mathbf{C_P}$ are averaged to obtain

$$\hat{\mathbf{C}}_P = \begin{bmatrix} \hat{C}_{aa} & \hat{C}_{ab} & \hat{C}_{ab} \\ \hat{C}_{ab} & \hat{C}_{aa} & \hat{C}_{ab} \\ \hat{C}_{ab} & \hat{C}_{ab} & \hat{C}_{aa} \end{bmatrix} \quad \text{F/m} \tag{4.11.13}$$

where

$$\hat{C}_{aa} = \tfrac{1}{3}(C_{aa} + C_{bb} + C_{cc}) \quad \text{F/m} \tag{4.11.14}$$

$$\hat{C}_{ab} = \tfrac{1}{3}(C_{ab} + C_{bc} + C_{ac}) \quad \text{F/m} \tag{4.11.15}$$

$\hat{\mathbf{C}}_P$ is a symmetrical capacitance matrix.

The shunt phase admittance matrix is given by

$$\mathbf{Y_P} = j\omega\mathbf{C_P} = j(2\pi f)\mathbf{C_P} \quad \text{S/m} \tag{4.11.16}$$

or, for a completely transposed line,

$$\hat{\mathbf{Y}}_P = j\omega\hat{\mathbf{C}}_P = j(2\pi f)\hat{\mathbf{C}}_P \quad \text{S/m} \tag{4.11.17}$$

4.12

ELECTRIC FIELD STRENGTH AT CONDUCTOR SURFACES AND AT GROUND LEVEL

When the electric field strength at a conductor surface exceeds the breakdown strength of air, current discharges occur. This phenomenon, called corona, causes additional line losses (corona loss), communications interference, and audible noise. Although breakdown strength depends on many factors, a rough value is 30 kV/cm in a uniform electric field for dry air at atmospheric pressure. The presence of water droplets or rain can lower this value significantly. To control corona, transmission lines are usually designed to maintain calculated values of conductor surface electric field strength below 20 kV_{rms}/cm.

When line capacitances are determined and conductor voltages are known, the conductor charges can be calculated from (4.9.3) for a single-phase line or from (4.11.10) for a three-phase line. Then the electric field strength at the surface of one phase conductor, neglecting the electric fields due to charges on other phase conductors and neutral wires, is, from (4.8.2),

$$E_r = \frac{q}{2\pi\varepsilon r} \quad \text{V/m} \tag{4.12.1}$$

where r is the conductor outside radius.

For bundled conductors with N_b conductors per bundle and with charge q C/m per phase, the charge per conductor is q/N_b and

$$E_{rave} = \frac{q/N_b}{2\pi\varepsilon r} \quad \text{V/m} \tag{4.12.2}$$

Equation (4.12.2) represents an average value for an individual conductor in a bundle. The maximum electric field strength at the surface of one conductor due to all charges in a bundle, obtained by the vector addition of electric fields (as shown in Figure 4.26), is as follows:

Two-conductor bundle ($N_b = 2$):

$$E_{rmax} = \frac{q/2}{2\pi\varepsilon r} + \frac{q/2}{2\pi\varepsilon d} = \frac{q/2}{2\pi\varepsilon r}\left(1 + \frac{r}{d}\right)$$

$$= E_{rave}\left(1 + \frac{r}{d}\right) \tag{4.12.3}$$

Three-conductor bundle ($N_b = 3$):

$$E_{rmax} = \frac{q/3}{2\pi\varepsilon}\left(\frac{1}{r} + \frac{2\cos 30°}{d}\right) = E_{rave}\left(1 + \frac{r\sqrt{3}}{d}\right) \tag{4.12.4}$$

Four-conductor bundle ($N_b = 4$):

$$E_{rmax} = \frac{q/4}{2\pi\varepsilon}\left(\frac{1}{r} + \frac{1}{d\sqrt{2}} + \frac{2\cos 45°}{d}\right) = E_{rave}\left[1 + \frac{r}{d}(2.1213)\right] \tag{4.12.5}$$

FIGURE 4.26

Vector addition of electric fields at the surface of one conductor in a bundle

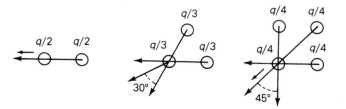

TABLE 4.5

Examples of maximum
ground-level electric
field strength versus
transmission-line voltage
[1] (© Copyright 1987.
Electric Power Research
Institute (EPRI),
Publication Number
FL-2500. Transmission
Line Reference Book,
345 kV and Above,
Second Edition,
Revised. Reprinted
with permission.)

Line Voltage (kV$_{rms}$)	Maximum Ground-Level Electric Field Strength (kV$_{rms}$/m)
23 (1ϕ)	0.01–0.025
23 (3ϕ)	0.01–0.05
115	0.1–0.2
345	2.3–5.0
345 (double circuit)	5.6
500	8.0
765	10.0

Although the electric field strength at ground level is much less than at conductor surfaces where corona occurs, there are still capacitive coupling effects. Charges are induced on ungrounded equipment such as vehicles with rubber tires located near a line. If a person contacts the vehicle and ground, a discharge current will flow to ground. Transmission-line heights are designed to maintain discharge currents below prescribed levels for any equipment that may be on the right-of-way. Table 4.5 shows examples of maximum ground-level electric field strength.

As shown in Figure 4.27, the ground-level electric field strength due to charged conductor k and its image conductor is perpendicular to the earth plane, with value

$$E_k(w) = \left(\frac{q_k}{2\pi\varepsilon}\right)\frac{2\cos\theta}{\sqrt{y_k^2 + (w - x_k)^2}}$$

$$= \left(\frac{q_k}{2\pi\varepsilon}\right)\frac{2y_k}{y_k^2 + (w - x_k)^2} \quad \text{V/m} \tag{4.12.6}$$

where (x_k, y_k) are the horizontal and vertical coordinates of conductor k with respect to reference point R, w is the horizontal coordinate of the ground-level point where the electric field strength is to be determined, and q_k is the charge on conductor k. The total ground-level electric field is the phasor sum of terms $E_k(w)$ for all overhead conductors. A lateral profile of ground-level

FIGURE 4.27

Ground-level electric
field strength due to an
overhead conductor and
its image

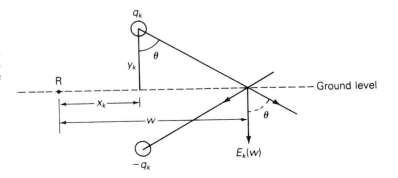

electric field strength is obtained by varying w from the center of the line to the edge of the right-of-way.

EXAMPLE 4.9 **Conductor surface and ground-level electric field strengths: single-phase line**

For the single-phase line of Example 4.8, calculate the conductor surface electric field strength in kV_{rms}/cm. Also calculate the ground-level electric field in kV_{rms}/m directly under conductor x. The line voltage is 20 kV.

SOLUTION From Example 4.8, $C_{xy} = 5.178 \times 10^{-12}$ F/m. Using (4.9.3) with $V_{xy} = 20\underline{/0°}$ kV,

$$q_x = -q_y = (5.178 \times 10^{-12})(20 \times 10^3\underline{/0°}) = 1.036 \times 10^{-7}\underline{/0°} \quad C/m$$

From (4.12.1), the conductor surface electric field strength is, with $r = 0.023$ ft $= 0.00701$ m,

$$E_r = \frac{1.036 \times 10^{-7}}{(2\pi)(8.854 \times 10^{-12})(0.00701)} \frac{V}{m} \times \frac{kV}{1000\ V} \times \frac{m}{100\ cm}$$

$$= 2.66 \quad kV_{rms}/cm$$

Selecting the center of the line as the reference point R, the coordinates (x_x, y_x) for conductor x are $(-2.5$ ft, 18 ft$)$ or $(-0.762$ m, 5.49 m$)$ and $(+0.762$ m, 5.49 m$)$ for conductor y. The ground-level electric field directly under conductor x, where $w = -0.762$ m, is, from (4.12.6),

$$E(-0.762) = E_x(-0.762) + E_y(-0.762)$$

$$= \frac{1.036 \times 10^{-7}}{(2\pi)(8.85 \times 10^{-12})} \left[\frac{(2)(5.49)}{(5.49)^2} - \frac{(2)(5.49)}{(5.49)^2 + (0.762 + 0.762)^2} \right]$$

$$= 1.862 \times 10^3 (0.364 - 0.338) = 48.5\underline{/0°} \ V/m = 0.0485 \ kV/m$$

For this 20-kV line, the electric field strengths at the conductor surface and at ground level are low enough to be of relatively small concern. For EHV lines, electric field strengths and the possibility of corona and shock hazard are of more concern. ■

4.13

PARALLEL CIRCUIT THREE-PHASE LINES

If two parallel three-phase circuits are close together, either on the same tower as in Figure 4.3, or on the same right-of-way, there are mutual inductive and capacitive couplings between the two circuits. When calculating the equivalent series impedance and shunt admittance matrices, these couplings should not be neglected unless the spacing between the circuits is large.

FIGURE 4.28

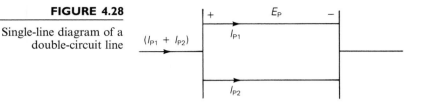

Consider the double-circuit line shown in Figure 4.28. For simplicity, assume that the lines are not transposed. Since both are connected in parallel, they have the same series-voltage drop for each phase. Following the same procedure as in Section 4.7, we can write $2(6 + N)$ equations similar to (4.7.6)–(4.7.9): six equations for the overhead phase conductors, N equations for the overhead neutral conductors, and $(6 + N)$ equations for the earth return conductors. After lumping the neutral voltage drop into the voltage drops across the phase conductors, and eliminating the neutral and earth return currents, we obtain

$$\begin{bmatrix} E_P \\ E_P \end{bmatrix} = Z_P \begin{bmatrix} I_{P1} \\ I_{P2} \end{bmatrix} \tag{4.13.1}$$

where E_P is the vector of phase-conductor voltage drops (including the neutral voltage drop), and I_{P1} and I_{P2} are the vectors of phase currents for lines 1 and 2. Z_P is a 6×6 impedance matrix. Solving (4.13.1)

$$\begin{bmatrix} I_{P1} \\ \hline I_{P2} \end{bmatrix} = Z_P^{-1} \begin{bmatrix} E_P \\ \hline E_P \end{bmatrix} = \begin{bmatrix} Y_A & Y_B \\ \hline Y_C & Y_D \end{bmatrix} \begin{bmatrix} E_P \\ \hline E_P \end{bmatrix} = \begin{bmatrix} (Y_A + Y_B) \\ (Y_C + Y_D) \end{bmatrix} E_P \tag{4.13.2}$$

where Y_A, Y_B, Y_C, and Y_D are obtained by partitioning Z_P^{-1} into four 3×3 matrices. Adding I_{P1} and I_{P2},

$$(I_{P1} + I_{P2}) = (Y_A + Y_B + Y_C + Y_D)E_P \tag{4.13.3}$$

and solving for E_P,

$$E_P = Z_{Peq}(I_{P1} + I_{P2}) \tag{4.13.4}$$

where

$$Z_{Peq} = (Y_A + Y_B + Y_C + Y_D)^{-1} \tag{4.13.5}$$

Z_{Peq} is the equivalent 3×3 series phase impedance matrix of the double-circuit line. Note that in (4.13.5) the matrices Y_B and Y_C account for the inductive coupling between the two circuits.

An analogous procedure can be used to obtain the shunt admittance matrix. Following the ideas of Section 4.11, we can write $(6 + N)$ equations similar to (4.11.4). After eliminating the neutral wire charges, we obtain

$$\begin{bmatrix} q_{P1} \\ \hline q_{P2} \end{bmatrix} = C_P \begin{bmatrix} V_P \\ \hline V_P \end{bmatrix} = \begin{bmatrix} C_A & C_B \\ \hline C_C & C_D \end{bmatrix} \begin{bmatrix} V_P \\ \hline V_P \end{bmatrix} = \begin{bmatrix} (C_A + C_B) \\ (C_C + C_D) \end{bmatrix} V_P \tag{4.13.6}$$

where V_P is the vector of phase-to-neutral voltages, and q_{P1} and q_{P2} are the vectors of phase-conductor charges for lines 1 and 2. $\mathbf{C}_P$ is a 6×6 capacitance matrix that is partitioned into four 3×3 matrices $\mathbf{C}_A$, $\mathbf{C}_B$, $\mathbf{C}_C$, and $\mathbf{C}_D$. Adding q_{P1} and q_{P2}

$$(q_{P1} + q_{P2}) = \mathbf{C}_{Peq} V_P \qquad (4.13.7)$$

where

$$\mathbf{C}_{Peq} = (\mathbf{C}_A + \mathbf{C}_B + \mathbf{C}_C + \mathbf{C}_D) \qquad (4.13.8)$$

Also,

$$Y_{Peq} = j\omega \mathbf{C}_{Peq} \qquad (4.13.9)$$

Y_{Peq} is the equivalent 3×3 shunt admittance matrix of the double-circuit line. The matrices $\mathbf{C}_B$ and $\mathbf{C}_C$ in (4.13.8) account for the capacitive coupling between the two circuits.

These ideas can be extended in a straightforward fashion to more than two parallel circuits.

PROBLEMS

SECTION 4.2

4.1 The *Aluminum Electrical Conductor Handbook* lists a dc resistance of 0.01552 ohm per 1000 ft at 20 °C and a 60-Hz resistance of 0.0951 ohm per mile at 50 °C for the all-aluminum Marigold conductor, which has 61 strands and whose size is 1113 kcmil. Assuming an increase in resistance of 1.6% for spiraling, calculate and verify the dc resistance. Then calculate the dc resistance at 50 °C, and determine the percentage increase due to skin effect.

4.2 The temperature dependence of resistance is also quantified by the relation $R_2 = R_1[1 + \alpha(T_2 - T_1)]$ where R_1 and R_2 are the resistances at temperatures T_1 and T_2, respectively, and α is known as the temperature coefficient of resistance. If a copper wire has a resistance of 50 Ω at 20 °C, find the maximum permissible operating temperature of the wire if its resistance is to increase by at most 10%. Take the temperature coefficient at 20 °C to be $\alpha = 0.00382$.

4.3 A transmission-line cable, of length 3 km, consists of 19 strands of identical copper conductors, each 1.5 mm in diameter. Because of the twist of the strands, the actual length of each conductor is increased by 5%. Determine the resistance of the cable, if the resistivity of copper is 1.72 $\mu\Omega$.cm at 20 °C.

4.4 One thousand circular mils or 1 kcmil is sometimes designated by the abbreviation MCM. Data for commercial bare aluminum electrical conductors lists a 60-Hz resistance of 0.0740 ohm per kilometer at 75 °C for a 954-MCM AAC conductor. (a) Determine the cross-sectional conducting area of this conductor in square meters. (b) Find the 60-Hz resistance of this conductor in ohms per kilometer at 45 °C.

4.5 A 60-Hz, 765-kV three-phase overhead transmission line has four ACSR 1113 kcmil 54/3 conductors per phase. Determine the 60-Hz resistance of this line in ohms per kilometer per phase at 50 °C.

4.6   A three-phase overhead transmission line is designed to deliver 190.5 MVA at 220 kV over a distance of 63 km, such that the total transmission line loss is not to exceed 2.5% of the rated line MVA. Given the resistivity of the conductor material to be 2.84×10^{-8} Ω-m, determine the required conductor diameter and the conductor size in circular mils. Neglect power losses due to insulator leakage currents and corona.

4.7   If the per-phase line loss in a 60-km-long transmission line is not to exceed 60 kW while it is delivering 100 A per phase, compute the required conductor diameter, if the resistivity of the conductor material is 1.72×10^{-8} Ωm.

SECTIONS 4.4 AND 4.5

4.8   A 60-Hz single-phase, two-wire overhead line has solid cylindrical copper conductors with 1.5 cm diameter. The conductors are arranged in a horizontal configuration with 0.5 m spacing. Calculate in mH/km (a) the inductance of each conductor due to internal flux linkages only, (b) the inductance of each conductor due to both internal and external flux linkages, and (c) the total inductance of the line.

4.9   Rework Problem 4.8 if the diameter of each conductor is: (a) increased by 20% to 1.8 cm, (b) decreased by 20% to 1.2 cm, without changing the phase spacing. Compare the results with those of Problem 4.8.

4.10   A 60-Hz three-phase, three-wire overhead line has solid cylindrical conductors arranged in the form of an equilateral triangle with 4 ft conductor spacing. Conductor diameter is 0.5 in. Calculate the positive-sequence inductance in H/m and the positive-sequence inductive reactance in Ω/km.

4.11   Rework Problem 4.10 if the phase spacing is: (a) increased by 20% to 4.8 ft, (b) decreased by 20% to 3.2 ft. Compare the results with those of Problem 4.10.

4.12   Find the inductive reactance per mile of a single-phase overhead transmission line operating at 60 Hz, given the conductors to be *Partridge* and the spacing between centers to be 20 ft.

4.13   A single-phase overhead transmission line consists of two solid aluminum conductors having a radius of 2.5 cm, with a spacing 3.6 m between centers. (a) Determine the total line inductance in mH/m. (b) Given the operating frequency to be 60 Hz, find the total inductive reactance of the line in Ω/km and in Ω/mi. (c) If the spacing is doubled to 7.2 m, how does the reactance change?

4.14   (a) In practice, one deals with the inductive reactance of the line per phase per mile and use the logarithm to the base 10.
Show that Eq. (4.5.9) of the text can be rewritten as

$$x = k \log \frac{D}{r'} \text{ ohms per mile per phase}$$

$$= x_d + x_a$$

Where $x_d = k \log D$ is the inductive reactance spacing factor in ohms per mile,

$$x_a = k \log \frac{1}{r'} \text{ is the inductive reactance at 1-ft spacing in ohms per mile.}$$

and $k = 4.657 \times 10^{-3} f = 0.2794$ at 60 Hz.

(b) Determine the inductive reactance per mile per phase at 60 Hz for a single-phase line with phase separation of 10 ft and conductor radius of 0.06677 ft.
If the spacing is doubled, how does the reactance change?

SECTION 4.6

4.15   Find the GMR of a stranded conductor consisting of six outer strands surrounding and touching one central strand, all strands having the same radius r.

4.16   A bundle configuration for UHV lines (above 1000 kV) has identical conductors equally spaced around a circle, as shown in Figure 4.29. N_b is the number of conductors in the bundle, A is the circle radius, and D_S is the conductor GMR. Using the distance D_{1n} between conductors 1 and n given by $D_{1n} = 2A \sin[(n-1)\pi/N_b]$ for $n = 1, 2, \ldots, N_b$, and the following trigonometric identity:

$$[2 \sin(\pi/N_b)][2 \sin(2\pi/N_b)][2 \sin(3\pi/N_b)] \cdots [2 \sin\{(N_b - 1)\pi/N_b\}] = N_b$$

show that the bundle GMR, denoted D_{SL}, is

$$D_{SL} = [N_b D_S A^{(N_b - 1)}]^{(1/N_b)}$$

Also show that the above formula agrees with (4.6.19)–(4.6.21) for EHV lines with $N_b = 2, 3$, and 4.

FIGURE 4.29

Bundle configuration for Problem 4.16

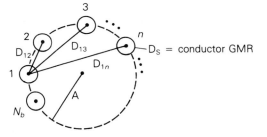

D_S = conductor GMR

4.17   Determine the GMR of each of the unconventional stranded conductors shown in Figure 4.30. All strands have the same radius r.

FIGURE 4.30

Unconventional stranded conductors for Problem 4.17

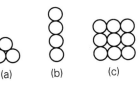

(a)   (b)   (c)

4.18   A 230-kV, 60-Hz, three-phase completely transposed overhead line has one ACSR 954-kcmil conductor per phase and flat horizontal phase spacing, with 8 m between adjacent conductors. Determine the inductance in H/m and the inductive reactance in Ω/km.

4.19   Rework Problem 4.18 if the phase spacing between adjacent conductors is: (a) increased by 10% to 8.8 m, (b) decreased by 10% to 7.2 m. Compare the results with those of Problem 4.18.

4.20 Calculate the inductive reactance in Ω/km of a bundled 500-kV, 60-Hz, three-phase completely transposed overhead line having three ACSR 1113-kcmil conductors per bundle, with 0.5 m between conductors in the bundle. The horizontal phase spacings between bundle centers are 10, 10, and 20 m.

4.21 Rework Problem 4.20 if the bundled line has: (a) three ACSR, 1351-kcmil conductors per phase, (b) three ACSR, 900-kcmil conductors per phase, without changing the bundle spacing or the phase spacings between bundle centers. Compare the results with those of Problem 4.20.

4.22 The conductor configuration of a bundled single-phase overhead transmission line is shown in Figure 4.31. Line X has its three conductors situated at the corners of an equilateral triangle with 10-cm spacing. Line Y has its three conductors arranged in a horizontal configuration with 10-cm spacing. All conductors are identical, solid-cylindrical conductors, each with a radius of 2 cm. (a) Find the equivalent representation in terms of the geometric mean radius of each bundle and a separation that is the geometric mean distance.

FIGURE 4.31

Problem 4.22

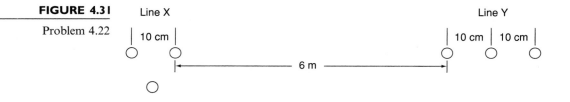

4.23 Figure 4.32 shows the conductor configuration of a completely transposed three-phase overhead transmission line with bundled phase conductors. All conductors have a radius of 0.74 cm with a 30-cm bundle spacing. (a) Determine the inductance per phase in mH/km and in mH/mi. (b) Find the inductive line reactance per phase in Ω/mi at 60 Hz.

FIGURE 4.32

Problem 4.23

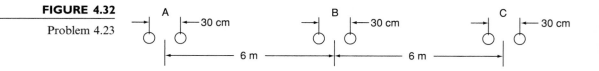

4.24 Consider a three-phase overhead line made up of three phase conductors, Linnet, 336.4 kcmil, ACSR 26/7. The line configuration is such that the horizontal separation between center of C and that of A is 40″, and between that of A and B is also 40″ in the same line; the vertical separation of A from the line of C–B is 16″. If the line is operated at 60 Hz at a conductor temperature of 75 °C, determine the inductive reactance per phase in Ω/mi,
(a) By using the formula given in Problem 4.14 (a), and
(b) By using (4.6.18) of the text.

4.25 For the overhead line of configuration shown in Figure 4.33, operating at 60 Hz, and a conductor temperature of 70 °C, determine the resistance per phase, inductive reactance in ohms/mile/phase and the current carrying capacity of the overhead line. The resistance of each conductor in the four-conductor bundle is 0.12 Ω/mi.

FIGURE 4.33

Line configuration for Problem 4.25

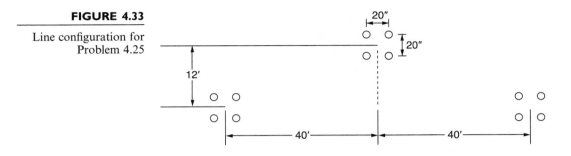

4.26 Consider a symmetrical bundle with N subconductors arranged in a circle of radius A. The inductance of a single-phase symmetrical bundle-conductor line is given by

$$L = 2 \times 10^{-7} \ln \frac{GMD}{GMR} \; H/m$$

Where GMR is given by $[Nr'(A)^{N-1}]^{1/N}$

$r' = (e^{-1/4}r)$, r being the subconductor radius, and GMD is approximately the distance D between the bundle centers. Note that A is related to the subconductor spacing S in the bundle circle by $S = 2A \sin(\Pi/N)$

Now consider a 965-kV, single-phase, bundle-conductor line with eight subconductors per phase, with phase spacing $D = 17$ m, and the subconductor spacing $S = 45.72$ cm. Each subconductor has a diameter of 4.572 cm. Determine the line inductance in H/m.

4.27 Figure 4.34 shows double-circuit conductors' relative positions in Segment 1 of transposition of a completely transposed three-phase overhead transmission line. The inductance is given by

$$L = 2 \times 10^{-7} \ln \frac{GMD}{GMR} \; H/m/phase$$

FIGURE 4.34

For Problem 4.27 (Double-circuit conductor configuration)

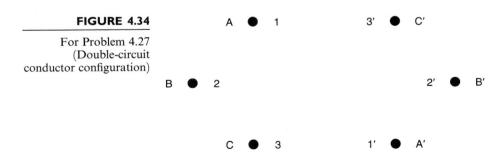

Where $\text{GMD} = (D_{AB_{eq}} D_{BC_{eq}} D_{AC_{eq}})^{1/3}$

With mean distances defined by equivalent spacings

$$D_{AB_{eq}} = (D_{12} D_{1'2'} D_{12'} D_{1'2})^{1/4}$$

$$D_{BC_{eq}} = (D_{23} D_{2'3'} D_{2'3} D_{23'})^{1/4}$$

$$D_{AC_{eq}} = (D_{13} D_{1'3'} D_{13'} D_{1'3})^{1/4}$$

And $\text{GMR} = [(\text{GMR})_A (\text{GMR})_B (\text{GMR})_C]^{1/3}$

with phase GMRs defined by

$$(\text{GMR})_A = [r' D_{11'}]^{1/2}; \quad (\text{GMR})_B = [r' D_{22'}]^{1/2}; \quad (\text{GMR})_C = [r' D_{33'}]^{1/2}$$

and r' is the GMR of phase conductors.

Now consider A 345-kV, three-phase, double-circuit line with phase-conductor's GMR of 0.0588 ft, and the horizontal conductor configuration shown in Figure 4.35.
(a) Determine the inductance per meter per phase in henries.
(b) Calculate the inductance of just one circuit and then divide by 2 to obtain the inductance of the double circuit.

FIGURE 4.35

For Problem 4.27

Find the relative error involved

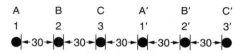

4.28   For the case of double-circuit, bundle-conductor lines, the same method indicated in Problem 4.27 applies with r' replaced by the bundle's GMR in the calculation of the overall GMR.

Now consider a double-circuit configuration shown in Figure 4.36, which belongs to a 500-kV, three-phase line with bundle conductors of three subconductors at 21-in. spacing. The GMR of each subconductor is given to be 0.0485 ft.

FIGURE 4.36

Configuration for Problem 4.28

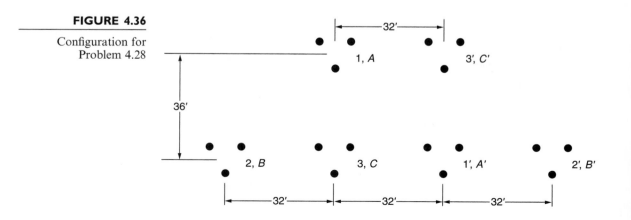

Determine the inductive reactance of the line in ohms per mile per phase. You may use

$$X_L = 0.2794 \log \frac{GMD}{GMR} \ \Omega/\text{mi/phase}.$$

4.29   Reconsider Problem 4.28 with an alternate phase placement given below:

Physical Position						
	1	2	3	1'	2'	3'
Phase Placement	A	B	B'	C	C'	A'

Calculate the inductive reactance of the line in $\Omega/\text{mi/phase}$.

4.30   Reconsider Problem 4.28 with still another alternate phase placement shown below.

Physical Position						
	1	2	3	1'	2'	3'
Phase Placement	C	A	B	B'	A'	C'

Find the inductive reactance of the line in $\Omega/\text{mi/phase}$.

4.31   Figure 4.37 shows the conductor configuration of a three-phase transmission line and a telephone line supported on the same towers. The power line carries a balanced current of 250 A/phase at 60 Hz, while the telephone line is directly located below phase b. Assume balanced three-phase currents in the power line. Calculate the voltage per kilometer induced in the telephone line.

FIGURE 4.37

Conductor layout for Problem 4.31

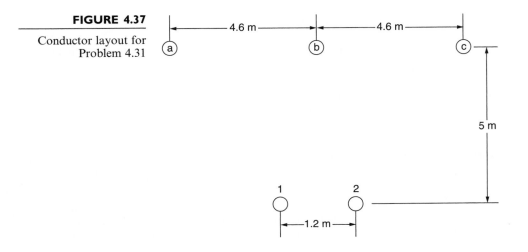

SECTION 4.9

4.32   Calculate the capacitance-to-neutral in F/m and the admittance-to-neutral in S/km for the single-phase line in Problem 4.8. Neglect the effect of the earth plane.

4.33   Rework Problem 4.32 if the diameter of each conductor is: (a) increased by 20% to 1.8 cm, (b) decreased by 20% to 1.2 cm. Compare the results with those of Problem 4.32.

4.34   Calculate the capacitance-to-neutral in F/m and the admittance-to-neutral in S/km for the three-phase line in Problem 4.10. Neglect the effect of the earth plane.

4.35   Rework Problem 4.34 if the phase spacing is: (a) increased by 20% to 4.8 ft, (b) decreased by 20% to 3.2 ft. Compare the results with those of Problem 4.34.

4.36   The line of Problem 4.23 as shown in Figure 4.32 is operating at 60 Hz. Determine (a) the line-to-neutral capacitance in nF/km per phase and in nF/mi per phase; (b) the capacitive reactance in Ω-km per phase and in Ω-mi per phase; and (c) the capacitive reactance in Ω per phase for a line length of 100 mi.

4.37   (a) In practice, one deals with the capacitive reactance of the line in ohms $\cdot$ mi to neutral. Show that Eq. (4.9.15) of the text can be rewritten as

$$X_C = k' \log \frac{D}{r} \text{ ohms} \cdot \text{mi to neutral}$$
$$= x_d' + x_a'$$

Where $x_d' = k' \log D$ is the capacitive reactance spacing factor,

$$x_a' = k' \log \frac{1}{r} \text{ is the capacitive reactance at 1-ft spacing,}$$

and $k' = (4.1 \times 10^6)/f = 0.06833 \times 10^6$ at $f = 60$ Hz.

(b) Determine the capacitive reactance in $\Omega \cdot$ mi. for a single-phase line of Problem 4.14. If the spacing is doubled, how does the reactance change?

4.38   The capacitance per phase of a balanced three-phase overhead line is given by

$$C = \frac{0.0389}{\log(\text{GMD}/r)} \mu f/\text{mi/phase}$$

For the line of Problem 4.24, determine the capacitive reactance per phase in $\Omega \cdot$ mi.

SECTION 4.10

4.39   Calculate the capacitance-to-neutral in F/m and the admittance-to-neutral in S/km for the three-phase line in Problem 4.18. Also calculate the line-charging current in kA/phase if the line is 100 km in length and is operated at 230 kV. Neglect the effect of the earth plane.

4.40   Rework Problem 4.39 if the phase spacing between adjacent conductors is: (a) increased by 10% to 8.8 m, (b) decreased by 10% to 7.2 m. Compare the results with those of Problem 4.39.

4.41   Calculate the capacitance-to-neutral in F/m and the admittance-to-neutral in S/km for the line in Problem 4.20. Also calculate the total reactive power in Mvar/km supplied by the line capacitance when it is operated at 500 kV. Neglect the effect of the earth plane.

4.42   Rework Problem 4.41 if the bundled line has: (a) three ACSR, 1351-kcmil conductors per phase, (b) three ACSR, 900-kcmil conductors per phase, without changing the bundle spacing or the phase spacings between bundle centers.

4.43   Three ACSR *Drake* conductors are used for a three-phase overhead transmission line operating at 60 Hz. The conductor configuration is in the form of an isosceles triangle with sides of 20, 20, and 38 ft. (a) Find the capacitance-to-neutral and capacitive reactance-to-neutral for each 1-mile length of line. (b) For a line length of 175 mi and a normal operating voltage of 220 kV, determine the capacitive reactance-to-neutral for the entire line length as well as the charging current per mile and total three-phase reactive power supplied by the line capacitance.

4.44   Consider the line of Problem 4.25. Calculate the capacitive reactance per phase in $\Omega \cdot$ mi.

SECTION 4.11

4.45   For an average line height of 10 m, determine the effect of the earth on capacitance for the single-phase line in Problem 4.32. Assume a perfectly conducting earth plane.

4.46   A three-phase 60-Hz, 125-km overhead transmission line has flat horizontal spacing with three identical conductors. The conductors have an outside diameter of 3.28 cm with 12 m between adjacent conductors. (a) Determine the capacitive reactance-to-neutral in Ω-m per phase and the capacitive reactance of the line in Ω per phase. Neglect the effect of the earth plane. (b) Assuming that the conductors are horizontally placed 20 m above ground, repeat (a) while taking into account the effect of ground. Consider the earth plane to be a perfect conductor.

4.47   For the single-phase line of Problem 4.14 (b), if the height of the conductor above ground is 80 ft., determine the line-to-line capacitance in F/m. Neglecting earth effect, evaluate the relative error involved.
If the phase separation is doubled, repeat the calculations.

4.48   The capacitance of a single-circuit, three-phase transposed line, and with configuration shown in Figure 4.38 including ground effect, with conductors not equilaterally spaced, is given by

FIGURE 4.38

Three-phase single-circuit line configuration including ground effect for Problem 4.48

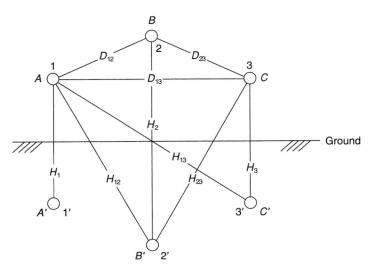

$$C_{a\eta} \frac{2\pi\varepsilon_0}{\ln \dfrac{D_{eq}}{r} - \ln \dfrac{H_m}{H_s}} \text{ F/m Line-to-neutral}$$

Where $D_{eq} = \sqrt[3]{D_{12}D_{23}D_{13}} = \text{GMD}$,

$$r = \text{conductor's outside radius,}$$

$$H_m = (H_{12}H_{23}H_{13})^{1/3}$$

$$\text{and} \quad H_s = (H_1 H_2 H_3)^{1/3}.$$

(a) Now consider Figure 4.39 in which the configuration of a three-phase, single circuit, 345-kV line, with conductors having an outside diameter of 1.065 in., is shown. Determine the capacitance to neutral in F/m, including the ground effect.
(b) Next, neglecting the effect of ground, see how the value changes.

FIGURE 4.39

Configuration for
Problem 4.48 (a)

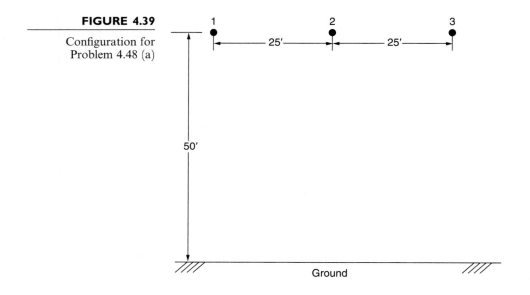

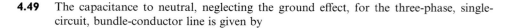
Ground

4.49   The capacitance to neutral, neglecting the ground effect, for the three-phase, single-circuit, bundle-conductor line is given by

$$C_{a\eta} = \frac{2\pi\varepsilon_0}{\ell\eta\left(\dfrac{\text{GMD}}{\text{GMR}}\right)} \text{ F/m Line-to-neutral}$$

Where $\text{GMD} = (D_{AB}D_{BC}D_{AC})^{1/3}$

$$\text{GMR} = [rN(A)^{N-1}]^{1/N}$$

in which N is the number of subconductors of the bundle conductor on a circle of radius A, and each subconductor has an outside radius of r,

The capacitive reactance in mega-ohms for 1 mi of line, at 60 Hz, can be shown to be

$$X_C = 0.0683 \log\left(\frac{GMD}{GMR}\right) = X'_a + X'_d$$

Where $X'_a = 0.0683 \log\left(\frac{1}{GMR}\right)$ and $X'_d = 0.0683 \log(GMD)$.

Note that A is related to the bundle spacing S given by

$$A = \frac{S}{2 \sin\left(\frac{\pi}{N}\right)}   \text{for } N > 1$$

Using the above information, for the configuration shown in Figure 4.40, compute the capacitance to neutral in F/m, and the capacitive reactance in $\Omega \cdot$ mi to neutral, for the three-phase, 765-kV, 60-Hz, single-circuit, bundle-conductor line $(N = 4)$, with subconductor's outside diameter of 1.16 in. and subconductor spacing (S) of 18 in.

FIGURE 4.40

Configuration for
Problem 4.49

SECTION 4.12

4.50   Calculate the conductor surface electric field strength in kV_{rms}/cm for the single-phase line in Problem 4.32 when the line is operating at 20 kV. Also calculate the ground-level electric field strength in kV_{rms}/m directly under one conductor. Assume a line height of 10 m.

4.51   Rework Problem 4.50 if the diameter of each conductor is: (a) increased by 25% to 1.875 cm, (b) decreased by 25% to 1.125 cm, without changing the phase spacings. Compare the results with those of Problem 4.50.

CASE STUDY QUESTIONS

a.   Why is aluminum today's choice of metal for conductors versus copper or some other metal? How does the use of steel together with aluminum as well as aluminum alloys and composite materials improve conductor performance?

b.   Are overhead transmission lines attractive, nondescript, or unsightly? What techniques are used to minimize the environmental impacts of transmission line construction and the visual impacts of transmission lines?

c.   Why is it so difficult to get a new transmission line built in the United States?

REFERENCES

1. Electric Power Research Institute (EPRI), *EPRI AC Transmission Line Reference Book—200 kV and Above* (Palo Alto, CA: EPRI, www.epri.com, December 2005).

2. Westinghouse Electric Corporation, *Electrical Transmission and Distribution Reference Book*, 4th ed. (East Pittsburgh, PA, 1964).

3. General Electric Company, *Electric Utility Systems and Practices*, 4th ed. (New York: Wiley, 1983).

4. John R. Carson, "Wave Propagation in Overhead Wires with Ground Return," *Bell System Tech. J.* 5 (1926): 539–554.

5. C. F. Wagner and R. D. Evans, *Symmetrical Components* (New York: McGraw-Hill, 1933).

6. Paul M. Anderson, *Analysis of Faulted Power Systems* (Ames, IA: Iowa State Press, 1973).

7. M. H. Hesse, "Electromagnetic and Electrostatic Transmission Line Parameters by Digital Computer," *Trans. IEEE* PAS-82 (1963): 282–291.

8. W. D. Stevenson, Jr., *Elements of Power System Analysis*, 4th ed. (New York: McGraw-Hill, 1982).

9. C. A. Gross, *Power System Analysis* (New York: Wiley, 1979).

10. A. J. Peterson, Jr. and S. Hoffmann, "Transmission Line Conductor Design Comes of Age," *Transmission & Distribution World Magazine* (www.tdworld.com, June 2003).

11. ANCI C2. *National Electrical Safety Code*, 2007 edition (New York: Institute of Electrical and Electronics Engineers).

12. J. Haunty, "Mammoth 765-kV Project," *Transmission & Distribution World Magazine* (www.tdworld.com, February 2006).

Series capacitor installation at Goshen Substation, Goshen, Idaho, USA rated at 395 kV, 965 Mvar (Courtesy of PacfiCorp)

5

TRANSMISSION LINES: STEADY-STATE OPERATION

In this chapter, we analyze the performance of single-phase and balanced three-phase transmission lines under normal steady-state operating conditions. Expressions for voltage and current at any point along a line are developed, where the distributed nature of the series impedance and shunt admittance is taken into account. A line is treated here as a two-port network for which the *ABCD* parameters and an equivalent π circuit are derived. Also, approximations are given for a medium-length line lumping the shunt admittance, for a short line neglecting the shunt admittance, and for a lossless line assuming zero series resistance and shunt conductance. The concepts of *surge impedance loading* and transmission-line *wavelength* are also presented.

An important issue discussed in this chapter is *voltage regulation*. Transmission-line voltages are generally high during light load periods and low during heavy load periods. Voltage regulation, defined in Section 5.1, refers to the change in line voltage as line loading varies from no-load to full load.

Another important issue discussed here is line loadability. Three major line-loading limits are: (1) the thermal limit, (2) the voltage-drop limit, and (3) the steady-state stability limit. Thermal and voltage-drop limits are discussed in Section 5.1. The theoretical steady-state stability limit, discussed in Section 5.4 for lossless lines and in Section 5.5 for lossy lines, refers to the ability of synchronous machines at the ends of a line to remain in synchronism. Practical line loadability is discussed in Section 5.6.

In Section 5.7 we discuss line compensation techniques for improving voltage regulation and for raising line loadings closer to the thermal limit.

CASE STUDY   Electric utilities have used inductors and capacitors on medium-length and long transmission lines for many years to provide increased line loadability and to maintain transmission voltages near rated values. The following article describes a flexible ac transmission systems (FACTS) technology based on power electronics and software-based information and control systems to provide dynamic control and compensation of voltage and power flow [6].

The FACTS on Resolving Transmission Gridlock

GREGORY REED, JOHN PASERBA, AND PETER SALAVANTIS

The challenges to providing a reliable and efficient stream of electricity to residential and commercial users in the digital age are great. Regulatory uncertainty, cost, and lengthy delays to transmission line construction are just a few of the barriers that have resulted in the serious deficiency in power transmission capacity that currently prevails in many regions of the United States. Solving these issues requires innovative thinking on the part of all involved. Increasing numbers of electricity stakeholders now recognize that low-environmental-impact technologies such as flexible ac transmission systems (FACTS) and dc links are a proven solution to rapidly enhancing reliability and upgrading trans-

("The FACTS on Resolving Transmission Gridlock" by Gregory Reed, John Paserba, and Peter Salavantis, IEEE Power & Energy (Sept/Oct 2003), pg. 41–46.)

mission capacity on a long-term and cost-effective basis.

A TRANSMISSION SYSTEM IN GRIDLOCK

While headlines of power shortages, rationing, and skyrocketing prices in California and other regions of the United States have faded from memory for much of the general public, energy professionals know that the state of the electrical transmission grid remains in crisis. The question is not whether but *when* the next major failure of the grid will occur absent government and private sector initiatives that enable significant infrastructure improvements to ensure reliable and affordable electricity. For the debate on electricity reliability and capacity to truly advance, however, electricity stakeholders must find solutions to the vexing problems of rapidly increasing demand, inadequate infrastructure, and the critical challenge of balancing energy growth with environmental protection.

Investment in new transmission facilities has declined steadily for the last 25 years while demand will grow by 45% over the next 20 years, according to U.S. Department of Energy (DOE) estimates. To meet these needs, it is estimated the United States must build between 1,300 and 1,900 power plants, yet there are no strategic plans to effectively deliver this amount of new generation to end users. As the DOE notes: "an antiquated and inadequate transmission grid prevents us from routing electricity over long distances and thereby avoiding regional blackouts, such as California's."

The costs of inaction are great. Recent EPRI studies reveal that direct economic losses attributed to transmission system stability and reliability deficiencies resulting in power interruptions and inadequate power quality in the United States conservatively exceed US$100 billion per year. Further, in order to reestablish adequately stable transmission system conditions nationwide, initial investments in overall transmission system infrastructure upgrades are estimated at up to US$30 billion in the western region of the United States alone. It is estimated that annual expenditures of up to US$3 billion or more would then be needed to maintain this condition in consideration of continued demand growth. Projecting these estimates conservatively nationwide yields an initial investment requirement in transmission facilities exceeding US$50 billion.

It is equally well documented that uncertainty about recovery of transmission system investments has played a major role in our current situation. Utilities and other owners of transmission infrastructure have little incentive to invest in new transmission assets given the present regulatory regime.

Finally, intense opposition to the construction of new power lines is often responsible for the interminable and often fatal roadblocks to new transmission capacity. NIMBY ("not in my backyard") challenges to needed transmission capacity amidst lacking infrastructure and growing demand round out the daunting mission of providing a reliable, ef-

ficient, and affordable stream of energy to complacent residential and industrial users of electricity.

AFFECTS OF DEFERRING TRANSMISSION INVESTMENT

It is universally recognized that to keep pace with present energy demand and to ensure reliability as electricity needs grow, the construction of new transmission lines must occur. Such modernization, however, presents a plethora of political, economic, and environmental issues that serve to indefinitely delay urgently needed upgrades of the U.S. transmission system. As the Federal Energy Regulatory Commission (FERC) notes, "the sluggishness of transmission construction is largely because siting transmission is a long and contentious process."

Indeed, the long-term trend of deferred investments in the electrical transmission grid continues to cause serious capacity and reliability issues that threaten the United State's economic security. Consider the case of the proposed Rainbow-Valley Transmission Project in Southern California. The proposal is for a 500-kV transmission line from Southern California Edison's Valley Substation to San Diego Gas and Electric's proposed new Rainbow Substation. The interconnect would greatly enhance system reliability in the area as well as provide 1,000 MW of import capacity into the San Diego region, which has had much in-city generation decommissioned in recent years. The case for the project not only entails transmission reliability and import capacity but is important for the continued economic growth and development of the region. Southern California-area businesses, industries, and residential customers alike have felt the adverse affects of the energy crisis in the region through periods of sky-rocketing prices and the potentially unstable supply during certain critical contingencies. Regardless, various NIMBY, political, and regulatory barriers continue to stifle the approval and development of this much-needed transmission corridor to provide electricity supply relief and increased reliability in the region. Further, citing absence of economic justification of the project, recent rulings have stalled all progress. Yet, no other adequate alternatives are available. Thus, this needed investment continues to be deferred at the expense of transmission system reliability and capacity deficits in one of the largest economic regions of the country. Both system and economic security remain exposed to risk under this current situation.

TRANSMISSION GRIDLOCK RESOLVED

A solution to this stalemate is the availability of proven and low-environmental-impact infrastructure technologies that are sound alternatives to the protracted process of power line construction, such as those identified in the DOE's National Transmission Grid Study. Two of these technology areas, FACTS and certain configurations of high-voltage direct current (HVDC) systems, represent long-term solutions to upgrading electrical transmission infrastructure in situations where power line construction is not feasible or attainable on a reasonable time basis to accommodate our urgent present-day needs.

Compelling financial and performance case studies exist for the wide-scale implementation of advanced, low-environmental impact technologies such as FACTS and HVDC. HVDC configurations in such analysis come in the form of back-to-back (BTB) dc links. Although detailed study of each individual application is required, these technologies can play a key role in alleviating transmission system constraints (i.e., bottlenecks) and provide rapidly implemented, cost-effective relief to the very serious, yet commonly experienced, congestion problems that occur throughout the grid.

FACTS technologies can essentially be defined as highly engineered power-electronics-based systems, integrating the control and operation of advanced power-semiconductor-based converters (or valves) with software-based information and control systems, which produce a compensated response to the transmission network that is interconnected via conventional switchgear and transformation equipment. FACTS technologies provide dynamic control and compensation of voltage and power flow and can be designed to coordinate the control of other transmission compensation devices, such as capaci-

tors, reactors, and transformer tap changers, in order to establish greater overall system operation improvements. In effect, FACTS "builds intelligence" into the grid by providing this type of enhanced system performance, optimization, and control.

FACTS are available in both "conventional" forms such as static VAR compensators (SVCs) and thyristor controlled series capacitors (TCSCs), as well as in "advanced" forms that are categorized as voltage-sourced converter (VSC)-based systems. The main VSC-based FACTS configurations are know as static reactive compensators (STATCOMs), static series synchronous compensators (SSSCs), unified power flow controllers (UPFCs), and VSC-BTBs.

The SVC and STATCOM designs are applied as shunt-connected devices providing dynamic reactive compensation for voltage control, system stabilization, and power quality improvement. The TCSC and SSSC designs provide dynamic real power flow control. The UPFC is an advanced configuration that combines the simultaneous operation of STATCOM and SSSC in one controller to allow increased power flow and dynamic voltage control and stability within the same device. The BTB systems are implemented to provide dc links for seamless interconnections as well as improved inter-tie reliability and control. All of these configurations of FACTS solutions allow for increased transmission capacity through the elimination of transmission system constraints (such as extending transient stability limits, improving power system damping, enhancing short-term and long-term voltage stability limits, providing loop flow control, etc.).

Numerous recent applications of FACTS in the United States and other parts of North America that have addressed localized issues have proven to be cost-effective, long-term solutions. Applying FACTS on a broad-scale basis for both local and regional solutions would result in numerous operational advantages.

FACTS Installation at the San Diego Gas and Electric Talega Substation—by Terry Snow

The FACTS technology recently installed in the SDG&E system at the Talega 138-kV substation is being applied to relieve transmission system con-

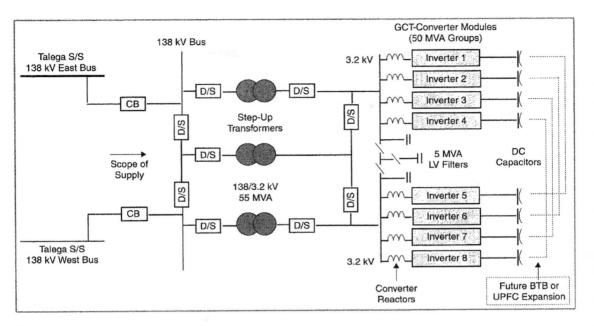

Figure 1
One-line diagram representation of Talega ±100-MVA, 138-kV STATCOM system (50 MW, 138 kV BTB dc-link or UPFC future expansion capability).

straints in the area through dynamic VAR control during peak load conditions. The transmission constraints addressed are related to voltage stability and dynamic VAR supply. The FACTS installation is operating as a STATCOM system and has a rated dynamic reactive capacity of ±100 MVA. As shown in Figure 1, the STATCOM system, as it is now configured, consists of two groups of voltage-sourced converters (50 MVA each). Each 50-MVA converter group consists of four sets of 12.5-MVA modules plus a 5-MVAR harmonic filter (plus one spare filter switchable to either group), with a nominal phase-to-phase ac voltage of 3.2 kV and a dc-link voltage of 6,000 V. The two 50-MVA VSC groups are connected to the 138-kV system via two three-phase step-up transformers each rated at 55 MVA, 3.2 kV/ 138 kV (plus one "hot" spare switchable via the motor operated disconnects). Either 50 MVA VSC group or both can be connected to each of the 138-kV buses via the various automatically controlled motor-operated disconnects.

The system is also designed for future operation as a BTB dc link connecting the Talega East and West buses or as a unified power flow controller (UPFC). The BTB dc-link system would have a power transfer rating of 50 MW and would be able to deliver power bidirectionally between the east and west buses at Talega. The dc links are physically in place for this future option, which would essentially only require software-based control adjustments for BTB operation. The potential expansion to a UPFC is also facilitated by the in-place dc links. Other expansion requirements for UPFC operation would be the connection of one group of 50-MVA converters to a transformer in series with a line emanating from the Talega Substation. The UPFC configuration would also require software adjustments and would allow for the simultaneous control of real power flow (MW) and reactive power (MVAR) at Talega. This type of "flexibility" in the design and operation of FACTS technologies is imperative in its ability to adapt to system topology changes or new operational requirements in the future.

The main power semiconductor devices incorporated in the converter design are 6-in gate commutated turn-off thyristors (GCTs), rated at 6 kV, 6 kA. These devices are arranged in each module, forming a three-level converter circuit, which reduces the harmonic current as compared to a two-level design. The control of the converter is achieved with a five-pulse PWM (pulse width modulation), which further decreases the harmonics as compared to three-pulse or one-pulse PWM control. Because of these two aforementioned features, only the small harmonic filter is required on the ac side. As part of the overall reactive compensation scheme at the Talega substation, there are also three 69-MVAR shunt capacitors that are connected directly at the 230-kV system. The STATCOM system is able to control the operation of the voltage-sourced converters and the three 69-MVAR-capacitor banks. It can be remotely operated via SDG&E's SCADA system or manually operated from the control building. A one-line diagram of the Talega FACTS installation is shown in Figure 1.

Some of the main benefits of this FACTS design are as follows:

- rapid response to system disturbances
- smooth voltage control over a wide range of operating conditions.

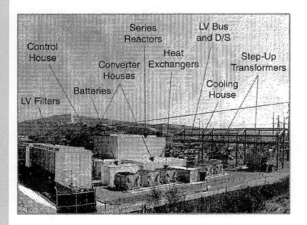

Figure 2
Photograph of the Talega ±100-MVA, 138-kV STATCOM (future BTB or UPFC) system.

- a significant amount of built-in redundancy (i.e., any one or more of the 12.5-MVA modules or 50-MVA groups can be out of service while all others remain in operation at their full rated capability)
- automatic reconfiguration to handle certain equipment failures (such as a transformer or filter) without shutting down the STATCOM
- expansion capability for future operation as a BTB dc link or UPFC system.

Figure 2 shows the FACTS installation, with various equipment and systems identified.

BENEFITS OF FACTS AND BTB DC-LINK TECHNOLOGIES

When implemented on a broad-scale basis, FACTS technologies deliver the following benefits.

- **Rapidly Implemented Installations:** FACTS projects are installed at existing substations and avoid the taking of public or private lands. They can be completed in less than 12 to 18 months—a substantially shorter time-frame than the process required for constructing new transmission lines.
- **Increased System Capacity:** FACTS provide increased capacity on the existing electrical transmission system infrastructure by allowing maximum operational efficiency of existing transmission lines and other equipment.
- **Enhanced System Reliability:** FACTS strengthen the operational integrity of transmission networks, allowing greater voltage stability and power flow control, which leads to enhanced system reliability and security.
- **Improved System Controllability:** FACTS allow improved system controllability by building "intelligence" into the transmission network via the ability to instantaneously respond to system disturbances and gridlock constraints and to enable redirection of power flows.
- **Seamless System Interconnections:** FACTS, in the form of BTB dc-link configurations, can establish "seamless" interconnections within and between regional and local

networks, allowing controlled power transfer and an increase in grid stability.

- **Fiscally Sound Investments:** FACTS are less expensive solutions to upgrading transmission system infrastructure as compared to conventional solutions such as the construction of new transmission lines. Strategic implementation of FACTS on a nationwide basis could reduce transmission system infrastructure expenditures by an estimated 30% overall. FACTS also provide transmission owners greater opportunity to realize profits through more efficient operation of existing networks.

A RECENT FACTS APPLICATION IN NORTH AMERICA

A project overview of a recent FACTS installation at the San Diego Gas and Electric Company's (SDG&E) Talega Substation, which was recently placed online, is described in the sidebar. The Talega FACTS project implements the advanced VSC-based technology, which provides superior performance along with design and control flexibility. The SDG&E FACTS project provides dynamic VAR support and control in order to compensate for transmission system needs arising from operational conditions due to deregulation, and it is the first of several other potential FACTS projects that SDG&E may complete in the future. The system is currently operating as a STATCOM, but it is also designed for future expansion to either a BTB dc link or a UPFC.

FACTS AND RENEWABLE ENERGY SOURCES

In addition to transmission applications and the benefits offered by FACTS for gridlock resolution and improved system operations, these technologies are also extremely beneficial when implemented for the interconnection of certain renewable energy sources to the grid, and they also have tremendous worth for industrial applications.

In the case of renewable energies, FACTS are especially advantageous when applied for wind generator interconnections. An increasing percentage of the United State's total generation supply is being produced from wind farm developments. However, wind generation, although beneficial in many aspects such as it pertains to economics, renewable sources of energy, and the environment, does not universally provide a steady and continuous interconnection to the electric power transmission grid.

Due to the nature of the source of wind power, a continuous and steady supply from a wind generation unit or wind farm is difficult to achieve. As such, the inherently unsteady nature of this type of generation source requires measures of stability and control on the interconnecting power transmission system. In addition, due to issues associated with voltage control, as well as both real power (MW) and reactive power (MVAR) dispatching, measures must be established for power system operators in order to adjust to wind generation output as base load, peak load, or other dispatching criteria.

As wind farms become a larger part of the total generation base and as the penetration levels increase, issues related to integration such as transients, stability, and voltage control are becoming increasingly important. In addition, due to the stochastic nature of wind, the integration of such renewable sources of generation into the transmission system is significantly different than conventional types of generation.

For wind generation applications, FACTS can be implemented for voltage control in the form of the shunt-connected SVC or STATCOM configurations. In addition to voltage support and control, there are also benefits that can be realized for allowing generating units to increase real power output by relieving the reactive power requirements through the application of these dynamic compensation technologies.

Even more advantageous, perhaps, is the application of the BTB dc link form of FACTS for wind interconnections. Such applications provide a seamless interconnection to the transmission system, allowing power flow control and at the same time providing the voltage control and stability required. In addition, wind generation facilities that are interconnected to the grid through dc links do not contribute to increased short circuit capacity on the transmission system, allowing for greater flexibility in the size and output range of wind generation, while allowing maximum real power output from the generating units.

By implementing FACTS technologies in coordination with wind generation applications (and other renewable applications), a reliable, steady, and secure interconnection to the power transmission grid is ensured. In addition, maximum output of wind capacity and efficient operation of wind generating units are realized through interconnection with FACTS controllers.

ADVANCING FACTS THROUGH ENERGY POLICY INITIATIVES

As indicated by the application areas described above, FACTS can provide significant enhancements to overall transmission system performance and at the same time are environmentally friendly and cost-effective solutions.

Government policies can assist in providing the vision and incentive for stakeholders to reap the significant benefit of lower-profile technologies such as FACTS that frequently compete for acceptance in a marketplace dominated by a traditional but problematic solution (i.e., power transmission lines).

Congress, the Department of Energy, and the Federal Energy Regulatory Commission have begun to recognize the solutions that low-environmental-impact technologies like FACTS afford in providing for rapid and reliable upgrades to transmission infrastructure. FERC has proposed incentives for utilities that invest in (currently undefined) "technologies that can be installed relatively quickly," bypass the "long siting process for procurement of new rights-of-way," and may be "environmentally benign." The proposed policy is presently structured such that these incentives apply to utilities that join a regional transmission organization (RTO).

Energy legislation approved by the U.S. House of Representatives includes a provision that directs FERC to undertake a rulemaking to encourage the deployment of economically efficient transmission technologies that increase the capacity and efficiency of existing transmission facilities. And in its National Transmission Grid Study, the DOE cites FACTS systems as a technology that increases electricity flows through existing transmission corridors without having to construct new transmission lines.

In light of the significant advantages that FACTS and certain HVDC technologies can provide for enhancement of transmission grid reliability, capacity, and control, regulatory policy should be structured to establish an environment with incentives for implementing FACTS and HVDC solutions. Such policy should put into place similar regulation and incentives that have previously been established for generator interconnections. FACTS technologies are indeed "generators" of reactive power (VARs) as applied on transmission networks, and BTB dc links are "controllers" of real power (megawatts).

Establishing policy that brings incentives toward a "merchant plant" approach for these technologies, with appropriate value given to VARs and controllable megawatts, would spawn the wide-scale applications that are critical to the future of transmission grid reliability. As such, regulatory policy should consider the following key points:

- Accelerated depreciation for investments in technologies that are, from a public policy and technical perspective, clear alternatives to the protracted process of power line construction.
- Increased rate of return on investment. The rate of return for transmission investment should be commensurate with the value to the system of having adequate transmission capacity. Compared to the costs of outages, congestion, and lack of access to low-cost electricity, the cost of this upgrade is minimal.
- Consistency between the regulations and incentives that have been established for generator interconnections with respect to var and megawatt value.

CONCLUSIONS

Over the past several years, FACTS and BTB dc-link applications have increased significantly as compared to the previous decade. There are now numerous FACTS applications, including recent installations in California and Texas, as well as in the New England region of the United States and in some areas of Canada. As more electricity stakeholders recognize the technical and public policy advantages that these technologies confer, additional applications will emerge. Advancements in the state of the art of FACTS technologies will continue and will further advance the case for breaking transmission gridlock with these innovative and proven systems.

ACKNOWLEDGMENTS

The authors would like to thank Terry Snow of San Diego Gas and Electric for his significant contributions to the Talega FACTS project and for his support on the article sidebar. Snow is manager of substation engineering and design at SDG&E and project manager for the Talega FACTS installation.

BIOGRAPHIES

Gregory Reed is vice president of T&D marketing and technology development at Mitsubishi Electric Power Products, Inc., located in Warrendale, Pennsylvania. His areas of expertise are in applications of power electronics technologies (including FACTS, HVDC, and custom power), information and control system technologies, and power system engineering and analysis. Dr. Reed received his Ph.D. in electric power engineering from the University of Pittsburgh in May 1997 and joined Mitsubishi Electric Power Products, Inc. (MEPPI) in June 1997. He received his B.S. in electrical engineering from Gannon University in 1985 and his M. Eng. in electric power engineering from Rensselaer Polytechnic Institute in 1986. Dr. Reed has been a Member of the IEEE Power Engineering Society since 1985 and is a contributing member to various committees and working groups related to FACTS and HVDC tech-

nologies. He is also an active participant in various government and industry forums related to energy policy and transmission technologies. Dr. Reed has authored or coauthored over 25 published papers and technical articles in the areas of electric power system analysis and the applications of power electronics technologies.

John Paserba earned his B.E.E. (1987) from Gannon University, Erie, Pennsylvania, and his M.E. (1988) from RPI, Troy, New York. He joined Mitsubishi Electric Power Products Inc. (MEPPI) in 1998 after working in GE's Power Systems Energy Consulting Department for over ten years. He is the secretary for the IEEE PES Power System Dynamic Performance Committee and was the chairman for the IEEE PES Power System Stability Subcommittee and the convenor of CIGRE Task Force 38.01.07 on Control of Power System Oscillations and is a contributing member to various committees and working groups related to FACTS and power system dynamic performance. He is also a member of the Editorial Board of the *IEEE Power & Energy Magazine* and was a member of the Editorial Board for *IEEE Transactions on Power Systems*. He is a Fellow (2003) of IEEE.

Peter Salavantis joined Mitsubishi Electric and Electronics, USA, in 1995 and is currently vice president for public affairs. Prior to joining the company, Salavantis consulted on public policy initiatives for multinational corporations involving international trade, finance, energy, telecommunications, and transportation issues. Salavantis has served in legislative and political campaign capacities in Vermont, Florida, and Washington, DC. He is a graduate of St. Michael's College.

5.1

MEDIUM AND SHORT LINE APPROXIMATIONS

In this section, we present short and medium-length transmission-line approximations as a means of introducing *ABCD* parameters. Some readers may prefer to start in Section 5.2, which presents the exact transmission-line equations.

FIGURE 5.1

Representation of two-port network

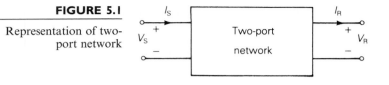

It is convenient to represent a transmission line by the two-port network shown in Figure 5.1, where V_S and I_S are the sending-end voltage and current, and V_R and I_R are the receiving-end voltage and current.

The relation between the sending-end and receiving-end quantities can be written as

$$V_S = AV_R + BI_R \quad \text{volts} \tag{5.1.1}$$

$$I_S = CV_R + DI_R \quad \text{A} \tag{5.1.2}$$

or, in matrix format,

$$\begin{bmatrix} V_S \\ I_S \end{bmatrix} = \begin{bmatrix} A & B \\ \hline C & D \end{bmatrix} \begin{bmatrix} V_R \\ I_R \end{bmatrix} \tag{5.1.3}$$

where A, B, C, and D are parameters that depend on the transmission-line constants R, L, C, and G. The $ABCD$ parameters are, in general, complex numbers. A and D are dimensionless. B has units of ohms, and C has units of siemens. Network theory texts [5] show that $ABCD$ parameters apply to linear, passive, bilateral two-port networks, with the following general relation:

$$AD - BC = 1 \tag{5.1.4}$$

The circuit in Figure 5.2 represents a short transmission line, usually applied to overhead 60-Hz lines less than 80 km long. Only the series resistance and reactance are included. The shunt admittance is neglected. The circuit applies to either single-phase or completely transposed three-phase lines operating under balanced conditions. For a completely transposed three-phase line, Z is the series impedance, V_S and V_R are positive-sequence line-to-neutral voltages, and I_S and I_R are positive-sequence line currents.

To avoid confusion between total series impedance and series impedance per unit length, we use the following notation:

$$z = \text{R} + j\omega\text{L} \quad \Omega/\text{m, series impedance per unit length}$$

$$y = \text{G} + j\omega\text{C} \quad \text{S/m, shunt admittance per unit length}$$

FIGURE 5.2

Short transmission line

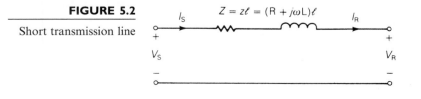

$Z = zl$ Ω, total series impedance

$Y = yl$ S, total shunt admittance

$l =$ line length   m

Recall that shunt conductance G is usually neglected for overhead transmission.

The *ABCD* parameters for the short line in Figure 5.2 are easily obtained by writing a KVL and KCL equation as

$$V_S = V_R + ZI_R \tag{5.1.5}$$

$$I_S = I_R \tag{5.1.6}$$

or, in matrix format,

$$\begin{bmatrix} V_S \\ I_S \end{bmatrix} = \left[\begin{array}{c|c} 1 & Z \\ \hline 0 & 1 \end{array} \right] \begin{bmatrix} V_R \\ I_R \end{bmatrix} \tag{5.1.7}$$

Comparing (5.1.7) and (5.1.3), the *ABCD* parameters for a short line are

$$A = D = 1 \quad \text{per unit} \tag{5.1.8}$$

$$B = Z \quad \Omega \tag{5.1.9}$$

$$C = 0 \quad \text{S} \tag{5.1.10}$$

For medium-length lines, typically ranging from 80 to 250 km at 60 Hz, it is common to lump the total shunt capacitance and locate half at each end of the line. Such a circuit, called a *nominal π circuit*, is shown in Figure 5.3.

To obtain the *ABCD* parameters of the nominal π circuit, note first that the current in the series branch in Figure 5.3 equals $I_R + \dfrac{V_R Y}{2}$. Then, writing a KVL equation,

$$V_S = V_R + Z\left(I_R + \frac{V_R Y}{2} \right)$$

$$= \left(1 + \frac{YZ}{2} \right) V_R + ZI_R \tag{5.1.11}$$

Also, writing a KCL equation at the sending end,

$$I_S = I_R + \frac{V_R Y}{2} + \frac{V_S Y}{2} \tag{5.1.12}$$

FIGURE 5.3

Medium-length transmission line— nominal π circuit

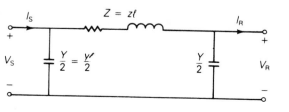

Using (5.1.11) in (5.1.12),

$$I_S = I_R + \frac{V_R Y}{2} + \left[\left(1 + \frac{YZ}{2}\right)V_R + ZI_R\right]\frac{Y}{2}$$

$$= Y\left(1 + \frac{YZ}{4}\right)V_R + \left(1 + \frac{YZ}{2}\right)I_R \qquad (5.1.13)$$

Writing (5.1.11) and (5.1.13) in matrix format,

$$\begin{bmatrix} V_S \\ \\ I_S \end{bmatrix} = \left[\begin{array}{c|c} \left(1 + \dfrac{YZ}{2}\right) & Z \\ \hline Y\left(1 + \dfrac{YZ}{4}\right) & \left(1 + \dfrac{YZ}{2}\right) \end{array}\right] \begin{bmatrix} V_R \\ \\ I_R \end{bmatrix} \qquad (5.1.14)$$

Thus, comparing (5.1.14) and (5.1.3)

$$A = D = 1 + \frac{YZ}{2} \quad \text{per unit} \qquad (5.1.15)$$

$$B = Z \quad \Omega \qquad (5.1.16)$$

$$C = Y\left(1 + \frac{YZ}{4}\right) \quad \text{S} \qquad (5.1.17)$$

Note that for both the short and medium-length lines, the relation $AD - BC = 1$ is verified. Note also that since the line is the same when viewed from either end, $A = D$.

Figure 5.4 gives the $ABCD$ parameters for some common networks, including a series impedance network that approximates a short line and a π circuit that approximates a medium-length line. A medium-length line could also be approximated by the T circuit shown in Figure 5.4, lumping half of the series impedance at each end of the line. Also given are the $ABCD$ parameters for networks in series, which are conveniently obtained by multiplying the $ABCD$ matrices of the individual networks.

$ABCD$ parameters can be used to describe the variation of line voltage with line loading. *Voltage regulation* is the change in voltage at the receiving end of the line when the load varies from no-load to a specified full load at a specified power factor, while the sending-end voltage is held constant. Expressed in percent of full-load voltage,

$$\text{percent VR} = \frac{|V_{RNL}| - |V_{RFL}|}{|V_{RFL}|} \times 100 \qquad (5.1.18)$$

where percent VR is the percent voltage regulation, $|V_{RNL}|$ is the magnitude of the no-load receiving-end voltage, and $|V_{RFL}|$ is the magnitude of the full-load receiving-end voltage.

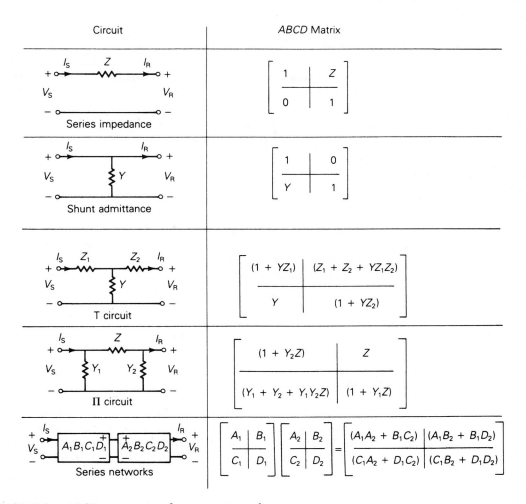

| Circuit | ABCD Matrix |

Series impedance

$$\begin{bmatrix} 1 & Z \\ 0 & 1 \end{bmatrix}$$

Shunt admittance

$$\begin{bmatrix} 1 & 0 \\ Y & 1 \end{bmatrix}$$

T circuit

$$\begin{bmatrix} (1 + YZ_1) & (Z_1 + Z_2 + YZ_1Z_2) \\ Y & (1 + YZ_2) \end{bmatrix}$$

Π circuit

$$\begin{bmatrix} (1 + Y_2Z) & Z \\ (Y_1 + Y_2 + Y_1Y_2Z) & (1 + Y_1Z) \end{bmatrix}$$

Series networks

$$\begin{bmatrix} A_1 & B_1 \\ C_1 & D_1 \end{bmatrix} \begin{bmatrix} A_2 & B_2 \\ C_2 & D_2 \end{bmatrix} = \begin{bmatrix} (A_1A_2 + B_1C_2) & (A_1B_2 + B_1D_2) \\ (C_1A_2 + D_1C_2) & (C_1B_2 + D_1D_2) \end{bmatrix}$$

FIGURE 5.4   *ABCD* parameters of common networks

FIGURE 5.5

Phasor diagrams for a short transmission line

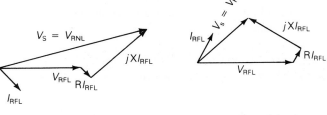

(a) Lagging p.f. load

(b) Leading p.f. load

The effect of load power factor on voltage regulation is illustrated by the phasor diagrams in Figure 5.5 for short lines. The phasor diagrams are graphical representations of (5.1.5) for lagging and leading power factor loads. Note that, from (5.1.5) at no-load, $I_{RNL} = 0$ and $V_S = V_{RNL}$ for a short line. As shown, the higher (worse) voltage regulation occurs for the lagging p.f. load, where V_{RNL} exceeds V_{RFL} by the larger amount. A smaller or even negative voltage regulation occurs for the leading p.f. load. In general, the no-load voltage is, from (5.1.1), with $I_{RNL} = 0$,

$$V_{RNL} = \frac{V_S}{A} \tag{5.1.19}$$

which can be used in (5.1.18) to determine voltage regulation.

In practice, transmission-line voltages decrease when heavily loaded and increase when lightly loaded. When voltages on EHV lines are maintained within $\pm 5\%$ of rated voltage, corresponding to about 10% voltage regulation, unusual operating problems are not encountered. Ten percent voltage regulation for lower voltage lines including transformer-voltage drops is also considered good operating practice.

In addition to voltage regulation, line loadability is an important issue. Three major line-loading limits are: (1) the thermal limit, (2) the voltage-drop limit, and (3) the steady-state stability limit.

The maximum temperature of a conductor determines its thermal limit. Conductor temperature affects the conductor sag between towers and the loss of conductor tensile strength due to annealing. If the temperature is too high, prescribed conductor-to-ground clearances may not be met, or the elastic limit of the conductor may be exceeded such that it cannot shrink to its original length when cooled. Conductor temperature depends on the current magnitude and its time duration, as well as on ambient temperature, wind velocity, and conductor surface conditions. Appendix Tables A.3 and A.4 give approximate current-carrying capacities of copper and ACSR conductors. The loadability of short transmission lines (less than 80 km in length for 60-Hz overhead lines) is usually determined by the conductor thermal limit or by ratings of line terminal equipment such as circuit breakers.

For longer line lengths (up to 300 km), line loadability is often determined by the voltage-drop limit. Although more severe voltage drops may be tolerated in some cases, a heavily loaded line with $V_R/V_S \geqslant 0.95$ is usually considered safe operating practice. For line lengths over 300 km, steady-state stability becomes a limiting factor. Stability, discussed in Section 5.4, refers to the ability of synchronous machines on either end of a line to remain in synchronism.

EXAMPLE 5.1   *ABCD parameters and the nominal π circuit: medium-length line*

A three-phase, 60-Hz, completely transposed 345-kV, 200-km line has two 795,000-cmil 26/2 ACSR conductors per bundle and the following positive-sequence line constants:

$$z = 0.032 + j0.35 \quad \Omega/\text{km}$$

$$y = j4.2 \times 10^{-6} \quad \text{S/km}$$

Full load at the receiving end of the line is 700 MW at 0.99 p.f. leading and at 95% of rated voltage. Assuming a medium-length line, determine the following:

a. $ABCD$ parameters of the nominal π circuit

b. Sending-end voltage V_S, current I_S, and real power P_S

c. Percent voltage regulation

d. Thermal limit, based on the approximate current-carrying capacity listed in Table A.4

e. Transmission-line efficiency at full load

SOLUTION

a. The total series impedance and shunt admittance values are

$$Z = zl = (0.032 + j0.35)(200) = 6.4 + j70 = 70.29\underline{/84.78^\circ} \quad \Omega$$

$$Y = yl = (j4.2 \times 10^{-6})(200) = 8.4 \times 10^{-4}\underline{/90^\circ} \quad \text{S}$$

From (5.1.15)–(5.1.17),

$$A = D = 1 + (8.4 \times 10^{-4}\underline{/90^\circ})(70.29\underline{/84.78^\circ})\left(\tfrac{1}{2}\right)$$

$$= 1 + 0.02952\underline{/174.78^\circ}$$

$$= 0.9706 + j0.00269 = 0.9706\underline{/0.159^\circ} \quad \text{per unit}$$

$$B = Z = 70.29\underline{/84.78^\circ} \quad \Omega$$

$$C = (8.4 \times 10^{-4}\underline{/90^\circ})(1 + 0.01476\underline{/174.78^\circ})$$

$$= (8.4 \times 10^{-4}\underline{/90^\circ})(0.9853 + j0.00134)$$

$$= 8.277 \times 10^{-4}\underline{/90.08^\circ} \quad \text{S}$$

b. The receiving-end voltage and current quantities are

$$V_R = (0.95)(345) = 327.8 \quad \text{kV}_{\text{LL}}$$

$$V_R = \frac{327.8}{\sqrt{3}}\underline{/0^\circ} = 189.2\underline{/0^\circ} \quad \text{kV}_{\text{LN}}$$

$$I_R = \frac{700\underline{/\cos^{-1} 0.99}}{(\sqrt{3})(0.95 \times 345)(0.99)} = 1.246\underline{/8.11^\circ} \quad \text{kA}$$

From (5.1.1) and (5.1.2), the sending-end quantities are

$$V_S = (0.9706\underline{/0.159^\circ})(189.2\underline{/0^\circ}) + (70.29\underline{/84.78^\circ})(1.246\underline{/8.11^\circ})$$

$$= 183.6\underline{/0.159^\circ} + 87.55\underline{/92.89^\circ}$$

$$= 179.2 + j87.95 = 199.6\underline{/26.14^\circ} \quad \text{kV}_{\text{LN}}$$

$$V_S = 199.6\sqrt{3} = 345.8 \text{ kV}_{LL} \approx 1.00 \quad \text{per unit}$$

$$I_S = (8.277 \times 10^{-4}\underline{/90.08°})(189.2\underline{/0°}) + (0.9706\underline{/0.159°})(1.246\underline{/8.11°})$$

$$= 0.1566\underline{/90.08°} + 1.209\underline{/8.27°}$$

$$= 1.196 + j0.331 = 1.241\underline{/15.5°} \quad \text{kA}$$

and the real power delivered to the sending end is

$$P_S = (\sqrt{3})(345.8)(1.241) \cos(26.14° - 15.5°)$$

$$= 730.5 \quad \text{MW}$$

c. From (5.1.19), the no-load receiving-end voltage is

$$V_{RNL} = \frac{V_S}{A} = \frac{345.8}{0.9706} = 356.3 \quad \text{kV}_{LL}$$

and, from (5.1.18),

$$\text{percent VR} = \frac{356.3 - 327.8}{327.8} \times 100 = 8.7\%$$

d. From Table A.4, the approximate current-carrying capacity of two 795,000-cmil 26/2 ACSR conductors is $2 \times 0.9 = 1.8$ kA.

e. The full-load line losses are $P_S - P_R = 730.5 - 700 = 30.5$ MW and the full-load transmission efficiency is

$$\text{percent EFF} = \frac{P_R}{P_S} \times 100 = \frac{700}{730.5} \times 100 = 95.8\%$$

Since $V_S = 1.00$ per unit, the full-load receiving-end voltage of 0.95 per unit corresponds to $V_R/V_S = 0.95$, considered in practice to be about the lowest operating voltage possible without encountering operating problems. Thus, for this 345-kV 200-km uncompensated line, voltage drop limits the full-load current to 1.246 kA at 0.99 p.f. leading, well below the thermal limit of 1.8 kA.   ■

5.2

TRANSMISSION-LINE DIFFERENTIAL EQUATIONS

The line constants R, L, and C are derived in Chapter 4 as per-length values having units of Ω/m, H/m, and F/m. They are not lumped, but rather are uniformly distributed along the length of the line. In order to account for the distributed nature of transmission-line constants, consider the circuit shown in Figure 5.6, which represents a line section of length Δx. $V(x)$ and $I(x)$ denote the voltage and current at position x, which is measured in meters from the right, or receiving end of the line. Similarly, $V(x + \Delta x)$ and $I(x + \Delta x)$ denote the voltage and current at position $(x + \Delta x)$. The circuit constants are

FIGURE 5.6

Transmission-line
section of length Δx

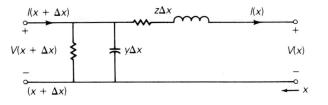

$$z = R + j\omega L \quad \Omega/m \tag{5.2.1}$$

$$y = G + j\omega C \quad S/m \tag{5.2.2}$$

where G is usually neglected for overhead 60-Hz lines. Writing a KVL equation for the circuit

$$V(x + \Delta x) = V(x) + (z\Delta x)I(x) \quad \text{volts} \tag{5.2.3}$$

Rearranging (5.2.3),

$$\frac{V(x + \Delta x) - V(x)}{\Delta x} = zI(x) \tag{5.2.4}$$

and taking the limit as Δx approaches zero,

$$\frac{dV(x)}{dx} = zI(x) \tag{5.2.5}$$

Similarly, writing a KCL equation for the circuit,

$$I(x + \Delta x) = I(x) + (y\Delta x)V(x + \Delta x) \quad \text{A} \tag{5.2.6}$$

Rearranging,

$$\frac{I(x + \Delta x) - I(x)}{\Delta x} = yV(x) \tag{5.2.7}$$

and taking the limit as Δx approaches zero,

$$\frac{dI(x)}{dx} = yV(x) \tag{5.2.8}$$

Equations (5.2.5) and (5.2.8) are two linear, first-order, homogeneous differential equations with two unknowns, $V(x)$ and $I(x)$. We can eliminate $I(x)$ by differentiating (5.2.5) and using (5.2.8) as follows:

$$\frac{d^2 V(x)}{dx^2} = z\frac{dI(x)}{dx} = zyV(x) \tag{5.2.9}$$

or

$$\frac{d^2 V(x)}{dx^2} - zyV(x) = 0 \tag{5.2.10}$$

Equation (5.2.10) is a linear, second-order, homogeneous differential equation with one unknown, $V(x)$. By inspection, its solution is

$$V(x) = A_1 e^{\gamma x} + A_2 e^{-\gamma x} \quad \text{volts} \tag{5.2.11}$$

where A_1 and A_2 are integration constants and

$$\gamma = \sqrt{zy} \quad \text{m}^{-1} \tag{5.2.12}$$

γ, whose units are m^{-1}, is called the *propagation constant*. By inserting (5.2.11) and (5.2.12) into (5.2.10), the solution to the differential equation can be verified.

Next, using (5.2.11) in (5.2.5),

$$\frac{dV(x)}{dx} = \gamma A_1 e^{\gamma x} - \gamma A_2 e^{-\gamma x} = zI(x) \tag{5.2.13}$$

Solving for $I(x)$,

$$I(x) = \frac{A_1 e^{\gamma x} - A_2 e^{-\gamma x}}{z/\gamma} \tag{5.2.14}$$

Using (5.2.12), $z/\gamma = z/\sqrt{zy} = \sqrt{z/y}$, (5.2.14) becomes

$$I(x) = \frac{A_1 e^{\gamma x} - A_2 e^{-\gamma x}}{Z_c} \tag{5.2.15}$$

where

$$Z_c = \sqrt{\frac{z}{y}} \quad \Omega \tag{5.2.16}$$

Z_c, whose units are Ω, is called the *characteristic impedance*.

Next, the integration constants A_1 and A_2 are evaluated from the boundary conditions. At $x = 0$, the receiving end of the line, the receiving-end voltage and current are

$$V_R = V(0) \tag{5.2.17}$$

$$I_R = I(0) \tag{5.2.18}$$

Also, at $x = 0$, (5.2.11) and (5.2.15) become

$$V_R = A_1 + A_2 \tag{5.2.19}$$

$$I_R = \frac{A_1 - A_2}{Z_c} \tag{5.2.20}$$

Solving for A_1 and A_2,

$$A_1 = \frac{V_R + Z_c I_R}{2} \tag{5.2.21}$$

$$A_2 = \frac{V_R - Z_c I_R}{2} \tag{5.2.22}$$

Substituting A_1 and A_2 into (5.2.11) and (5.2.15),

$$V(x) = \left(\frac{V_R + Z_c I_R}{2}\right)e^{\gamma x} + \left(\frac{V_R - Z_c I_R}{2}\right)e^{-\gamma x} \qquad (5.2.23)$$

$$I(x) = \left(\frac{V_R + Z_c I_R}{2Z_c}\right)e^{\gamma x} - \left(\frac{V_R - Z_c I_R}{2Z_c}\right)e^{-\gamma x} \qquad (5.2.24)$$

Rearranging (5.2.23) and (5.2.24),

$$V(x) = \left(\frac{e^{\gamma x} + e^{-\gamma x}}{2}\right)V_R + Z_c\left(\frac{e^{\gamma x} - e^{-\gamma x}}{2}\right)I_R \qquad (5.2.25)$$

$$I(x) = \frac{1}{Z_c}\left(\frac{e^{\gamma x} - e^{-\gamma x}}{2}\right)V_R + \left(\frac{e^{\gamma x} + e^{-\gamma x}}{2}\right)I_R \qquad (5.2.26)$$

Recognizing the hyperbolic functions cosh and sinh,

$$V(x) = \cosh(\gamma x)V_R + Z_c \sinh(\gamma x)I_R \qquad (5.2.27)$$

$$I(x) = \frac{1}{Z_c} \sinh(\gamma x)V_R + \cosh(\gamma x)I_R \qquad (5.2.28)$$

Equations (5.2.27) and (5.2.28) give the *ABCD* parameters of the distributed line. In matrix format,

$$\begin{bmatrix} V(x) \\ I(x) \end{bmatrix} = \left[\begin{array}{c|c} A(x) & B(x) \\ \hline C(x) & D(x) \end{array}\right]\begin{bmatrix} V_R \\ I_R \end{bmatrix} \qquad (5.2.29)$$

where

$$A(x) = D(x) = \cosh(\gamma x) \quad \text{per unit} \qquad (5.2.30)$$

$$B(x) = Z_c \sinh(\gamma x) \quad \Omega \qquad (5.2.31)$$

$$C(x) = \frac{1}{Z_c} \sinh(\gamma x) \quad \text{S} \qquad (5.2.32)$$

Equation (5.2.29) gives the current and voltage at any point x along the line in terms of the receiving-end voltage and current. At the sending end, where $x = l$, $V(l) = V_S$ and $I(l) = I_S$. That is,

$$\begin{bmatrix} V_S \\ I_S \end{bmatrix} = \left[\begin{array}{c|c} A & B \\ \hline C & D \end{array}\right]\begin{bmatrix} V_R \\ I_R \end{bmatrix} \qquad (5.2.33)$$

where

$$A = D = \cosh(\gamma l) \quad \text{per unit} \qquad (5.2.34)$$

$$B = Z_c \sinh(\gamma l) \quad \Omega \qquad (5.2.35)$$

$$C = \frac{1}{Z_c} \sinh(\gamma l) \quad \text{S} \qquad (5.2.36)$$

Equations (5.2.34)–(5.2.36) give the *ABCD* parameters of the distributed line. In these equations, the propagation constant γ is a complex quantity with real and imaginary parts denoted α and β. That is,

$$\gamma = \alpha + j\beta \quad \text{m}^{-1} \tag{5.2.37}$$

The quantity γl is dimensionless. Also

$$e^{\gamma l} = e^{(\alpha l + j\beta l)} = e^{\alpha l} e^{j\beta l} = e^{\alpha l} \underline{/\beta l} \tag{5.2.38}$$

Using (5.2.38) the hyperbolic functions cosh and sinh can be evaluated as follows:

$$\cosh(\gamma l) = \frac{e^{\gamma l} + e^{-\gamma l}}{2} = \frac{1}{2}\left(e^{\alpha l} \underline{/\beta l} + e^{-\alpha l} \underline{/-\beta l}\right) \tag{5.2.39}$$

and

$$\sinh(\gamma l) = \frac{e^{\gamma l} - e^{-\gamma l}}{2} = \frac{1}{2}\left(e^{\alpha l} \underline{/\beta l} - e^{-\alpha l} \underline{/-\beta l}\right) \tag{5.2.40}$$

Alternatively, the following identities can be used:

$$\cosh(\alpha l + j\beta l) = \cosh(\alpha l)\cos(\beta l) + j\sinh(\alpha l)\sin(\beta l) \tag{5.2.41}$$

$$\sinh(\alpha l + j\beta l) = \sinh(\alpha l)\cos(\beta l) + j\cosh(\alpha l)\sin(\beta l) \tag{5.2.42}$$

Note that in (5.2.39)–(5.2.42), the dimensionless quantity βl is in radians, not degrees.

The *ABCD* parameters given by (5.2.34)–(5.2.36) are exact parameters valid for any line length. For accurate calculations, these equations must be used for overhead 60-Hz lines longer than 250 km. The *ABCD* parameters derived in Section 5.1 are approximate parameters that are more conveniently used for hand calculations involving short and medium-length lines. Table 5.1 summarizes the *ABCD* parameters for short, medium, long, and lossless (see Section 5.4) lines.

TABLE 5.1

Summary: Transmission-line *ABCD* parameters

Parameter	$A = D$	B	C
Units	per Unit	Ω	S
Short line (less than 80 km)	1	Z	0
Medium line—nominal π circuit (80 to 250 km)	$1 + \dfrac{YZ}{2}$	Z	$Y\left(1 + \dfrac{YZ}{4}\right)$
Long line—equivalent π circuit (more than 250 km)	$\cosh(\gamma\ell) = 1 + \dfrac{Y'Z'}{2}$	$Z_c \sinh(\gamma\ell) = Z'$	$\begin{array}{l}(1/Z_c)\sinh(\gamma\ell)\\ = Y'\left(1 + \dfrac{Y'Z'}{4}\right)\end{array}$
Lossless line (R = G = 0)	$\cos(\beta\ell)$	$jZ_c \sin(\beta\ell)$	$\dfrac{j\sin(\beta\ell)}{Z_c}$

EXAMPLE 5.2   **Exact *ABCD* parameters: long line**

A three-phase 765-kV, 60-Hz, 300-km, completely transposed line has the following positive-sequence impedance and admittance:

$$z = 0.0165 + j0.3306 = 0.3310\underline{/87.14°}\quad \Omega/\text{km}$$

$$y = j4.674 \times 10^{-6}\quad \text{S/km}$$

Assuming positive-sequence operation, calculate the exact *ABCD* parameters of the line. Compare the exact *B* parameter with that of the nominal π circuit.

SOLUTION   From (5.2.12) and (5.2.16):

$$Z_c = \sqrt{\frac{0.3310\underline{/87.14°}}{4.674 \times 10^{-6}\underline{/90°}}} = \sqrt{7.082 \times 10^4\underline{/-2.86°}}$$

$$= 266.1\underline{/-1.43°}\quad \Omega$$

and

$$\gamma l = \sqrt{(0.3310\underline{/87.14°})(4.674 \times 10^{-6}\underline{/90°})} \times (300)$$

$$= \sqrt{1.547 \times 10^{-6}\underline{/177.14°}} \times (300)$$

$$= 0.3731\underline{/88.57°} = 0.00931 + j0.3730\quad \text{per unit}$$

From (5.2.38),

$$e^{\gamma l} = e^{0.00931}e^{+j0.3730} = 1.0094\underline{/0.3730}\quad \text{radians}$$

$$= 0.9400 + j0.3678$$

and

$$e^{-\gamma l} = e^{-0.00931}e^{-j0.3730} = 0.9907\underline{/-0.3730}\quad \text{radians}$$

$$= 0.9226 - j0.3610$$

Then, from (5.2.39) and (5.2.40),

$$\cosh(\gamma l) = \frac{(0.9400 + j0.3678) + (0.9226 - j0.3610)}{2}$$

$$= 0.9313 + j0.0034 = 0.9313\underline{/0.209°}$$

$$\sinh(\gamma l) = \frac{(0.9400 + j0.3678) - (0.9226 - j0.3610)}{2}$$

$$= 0.0087 + j0.3644 = 0.3645\underline{/88.63°}$$

Finally, from (5.2.34)–(5.2.36),

$$A = D = \cosh(\gamma l) = 0.9313\underline{/0.209^\circ} \quad \text{per unit}$$

$$B = (266.1\underline{/-1.43^\circ})(0.3645\underline{/88.63^\circ}) = 97.0\underline{/87.2^\circ} \quad \Omega$$

$$C = \frac{0.3645\underline{/88.63^\circ}}{266.1\underline{/-1.43^\circ}} = 1.37 \times 10^{-3}\underline{/90.06^\circ} \quad S$$

Using (5.1.16), the B parameter for the nominal π circuit is

$$B_{\text{nominal }\pi} = Z = (0.3310\underline{/87.14^\circ})(300) = 99.3\underline{/87.14^\circ} \quad \Omega$$

which is 2% larger than the exact value. ∎

5.3

EQUIVALENT π CIRCUIT

Many computer programs used in power system analysis and design assume circuit representations of components such as transmission lines and transformers (see the power-flow program described in Chapter 6 as an example). It is therefore convenient to represent the terminal characteristics of a transmission line by an equivalent circuit instead of its $ABCD$ parameters.

The circuit shown in Figure 5.7 is called an *equivalent π circuit*. It is identical in structure to the nominal π circuit of Figure 5.3, except that Z' and Y' are used instead of Z and Y. Our objective is to determine Z' and Y' such that the equivalent π circuit has the same $ABCD$ parameters as those of the distributed line, (5.2.34)–(5.2.36). The $ABCD$ parameters of the equivalent π circuit, which has the same structure as the nominal π, are

$$A = D = 1 + \frac{Y'Z'}{2} \quad \text{per unit} \tag{5.3.1}$$

$$B = Z' \quad \Omega \tag{5.3.2}$$

FIGURE 5.7

Transmission-line equivalent π circuit

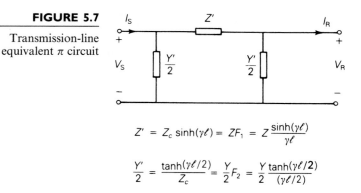

$$Z' = Z_c \sinh(\gamma\ell) = ZF_1 = Z\frac{\sinh(\gamma\ell)}{\gamma\ell}$$

$$\frac{Y'}{2} = \frac{\tanh(\gamma\ell/2)}{Z_c} = \frac{Y}{2}F_2 = \frac{Y}{2}\frac{\tanh(\gamma\ell/2)}{(\gamma\ell/2)}$$

$$C = Y'\left(1 + \frac{Y'Z'}{4}\right) \quad \text{S} \tag{5.3.3}$$

where we have replaced Z and Y in (5.1.15)–(5.1.17) with Z' and Y' in (5.3.1)–(5.3.3). Equating (5.3.2) to (5.2.35),

$$Z' = Z_c \sinh(\gamma l) = \sqrt{\frac{z}{y}} \sinh(\gamma l) \tag{5.3.4}$$

Rewriting (5.3.4) in terms of the nominal π circuit impedance $Z = zl$,

$$Z' = zl\left[\sqrt{\frac{z}{y}} \frac{\sinh(\gamma l)}{zl}\right] = zl\left[\frac{\sinh(\gamma l)}{\sqrt{zy}\, l}\right]$$

$$= ZF_1 \quad \Omega \tag{5.3.5}$$

where

$$F_1 = \frac{\sinh(\gamma l)}{\gamma l} \quad \text{per unit} \tag{5.3.6}$$

Similarly, equating (5.3.1) to (5.2.34),

$$1 + \frac{Y'Z'}{2} = \cosh(\gamma l)$$

$$\frac{Y'}{2} = \frac{\cosh(\gamma l) - 1}{Z'} \tag{5.3.7}$$

Using (5.3.4) and the identity $\tanh\left(\dfrac{\gamma l}{2}\right) = \dfrac{\cosh(\gamma l) - 1}{\sinh(\gamma l)}$, (5.3.7) becomes

$$\frac{Y'}{2} = \frac{\cosh(\gamma l) - 1}{Z_c \sinh(\gamma l)} = \frac{\tanh(\gamma l/2)}{Z_c} = \frac{\tanh(\gamma l/2)}{\sqrt{\dfrac{z}{y}}} \tag{5.3.8}$$

Rewriting (5.3.8) in terms of the nominal π circuit admittance $Y = yl$,

$$\frac{Y'}{2} = \frac{yl}{2}\left[\frac{\tanh(\gamma l/2)}{\sqrt{\dfrac{z}{y}}\dfrac{yl}{2}}\right] = \frac{yl}{2}\left[\frac{\tanh(\gamma l/2)}{\sqrt{zy}\, l/2}\right]$$

$$= \frac{Y}{2}F_2 \quad \text{S} \tag{5.3.9}$$

where

$$F_2 = \frac{\tanh(\gamma l/2)}{\gamma l/2} \quad \text{per unit} \tag{5.3.10}$$

Equations (5.3.6) and (5.3.10) give the correction factors F_1 and F_2 to convert Z and Y for the nominal π circuit to Z' and Y' for the equivalent π circuit.

EXAMPLE 5.3   **Equivalent π circuit: long line**

Compare the equivalent and nominal π circuits for the line in Example 5.2.

SOLUTION   For the nominal π circuit,

$$Z = zl = (0.3310\underline{/87.14°})(300) = 99.3\underline{/87.14°} \quad \Omega$$

$$\frac{Y}{2} = \frac{yl}{2} = \left(\frac{j4.674 \times 10^{-6}}{2}\right)(300) = 7.011 \times 10^{-4}\underline{/90°} \quad \text{S}$$

From (5.3.6) and (5.3.10), the correction factors are

$$F_1 = \frac{0.3645\underline{/88.63°}}{0.3731\underline{/88.57°}} = 0.9769\underline{/0.06°} \quad \text{per unit}$$

$$F_2 = \frac{\tanh(\gamma l/2)}{\gamma l/2} = \frac{\cosh(\gamma l) - 1}{(\gamma l/2)\sinh(\gamma l)}$$

$$= \frac{0.9313 + j0.0034 - 1}{\left(\dfrac{0.3731}{2}\underline{/88.57°}\right)(0.3645\underline{/88.63°})}$$

$$= \frac{-0.0687 + j0.0034}{0.06800\underline{/177.20°}}$$

$$= \frac{0.06878\underline{/177.17°}}{0.06800\underline{/177.20°}} = 1.012\underline{/-0.03°} \quad \text{per unit}$$

Then, from (5.3.5) and (5.3.9), for the equivalent π circuit,

$$Z' = (99.3\underline{/87.14°})(0.9769\underline{/0.06°}) = 97.0\underline{/87.2°} \quad \Omega$$

$$\frac{Y'}{2} = (7.011 \times 10^{-4}\underline{/90°})(1.012\underline{/-0.03°}) = 7.095 \times 10^{-4}\underline{/89.97°} \quad \text{S}$$

$$= 3.7 \times 10^{-7} + j7.095 \times 10^{-4} \quad \text{S}$$

Comparing these nominal and equivalent π circuit values, Z' is about 2% smaller than Z, and $Y'/2$ is about 1% larger than $Y/2$. Although the circuit values are approximately the same for this line, the equivalent π circuit should be used for accurate calculations involving long lines. Note the small shunt conductance, $G' = 3.7 \times 10^{-7}$ S, introduced in the equivalent π circuit. G' is often neglected.   ∎

5.4

LOSSLESS LINES

In this section, we discuss the following concepts for lossless lines: surge impedance, $ABCD$ parameters, equivalent π circuit, wavelength, surge impedance loading, voltage profiles, and steady-state stability limit.

When line losses are neglected, simpler expressions for the line parameters are obtained and the above concepts are more easily understood. Since transmission and distribution lines for power transfer generally are designed to have low losses, the equations and concepts developed here can be used for quick and reasonably accurate hand calculations leading to seat-of-the-pants analyses and to initial designs. More accurate calculations can then be made with computer programs for follow-up analysis and design.

SURGE IMPEDANCE

For a lossless line, $R = G = 0$, and

$$z = j\omega L \quad \Omega/m \tag{5.4.1}$$

$$y = j\omega C \quad S/m \tag{5.4.2}$$

From (5.2.12) and (5.2.16),

$$Z_c = \sqrt{\frac{z}{y}} = \sqrt{\frac{j\omega L}{j\omega C}} = \sqrt{\frac{L}{C}} \quad \Omega \tag{5.4.3}$$

and

$$\gamma = \sqrt{zy} = \sqrt{(j\omega L)(j\omega C)} = j\omega\sqrt{LC} = j\beta \quad m^{-1} \tag{5.4.4}$$

where

$$\beta = \omega\sqrt{LC} \quad m^{-1} \tag{5.4.5}$$

The characteristic impedance $Z_c = \sqrt{L/C}$, commonly called *surge* impedance for a lossless line, is pure real—that is, resistive. The propagation constant $\gamma = j\beta$ is pure imaginary.

ABCD PARAMETERS

The *ABCD* parameters are, from (5.2.30)–(5.2.32),

$$A(x) = D(x) = \cosh(\gamma x) = \cosh(j\beta x)$$

$$= \frac{e^{j\beta x} + e^{-j\beta x}}{2} = \cos(\beta x) \quad \text{per unit} \tag{5.4.6}$$

$$\sinh(\gamma x) = \sinh(j\beta x) = \frac{e^{j\beta x} - e^{-j\beta x}}{2} = j\sin(\beta x) \quad \text{per unit} \tag{5.4.7}$$

$$B(x) = Z_c \sinh(\gamma x) = jZ_c \sin(\beta x) = j\sqrt{\frac{L}{C}}\sin(\beta x) \quad \Omega \tag{5.4.8}$$

$$C(x) = \frac{\sinh(\gamma x)}{Z_c} = \frac{j\sin(\beta x)}{\sqrt{\frac{L}{C}}} \quad S \tag{5.4.9}$$

$A(x)$ and $D(x)$ are pure real; $B(x)$ and $C(x)$ are pure imaginary.

A comparison of lossless versus lossy $ABCD$ parameters is shown in Table 5.1.

EQUIVALENT π CIRCUIT

For the equivalent π circuit, using (5.3.4),

$$Z' = jZ_c \sin(\beta l) = jX' \quad \Omega \tag{5.4.10}$$

or, from (5.3.5) and (5.3.6),

$$Z' = (j\omega Ll)\left(\frac{\sin(\beta l)}{\beta l}\right) = jX' \quad \Omega \tag{5.4.11}$$

Also, from (5.3.9) and (5.3.10),

$$\frac{Y'}{2} = \frac{Y}{2}\frac{\tanh(j\beta l/2)}{j\beta l/2} = \frac{Y}{2}\frac{\sinh(j\beta l/2)}{(j\beta l/2)\cosh(j\beta l/2)}$$

$$= \left(\frac{j\omega Cl}{2}\right)\frac{j\sin(\beta l/2)}{(j\beta l/2)\cos(\beta l/2)} = \left(\frac{j\omega Cl}{2}\right)\frac{\tan(\beta l/2)}{\beta l/2}$$

$$= \left(\frac{j\omega C'l}{2}\right) \quad S \tag{5.4.12}$$

Z' and Y' are both pure imaginary. Also, for βl less than π radians, Z' is pure inductive and Y' is pure capacitive. Thus the equivalent π circuit for a lossless line, shown in Figure 5.8, is also lossless.

WAVELENGTH

A *wavelength* is the distance required to change the phase of the voltage or current by 2π radians or 360°. For a lossless line, using (5.2.29),

$$V(x) = A(x)V_R + B(x)I_R$$

$$= \cos(\beta x)V_R + jZ_c\sin(\beta x)I_R \tag{5.4.13}$$

FIGURE 5.8

Equivalent π circuit for a lossless line ($\beta\ell$ less than π)

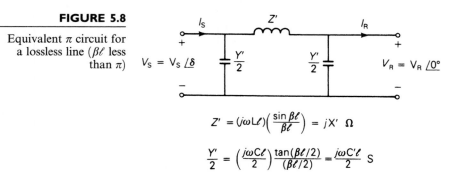

$$Z' = (j\omega L\ell)\left(\frac{\sin\beta\ell}{\beta\ell}\right) = jX' \;\; \Omega$$

$$\frac{Y'}{2} = \left(\frac{j\omega C\ell}{2}\right)\frac{\tan(\beta\ell/2)}{(\beta\ell/2)} = \frac{j\omega C'\ell}{2} \;\; S$$

and

$$I(x) = C(x)V_R + D(x)I_R$$

$$= \frac{j \sin(\beta x)}{Z_c} V_R + \cos(\beta x)I_R \qquad (5.4.14)$$

From (5.4.13) and (5.4.14), $V(x)$ and $I(x)$ change phase by 2π radians when $x = 2\pi/\beta$. Denoting wavelength by λ, and using (5.4.5),

$$\lambda = \frac{2\pi}{\beta} = \frac{2\pi}{\omega\sqrt{LC}} = \frac{1}{f\sqrt{LC}} \quad \text{m} \qquad (5.4.15)$$

or

$$f\lambda = \frac{1}{\sqrt{LC}} \qquad (5.4.16)$$

We will show in Chapter 12 that the term $(1/\sqrt{LC})$ in (5.4.16) is the velocity of propagation of voltage and current waves along a lossless line. For overhead lines, $(1/\sqrt{LC}) \approx 3 \times 10^8$ m/s, and for $f = 60$ Hz, (5.4.14) gives

$$\lambda \approx \frac{3 \times 10^8}{60} = 5 \times 10^6 \text{ m} = 5000 \text{ km} = 3100 \text{ mi}$$

Typical power-line lengths are only a small fraction of the above 60-Hz wavelength.

SURGE IMPEDANCE LOADING

Surge impedance loading (SIL) is the power delivered by a lossless line to a load resistance equal to the surge impedance $Z_c = \sqrt{L/C}$. Figure 5.9 shows a lossless line terminated by a resistance equal to its surge impedance. This line represents either a single-phase line or one phase-to-neutral of a balanced three-phase line. At SIL, from (5.4.13),

$$V(x) = \cos(\beta x)V_R + jZ_c \sin(\beta x)I_R$$

$$= \cos(\beta x)V_R + jZ_c \sin(\beta x)\left(\frac{V_R}{Z_c}\right)$$

$$= (\cos \beta x + j \sin \beta x)V_R$$

$$= e^{j\beta x} V_R \quad \text{volts} \qquad (5.4.17)$$

$$|V(x)| = |V_R| \quad \text{volts} \qquad (5.4.18)$$

FIGURE 5.9

Lossless line terminated by its surge impedance

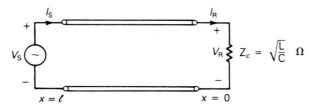

TABLE 5.2	V_{rated} (kV)	$Z_c = \sqrt{L/C}$ (Ω)	SIL$= V^2_{rated}/Z_c$ (MW)
Surge impedance and SIL values for typical 60-Hz overhead lines [1, 2]	69	366–400	12–13
	138	366–405	47–52
	230	365–395	134–145
	345	280–366	325–425
	500	233–294	850–1075
	765	254–266	2200–2300

Thus, at SIL, the voltage profile is flat. That is, the voltage magnitude at any point x along a lossless line at SIL is constant.

Also from (5.4.14) at SIL,

$$
\begin{aligned}
I(x) &= \frac{j\,\sin(\beta x)}{Z_c}\,V_R + (\cos\beta x)\frac{V_R}{Z_c} \\
&= (\cos\beta x + j\,\sin\beta x)\frac{V_R}{Z_c} \\
&= (e^{j\beta x})\frac{V_R}{Z_c} \quad \text{A}
\end{aligned}
\tag{5.4.19}
$$

Using (5.4.17) and (5.4.19), the complex power flowing at any point x along the line is

$$
\begin{aligned}
S(x) &= P(x) + jQ(x) = V(x)I^*(x) \\
&= (e^{j\beta x}V_R)\left(\frac{e^{j\beta x}V_R}{Z_c}\right)^* \\
&= \frac{|V_R|^2}{Z_c}
\end{aligned}
\tag{5.4.20}
$$

Thus the real power flow along a lossless line at SIL remains constant from the sending end to the receiving end. The reactive power flow is zero.

At rated line voltage, the real power delivered, or SIL, is, from (5.4.20),

$$
\text{SIL} = \frac{V^2_{rated}}{Z_c}
\tag{5.4.21}
$$

where rated voltage is used for a single-phase line and rated line-to-line voltage is used for the total real power delivered by a three-phase line. Table 5.2 lists surge impedance and SIL values for typical overhead 60-Hz three-phase lines.

VOLTAGE PROFILES

In practice, power lines are not terminated by their surge impedance. Instead, loadings can vary from a small fraction of SIL during light load conditions

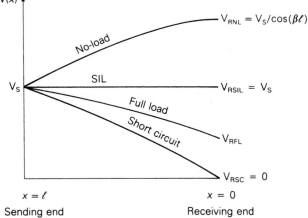

Voltage profiles of an uncompensated lossless line with fixed sending-end voltage for line lengths up to a quarter wavelength

up to multiples of SIL, depending on line length and line compensation, during heavy load conditions. If a line is not terminated by its surge impedance, then the voltage profile is not flat. Figure 5.10 shows voltage profiles of lines with a fixed sending-end voltage magnitude V_S for line lengths l up to a quarter wavelength. This figure shows four loading conditions: (1) no-load, (2) SIL, (3) short circuit, and (4) full load, which are described as follows:

1. At no-load, $I_{RNL} = 0$ and (5.4.13) yields

$$V_{NL}(x) = (\cos \beta x) V_{RNL} \qquad (5.4.22)$$

The no-load voltage increases from $V_S = (\cos \beta l) V_{RNL}$ at the sending end to V_{RNL} at the receiving end (where $x = 0$).

2. From (5.4.18), the voltage profile at SIL is flat.

3. For a short circuit at the load, $V_{RSC} = 0$ and (5.4.13) yields

$$V_{SC}(x) = (Z_c \sin \beta x) I_{RSC} \qquad (5.4.23)$$

The voltage decreases from $V_S = (\sin \beta l)(Z_c I_{RSC})$ at the sending end to $V_{RSC} = 0$ at the receiving end.

4. The full-load voltage profile, which depends on the specification of full-load current, lies above the short-circuit voltage profile.

Figure 5.10 summarizes these results, showing a high receiving-end voltage at no-load and a low receiving-end voltage at full load. This voltage regulation problem becomes more severe as the line length increases. In Section 5.6, we discuss shunt compensation methods to reduce voltage fluctuations.

STEADY-STATE STABILITY LIMIT

The equivalent π circuit of Figure 5.8 can be used to obtain an equation for the real power delivered by a lossless line. Assume that the voltage magnitudes V_S and V_R at the ends of the line are held constant. Also, let δ denote the voltage-phase angle at the sending end with respect to the receiving end. From KVL, the receiving-end current I_R is

$$
\begin{aligned}
I_R &= \frac{V_S - V_R}{Z'} - \frac{Y'}{2} V_R \\
&= \frac{V_S e^{j\delta} - V_R}{jX'} - \frac{j\omega C' l}{2} V_R
\end{aligned}
\tag{5.4.24}
$$

and the complex power S_R delivered to the receiving end is

$$
\begin{aligned}
S_R &= V_R I_R^* = V_R \left(\frac{V_S e^{j\delta} - V_R}{jX'} \right)^* + \frac{j\omega C' l}{2} V_R^2 \\
&= V_R \left(\frac{V_S e^{-j\delta} - V_R}{-jX'} \right) + \frac{j\omega Cl}{2} V_R^2 \\
&= \frac{jV_R V_S \cos \delta + V_R V_S \sin \delta - jV_R^2}{X'} + \frac{j\omega Cl}{2} V_R^2
\end{aligned}
\tag{5.4.25}
$$

The real power delivered is

$$
P = P_S = P_R = \text{Re}(S_R) = \frac{V_R V_S}{X'} \sin \delta \quad \text{W}
\tag{5.4.26}
$$

Note that since the line is lossless, $P_S = P_R$.

Equation (5.4.26) is plotted in Figure 5.11. For fixed voltage magnitudes V_S and V_R, the phase angle δ increases from 0 to 90° as the real power delivered increases. The maximum power that the line can deliver, which occurs when $\delta = 90°$, is given by

$$
P_{\max} = \frac{V_S V_R}{X'} \quad \text{W}
\tag{5.4.27}
$$

FIGURE 5.11

Real power delivered by a lossless line versus voltage angle across the line

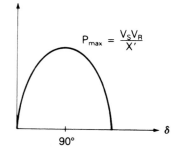

Real power P

$P_{\max} = \dfrac{V_S V_R}{X'}$

90°

δ

P_{max} represents the theoretical *steady-state stability* limit of a lossless line. If an attempt were made to exceed this steady-state stability limit, then synchronous machines at the sending end would lose synchronism with those at the receiving end. Stability is further discussed in Chapter 13.

It is convenient to express the steady-state stability limit in terms of SIL. Using (5.4.10) in (5.4.26),

$$P = \frac{V_S V_R \sin \delta}{Z_c \sin \beta l} = \left(\frac{V_S V_R}{Z_c}\right) \frac{\sin \delta}{\sin\left(\dfrac{2\pi l}{\lambda}\right)} \tag{5.4.28}$$

Expressing V_S and V_R in per-unit of rated line voltage,

$$P = \left(\frac{V_S}{V_{rated}}\right)\left(\frac{V_R}{V_{rated}}\right)\left(\frac{V_{rated}^2}{Z_c}\right) \frac{\sin \delta}{\sin\left(\dfrac{2\pi l}{\lambda}\right)}$$

$$= V_{S.p.u.} V_{R.p.u.} (SIL) \frac{\sin \delta}{\sin\left(\dfrac{2\pi l}{\lambda}\right)} \quad W \tag{5.4.29}$$

And for $\delta = 90°$, the theoretical steady-state stability limit is

$$P_{max} = \frac{V_{S.p.u.} V_{R.p.u.} (SIL)}{\sin\left(\dfrac{2\pi l}{\lambda}\right)} \quad W \tag{5.4.30}$$

Equations (5.4.27)–(5.4.30) reveal two important factors affecting the steady-state stability limit. First, from (5.4.27), it increases with the square of the line voltage. For example, a doubling of line voltage enables a fourfold increase in maximum power flow. Second, it decreases with line length. Equation (5.4.30) is plotted in Figure 5.12 for $V_{S.p.u.} = V_{R.p.u.} = 1$, $\lambda = 5000$ km, and line lengths up to 1100 km. As shown, the theoretical steady-state stability limit decreases from 4(SIL) for a 200-km line to about 2(SIL) for a 400-km line.

FIGURE 5.12

Transmission-line loadability curve for 60-Hz overhead lines—no series or shunt compensation

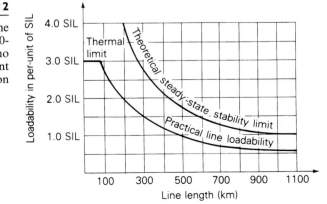

EXAMPLE 5.4 Theoretical steady-state stability limit: long line

Neglecting line losses, find the theoretical steady-state stability limit for the 300-km line in Example 5.2. Assume a 266.1-Ω surge impedance, a 5000-km wavelength, and $V_S = V_R = 765$ kV.

SOLUTION From (5.4.21),

$$SIL = \frac{(765)^2}{266.1} = 2199 \quad MW$$

From (5.4.30) with $l = 300$ km and $\lambda = 5000$ km,

$$P_{max} = \frac{(1)(1)(2199)}{\sin\left(\dfrac{2\pi \times 300}{5000}\right)} = (2.716)(2199) = 5974 \quad MW$$

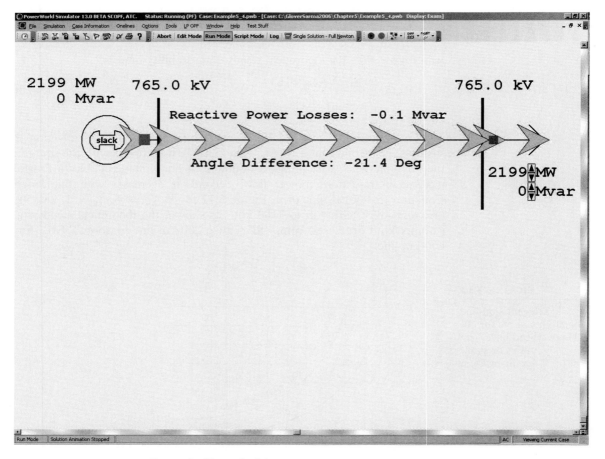

Screen for Example 5.4

Alternatively, from Figure 5.12, for a 300-km line, the theoretical steady-state stability limit is $(2.72)\text{SIL} = (2.72)(2199) = 5980$ MW, about the same as the above result.

Open PowerWorld Simulator case Example 5_4 and select **Simulation, Solve and Animate** to see an animated view of this example. When the load on a line is equal to the SIL, the voltage profile across the line is flat and the line's net reactive power losses are zero. For loads above the SIL, the line consumes reactive power and the load's voltage magnitude is below the sending-end value. Conversely, for loads below the SIL, the line actually generates reactive power and the load's voltage magnitude is above the sending-end value. Use the load arrow button to vary the load to see the changes in the receiving-end voltage and the line's reactive power consumption. ∎

5.5

MAXIMUM POWER FLOW

Maximum power flow, discussed in Section 5.4 for lossless lines, is derived here in terms of the *ABCD* parameters for lossy lines. The following notation is used:

$$A = \cosh(\gamma l) = A \underline{/\theta_A}$$

$$B = Z' = Z' \underline{/\theta_Z}$$

$$V_S = V_S \underline{/\delta} \qquad V_R = V_R \underline{/0°}$$

Solving (5.2.33) for the receiving-end current,

$$I_R = \frac{V_S - AV_R}{B} = \frac{V_S e^{j\delta} - AV_R e^{j\theta_A}}{Z' e^{j\theta_Z}} \tag{5.5.1}$$

The complex power delivered to the receiving end is

$$S_R = P_R + jQ_R = V_R I_R^* = V_R \left[\frac{V_S e^{j(\delta - \theta_Z)} - AV_R e^{j(\theta_A - \theta_Z)}}{Z'} \right]^*$$

$$= \frac{V_R V_S}{Z'} e^{j(\theta_Z - \delta)} - \frac{AV_R^2}{Z'} e^{j(\theta_Z - \theta_A)} \tag{5.5.2}$$

The real and reactive power delivered to the receiving end are thus

$$P_R = \text{Re}(S_R) = \frac{V_R V_S}{Z'} \cos(\theta_Z - \delta) - \frac{AV_R^2}{Z'} \cos(\theta_Z - \theta_A) \tag{5.5.3}$$

$$Q_R = \text{Im}(S_R) = \frac{V_R V_S}{Z'} \sin(\theta_Z - \delta) - \frac{AV_R^2}{Z'} \sin(\theta_Z - \theta_A) \tag{5.5.4}$$

Note that for a lossless line, $\theta_A = 0°$, $B = Z' = jX'$, $Z' = X'$, $\theta_Z = 90°$, and (5.5.3) reduces to

$$P_R = \frac{V_R V_S}{X'} \cos(90 - \delta) - \frac{A V_R^2}{X'} \cos(90°)$$

$$= \frac{V_R V_S}{X'} \sin \delta \tag{5.5.5}$$

which is the same as (5.4.26).

The theoretical maximum real power delivered (or steady-state stability limit) occurs when $\delta = \theta_Z$ in (5.5.3):

$$P_{Rmax} = \frac{V_R V_S}{Z'} - \frac{A V_R^2}{Z'} \cos(\theta_Z - \theta_A) \tag{5.5.6}$$

The second term in (5.5.6), and the fact that Z' is larger than X', reduce P_{Rmax} to a value somewhat less than that given by (5.4.27) for a lossless line.

EXAMPLE 5.5   Theoretical maximum power delivered: long line

Determine the theoretical maximum power, in MW and in per-unit of SIL, that the line in Example 5.2 can deliver. Assume $V_S = V_R = 765$ kV.

SOLUTION   From Example 5.2,

$$A = 0.9313 \quad \text{per unit}; \qquad \theta_A = 0.209°$$
$$B = Z' = 97.0 \quad \Omega; \qquad \theta_Z = 87.2°$$
$$Z_c = 266.1 \quad \Omega$$

From (5.5.6) with $V_S = V_R = 765$ kV,

$$P_{Rmax} = \frac{(765)^2}{97} - \frac{(0.9313)(765)^2}{97} \cos(87.2° - 0.209°)$$

$$= 6033 - 295 = 5738 \quad \text{MW}$$

From (5.4.20),

$$SIL = \frac{(765)^2}{266.1} = 2199 \quad \text{MW}$$

Thus

$$P_{Rmax} = \frac{5738}{2199} = 2.61 \quad \text{per unit}$$

This value is about 4% less than that found in Example 5.4, where losses were neglected. ∎

5.6

LINE LOADABILITY

In practice, power lines are not operated to deliver their theoretical maximum power, which is based on rated terminal voltages and an angular displacement $\delta = 90°$ across the line. Figure 5.12 shows a practical line loadability curve plotted below the theoretical steady-state stability limit. This curve is based on the voltage-drop limit $V_R/V_S \geqslant 0.95$ and on a maximum angular displacement of 30 to 35° across the line (or about 45° across the line and equivalent system reactances), in order to maintain stability during transient disturbances [1, 3]. The curve is valid for typical overhead 60-Hz lines with no compensation. Note that for short lines less than 80 km long, loadability is limited by the thermal rating of the conductors or by terminal equipment ratings, not by voltage drop or stability considerations. In Section 5.7, we investigate series and shunt compensation techniques to increase the loadability of longer lines toward their thermal limit.

EXAMPLE 5.6 **Practical line loadability and percent voltage regulation: long line**

The 300-km uncompensated line in Example 5.2 has four 1,272,000-cmil 54/3 ACSR conductors per bundle. The sending-end voltage is held constant at 1.0 per-unit of rated line voltage. Determine the following:

  a. The practical line loadability (Assume an approximate receiving-end voltage $V_R = 0.95$ per unit and $\delta = 35°$ maximum angle across the line.)

  b. The full-load current at 0.986 p.f. leading based on the above practical line loadability

  c. The exact receiving-end voltage for the full-load current found in part (b)

  d. Percent voltage regulation for the above full-load current

  e. Thermal limit of the line, based on the approximate current-carrying capacity given in Table A.4

SOLUTION

  a. From (5.5.3), with $V_S = 765$, $V_R = 0.95 \times 765$ kV, and $\delta = 35°$, using the values of Z', θ_Z, A, and θ_A from Example 5.5,

$$P_R = \frac{(765)(0.95 \times 765)}{97.0} \cos(87.2° - 35°)$$

$$- \frac{(0.9313)(0.95 \times 765)^2}{97.0} \cos(87.2° - 0.209°)$$

$$= 3513 - 266 = 3247 \quad \text{MW}$$

$P_R = 3247$ MW is the practical line loadability, provided the thermal and voltage-drop limits are not exceeded. Alternatively, from Figure 5.12 for a 300-km line, the practical line loadability is $(1.49)\text{SIL} = (1.49)(2199) = 3277$ MW, about the same as the above result.

b. For the above loading at 0.986 p.f. leading and at 0.95×765 kV, the full-load receiving-end current is

$$I_{RFL} = \frac{P}{\sqrt{3}V_R(\text{p.f.})} = \frac{3247}{(\sqrt{3})(0.95 \times 765)(0.986)} = 2.616 \quad \text{kA}$$

c. From (5.1.1) with $I_{RFL} = 2.616\underline{/\cos^{-1} 0.986} = 2.616\underline{/9.599°}$ kA, using the A and B parameters from Example 5.2,

$$V_S = AV_{RFL} + BI_{RFL}$$

$$\frac{765}{\sqrt{3}}\underline{/\delta} = (0.9313\underline{/0.209°})(V_{RFL}\underline{/0°}) + (97.0\underline{/87.2°})(2.616\underline{/9.599°})$$

$$441.7\underline{/\delta} = (0.9313V_{RFL} - 30.04) + j(0.0034V_{RFL} + 251.97)$$

Taking the squared magnitude of the above equation,

$$(441.7)^2 = 0.8673V_{RFL}^2 - 54.24V_{RFL} + 64{,}391$$

Solving,

$$V_{RFL} = 420.7 \quad \text{kV}_{LN}$$

$$= 420.7\sqrt{3} = 728.7 \quad \text{kV}_{LL} = 0.953 \quad \text{per unit}$$

d. From (5.1.19), the receiving-end no-load voltage is

$$V_{RNL} = \frac{V_S}{A} = \frac{765}{0.9313} = 821.4 \quad \text{kV}_{LL}$$

And from (5.1.18),

$$\text{percent VR} = \frac{821.4 - 728.7}{728.7} \times 100 = 12.72\%$$

e. From Table A.4, the approximate current-carrying capacity of four 1,272,000-cmil 54/3 ACSR conductors is $4 \times 1.2 = 4.8$ kA.

Since the voltages $V_S = 1.0$ and $V_{RFL} = 0.953$ per unit satisfy the voltage-drop limit $V_R/V_S \geqslant 0.95$, the factor that limits line loadability is steady-state stability for this 300-km uncompensated line. The full-load current of 2.616 kA corresponding to loadability is also well below the thermal limit of 4.8 kA. The 12.7% voltage regulation is too high because the no-load voltage is too high. Compensation techniques to reduce no-load voltages are discussed in Section 5.7. ∎

EXAMPLE 5.7   **Selection of transmission line voltage and number of lines for power transfer**

From a hydroelectric power plant 9000 MW are to be transmitted to a load center located 500 km from the plant. Based on practical line loadability criteria, determine the number of three-phase, 60-Hz lines required to transmit this power, with one line out of service, for the following cases: (a) 345-kV lines with $Z_c = 297$ Ω; (b) 500-kV lines with $Z_c = 277$ Ω; (c) 765-kV lines with $Z_c = 266$ Ω. Assume $V_S = 1.0$ per unit, $V_R = 0.95$ per unit, and $\delta = 35°$. Also assume that the lines are uncompensated and widely separated such that there is negligible mutual coupling between them.

SOLUTION

a. For 345-kV lines, (5.4.21) yields

$$SIL = \frac{(345)^2}{297} = 401 \quad MW$$

Neglecting losses, from (5.4.29), with $l = 500$ km and $\delta = 35°$,

$$P = \frac{(1.0)(0.95)(401)\ \sin(35°)}{\sin\left(\dfrac{2\pi \times 500}{5000}\right)} = (401)(0.927) = 372 \quad MW/line$$

Alternatively, the practical line loadability curve in Figure 5.12 can be used to obtain $P = (0.93)SIL$ for typical 500-km overhead 60-Hz uncompensated lines.

In order to transmit 9000 MW with one line out of service,

$$\#345\text{-kV lines} = \frac{9000\ MW}{372\ MW/line} + 1 = 24.2 + 1 \approx 26$$

b. For 500-kV lines,

$$SIL = \frac{(500)^2}{277} = 903 \quad MW$$

$$P = (903)(0.927) = 837 \quad MW/line$$

$$\#500\text{-kV lines} = \frac{9000}{837} + 1 = 10.8 + 1 \approx 12$$

c. For 765-kV lines,

$$SIL = \frac{(765)^2}{266} = 2200 \quad MW$$

$$P = (2200)(0.927) = 2039 \quad MW/line$$

$$\#765\text{-kV lines} = \frac{9000}{2039} + 1 = 4.4 + 1 \approx 6$$

Increasing the line voltage from 345 to 765 kV, a factor of 2.2, reduces the required number of lines from 26 to 6, a factor of 4.3. ∎

EXAMPLE 5.8 **Effect of intermediate substations on number of lines required for power transfer**

Can five instead of six 765-kV lines transmit the required power in Example 5.7 if there are two intermediate substations that divide each line into three 167-km line sections, and if only one line section is out of service?

SOLUTION The lines are shown in Figure 5.13. For simplicity, we neglect line losses. The equivalent π circuit of one 500-km, 765-kV line has a series reactance, from (5.4.10) and (5.4.15),

$$X' = (266) \sin\left(\frac{2\pi \times 500}{5000}\right) = 156.35 \quad \Omega$$

Combining series/parallel reactances in Figure 5.13, the equivalent reactance of five lines with one line section out of service is

$$X_{eq} = \frac{1}{5}\left(\frac{2}{3}X'\right) + \frac{1}{4}\left(\frac{X'}{3}\right) = 0.2167X' = 33.88 \quad \Omega$$

Then, from (5.4.26) with $\delta = 35°$,

$$P = \frac{(765)(765 \times 0.95) \sin(35°)}{33.88} = 9412 \quad MW$$

Inclusion of line losses would reduce the above value by 3 or 4% to about 9100 MW. Therefore, the answer is yes. Five 765-kV, 500-km uncompensated lines with two intermediate substations and with one line section out of service will transmit 9000 MW. Intermediate substations are often economical if their costs do not outweigh the reduction in line costs.

This example is modeled in PowerWorld Simulator case Example 5_8. Each line segment is represented with the lossless line model from Example 5.4 with the π circuit parameters modified to exactly match those for a 167 km distributed line. The pie charts on each line segment show the percentage loading of the line, assuming a rating of 3500 MVA. The solid red squares on the lines represent closed circuit breakers, and the green squares correspond

FIGURE 5.13

Transmission-line configuration for Example 5.8

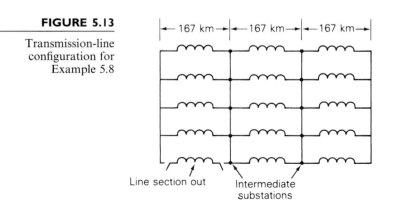

to open circuit breakers. Clicking on a circuit breaker toggles its status. The simulation results differ slightly from the simplified analysis done earlier in the example because the simulation includes the charging capacitance of the transmission lines. With all line segments in-service, use the load's arrow to verify that the SIL for this system is 11,000 MW, five times that of the single circuit line in Example 5.4.

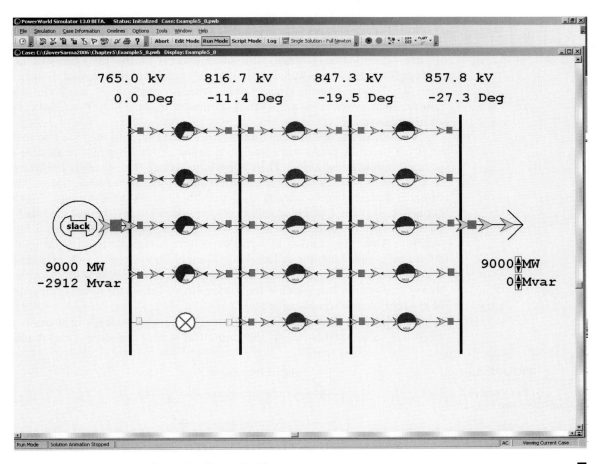

Screen for Example 5.8   ∎

5.7

REACTIVE COMPENSATION TECHNIQUES

Inductors and capacitors are used on medium-length and long transmission lines to increase line loadability and to maintain voltages near rated values.

Shunt reactors (inductors) are commonly installed at selected points along EHV lines from each phase to neutral. The inductors absorb reactive power and reduce overvoltages during light load conditions. They also reduce transient overvoltages due to switching and lightning surges. However, shunt reactors can reduce line loadability if they are not removed under full-load conditions.

In addition to shunt reactors, shunt capacitors are sometimes used to deliver reactive power and increase transmission voltages during heavy load conditions. Another type of shunt compensation includes thyristor-switched reactors in parallel with capacitors. These devices, called *static var compensators*, can absorb reactive power during light loads and deliver reactive power during heavy loads. Through automatic control of the thyristor switches, voltage fluctuations are minimized and line loadability is increased. Synchronous condensors (synchronous motors with no mechanical load) can also control their reactive power output, although more slowly than static var compensators.

Series capacitors are sometimes used on long lines to increase line loadability. Capacitor banks are installed in series with each phase conductor at selected points along a line. Their effect is to reduce the net series impedance of the line in series with the capacitor banks, thereby reducing line-voltage drops and increasing the steady-state stability limit. A disadvantage of series capacitor banks is that automatic protection devices must be installed to by-pass high currents during faults and to reinsert the capacitor banks after fault clearing. Also, the addition of series capacitors can excite low-frequency oscillations, a phenomenon called *subsynchronous resonance*, which may damage turbine-generator shafts. Studies have shown, however, that series capacitive compensation can increase the loadability of long lines at only a fraction of the cost of new transmission [1].

Figure 5.14 shows a schematic and an equivalent circuit for a compensated line section, where N_C is the amount of series capacitive compensation

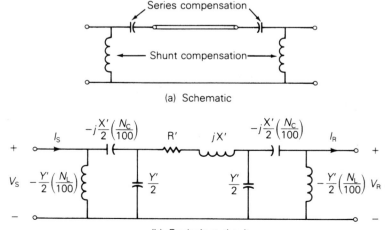

FIGURE 5.14

Compensated transmission-line section

(a) Schematic

(b) Equivalent circuit

expressed in percent of the positive-sequence line impedance and N_L is the amount of shunt reactive compensation in percent of the positive-sequence line admittance. It is assumed in Figure 5.14 that half of the compensation is installed at each end of the line section. The following two examples illustrate the effect of compensation.

EXAMPLE 5.9 Shunt reactive compensation to improve transmission-line voltage regulation

Identical shunt reactors (inductors) are connected from each phase conductor to neutral at both ends of the 300-km line in Example 5.2 during light load conditions, providing 75% compensation. The reactors are removed during heavy load conditions. Full load is 1.90 kA at unity p.f. and at 730 kV. Assuming that the sending-end voltage is constant, determine the following:

a. Percent voltage regulation of the uncompensated line

b. The equivalent shunt admittance and series impedance of the compensated line

c. Percent voltage regulation of the compensated line

SOLUTION

a. From (5.1.1) with $I_{RFL} = 1.9\underline{/0°}$ kA, using the A and B parameters from Example 5.2,

$$V_S = AV_{RFL} + BI_{RFL}$$

$$= (0.9313\underline{/0.209°})\left(\frac{730}{\sqrt{3}}\underline{/0°}\right) + (97.0\underline{/87.2°})(1.9\underline{/0°})$$

$$= 392.5\underline{/0.209°} + 184.3\underline{/87.2°}$$

$$= 401.5 + j185.5$$

$$= 442.3\underline{/24.8°} \quad kV_{LN}$$

$$V_S = 442.3\sqrt{3} = 766.0 \quad kV_{LL}$$

The no-load receiving-end voltage is, from (5.1.19),

$$V_{RNL} = \frac{766.0}{0.9313} = 822.6 \quad kV_{LL}$$

and the percent voltage regulation for the uncompensated line is, from (5.1.18),

$$\text{percent VR} = \frac{822.6 - 730}{730} \times 100 = 12.68\%$$

b. From Example 5.3, the shunt admittance of the equivalent π circuit without compensation is

$$Y' = 2(3.7 \times 10^{-7} + j7.094 \times 10^{-4})$$

$$= 7.4 \times 10^{-7} + j14.188 \times 10^{-4} \quad \text{S}$$

With 75% shunt compensation, the equivalent shunt admittance is

$$Y_{eq} = 7.4 \times 10^{-7} + j14.188 \times 10^{-4}\left(1 - \tfrac{75}{100}\right)$$

$$= 3.547 \times 10^{-4}\underline{/89.88^\circ} \quad \text{S}$$

Since there is no series compensation, the equivalent series impedance is the same as without compensation:

$$Z_{eq} = Z' = 97.0\underline{/87.2^\circ} \quad \Omega$$

c. The equivalent A parameter for the compensated line is

$$A_{eq} = 1 + \frac{Y_{eq}Z_{eq}}{2}$$

$$= 1 + \frac{(3.547 \times 10^{-4}\underline{/89.88^\circ})(97.0\underline{/87.2^\circ})}{2}$$

$$= 1 + 0.0172\underline{/177.1^\circ}$$

$$= 0.9828\underline{/0.05^\circ} \quad \text{per unit}$$

Then, from (5.1.19),

$$V_{RNL} = \frac{766}{0.9828} = 779.4 \quad \text{kV}_{LL}$$

Since the shunt reactors are removed during heavy load conditions, $V_{RFL} = 730$ kV is the same as without compensation. Therefore

$$\text{percent VR} = \frac{779.4 - 730}{730} \times 100 = 6.77\%$$

The use of shunt reactors at light loads improves the voltage regulation from 12.68% to 6.77% for this line. ∎

EXAMPLE 5.10   Series capacitive compensation to increase transmission-line loadability

Identical series capacitors are installed in each phase at both ends of the line in Example 5.2, providing 30% compensation. Determine the theoretical maximum power that this compensated line can deliver and compare with that of the uncompensated line. Assume $V_S = V_R = 765$ kV.

SOLUTION   From Example 5.3, the equivalent series reactance without compensation is

$$X' = 97.0 \sin 87.2^\circ = 96.88 \quad \Omega$$

Based on 30% series compensation, half at each end of the line, the impedance of each series capacitor is

$$Z_{cap} = -jX_{cap} = -j(\tfrac{1}{2})(0.30)(96.88) = -j14.53 \quad \Omega$$

From Figure 5.4, the $ABCD$ matrix of this series impedance is

$$\begin{bmatrix} 1 & -j14.53 \\ 0 & 1 \end{bmatrix}$$

As also shown in Figure 5.4, the equivalent $ABCD$ matrix of networks in series is obtained by multiplying the $ABCD$ matrices of the individual networks. For this example there are three networks: the series capacitors at the sending end, the line, and the series capacitors at the receiving end. Therefore the equivalent $ABCD$ matrix of the compensated line is, using the $ABCD$ parameters, from Example 5.2,

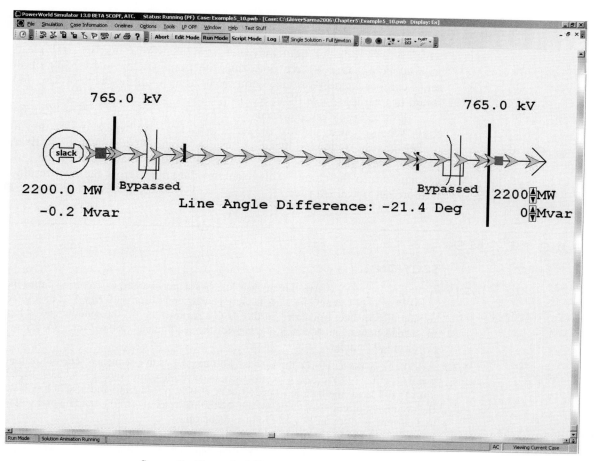

Screen for Example 5.10

$$\begin{bmatrix} 1 & -j14.53 \\ \hline 0 & 1 \end{bmatrix} \begin{bmatrix} 0.9313\underline{/0.209°} & 97.0\underline{/87.2°} \\ \hline 1.37 \times 10^{-3}\underline{/90.06°} & 0.9313\underline{/0.209°} \end{bmatrix} \begin{bmatrix} 1 & -j14.53 \\ \hline 0 & 1 \end{bmatrix}$$

After performing these matrix multiplications, we obtain

$$\begin{bmatrix} A_{eq} & B_{eq} \\ \hline C_{eq} & D_{eq} \end{bmatrix} = \begin{bmatrix} 0.9512\underline{/0.205°} & 69.70\underline{/86.02°} \\ \hline 1.37 \times 10^{-3}\underline{/90.06°} & 0.9512\underline{/0.205°} \end{bmatrix}$$

Therefore

$$A_{eq} = 0.9512 \quad \text{per unit} \qquad \theta_{Aeq} = 0.205°$$
$$B_{eq} = Z'_{eq} = 69.70 \quad \Omega \qquad \theta_{Zeq} = 86.02°$$

From (5.5.6) with $V_S = V_R = 765$ kV,

$$P_{Rmax} = \frac{(765)^2}{69.70} - \frac{(0.9512)(765)^2}{69.70} \cos(86.02° - 0.205°)$$
$$= 8396 - 583 = 7813 \quad \text{MW}$$

which is 36.2% larger than the value of 5738 MW found in Example 5.5 without compensation. We note that the practical line loadability of this series compensated line is also about 35% larger than the value of 3247 MW found in Example 5.6 without compensation.

This example is modeled in PowerWorld Simulator case Example 5_10. When opened, both of the series capacitors are bypassed (i.e., they are modeled as short circuits) meaning this case is initially identical to the Example 5.4 case. Click on the blue "Bypassed" field to place each of the series capacitors into the circuit. This decreases the angle across the line, resulting in more net power transfer. ∎

PROBLEMS

SECTION 5.1

5.1 A 30-km, 34.5-kV, 60-Hz three-phase line has a positive-sequence series impedance $z = 0.19 + j0.34$ Ω/km. The load at the receiving end absorbs 10 MVA at 33 kV. Assuming a short line, calculate: (a) the *ABCD* parameters, (b) the sending-end voltage for a load power factor of 0.9 lagging, (c) the sending-end voltage for a load power factor of 0.9 leading.

5.2 A 150-km, 230-kV, 60-Hz three-phase line has a positive-sequence series impedance $z = 0.08 + j0.48$ Ω/km and a positive-sequence shunt admittance $y = j3.33 \times 10^{-6}$ S/km. At full load, the line delivers 250 MW at 0.99 p.f. lagging and at 220 kV. Using the nominal π circuit, calculate: (a) the *ABCD* parameters, (b) the sending-end voltage and current, and (c) the percent voltage regulation.

5.3 Rework Problem 5.2 in per-unit using 100-MVA (three-phase) and 230-kV (line-to-line) base values. Calculate: (a) the per-unit *ABCD* parameters, (b) the per-unit sending-end voltage and current, and (c) the percent voltage regulation.

5.4 Derive the *ABCD* parameters for the two networks in series, as shown in Figure 5.4.

5.5 Derive the *ABCD* parameters for the T circuit shown in Figure 5.4.

5.6 (a) Consider a medium-length transmission line represented by a nominal π circuit shown in Figure 5.3 of the text. Draw a phasor diagram for lagging power-factor condition at the load (receiving end).

(b) Now consider a nominal *T*-circuit of the medium-length transmission line shown in Figure 5.15.

  (i) Draw the corresponding phasor diagram for lagging power-factor load condition

  (ii) Determine the *ABCD* parameters in terms of *Y* and *Z*, for the nominal *T*-circuit and for the nominal π-circuit of part (a).

FIGURE 5.15

Nominal T-circuit for Problem 5.6

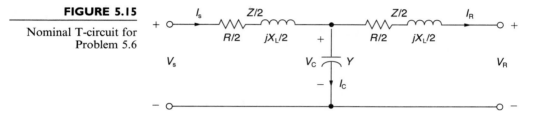

5.7 The per-phase impedance of a short three—phase transmission line is $0.5\underline{/53.15^\circ}\,\Omega$. The three-phase load at the receiving end is 900 kW at 0.8 PF lagging. If the line-to-line sending-end voltage is 3.3 kV, determine (a) the receiving-end line-to-line voltage in kV, and (b) the line current.
Draw the phasor diagram with the line current *I*, as reference.

5.8 Reconsider Problem 5.7 and find the following: (a) sending-end power factor, (b) sending-end three-phase power, and (c) the three-phase line loss.

5.9 The 100-km, 230-kV, 60-Hz three-phase line in Problems 4.18 and 4.39 delivers 300 MVA at 218 kV to the receiving end at full load. Using the nominal π circuit, calculate the: (a) *ABCD* parameters, sending-end voltage, and percent voltage regulation when the receiving-end power factor is (b) 0.9 lagging, (c) unity, and (d) 0.9 leading. Assume a $50\,^\circ$C conductor temperature to determine the resistance of this line.

5.10 The 500-kV, 60-Hz three-phase line in Problems 4.20 and 4.41 has a 180-km length and delivers 1600 MW at 475 kV and at 0.95 power factor leading to the receiving end at full load. Using the nominal π circuit, calculate the: (a) *ABCD* parameters, (b) sending-end voltage and current, (c) sending-end power and power factor, (d) full-load line losses and efficiency, and (e) percent voltage regulation. Assume a $50\,^\circ$C conductor temperature to determine the resistance of this line.

5.11 A 40-km, 220-kV, 60-Hz three-phase overhead transmission line has a per-phase resistance of $0.15\,\Omega$/km, a per-phase inductance of 1.3263 mH/km, and negligible shunt capacitance. Using the short line model, find the sending-end voltage, voltage regulation, sending-end power, and transmission line efficiency when the line is supplying a three-phase load of: (a) 381 MVA at 0.8 power factor lagging and at 220 kV, (b) 381 MVA at 0.8 power factor leading and at 220 kV.

5.12   A 60-Hz, 100-mile, three-phase overhead transmission line, constructed of ACSR conductors, has a series impedance of $(0.1826 + j0.784)$ Ω/mi per phase and a shunt capacitive reactance-to-neutral of $185.5 \times 10^3 \underline{/-90°}$ Ω-mi per phase. Using the nominal π circuit for a medium-length transmission line, (a) determine the total series impedance and shunt admittance of the line. (b) Compute the voltage, the current, and the real and reactive power at the sending end if the load at the receiving end draws 200 MVA at unity power factor and at a line-to-line voltage of 230 kV. (c) Find the percent voltage regulation of the line.

SECTION 5.2

5.13   Evaluate $\cosh(\gamma l)$ and $\tanh(\gamma l/2)$ for $\gamma l = 0.45 \underline{/87°}$ per unit.

5.14   A 500-km, 500-kV, 60-Hz uncompensated three-phase line has a positive-sequence series impedance $z = 0.03 + j0.35$ Ω/km and a positive-sequence shunt admittance $y = j4.4 \times 10^{-6}$ S/km. Calculate: (a) Z_c, (b) (γl), and (c) the exact $ABCD$ parameters for this line.

5.15   At full load the line in Problem 5.14 delivers 1000 MW at unity power factor and at 480 kV. Calculate: (a) the sending-end voltage, (b) the sending-end current, (c) the sending-end power factor, (d) the full-load line losses, and (e) the percent voltage regulation.

5.16   The 500-kV, 60-Hz three-phase line in Problems 4.20 and 4.41 has a 300-km length. Calculate: (a) Z_c, (b) (γl), and (c) the exact $ABCD$ parameters for this line. Assume a 50 °C conductor temperature.

5.17   At full load, the line in Problem 5.16 delivers 1500 MVA at 480 kV to the receiving-end load. Calculate the sending-end voltage and percent voltage regulation when the receiving-end power factor is (a) 0.9 lagging, (b) unity, and (c) 0.9 leading.

5.18   A 60-Hz, 230-mile, three-phase overhead transmission line has a series impedance $z = 0.8431 \underline{/79.04°}$ Ω/mi and a shunt admittance $y = 5.105 \times 10^{-6} \underline{/90°}$ S/mi. The load at the receiving end is 125 MW at unity power factor and at 215 kV. Determine the voltage, current, real and reactive power at the sending end and the percent voltage regulation of the line. Also find the wavelength and velocity of propagation of the line.

5.19   Using per-unit calculations, rework Problem 5.18 to determine the sending-end voltage and current.

5.20   (a) The series expansions of the hyperbolic functions are given by

$$\cosh\theta = 1 + \frac{\theta^2}{2} + \frac{\theta^4}{24} + \frac{\theta^6}{720} + \cdots.$$

$$\sinh\theta = 1 + \frac{\theta^2}{6} + \frac{\theta^4}{120} + \frac{\theta^6}{5040} + \cdots.$$

For the $ABCD$ parameters of a long transmission line represented by an equivalent π circuit, apply the above expansion and consider only the first two terms, and express the result in terms of Y and Z.

(b) For the nominal π and equivalent π circuits shown in Figures 5.3 and 5.7 of the text, show that

$$\frac{A-1}{B} = \frac{Y}{2} \quad \text{and} \quad \frac{A-1}{B} = \frac{Y'}{2}$$

hold good, respectively.

5.21 Starting with (5.1.1) of the text, show that

$$A = \frac{V_S I_S + V_R I_R}{V_R I_S + V_S I_R} \quad \text{and} \quad B = \frac{V_S^2 - V_R^2}{V_R I_S + V_S I_R}$$

5.22 Consider the A parameter of the long line given by $\cosh\theta$, where $\theta = \sqrt{ZY}$. With $x = e^{-\theta} = x_1 + jx_2$, and $A = A_1 + jA_2$, show that x_1 and x_2 satisfy the following:

$$x_1^2 - x_2^2 - 2(A_1 x_1 - A_2 x_2) + 1 = 0$$

and $x_1 x_2 - (A_2 x_1 + A_1 x_2) = 0$.

SECTION 5.3

5.23 Determine the equivalent π circuit for the line in Problem 5.14 and compare it with the nominal π circuit.

5.24 Determine the equivalent π circuit for the line in Problem 5.16. Compare the equivalent π circuit with the nominal π circuit.

5.25 Let the transmission line of Problem 5.12 be extended to cover a distance of 200 miles. Assume conditions at the load to be the same as in Problem 5.12. Determine the: (a) sending-end voltage, (b) sending-end current, (c) sending-end real and reactive powers, and (d) percent voltage regulation.

SECTION 5.4

5.26 A 320-km 500-kV, 60-Hz three-phase uncompensated line has a positive-sequence series reactance $x = 0.34$ Ω/km and a positive-sequence shunt admittance $y = j4.5 \times 10^{-6}$ S/km. Neglecting losses, calculate: (a) Z_c, (b) (γl), (c) the $ABCD$ parameters, (d) the wavelength λ of the line, in kilometers, and (e) the surge impedance loading in MW.

5.27 Determine the equivalent π circuit for the line in Problem 5.26.

5.28 Rated line voltage is applied to the sending end of the line in Problem 5.26. Calculate the receiving-end voltage when the receiving end is terminated by (a) an open circuit, (b) the surge impedance of the line, and (c) one-half of the surge impedance. (d) Also calculate the theoretical maximum real power that the line can deliver when rated voltage is applied to both ends of the line.

5.29 Rework Problems 5.9 and 5.16 neglecting the conductor resistance. Compare the results with and without losses.

5.30 From (4.6.22) and (4.10.4), the series inductance and shunt capacitance of a three-phase overhead line are

$$L_a = 2 \times 10^{-7} \ln(D_{eq}/D_{SL}) = \frac{\mu_0}{2\pi} \ln(D_{eq}/D_{SL}) \quad \text{H/m}$$

$$C_{an} = \frac{2\pi\varepsilon_0}{\ln(D_{eq}/D_{SC})} \quad \text{F/m}$$

where $\mu_0 = 4\pi \times 10^{-7}$ H/m and $\varepsilon_0 = \left(\frac{1}{36\pi}\right) \times 10^{-9}$ F/m

Using these equations, determine formulas for surge impedance and velocity of propagation of an overhead lossless line. Then determine the surge impedance and velocity of propagation for the three-phase line given in Example 4.5. Assume positive-sequence operation. Neglect line losses as well as the effects of the overhead neutral wires and the earth plane.

5.31 A 500-kV, 300-km, 60-Hz three-phase overhead transmission line, assumed to be loss-less, has a series inductance of 0.97 mH/km per phase and a shunt capacitance of 0.115 μF/km per phase. (a) Determine the phase constant β, the surge impedance Z_C, velocity of propagation v, and the wavelength λ of the line. (b) Determine the voltage, current, real and reactive power at the sending end, and the percent voltage regulation of the line if the receiving-end load is 800 MW at 0.8 power factor lagging and at 500 kV.

5.32 The following parameters are based on a preliminary line design: $V_S = 1.0$ per unit, $V_R = 0.9$ per unit, $\lambda = 5000$ km, $Z_C = 320$ Ω, $\delta = 36.8°$. A three-phase power of 700 MW is to be transmitted to a substation located 315 km from the source of power. (a) Determine a nominal voltage level for the three-phase transmission line, based on the practical line-loadability equation. (b) For the voltage level obtained in (a), determine the theoretical maximum power that can be transferred by the line.

5.33 Consider a long radial line terminated in its characteristic impedance Z_C. Determine the following:

(a) V_1/I_1, known as the driving point impedance.

(b) $|V_2|/|V_1|$, known as the voltage gain, in terms of $\alpha\ell$.

(c) $|I_2|/|I_1|$, known as the current gain, in terms of $\alpha\ell$.

(d) The complex power gain, $-S_{21}/S_{12}$, in terms of $\alpha\ell$.

(e) The real power efficiency, $(-P_{21}/P_{12}) = \eta$, in terms of $\alpha\ell$.

[Note: 1 refers to sending end and 2 refers to receiving end. (S_{21}) is the complex power received at 2; S_{12} is sent from 1.]

5.34 For the case of a lossless line, how would the results of Problem 5.33 change? In terms of Z_C, which will be a real quantity for this case, Express P_{12} in terms $|I_1|$ and $|V_1|$.

5.35 For a lossless open-circuited line, express the sending-end voltage, V_1, in terms of the receiving-end voltage, V_2, for the three cases of short-line model, medium-length line model, and long-line model. Is it true that the voltage at the open receiving end of a long line is higher than that at the sending end, for small $\beta\ell$.

5.36 For a short transmission line of impedance $(R + jX)$ ohms per phase, show that the maximum power that can be transmitted over the line is

$$P_{max} = \frac{V_R^2}{Z^2}\left(\frac{ZV_S}{V_R} - R\right), \quad \text{where } Z = \sqrt{R^2 + X^2},$$

when the sending-end and receiving-end voltages are fixed, and for the condition

$$Q = \frac{-V_R^2 X}{R^2 + X^2} \quad \text{when } dP/dQ = 0$$

5.37 (a) Consider complex power transmission via the three-phase short line for which the per-phase circuit is shown in Figure 5.16. Express S_{12}, the complex power sent by bus 1 (or V_1), and $(-S_{21})$, the complex power received by bus 2 (or V_2), in terms of V_1, V_2, Z, $\underline{/Z}$, and $\theta_{12} = \theta_1 - \theta_2$, the power angle.

(b) For a balanced three-phase transmission line, in per-unit notation, with $Z = 1\underline{/85°}$, $\theta_{12} = 10°$, determine S_{12} and $(-S_{21})$ for

(i) $V_1 = V_2 = 1.0$

(ii) $V_1 = 1.1$ and $V_2 = 0.9$

Comment on the changes of Real and Reactive powers from (i) to (ii).

FIGURE 5.16

Per-phase circuit for
Problem 5.37

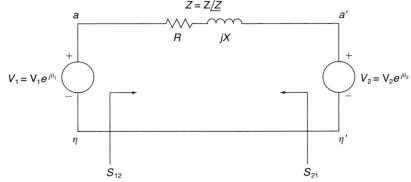

SECTION 5.5

5.38   The line in Problem 5.14 has three ACSR 1113-kcmil conductors per phase. Calculate the theoretical maximum real power that this line can deliver and compare with the thermal limit of the line. Assume $V_S = V_R = 1.0$ per unit and unity power factor at the receiving end.

5.39   Repeat Problems 5.14 and 5.38 if the line length is (a) 200 km, (b) 550 km.

5.40   For the 500-kV line given in Problem 5.16, (a) calculate the theoretical maximum real power that the line can deliver to the receiving end when rated voltage is applied to both ends. (b) Calculate the receiving-end reactive power and power factor at this theoretical loading.

5.41   A 230-kV, 100-km, 60-Hz three-phase overhead transmission line with a rated current of 900 A/phase has a series impedance $z = 0.088 + j0.465$ Ω/km and a shunt admittance $y = j3.524$ μS/km. (a) Obtain the nominal π equivalent circuit in normal units and in per unit on a base of 100 MVA (three phase) and 230 kV (line-to-line). (b) Determine the three-phase rated MVA of the line. (c) Compute the ABCD parameters. (d) Calculate the SIL.

5.42   A three-phase power of 460 MW is to the transmitted to a substation located 500 km from the source of power. With $V_S = 1$ per unit, $V_R = 0.9$ per unit, $\lambda = 5000$ km, $Z_C = 500$ Ω, and $\delta = 36.87°$, determine a nominal voltage level for the lossless transmission line, based on Eq. (5.4.29) of the text.

Using this result, find the theoretical three-phase maximum power that can be transferred by the lossless transmission line.

PW   **5.43**   Open Power World Simulator case Example 5_4 and graph the load bus voltage as a function of load real power (assuming unity power factor at the load). What is the maximum amount of real power that can be transferred to the load at unity power factor if we require the load voltage always be greater than 0.9 per unit?

PW **5.44** Repeat Problem 5.43, but now vary the load reactive power, assuming the load real power is fixed at 1000 MW.

SECTION 5.6

5.45 For the line in Problems 5.14 and 5.38, determine: (a) the practical line loadability in MW, assuming $V_S = 1.0$ per unit, $V_R \approx 0.95$ per unit, and $\delta_{max} = 35°$; (b) the full-load current at 0.99 p.f. leading, based on the above practical line loadability; (c) the exact receiving-end voltage for the full-load current in (b) above; and (d) the percent voltage regulation. For this line, is loadability determined by the thermal limit, the voltage-drop limit, or steady-state stability?

5.46 Repeat Problem 5.45 for the 500-kV line given in Problem 5.10.

5.47 Determine the practical line loadability in MW and in per-unit of SIL for the line in Problem 5.14 if the line length is (a) 200 km, (b) 600 km. Assume $V_S = 1.0$ per unit, $V_R = 0.95$ per unit, $\delta_{max} = 35°$, and 0.99 leading power factor at the receiving end.

5.48 It is desired to transmit 2200 MW from a power plant to a load center located 300 km from the plant. Determine the number of 60-Hz three-phase, uncompensated transmission lines required to transmit this power with one line out of service for the following cases: (a) 345-kV lines, $Z_c = 300\ \Omega$, (b) 500-kV lines, $Z_c = 275\ \Omega$, (c) 765-kV lines, $Z_c = 260\ \Omega$. Assume that $V_S = 1.0$ per unit, $V_R = 0.95$ per unit, and $\delta_{max} = 35°$.

5.49 Repeat Problem 5.48 if it is desired to transmit: (a) 3200 MW to a load center located 300 km from the plant, (b) 2000 MW to a load center located 400 km from the plant.

5.50 A three-phase power of 3600 MW is to be transmitted through four identical 60-Hz overhead transmission lines over a distance of 300 km. Based on a preliminary design, the phase constant and surge impedance of the line are $\beta = 9.46 \times 10^{-4}$ rad/km and $Z_C = 343\ \Omega$, respectively. Assuming $V_S = 1.0$ per unit, $V_R = 0.9$ per unit, and a power angle $\delta = 36.87°$, determine a suitable nominal voltage level in kV, based on the practical line-loadability criteria.

5.51 The power flow at any point on a transmission line can be calculated in terms of the $ABCD$ parameters. By letting $A = |A|\underline{/\alpha}$, $B = |B|\underline{/\beta}$, $V_R = |V_R|\underline{/0°}$, and $V_S = |V_S|\underline{/\delta}$, the complex power at the receiving end can be shown to be

$$P_R + jQ_R = \frac{|V_R||V_S|\underline{/\beta - \alpha}}{|B|} - \frac{|A||V_R^2|\underline{/\beta - \alpha}}{|B|}$$

(a) Draw a phasor diagram corresponding to the above equation. Let it be represented by a triangle $O'OA$ with O' as the origin and OA representing $P_R + jQ_R$.

(b) By shifting the origin from O' to O, turn the result of (a) into a power diagram, redrawing the phasor diagram. For a given fixed value of $|V_R|$ and a set of values for $|V_S|$, draw the loci of point A, thereby showing the so-called receiving-end circles.

(c) From the result of (b) for a given load with a lagging power factor angle θ_R, determine the amount of reactive power that must be supplied to the receiving end to maintain a constant receiving-end voltage, if the sending-end voltage magnitude decreases from $|V_{S1}|$ to $|V_{S2}|$.

5.52 (a) Consider complex power transmission via the three-phase long line for which the per-phase circuit is shown in Figure 5.17. See Problem 5.37 in which the short-line case was considered. Show that

$$\text{sending-end power} = S_{12} = \frac{Y'^*}{2}V_1^2 + \frac{V_1^2}{Z'^*} - \frac{V_1 V_2}{Z'^*}e^{j\theta_{12}}$$

$$\text{and received power} = -S_{21} = -\frac{Y'^*}{2}V_2^2 - \frac{V_2^2}{Z'^*} + \frac{V_1 V_2}{Z'^*}e^{-j\theta_{12}}$$

where $\theta_{12} = \theta_1 - \theta_2$.

(b) For a lossless line with equal voltage magnitudes at each end, show that

$$P_{12} = -P_{21} = \frac{V_1^2 \sin\theta_{12}}{Z_C \sin\beta\ell} = P_{SIL}\frac{\sin\theta_{12}}{\sin\beta\ell}$$

(c) For $\theta_{12} = 45°$, and $\beta = 0.002$ rad/km, find (P_{12}/P_{SIL}) as a function of line length in km, and sketch it.

(d) If a thermal limit of $(P_{12}/P_{SIL}) = 2$ is set, which limit governs for short lines and long lines?

FIGURE 5.17

Per-phase circuit for
Problem 5.52

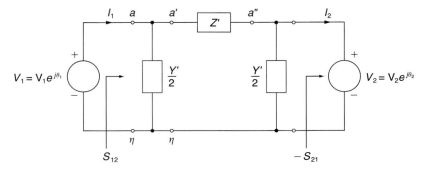

5.53 Open PowerWorld Simulator case Example 5_8. If we require the load bus voltage to be greater than or equal to 730 kV even with any line segment out of service, what is the maximum amount of real power that can be delivered to the load?

5.54 Repeat Problem 5.53, but now assume any two line segments may be out of service.

SECTION 5.7

5.55 Recalculate the percent voltage regulation in Problem 5.15 when identical shunt reactors are installed at both ends of the line during light loads, providing 70% total shunt compensation. The reactors are removed at full load. Also calculate the impedance of each shunt reactor.

5.56 Rework Problem 5.17 when identical shunt reactors are installed at both ends of the line, providing 50% total shunt compensation. The reactors are removed at full load.

5.57 Identical series capacitors are installed at both ends of the line in Problem 5.14, providing 40% total series compensation. Determine the equivalent $ABCD$ parameters of this compensated line. Also calculate the impedance of each series capacitor.

5.58   Identical series capacitors are installed at both ends of the line in Problem 5.16, providing 30% total series compensation. (a) Determine the equivalent *ABCD* parameters for this compensated line. (b) Determine the theoretical maximum real power that this series-compensated line can deliver when $V_S = V_R = 1.0$ per unit. Compare your result with that of Problem 5.40.

5.59   Determine the theoretical maximum real power that the series-compensated line in Problem 5.57 can deliver when $V_S = V_R = 1.0$ per unit. Compare your result with that of Problem 5.38.

5.60   What is the minimum amount of series capacitive compensation N_C in percent of the positive-sequence line reactance needed to reduce the number of 765-kV lines in Example 5.8 from five to four. Assume two intermediate substations with one line section out of service. Also, neglect line losses and assume that the series compensation is sufficiently distributed along the line so as to effectively reduce the series reactance of the equivalent π circuit to $X'(1 - N_C/100)$.

5.61   Determine the equivalent *ABCD* parameters for the line in Problem 5.14 if it has 70% shunt reactive (inductors) compensation and 40% series capacitive compensation. Half of this compensation is installed at each end of the line, as in Figure 5.14.

5.62   Consider the transmission line of Problem 5.18. (a) Find the *ABCD* parameters of the line when uncompensated. (b) For a series capacitive compensation of 70% (35% at the sending end and 35% at the receiving end), determine the *ABCD* parameters. Comment on the relative change in the magnitude of the *B* parameter with respect to the relative changes in the magnitudes of the *A*, *C*, and *D* parameters. Also comment on the maximum power that can be transmitted when series compensated.

5.63   Given the uncompensated line of Problem 5.18, let a three-phase shunt reactor (inductor) that compensates for 70% of the total shunt admittance of the line be connected at the receiving end of the line during no-load conditions. Determine the effect of voltage regulation with the reactor connected at no load. Assume that the reactor is removed under full-load conditions.

5.64   Let the three-phase lossless transmission line of Problem 5.31 supply a load of 1000 MVA at 0.8 power factor lagging and at 500 kV. (a) Determine the capacitance/phase and total three-phase Mvars supplied by a three-phase, Δ-connected shunt-capacitor bank at the receiving end to maintain the receiving-end voltage at 500 kV when the sending end of the line is energized at 500 kV. (b) If series capacitive compensation of 40% is installed at the midpoint of the line, without the shunt capacitor bank at the receiving end, compute the sending-end voltage and percent voltage regulation.

PW   **5.65**   Open PowerWorld Simulator case Example 5_10 with the series capacitive compensation at both ends of the line in service. Graph the load bus voltage as a function of load real power (assuming unity power factor at the load). What is the maximum amount of real power that can be transferred to the load at unity power factor if we require the load voltage always be greater than 0.85 per unit?

PW   **5.66**   Problem 5.66 Open PowerWorld Simulator case Example 5.10 with the series capacitive compensation at both ends of the line in service. With the reactive power load fixed at 500 Mvar, graph the load bus voltage as the MW load is varied between 0 and 2600 MW in 200 MW increments. Then repeat with both of the series compensation elements out of service.

CASE STUDY QUESTIONS

A. Dispatchers at power system control centers have limited control of power flows in today's transmission and distribution networks. Dispatch options include rescheduling of generation and intertie flows, adjusting generator excitation and transformer taps, removing lines or equipment from service, and bringing standby equipment on-line. However, Ohm's law seems to have the greatest influence on network flows. That is, power (more correctly, energy) flows from generators to customer loads through paths of lowest impedance. Do FACTS installations provide the opportunity for central dispatch (control) of flow on each compensated line?

B. What are the benefits of controlling power flows on individual lines? What are the risks?

REFERENCES

1. Electric Power Research Institute (EPRI), *EPRI AC Transmission Line Reference Book—200 kV and Above* (Palo Alto, CA: EPRI, www.epri.com, December 2005).

2. Westinghouse Electric Corporation, *Electrical Transmission and Distribution Reference Book*, 4th ed. (East Pittsburgh, PA, 1964).

3. R. D. Dunlop, R. Gutman, and P. P. Marchenko, "Analytical Development of Loadability Characteristics for EHV and UHV Lines," *IEEE Trans. PAS*, Vol. PAS-98, No. 2 (March/April 1979): pp. 606–607.

4. W. D. Stevenson, Jr., *Elements of Power System Analysis*, 4th ed. (New York: McGraw-Hill, 1982).

5. W. H. Hayt, Jr., and J. E. Kemmerly, *Engineering Circuit Analysis*, 3rd ed. (New York: McGraw-Hill, 1978).

6. G. Reed, J. Paserba, and P. Salavantis, "The FACTS on Resolving Transmission Gridlock," *IEEE Power & Energy Magazine, 1*, 5 (September/October 2003), pp. 41–46.

6

POWER FLOWS

Successful power system operation under normal balanced three-phase steady-state conditions requires the following:

1. Generation supplies the demand (load) plus losses.

2. Bus voltage magnitudes remain close to rated values.

3. Generators operate within specified real and reactive power limits.

4. Transmission lines and transformers are not overloaded.

The power-flow computer program (sometimes called *load flow*) is the basic tool for investigating these requirements. This program computes the voltage magnitude and angle at each bus in a power system under balanced three-phase steady-state conditions. It also computes real and reactive power flows for all equipment interconnecting the buses, as well as equipment losses.

Both existing power systems and proposed changes including new generation and transmission to meet projected load growth are of interest.

Conventional nodal or loop analysis is not suitable for power-flow studies because the input data for loads are normally given in terms of power, not impedance. Also, generators are considered as power sources, not voltage or current sources. The power-flow problem is therefore formulated as a set of nonlinear algebraic equations suitable for computer solution.

In Sections 6.1–6.3 we review some basic methods, including direct and iterative techniques for solving algebraic equations. Then in Sections 6.4–6.6 we formulate the power-flow problem, specify computer input data, and present two solution methods, Gauss–Seidel and Newton–Raphson. Means for controlling power flows are discussed in Section 6.7. Sections 6.8 and 6.9 introduce sparsity techniques and a fast decoupled power-flow method, while Section 6.10 discusses the DC power flow.

Since balanced three-phase steady-state conditions are assumed, we use only positive-sequence networks in this chapter. Also, all power-flow equations and input/output data are given in per-unit.

CASE STUDY Power-flow programs are used to analyze large transmission grids and the complex interaction between transmission grids and the power market. The following article describes visualization tools that are integrated with power-flow programs, enabling users to interpret large volumes of power-flow data more rapidly, more accurately, and more intuitively [9].

Visualizing the Electric Grid

THOMAS J. OVERBYE, JAMES D. WEBER

Visualization software packs a large amount of information into a single computer-generated image, enabling viewers to interpret the data more rapidly and more accurately than ever before. This kind of software will become still more useful, even indispensable, as electricity grids are integrated over ever-larger areas, as transmission and generation become competitive markets, and as transactions grow in number and complexity.

Tracking and managing these burgeoning transaction flows puts operating authorities on their mettle. While the electric power system was designed as the ultimate in plug-and-play convenience,

the humble wall outlet has become a gateway to one of the largest and most complex of man-made objects. For example, barring a few islands and other small isolated systems, the grid in most of North America is just one big electric circuit. It encompasses billions of components, tens of millions of kilometers of transmission line, and thousands of generators with power outputs ranging from less than 100 kW to 1000 MW and beyond. Grids on other continents are similarly interconnected.

In recent years, a further complicating factor has emerged. Along with the broadening integration of power systems has come the increased transfer of large blocks of power from one region to another. In the United States, because of varying local power loads and availability, utilities purchase electricity from distant counterparts and independent suppliers, exploiting price differentials to economize on costs. For one, the Tennessee Valley Authority, which provides power to more than 8 million residents in seven states using over 27,000 km of transmission lines, handled a mere 20,000 transaction requests through its service territory in 1996, compared to the 300,000 in 1999.

The net effect is that data once of interest mainly to small cadres of utilities now must be communicated to the new entities being established to manage restructured grids. In the United States, that means independent system operators (ISOs) and regional transmission organizations (RTOs), which have to be able to grasp fast-changing situations instantaneously and evaluate corrective strategies nearly as fast.

Power marketers' needs, too, become more urgent, as access to the grid is opened and competition among generators is introduced across the United States and elsewhere. They must be able to see just how much existing and proposed transactions will cost, and the availability of electricity at any time and any point in the system.

Finally, concepts like power flow, loop flow, and reactive power, which once mattered only to the engineers directly involved in grid operations, now must be made intuitive. This is because they must be communicated to public service commissions and the consumer-voters to whom such boards are answerable.

In short, whether the client/user is a power marketer, a grid operator or manager, a public authority, or a member of the public, power system visualization tools can aid their comprehension by lifting the truly significant above background noise. Such tools can expedite decision-making for congestion management, power trading, market organization, and investment planning for the long term.

The visualization tools illustrated here are available from PowerWorld Corp., Urbana, Illinois. Visualization tools offered by others rely on updated text. ABB, Alstom ESCA, GE Harris, and Siemens, for example, offer tools that are part of larger energy management systems packages.

HOW FLOWS ARE MANAGED

The usual reason that a large transfer of power can be hard to handle is that there are few mechanisms to control its route through the transmission system from generator to distant load. Often that route is indirect, dictated by the impedances of the lines and places where power enters or leaves the system. In effect, a single transaction between a generator and a utility spreads throughout a large portion of the grid—a phenomenon termed loop flow.

(To be sure, current can be and is directly guided during high-voltage direct-current [HVDC] transmission. And ac current is being nudged in desired directions by devices like phase-shifting transformers and series compensation capacitors, often lumped together as flexible ac transmission (FACT) devices. However, very few of these devices are available in most large power systems, so in effect transmission flows are not controllable.)

The percentage of a transfer that flows on any component in the grid—a transformer, say—is known, in language developed for the U.S. Eastern Interconnect, as the power transfer distribution factor (PTDF). A transaction that would send power through an overloaded component, in a direction to increase the loading, may not be allowed, or if already under way, may have to be curtailed. The U.S. procedure for ordering such curtailments is known as transmission-line loading relief (TLR). Its developer was the North American Electric Reliability Council (NERC), the utilities' voluntary reliability organization in Princeton, New Jersey.

To reiterate, a grid component owner that detects overloading serves notice with the relevant authority—an ISO or RTO, for example—and asks for relief. The independent operator, or whoever, thereupon orders loading relief measures. For the component in question, any transaction involving a distribution factor higher than a predetermined level—set by NERC at 5% of the transaction—is a candidate for curtailment. If more than 5% of the power transferred as part of a transaction will go over a grid component subject to a TLR, the transaction may be scaled back or canceled.

Those TLR measures in turn will affect other existing and proposed transactions, requiring further near-instantaneous analysis by utilities, grid supervisors, and power marketers. The need at every level for state-of-the-art visualization tools is obvious, since any bottleneck in this complex system can

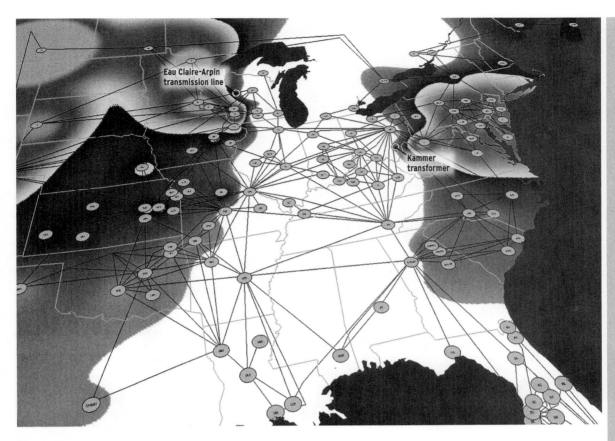

Figure 1
Seeing Potential Transmission Overloads. Color contour map highlights salient features of hypothetical grid activity during a price spike in the U.S. Midwest. A transmission line in Wisconsin (Eau Claire-Arpin) and a transformer in Ohio (Kammer) were chokepoints because they were at their maximum loading levels. Regulatory constraints then prevented further power transfers, which would only have increased the loading on either device.

Because of loop flow, the two chokepoints vastly reduced the market of potential additional generation to Midwest utilities. The colored regions, ranging from green to red for degrees of escalating percentage, exceeded the 5% regulatory cutoff and hence were prohibited from handling new transactions.

quickly cause brownouts, blackouts, or nasty price spikes.

AVERTING PRICE SPIKES, ISLANDING

Problems with grid management are not necessarily the cause of electricity outages or price spikes—California's current electricity crisis seems to have been induced primarily by unforeseen generating shortages and misguided public policy. Here, visualization can help only indirectly, by better showing policymakers the potential impact policy decisions can have on grid operation.

But when grid congestion is at the root of problems and floods of data are involved, visualization tools like conttouring, dynamic pie charts, animated diagrams, and two- and three-dimensional outlines have much more to offer.

Congestion played a pivotal role, for example, in the notorious U.S. midwestern price spikes of June 1998. That month, spot market prices for electricity soared three-hundredfold from US $25 to $7500

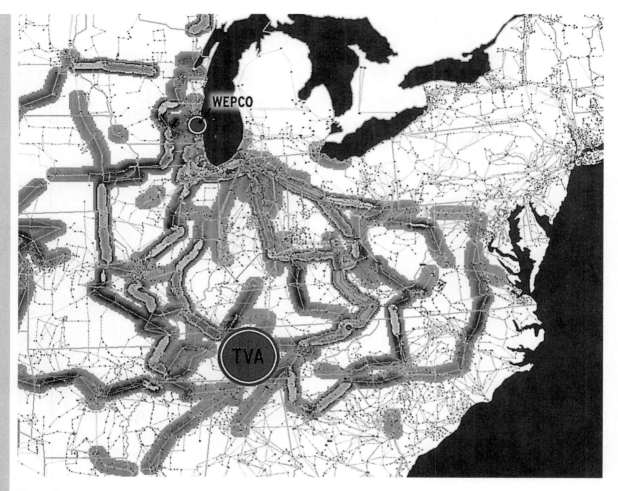

Figure 2
Assessing the Impact of a Power Transfer. The power transfer distribution factor (PTDF) is the percentage of a transaction that loops over a grid component. Here, superimposed on a map of the U.S. high-voltage transmission grid, a color contour indicates PTDF values for a hypothetical transfer between a utility in eastern Wisconsin (WEPCO) and the Tennessee Valley Authority (TVA). A value of 2% [green] is acceptable, 15% [red] unacceptable.

per megawatt-hour. Though there were many contributing factors, the most important were barriers to importing electricity from outside the region. Electricity was available elsewhere on the grid to the east and west, but could not be transferred because of overloads (congestion) on just two elements: a transmission line in northwest Wisconsin and a transformer in southeast Ohio.

The situation at the time of the June 1998 price spikes is diagrammed on the preceding page, where the small ovals represent operating areas in the Eastern Interconnect, each a potential seller. In the transaction illustrated, the buyer was a utility in northern Illinois. The contour indicates what percentage of the power transfer requested would have flowed through overloaded devices; shaded areas on the left could not sell because of the overload in northwest Wisconsin, those on the right because of the overload in southeast Ohio.

The visualization provides a picture of the complex interaction between the grid and the power market, allowing market participants to respond more quickly to changing conditions. With the market segmentation visualized on the prior page, power buyers in the affected areas could move quickly to procure long-term power capacity contracts, rather than having to buy at the astronomical spot market prices.

In the past, to form a mental picture of how line-loading relief measures might affect a market or reliability area, marketers or operators would have had to scan a long numerical list of distribution factors—no easy task once the list grows beyond a hundred or so entries. This is because in any large grid system, there are huge numbers of distribution factor sets, each dependent on pairs of buyers and sellers. Contouring provides a good solution, making the impact of loop flow apparent at a glance.

Another way of mapping the implications of TLRs is illustrated above: the map shows the distribution factors for a hypothetical power transfer from a utility in eastern Wisconsin and the Tennessee Valley Authority. Note that the transfer affects lines as far away as Nebraska and eastern Virginia. Of the 45 000 lines modeled in the case, 171 had PTDFs above 5%, while for 578 the PTDFs were above 2%.

With the aid of such tools, a marketer can easily start considering a host of WHAT IF scenarios. How might a loading relief on a transmission line affect market participants other than those directly involved in a transaction? What if there is an outage of a major transmission line? What is the outlook for other potential buyers?

VISUALIZING VOLUMINOUS FLOWS

To determine how power moves through a transmission network from generators to loads, it is necessary to calculate the real and reactive power flow on each and every transmission line or transformer, along with associated bus voltages (in other words, the voltages at each node). With networks containing tens of thousands of buses and branches, such calculations yield a lot of numbers. Traditionally they were presented either in reams of tabular output showing the power flows at each bus or else as data in a static so-called one-line diagram. (One-line diagrams are so named because they represent the actual three conductors of the underlying three-phase electric system with a single equivalent line.)

The visualization challenge is to make these concepts intuitive. One simple yet effective technique to depict the flow of power in an electricity network is to use animated line flow [see Figure 3, and link to PowerWorld site]. Here, the size, orientation, and speed of the arrows indicate the direction of power flow on the line, bringing the system almost literally to life.

Dynamically sized pie charts are another visualization idea that has proven useful for quickly detecting overloads in a large network. On the one-line, the percentage fill in each pie chart indicates how close each transmission line is to its thermal limit.

When thousands of lines must be considered, however, checking each and every value is not an option. Of course, tabular displays can be used to sort the values by loading percentage, but with a loss of geographical relevance. Because engineers and traders are mostly concerned with transmission lines near or above their limits, low-loaded lines can be eliminated by dynamically sizing the pie charts to become visible only when the loading is above a certain threshold.

CONTOURING THE GRID

Using pie charts to visualize these values is helpful, unless a whole host of them appear on the screen. Here, an entirely different visualization approach is useful—contouring.

For decades, power system engineers have represented bus-based values by drawing one-line diagrams embellished with digital numerical displays of the nearest bus's values. The results, being numerical, are precise and displayed next to the bus to which they refer. But for more than a handful of

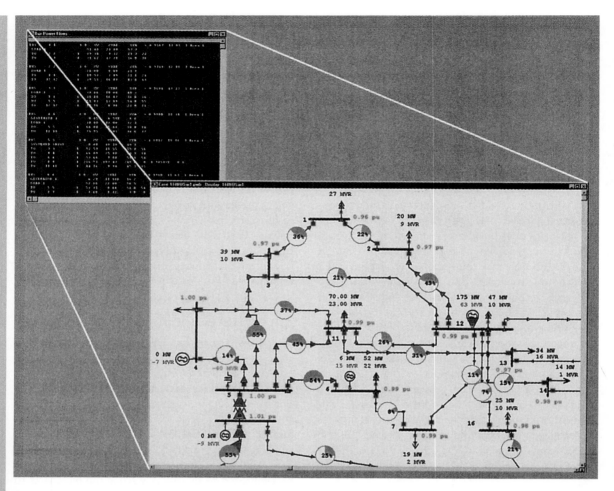

Figure 3
Animating Tabular Flow Information. Output that once had to be scanned in unwieldy tables [upper left] now can be absorbed at a glance from an enhanced diagram. For an animated version go to: http://www.powerworld.com/spectrum.

buses, it takes a lot of time to find a pattern. Contours are a familiar way of displaying continuous, spatially distributed data. The equal-temperature contours provided in a newspaper's weather forecast form a well-known example.

The trouble with contouring power system data is that it is not spatially continuous. Bus voltage magnitudes exist only at buses, and power only as flows on the lines, yet the spaces between buses and lines appear in contour maps as continuous gradients, not as gaps.

In practice the artificially blended spaces between nodes and lines do not matter much, as the main purpose of a contour is to show trends in data. Values are exact only at the buses or on the lines. Colors can be used to represent a weighted average of nearby data-points. This color gradation brings out the spatial relationships in the data.

MARKET POWER

Power flows matter not only to operations engineers and power traders, but also to the authorities charged with deciding whether two utilities should be allowed to merge, whether a new combustion turbine is needed in a trendy suburb, or whether the absence of a single transmission line could send electricity prices soaring. Overloads on just a few transmission lines can segment, or separate, even large power markets, causing prices to spike, with some players reaping huge windfall profits. The end result: serious misgivings about the whole process of restructuring and deregulation.

The central concern is that benefits from breaking up the old vertically integrated utilities will be for nought if the newly unbundled generation and transmission companies are able to exercise quasi-monopolistic power over local and regional markets. Collusion is one method, and another is "gaming" the system—taking advantage of legal loopholes and operational quirks to create or exploit bottlenecks and chokepoints.

Such abusive power, dubbed market power in the electricity context, refers to the ability of one seller or a group of sellers to maintain prices above competitive levels for a significant period of time. This can be done in various ways, depending on how markets have been organized to set prices in jurisdictions adopting market mechanisms—notably the United Kingdom, Norway, New Zealand, and, in the United States, the California Power Exchange, the PJM (Pennsylvania–New Jersey—Maryland) Interconnection, the New England Power Pool and New York ISO.

In most jurisdictions introducing competition, markets are organized so that spot prices can be determined at every node (or bus) in the system. The U.S. name for this is the locational marginal price (LMP). Under truly competitive conditions, it equals the marginal cost of providing electricity to that point in the transmission system, where the provider is any generator bidding into the system.

In the absence of overloads, spot marginal prices are about equal across an entire power market (though this depends somewhat on how resistive line losses are taken into account). But when overloads occur, spot prices can rapidly diverge. Because LMPs are busbased values, contouring is again extremely useful for showing market-wide patterns.

As an example, the figure at the left shows a contour of the LMPs generated by an optimal power flow (OPF) study using a 9270-bus system to model those in the northeast [see "A Brief History of the Power Flow, page 260]. An OPF sets the outputs of generators to minimize the total cost of operating the power system, while at the same time ensuring that no transmission system elements are overloaded. In the study, marginal prices were calculated for 5774 buses, and some 2000 of these values were used in creating the contour. The contour is superimposed on a map of the high-voltage transmission lines in the Northeast, and the pie charts indicate which transmission lines or corridors are congested.

Note the price differential between New York and New England caused by a congested line on the boundary between northern New York and New England. The pocket of high prices in western New York is due to a constraint on a single 230/115-kV transformer. A transmission element is said to constrain the power system when generation must be moved from the most economical operating point in order to reduce the loading on the element. This constraint can be eliminated by bringing on stream a relatively small 85-MW generator on the constrained side of the transformer.

To be truly effective, however, computer visualization must be interactive and it must be fast. Using a standard desktop computer, the contour on the opposite page can be re-created, with a reasonable resolution, within a few seconds. Fast contouring, coupled with easy zooming and panning, equips the market analyst with an interactive tool with which to quickly explore a power system data set. For instance, zooming could be used to provide more details about pricing across Massachusetts, while dynamically sized pie charts could be show lines that are close to but not yet exceeding their limits.

THE GRID GOES 3-D

Contouring can be quite helpful when one is primarily concerned with the visualization of a single type of spatially oriented data, such as bus voltages or transmission line flows. But the data of interest in

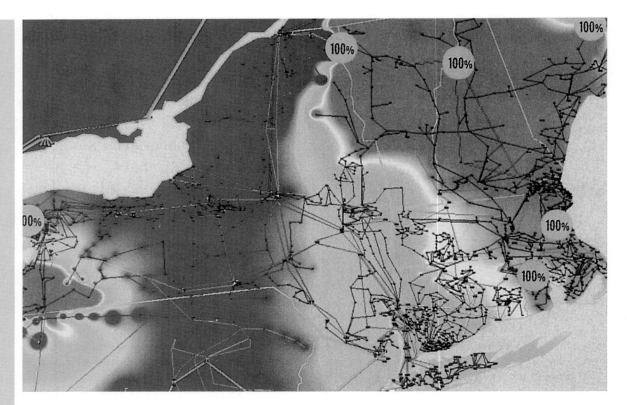

Figure 4
Dynamically Sized Data Displays Provide Early Warning of Possible Price Spikes. The real cost of supplying power fluctuates widely as capacity, load, and other conditions on the grid change from moment to moment. Congestion on just a handful of lines can result in big regional variations in prices. This contour of the price variation across a potential power market in the U.S. Northeast includes circles showing transmission lines that are fully loaded.

a power system could include a long list of independent and dependent variables. Bus voltage magnitudes and prices, transmission line loadings and PTDFs, generator reserves and bids, and scheduled flows between areas all come to mind.

In more advanced applications involving OPF and available transfer capability (ATC) calculations, this list of variables is even longer. ATC calculations determine maximum amounts of megawatt transfers that can occur across the transmission system. One solution is to leave the two-dimensional views behind and enter the third dimension.

Interactive, 3-D visualization is certainly nothing new. Nevertheless, in designing it for power system visualization several issues arise. First and foremost,

in visualizing power system data there is usually no corresponding "physical" representation for the variables. For example, there is no physical correlate to the reactive power output of a generator, or for the marginal cost of enforcing a transmission line constraint. These are abstract calculated values, to be added as desired to diagrams in which physical flows are represented in the first two dimensions.

The abstract nature of the data makes this kind of visualization different from the characteristic use of interactive 3-D for some types of scientific visualization, in which the purpose of the environment is to visualize physical phenomena, such as flows in a wind tunnel or molecular interactions. To address this issue, an environment based upon the traditional one-line representation (in which the three ac phases are represented by single lines) serves as a good starting point. Those concerned with power systems are familiar with it. The new environment differs from the old in that a traditional one-line is a two-dimensional representation, whereas the new is 3-D, opening a world of possibilities of how to use this additional dimension.

The 3-D environment must also be highly interactive. In power systems, there is too much data for everything of interest to be displayed. Rather the user should be able to access the data of interest quickly and intuitively.

This leads to the question of how to interact with the 3-D environment. With a 2-D one-line, there are just three degrees of freedom associated with viewing: panning in either the x or y directions, and zooming. The 2-D one-line could be thought of as lying in the xy-plane, with the viewing "camera" poised above. Panning the one-line can then be thought of as moving the camera in the x and y directions. And zooming is simply changing the height of the camera above the one-line. A 3-D environment has the same three degrees of freedom, but adds three more, because now the camera itself can be rotated about each of its three axes. Navigation can, however, be simplified by restricting camera rotation.

One useful approach is to allow rotation about two axes only. If the camera rotates about the axis passing through its sides, it can change its angle with respect to the horizon (elevation). Alternatively, it can rotate about the axis passing through it from top to bottom. Rotation about the axis passing through the camera from front to back (twist) is not allowed.

The results can be stunning. Suddenly the one-lines come to life. The 3-D environment gives the viewer a greater sense of involvement with the system, making important information harder to overlook and otherwise hidden relationships easier to see.

For example, one limitation on the transmission system is the need to maintain high enough bus voltages—values that contouring can display quite effectively but minus information about controlling factors, such as the reactive power output of generators, that could correct problems with the voltages. Voltage security analysis requires a simultaneous awareness of both the bus voltage magnitudes and the generator reactive characteristics, including the generator reactive reserves. Showing all this information numerically on a 2-D one-line would only be effective for a very small system.

PERSPECTIVE PROJECTION

The 3-D alternative is to draw the 2-D one-line in the xy-plane using a perspective projection, in which closer objects appear larger. The generator reactive output and reserves are then shown using cylinders in the third dimension. In Figure 5, the height of each cylinder is proportional to the maximum reactive capacity of the generator; the dark lower part of the cylinder indicates the present reactive output, while the lighter part indicates the reactive reserves. The bus voltage values are indicated by a contour in the xy-plane; voltages below 98% of desired values are shaded.

Thus the 3-D one-line now shows, at one and the same time, the location of low voltages, the present generator reactive power outputs, and the reserves. It does a good job of conveying qualitative information about the magnitude of these values, but not exact numerical values. Thus the figure only shows that the reactive power at bus 20 is about 50% of its maximum, not its actual value. In some situations this could be a serious limitation. Hence the 3-D one-lines are meant to supplement, rather than to replace, existing display formats.

The last issue to address is performance. For effective use of the 3-D environment, fast display refresh is crucial. This in turn depends on, among other things, the speed of the computer's processor, the speed of the graphics card, whether the graphics card has hardware support for 3-D, and software considerations such as the level of display detail. At present, PCs with the best mainstream display cards are approaching display rates of 10 million polygons per second, enough for good refresh rates plus a fair amount of detail.

Performance is driven strongly by computer games—their popularity, indeed, could not come at a better time for power visualization.

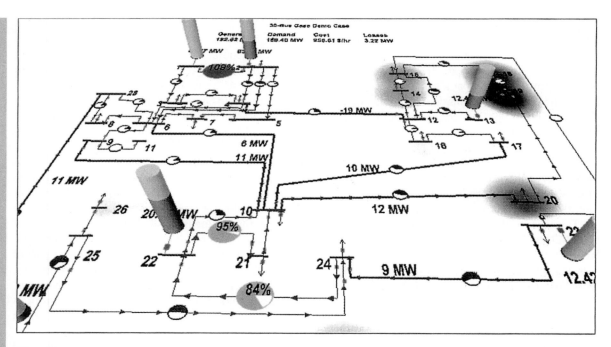

Figure 5
Introducing a Third Dimension for Abstract Data. A three-dimensional environment allows quick comparison of a number of system variables, as shown for this small 30-bus power system. Actual flows are diagrammed in the horizontal plane, while the vertical cylinders show the generators' reactive-power outputs [shaded gray] and their reactive reserves [shaded orange]. The pie charts represent line loadings.

A Brief History of the Power Flow

FERNANDO L. ALVARADO, ROBERT J. THOMAS

Improved economy and reliability were recognized well over half a century ago as benefits of using an interconnected network for the transport of electric power. But critical to its realization was (and still is) the ability to predict the voltages and flows on network components. As the networks evolved, the challenge was to develop a tool that would produce this critical information. The load-flow (or power flow), as the tool came to be known, predicts all flows and voltages in the network when given the status of generators and loads. It is the tool most heavily used by power engineers.

Early load-flows were solved using what were called calculator boards. These boards were a kind of analog computer, in that they emulated a specific system by using a physical lumped-parameter resistor-inductor-capacitor realization of the actual system, the components being connected in the same topology. For a realistic system, these boards filled several rooms, consumed substantial power, and had to be rewired when any modification was desired. As studies often required teams of engineers working in unison adjusting knobs and settings and reading out results aloud, the need for a flexible alternative was clear.

Enter the modern digital computer, which, in fact, owes much of the impetus behind its original development to power engineers and their need for a better way to solve load-flows. In the early days of computing, the electric power business was by far the largest commercial user (and even developer) of digital machines. It was not unusual for a utility to spend several million dollars (not adjusted for inflation) on the development of digital hardware and software. While IBM Corp. was advancing mainframe machine architectures, theorists were publishing the first papers on load-flow algorithms.

The earliest algorithms were based on the Gauss–Seidel method, which made it possible, for the first time, to solve the load-flow problem for relatively large systems. It suffered, however, from relatively poor convergence characteristics. Then the Newton algorithm was developed to improve the convergence of the Gauss–Seidel method, but was initially thought to be impractical for realistically sized systems because of computational problems with large networks. The underlying problem for the iterative Newton method is the solution of a matrix equation of large dimension.

In the 1960s Bill Tinney and his colleagues at the Bonneville Power Administration observed that,

although the main system matrix was very large, it was also very sparse (meaning it had a very small proportion of nonzero values). This observation gave rise to the development of sparsity methods. The concept made it possible to apply the Newton method to systems of arbitrary size, to attain for the first time both speed and excellent convergence characteristics.

Since the '60s, numerous advances and extensions have been made in load-flow methods. In the early '70s came the fast-decoupled load-flow, which enhanced computational speed. Extensions to the load-flow itself included the representation of components such as high-voltage direct-current (HVDC) transmission lines, better methods for loss calculation, solution of the optimum power flow and state estimation problems, the continuation power flow, and the determination of spot prices of electricity in the presence of constraints—plus, of course, the development of better ways of visualizing and presenting load-flow results.

6.1

DIRECT SOLUTIONS TO LINEAR ALGEBRAIC EQUATIONS: GAUSS ELIMINATION

Consider the following set of linear algebraic equations in matrix format:

$$
\begin{bmatrix}
A_{11} & A_{12} & \cdots & A_{1N} \\
A_{21} & A_{22} & \cdots & A_{2N} \\
\vdots & & \vdots & \\
A_{N1} & A_{N2} & \cdots & A_{NN}
\end{bmatrix}
\begin{bmatrix}
x_1 \\
x_2 \\
\vdots \\
x_N
\end{bmatrix}
=
\begin{bmatrix}
y_1 \\
y_2 \\
\vdots \\
y_N
\end{bmatrix}
\tag{6.1.1}
$$

or

$$
\mathbf{A}\mathbf{x} = \mathbf{y} \tag{6.1.2}
$$

where $\mathbf{x}$ and $\mathbf{y}$ are N vectors and $\mathbf{A}$ is an $N \times N$ square matrix. The components of $\mathbf{x}$, $\mathbf{y}$, and $\mathbf{A}$ may be real or complex. Given $\mathbf{A}$ and $\mathbf{y}$, we want to solve for $\mathbf{x}$. We assume the $\det(\mathbf{A})$ is nonzero, so a unique solution to (6.1.1) exists.

The solution $\mathbf{x}$ can easily be obtained when $\mathbf{A}$ is an upper triangular matrix with nonzero diagonal elements. Then (6.1.1) has the form

$$
\begin{bmatrix}
A_{11} & A_{12}\dots & & A_{1N} \\
0 & A_{22}\dots & & A_{2N} \\
\vdots & & & \\
0 & 0\dots & A_{N-1,N-1} & A_{N-1,N} \\
0 & 0\dots 0 & & A_{NN}
\end{bmatrix}
\begin{bmatrix}
x_1 \\ x_2 \\ \vdots \\ x_{N-1} \\ x_N
\end{bmatrix}
=
\begin{bmatrix}
y_1 \\ y_2 \\ \vdots \\ y_{N-1} \\ y_N
\end{bmatrix}
\tag{6.1.3}
$$

Since the last equation in (6.1.3) involves only x_N,

$$
x_N = \frac{y_N}{A_{NN}}
\tag{6.1.4}
$$

After x_N is computed, the next-to-last equation can be solved:

$$
x_{N-1} = \frac{y_{N-1} - A_{N-1,N}x_N}{A_{N-1,N-1}}
\tag{6.1.5}
$$

In general, with $x_N, x_{N-1}, \dots, x_{k+1}$ already computed, the kth equation can be solved

$$
x_k = \frac{y_k - \displaystyle\sum_{n=k+1}^{N} A_{kn}x_n}{A_{kk}} \qquad k = N, N-1, \dots, 1
\tag{6.1.6}
$$

This procedure for solving (6.1.3) is called *back substitution*.

If $\mathbf{A}$ is not upper triangular, (6.1.1) can be transformed to an equivalent equation with an upper triangular matrix. The transformation, called *Gauss elimination*, is described by the following $(N-1)$ steps. During step 1, we use the first equation in (6.1.1) to eliminate x_1 from the remaining equations. That is, equation 1 is multiplied by A_{n1}/A_{11} and then subtracted from equation n, for $n = 2, 3, \dots, N$. After completing step 1, we have

$$
\begin{bmatrix}
A_{11} & A_{12} & \cdots & A_{1N} \\
0 & \left(A_{22} - \dfrac{A_{21}}{A_{11}}A_{12}\right) & \cdots & \left(A_{2N} - \dfrac{A_{21}}{A_{11}}A_{1N}\right) \\
0 & \left(A_{32} - \dfrac{A_{31}}{A_{11}}A_{12}\right) & \cdots & \left(A_{3N} - \dfrac{A_{31}}{A_{11}}A_{1N}\right) \\
\vdots & \vdots & & \vdots \\
0 & \left(A_{N2} - \dfrac{A_{N1}}{A_{11}}A_{12}\right) & \cdots & \left(A_{NN} - \dfrac{A_{N1}}{A_{11}}A_{1N}\right)
\end{bmatrix}
\begin{bmatrix}
x_1 \\ x_2 \\ x_3 \\ \vdots \\ x_N
\end{bmatrix}
$$

$$
=
\begin{bmatrix}
y_1 \\
y_2 - \dfrac{A_{21}}{A_{11}}y_1 \\
y_3 - \dfrac{A_{31}}{A_{11}}y_1 \\
\vdots \\
y_N - \dfrac{A_{N1}}{A_{11}}y_1
\end{bmatrix}
\tag{6.1.7}
$$

Equation (6.1.7) has the following form:

$$
\begin{bmatrix}
A_{11}^{(1)} & A_{12}^{(1)} & \cdots & A_{1N}^{(1)} \\
0 & A_{22}^{(1)} & \cdots & A_{2N}^{(1)} \\
0 & A_{32}^{(1)} & \cdots & A_{3N}^{(1)} \\
\vdots & \vdots & & \vdots \\
0 & A_{N2}^{(1)} & \cdots & A_{NN}^{(1)}
\end{bmatrix}
\begin{bmatrix}
x_1 \\ x_2 \\ x_3 \\ \vdots \\ x_N
\end{bmatrix}
=
\begin{bmatrix}
y_1^{(1)} \\ y_2^{(1)} \\ y_3^{(1)} \\ \vdots \\ y_N^{(1)}
\end{bmatrix}
\qquad (6.1.8)
$$

where the superscript (1) denotes step 1 of Gauss elimination.

During step 2 we use the second equation in (6.1.8) to eliminate x_2 from the remaining (third, fourth, fifth, and so on) equations. That is, equation 2 is multiplied by $A_{n2}^{(1)}/A_{22}^{(1)}$ and subtracted from equation n, for $n = 3, 4, \ldots, N$. After step 2, we have

$$
\begin{bmatrix}
A_{11}^{(2)} & A_{12}^{(2)} & A_{13}^{(2)} & \cdots & A_{1N}^{(2)} \\
0 & A_{22}^{(2)} & A_{23}^{(2)} & \cdots & A_{2N}^{(2)} \\
0 & 0 & A_{33}^{(2)} & \cdots & A_{3N}^{(2)} \\
0 & 0 & A_{43}^{(2)} & \cdots & A_{4N}^{(2)} \\
\vdots & \vdots & \vdots & & \vdots \\
0 & 0 & A_{N3}^{(2)} & \cdots & A_{NN}^{(2)}
\end{bmatrix}
\begin{bmatrix}
x_1 \\ x_2 \\ x_3 \\ x_4 \\ \vdots \\ x_N
\end{bmatrix}
=
\begin{bmatrix}
y_1^{(2)} \\ y_2^{(2)} \\ y_3^{(2)} \\ y_4^{(2)} \\ \vdots \\ y_N^{(2)}
\end{bmatrix}
\qquad (6.1.9)
$$

During step k, we start with $\mathbf{A}^{(k-1)}\mathbf{x} = \mathbf{y}^{(k-1)}$. The first k of these equations, already triangularized, are left unchanged. Also, equation k is multiplied by $A_{nk}^{(k-1)}/A_{kk}^{(k-1)}$ and then subtracted from equation n, for $n = k+1, k+2, \ldots, N$.

After $(N-1)$ steps, we arrive at the equivalent equation $\mathbf{A}^{(N-1)}\mathbf{x} = \mathbf{y}^{(N-1)}$, where $\mathbf{A}^{(N-1)}$ is upper triangular.

EXAMPLE 6.1   Gauss elimination and back substitution: direct solution to linear algebraic equations

Solve

$$
\begin{bmatrix}
10 & 5 \\
2 & 9
\end{bmatrix}
\begin{bmatrix}
x_1 \\ x_2
\end{bmatrix}
=
\begin{bmatrix}
6 \\ 3
\end{bmatrix}
$$

using Gauss elimination and back substitution.

SOLUTION   Since $N = 2$ for this example, there is $(N-1) = 1$ Gauss elimination step. Multiplying the first equation by $A_{21}/A_{11} = 2/10$ and then subtracting from the second,

$$\begin{bmatrix} 10 & 5 \\ 0 & 9 - \dfrac{2}{10}(5) \end{bmatrix} \begin{bmatrix} x_1 \\ x_2 \end{bmatrix} = \begin{bmatrix} 6 \\ 3 - \dfrac{2}{10}(6) \end{bmatrix}$$

or

$$\begin{bmatrix} 10 & 5 \\ 0 & 8 \end{bmatrix} \begin{bmatrix} x_1 \\ x_2 \end{bmatrix} = \begin{bmatrix} 6 \\ 1.8 \end{bmatrix}$$

which has the form $\mathbf{A}^{(1)}\mathbf{x} = \mathbf{y}^{(1)}$, where $\mathbf{A}^{(1)}$ is upper triangular. Now, using back substitution, (6.1.6) gives, for $k = 2$:

$$x_2 = \frac{y_2^{(1)}}{A_{22}^{(1)}} = \frac{1.8}{8} = 0.225$$

and, for $k = 1$,

$$x_1 = \frac{y_1^{(1)} - A_{12}^{(1)} x_2}{A_{11}^{(1)}} = \frac{6 - (5)(0.225)}{10} = 0.4875$$ ∎

EXAMPLE 6.2   Gauss elimination: triangularizing a matrix

Use Gauss elimination to triangularize

$$\begin{bmatrix} 2 & 3 & -1 \\ -4 & 6 & 8 \\ 10 & 12 & 14 \end{bmatrix} \begin{bmatrix} x_1 \\ x_2 \\ x_3 \end{bmatrix} = \begin{bmatrix} 5 \\ 7 \\ 9 \end{bmatrix}$$

SOLUTION   There are $(N - 1) = 2$ Gauss elimination steps. During step 1, we subtract $A_{21}/A_{11} = -4/2 = -2$ times equation 1 from equation 2, and we subtract $A_{31}/A_{11} = 10/2 = 5$ times equation 1 from equation 3, to give

$$\begin{bmatrix} 2 & 3 & -1 \\ 0 & 6 - (-2)(3) & 8 - (-2)(-1) \\ 0 & 12 - (5)(3) & 14 - (5)(-1) \end{bmatrix} \begin{bmatrix} x_1 \\ x_2 \\ x_3 \end{bmatrix} = \begin{bmatrix} 5 \\ 7 - (-2)(5) \\ 9 - (5)(5) \end{bmatrix}$$

or

$$\begin{bmatrix} 2 & 3 & -1 \\ 0 & 12 & 6 \\ 0 & -3 & 19 \end{bmatrix} \begin{bmatrix} x_1 \\ x_2 \\ x_3 \end{bmatrix} = \begin{bmatrix} 5 \\ 17 \\ -16 \end{bmatrix}$$

which is $\mathbf{A}^{(1)}\mathbf{x} = \mathbf{y}^{(1)}$. During step 2, we subtract $A_{32}^{(1)}/A_{22}^{(1)} = -3/12 = -0.25$ times equation 2 from equation 3, to give

$$
\begin{bmatrix}
2 & 3 & -1 \\
0 & 12 & 6 \\
0 & 0 & 19 - (-.25)(6)
\end{bmatrix}
\begin{bmatrix}
x_1 \\
x_2 \\
x_3
\end{bmatrix}
=
\begin{bmatrix}
5 \\
17 \\
-16 - (-.25)(17)
\end{bmatrix}
$$

or

$$
\begin{bmatrix}
2 & 3 & -1 \\
0 & 12 & 6 \\
0 & 0 & 20.5
\end{bmatrix}
\begin{bmatrix}
x_1 \\
x_2 \\
x_3
\end{bmatrix}
=
\begin{bmatrix}
5 \\
17 \\
-11.75
\end{bmatrix}
$$

which is triangularized. The solution $\mathbf{x}$ can now be easily obtained via back substitution. ∎

Computer storage requirements for Gauss elimination and back substitution include N^2 memory locations for $\mathbf{A}$ and N locations for $\mathbf{y}$. If there is no further need to retain $\mathbf{A}$ and $\mathbf{y}$, then $\mathbf{A}^{(k)}$ can be stored in the location of $\mathbf{A}$, and $\mathbf{y}^{(k)}$, as well as the solution $\mathbf{x}$, can be stored in the location of $\mathbf{y}$. Additional memory is also required for iterative loops, arithmetic statements, and working space.

Computer time requirements can be evaluated by determining the number of arithmetic operations required for Gauss elimination and back substitution. One can show that Gauss elimination requires $(N^3 - N)/3$ multiplications, $(N)(N-1)/2$ divisions, and $(N^3 - N)/3$ subtractions. Also, back substitution requires $(N)(N-1)/2$ multiplications, N divisions, and $(N)(N-1)/2$ subtractions. Therefore, for very large N, the approximate computer time for solving (6.1.1) by Gauss elimination and back substitution is the time required to perform $N^3/3$ multiplications and $N^3/3$ subtractions.

For example, consider a digital computer with a 2×10^{-9} s multiplication time and 1×10^{-9} s addition or subtraction time. Solving $N = 10,000$ equations would require approximately

$$
\tfrac{1}{3}N^3(2 \times 10^{-9}) + \tfrac{1}{3}N^3(1 \times 10^{-9}) = \tfrac{1}{3}(10,000)^3(3 \times 10^{-9}) = 1000 \quad \text{s}
$$

plus some additional bookkeeping time for indexing and managing loops.

Since the power flow problem often involves solving power systems with tens of thousands of equations, by itself Gauss elimination would not be a good solution. However, for matrixes that have relatively few nonzero elements, known as sparse matrices, special techniques can be employed to significantly reduce computer storage and time requirements. Since all large power systems can be modeled using sparse matrices, these techniques are briefly introduced in Section 6.8.

6.2

ITERATIVE SOLUTIONS TO LINEAR ALGEBRAIC EQUATIONS: JACOBI AND GAUSS–SEIDEL

A general iterative solution to (6.1.1) proceeds as follows. First select an initial guess $\mathbf{x}(0)$. Then use

$$\mathbf{x}(i+1) = \mathbf{g}[\mathbf{x}(i)] \qquad i = 0, 1, 2, \ldots \tag{6.2.1}$$

where $\mathbf{x}(i)$ is the ith guess and $\mathbf{g}$ is an N vector of functions that specify the iteration method. Continue the procedure until the following stopping condition is satisfied:

$$\left| \frac{x_k(i+1) - x_k(i)}{x_k(i)} \right| < \varepsilon \qquad \text{for all } k = 1, 2, \ldots, N \tag{6.2.2}$$

where $x_k(i)$ is the kth component of $\mathbf{x}(i)$ and ε is a specified *tolerance level*.

The following questions are pertinent:

1. Will the iteration procedure converge to the unique solution?

2. What is the convergence rate (how many iterations are required)?

3. When using a digital computer, what are the computer storage and time requirements?

These questions are addressed for two specific iteration methods: *Jacobi* and *Gauss–Seidel*.* The Jacobi method is obtained by considering the kth equation of (6.1.1), as follows:

$$y_k = A_{k1}x_1 + A_{k2}x_2 + \cdots + A_{kk}x_k + \cdots + A_{kN}x_N \tag{6.2.3}$$

Solving for x_k,

$$x_k = \frac{1}{A_{kk}} [y_k - (A_{k1}x_1 + \cdots + A_{k,k-1}x_{k-1} + A_{k,k+1}x_{k+1} + \cdots + A_{kN}x_N)]$$

$$= \frac{1}{A_{kk}} \left[y_k - \sum_{n=1}^{k-1} A_{kn}x_n - \sum_{n=k+1}^{N} A_{kn}x_n \right] \tag{6.2.4}$$

The Jacobi method uses the "old" values of $\mathbf{x}(i)$ at iteration i on the right side of (6.2.4) to generate the "new" value $x_k(i+1)$ on the left side of (6.2.4). That is,

$$x_k(i+1) = \frac{1}{A_{kk}} \left[y_k - \sum_{n=1}^{k-1} A_{kn}x_n(i) - \sum_{n=k+1}^{N} A_{kn}x_n(i) \right] \qquad k = 1, 2, \ldots, N \tag{6.2.5}$$

The Jacobi method given by (6.2.5) can also be written in the following matrix format:

---

*The Jacobi method is also called the Gauss method.

$$\mathbf{x}(i+1) = \mathbf{M}\mathbf{x}(i) + \mathbf{D}^{-1}\mathbf{y} \tag{6.2.6}$$

where

$$\mathbf{M} = \mathbf{D}^{-1}(\mathbf{D} - \mathbf{A}) \tag{6.2.7}$$

and

$$\mathbf{D} = \begin{bmatrix} A_{11} & 0 & 0 & \cdots & 0 \\ 0 & A_{22} & 0 & \cdots & 0 \\ 0 & \vdots & \vdots & & \vdots \\ \vdots & & & & 0 \\ 0 & 0 & 0 & \cdots & A_{NN} \end{bmatrix} \tag{6.2.8}$$

For Jacobi, $\mathbf{D}$ consists of the diagonal elements of the $\mathbf{A}$ matrix.

EXAMPLE 6.3   Jacobi method: iterative solution to linear algebraic equations

Solve Example 6.1 using the Jacobi method. Start with $x_1(0) = x_2(0) = 0$ and continue until (6.2.2) is satisfied for $\varepsilon = 10^{-4}$.

SOLUTION   From (6.2.5) with $N = 2$,

$$k = 1 \qquad x_1(i+1) = \frac{1}{A_{11}}[y_1 - A_{12}x_2(i)] = \frac{1}{10}[6 - 5x_2(i)]$$

$$k = 2 \qquad x_2(i+1) = \frac{1}{A_{22}}[y_2 - A_{21}x_1(i)] = \frac{1}{9}[3 - 2x_1(i)]$$

Alternatively, in matrix format using (6.2.6)–(6.2.8),

$$\mathbf{D}^{-1} = \begin{bmatrix} 10 & 0 \\ \hline 0 & 9 \end{bmatrix}^{-1} = \begin{bmatrix} \dfrac{1}{10} & 0 \\ \hline 0 & \dfrac{1}{9} \end{bmatrix}$$

$$\mathbf{M} = \begin{bmatrix} \dfrac{1}{10} & 0 \\ \hline 0 & \dfrac{1}{9} \end{bmatrix} \begin{bmatrix} 0 & -5 \\ \hline -2 & 0 \end{bmatrix} = \begin{bmatrix} 0 & -\dfrac{5}{10} \\ \hline -\dfrac{2}{9} & 0 \end{bmatrix}$$

$$\begin{bmatrix} x_1(i+1) \\ x_2(i+1) \end{bmatrix} = \begin{bmatrix} 0 & -\dfrac{5}{10} \\ \hline -\dfrac{2}{9} & 0 \end{bmatrix} \begin{bmatrix} x_1(i) \\ x_2(i) \end{bmatrix} + \begin{bmatrix} \dfrac{1}{10} & 0 \\ \hline 0 & \dfrac{1}{9} \end{bmatrix} \begin{bmatrix} 6 \\ 3 \end{bmatrix}$$

The above two formulations are identical. Starting with $x_1(0) = x_2(0) = 0$, the iterative solution is given in the following table:

JACOBI

i	0	1	2	3	4	5	6	7	8	9	10
$x_1(i)$	0	0.60000	0.43334	0.50000	0.48148	0.48889	0.48683	0.48766	0.48743	0.48752	0.48749
$x_2(i)$	0	0.33333	0.20000	0.23704	0.22222	0.22634	0.22469	0.22515	0.22496	0.22502	0.22500

As shown, the Jacobi method converges to the unique solution obtained in Example 6.1. The convergence criterion is satisfied at the 10th iteration, since

$$\left|\frac{x_1(10) - x_1(9)}{x_1(9)}\right| = \left|\frac{0.48749 - 0.48752}{0.48749}\right| = 6.2 \times 10^{-5} < \varepsilon$$

and

$$\left|\frac{x_2(10) - x_2(9)}{x_2(9)}\right| = \left|\frac{0.22500 - 0.22502}{0.22502}\right| = 8.9 \times 10^{-5} < \varepsilon \qquad \blacksquare$$

The Gauss–Seidel method is given by

$$x_k(i+1) = \frac{1}{A_{kk}}\left[y_k - \sum_{n=1}^{k-1} A_{kn}x_n(i+1) - \sum_{n=k+1}^{N} A_{kn}x_n(i) \right] \qquad (6.2.9)$$

Comparing (6.2.9) with (6.2.5), note that Gauss–Seidel is similar to Jacobi except that during each iteration, the "new" values, $x_n(i+1)$, for $n < k$ are used on the right side of (6.2.9) to generate the "new" value $x_k(i+1)$ on the left side.

The Gauss–Seidel method of (6.2.9) can also be written in the matrix format of (6.2.6) and (6.2.7), where

$$\mathbf{D} = \begin{bmatrix} A_{11} & 0 & 0 & \cdots & 0 \\ A_{21} & A_{22} & 0 & \cdots & 0 \\ \vdots & \vdots & & & \vdots \\ A_{N1} & A_{N2} & \cdots & & A_{NN} \end{bmatrix} \qquad (6.2.10)$$

For Gauss–Seidel, $\mathbf{D}$ in (6.2.10) is the lower triangular portion of $\mathbf{A}$, whereas for Jacobi, $\mathbf{D}$ in (6.2.8) is the diagonal portion of $\mathbf{A}$.

EXAMPLE 6.4   Gauss–Seidel method: iterative solution to linear algebraic equations

Rework Example 6.3 using the Gauss–Seidel method.

SOLUTION   From (6.2.9),

$$k = 1 \qquad x_1(i+1) = \frac{1}{A_{11}}[y_1 - A_{12}x_2(i)] = \frac{1}{10}[6 - 5x_2(i)]$$

$$k = 2 \qquad x_2(i+1) = \frac{1}{A_{22}}[y_2 - A_{21}x_1(i+1)] = \frac{1}{9}[3 - 2x_1(i+1)]$$

Using this equation for $x_1(i+1)$, $x_2(i+1)$ can also be written as

$$x_2(i+1) = \frac{1}{9}\left\{3 - \frac{2}{10}[6 - 5x_2(i)]\right\}$$

Alternatively, in matrix format, using (6.2.10), (6.2.6), and (6.2.7):

$$\mathbf{D}^{-1} = \left[\begin{array}{c|c} 10 & 0 \\ \hline 2 & 9 \end{array}\right]^{-1} = \left[\begin{array}{c|c} \dfrac{1}{10} & 0 \\ \hline -\dfrac{2}{90} & \dfrac{1}{9} \end{array}\right]$$

$$\mathbf{M} = \left[\begin{array}{c|c} \dfrac{1}{10} & 0 \\ \hline -\dfrac{2}{90} & \dfrac{1}{9} \end{array}\right]\left[\begin{array}{c|c} 0 & -5 \\ \hline 0 & 0 \end{array}\right] = \left[\begin{array}{c|c} 0 & -\dfrac{1}{2} \\ \hline 0 & \dfrac{1}{9} \end{array}\right]$$

$$\left[\begin{array}{c} x_1(i+1) \\ x_2(i+1) \end{array}\right] = \left[\begin{array}{c|c} 0 & -\dfrac{1}{2} \\ \hline 0 & \dfrac{1}{9} \end{array}\right]\left[\begin{array}{c} x_1(i) \\ x_2(i) \end{array}\right] + \left[\begin{array}{c|c} \dfrac{1}{10} & 0 \\ \hline -\dfrac{2}{90} & \dfrac{1}{9} \end{array}\right]\left[\begin{array}{c} 6 \\ 3 \end{array}\right]$$

These two formulations are identical. Starting with $x_1(0) = x_2(0) = 0$, the solution is given in the following table:

GAUSS–SEIDEL	i	0	1	2	3	4	5	6
	$x_1(i)$	0	0.60000	0.50000	0.48889	0.48765	0.48752	0.48750
	$x_2(i)$	0	0.20000	0.22222	0.22469	0.22497	0.22500	0.22500

For this example, Gauss–Seidel converges in 6 iterations, compared to 10 iterations with Jacobi. ∎

The convergence rate is faster with Gauss–Seidel for some **A** matrices, but faster with Jacobi for other **A** matrices. In some cases, one method diverges while the other converges. In other cases both methods diverge, as illustrated by the next example.

EXAMPLE 6.5 Divergence of Gauss–Seidel method

Using the Gauss–Seidel method with $x_1(0) = x_2(0) = 0$, solve

$$\left[\begin{array}{c|c} 5 & 10 \\ \hline 9 & 2 \end{array}\right]\left[\begin{array}{c} x_1 \\ x_2 \end{array}\right] = \left[\begin{array}{c} 6 \\ 3 \end{array}\right]$$

SOLUTION Note that these equations are the same as those in Example 6.1, except that x_1 and x_2 are interchanged. Using (6.2.9),

$$k = 1 \qquad x_1(i+1) = \frac{1}{A_{11}}[y_1 - A_{12}x_2(i)] = \frac{1}{5}[6 - 10x_2(i)]$$

$$k = 2 \qquad x_2(i+1) = \frac{1}{A_{22}}[y_2 - A_{21}x_1(i+1)] = \frac{1}{2}[3 - 9x_1(i+1)]$$

Successive calculations of x_1 and x_2 are shown in the following table:

<table>
<tr><td>**GAUSS–SEIDEL**</td><td>i</td><td>0</td><td>1</td><td>2</td><td>3</td><td>4</td><td>5</td></tr>
<tr><td></td><td>$x_1(i)$</td><td>0</td><td>1.2</td><td>9</td><td>79.2</td><td>711</td><td>6397</td></tr>
<tr><td></td><td>$x_2(i)$</td><td>0</td><td>−3.9</td><td>−39</td><td>−354.9</td><td>−3198</td><td>−28786</td></tr>
</table>

The unique solution by matrix inversion is

$$\begin{bmatrix} x_1 \\ x_2 \end{bmatrix} = \begin{bmatrix} 5 & 10 \\ 9 & 2 \end{bmatrix}^{-1} \begin{bmatrix} 6 \\ 3 \end{bmatrix} = \frac{-1}{80} \begin{bmatrix} 2 & -10 \\ -9 & 5 \end{bmatrix} \begin{bmatrix} 6 \\ 3 \end{bmatrix} = \begin{bmatrix} 0.225 \\ 0.4875 \end{bmatrix}$$

As shown, Gauss–Seidel does not converge to the unique solution; instead it diverges. We could show that Jacobi also diverges for this example. ∎

If any diagonal element A_{kk} equals zero, then Jacobi and Gauss–Seidel are undefined, because the right-hand sides of (6.2.5) and (6.2.9) are divided by A_{kk}. Also, if any one diagonal element has too small a magnitude, these methods will diverge. In Examples 6.3 and 6.4, Jacobi and Gauss–Seidel converge, since the diagonals (10 and 9) are both large; in Example 6.5, however, the diagonals (5 and 2) are small compared to the off-diagonals, and the methods diverge.

In general, convergence of Jacobi or Gauss–Seidel can be evaluated by recognizing that (6.2.6) represents a digital filter with input $\mathbf{y}$ and output $\mathbf{x}(i)$. The z-transform of (6.2.6) may be employed to determine the filter transfer function and its poles. The output $\mathbf{x}(i)$ converges if and only if all the filter poles have magnitudes less than 1 (see Problems 6.16, 6.17, and 6.18).

Rate of convergence is also established by the filter poles. Fast convergence is obtained when the magnitudes of all the poles are small. In addition, experience with specific $\mathbf{A}$ matrices has shown that more iterations are required for Jacobi and Gauss–Seidel as the dimension N increases.

Computer storage requirements for Jacobi include N^2 memory locations for the $\mathbf{A}$ matrix and $3N$ locations for the vectors $\mathbf{y}$, $\mathbf{x}(i)$, and $\mathbf{x}(i+1)$. Storage space is also required for loops, arithmetic statements, and working space to compute (6.2.5). Gauss–Seidel requires N fewer memory locations, since for (6.2.9) the new value $x_k(i+1)$ can be stored in the location of the old value $x_k(i)$.

Computer time per iteration is relatively small for Jacobi and Gauss–Seidel. Inspection of (6.2.5) or (6.2.9) shows that N^2 multiplications/divisions and $N(N-1)$ subtractions per iteration are required [one division, $(N-1)$ multiplications, and $(N-1)$ subtractions for each $k = 1, 2, \ldots, N$]. But as was the case with Gauss elimination, if the matrix is sparse (i.e., most of the elements are zero), special sparse matrix algorithms can be used to substantially decrease both the storage requirements and the computation time.

6.3

ITERATIVE SOLUTIONS TO NONLINEAR ALGEBRAIC EQUATIONS: NEWTON–RAPHSON

A set of nonlinear algebraic equations in matrix format is given by

$$\mathbf{f(x)} = \begin{bmatrix} f_1(\mathbf{x}) \\ f_2(\mathbf{x}) \\ \vdots \\ f_N(\mathbf{x}) \end{bmatrix} = \mathbf{y} \tag{6.3.1}$$

where $\mathbf{y}$ and $\mathbf{x}$ are N vectors and $\mathbf{f(x)}$ is an N vector of functions. Given $\mathbf{y}$ and $\mathbf{f(x)}$, we want to solve for $\mathbf{x}$. The iterative methods described in Section 6.2 can be extended to nonlinear equations as follows. Rewriting (6.3.1),

$$\mathbf{0} = \mathbf{y} - \mathbf{f(x)} \tag{6.3.2}$$

Adding $\mathbf{Dx}$ to both sides of (6.3.2), where $\mathbf{D}$ is a square $N \times N$ invertible matrix,

$$\mathbf{Dx} = \mathbf{Dx} + \mathbf{y} - \mathbf{f(x)} \tag{6.3.3}$$

Premultiplying by $\mathbf{D}^{-1}$,

$$\mathbf{x} = \mathbf{x} + \mathbf{D}^{-1}[\mathbf{y} - \mathbf{f(x)}] \tag{6.3.4}$$

The old values $\mathbf{x}(i)$ are used on the right side of (6.3.4) to generate the new values $\mathbf{x}(i+1)$ on the left side. That is,

$$\mathbf{x}(i+1) = \mathbf{x}(i) + \mathbf{D}^{-1}\{\mathbf{y} - \mathbf{f}[\mathbf{x}(i)]\} \tag{6.3.5}$$

For linear equations, $\mathbf{f(x)} = \mathbf{Ax}$ and (6.3.5) reduces to

$$\mathbf{x}(i+1) = \mathbf{x}(i) + \mathbf{D}^{-1}[\mathbf{y} - \mathbf{Ax}(i)] = \mathbf{D}^{-1}(\mathbf{D} - \mathbf{A})\mathbf{x}(i) + \mathbf{D}^{-1}\mathbf{y} \tag{6.3.6}$$

which is identical to the Jacobi and Gauss–Seidel methods of (6.2.6). For nonlinear equations, the matrix $\mathbf{D}$ in (6.3.5) must be specified.

One method for specifying $\mathbf{D}$, called *Newton–Raphson*, is based on the following Taylor series expansion of $\mathbf{f(x)}$ about an operating point $\mathbf{x}_0$.

$$y = f(x_0) + \frac{d\mathbf{f}}{d\mathbf{x}}\bigg|_{\mathbf{x}=\mathbf{x}_0} (\mathbf{x} - \mathbf{x}_0) \cdots \tag{6.3.7}$$

Neglecting the higher order terms in (6.3.7) and solving for $\mathbf{x}$,

$$\mathbf{x} = \mathbf{x}_0 + \left[\frac{d\mathbf{f}}{d\mathbf{x}}\bigg|_{\mathbf{x}=\mathbf{x}_0} \right]^{-1} [\mathbf{y} - \mathbf{f}(\mathbf{x}_0)] \tag{6.3.8}$$

The Newton–Raphson method replaces $\mathbf{x}_0$ by the old value $\mathbf{x}(i)$ and $\mathbf{x}$ by the new value $\mathbf{x}(i + 1)$ in (6.3.8). Thus,

$$\mathbf{x}(i + 1) = \mathbf{x}(i) + \mathbf{J}^{-1}(i)\{\mathbf{y} - \mathbf{f}[\mathbf{x}(i)]\} \tag{6.3.9}$$

where

$$\mathbf{J}(i) = \frac{d\mathbf{f}}{d\mathbf{x}}\bigg|_{\mathbf{x}=\mathbf{x}(i)} = \begin{bmatrix} \dfrac{\partial f_1}{\partial x_1} & \dfrac{\partial f_1}{\partial x_2} & \cdots & \dfrac{\partial f_1}{\partial x_N} \\[2mm] \dfrac{\partial f_2}{\partial x_1} & \dfrac{\partial f_2}{\partial x_2} & \cdots & \dfrac{\partial f_2}{\partial x_N} \\[2mm] \vdots & \vdots & & \vdots \\[2mm] \dfrac{\partial f_N}{\partial x_1} & \dfrac{\partial f_N}{\partial x_2} & \cdots & \dfrac{\partial f_N}{\partial x_N} \end{bmatrix}_{\mathbf{x}=\mathbf{x}(i)} \tag{6.3.10}$$

The $N \times N$ matrix $\mathbf{J}(i)$, whose elements are the partial derivatives shown in (6.3.10), is called the Jacobian matrix. The Newton–Raphson method is similar to extended Gauss–Seidel, except that $\mathbf{D}$ in (6.3.5) is replaced by $\mathbf{J}(i)$ in (6.3.9).

EXAMPLE 6.6   Newton–Raphson method: solution to polynomial equations

---

Solve the scalar equation $f(x) = y$, where $y = 9$ and $f(x) = x^2$. Starting with $x(0) = 1$, use (a) Newton–Raphson and (b) extended Gauss–Seidel with $D = 3$ until (6.2.2) is satisfied for $\varepsilon = 10^{-4}$. Compare the two methods.

SOLUTION

a. Using (6.3.10) with $f(x) = x^2$,

$$\mathbf{J}(i) = \frac{d}{dx}(x^2)\bigg|_{x=x(i)} = 2x\bigg|_{x=x(i)} = 2x(i)$$

Using $\mathbf{J}(i)$ in (6.3.9),

$$x(i + 1) = x(i) + \frac{1}{2x(i)}[9 - x^2(i)]$$

Starting with $x(0) = 1$, successive calculations of the Newton–Raphson equation are shown in the following table:

NEWTON– RAPHSON	i	0	1	2	3	4	5
	$x(i)$	1	5.00000	3.40000	3.02353	3.00009	3.00000

b. Using (6.3.5) with $D = 3$, the Gauss–Seidel method is

$$x(i+1) = x(i) + \tfrac{1}{3}[9 - x^2(i)]$$

The corresponding Gauss–Seidel calculations are as follows:

GAUSS–SEIDEL (D = 3)	i	0	1	2	3	4	5	6
	$x(i)$	1	3.66667	2.18519	3.59351	2.28908	3.54245	2.35945

As shown, Gauss–Seidel oscillates about the solution, slowly converging, whereas Newton–Raphson converges in five iterations to the solution $x = 3$. Note that if $x(0)$ is negative, Newton–Raphson converges to the negative solution $x = -3$. Also, it is assumed that the matrix inverse $\mathbf{J}^{-1}$ exists. Thus the initial value $x(0) = 0$ should be avoided for this example. ∎

EXAMPLE 6.7 Newton–Raphson method: solution to nonlinear algebraic equations

Solve

$$\begin{bmatrix} x_1 + x_2 \\ x_1 x_2 \end{bmatrix} = \begin{bmatrix} 15 \\ 50 \end{bmatrix} \qquad \mathbf{x}(0) = \begin{bmatrix} 4 \\ 9 \end{bmatrix}$$

Use the Newton–Raphson method starting with the above $\mathbf{x}(0)$ and continue until (6.2.2) is satisfied with $\varepsilon = 10^{-4}$.

SOLUTION Using (6.3.10) with $f_1 = (x_1 + x_2)$ and $f_2 = x_1 x_2$,

$$\mathbf{J}(i)^{-1} = \left[\begin{array}{c|c} \dfrac{\partial f_1}{\partial x_1} & \dfrac{\partial f_1}{\partial x_2} \\ \hline \dfrac{\partial f_2}{\partial x_1} & \dfrac{\partial f_2}{\partial x_2} \end{array} \right]^{-1}_{\mathbf{x}=\mathbf{x}(i)} = \left[\begin{array}{c|c} 1 & 1 \\ \hline x_2(i) & x_1(i) \end{array} \right]^{-1} = \dfrac{\left[\begin{array}{c|c} x_1(i) & -1 \\ \hline -x_2(i) & 1 \end{array} \right]}{x_1(i) - x_2(i)}$$

Using $\mathbf{J}(i)^{-1}$ in (6.3.9),

$$\begin{bmatrix} x_1(i+1) \\ x_2(i+1) \end{bmatrix} = \begin{bmatrix} x_1(i) \\ x_2(i) \end{bmatrix} + \dfrac{\left[\begin{array}{c|c} x_1(i) & -1 \\ \hline -x_2(i) & 1 \end{array} \right]}{x_1(i) - x_2(i)} \begin{bmatrix} 15 - x_1(i) - x_2(i) \\ 50 - x_1(i)x_2(i) \end{bmatrix}$$

Writing the preceding as two separate equations,

$$x_1(i+1) = x_1(i) + \frac{x_1(i)[15 - x_1(i) - x_2(i)] - [50 - x_1(i)x_2(i)]}{x_1(i) - x_2(i)}$$

$$x_2(i+1) = x_2(i) + \frac{-x_2(i)[15 - x_1(i) - x_2(i)] + [50 - x_1(i)x_2(i)]}{x_1(i) - x_2(i)}$$

Successive calculations of these equations are shown in the following table:

i	0	1	2	3	4
$x_1(i)$	4	5.20000	4.99130	4.99998	5.00000
$x_2(i)$	9	9.80000	10.00870	10.00002	10.00000

Newton–Raphson converges in four iterations for this example. ∎

Equation (6.3.9) contains the matrix inverse $\mathbf{J}^{-1}$. Instead of computing $\mathbf{J}^{-1}$, (6.3.9) can be rewritten as follows:

$$\mathbf{J}(i)\Delta\mathbf{x}(i) = \Delta\mathbf{y}(i) \tag{6.3.11}$$

where

$$\Delta\mathbf{x}(i) = \mathbf{x}(i+1) - \mathbf{x}(i) \tag{6.3.12}$$

and

$$\Delta\mathbf{y}(i) = \mathbf{y} - \mathbf{f}[\mathbf{x}(i)] \tag{6.3.13}$$

Then, during each iteration, the following four steps are completed:

STEP 1 Compute $\Delta\mathbf{y}(i)$ from (6.3.13).

STEP 2 Compute $\mathbf{J}(i)$ from (6.3.10).

STEP 3 Using Gauss elimination and back substitution, solve (6.3.11) for $\Delta\mathbf{x}(i)$.

STEP 4 Compute $\mathbf{x}(i+1)$ from (6.3.12).

EXAMPLE 6.8   Newton–Raphson method in four steps

Complete the above four steps for the first iteration of Example 6.7.

SOLUTION

STEP 1 $\Delta\mathbf{y}(0) = \mathbf{y} - \mathbf{f}[\mathbf{x}(0)] = \begin{bmatrix} 15 \\ 50 \end{bmatrix} - \begin{bmatrix} 4+9 \\ (4)(9) \end{bmatrix} = \begin{bmatrix} 2 \\ 14 \end{bmatrix}$

STEP 2 $\mathbf{J}(0) = \begin{bmatrix} 1 & 1 \\ x_2(0) & x_1(0) \end{bmatrix} = \begin{bmatrix} 1 & 1 \\ 9 & 4 \end{bmatrix}$

STEP 3 Using $\Delta\mathbf{y}(0)$ and $\mathbf{J}(0)$, (6.3.11) becomes

$$\left[\begin{array}{c|c} 1 & 1 \\ \hline 9 & 4 \end{array}\right] \begin{bmatrix} \Delta x_1(0) \\ \Delta x_2(0) \end{bmatrix} = \begin{bmatrix} 2 \\ 14 \end{bmatrix}$$

Using Gauss elimination, subtract $\mathbf{J}_{21}/\mathbf{J}_{11} = 9/1 = 9$ times the first equation from the second equation, giving

$$\left[\begin{array}{c|c} 1 & 1 \\ \hline 0 & -5 \end{array}\right] \begin{bmatrix} \Delta x_1(0) \\ \Delta x_2(0) \end{bmatrix} = \begin{bmatrix} 2 \\ -4 \end{bmatrix}$$

Solving by back substitution,

$$\Delta x_2(0) = \frac{-4}{-5} = 0.8$$

$$\Delta x_1(0) = 2 - 0.8 = 1.2$$

STEP 4 $\quad \mathbf{x}(1) = \mathbf{x}(0) + \Delta\mathbf{x}(0) = \begin{bmatrix} 4 \\ 9 \end{bmatrix} + \begin{bmatrix} 1.2 \\ 0.8 \end{bmatrix} = \begin{bmatrix} 5.2 \\ 9.8 \end{bmatrix}$

This is the same as computed in Example 6.7. ∎

Experience from power-flow studies has shown that Newton–Raphson converges in many cases where Jacobi and Gauss–Seidel diverge. Furthermore, the number of iterations required for convergence is independent of the dimension N for Newton–Raphson, but increases with N for Jacobi and Gauss–Seidel. Most Newton–Raphson power-flow problems converge in fewer than ten iterations [1].

6.4

THE POWER-FLOW PROBLEM

The power-flow problem is the computation of voltage magnitude and phase angle at each bus in a power system under balanced three-phase steady-state conditions. As a by-product of this calculation, real and reactive power flows in equipment such as transmission lines and transformers, as well as equipment losses, can be computed.

The starting point for a power-flow problem is a single-line diagram of the power system, from which the input data for computer solutions can be obtained. Input data consist of bus data, transmission line data, and transformer data.

As shown in Figure 6.1, the following four variables are associated with each bus k: voltage magnitude V_k, phase angle δ_k, net real power P_k, and reactive power Q_k supplied to the bus. At each bus, two of these variables are specified as input data, and the other two are unknowns to be computed by

FIGURE 6.1

Bus variables V_k, δ_k, P_k, and Q_k

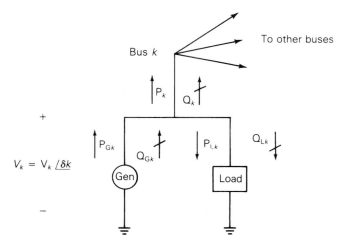

the power-flow program. For convenience, the power delivered to bus k in Figure 6.1 is separated into generator and load terms. That is,

$$P_k = P_{Gk} - P_{Lk}$$

$$Q_k = Q_{Gk} - Q_{Lk} \tag{6.4.1}$$

Each bus k is categorized into one of the following three bus types:

1. **Swing bus (or slack bus)**—There is only one swing bus, which for convenience is numbered bus 1 in this text. The swing bus is a reference bus for which V_1/δ_1, typically $1.0/0°$ per unit, is input data. The power-flow program computes P_1 and Q_1.

2. **Load bus**—P_k and Q_k are input data. The power-flow program computes V_k and δ_k. Most buses in a typical power-flow program are load buses.

3. **Voltage controlled bus**—P_k and V_k are input data. The power-flow program computes Q_k and δ_k. Examples are buses to which generators, switched shunt capacitors, or static var systems are connected. Maximum and minimum var limits Q_{Gkmax} and Q_{Gkmin} that this equipment can supply are also input data. Another example is a bus to which a tap-changing transformer is connected; the power-flow program then computes the tap setting.

Note that when bus k is a load bus with no generation, $P_k = -P_{Lk}$ is negative; that is, the real power supplied to bus k in Figure 6.1 is negative. If the load is inductive, $Q_k = -Q_{Lk}$ is negative.

Transmission lines are represented by the equivalent π circuit, shown in Figure 5.7. Transformers are also represented by equivalent circuits, as shown in Figure 3.9 for a two-winding transformer, Figure 3.20 for a three-winding transformer, or Figure 3.25 for a tap-changing transformer.

Input data for each transmission line include the per-unit equivalent π circuit series impedance Z' and shunt admittance Y', the two buses to which the line is connected, and maximum MVA rating. Similarly, input data for each transformer include per-unit winding impedances Z, the per-unit exciting branch admittance Y, the buses to which the windings are connected, and maximum MVA ratings. Input data for tap-changing transformers also include maximum tap settings.

The bus admittance matrix Y_{bus} can be constructed from the line and transformer input data. From (2.4.3) and (2.4.4), the elements of Y_{bus} are:

Diagonal elements: Y_{kk} = sum of admittances connected to bus k

Off-diagonal elements: $Y_{kn} = -($sum of admittances connected between buses k and $n)$

$$k \neq n \tag{6.4.2}$$

EXAMPLE 6.9   Power-flow input data and Y_{bus}

Figure 6.2 shows a single-line diagram of a five-bus power system. Input data are given in Tables 6.1, 6.2, and 6.3. As shown in Table 6.1, bus 1, to which a generator is connected, is the swing bus. Bus 3, to which a generator and a load are connected, is a voltage-controlled bus. Buses 2, 4, and 5 are load buses. Note that the loads at buses 2 and 3 are inductive since $Q_2 = -Q_{L2} = -2.8$ and $-Q_{L3} = -0.4$ are negative.

For each bus k, determine which of the variables V_k, δ_k, P_k, and Q_k are input data and which are unknowns. Also, compute the elements of the second row of Y_{bus}.

SOLUTION   The input data and unknowns are listed in Table 6.4. For bus 1, the swing bus, P_1 and Q_1 are unknowns. For bus 3, a voltage-controlled bus,

FIGURE 6.2

Single-line diagram for
Example 6.9

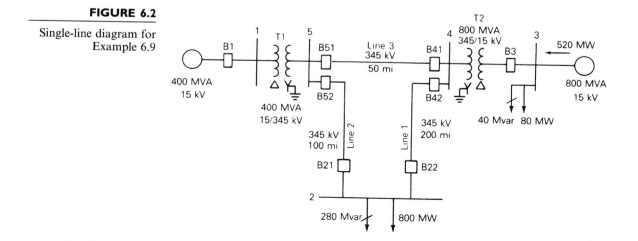

TABLE 6.1

Bus input data for Example 6.9*

Bus	Type	V per unit	δ degrees	P_G per unit	Q_G per unit	P_L per unit	Q_L per unit	Q_{Gmax} per unit	Q_{Gmin} per unit
1	Swing	1.0	0	—	—	0	0	—	—
2	Load	—	—	0	0	8.0	2.8	—	—
3	Constant voltage	1.05	—	5.2	—	0.8	0.4	4.0	−2.8
4	Load	—	—	0	0	0	0	—	—
5	Load	—	—	0	0	0	0	—	—

* $S_{base} = 100$ MVA, $V_{base} = 15$ kV at buses 1, 3, and 345 kV at buses 2, 4, 5

TABLE 6.2

Line input data for Example 6.9

Bus-to-Bus	R′ per unit	X′ per unit	G′ per unit	B′ per unit	Maximum MVA per unit
2–4	0.0090	0.100	0	1.72	12.0
2–5	0.0045	0.050	0	0.88	12.0
4–5	0.00225	0.025	0	0.44	12.0

TABLE 6.3

Transformer input data for Example 6.9

Bus-to-Bus	R per unit	X per unit	G_c per unit	B_m per unit	Maximum MVA per unit	Maximum TAP Setting per unit
1–5	0.00150	0.02	0	0	6.0	—
3–4	0.00075	0.01	0	0	10.0	—

TABLE 6.4

Input data and unknowns for Example 6.9

Bus	Input Data	Unknowns
1	$V_1 = 1.0$, $\delta_1 = 0$	P_1, Q_1
2	$P_2 = P_{G2} - P_{L2} = -8$	V_2, δ_2
	$Q_2 = Q_{G2} - Q_{L2} = -2.8$	
3	$V_3 = 1.05$	Q_3, δ_3
	$P_3 = P_{G3} - P_{L3} = 4.4$	
4	$P_4 = 0$, $Q_4 = 0$	V_4, δ_4
5	$P_5 = 0$, $Q_5 = 0$	V_5, δ_5

Q_3 and δ_3 are unknowns. For buses 2, 4, and 5, load buses, V_2, V_4, V_5 and δ_2, δ_4, δ_5 are unknowns.

The elements of Y_{bus} are computed from (6.4.2). Since buses 1 and 3 are not directly connected to bus 2,

$$Y_{21} = Y_{23} = 0$$

Using (6.4.2),

$$Y_{24} = \frac{-1}{R'_{24} + jX'_{24}} = \frac{-1}{0.009 + j0.1} = -0.89276 + j9.91964 \quad \text{per unit}$$

$$= 9.95972\underline{/95.143°} \quad \text{per unit}$$

$$Y_{25} = \frac{-1}{R'_{25} + jX'_{25}} = \frac{-1}{0.0045 + j0.05} = -1.78552 + j19.83932 \quad \text{per unit}$$

$$= 19.9195\underline{/95.143°} \quad \text{per unit}$$

$$Y_{22} = \frac{1}{R'_{24} + jX'_{24}} + \frac{1}{R'_{25} + jX'_{25}} + j\frac{B'_{24}}{2} + j\frac{B'_{25}}{2}$$

$$= (0.89276 - j9.91964) + (1.78552 - j19.83932) + j\frac{1.72}{2} + j\frac{0.88}{2}$$

$$= 2.67828 - j28.4590 = 28.5847\underline{/-84.624°} \quad \text{per unit}$$

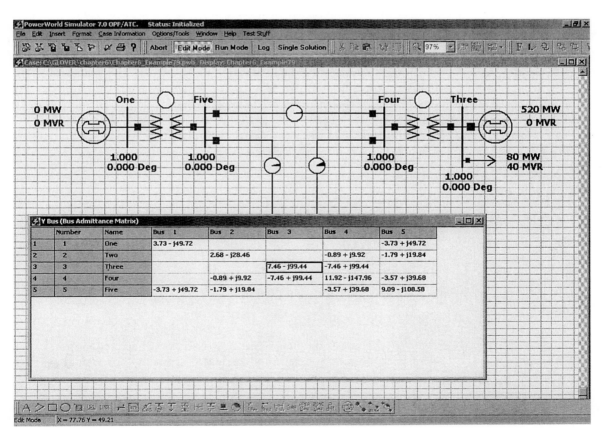

Screen for Example 6.9

where half of the shunt admittance of each line connected to bus 2 is included in Y_{22} (the other half is located at the other ends of these lines).

This five-bus power system is modeled in PowerWorld Simulator case Example 6_9. To view the input data, first click on the **Edit Mode** button to switch into the Edit mode (the Edit mode is used for modifying system parameters). Using the **Case Information** menu, you can view tabular displays showing the various parameters for the system. For example, Select **Case Information, Buses...** to view the parameters for each bus, and **Case Information, Lines and Transformers...** to view the parameters for the lines and transformer values. Fields shown in blue can be directly changed simply by typing over them, and those shown in green can be toggled by clicking on them.

The elements of the $\mathbf{Y}_{\text{bus}}$ can also be display by selecting **Case Information, Solution Details, $\mathbf{Y}_{\text{bus}}$**. Since the $\mathbf{Y}_{\text{bus}}$ entries are derived from other system parameters, they cannot be changed directly. Notice that several of the entries are blank, indicating that there is no line directly connecting these two buses (a blank entry is equivalent to zero). For larger networks, most of the elements of the $\mathbf{Y}_{\text{bus}}$ are zero since any single bus usually only has a few incident lines. The elements of the $\mathbf{Y}_{\text{bus}}$ can be saved in a Matlab format file by first right-clicking within the Ybus matrix to display the local menu, and then selecting "Save Ybus in Matlab Format" from the menu.

Finally, notice that no flows are shown on the one-line, because the nonlinear power-flow equations have not yet been solved. We cover the solution of these equations next. ∎

Using $\mathbf{Y}_{\text{bus}}$, we can write nodal equations for a power system network, as follows:

$$I = Y_{\text{bus}} V \tag{6.4.3}$$

where I is the N vector of source currents injected into each bus and V is the N vector of bus voltages. For bus k, the kth equation in (6.4.3) is

$$I_k = \sum_{n=1}^{N} Y_{kn} V_n \tag{6.4.4}$$

The complex power delivered to bus k is

$$S_k = \mathrm{P}_k + j\mathrm{Q}_k = V_k I_k^* \tag{6.4.5}$$

Power-flow solutions by Gauss–Seidel are based on nodal equations, (6.4.4), where each current source I_k is calculated from (6.4.5). Using (6.4.4) in (6.4.5),

$$\mathrm{P}_k + j\mathrm{Q}_k = V_k \left[\sum_{n=1}^{N} Y_{kn} V_n \right]^* \qquad k = 1, 2, \ldots, N \tag{6.4.6}$$

With the following notation,

$$V_n = V_n e^{j\delta_n} \tag{6.4.7}$$

$$Y_{kn} = Y_{kn} e^{j\theta_{kn}} = G_{kn} + jB_{kn} \qquad k, n = 1, 2, \ldots, N \tag{6.4.8}$$

(6.4.6) becomes

$$P_k + jQ_k = V_k \sum_{n=1}^{N} Y_{kn} V_n e^{j(\delta_k - \delta_n - \theta_{kn})} \tag{6.4.9}$$

Taking the real and imaginary parts of (6.4.9), we can write the power balance equations as either

$$P_k = V_k \sum_{n=1}^{N} Y_{kn} V_n \cos(\delta_k - \delta_n - \theta_{kn}) \tag{6.4.10}$$

$$Q_k = V_k \sum_{n=1}^{N} Y_{kn} V_n \sin(\delta_k - \delta_n - \theta_{kn}) \qquad k = 1, 2, \ldots, N \tag{6.4.11}$$

or when the Y_{kn} is expressed in rectangular coordinates by

$$P_K = V_K \sum_{n=1}^{N} V_n [G_{kn} \cos(\delta_k - \delta_n) + B_{kn} \sin(\delta_k - \delta_n)] \tag{6.4.12}$$

$$Q_K = V_K \sum_{n=1}^{N} V_n [G_{kn} \sin(\delta_k - \delta_n) - B_{kn} \cos(\delta_k - \delta_n)] \qquad k = 1, 2, \ldots, N \tag{6.4.13}$$

Power-flow solutions by Newton–Raphson are based on the nonlinear power-flow equations given by (6.4.10) and (6.4.11) [or alternatively by (6.4.12) and (6.4.13)].

6.5

POWER-FLOW SOLUTION BY GAUSS–SEIDEL

Nodal equations $I = Y_{bus} V$ are a set of linear equations analogous to $y = Ax$, solved in Section 6.2 using Gauss–Seidel. Since power-flow bus data consists of P_k and Q_k for load buses or P_k and V_k for voltage-controlled buses, nodal equations do not directly fit the linear equation format; the current source vector I is unknown and the equations are actually nonlinear. For each load bus, I_k can be calculated from (6.4.5), giving

$$I_k = \frac{P_k - jQ_k}{V_k^*} \tag{6.5.1}$$

Applying the Gauss–Seidel method, (6.2.9), to the nodal equations, with I_k given above, we obtain

$$V_k(i+1) = \frac{1}{Y_{kk}} \left[\frac{P_k - jQ_k}{V_k^*(i)} - \sum_{n=1}^{k-1} Y_{kn} V_n(i+1) - \sum_{n=k+1}^{N} Y_{kn} V_n(i) \right] \quad (6.5.2)$$

Equation (6.5.2) can be applied twice during each iteration for load buses, first using $V_k^*(i)$, then replacing $V_k^*(i)$, by $V_k^*(i+1)$ on the right side of (6.5.2).

For a voltage-controlled bus, Q_k is unknown, but can be calculated from (6.4.11), giving

$$Q_k = V_k(i) \sum_{n=1}^{N} Y_{kn} V_n(i) \, \sin[\delta_k(i) - \delta_n(i) - \theta_{kn}] \quad (6.5.3)$$

Also,

$$Q_{Gk} = Q_k + Q_{Lk}$$

If the calculated value of Q_{Gk} does not exceed its limits, then Q_k is used in (6.5.2) to calculate $V_k(i+1) = V_k(i+1)/\delta_k(i+1)$. Then the magnitude $V_k(i+1)$ is changed to V_k, which is input data for the voltage-controlled bus. Thus we use (6.5.2) to compute only the angle $\delta_k(i+1)$ for voltage-controlled buses.

If the calculated value exceeds its limit $Q_{Gk\text{max}}$ or $Q_{Gk\text{min}}$ during any iteration, then the bus type is changed from a voltage-controlled bus to a load bus, with Q_{Gk} set to its limit value. Under this condition, the voltage-controlling device (capacitor bank, static var system, and so on) is not capable of maintaining V_k as specified by the input data. The power-flow program then calculates a new value of V_k.

For the swing bus, denoted bus 1, V_1 and δ_1 are input data. As such, no iterations are required for bus 1. After the iteration process has converged, one pass through (6.4.10) and (6.4.11) can be made to compute P_1 and Q_1.

EXAMPLE 6.10   Power-flow solution by Gauss–Seidel

For the power system of Example 6.9, use Gauss–Seidel to calculate $V_2(1)$, the phasor voltage at bus 2 after the first iteration. Use zero initial phase angles and 1.0 per-unit initial voltage magnitudes (except at bus 3, where $V_3 = 1.05$) to start the iteration procedure.

SOLUTION   Bus 2 is a load bus. Using the input data and bus admittance values from Example 6.9 in (6.5.2),

$$V_2(1) = \frac{1}{Y_{22}} \left\{ \frac{P_2 - jQ_2}{V_2^*(0)} - [Y_{21}V_1(1) + Y_{23}V_3(0) + Y_{24}V_4(0) + Y_{25}V_5(0)] \right\}$$

$$= \frac{1}{28.5847\underline{/-84.624°}} \left\{ \frac{-8 - j(-2.8)}{1.0\underline{/0°}} \right.$$

$$\left. - [(-1.78552 + j19.83932)(1.0) + (-0.89276 + j9.91964)(1.0)] \right\}$$

$$= \frac{(-8 + j2.8) - (-2.67828 + j29.7589)}{28.5847\underline{/-84.624°}}$$

$$= 0.96132\underline{/-16.543°} \quad \text{per unit}$$

Next, the above value is used in (6.5.2) to recalculate $V_2(1)$:

$$V_2(1) = \frac{1}{28.5847\underline{/-84.624°}} \left\{ \frac{-8 + j2.8}{0.96132\underline{/16.543°}} \right.$$

$$\left. - [-2.67828 + j29.75829] \right\}$$

$$= \frac{-4.4698 - j24.5973}{28.5847\underline{/-84.624°}} = 0.87460\underline{/-15.675°} \quad \text{per unit}$$

Computations are next performed at buses 3, 4, and 5 to complete the first Gauss–Seidel iteration.

To see the complete convergence of this case, open PowerWorld Simulator case Example 6_10. By default, PowerWorld Simulator uses the Newton–Raphson method described in the next section. However, the case can be solved with the Gauss–Seidel approach by selecting **Simulation, Gauss– Seidel Power Flow**. To avoid getting stuck in an infinite loop if a case does not converge, PowerWorld Simulator places a limit on the maximum number of iterations. Usually for a Gauss–Seidel procedure this number is quite high, perhaps equal to 100 iterations. However, in this example to demonstrate the convergence characteristics of the Gauss–Seidel method it has been set to a single iteration, allowing the voltages to be viewed after each iteration. To step through the solution one iteration at a time, just repeatedly select **Simulation, Gauss–Seidel Power Flow**.

A common stopping criteria for the Gauss–Seidel is to use the scaled difference in the voltage from one iteration to the next (6.2.2). When this difference is below a specified convergence tolerance ε for each bus, the problem is considered solved. An alternative approach, implemented in PowerWorld Simulator, is to examine the real and reactive mismatch equations, defined as the difference between the right- and left-hand sides of (6.4.10) and (6.4.11). PowerWorld Simulator continues iterating until all the bus mismatches are below an MVA (or kVA) tolerance. When single-stepping through the solution, the bus mismatches can be viewed after each iteration on the **Case**

Information, Mismatches display. The solution mismatch tolerance can be changed on the Power Flow Solution page of the PowerWorld Simulator Options dialog (select **Options, Simulator Options**, then select the **Power Flow Solution** category to view this dialog); the maximum number of iterations can also be changed from this page. A typical convergence tolerance is about 0.5 MVA.

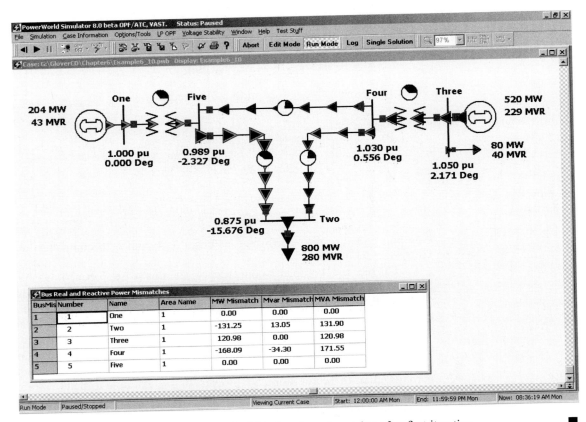

Screen for Example 6.10, showing mismatches after first iteration ∎

6.6

POWER-FLOW SOLUTION BY NEWTON–RAPHSON

Equations (6.4.10) and (6.4.11) are analogous to the nonlinear equation $y = f(x)$, solved in Section 6.3 by Newton–Raphson. We define the x, y, and f vectors for the power-flow problem as

$$\mathbf{x} = \begin{bmatrix} \boldsymbol{\delta} \\ \mathbf{V} \end{bmatrix} = \begin{bmatrix} \delta_2 \\ \vdots \\ \delta_N \\ V_2 \\ \vdots \\ V_N \end{bmatrix}; \quad \mathbf{y} = \begin{bmatrix} \mathbf{P} \\ \mathbf{Q} \end{bmatrix} = \begin{bmatrix} P_2 \\ \vdots \\ P_N \\ Q_2 \\ \vdots \\ Q_N \end{bmatrix};$$

$$\mathbf{f}(\mathbf{x}) = \begin{bmatrix} \mathbf{P}(\mathbf{x}) \\ \mathbf{Q}(\mathbf{x}) \end{bmatrix} = \begin{bmatrix} P_2(\mathbf{x}) \\ \vdots \\ P_N(\mathbf{x}) \\ Q_2(\mathbf{x}) \\ \vdots \\ Q_N(\mathbf{x}) \end{bmatrix} \tag{6.6.1}$$

where all V, P, and Q terms are in per-unit and δ terms are in radians. The swing bus variables δ_1 and V_1 are omitted from (6.6.1), since they are already known. Equations (6.4.10) and (6.4.11) then have the following form:

$$y_k = P_k = P_k(\mathbf{x}) = V_k \sum_{n=1}^{N} Y_{kn} V_n \cos(\delta_k - \delta_n - \theta_{kn}) \tag{6.6.2}$$

$$y_{k+N} = Q_k = Q_k(\mathbf{x}) = V_k \sum_{n=1}^{N} Y_{kn} V_n \sin(\delta_k - \delta_n - \theta_{kn})$$

$$k = 2, 3, \ldots, N \tag{6.6.3}$$

The Jacobian matrix of (6.3.10) has the form

$$\mathbf{J} = \begin{bmatrix}
\dfrac{\partial P_2}{\partial \delta_2} & \cdots & \dfrac{\partial P_2}{\partial \delta_N} & \dfrac{\partial P_2}{\partial V_2} & \cdots & \dfrac{\partial P_2}{\partial V_N} \\
\vdots & & & \vdots & & \\
\dfrac{\partial P_N}{\partial \delta_2} & \cdots & \dfrac{\partial P_N}{\partial \delta_N} & \dfrac{\partial P_N}{\partial V_2} & \cdots & \dfrac{\partial P_N}{\partial V_N} \\
\dfrac{\partial Q_2}{\partial \delta_2} & \cdots & \dfrac{\partial Q_2}{\partial \delta_N} & \dfrac{\partial Q_2}{\partial V_2} & \cdots & \dfrac{\partial Q_2}{\partial V_N} \\
\vdots & & & \vdots & & \\
\dfrac{\partial Q_N}{\partial \delta_2} & \cdots & \dfrac{\partial Q_N}{\partial \delta_N} & \dfrac{\partial Q_N}{\partial V_2} & \cdots & \dfrac{\partial Q_N}{\partial V_N}
\end{bmatrix} \tag{6.6.4}$$

where the blocks are labeled **J1**, **J2**, **J3**, and **J4**.

TABLE 6.5

Elements of the
Jacobian matrix

$n \neq k$

$$J1_{kn} = \frac{\partial P_k}{\partial \delta_n} = V_k Y_{kn} V_n \sin(\delta_k - \delta_n - \theta_{kn})$$

$$J2_{kn} = \frac{\partial P_k}{\partial V_n} = V_k Y_{kn} \cos(\delta_k - \delta_n - \theta_{kn})$$

$$J3_{kn} = \frac{\partial Q_k}{\partial \delta_n} = -V_k Y_{kn} V_n \cos(\delta_k - \delta_n - \theta_{kn})$$

$$J4_{kn} = \frac{\partial Q_k}{\partial V_n} = V_k Y_{kn} \sin(\delta_k - \delta_n - \theta_{kn})$$

$n = k$

$$J1_{kk} = \frac{\partial P_k}{\partial \delta_k} = -V_k \sum_{\substack{n=1 \\ n \neq k}}^{N} Y_{kn} V_n \sin(\delta_k - \delta_n - \theta_{kn})$$

$$J2_{kk} = \frac{\partial P_k}{\partial V_k} = V_k Y_{kk} \cos\theta_{kk} + \sum_{n=1}^{N} Y_{kn} V_n \cos(\delta_k - \delta_n - \theta_{kn})$$

$$J3_{kk} = \frac{\partial Q_k}{\partial \delta_k} = V_k \sum_{\substack{n=1 \\ n \neq k}}^{N} Y_{kn} V_n \cos(\delta_k - \delta_n - \theta_{kn})$$

$$J4_{kk} = \frac{\partial Q_k}{\partial V_k} = -V_k Y_{kk} \sin\theta_{kk} + \sum_{n=1}^{N} Y_{kn} V_n \sin(\delta_k - \delta_n - \theta_{kn})$$

$$k, n = 2, 3, \dots, N$$

Equation (6.6.4) is partitioned into four blocks. The partial derivatives in each block, derived from (6.6.2) and (6.6.3), are given in Table 6.5.

We now apply to the power-flow problem the four Newton–Raphson steps outlined in Section 6.3, starting with $\mathbf{x}(i) = \begin{bmatrix} \boldsymbol{\delta}(i) \\ \mathbf{V}(i) \end{bmatrix}$ at the ith iteration.

STEP 1   Use (6.6.2) and (6.6.3) to compute

$$\Delta \mathbf{y}(i) = \begin{bmatrix} \Delta \mathbf{P}(i) \\ \Delta \mathbf{Q}(i) \end{bmatrix} = \begin{bmatrix} \mathbf{P} - \mathbf{P}[\mathbf{x}(i)] \\ \mathbf{Q} - \mathbf{Q}[\mathbf{x}(i)] \end{bmatrix} \tag{6.6.5}$$

STEP 2   Use the equations in Table 6.5 to calculate the Jacobian matrix.

STEP 3   Use Gauss elimination and back substitution to solve

$$\left[\begin{array}{c|c} \mathbf{J1}(i) & \mathbf{J2}(i) \\ \hline \mathbf{J3}(i) & \mathbf{J4}(i) \end{array} \right] \begin{bmatrix} \Delta \boldsymbol{\delta}(i) \\ \Delta \mathbf{V}(i) \end{bmatrix} = \begin{bmatrix} \Delta \mathbf{P}(i) \\ \Delta \mathbf{Q}(i) \end{bmatrix} \tag{6.6.6}$$

STEP 4   Compute

$$\mathbf{x}(i+1) = \begin{bmatrix} \boldsymbol{\delta}(i+1) \\ \mathbf{V}(i+1) \end{bmatrix} = \begin{bmatrix} \boldsymbol{\delta}(i) \\ \mathbf{V}(i) \end{bmatrix} + \begin{bmatrix} \Delta \boldsymbol{\delta}(i) \\ \Delta \mathbf{V}(i) \end{bmatrix} \tag{6.6.7}$$

Starting with initial value $\mathbf{x}(0)$, the procedure continues until convergence is obtained or until the number of iterations exceeds a specified maximum. Convergence criteria are often based on $\Delta\mathbf{y}(i)$ (called *power mismatches*) rather than on $\Delta\mathbf{x}(i)$ (phase angle and voltage magnitude mismatches).

For each voltage-controlled bus, the magnitude V_k is already known, and the function $Q_k(\mathbf{x})$ is not needed. Therefore, we could omit V_k from the $\mathbf{x}$ vector and Q_k from the $\mathbf{y}$ vector. We could also omit from the Jacobian matrix the column corresponding to partial derivatives with respect to V_k and the row corresponding to partial derivatives of $Q_k(\mathbf{x})$. Alternatively, rows and corresponding columns for voltage-controlled buses can be retained in the Jacobian matrix. Then during each iteration, the voltage magnitude $V_k(i+1)$ of each voltage-controlled bus is reset to V_k, which is input data for that bus.

At the end of each iteration, we compute $Q_k(\mathbf{x})$ from (6.6.3) and $Q_{Gk} = Q_k(\mathbf{x}) + Q_{Lk}$ for each voltage-controlled bus. If the computed value of Q_{Gk} exceeds its limits, then the bus type is changed to a load bus with Q_{Gk} set to its limit value. The power-flow program also computes a new value for V_k.

EXAMPLE 6.11   **Jacobian matrix and power-flow solution by Newton–Raphson**

Determine the dimension of the Jacobian matrix for the power system in Example 6.9. Also calculate $\Delta P_2(0)$ in Step 1 and $J1_{24}(0)$ is Step 2 of the first Newton–Raphson iteration. Assume zero initial phase angles and 1.0 per-unit initial voltage magnitudes (except $V_3 = 1.05$).

SOLUTION   Since there are $N = 5$ buses for Example 6.9, (6.6.2) and (6.6.3) constitute $2(N-1) = 8$ equations, for which $\mathbf{J}(i)$ has dimension 8×8. However, there is one voltage-controlled bus, bus 3. Therefore, V_3 and the equation for $Q_3(\mathbf{x})$ could be eliminated, with $\mathbf{J}(i)$ reduced to a 7×7 matrix.

From Step 1 and (6.6.2),

$$\Delta P_2(0) = P_2 - P_2(\mathbf{x}) = P_2 - V_2(0)\{Y_{21}V_1 \cos[\delta_2(0) - \delta_1(0) - \theta_{21}]$$
$$+ Y_{22}V_2 \cos[-\theta_{22}] + Y_{23}V_3 \cos[\delta_2(0) - \delta_3(0) - \theta_{23}]$$
$$+ Y_{24}V_4 \cos[\delta_2(0) - \delta_4(0) - \theta_{24}]$$
$$+ Y_{25}V_5 \cos[\delta_2(0) - \delta_5(0) - \theta_{25}]\}$$

$$\Delta P_2(0) = -8.0 - 1.0\{28.5847(1.0) \cos(84.624°)$$
$$+ 9.95972(1.0) \cos(-95.143°)$$
$$+ 19.9159(1.0) \cos(-95.143°)\}$$

$$= -8.0 - (-2.89 \times 10^{-4}) = -7.99972 \quad \text{per unit}$$

From Step 2 and J1 given in Table 6.5

$$J1_{24}(0) = V_2(0)Y_{24}V_4(0) \sin[\delta_2(0) - \delta_4(0) - \theta_{24}]$$
$$= (1.0)(9.95972)(1.0) \sin[-95.143°]$$
$$= -9.91964 \quad \text{per unit}$$

To see the complete convergence of this case, open PowerWorld Simulator case Example 6_11. Select **Case Information, Mismatches** to see the initial mismatches, and **Case Information, Solution Details, Power Flow Jacobian** to view the initial Jacobian matrix. As is common in commercial power flows, PowerWorld Simulator actually includes rows in the Jacobian for voltage-controlled buses. When a generator is regulating its terminal voltage, this row corresponds to the equation setting the bus voltage magnitude equal to the generator voltage setpoint. However, if the generator hits a reactive power limit, the bus type is switched to a load bus.

To step through the Newton–Raphson solution, select **Simulation, Polar NR Power Flow**. As in Example 6.10, the maximum number of iterations has been set to 1, allowing the voltages, mismatches, and Jacobian to be viewed after each iteration. The power flow should converge to the Tables 6.6, 6.7, and 6.8 solution in three iterations.

TABLE 6.6

Bus output data for the power system given in Example 6.9

Bus #	Voltage Magnitude (per unit)	Phase Angle (degrees)	Generation PG (per unit)	Generation QG (per unit)	Load PL (per unit)	Load QL (per unit)
1	1.000	0.000	3.948	1.144	0.000	0.000
2	0.834	−22.407	0.000	0.000	8.000	2.800
3	1.050	−0.597	5.200	3.376	0.800	0.400
4	1.019	−2.834	0.000	0.000	0.000	0.000
5	0.974	−4.548	0.000	0.000	0.000	0.000
		TOTAL	9.148	4.516	8.800	3.200

TABLE 6.7

Line output data for the power system given in Example 6.9

Line #	Bus to Bus		P	Q	S
1	2	4	−2.920	−1.392	3.232
	4	2	3.036	1.216	3.272
2	2	5	−5.080	−1.408	5.272
	5	2	5.256	2.632	5.876
3	4	5	1.344	1.504	2.016
	5	4	−1.332	−1.824	2.260

TABLE 6.8

Transformer output data for the power system given in Example 6.9

Tran. #	Bus to Bus		P	Q	S
1	1	5	3.948	1.144	4.112
	5	1	−3.924	−0.804	4.004
2	3	4	4.400	2.976	5.312
	4	3	−4.380	−2.720	5.156

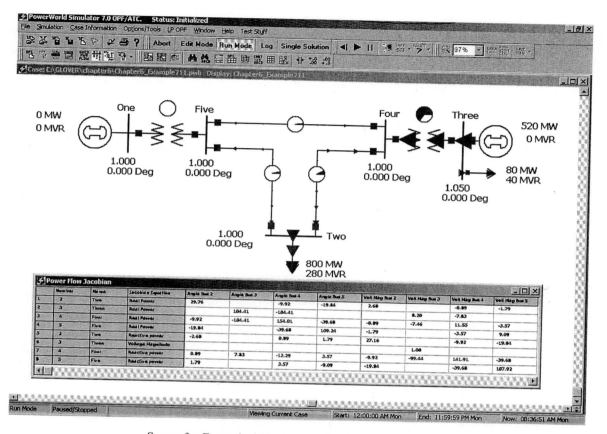

Screen for Example 6.11, showing Jacobian matrix at first iteration ∎

EXAMPLE 6.12   Power-flow program: change in generation

Using the power-flow system given in Example 6_9, determine the acceptable generation range at bus 3, keeping each line and transformer loaded at or below 100% of its MVA limit.

SOLUTION   Load PowerWorld Simulator case Example 6_9. Select **Single Solution** to perform a single power-flow solution using the Newton–Raphson approach. Then view the **Case Information** displays to verify that the Power-World Simulator solution matches the solution shown in Tables 6.6, 6.7, and 6.8. Additionally, the pie charts on the one-lines show the percentage line and transformer loadings. Initially transformer T1, between buses 1 and 5, is loaded at about 68% of its maximum MVA limit, while transformer T2, between buses 3 and 4, is loaded at about 53%.

Next, the bus 3 generation needs to be varied. This can be done a number of different ways in PowerWorld Simulator. The easiest (for this example) is to use the bus 3 generator MW one-line field to manually change the generation. Right-click on the "520 MW" field to the right of the bus 3 generator and select 'Generator Field Information' dialog to view the 'Generator Field Options' dialog. Set the "Delta Per Mouse Click" field to 10 and select OK. Small arrows are now visible next to this field on the one-line; clicking on the up arrow increases the generator's MW output by 10 MW, while clicking on the down arrow decreases the generation by 10 MW. Select **Simulation, Solve and Animate** to begin the simulation. Increase the generation until the pie chart for the transformer from bus 3 to 4 is loaded to 100%. This occurs at about 1000 MW. Notice that as the bus 3 generation is increased the bus 1 slack generation decreases by a similar amount. Repeat the process, except now decreasing the generation. This unloads the transformer from bus 3 to 4, but increases the loading on the transformer from bus 1 to bus 5. The bus 1 to 5 transformer should reach 100% loading with the bus 3 generation equal to about 330 MW.

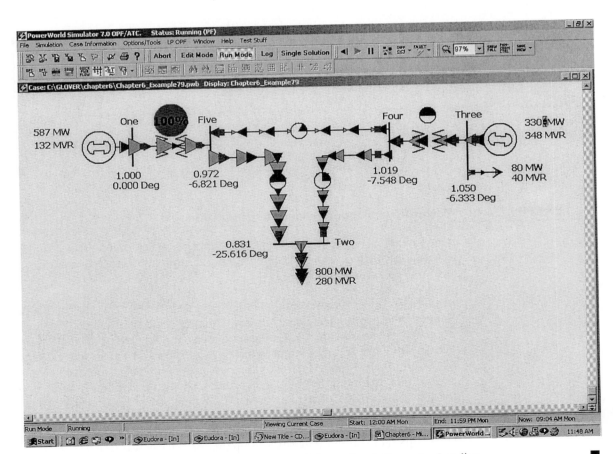

Screen for Example 6.12, Minimum Bus 3 Generator Loading

Voltage-controlled buses to which tap-changing or voltage-regulating transformers are connected can be handled by various methods. One method is to treat each of these buses as a load bus. The equivalent π circuit parameters (Figure 3.25) are first calculated with tap setting $c = 1.0$ for starting. During each iteration, the computed bus voltage magnitude is compared with the desired value specified by the input data. If the computed voltage is low (or high), c is increased (or decreased) to its next setting, and the parameters of the equivalent π circuit as well as Y_{bus} are recalculated. The procedure continues until the computed bus voltage magnitude equals the desired value within a specified tolerance, or until the high or low tap-setting limit is reached. Phase-shifting transformers can be handled in a similar way by using a complex turns ratio $c = 1.0 \underline{/\alpha}$, and by varying the phase-shift angle α.

A method with faster convergence makes c a variable and includes it in the **x** vector of (6.6.1). An equation is then derived to enter into the Jacobian matrix [4].

In comparing the Gauss-Seidel and Newton-Raphson algorithms, experience from power-flow studies has shown that Newton-Raphson converges in many cases where Jacobi and Gauss-Seidel diverge. Furthermore, the number of iterations required for convergence is independent of the number of buses N for Newton-Raphson, but increases with N for Jacobi and Gauss-Seidel. The principal advantage of the Jacobi and Gauss-Seidel methods had been their more modest memory storage requirements and their lower computational requirements per iteration. However, with the vast increases in low-cost computer memory over the last several decades, coupled with the need to solve power-flow problems with tens of thousands of buses, these advantages have been essentially eliminated. Therefore the Newton-Raphson, or one of the derivative methods discussed in Sections 6.9 and 6.10, are the preferred power-flow solution approaches.

EXAMPLE 6.13	**Power-flow program: 37-bus system**

To see a power-flow example of a larger system, open PowerWorld Simulator case Example 6_13. This case models a 37-bus, 9-generator power system containing three different voltage levels (345 kV, 138 kV, and 69 kV) with 57 transmission lines or transformers. The one-line can be panned by pressing the arrow keys, and it can be zoomed by pressing the ⟨ctrl⟩ with the up arrow key to zoom in or with the down arrow key to zoom out. Use **Simulation, Solve and Animate** to animate the one-line and **Simulation, Stop Solution/ Animation** to stop the animation.

Determine the lowest per-unit voltage and the maximum line/transformer loading both for the initial case and for the case with the line from bus TIM69 to HANNAH69 out of service.

SOLUTION Select **Single Solution** to initially solve the power flow, and then **Case Information, Buses...** to view a listing of all the buses in the case. To quickly determine the lowest per-unit voltage magnitude, left-click on the PU Volt column header to sort the column (clicking a second time reverses the

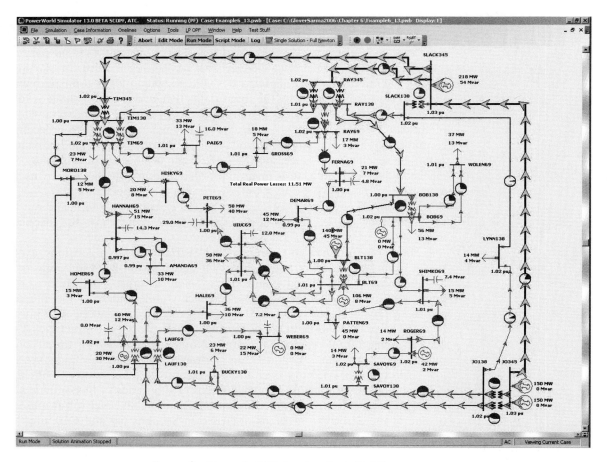

Screen for Example 6.13 showing the initial flows

sort). The lowest initial voltage magnitude is 0.9902 at bus DEMAR69. Next, select **Case Information, Lines and Transformers...** to view the Line and Transformer Records display. Left-click on % of Max Limit to sort the lines by percentage loading. Initially the highest percentage loading is 64.9% on the line from UIUC69 to BLT69 circuit 1.

There are several ways to remove the TIM69 to HANA69 line. One approach is to locate the line on the Line and Transformer Records display and then double-click on the Status field to change its value. An alternative approach is to find the line on the one-line (it is in the upper-lefthand portion) and then click on one of its circuit breakers. Once the line is removed, use **Single Solution** to resolve the power flow. The lowest per-unit voltage is now 0.9104 at AMANDA69 and the highest percentage line loading is 134.8%, on the line from HOMER69 to LAUF69. Since there are now several bus and line violations, the power system is no longer at a secure operating point. Control actions and/or design improvements are needed to correct these problems. Design Project 1 discusses these options. ∎

6.7

CONTROL OF POWER FLOW

The following means are used to control system power flows:

1. Prime mover and excitation control of generators.

2. Switching of shunt capacitor banks, shunt reactors, and static var systems.

3. Control of tap-changing and regulating transformers.

A simple model of a generator operating under balanced steady-state conditions is the Thévenin equivalent shown in Figure 6.3. V_t is the generator terminal voltage, E_g is the excitation voltage, δ is the power angle, and X_g is the positive-sequence synchronous reactance. From the figure, the generator current is

$$I = \frac{E_g e^{j\delta} - V_t}{jX_g} \tag{6.7.1}$$

and the complex power delivered by the generator is

$$S = P + jQ = V_t I^* = V_t \left(\frac{E_g e^{-j\delta} - V_t}{-jX_g} \right)$$

$$= \frac{V_t E_g (j \cos \delta + \sin \delta) - jV_t^2}{X_g} \tag{6.7.2}$$

The real and reactive powers delivered are then

$$P = \text{Re } S = \frac{V_t E_g}{X_g} \sin \delta \tag{6.7.3}$$

$$Q = \text{Im } S = \frac{V_t}{X_g} (E_g \cos \delta - V_t) \tag{6.7.4}$$

Equation (6.7.3) shows that the real power P increases when the power angle δ increases. From an operational standpoint, when the prime mover increases the power input to the generator while the excitation voltage is held constant, the rotor speed increases. As the rotor speed increases, the power angle δ also increases, causing an increase in generator real power output P. There is also a decrease in reactive power output Q, given by (6.7.4). However, when δ is

FIGURE 6.3

Generator Thévenin
equivalent

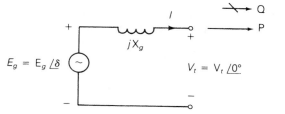

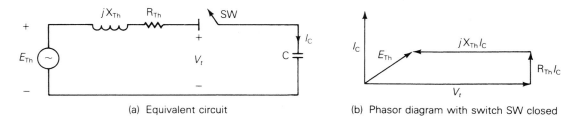

FIGURE 6.4   Effect of adding a shunt capacitor bank to a power system bus

less than $15°$, the increase in P is much larger than the decrease in Q. From the power-flow standpoint, an increase in prime-move power corresponds to an increase in P at the constant-voltage bus to which the generator is connected. The power-flow program computes the increase in δ along with the small change in Q.

Equation (6.7.4) shows that reactive power output Q increases when the excitation voltage E_g increases. From the operational standpoint, when the generator exciter output increases while holding the prime-mover power constant, the rotor current increases. As the rotor current increases, the excitation voltage E_g also increases, causing an increase in generator reactive power output Q. There is also a small decrease in δ required to hold P constant in (6.7.3). From the power-flow standpoint, an increase in generator excitation corresponds to an increase in voltage magnitude at the constant-voltage bus to which the generator is connected. The power-flow program computes the increase in reactive power Q supplied by the generator along with the small change in δ.

Figure 6.4 shows the effect of adding a shunt capacitor bank to a power system bus. The system is modeled by its Thévenin equivalent. Before the capacitor bank is connected, the switch SW is open and the bus voltage equals E_{Th}. After the bank is connected, SW is closed, and the capacitor current I_C leads the bus voltage V_t by $90°$. The phasor diagram shows that V_t is larger than E_{Th} when SW is closed. From the power-flow standpoint, the addition of a shunt capacitor bank to a load bus corresponds to the addition of a negative reactive load, since a capacitor absorbs negative reactive power. The power-flow program computes the increase in bus voltage magnitude along with the small change in δ. Similarly, the addition of a shunt reactor corresponds to the addition of a positive reactive load, wherein the power-flow program computes the decrease in voltage magnitude.

Tap-changing and voltage-magnitude–regulating transformers are used to control bus voltages as well as reactive power flows on lines to which they are connected. Similarly, phase-angle regulating transformers are used to control bus angles as well as real power flows on lines to which they are connected. Both tap-changing and regulating transformers are modeled by a transformer with an off-nominal turns ratio c (Figure 3.25). From the power-flow standpoint, a change in tap setting or voltage regulation corresponds to a change in c. The power-flow program computes the changes in Y_{bus}, bus voltage magnitudes and angles, and branch flows.

Besides the above controls, the power-flow program can be used to investigate the effect of switching in or out lines, transformers, loads, and generators. Proposed system changes to meet future load growth, including new transmission, new transformers, and new generation can also be investigated. Power-flow design studies are normally conducted by trial and error. Using engineering judgment, adjustments in generation levels and controls are made until the desired equipment loadings and voltage profile are obtained.

EXAMPLE 6.14   **Power-flow program: effect of shunt capacitor banks**

Determine the effect of adding a 200-Mvar shunt capacitor bank at bus 2 on the power system in Example 6.9.

SOLUTION   Open PowerWorld Simulator case Example 6_14. This case is identical to Example 6.9 except that a 200-Mvar shunt capacitor bank has

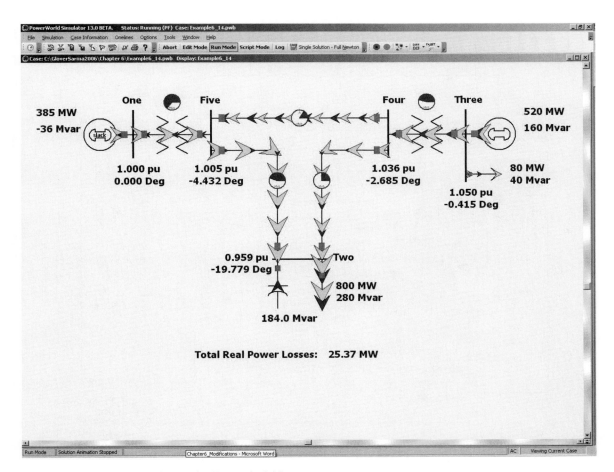

Screen for Example 6.14

been added at bus 2. Initially this capacitor is open. Click on the capacitor's circuit to close the capacitor and then select **Single Solution** to solve the case. The capacitor increases the bus 2 voltage from 0.834 per unit to a more acceptable 0.959 per unit. The insertion of the capacitor has also substantially decreased the losses, from 34.84 to 25.37 MW.

Notice that the amount of reactive power actually supplied by the capacitor is only 184 Mvar. This discrepancy arises because a capacitor's reactive varies with the square of the terminal voltage, $Q_{cap} = V_{cap}^2/X_c$ (see 2.3.5). A capacitor's Mvar rating is based on an assumed voltage of 1.0 per unit. ■

EXAMPLE 6.15

PowerWorld Simulator Case Example 6_15, which modifies the Example 6.13 case by (1) opening one of the 138/69 kV transformers at the LAUF substation, and (2) opening the 69 kV transmission line between PATTEN69

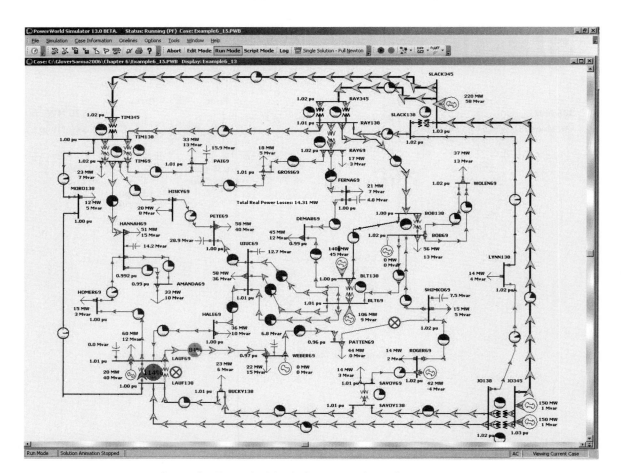

Screen for Example 6.15, before generation adjustment

and SHIMKO69. This causes a flow of 116.2 MVA on the remaining 138/69 kV transformer at LAUF. Since this transformer has a limit of 101 MVA, it results in an overload at 115%. Redispatch the generators in order to remove this overload.

SOLUTION There are a number of solutions to this problem, and several solution techniques. One solution technique would be to use engineering intuition, along with a trial and error approach. Since the overload is from the 138 kV level to the 69 kV level, and there is a generator directly connected to at the LAUF 69 kV bus, it stands to reason that increasing this generation would decrease the overload. Using this approach, we can remove the overload by increasing the Lauf generation until the transformer flow is reduced to 100%. This occurs when the generation is increased from 20 MW to 51 MW. Notice that as the generation is increased, the swing bus generation automatically decreases in order to satisfy the requirement that total system load plus losses must be equal to total generation.

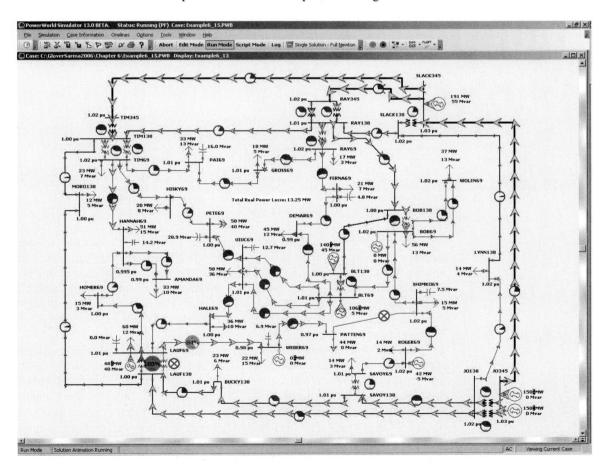

A solution to Example 6.15

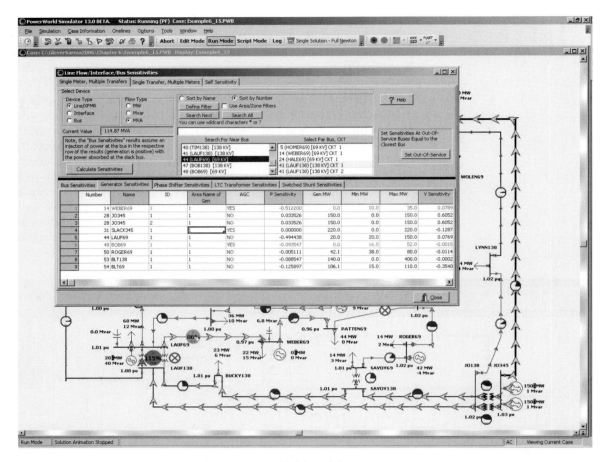

Example 6.15 Flow Sensitivities Dialog

An alternative possible solution is seen by noting that since the overload is caused by power flowing from the 138 kV bus, decreasing the generation at JO345 might also decrease this flow. This is indeed the case, but now the trial and error approach requires a substantial amount of work, and ultimately doesn't solve the problem. Even when we decrease the total JO345 generation from 300 MW to 0 MW, the overload is still present, albeit with its percentage decreased to 105%.

An alternative solution approach would be to first determine the generators with the most sensitivity to this violation and then adjust these. This can be done in PowerWorld Simulator by selecting **Tools, Flows and Voltage Sensitivities**. Select the Lauf 138/69 kV transformer, click on the **Calculate Sensitivities** button, and select the Generator Sensitivities tab towards the bottom of the dialog. The "P Sensitivity" field tells how increasing the output of each generator by one MW would affect the MVA flow on this transformer. Note that the sensitivity for the Lauf generator is −0.494, indicating that if we increase this generation by 1 MW the transformer MVA flow would decrease by 0.494 MVA. Hence, in order to decrease the flow by 15.2 MVA we would expect to increase the LAUF69 generator by 31 MW, exactly what we got by the trial and error approach. It is also clear that the JO345 generators, with a sensitivity of just 0.0335, would be relatively ineffective. In actual power system operation these sensitivities, known as generator shift factors, are used extensively. These sensitivities are also used in the Optimal Power Flow (introduced in Section 11.5). ■

6.8

SPARSITY TECHNIQUES

A typical power system has an average of fewer than three lines connected to each bus. As such, each row of Y_{bus} has an average of fewer than four nonzero elements, one off-diagonal for each line and the diagonal. Such a matrix, which has only a few nonzero elements, is said to be *sparse*.

Newton–Raphson power-flow programs employ sparse matrix techniques to reduce computer storage and time requirements [2]. These techniques include compact storage of Y_{bus} and $\mathbf{J}(i)$ and reordering of buses to avoid fill-in of $\mathbf{J}(i)$ during Gauss elimination steps. Consider the following matrix:

$$\mathbf{S} = \begin{bmatrix} 1.0 & -1.1 & -2.1 & -3.1 \\ -4.1 & 2.0 & 0 & -5.1 \\ -6.1 & 0 & 3.0 & 0 \\ -7.1 & 0 & 0 & 4.0 \end{bmatrix} \qquad (6.8.1)$$

One method for compact storage of $\mathbf{S}$ consists of the following four vectors:

$$\mathbf{DIAG} = \begin{bmatrix} 1.0 & 2.0 & 3.0 & 4.0 \end{bmatrix} \qquad (6.8.2)$$

$$\mathbf{OFFDIAG} = \begin{bmatrix} -1.1 & -2.1 & -3.1 & -4.1 & -5.1 & -6.1 & -7.1 \end{bmatrix} \quad (6.8.3)$$

$$\mathbf{COL} = \begin{bmatrix} 2 & 3 & 4 & 1 & 4 & 1 & 1 \end{bmatrix} \qquad (6.8.4)$$

$$\mathbf{ROW} = \begin{bmatrix} 3 & 2 & 1 & 1 \end{bmatrix} \qquad (6.8.5)$$

DIAG contains the ordered diagonal elements and **OFFDIAG** contains the nonzero off-diagonal elements of **S**. **COL** contains the column number of each off-diagonal element. For example, the *fourth* element in **COL** is 1, indicating that the *fourth* element of **OFFDIAG**, -4.1, is located in column 1. **ROW** indicates the number of off-diagonal elements in each row of **S**. For example, the *first* element of **ROW** is 3, indicating the *first* three elements of **OFFDIAG**, -1.1, -2.1, and -3.1, are located in the *first* row. The *second* element of **ROW** is 2, indicating the next two elements of **OFFDIAG**, -4.1 and -5.1, are located in the *second* row. The **S** matrix can be completely reconstructed from these four vectors. Note that the dimension of **DIAG** and

ROW equals the number of diagonal elements of **S**, whereas the dimension of **OFFDIAG** and **COL** equals the number of nonzero off-diagonals.

Now assume that computer storage requirements are 4 bytes to store each magnitude and 4 bytes to store each phase of Y_{bus} in an N-bus power system. Also assume Y_{bus} has an average of $3N$ nonzero off-diagonals (three lines per bus) along with its N diagonals. Using the preceding compact storage technique, we need $(4+4)3N = 24N$ bytes for **OFFDIAG** and $(4+4)N = 8N$ bytes for **DIAG**. Also, assuming 2 bytes to store each integer, we need $6N$ bytes for **COL** and $2N$ bytes for **ROW**. Total computer memory required is then $(24+8+6+2)N = 40N$ bytes with compact storage of Y_{bus}, compared to $8N^2$ bytes without compact storage. For a 1000-bus power system, this means 40 instead of 8000 kilobytes to store Y_{bus}. Further storage reduction could be obtained by storing only the upper triangular portion of the symmetric Y_{bus} matrix.

The Jacobian matrix is also sparse. From Table 6.5, whenever $Y_{kn} = 0$, $J1_{kn} = J2_{kn} = J3_{kn} = J4_{kn} = 0$. Compact storage of **J** for a 30,000-bus power system requires less than 10 megabytes with the above assumptions.

The other sparsity technique is to reorder buses. Suppose Gauss elimination is used to triangularize **S** in (6.8.1). After one Gauss elimination step, as described in Section 6.1, we have

$$S^{(1)} = \begin{bmatrix} 1.0 & -1.1 & -2.1 & -3.1 \\ 0 & -2.51 & -8.61 & -7.61 \\ 0 & -6.71 & -9.81 & -18.91 \\ 0 & -7.81 & -14.91 & -18.01 \end{bmatrix} \tag{6.8.6}$$

We can see that the zeros in columns 2, 3, and 4 of **S** are filled in with non-zero elements in $S^{(1)}$. The original degree of sparsity is lost.

One simple reordering method is to start with those buses having the fewest connected branches and to end with those having the most connected branches. For example, **S** in (6.8.1) has three branches connected to bus 1 (three off-diagonals in row 1), two branches connected to bus 2, and one branch connected to buses 3 and 4. Reordering the buses 4, 3, 2, 1 instead of 1, 2, 3, 4 we have

$$S_{reordered} = \begin{bmatrix} 4.0 & 0 & 0 & -7.1 \\ 0 & 3.0 & 0 & -6.1 \\ -5.1 & 0 & 2.0 & -4.1 \\ -3.1 & -2.1 & -1.1 & 1.0 \end{bmatrix} \tag{6.8.7}$$

Now, after one Gauss elimination step,

$$S_{reordered}^{(1)} = \begin{bmatrix} 4.0 & 0 & 0 & -7.1 \\ 0 & 3.0 & 0 & -6.1 \\ 0 & 0 & 2.0 & -13.15 \\ 0 & -2.1 & -1.1 & -4.5025 \end{bmatrix} \tag{6.8.8}$$

Note that the original degree of sparsity is not lost in (6.8.8).

Reordering buses according to the fewest connected branches can be performed once, before the Gauss elimination process begins. Alternatively, buses can be renumbered during each Gauss elimination step in order to account for changes during the elimination process.

Sparsity techniques similar to those described in this section are a standard feature of today's Newton–Raphson power-flow programs. As a result of these techniques, typical 30,000-bus power-flow solutions require less than 10 megabytes of storage, less than one second per iteration of computer time, and less than 10 iterations to converge.

EXAMPLE 6.16   Sparsity in a 37-bus system

To see a visualization of the sparsity of the power-flow Ybus and Jacobian matrices in a 37-bus system, open PowerWorld Simulator case Example 6_13.

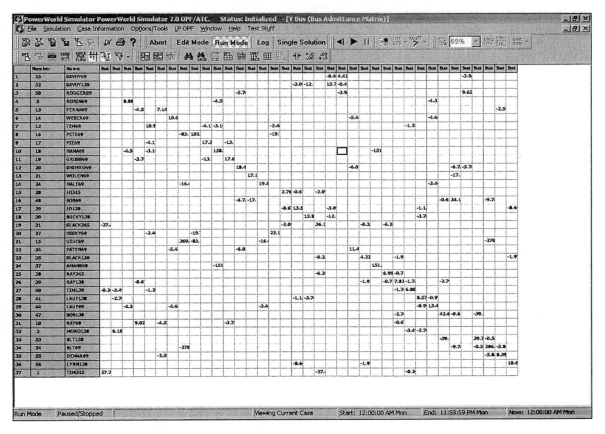

Screen for Example 6.16

Select **Case Information, Solution Details, Ybus** to view the bus admittance matrix. Then press ⟨ctrl⟩ Page Down to zoom the display out. Blank entries in the matrix correspond to zero entries. The 37×37 Ybus has a total of 1369 entries, with only about 10% nonzero. Select **Case Information, Solution Details, Power Flow Jacobian** to view the Jacobian matrix.   ■

6.9

FAST DECOUPLED POWER FLOW

Contingencies are a major concern in power system operations. For example, operating personnel need to know what power-flow changes will occur due to a particular generator outage or transmission-line outage. Contingency information, when obtained in real time, can be used to anticipate problems caused by such outages, and can be used to develop operating strategies to overcome the problems.

Fast power-flow algorithms have been developed to give power-flow solutions in seconds or less [8]. These algorithms are based on the following simplification of the Jacobian matrix. Neglecting $\mathbf{J}_2(i)$ and $\mathbf{J}_3(i)$, (6.6.6) reduces to two sets of decoupled equations:

$$\mathbf{J}_1(i)\Delta\boldsymbol{\delta}(i) = \Delta\mathbf{P}(i) \tag{6.9.1}$$

$$\mathbf{J}_4(i)\Delta\mathbf{V}(i) = \Delta\mathbf{Q}(i) \tag{6.9.2}$$

The computer time required to solve (6.9.1) and (6.9.2) is significantly less than that required to solve (6.6.6). Further reduction in computer time can be obtained from additional simplification of the Jacobian matrix. For example, assume $V_k \approx V_n \approx 1.0$ per unit and $\delta_k \approx \delta_n$. Then $\mathbf{J}_1$ and $\mathbf{J}_4$ are constant matrices whose elements in Table 6.5 are the negative of the imaginary components of $\mathbf{Y}_{\text{bus}}$. As such, $\mathbf{J}_1$ and $\mathbf{J}_4$ do not have to be recalculated during successive iterations.

The above simplifications can result in rapid power flow solutions for most systems. While the fast decoupled power flow usually takes more iterations to converge, it is usually significantly faster then the Newton-Raphson algorithm since the Jacobian does not need to be recomputed each iteration. And since the mismatch equations themselves have not been modified, the solution obtained by the fast decoupled algorithm is the same as that found with the Newton-Raphson algorithm. However, in some situations in which only an approximate power flow solution is needed the fast decoupled approach can be used with a fixed number of iterations (typically one) to give an extremely fast, albeit approximate solution.

6.10

THE "DC" POWER FLOW

The power flow problem can be further simplified by extending the fast decoupled power flow to completely neglect the Q-V equation, assuming that the voltage magnitudes are constant at 1.0 per unit. With these simplifications the power flow on the line from bus j to bus k with reactive X_{jk} becomes

$$P_{jk} = \frac{\delta_j - \delta_k}{X_{jk}} \tag{6.10.1}$$

and the real power balance equations reduce to a completely linear problem

$$-\mathbf{B}\boldsymbol{\delta} = \mathbf{P} \tag{6.10.2}$$

where $\mathbf{B}$ is the imaginary component of the of $\mathbf{Y}_{\text{bus}}$ calculated neglecting line resistance and excepting the slack bus row and column.

Because (6.10.2) is a linear equation with a form similar to that found in solving dc resistive circuits, this technique is referred to as the DC power flow. However, in contrast to the previous power flow algorithms, the DC power flow only gives an approximate solution, with the degree of approximation system dependent. Nevertheless, with the advent of power system restructuring the DC power flow has become a commonly used analysis technique.

EXAMPLE 6.17

Determine the dc power flow solution for the five bus from Example 6.9.

SOLUTION With bus 1 as the system slack, the $\mathbf{B}$ matrix and $\mathbf{P}$ vector for this system are

$$\mathbf{B} = \begin{bmatrix} -30 & 0 & 10 & 20 \\ 0 & -100 & 100 & 0 \\ 10 & 100 & -150 & 40 \\ 20 & 0 & 40 & -110 \end{bmatrix} \quad \mathbf{P} = \begin{bmatrix} -8.0 \\ 4.4 \\ 0 \\ 0 \end{bmatrix}$$

$$\boldsymbol{\delta} = -\mathbf{B}^{-1}\mathbf{P} = \begin{bmatrix} -0.3263 \\ 0.0091 \\ -0.0349 \\ -0.0720 \end{bmatrix} \text{radians} = \begin{bmatrix} -18.70 \\ 0.5214 \\ -2.000 \\ -4.125 \end{bmatrix} \text{degrees}$$

To view this example in PowerWorld Simulator open case Example 6_17 which has this example solved using the DC power flow. To view the DC power flow options select **Options, Simulator Options** to show the PowerWorld Simulator Options dialog. Then select the Power Flow Solution category, and the DC Options page.

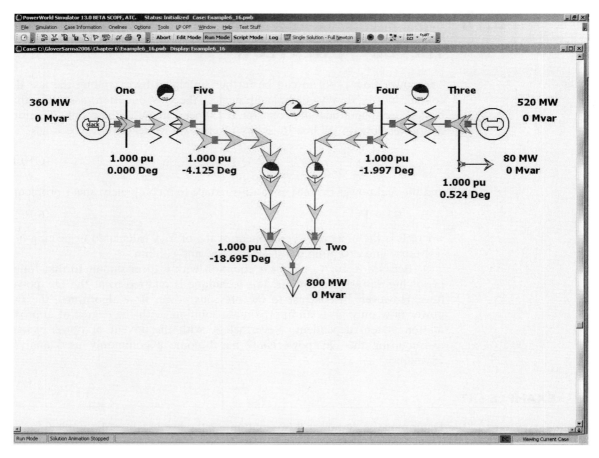

Screen for Example 6.17   ■

PROBLEMS

SECTION 6.1

6.1   Using Gauss elimination, solve the following linear algebraic equations:

$$5x_1 - 2x_2 - 3x_3 = 4$$
$$-5x_1 + 7x_2 - 2x_3 = -10$$
$$-3x_1 - 3x_2 + 8x_3 = 6$$

Find the three unknowns $x_1, x_2,$ and x_3. Check your answers using Cramer's rule. Also check by using matrix solution of linear equations.

6.2 Using Gauss elimination and back substitution, solve

$$\begin{bmatrix} 6 & 2 & 1 \\ 4 & 10 & 2 \\ 3 & 4 & 14 \end{bmatrix} \begin{bmatrix} x_1 \\ x_2 \\ x_3 \end{bmatrix} = \begin{bmatrix} 3 \\ 4 \\ 2 \end{bmatrix}$$

6.3 Rework Problem 6.2 with the value of A_{33} change from 14 to 1.4.

6.4 Solve for x_1 and x_2 in the system of equations given by

$$x_2 - 3x_1 + 1.9 = 0$$

$$x_2 + x_1^2 - 1.8 = 0$$

by Gauss method with an initial guess of $x_1 = 1$ and $x_2 = 1$.

6.5 Show that the Gauss elimination method, which transforms the set of N linear equations $\mathbf{Ax} = \mathbf{y}$ to $\mathbf{A}^{(N-1)}\mathbf{x} = \mathbf{y}^{(N-1)}$, where $\mathbf{A}^{(N-1)}$ is triangular, requires $(N^3 - N)/3$ multiplications, $N(N-1)/2$ divisions, and $(N^3 - N)/3$ subtractions. Assume that all the elements of $\mathbf{A}$ and $\mathbf{y}$ are nonzero and real. (*Hint:* Investigate (6.1.7). Note that during the first Gauss elimination step, *each* of the $(N-1)$ rows that are changed requires *one* division, N multiplications, and N subtractions.)

6.6 Show that, after triangularizing $\mathbf{Ax} = \mathbf{y}$, the back substitution method of solving $\mathbf{A}^{(N-1)}\mathbf{x} = \mathbf{y}^{(N-1)}$ requires N divisions, $N(N-1)/2$ multiplications, and $N(N-1)/2$ subtractions. Assume that all the elements of $\mathbf{A}^{(N-1)}$ and $\mathbf{y}^{(N-1)}$ are nonzero and real.

SECTION 6.2

6.7 Solve $x^2 - 3x + 1 = 0$ using the Jacobi iterative method with $x(0) = 1$. Continue until (Eq. 6.2.2) is satisfied with $\varepsilon = 0.01$. Check using the quadratic formula.

6.8 Solve the following complex number equation by the Jacobi iterative method:

$$x^2 - (3 + 5j)x = 4 + 3j$$

Use an initial guess $x(0) = 1 + j$, and a stopping criteria of $\varepsilon = 0.05$.

6.9 Solve Problem 6.2 using the Jacobi iterative method. Start with $x_1(0) = x_2(0) = x_3(0) = 0$, and continue until (6.2.2) is satisfied with $\varepsilon = 0.01$.

6.10 Repeat Problem 6.9 using the Gauss–Seidel iterative method. Which method converges more rapidly?

6.11 Try to solve Problem 6.2 using the Jacobi and Gauss–Seidel iterative methods with the value of A_{33} changed from 14 to 0.14 and with $x_1(0) = x_2(0) = x_3(0) = 0$. Show that neither method converges to the unique solution.

6.12 Using the Jacobi method (also known as the Gauss method), solve for x_1 and x_2 in the system of equations

$$x_2 - 3x_1 + 1.9 = 0$$

$$x_2 + x_1^2 - 1.8 = 0$$

Use an initial guess $x_1(0) = 1.0 = x_2(0) = 1.0$. Also, see what happens when you choose an uneducated initial guess $x_1(0) = x_2(0) = 100$.

6.13   Use the Gauss-Seidel method to solve the following equations that contain terms that are often found in power flow equations

$$x_1 = (1/(-20j)) * [(-1 + 0.5j)/(x_1)^* - (j10) * x_2 - (j10)]$$

$$x_2 = (1/(-20j)) * [(-2 + j)/(x_2)^* - (j10) * x_1 - (j10)]$$

Use an initial estimate of $x_1(0) = 1$ and $x_2(0) = 1$, and a stopping of $\varepsilon = 0.05$.

6.14   Find a root of the following equation by using the Gauss-Seidel method: (use an initial estimate of $x = 2$) $f(x) = x^3 - 6x^2 + 9x - 4 = 0$.

6.15   Consider the following three linear algebraic equations:

$$8x_1 - 4x_2 = 24$$

$$-4x_1 + 7x_2 + 2x_3 = 0$$

$$2x_2 + 8x_3 = 12$$

Solve these for x_1, x_2, and x_3 within a tolerance of $\varepsilon = 0.01$. Start with $x_1(0) = x_2(0) = x_3(0) = 1$.

(a) by the Jacobi iterative method.
(b) by Gauss-Seidel Iteration.

6.16   Take the z-transform of (6.2.6) and show that $\mathbf{X}(z) = \mathbf{G}(z)\mathbf{Y}(z)$, where $\mathbf{G}(z) = (z\mathbf{U} - \mathbf{M})^{-1}\mathbf{D}^{-1}$ and $\mathbf{U}$ is the unit matrix.

$\mathbf{G}(z)$ is the matrix transfer function of a digital filter that represents the Jacobi or Gauss–Seidel methods. The filter poles are obtained by solving $\det(z\mathbf{U} - \mathbf{M}) = 0$. The filter is stable if and only if all the poles have magnitudes less than 1.

6.17   Using the results of Problem 6.16, determine the filter poles for Examples 6.3 and 6.5. Note that in Example 6.3 both poles have magnitudes less than 1, which means the filter is stable and Jacobi converges for this example. However, in Example 6.5, one pole has a magnitude greater than 1, which means the filter is unstable and Gauss–Seidel diverges for this example.

6.18   Determine the poles of the Jacobi and Gauss–Seidel digital filters for the general two-dimensional problem ($N = 2$):

$$\left[\begin{array}{c|c} A_{11} & A_{12} \\ \hline A_{21} & A_{22} \end{array}\right] \left[\begin{array}{c} x_1 \\ x_2 \end{array}\right] = \left[\begin{array}{c} y_1 \\ y_2 \end{array}\right]$$

Then determine a necessary and sufficient condition for convergence of these filters when $N = 2$.

SECTION 6.3

6.19   Use Newton-Raphson to find a solution to the polynomial equation $f(x) = y$ where $y = 0$ and $f(x) = x^3 + 9x^2 + 2x - 48$. Start with $x(0) = 1$ and continue until (6.2.2) is satisfied with $\varepsilon = 0.001$.

6.20   Repeat 6.19 using $x(0) = -1$.

6.21   Use Newton–Raphson to find one solution to the polynomial equation $f(x) = y$, where $y = 0$ and $f(x) = x^4 + 12x^3 + 54x^2 + 108x + 81$. Start with $x(0) = -1$ and continue until (6.2.2) is satisfied with $\varepsilon = 0.001$.

6.22 Use Newton–Raphson to find a solution to

$$\begin{bmatrix} e^{x_1 x_2} \\ \cos(x_1 + x_2) \end{bmatrix} = \begin{bmatrix} 1.2 \\ 0.5 \end{bmatrix}$$

where x_1 and x_2 are in radians. (a) Start with $x_1(0) = 1.0$ and $x_2(0) = 0.5$ and continue until (6.2.2) is satisfied with $\varepsilon = 0.005$. (b) Show that Newton–Raphson diverges for this example if $x_1(0) = 1.0$ and $x_2(0) = 2.0$.

6.23 Solve the following equations by the Newton-Raphson method:

$$2x_1^2 + x_2^2 - 8 = 0$$
$$x_1^2 - x_2^2 + x_1 x_2 - 4 = 0$$

Start with an initial guess of $x_1 = 1$ and $x_2 = 1$.

6.24 The following nonlinear equation contains terms that are often found in power-flow equations:

$$y \sin y + 4 = 0$$

Find a solution by using the Newton–Raphson method with an initial guess $y(0) = 4$ radians.

6.25 Reconsider Problem 6.14 and use the Newton-Raphson method, with an initial estimate of $x = 6$.

6.26 With an initial estimate of $x = 0$, obtain a root of the equation

$$x = 2 - \sin x$$

by the Newton-Raphson method within a tolerance of $\varepsilon = 0.001$.

SECTION 6.4

6.27 Consider the simplified electric power system shown in Figure 6.5 for which the power-flow solution can be obtained without resorting to iterative techniques. (a) Compute the elements of the bus admittance matrix Y_{bus}. (b) Calculate the phase angle δ_2 by using the real power equation at bus 2 (voltage-controlled bus). (c) Determine $|V_3|$ and δ_3 by using both the real and reactive power equations at bus 3 (load bus). (d) Find the real power generated at bus 1 (swing bus). (e) Evaluate the total real power losses in the system.

FIGURE 6.5

Problem 6.27

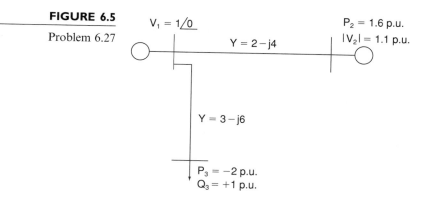

6.28 Compute the elements of the second row of Y_{bus} for the power system in Example 6.9.

6.29 In Example 6.9, double the impedance of the line from bus 2 to bus 4. Determine the new values for the second row of Y_{bus}. Verify your result using PowerWorld Simulator case Example 6.9.

6.30 Figure 6.6 shows a single-line diagram of a three-bus power system. Power-flow input data are given in Tables 6.9 and 6.10. (a) Determine the 3×3 per-unit bus admittance matrix Y_{bus}. (b) For each bus $k = 1, 2, 3$ determine which of the variables $V_k, \partial_k, P_k,$ and Q_k are input data and which are unknowns.

FIGURE 6.6

Single-line diagram for Problem 6.30 (per unit impedances and per unit real and reactive powers are shown)

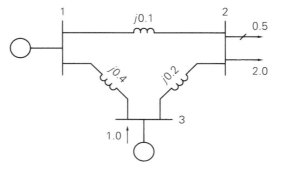

TABLE 6.9

Bus input data for Problem 6.30

Bus	Type	V per Unit	∂ Degrees	P_G per Unit	Q_G per Unit	P_L per Unit	Q_L per Unit	Q_{Gmin} per Unit	Q_{Gmax} per Unit
1	Swing	1.0	0	—	—	0	0	—	—
2	Load	—	—	0	0	2.0	0.5	—	—
3	Constant voltage	1.0	—	1.0	—	0	0	−5.0	+5.0

TABLE 6.10

Line input data for Problem 6.30

Line	Bus-to-Bus	R' per Unit	X' per Unit	G' per Unit	B' per Unit	Maximum MVA per Unit
1	1–2	0	0.1	0	0	3.0
2	2–3	0	0.2	0	0	3.0
3	1–3	0	0.4	0	0	3.0

(Note: There are no transformers)

SECTION 6.5

6.31 Assume a $1 + 0.5j$ per unit load at bus 2 is being supplied by a generator at bus 1 through a transmission line with series impedance of $0.05 + j0.1$ per unit. Assuming bus 1 is the swing bus with a fixed per unit voltage of $1.0\underline{/0}$, use Gauss-Seidel method to calculate the voltage at bus 2 after three iterations.

6.32   Repeat the above problem with the swing bus voltage changed to $1.0\underline{/30°}$ per unit.

6.33   For the power system in Example 6.9, use Gauss–Seidel to calculate $V_3(1)$, the phasor voltage at bus 3 after the first iteration. Note that bus 3 is a voltage-controlled bus.

6.34   For the power system given in Problem 6.30, use Gauss–Seidel to compute $V_2(1)$ and $V_3(1)$, the phasor voltages at bus 2 and 3 after the first iteration. Use zero initial phase angles and 1.0 per-unit initial bus voltage magnitudes.

6.35   The bus admittance matrix for the power system shown in Figure 6.7 is given by

$$Y_{bus} = \begin{bmatrix} 3-j9 & -2+j6 & -1+j3 & 0 \\ -2+j6 & 3.666-j11 & -0.666+j2 & -1+j3 \\ -1+j3 & -0.666+j2 & 3.666-j11 & -2+j6 \\ 0 & -1+j3 & -2+j6 & 3-j9 \end{bmatrix} \text{ per unit}$$

With the complex powers on load buses 2, 3, and 4 as shown in Figure 6.7, determine the value for V_2 that is produced by the first and second iterations of the Gauss–Seidel procedure. Choose the initial guess $V_2(0) = V_3(0) = V_4(0) = 1.0\underline{/0°}$ per unit.

FIGURE 6.7

Problem 6.35

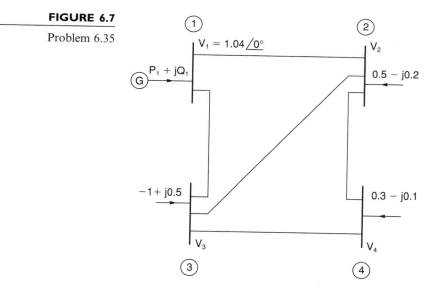

6.36   The bus admittance matrix of a three-bus power system is given by

$$Y_{bus} = -j \begin{bmatrix} 7 & -2 & -5 \\ -2 & 6 & -4 \\ -5 & -4 & 9 \end{bmatrix} \text{ per unit}$$

with $V_1 = 1.0\underline{/0°}$ per unit; $V_2 = 1.0$ per unit; $P_2 = 60$ MW; $P_3 = -80$ MW; $Q_3 = -60$ MVAR (lagging) as a part of the power-flow solution of the system, find V_2 and V_3 within a tolerance of 0.01 per unit, by using Gauss-Seidel iteration method. Start with $\delta_2 = 0$, $V_3 = 1.0$ per unit, and $\delta_3 = 0$.

PW   **6.37**   Using PowerWorld Simulator, determine the maximum mismatch after the first, second, and third iterations for the Example 6.10 case using Gauss–Seidel. How many iterations does it take to converge to a mismatch less than 0.5 MVA?

PW   **6.38**   Repeat Problem 6.37, except first decrease the load at bus 2 to 500 MW and 100 Mvar.

PW   **6.39**   Open the PowerWorld Simulator Problem 6_39 case. This case is similar to the Example 6.10 case, except that the maximum number of iterations has been increased, and the case has been set to automatically initialize from a flat start solution (i.e., one with all the voltage angles equal to zero and the voltage magnitudes at load buses equal to 1.0 per unit). Increase the load at bus 2 in 10-MW steps, keeping the load power factor constant. For each load increase, how many iterations are required to converge using Gauss–Seidel? What is the maximum load level before the iterations diverge?

SECTION 6.6

6.40   For the power system in Example 6.9, calculate $\Delta P_4(0)$ in Step 1 and $J1_{44}(0)$ in Step 2 of the first Newton–Raphson iteration. Assume zero initial phase angles and 1.0 per-unit initial voltage magnitudes (except $V_3 = 1.05$).

6.41   For the power system given in Problem 6.30, use (6.6.2) and (6.6.3) to write the three power-flow equations to be solved by the Newton–Raphson method. Also, identify the three unknown variables to be solved. Do not solve the equations.

6.42   For the power system given in Problem 6.30, use Newton–Raphson to compute $V_2(1)$ and $V_3(1)$, the phasor voltages at bus 2 and 3 after the first iteration, as follows. (a) Step 1: use (6.6.2) and (6.6.3) to compute $\Delta \mathbf{y}(0)$. (b) Step 2: compute the 3×3 Jacobian matrix $\mathbf{J}(0)$ using the equations in Table 6.5. (c) Step 3: use Gauss elimination and back substitution to solve (6.6.6). (d) Step 4: compute $\mathbf{x}(1)$ in (6.6.7). Also, use (6.5.3) to compute Q_{G3} and verify that it is within the limits shown in Table 6.12. In Steps 1 and 2, use zero initial phase angles and 1.0 per-unit initial bus voltage magnitudes.

6.43   For the transmission system shown in Figure 6.8, all shunt elements are capacitors with an admittance $y_C = j0.01$ per unit, and all series elements are inductors with an impedance $z_L = j0.08$ per unit. Determine δ_2, $|V_3|$, δ_3, P_{G1}, Q_{G1}, and Q_{G2} for the system.

FIGURE 6.8

Problem 6.43

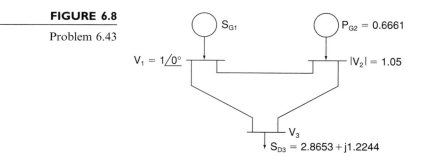

PW   **6.44**   Load PowerWorld Simulator case Example 6.11; this case is set to perform a single iteration of the Newton–Raphson power flow each time **Single Solution** is selected. Verify that initially the Jacobian element J_{33} is 104.41. Then, give and verify the value of this element after each of the next three iterations (until the case converges).

SECTION 6.7

PW   **6.45**   Use PowerWorld Simulator to determine the Mvar rating of the shunt capacitor bank in the Example 6.14 case that increases V_2 to 1.02 per unit. Also determine the effect of this capacitor bank on line loadings and total real power losses (total real and reactive losses are shown on the **Case Information, Case Summary** dialog). To vary the capacitor's nominal Mvar rating, right-click on the capacitor symbol to view the **Switched Shunt** dialog and then change the Nominal Mvar field.

PW   **6.46**   Use PowerWorld Simulator to modify the Example 6.9 case by inserting a second line between bus 2 and bus 5. Give the new line a circuit identifier of "2" to distinguish it from the existing line. The line parameters of the added line should be identical to those of the existing line 2–5. Determine the new line's effect on V_2, the line loadings, and on the total real power losses.

PW   **6.47**   Using PowerWorld Simulator with the Example 6.9 case, change the generator 1 voltage setpoint between 1.00 and 1.08 per unit in 0.005 per-unit steps. Show the variation in the reactive power output of generator 1, V_2, and total real power losses.

PW   **6.48**   Open PowerWorld Simulator case Problem 6_48. This case is identical to Example 6.9 except that the transformer between buses 1 and 5 is now a tap-changing transformer with a tap range between 0.9 and 1.1 and a tap step size of 0.00625. The tap is on the high side of the transformer. As the tap is varied between 0.975 and 1.1, show the variation in the reactive power output of generator 1, V_5, V_2, and the total real power losses.

PW   **6.49**   Open PowerWorld Simulator case Example 6_13. As in Example 6.13, remove the TIM69 to HANNAH69 line. Determine the Mvar rating of the shunt capacitor bank at Bus HANNAH69 necessary to correct the HANNAH69 voltage back to 0.96 per unit. Use ⟨ctrl⟩ up arrow to zoom the one-line to better see the one-line values.

PW   **6.50**   Open PowerWorld Simulator case Example 6_13. Plot the variation in the total system real power losses as the generation at bus BLT138 is varied in 20-MW blocks between 0 MW and 400 MW. What value of BLT138 generation minimizes system losses?

PW   **6.51**   Repeat Problem 6.50, except first remove the 138-kV line from RAY138 to BOB138.

PW   **6.52**   Open PowerWorld Simulator case Example 6_13. Sequentially open (remove from service) each of the case's twelve 138 kV transmission lines (always closing the previous device). Record the impact each outage has on system losses. Which device has the largest impact on system losses?

PW   **6.53**   Open PowerWorld Simulator case Problem 6_53. This case contains a modified version of the 37 bus case from Example 6.13, except now with three lines out of service. This causes an overload on the BLT69 to UIUC69 transmission line. Using the techniques discussed in Example 6.15, determine a solution of how the system generation can be adjusted to remove this overload. *Note:* you may need to adjust more than one generator.

SECTION 6.8

6.54 Using the compact storage technique described in Section 6.8, determine the vectors **DIAG**, **OFFDIAG**, **COL**, and **ROW** for the following matrix:

$$
S = \begin{bmatrix}
17 & -9.1 & 0 & 0 & -2.1 & -7.1 \\
-9.1 & 25 & -8.1 & -1.1 & -6.1 & 0 \\
0 & -8.1 & 9 & 0 & 0 & 0 \\
0 & -1.1 & 0 & 2 & 0 & 0 \\
-2.1 & -6.1 & 0 & 0 & 14 & -5.1 \\
-7.1 & 0 & 0 & 0 & -5.1 & 15
\end{bmatrix}
$$

6.55 If 4 bytes of computer storage are used for each floating-point number and 2 bytes for each integer, determine the total number of bytes required to store the **S** matrix in Problem 6.52 (a) with compact storage and (b) without compact storage.

6.56 For the triangular factorization of the corresponding Y_{bus}, number the nodes of the graph shown in Figure 6.9 in an optimal order.

FIGURE 6.9

Problem 6.56

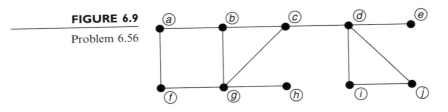

SECTION 6.9

6.57 Compare the angles and line flows between the Example 6.17 case and results shown in Tables 6.6, 6.7, and 6.8.

6.58 Resolve the Example 6.17 case except with the line between bus 2 and 4 removed.

6.59 Discuss the power system parameters and operating conditions that would significantly affect the accuracy of a DC power flow.

CASE STUDY QUESTIONS

A. How does transmission congestion contribute to spikes in electricity prices?

B. Transmission congestion is one phenomenon that can be studied via power-flow programs. What other studies can be performed with power-flow programs?

C. How can advances in computer visualization be applied to displaying power flow results?

D. One of the causes of the August 14th, 2003 blackout was lack of situational awareness by the utility operators. Discuss how better power system visualization could help to reduce blackout risk. You may wish to consult the Final Report from this blackout, which is available online at http://www.nerc.com/~filez/blackout.html.

DESIGN PROJECT I: SYSTEM PLANNING

After months of negotiation the AMA company is close to committing to build a new plant in the eastern portion of the city of Metropolis. The new jobs this plant will bring are much anticipated by city officials. With an anticipated peak load of about 48 MW and 10 Mvar, the plant will also bring additional revenue to the local utility, Metropolis Light and Power (MLP). However, in order to accommodate the new AMA load a new substation will need to be constructed at the AMA plant location. While AMA needs to receive electricity at the 69 kV level, the new substation location is large enough to accommodate a 138/69 kV transformer if needed. Additionally, for reliability purposes AMA needs to have at least two separate feeds into their substation.

As a planning engineer for MLP your job is to make recommendations to insure that with AMA's new load under peak loading conditions the transmission system in the eastern region is adequate for any base case or first contingency loading situation. This is also a good opportunity not only to meet the new AMA load, but also to fix some existing first contingency violations. The below table shows the available right-of-way distances that are available for the construction of new 69 kV and/or new 138 kV lines. All existing 69 kV only substations are large enough to accommodate 138 kV as well.

Design Procedure

1. DesignCase1 into PowerWorld Simulator. Perform an initial power-flow solution to verify the base case system operation without the **AMA load**. Note that the entire line flows and bus voltage magnitudes are within their limits. Assume all line MVA flows must be at or below 100% of the limit A values, and all voltages must be between 0.95 and 1.10 per unit.

2. Repeat the above analysis considering the impact of any single transmission line or transformer outage. This is known as contingency analysis. To simplify this analysis, PowerWorld Simulator has the ability to automatically perform a contingency analysis study. Select **Tools, Contingency Analysis** to show the Contingency Analysis display. Note that the 57 single line/transformer contingencies are already defined. Select **Start Run** to automatically see the impact of removing any single element. This system has line violations for three different contingencies.

3. Using the available rights-of-ways given in the table at the end of this design project, and the transmission line parameters/costs given in the table at the end of design project 2, iteratively determine the least expensive system additions so that the base case and all the contingences result in secure operation points (i.e., one with no vio-

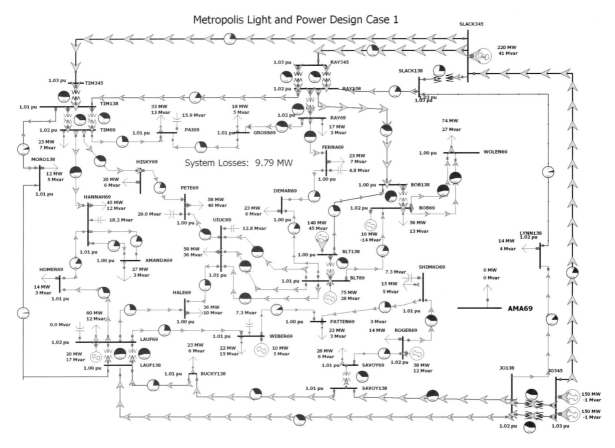

FIGURE 6.10

lations) with the new *AMA load*. The parameters of the new transmission lines(s) need to be derived using the tower configurations and conductor types provided by your instructor. The total cost of an addition is defined as the construction costs minus the savings associated with any decrease in system losses over the next 5 years.

4. Write a detailed report discussing the initial system problems, your approach to optimally solving the system problems, and the justification for your final recommendation.

Simplifying Assumptions

To simplify the analysis, several assumptions are made:

1. You need only consider the base case loading level given in Design-Case1. In a real design, typically a number of different operating points/loading levels must be considered.

2. You should consider the generator outputs as fixed values; any changes in the losses and the new *AMA load* are always picked up by the system slack generator.

3. You should not modify the status of the capacitors or the transformer taps.

4. You should assume that the system losses remain constant over the 5-year period and need only consider the impact and new design has on the base case losses. Electricity is priced at **$55/MWh**.

Available New Rights-of-Ways

Right-of-Way/Substation	Right-of-Way Mileage (miles)
AMA to JO	4.5
AMA to LYNN	5
AMA to ROGER	8
AMA to WOLEN	11
AMA to SHIMKO	4
BOB to SHIMKO	6.2
WOLEN to SHIMKO	12
SAVOY TO SHIMKO	6

Note: each substation is physically large enough to accommodate the addition of 138 kV bus work, and a 138/69 kV transformer.

DESIGN PROJECT 2: SYSTEM PLANNING

After months of negotiation the Kyle Aluminum company is close to committing to build a new plant in the western portion of Metropolis, a city whose electricity is provided by the Metropolis Light and Power Company (MLP). To meet the new load, a new substation will be constructed at this location called KYLE69, which will be served at 69 kV.

MLP anticipates a peak load at the KYLE69 substation of 45 MW and 10 Mvar. Since aluminum production is very sensitive to blackouts, Kyle Aluminum is requiring that MLP provide at least two separate feeds into this new substation. While 69 kV service is required the new substation is large enough to accommodate a 138/69-kV transformer as well.

You are currently interning with MLP as a planning engineer. Your job is to develop recommendations for management on the least cost way to supply this new substation. At the same time, your boss has asked you to use this opportunity to fix some existing contingency problems in the western portion of the MLP system. Specifically your task is to make recommendations for the best transmission system additions to strengthen the western portion of the system so that there will be no violations for either the base case loading or during any single transmission element contingency with the inclusion of the new load. Secure system operations require that no lines or transformers be loaded higher than 100% of their ratings and that all bus voltage magnitudes are between 0.95 and 1.08 per unit. The table that follows summarizes various rights-of-way that can be used for the installation of new lines. Design costs are also provided.

Design Procedure

1. Load DesignCase2 into PowerWorld Simulator. Perform an initial power-flow solution to verify the base case system operation without the new load. Note that all the line flows and bus voltage magnitudes are within their limits.

2. Repeat the above analysis considering the impact of any single transmission line or transformer outage. This is known as contingency analysis. To simplify this analysis, PowerWorld Simulator has

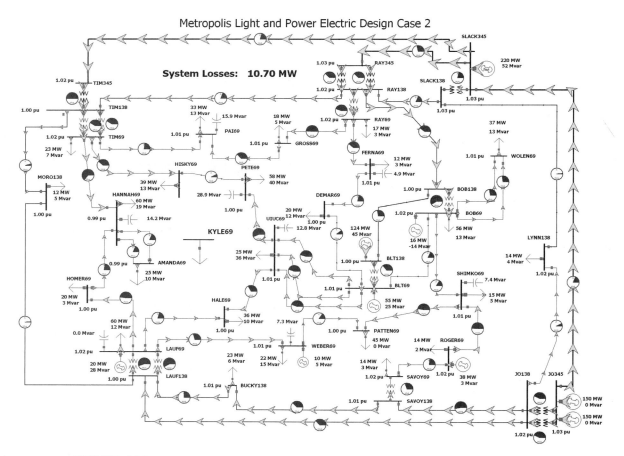

FIGURE 6.11

the ability to automatically perform a contingency analysis study. Select **Tools, Contingency Analysis** to show the Contingency Analysis display. Note that the 57 single line/transformer contingencies are already defined. Select **Start Run** to automatically see the impact of removing any single element. This system is insecure for several contingencies.

3. Using the rights-of-ways given in the table at the end of this design project, and the transmission line parameters/costs given in the table after the right-of-ways, iteratively determine the least expensive system addition (i.e., best) so that the base case and all the contingences result in secure operation points with the new KYLE69 load. The parameters of the new transmission lines(s) need to be derived using the tower configurations and conductor types provided by your instructor. The total cost of an addition is defined as the construction costs plus the cost (or minus the savings) from changes in system losses over the next 5 *years*.

4. Write a detailed report discussing the initial system problems, your approach to optimally solving the system problems, and the justification for your final recommendation.

Simplifying Assumptions

To simplify the analysis, several assumptions are made:

1. You need only consider the base case loading level given in Design-Case2. In a real design, typically a number of different operating points/loading levels must be considered.

2. You should consider the generator outputs as fixed values; any changes in the losses and the new KYLE69 load are always picked up by the system slack.

3. You should not modify the status of the capacitors or the transformer taps.

4. You should assume that the system losses remain constant over the 5-year period and need only consider the impact and new design has on the base case losses. Electricity is priced at **$55/MWh**.

Available New Rights-of-Ways

Right-of-Way/Substation	Right-of-Way/Mileage (miles)
KYLE to HISKY	5
KYLE to AMANDA	5.2
KYLE to PETE	6
KYLE to TIM	10.5
KYLE to MORO	8

KYLE to LAUF	10
KYLE to UIUC	13
KYLE to HOMER	7

Note: each substation is physically large enough to accommodate the addition of 138 kV bus work, and a 138/69 kV transformer.

Design Projects 1 and 2: Sample Transmission System Design Costs

Transmission lines (69 kV and 138 kV)   New transmission lines include a fixed cost and a variable cost. The fixed cost is for the purchase/installation of the three-phase circuit breakers, associated relays, and changes to the substation bus structure. The fixed costs are $100,000 for a 138-kV line and $50,000 for a 69-kV line.

The variable costs depend on the type of conductor and the length of the line. The assumed costs in $/mile are given here.

Conductor Type	Current Rating (Amps)	138-kV Lines	69-kV Lines
Partridge	460		$90,000/mi
Lark	600	$170,000/mi	$105,000/mi
Rook	770	$180,000/mi	$120,000/mi
Condor	990	$190,000/mi	

Line impedance data and MVA ratings are determined based on the conductor type and tower configuration. The conductor characteristics are given in Table A.4. For these design problems, use the tower configurations provided by the instructor.

Transformers (138 kV/69 kV)   Transformer costs include associated circuit breakers, relaying and installation.

101 MVA	$870,000
187 MVA	$1,150,000

Assume any new 138/69 kV transformer has 0.0025 per unit resistance and 0.04 per unit reactance on a 100-MVA base.

Bus work

New 69-kV substation:	$250,000
New 138/69-kV substation:	$400,000
Upgrade 69-kV substation to 138/69 kV	$200,000

DESIGN PROJECT 3: SYSTEM PLANNING*

Time given: 11 weeks
Approximate time required: 40 hours
Additional references: [10, 11]

Figure 6.12 shows a single-line diagram of four interconnected power systems identified by different graphic bus designations. The following data are given:

1. There are 31 buses, 21 lines, and 13 transformers.

2. Generation is present at buses 1, 16, 17, 22, and 23.

3. Total load of the four systems is 400 MW.

4. Bus 1 is the swing bus.

5. The system base is 100 MVA.

6. Additional information on transformers and transmission lines is provided in [10, 11].

Based on the data given:

1. Allocate the total 400-MW system load among the four systems.

2. For each system, allocate the load to buses that you want to represent as load buses. Select reasonable load power factors.

FIGURE 6.12

Design Project 3:
Single-line diagram for
31-bus interconnected
power system

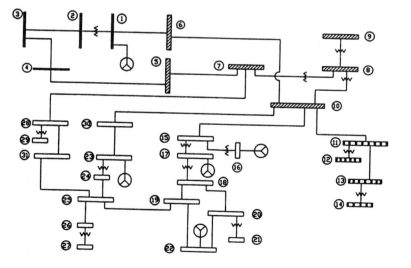

*This is based on a project assigned by Adjunct Professor Leonard Dow at Northeastern University, Boston, Massachusetts.

3. Taking into consideration the load you allocated above, select appropriate transmission-line voltage ratings, MVA ratings, and distances necessary to supply these loads. Then determine per-unit transmission-line impedances for the lines shown on the single-line diagram (show your calculations).

4. Also select appropriate transformer voltage and MVA ratings, and determine per-unit transformer leakage impedances for the transformers shown on the single-line diagram.

5. Develop a generation schedule for the 5 generator buses.

6. Show on a copy of the single-line diagram per-unit line impedances, transformer impedances, generator outputs, and loads that you selected above.

7. Using PowerWorld Simulator, run a base case power flow. In addition to the printed input/output data files, show on a separate copy of the single-line diagram per-unit bus voltages as well as real and reactive line flows, generator outputs, and loads. Flag any high/low bus voltages for which $0.95 \le V \le 1.05$ per unit and any line or transformer flows that exceed normal ratings.

8. If the base case shows any high/low voltages or ratings exceeded, then correct the base case by making changes. Explain the changes you have made.

9. Repeat (7). Rerun the power-flow program and show your changes on a separate copy of the single-line diagram.

10. Provide a typed summary of your results along with your above calculations, printed power-flow input/output data files, and copies of the single-line diagram.

DESIGN PROJECT 4: POWER FLOW/SHORT CIRCUITS

Time given: 3 weeks
Approximate time required: 15 hours

Each student is assigned one of the single-line diagrams shown in Figures 6.13 and 6.14. Also, the length of line 2 in these figures is varied for each student.

Assignment 1: Power-Flow Preparation

For the single-line diagram that you have been assigned (Figure 6.13 or 6.14), convert all positive-sequence impedance, load, and voltage data to per unit

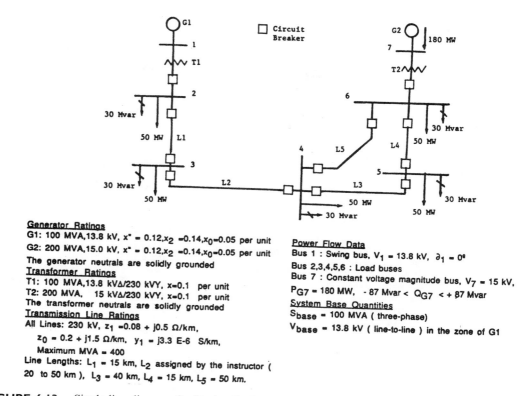

Generator Ratings
G1: 100 MVA,13.8 kV, $x'' = 0.12, x_2 = 0.14, x_0 = 0.05$ per unit
G2: 200 MVA,15.0 kV, $x'' = 0.12, x_2 = 0.14, x_0 = 0.05$ per unit
The generator neutrals are solidly grounded
Transformer Ratings
T1: 100 MVA,13.8 kVΔ/230 kVY, $x=0.1$ per unit
T2: 200 MVA, 15 kVΔ/230 kVY, $x=0.1$ per unit
The transformer neutrals are solidly grounded
Transmission Line Ratings
All Lines: 230 kV, $z_1 = 0.08 + j0.5$ Ω/km,

$z_0 = 0.2 + j1.5$ Ω/km, $y_1 = j3.3$ E-6 S/km,

Maximum MVA = 400
Line Lengths: $L_1 = 15$ km, L_2 assigned by the instructor (
20 to 50 km), $L_3 = 40$ km, $L_4 = 15$ km, $L_5 = 50$ km.

Power Flow Data
Bus 1 : Swing bus, $V_1 = 13.8$ kV, $\partial_1 = 0^\circ$
Bus 2,3,4,5,6 : Load buses
Bus 7 : Constant voltage magnitude bus, $V_7 = 15$ kV,
$P_{G7} = 180$ MW, -87 Mvar $< Q_{G7} < +87$ Mvar
System Base Quantities
$S_{base} = 100$ MVA (three-phase)
$V_{base} = 13.8$ kV (line-to-line) in the zone of G1

FIGURE 6.13 Single-line diagram for Design Project 4—transmission loop

using the given system base quantities. Then using PowerWorld Simulator, create three input data files: bus input data, line input data, and transformer input data. Note that bus 1 is the swing bus. Your output for this assignment consists of three power-flow input data files.

The purpose of this assignment is to get started and to correct errors before going to the next assignment. It requires a knowledge of the per-unit system, which was covered in Chapter 3, but may need review.

Assignment 2: Power Flow

Case 1. Run the power flow program and obtain the bus, line, and transformer input/output data files that you prepared in Assignment 1.

Case 2. Suggest one method of increasing the voltage magnitude at bus 4 by 5%. Demonstrate the effectiveness of your method by making appropriate changes to the input data of case 1 and by running the power flow program.

Your output for this assignment consists of 12 data files, 3 input and 3 output data files for each case, along with a one-paragraph explanation of your method for increasing the voltage at bus 4 by 5%.

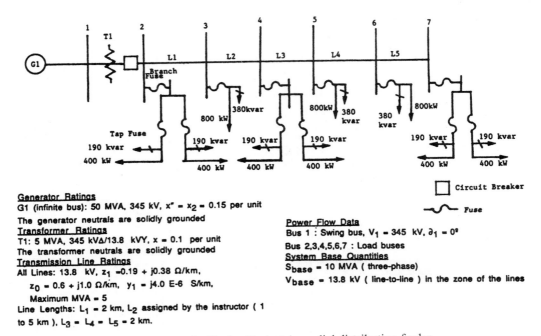

Generator Ratings
G1 (infinite bus): 50 MVA, 345 kV, $x'' = x_2 = 0.15$ per unit
The generator neutrals are solidly grounded
Transformer Ratings
T1: 5 MVA, 345 kVΔ/13.8 kVY, $x = 0.1$ per unit
The transformer neutrals are solidly grounded
Transmission Line Ratings
All Lines: 13.8 kV, $z_1 = 0.19 + j0.38$ Ω/km,
$z_0 = 0.6 + j1.0$ Ω/km, $y_1 = j4.0$ E-6 S/km,
Maximum MVA = 5
Line Lengths: $L_1 = 2$ km, L_2 assigned by the instructor (1
to 5 km), $L_3 = L_4 = L_5 = 2$ km.

Power Flow Data
Bus 1 : Swing bus, $V_1 = 345$ kV, $\partial_1 = 0°$
Bus 2,3,4,5,6,7 : Load buses
System Base Quantities
$S_{base} = 10$ MVA (three-phase)
$V_{base} = 13.8$ kV (line-to-line) in the zone of the lines

FIGURE 6.14 Single-line diagram for Design Project 4—radial distribution feeder

During this assignment, course material contains voltage control methods, including use of generator excitation control, tap changing and regulating transformers, static capacitors, static var systems, and parallel transmission lines.

This project continues in Chapters 7 and 9.

DESIGN PROJECT 5: POWER FLOW*

Time given: 4 weeks
Approximate time required: 25 hours

Figure 6.15 shows the single-line diagram of a 10-bus power system with 7 generating units, 2 345-kV lines, 7 230-kV lines, and 5 transformers. Per-unit transformer leakage reactances, transmission-line series impedances and shunt susceptances, real power generation, and real and reactive loads during heavy load periods, all on a 100-MVA system base, are given on the diagram. Fixed transformer tap settings are also shown. During light load periods, the

---

*This is based on a project assigned by Adjunct Professor Richard Farmer at Arizona State University, Tempe, Arizona.

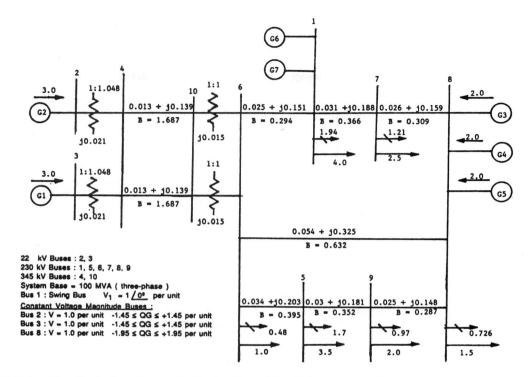

FIGURE 6.15   Single-line diagram for Design Project 5—10-bus power system

real and reactive loads (and generation) are 25% of those shown. Note that bus 1 is the swing bus.

Design Procedure

Using PowerWorld Simulator (convergence can be achieved by changing load buses to constant voltage magnitude buses with wide var limits), determine:

1. The amount of shunt compensation required at 230- and 345-kV buses such that the voltage magnitude $0.99 \leq V \leq 1.02$ per unit at all buses during both light and heavy loads. Find two settings for the compensation, one for light and one for heavy loads.

2. The amount of series compensation required during heavy loads on each 345-kV line such that there is a maximum of 40° angular displacement between bus 4 and bus 10. Assume that one 345-kV line is out of service. Also assume that the series compensation is effectively distributed such that the net series reactance of each 345-kV line is reduced by the percentage compensation. Determine the percentage series compensation to within ±10%.

REFERENCES

1. W. F. Tinney and C. E. Hart, "Power Flow Solutions by Newton's Method," *IEEE Trans. PAS*, *86* (November 1967), p. 1449.

2. W. F. Tinney and J. W. Walker, "Direct Solution of Sparse Network Equations by Optimally Ordered Triangular Factorization," *Proc. IEEE*, *55* (November 1967), pp. 1801–1809.

3. Glenn W. Stagg and Ahmed H. El-Abiad, *Computer Methods in Power System Analysis* (New York: McGraw-Hill, 1968).

4. N. M. Peterson and W. S. Meyer, "Automatic Adjustment of Transformer and Phase Shifter Taps in Newton Power Flow," *IEEE Trans. PAS*, *90* (January–February 1971), pp. 103–108.

5. W. D. Stevenson, Jr., *Elements of Power Systems Analysis*, 4th ed. (New York: McGraw-Hill, 1982).

6. A. Bramellar and R. N. Allan, *Sparsity* (London: Pitman, 1976).

7. C. A. Gross, *Power Systems Analysis* (New York: Wiley, 1979).

8. B. Stott, "Fast Decoupled Load Flow," *IEEE Trans. PAS*, Vol. PAS 91 (September–October 1972), pp. 1955–1959.

9. T. Overbye and J. Weber, "Visualizing the Electric Grid," *IEEE Spectrum*, 38, 2 (February 2001), pp. 52–58.

10. Westinghouse Electric Corporation, *Transmission and Distribution Reference Book*, 4th edition (Pittsburgh: Westinghouse, 1964).

11. Aluminum Association, *The Aluminum Electrical Conductor Handbook* (Washington DC: Aluminum Association).

12. A. J. Wood and B. F. Wollenberg, *Power Generation, Operation and Control*, 2nd ed. (New York: John Wiley & Sons, 1996).

13. A. R. Bergen and V. Vittal, *Power System Analysis*, 2nd ed. (Upper Saddle River, NJ: Prentice Hall, 2000).

SF6 circuit breaker installation (Courtesy of PacifiCorp).

7

SYMMETRICAL FAULTS

Short circuits occur in power systems when equipment insulation fails, due to system overvoltages caused by lightning or switching surges, to insulation contamination (salt spray or pollution), or to other mechanical causes. The resulting short circuit or "fault" current is determined by the internal voltages of the synchronous machines and by the system impedances between the machine voltages and the fault. Short-circuit currents may be several orders of magnitude larger than normal operating currents and, if allowed to persist, may cause thermal damage to equipment. Windings and busbars may also suffer mechanical damage due to high magnetic forces during faults. It is therefore necessary to remove faulted sections of a power system from service as soon as possible. Standard EHV protective equipment is designed to clear faults within 3 cycles (50 ms at 60 Hz). Lower voltage protective equipment operates more slowly (for example, 5 to 20 cycles).

We begin this chapter by reviewing series R–L circuit transients in Section 7.1, followed in Section 7.2 by a description of three-phase short-circuit currents at unloaded synchronous machines. We analyze both the ac component, including subtransient, transient, and steady-state currents, and the dc component of fault current. We then extend these results in Sections 7.3 and 7.4 to power system three-phase short circuits by means of the superposition principle. We observe that the bus impedance matrix is the key to calculating fault currents. The SHORT CIRCUITS computer program that accompanies this text may be utilized in power system design to select, set, and coordinate protective equipment such as circuit breakers, fuses, relays, and instrument transformers. We discuss circuit breaker and fuse selection in Section 7.5.

Balanced three-phase power systems are assumed throughout this chapter. We also work in per-unit.

CASE STUDY Short circuits can cause severe damage when not interrupted promptly. In some cases, high-impedance fault currents may be insufficient to operate protective relays or blow fuses. Standard overcurrent protection schemes utilized on secondary distribution at some industrial, commercial, and large residential buildings may not detect high-impedance faults, commonly called arcing faults. In these cases, more careful design techniques, such as the use of ground fault circuit interruption, are required to detect arcing faults and prevent burndown. The following case histories [11] give examples of the destructive effects of arcing faults.

The Problem of Arcing Faults in Low-Voltage Power Distribution Systems

FRANCIS J. SHIELDS

ABSTRACT

Many cases of electrical equipment burndown arising from low-level arcing-fault currents have occurred in recent years in low-voltage power distribution systems. Burndown, which is the severe damage or complete destruction of conductors, insulation systems and metallic enclosures, is caused by the concentrated release of energy in the fault arc. Both grounded and ungrounded electrical distribution systems have experienced burndown, and

("The Problem of Arcing Faults in Low-Voltage Power Distribution Systems", Francis J. Shields © 1967 IEEE. Reprinted, with permission, from IEEE Transactions on Industry and General Applications, Vol. IGA-3, No. 1, Jan./Feb. 1967, pp. 16–17.)

the reported incidents have involved both industrial and commercial building distribution equipment, without regard to manufacturer, geographical location, or operating environment.

BURNDOWN CASE HISTORIES

The reported incidents of equipment burndown are many. One of the most publicized episodes involved a huge apartment building complex in New York City (Fig. 1), in which two main 480Y/277-volt switchboards were completely destroyed, and two 5000-ampere service entrance buses were burned-off right back to the utility vault. This arcing fault blazed and sputtered for over an hour, and inconvenienced some 10,000 residents of the development through loss of service to building water

Figure 1
Burndown damage caused by arcing fault. View shows low-voltage cable compartments of secondary unit substation.

Figure 2
Service entrance switch and current-limiting fuses completely destroyed by arcing fault in main low-voltage switchboard.

Figure 3
Fused feeder switch consumed by arcing fault in high-rise apartment main switchboard. No intermediate segregating barriers had been used in construction.

pumps, hall and stair lighting, elevators, appliances, and apartment lights. Several days elapsed before service resembling normal was restored through temporary hookups. Illustrations of equipment damage in this burndown are shown in Figs. 2 and 3.

Another example of burndown occurred in the Midwest, and resulted in completely gutting a service entrance switchboard and burning up two 1000-kVA supply transformers. This burndown arc current flowed for about 15 minutes.

In still other reported incidents, a Maryland manufacturer experienced four separate burndowns of secondary unit substations in a little over a year; on the West Coast a unit substation at an industrial process plant burned for more than eight minutes, resulting in destruction of the low-voltage switchgear equipment; and this year [1966] several burndowns have occurred in government office buildings at scattered locations throughout the country.

An example of the involvement of the latter type of equipment in arcing-fault burndowns is shown in

Fig. 4. The arcing associated with this fault continued for over 20 minutes, and the fault was finally extinguished only when the relays on the primary system shut down the whole plant.

The electrical equipment destruction shown in the sample photographs is quite startling, but it is only one aspect of this type of fault. Other less graphic but no less serious effects of electrical

Figure 4
Remains of main secondary circuit breaker burned down during arcing fault in low-voltage switchgear section of unit substation.

equipment burndown may include personnel fatalities or serious injury, contingent fire damage, loss of vital services (lighting, elevators, ventilation, fire pumps, etc.), shutdown of critical loads, and loss of product revenue. It should be pointed out that the cases reported have involved both industrial and commercial building distribution equipment, without regard to manufacturer, geographical location, operating environment, or the presence or absence of electrical system neutral grounding. Also, the reported burndowns have included a variety of distribution equipment—load center unit substations, switchboards, busway, panelboards, service-entrance equipment, motor control centers, and cable in conduit, for example.

It is obvious, therefore, when all the possible effects of arcing-fault burndowns are taken into consideration, that engineers responsible for electrical power system layout and operation should be anxious both to minimize the probability of arcing faults in electrical systems and to alleviate or mitigate the destructive effects of such faults if they should inadvertently occur despite careful design and the use of quality equipment.

7.1

SERIES R–L CIRCUIT TRANSIENTS

Consider the series R–L circuit shown in Figure 7.1. The closing of switch SW at $t = 0$ represents to a first approximation a three-phase short circuit at the terminals of an unloaded synchronous machine. For simplicity, assume zero fault impedance; that is, the short circuit is a solid or "bolted" fault. The current is assumed to be zero before SW closes, and the source angle α determines the source voltage at $t = 0$. Writing a KVL equation for the circuit,

$$\frac{\mathrm{L}di(t)}{dt} + \mathrm{R}i(t) = \sqrt{2}\mathrm{V}\sin(\omega t + \alpha) \quad t \geqslant 0 \tag{7.1.1}$$

The solution to (7.1.1) is

$$i(t) = i_{\mathrm{ac}}(t) + i_{\mathrm{dc}}(t)$$

$$= \frac{\sqrt{2}\mathrm{V}}{\mathrm{Z}}[\sin(\omega t + \alpha - \theta) - \sin(\alpha - \theta)e^{-t/\mathrm{T}}] \quad \mathrm{A} \tag{7.1.2}$$

FIGURE 7.1

Current in a series R–L
circuit with ac voltage
source

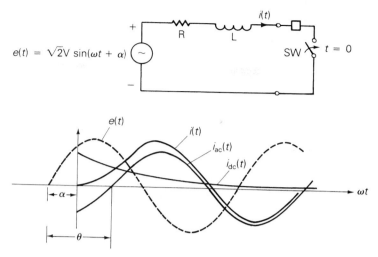

where

$$i_{ac}(t) = \frac{\sqrt{2}V}{Z} \sin(\omega t + \alpha - \theta) \quad A \tag{7.1.3}$$

$$i_{dc}(t) = -\frac{\sqrt{2}V}{Z} \sin(\alpha - \theta)e^{-t/T} \quad A \tag{7.1.4}$$

$$Z = \sqrt{R^2 + (\omega L)^2} = \sqrt{R^2 + X^2} \quad \Omega \tag{7.1.5}$$

$$\theta = \tan^{-1} \frac{\omega L}{R} = \tan^{-1} \frac{X}{R} \tag{7.1.6}$$

$$T = \frac{L}{R} = \frac{X}{\omega R} = \frac{X}{2\pi f R} \quad s \tag{7.1.7}$$

The total fault current in (7.1.2), called the *asymmetrical fault current*, is plotted in Figure 7.1 along with its two components. The ac fault current (also called *symmetrical* or *steady-state fault current*), given by (7.1.3), is a sinusoid. The *dc offset current*, given by (7.1.4), decays exponentially with time constant $T = L/R$.

The rms ac fault current is $I_{ac} = V/Z$. The magnitude of the dc offset, which depends on α, varies from 0 when $\alpha = \theta$ to $\sqrt{2}I_{ac}$ when $\alpha = (\theta \pm \pi/2)$. Note that a short circuit may occur at any instant during a cycle of the ac source; that is, α can have any value. Since we are primarily interested in the largest fault current, we choose $\alpha = (\theta - \pi/2)$. Then (7.1.2) becomes

$$i(t) = \sqrt{2}I_{ac}[\sin(\omega t - \pi/2) + e^{-t/T}] \quad A \tag{7.1.8}$$

where

TABLE 7.1

Short-circuit current—
series R–L circuit*

Component	Instantaneous Current (A)	rms Current (A)
Symmetrical (ac)	$i_{ac}(t) = \dfrac{\sqrt{2}V}{Z}\sin(\omega t + \alpha - \theta)$	$I_{ac} = \dfrac{V}{Z}$
dc offset	$i_{dc}(t) = \dfrac{-\sqrt{2}V}{Z}\sin(\alpha - \theta)e^{-t/T}$	
Asymmetrical (total)	$i(t) = i_{ac}(t) + i_{dc}(t)$	$I_{rms}(t) = \sqrt{I_{ac}^2 + i_{dc}(t)^2}$ with maximum dc offset: $I_{rms}(\tau) = K(\tau)I_{ac}$

*See Figure 7.1 and (7.1.1)–(7.1.12).

$$I_{ac} = \frac{V}{Z} \quad A \tag{7.1.9}$$

The rms value of $i(t)$ is of interest. Since $i(t)$ in (7.1.8) is not strictly periodic, its rms value is not strictly defined. However, treating the exponential term as a constant, we stretch the rms concept to calculate the rms asymmetrical fault current with maximum dc offset, as follows:

$$\begin{aligned} I_{rms}(t) &= \sqrt{[I_{ac}]^2 + [I_{dc}(t)]^2} \\ &= \sqrt{[I_{ac}]^2 + [\sqrt{2}I_{ac}e^{-t/T}]^2} \\ &= I_{ac}\sqrt{1 + 2e^{-2t/T}} \quad A \end{aligned} \tag{7.1.10}$$

It is convenient to use $T = X/(2\pi fR)$ and $t = \tau/f$, where τ is time in cycles, and write (7.1.10) as

$$I_{rms}(\tau) = K(\tau)I_{ac} \quad A \tag{7.1.11}$$

where

$$K(\tau) = \sqrt{1 + 2e^{-4\pi\tau/(X/R)}} \quad \text{per unit} \tag{7.1.12}$$

From (7.1.11) and (7.1.12), the rms asymmetrical fault current equals the rms ac fault current times an "asymmetry factor," $K(\tau)$. $I_{rms}(\tau)$ decreases from $\sqrt{3}I_{ac}$ when $\tau = 0$ to I_{ac} when τ is large. Also, higher X to R ratios (X/R) give higher values of $I_{rms}(\tau)$. The above series R–L short-circuit currents are summarized in Table 7.1.

EXAMPLE 7.1 Fault currents: R–L circuit with ac source

A bolted short circuit occurs in the series R–L circuit of Figure 7.1 with $V = 20$ kV, $X = 8\ \Omega$, $R = 0.8\ \Omega$, and with maximum dc offset. The circuit breaker opens 3 cycles after fault inception. Determine (a) the rms ac fault current, (b) the rms "momentary" current at $\tau = 0.5$ cycle, which passes through the breaker before it opens, and (c) the rms asymmetrical fault current that the breaker interrupts.

SOLUTION

a. From (7.1.9),

$$I_{ac} = \frac{20 \times 10^3}{\sqrt{(8)^2 + (0.8)^2}} = \frac{20 \times 10^3}{8.040} = 2.488 \quad kA$$

b. From (7.1.11) and (7.1.12) with $(X/R) = 8/(0.8) = 10$ and $\tau = 0.5$ cycle,

$$K(0.5 \text{ cycle}) = \sqrt{1 + 2e^{-4\pi(0.5)/10}} = 1.438$$

$$I_{momentary} = K(0.5 \text{ cycle})I_{ac} = (1.438)(2.488) = 3.576 \quad kA$$

c. From (7.1.11) and (7.1.12) with $(X/R) = 10$ and $\tau = 3$ cycles,

$$K(3 \text{ cycles}) = \sqrt{1 + 2e^{-4\pi(3)/10}} = 1.023$$

$$I_{rms}(3 \text{ cycles}) = (1.023)(2.488) = 2.544 \quad kA \qquad \blacksquare$$

7.2

THREE-PHASE SHORT CIRCUIT—UNLOADED SYNCHRONOUS MACHINE

One way to investigate a three-phase short circuit at the terminals of a synchronous machine is to perform a test on an actual machine. Figure 7.2 shows an oscillogram of the ac fault current in one phase of an unloaded synchronous machine during such a test. The dc offset has been removed

FIGURE 7.2

ac fault current in one phase of an unloaded synchronous machine during a three-phase short circuit (the dc offset current is removed)

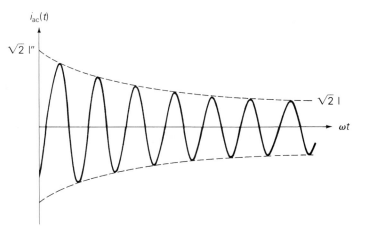

from the oscillogram. As shown, the amplitude of the sinusoidal waveform decreases from a high initial value to a lower steady-state value.

A physical explanation for this phenomenon is that the magnetic flux caused by the short-circuit armature currents (or by the resultant armature MMF) is initially forced to flow through high reluctance paths that do not link the field winding or damper circuits of the machine. This is a result of the theorem of constant flux linkages, which states that the flux linking a closed winding cannot change instantaneously. The armature inductance, which is inversely proportional to reluctance, is therefore initially low. As the flux then moves toward the lower reluctance paths, the armature inductance increases.

The ac fault current in a synchronous machine can be modeled by the series R–L circuit of Figure 7.1 if a time-varying inductance $L(t)$ or reactance $X(t) = \omega L(t)$ is employed. In standard machine theory texts [3, 4], the following reactances are defined:

X''_d = direct axis subtransient reactance

X'_d = direct axis transient reactance

X_d = direct axis synchronous reactance

where $X''_d < X'_d < X_d$. The subscript d refers to the direct axis. There are similar quadrature axis reactances X''_q, X'_q, and X_q [3, 4]. However, if the armature resistance is small, the quadrature axis reactances do not significantly affect the short-circuit current. Using the above direct axis reactances, the instantaneous ac fault current can be written as

$$i_{ac}(t) = \sqrt{2}E_g \left[\left(\frac{1}{X''_d} - \frac{1}{X'_d} \right) e^{-t/T''_d} \right.$$
$$\left. + \left(\frac{1}{X'_d} - \frac{1}{X_d} \right) e^{-t/T'_d} + \frac{1}{X_d} \right] \sin\left(\omega t + \alpha - \frac{\pi}{2} \right) \quad (7.2.1)$$

where E_g is the rms line-to-neutral prefault terminal voltage of the unloaded synchronous machine. Armature resistance is neglected in (7.2.1). Note that at $t = 0$, when the fault occurs, the rms value of $i_{ac}(t)$ in (7.2.1) is

$$I_{ac}(0) = \frac{E_g}{X''_d} = I'' \quad (7.2.2)$$

which is called the rms *subtransient fault current*, I''. The duration of I'' is determined by the time constant T''_d, called the *direct axis short-circuit subtransient time constant*.

At a later time, when t is large compared to T''_d but small compared to the *direct axis short-circuit transient time constant* T'_d, the first exponential term in (7.2.1) has decayed almost to zero, but the second exponential has not decayed significantly. The rms ac fault current then equals the rms *transient fault current*, given by

$$I' = \frac{E_g}{X'_d} \quad (7.2.3)$$

	Instantaneous Current (A)	rms Current (A)
TABLE 7.2 Short-circuit current— unloaded synchronous machine*	**Component**	
Symmetrical (ac)	(7.2.1)	$I_{ac}(t) = E_g \left[\left(\dfrac{1}{X_d''} - \dfrac{1}{X_d'} \right) e^{-t/T_d''} \right.$
		$\left. + \left(\dfrac{1}{X_d'} - \dfrac{1}{X_d} \right) e^{-t/T_d'} + \dfrac{1}{X_d} \right]$
Subtransient		$I'' = E_g / X_d''$
Transient		$I' = E_g / X_d'$
Steady-state		$I = E_g / X_d$
Maximum dc offset	$i_{dc}(t) = \sqrt{2}\,I'' e^{-t/T_A}$	
Asymmetrical (total)	$i(t) = i_{ac}(t) + i_{dc}(t)$	$I_{rms}(t) = \sqrt{I_{ac}(t)^2 + i_{dc}(t)^2}$
		with maximum dc offset:
		$I_{rms}(t) = \sqrt{I_{ac}(t)^2 + [\sqrt{2}\,I'' e^{-t/T_A}]^2}$

*See Figure 7.2 and (7.2.1)–(7.2.5).

When t is much larger than T_d', the rms ac fault current approaches its steady-state value, given by

$$ I_{ac}(\infty) = \frac{E_g}{X_d} = I \tag{7.2.4} $$

Since the three-phase no-load voltages are displaced 120° from each other, the three-phase ac fault currents are also displaced 120° from each other. In addition to the ac fault current, each phase has a different dc offset. The maximum dc offset in any one phase, which occurs when $\alpha = 0$ in (7.2.1), is

$$ i_{dcmax}(t) = \frac{\sqrt{2} E_g}{X_d''} e^{-t/T_A} = \sqrt{2}\,I'' e^{-t/T_A} \tag{7.2.5} $$

where T_A is called the *armature time constant*. Note that the magnitude of the maximum dc offset depends only on the rms subtransient fault current I''. The above synchronous machine short-circuit currents are summarized in Table 7.2.

Machine reactances X_d'', X_d', and X_d as well as time constants T_d'', T_d', and T_A are usually provided by synchronous machine manufacturers. They can also be obtained from a three-phase short-circuit test, by analyzing an oscillogram such as that in Figure 7.2 [2]. Typical values of synchronous machine reactances and time constants are given in Appendix Table A.1.

EXAMPLE 7.2   **Three-phase short-circuit currents, unloaded synchronous generator**

A 500-MVA 20-kV, 60-Hz synchronous generator with reactances $X_d'' = 0.15$, $X_d' = 0.24$, $X_d = 1.1$ per unit and time constants $T_d'' = 0.035$, $T_d' = 2.0$, $T_A = 0.20$ s is connected to a circuit breaker. The generator is operating at 5% above rated voltage and at no-load when a bolted three-phase short circuit occurs on the load side of the breaker. The breaker interrupts the fault

3 cycles after fault inception. Determine (a) the subtransient fault current in per-unit and kA rms; (b) maximum dc offset as a function of time; and (c) rms asymmetrical fault current, which the breaker interrupts, assuming maximum dc offset.

SOLUTION

a. The no-load voltage before the fault occurs is $E_g = 1.05$ per unit. From (7.2.2), the subtransient fault current that occurs in each of the three phases is

$$I'' = \frac{1.05}{0.15} = 7.0 \quad \text{per unit}$$

The generator base current is

$$I_{base} = \frac{S_{rated}}{\sqrt{3}V_{rated}} = \frac{500}{(\sqrt{3})(20)} = 14.43 \quad \text{kA}$$

The rms subtransient fault current in kA is the per-unit value multiplied by the base current:

$$I'' = (7.0)(14.43) = 101.0 \quad \text{kA}$$

b. From (7.2.5), the maximum dc offset that may occur in any one phase is

$$i_{dcmax}(t) = \sqrt{2}(101.0)e^{-t/0.20} = 142.9e^{-t/0.20} \quad \text{kA}$$

c. From (7.2.1), the rms ac fault current at $t = 3$ cycles $= 0.05$ s is

$$I_{ac}(0.05\ \text{s}) = 1.05\left[\left(\frac{1}{0.15} - \frac{1}{0.24}\right)e^{-0.05/0.035}\right.$$
$$\left. + \left(\frac{1}{0.24} - \frac{1}{1.1}\right)e^{-0.05/2.0} + \frac{1}{1.1}\right]$$
$$= 4.920 \quad \text{per unit}$$
$$= (4.920)(14.43) = 71.01 \quad \text{kA}$$

Modifying (7.1.10) to account for the time-varying symmetrical component of fault current, we obtain

$$I_{rms}(0.05) = \sqrt{[I_{ac}(0.05)]^2 + [\sqrt{2}I''e^{-t/T_a}]^2}$$
$$= I_{ac}(0.05)\sqrt{1 + 2\left[\frac{I''}{I_{ac}(0.05)}\right]^2 e^{-2t/T_a}}$$
$$= (71.01)\sqrt{1 + 2\left[\frac{101}{71.01}\right]^2 e^{-2(0.05)/0.20}}$$
$$= (71.01)(1.8585)$$
$$= 132 \quad \text{kA}$$

7.3

POWER SYSTEM THREE-PHASE SHORT CIRCUITS

In order to calculate the subtransient fault current for a three-phase short circuit in a power system, we make the following assumptions:

1. **Transformers** are represented by their leakage reactances. Winding resistances, shunt admittances, and Δ–Y phase shifts are neglected.

2. **Transmission lines** are represented by their equivalent series reactances. Series resistances and shunt admittances are neglected.

3. **Synchronous machines** are represented by constant-voltage sources behind subtransient reactances. Armature resistance, saliency, and saturation are neglected.

4. All nonrotating impedance loads are neglected.

5. **Induction motors** are either neglected (especially for small motors rated less than 50 hp) or represented in the same manner as synchronous machines.

These assumptions are made for simplicity in this text, and in practice they should not be made for all cases. For example, in distribution systems, resistances of primary and secondary distribution lines may in some cases significantly reduce fault current magnitudes.

Figure 7.3 shows a single-line diagram consisting of a synchronous generator feeding a synchronous motor through two transformers and a transmission line. We shall consider a three-phase short circuit at bus 1. The positive-sequence equivalent circuit is shown in Figure 7.4(a), where the voltages E_g'' and E_m'' are the prefault internal voltages behind the subtransient reactances of the machines, and the closing of switch SW represents the fault. For purposes of calculating the subtransient fault current, E_g'' and E_m'' are assumed to be constant-voltage sources.

In Figure 7.4(b) the fault is represented by two opposing voltage sources with equal phasor values V_F. Using superposition, the fault current can then be calculated from the two circuits shown in Figure 7.4(c). However, if V_F equals the prefault voltage at the fault, then the second circuit in Figure 7.4(c) represents the system before the fault occurs. As such, $I_{F2}'' = 0$ and V_F,

FIGURE 7.3

Single-line diagram of a synchronous generator feeding a synchronous motor

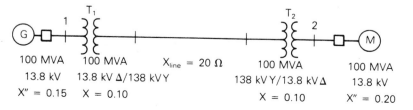

100 MVA	100 MVA	$X_{line} = 20\ \Omega$	100 MVA	100 MVA
13.8 kV	13.8 kV Δ/138 kV Y		138 kV Y/13.8 kV Δ	13.8 kV
X″ = 0.15	X = 0.10		X = 0.10	X″ = 0.20

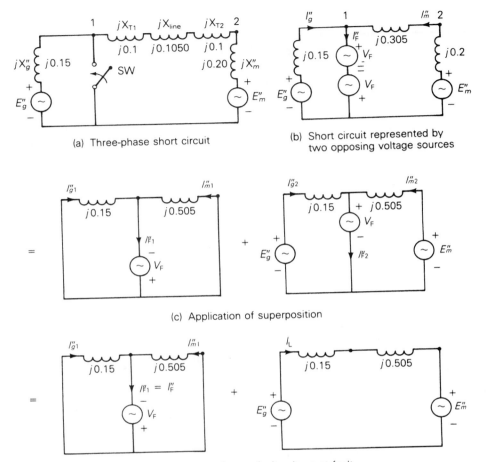

(a) Three-phase short circuit

(b) Short circuit represented by two opposing voltage sources

(c) Application of superposition

(d) V_F set equal to prefault voltage at fault

FIGURE 7.4 Application of superposition to a power system three-phase short circuit

which has no effect, can be removed from the second circuit, as shown in Figure 7.4(d). The subtransient fault current is then determined from the first circuit in Figure 7.4(d), $I_F'' = I_{F1}''$. The contribution to the fault from the generator is $I_g'' = I_{g1}'' + I_{g2}'' = I_{g1}'' + I_L$, where I_L is the prefault generator current. Similarly, $I_m'' = I_{m1}'' - I_L$.

EXAMPLE 7.3 Three-phase short-circuit currents, power system

The synchronous generator in Figure 7.3 is operating at rated MVA, 0.95 p.f. lagging and at 5% above rated voltage when a bolted three-phase short circuit occurs at bus 1. Calculate the per-unit values of (a) subtransient fault current; (b) subtransient generator and motor currents, neglecting prefault current; and (c) subtransient generator and motor currents including prefault current.

SOLUTION

a. Using a 100-MVA base, the base impedance in the zone of the transmission line is

$$Z_{\text{base, line}} = \frac{(138)^2}{100} = 190.44 \quad \Omega$$

and

$$X_{\text{line}} = \frac{20}{190.44} = 0.1050 \quad \text{per unit}$$

The per-unit reactances are shown in Figure 7.4. From the first circuit in Figure 7.4(d), the Thévenin impedance as viewed from the fault is

$$Z_{\text{Th}} = jX_{\text{Th}} = j\frac{(0.15)(0.505)}{(0.15 + 0.505)} = j0.11565 \quad \text{per unit}$$

and the prefault voltage at the generator terminals is

$$V_{\text{F}} = 1.05\underline{/0°} \quad \text{per unit}$$

The subtransient fault current is then

$$I''_{\text{F}} = \frac{V_{\text{F}}}{Z_{\text{Th}}} = \frac{1.05\underline{/0°}}{j0.11565} = -j9.079 \quad \text{per unit}$$

b. Using current division in the first circuit of Figure 7.4(d),

$$I''_{g1} = \left(\frac{0.505}{0.505 + 0.15}\right)I''_{\text{F}} = (0.7710)(-j9.079) = -j7.000 \quad \text{per unit}$$

$$I''_{m1} = \left(\frac{0.15}{0.505 + 0.15}\right)I''_{\text{F}} = (0.2290)(-j9.079) = -j2.079 \quad \text{per unit}$$

c. The generator base current is

$$I_{\text{base, gen}} = \frac{100}{(\sqrt{3})(13.8)} = 4.1837 \quad \text{kA}$$

and the prefault generator current is

$$I_{\text{L}} = \frac{100}{(\sqrt{3})(1.05 \times 13.8)}\underline{/-\cos^{-1} 0.95} = 3.9845\underline{/-18.19°} \quad \text{kA}$$

$$= \frac{3.9845\underline{/-18.19°}}{4.1837} = 0.9524\underline{/-18.19°}$$

$$= 0.9048 - j0.2974 \quad \text{per unit}$$

The subtransient generator and motor currents, including prefault current, are then

$$I_g'' = I_{g1}'' + I_L = -j7.000 + 0.9048 - j0.2974$$

$$= 0.9048 - j7.297 = 7.353\underline{/-82.9°} \quad \text{per unit}$$

$$I_m'' = I_{m1}'' - I_L = -j2.079 - 0.9048 + j0.2974$$

$$= -0.9048 - j1.782 = 1.999\underline{/243.1°} \quad \text{per unit}$$

An alternate method of solving Example 7.3 is to first calculate the internal voltages E_g'' and E_m'' using the prefault load current I_L. Then, instead of using superposition, the fault currents can be resolved directly from the circuit in Figure 7.4(a) (see Problem 7.11). However, in a system with many synchronous machines, the superposition method has the advantage that all machine voltage sources are shorted, and the prefault voltage is the only source required to calculate the fault current. Also, when calculating the contributions to fault current from each branch, prefault currents are usually small, and hence can be neglected. Otherwise, prefault load currents could be obtained from a power-flow program. ∎

7.4

BUS IMPEDANCE MATRIX

We now extend the results of the previous section to calculate subtransient fault currents for three-phase faults in an N-bus power system. The system is modeled by its positive-sequence network, where lines and transformers are represented by series reactances and synchronous machines are represented by constant-voltage sources behind subtransient reactances. As before, all resistances, shunt admittances, and nonrotating impedance loads are neglected. For simplicity, we also neglect prefault load currents.

Consider a three-phase short circuit at any bus n. Using the superposition method described in Section 7.3, we analyze two separate circuits. (For example, see Figure 7.4d.) In the first circuit, all machine-voltage sources are short-circuited and the only source is due to the prefault voltage at the fault. Writing nodal equations for the first circuit,

$$Y_{\text{bus}} E^{(1)} = I^{(1)} \tag{7.4.1}$$

where Y_{bus} is the positive-sequence bus admittance matrix, $E^{(1)}$ is the vector of bus voltages, and $I^{(1)}$ is the vector of current sources. The superscript (1) denotes the first circuit. Solving (7.4.1),

$$Z_{\text{bus}} I^{(1)} = E^{(1)} \tag{7.4.2}$$

where

$$\mathbf{Z}_{\text{bus}} = \mathbf{Y}_{\text{bus}}^{-1} \tag{7.4.3}$$

$\mathbf{Z}_{\text{bus}}$, the inverse of $\mathbf{Y}_{\text{bus}}$, is called the positive-sequence *bus impedance matrix*. Both $\mathbf{Z}_{\text{bus}}$ and $\mathbf{Y}_{\text{bus}}$ are symmetric matrices.

Since the first circuit contains only one source, located at faulted bus n, the current source vector contains only one nonzero component, $I_n^{(1)} = -I_{Fn}''$. Also, the voltage at faulted bus n in the first circuit is $E_n^{(1)} = -V_F$. Rewriting (7.4.2),

$$\begin{bmatrix} Z_{11} & Z_{12} & \cdots & Z_{1n} & \cdots & Z_{1N} \\ Z_{21} & Z_{22} & \cdots & Z_{2n} & \cdots & Z_{2N} \\ \vdots & & & & & \\ Z_{n1} & Z_{n2} & \cdots & Z_{nn} & \cdots & Z_{nN} \\ \vdots & & & & & \\ Z_{N1} & Z_{N2} & \cdots & Z_{Nn} & \cdots & Z_{NN} \end{bmatrix} \begin{bmatrix} 0 \\ 0 \\ \vdots \\ -I_{Fn}'' \\ \vdots \\ 0 \end{bmatrix} = \begin{bmatrix} E_1^{(1)} \\ E_2^{(1)} \\ \vdots \\ -V_F \\ \vdots \\ E_N^{(1)} \end{bmatrix} \tag{7.4.4}$$

The minus sign associated with the current source in (7.4.4) indicates that the current injected into bus n is the negative of I_{Fn}'', since I_{Fn}'' flows away from bus n to the neutral. From (7.4.4), the subtransient fault current is

$$I_{Fn}'' = \frac{V_F}{Z_{nn}} \tag{7.4.5}$$

Also from (7.4.4) and (7.4.5), the voltage at any bus k in the first circuit is

$$E_k^{(1)} = Z_{kn}(-I_{Fn}'') = \frac{-Z_{kn}}{Z_{nn}} V_F \tag{7.4.6}$$

The second circuit represents the prefault conditions. Neglecting prefault load current, all voltages throughout the second circuit are equal to the prefault voltage; that is, $E_k^{(2)} = V_F$ for each bus k. Applying superposition,

$$E_k = E_k^{(1)} + E_k^{(2)} = \frac{-Z_{kn}}{Z_{nn}} V_F + V_F$$

$$= \left(1 - \frac{Z_{kn}}{Z_{nn}}\right) V_F \qquad k = 1, 2, \ldots, N \tag{7.4.7}$$

EXAMPLE 7.4 **Using $\mathbf{Z}_{bus}$ to compute three-phase short-circuit currents in a power system**

Faults at bus 1 and 2 in Figure 7.3 are of interest. The prefault voltage is 1.05 per unit and prefault load current is neglected. (a) Determine the 2×2 positive-sequence bus impedance matrix. (b) For a bolted three-phase short circuit at bus 1, use $\mathbf{Z}_{\text{bus}}$ to calculate the subtransient fault current and the contribution to the fault current from the transmission line. (c) Repeat part (b) for a bolted three-phase short circuit at bus 2.

FIGURE 7.5

Circuit of Figure 7.4(a)
showing per-unit
admittance values

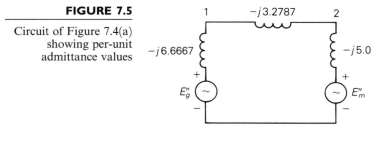

SOLUTION

a. The circuit of Figure 7.4(a) is redrawn in Figure 7.5 showing per-unit admittance rather than per-unit impedance values. Neglecting prefault load current, $E_g'' = E_m'' = V_F = 1.05\underline{/0°}$ per unit. From Figure 7.5, the positive-sequence bus admittance matrix is

$$Y_{bus} = -j \begin{bmatrix} 9.9454 & -3.2787 \\ -3.2787 & 8.2787 \end{bmatrix} \quad \text{per unit}$$

Inverting Y_{bus},

$$Z_{bus} = Y_{bus}^{-1} = +j \begin{bmatrix} 0.11565 & 0.04580 \\ 0.04580 & 0.13893 \end{bmatrix} \quad \text{per unit}$$

b. Using (7.4.5) the subtransient fault current at bus 1 is

$$I_{F1}'' = \frac{V_F}{Z_{11}} = \frac{1.05\underline{/0°}}{j0.11565} = -j9.079 \quad \text{per unit}$$

which agrees with the result in Example 7.3, part (a). The voltages at buses 1 and 2 during the fault are, from (7.4.7),

$$E_1 = \left(1 - \frac{Z_{11}}{Z_{11}}\right) V_F = 0$$

$$E_2 = \left(1 - \frac{Z_{21}}{Z_{11}}\right) V_F = \left(1 - \frac{j0.04580}{j0.11565}\right) 1.05\underline{/0°} = 0.6342\underline{/0°}$$

The current to the fault from the transmission line is obtained from the voltage drop from bus 2 to 1 divided by the impedance of the line and transformers T_1 and T_2:

$$I_{21} = \frac{E_2 - E_1}{j(X_{line} + X_{T1} + X_{T2})} = \frac{0.6342 - 0}{j0.3050} = -j2.079 \quad \text{per unit}$$

which agrees with the motor current calculated in Example 7.3, part (b), where prefault load current is neglected.

c. Using (7.4.5), the subtransient fault current at bus 2 is

$$I_{F2}'' = \frac{V_F}{Z_{22}} = \frac{1.05\underline{/0^\circ}}{j0.13893} = -j7.558 \quad \text{per unit}$$

and from (7.4.7),

$$E_1 = \left(1 - \frac{Z_{12}}{Z_{22}}\right)V_F = \left(1 - \frac{j0.04580}{j0.13893}\right)1.05\underline{/0^\circ} = 0.7039\underline{/0^\circ}$$

$$E_2 = \left(1 - \frac{Z_{22}}{Z_{22}}\right)V_F = 0$$

The current to the fault from the transmission line is

$$I_{12} = \frac{E_1 - E_2}{j(X_{\text{line}} + X_{T1} + X_{T2})} = \frac{0.7039 - 0}{j0.3050} = -j2.308 \quad \text{per unit} \qquad \blacksquare$$

Figure 7.6 shows a bus impedance equivalent circuit that illustrates the short-circuit currents in an N-bus system. This circuit is given the name *rake equivalent* in Neuenswander [5] due to its shape, which is similar to a garden rake.

The diagonal elements $Z_{11}, Z_{22}, \ldots, Z_{NN}$ of the bus impedance matrix, which are the *self-impedances*, are shown in Figure 7.6. The off-diagonal elements, or the *mutual impedances*, are indicated by the brackets in the figure.

Neglecting prefault load currents, the internal voltage sources of all synchronous machines are equal both in magnitude and phase. As such, they can be connected, as shown in Figure 7.7, and replaced by one equivalent source V_F from neutral bus 0 to a references bus, denoted r. This equivalent source is also shown in the rake equivalent of Figure 7.6.

FIGURE 7.6

Bus impedance equivalent circuit (*rake equivalent*)

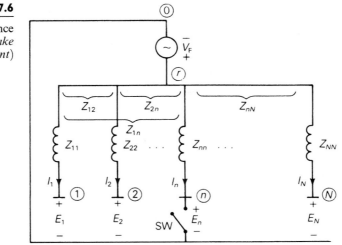

FIGURE 7.7

Parallel connection of
unloaded synchronous
machine internal-voltage
sources

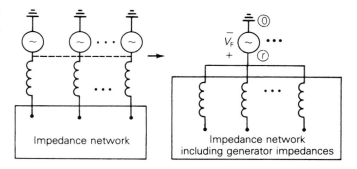

Using Z_{bus}, the fault currents in Figure 7.6 are given by

$$
\begin{bmatrix}
Z_{11} & Z_{12} & \cdots & Z_{1n} & \cdots & Z_{1N} \\
Z_{21} & Z_{22} & \cdots & Z_{2n} & \cdots & Z_{2N} \\
\vdots & & & & & \\
Z_{n1} & Z_{n2} & \cdots & Z_{nn} & \cdots & Z_{nN} \\
\vdots & & & & & \\
Z_{N1} & Z_{N2} & \cdots & Z_{Nn} & \cdots & Z_{NN}
\end{bmatrix}
\begin{bmatrix}
I_1 \\ I_2 \\ \vdots \\ I_n \\ \vdots \\ I_N
\end{bmatrix}
=
\begin{bmatrix}
V_F - E_1 \\ V_F - E_2 \\ \vdots \\ V_F - E_n \\ \vdots \\ V_F - E_N
\end{bmatrix}
\tag{7.4.8}
$$

where $I_1, I_2, \ldots$ are the branch currents and $(V_F - E_1), (V_F - E_2), \ldots$ are the voltages across the branches.

If switch SW in Figure 7.6 is open, all currents are zero and the voltage at each bus with respect to the neutral equals V_F. This corresponds to pre-fault conditions, neglecting prefault load currents. If switch SW is closed, corresponding to a short circuit at bus n, $E_n = 0$ and all currents except I_n remain zero. The fault current is $I_{Fn}'' = I_n = V_F/Z_{nn}$, which agrees with (7.4.5). This fault current also induces a voltage drop $Z_{kn}I_n = (Z_{kn}/Z_{nn})V_F$ across each branch k. The voltage at bus k with respect to the neutral then equals V_F minus this voltage drop, which agrees with (7.4.7).

As shown by Figure 7.6 as well as (7.4.5), subtransient fault currents throughout an N-bus system can be determined from the bus impedance matrix and the prefault voltage. Z_{bus} can be computed by first constructing Y_{bus}, via nodal equations, and then inverting Y_{bus}. Once Z_{bus} has been obtained, these fault currents are easily computed.

EXAMPLE 7.5

PowerWorld Simulator case Example 7_5 models the 5-bus power system whose one-line diagram is shown in Figure 6.2. Machine, line, and transformer data are given in Tables 7.3, 7.4, and 7.5. This system is initially unloaded. Prefault voltages at all the buses are 1.05 per unit. Use PowerWorld Simulator to determine the fault current for three-phase faults at each of the buses.

		Machine Subtransient Reactance—X_d'' (per unit)
TABLE 7.3	**Bus**	
Synchronous machine data for SYMMETRICAL SHORT CIRCUITS program*	1	0.045
	3	0.0225

*S_{base} = 100 MVA
V_{base} = 15 kV at buses 1, 3
= 345 kV at buses 2, 4, 5

		Equivalent Positive-Sequence Series Reactance (per unit)
TABLE 7.4	**Bus-to-Bus**	
Line data for SYMMETRICAL SHORT CIRCUITS program	2–4	0.1
	2–5	0.05
	4–5	0.025

		Leakage Reactance—X (per unit)
TABLE 7.5	**Bus-to-Bus**	
Transformer data for SYMMETRICAL SHORT CIRCUITS program	1–5	0.02
	3–4	0.01

SOLUTION   To fault a bus from the one-line, first right-click on the bus symbol to display the local menu, and then select "Fault." This displays the **Fault** dialog. The selected bus will be automatically selected as the fault location. Verify that the Fault Location is "Bus Fault" and the Fault Type is "3 Phase Balanced" (unbalanced faults are covered in Chapter 9). Then select "Calculate" to determine the fault currents and voltages. The results are shown in the tables at the bottom of the dialog. Additionally, the values can be animated on the one-line by changing the Oneline Display Field value. Since with a three-phase fault the system remains balanced, the magnitudes of the a phase, b phase and c phase values are identical. The 5 × 5 Z_{bus} matrix for this system is shown in Table 7.6, and the fault currents and bus voltages for faults at each of the buses are given in Table 7.7. Note that these fault currents are subtransient fault currents, since the machine reactance input data consist of direct axis subtransient reactances.

TABLE 7.6

Z_{bus} for Example 7.5

$$j\begin{bmatrix} 0.0279725 & 0.0177025 & 0.0085125 & 0.0122975 & 0.020405 \\ 0.0177025 & 0.0569525 & 0.0136475 & 0.019715 & 0.02557 \\ 0.0085125 & 0.0136475 & 0.0182425 & 0.016353 & 0.012298 \\ 0.0122975 & 0.019715 & 0.016353 & 0.0236 & 0.017763 \\ 0.020405 & 0.02557 & 0.012298 & 0.017763 & 0.029475 \end{bmatrix}$$

TABLE 7.7

Fault currents and bus voltages for Example 7.5

Fault Bus	Fault Current (per unit)	Contributions to Fault Current		
		Gen Line or TRSF	Bus-to-Bus	Current (per unit)
1	37.536			
		G 1	GRND–1	23.332
		T 1	5–1	14.204
2	18.436			
		L 1	4–2	6.864
		L 2	5–2	11.572
3	57.556			
		G 2	GRND–3	46.668
		T 2	4–3	10.888
4	44.456			
		L 1	2–4	1.736
		L 3	5–4	10.412
		T 2	3–4	32.308
5	35.624			
		L 2	2–5	2.78
		L 3	4–5	16.688
		T 1	1–5	16.152

$V_F = 1.05$ Fault Bus:	Per-Unit Bus Voltage Magnitudes during the Fault				
	Bus 1	Bus 2	Bus 3	Bus 4	Bus 5
1	0.0000	0.7236	0.5600	0.5033	0.3231
2	0.3855	0.0000	0.2644	0.1736	0.1391
3	0.7304	0.7984	0.0000	0.3231	0.6119
4	0.5884	0.6865	0.1089	0.0000	0.4172
5	0.2840	0.5786	0.3422	0.2603	0.0000

Fault Analysis Dialog for Example 7.5—fault at bus 1

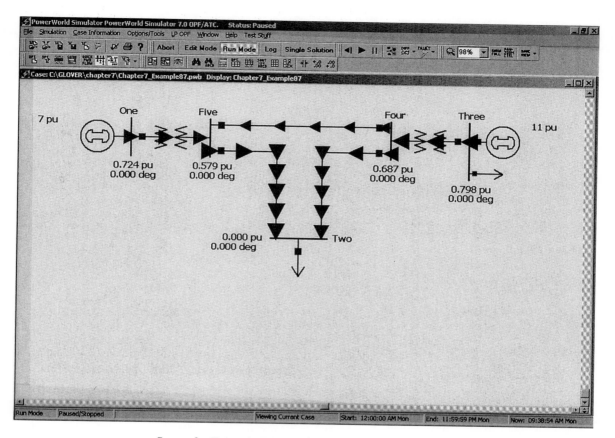

Screen for Example 7.5—fault at bus 2

EXAMPLE 7.6

Redo Example 7.5 with an additional line installed between buses 2 and 4. This line, whose reactance is 0.075 per unit, is not mutually coupled to any other line.

SOLUTION   The modified system is contained in PowerWorld Simulator case Example 7_6. Z_{bus} along with the fault currents and bus voltages are shown in Tables 7.8 and 7.9.

TABLE 7.8	
Z_{bus} for Example 7.6	

$$
j \begin{bmatrix}
0.027723 & 0.01597 & 0.00864 & 0.01248 & 0.02004 \\
0.01597 & 0.04501 & 0.01452 & 0.02097 & 0.02307 \\
0.00864 & 0.01452 & 0.01818 & 0.01626 & 0.01248 \\
0.01248 & 0.02097 & 0.01626 & 0.02349 & 0.01803 \\
0.02004 & 0.02307 & 0.01248 & 0.01803 & 0.02895
\end{bmatrix}
$$

TABLE 7.9

Fault currents and bus voltages for Example 7.6

Fault Bus	Fault Current (per unit)	Contributions to Fault Current		
		Gen Line or TRSF	Bus-to-Bus	Current (per unit)
1	37.872			
		G 1	GRND–1	23.332
		T 1	5–1	14.544
2	23.328			
		L 1	4–2	5.608
		L 2	5–2	10.24
		L 4	4–2	7.48
3	57.756			
		G 2	GRND–3	46.668
		T 2	4–3	11.088
4	44.704			
		L 1	2–4	1.128
		L 3	5–4	9.768
		L 4	2–4	1.504
		T 2	3–4	32.308
5	36.268			
		L 2	2–5	4.268
		L 3	4–5	15.848
		T 1	1–5	16.152

$V_F = 1.05$ Fault Bus:	Per-Unit Bus Voltage Magnitudes during the Fault				
	Bus 1	Bus 2	Bus 3	Bus 4	Bus 5
1	0.0000	0.6775	0.5510	0.4921	0.3231
2	0.4451	0.0000	0.2117	0.1127	0.2133
3	0.7228	0.7114	0.0000	0.3231	0.5974
4	0.5773	0.5609	0.1109	0.0000	0.3962
5	0.2909	0.5119	0.3293	0.2442	0.0000

∎

7.5

CIRCUIT BREAKER AND FUSE SELECTION

A SHORT CIRCUITS computer program may be utilized in power system design to select, set, and coordinate protective equipment such as circuit breakers, fuses, relays, and instrument transformers. In this section we discuss basic principles of circuit breaker and fuse selection.

AC CIRCUIT BREAKERS

A *circuit breaker* is a mechanical switch capable of interrupting fault currents and of reclosing. When circuit-breaker contacts separate while carrying current, an arc forms. The breaker is designed to extinguish the arc by elongating and cooling it. The fact that ac arc current naturally passes through zero twice during its 60-Hz cycle aids the arc extinction process.

Circuit breakers are classified as *power* circuit breakers when they are intended for service in ac circuits above 1500 V, and as *low-voltage* circuit breakers in ac circuits up to 1500 V. There are different types of circuit breakers depending on the medium—air, oil, SF_6 gas, or vacuum—in which the arc is elongated. Also, the arc can be elongated either by a magnetic force or by a blast of air.

Some circuit breakers are equipped with a high-speed automatic reclosing capability. Since most faults are temporary and self-clearing, reclosing is based on the idea that if a circuit is deenergized for a short time, it is likely that whatever caused the fault has disintegrated and the ionized arc in the fault has dissipated.

When reclosing breakers are employed in EHV systems, standard practice is to reclose only once, approximately 15 to 50 cycles (depending on operating voltage) after the breaker interrupts the fault. If the fault persists and the EHV breaker recloses into it, the breaker reinterrupts the fault current and then "locks out," requiring operator resetting. Multiple-shot reclosing in EHV systems is not standard practice because transient stability (Chapter 13) may be compromised. However, for distribution systems (2.4–46 kV) where customer outages are of concern, standard reclosers are equipped for two or more reclosures.

For low-voltage applications, molded case circuit breakers with dual trip capability are available. There is a magnetic instantaneous trip for large fault currents above a specified threshold, and a thermal trip with time delay for smaller fault currents.

Modern circuit-breaker standards are based on symmetrical interrupting current. It is usually necessary to calculate only symmetrical fault current at a system location, and then select a breaker with a symmetrical interrupting capability equal to or above the calculated current. The breaker has the additional capability to interrupt the asymmetrical (or total) fault current if the dc offset is not too large.

Recall from Section 7.1 that the maximum asymmetry factor K ($\tau = 0$) is $\sqrt{3}$, which occurs at fault inception ($\tau = 0$). After fault inception, the dc fault current decays exponentially with time constant $T = (L/R) = (X/\omega R)$, and the asymmetry factor decreases. Power circuit breakers with a 2-cycle rated interruption time are designed for an asymmetrical interrupting capability up to 1.4 times their symmetrical interrupting capability, whereas slower circuit breakers have a lower asymmetrical interrupting capability.

A simplified method for breaker selection is called the "E/X simplified method" [1, 7]. The maximum symmetrical short-circuit current at the system location in question is calculated from the prefault voltage and system reactance characteristics, using computer programs. Resistances, shunt admittances, nonrotating impedance loads, and prefault load currents are neglected. Then, if the X/R ratio at the system location is less than 15, a breaker with a symmetrical interrupting capability equal to or above the cal-

	Rated Values						
				Insulation Level		Current	
Identification		Voltage		Rated Withstand Test Voltage		Rated Continuous Current at 60 Hz (Amperes, rms)	Rated Short-Circuit Current (at Rated Max kV) (kA, rms)
Nominal Voltage Class (kV, rms)	Nominal 3-Phase MVA Class	Rated Max Voltage (kV, rms)	Rated Voltage Range Factor (K)	Low Frequency (kV, rms)	Impulse (kV, Crest)		
Col 1	Col 2	Col 3	Col 4	Col 5	Col 6	Col 7	Col 8
14.4	250	15.5	2.67			600	8.9
14.4	500	15.5	1.29			1200	18
23	500	25.8	2.15			1200	11
34.5	1500	38	1.65			1200	22
46	1500	48.3	1.21			1200	17
69	2500	72.5	1.21			1200	19
115		121	1.0			1200	20
115		121	1.0			1600	40
115		121	1.0			2000	40
115		121	1.0			2000	63
115		121	1.0			3000	40
115		121	1.0			3000	63
138		145	1.0			1200	20
138	Not	145	1.0			1600	40
138		145	1.0			2000	40
138		145	1.0			2000	63
138		145	1.0			2000	80
138	Applicable	145	1.0			3000	40
138		145	1.0			3000	63
138		145	1.0			3000	80
161		169	1.0			1200	16
161		169	1.0			1600	31.5
161		169	1.0			2000	40
161		169	1.0			2000	50
230		242	1.0			1600	31.5
230		242	1.0			2000	31.5
230		242	1.0			3000	31.5
230		242	1.0			2000	40
230		242	1.0			3000	40
230		242	1.0			3000	63
345		362	1.0			2000	40
345		362	1.0			3000	40
500		550	1.0			2000	40
500		550	1.0			3000	40
700		765	1.0			2000	40
700		765	1.0			3000	40

TABLE 7.10

(continued)

			Related Required Capabilities		
			Current Values		
Rated Values		Rated Max Voltage Divided by K (kV, rms)	Max Symmetrical Interrupting Capability	3-Second Short-Time Current Carrying Capability	Closing and Latching Capability 1.6K Times Rated Short-Circuit Current (kA, rms)
Rated Interrupting Time (Cycles)	Rated Permissible Tripping Delay (Seconds)		K Times Rated Short-Circuit Current		
			(kA, rms)	(kA, rms)	
Col 9	Col 10	Col 11	Col 12	Col 13	Col 14
5	2	5.8	24	24	38
5	2	12	23	23	37
5	2	12	24	24	38
5	2	23	36	36	58
5	2	40	21	21	33
5	2	60	23	23	37
3	1	121	20	20	32
3	1	121	40	40	64
3	1	121	40	40	64
3	1	121	63	63	101
3	1	121	40	40	64
3	1	121	63	63	101
3	1	145	20	20	32
3	1	145	40	40	64
3	1	145	40	40	64
3	1	145	63	63	101
3	1	145	80	80	128
3	1	145	40	40	64
3	1	145	63	63	101
3	1	145	80	80	128
3	1	169	16	16	26
3	1	169	31.5	31.5	50
3	1	169	40	40	64
3	1	169	50	50	80
3	1	242	31.5	31.5	50
3	1	242	31.5	31.5	50
3	1	242	31.5	31.5	50
3	1	242	40	40	64
3	1	242	40	40	64
3	1	242	63	63	101
3	1	362	40	40	64
3	1	362	40	40	64
2	1	550	40	40	64
2	1	550	40	40	64
2	1	765	40	40	64
2	1	765	40	40	64

culated current at the given operating voltage is satisfactory. However, if X/R is greater than 15, the dc offset may not have decayed to a sufficiently low value. In this case, a method for correcting the calculated fault current to account for dc and ac time constants as well as breaker speed can be used [10]. If X/R is unknown, the calculated fault current should not be greater than 80% of the breaker interrupting capability.

When selecting circuit breakers for generators, two cycle breakers are employed in practice, and the subtransient fault current is calculated; therefore subtransient machine reactances X''_d are used in fault calculations. For synchronous motors, subtransient reactances X''_d or transient reactances X'_d are used, depending on breaker speed. Also, induction motors can momentarily contribute to fault current. Large induction motors are usually modeled as sources in series with X''_d or X'_d, depending on breaker speed. Smaller induction motors (below 50 hp) are often neglected entirely.

Table 7.10 shows a schedule of preferred ratings for outdoor power circuit breakers. We describe some of the more important ratings shown next.

Voltage ratings

Rated maximum voltage: Designates the maximum rms line-to-line operating voltage. The breaker should be used in systems with an operating voltage less than or equal to this rating.

Rated low frequency withstand voltage: The maximum 60-Hz rms line-to-line voltage that the circuit breaker can withstand without insulation damage.

Rated impulse withstand voltage: The maximum crest voltage of a voltage pulse with standard rise and delay times that the breaker insulation can withstand.

Rated voltage range factor K: The range of voltage for which the symmetrical interrupting capability times the operating voltage is constant.

Current ratings

Rated continuous current: The maximum 60-Hz rms current that the breaker can carry continuously while it is in the closed position without overheating.

Rated short-circuit current: The maximum rms symmetrical current that the breaker can safely interrupt at rated maximum voltage.

Rated momentary current: The maximum rms asymmetrical current that the breaker can withstand while in the closed position without damage. Rated momentary current for standard breakers is 1.6 times the symmetrical interrupting capability.

Rated interrupting time: The time in cycles on a 60-Hz basis from the instant the trip coil is energized to the instant the fault current is cleared.

FIGURE 7.8

Symmetrical interrupting
capability of a 69-kV
class breaker

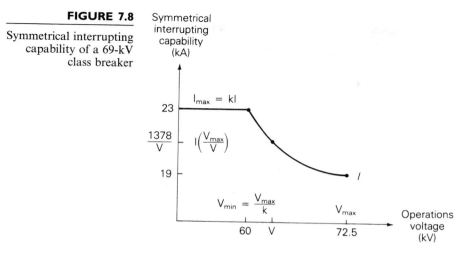

Rated interrupting MVA: For a three-phase circuit breaker, this is $\sqrt{3}$ times the rated maximum voltage in kV times the rated short-circuit current in kA. It is more common to work with current and voltage ratings than with MVA rating.

As an example, the symmetrical interrupting capability of the 69-kV class breaker listed in Table 7.10 is plotted versus operating voltage in Figure 7.8. As shown, the symmetrical interrupting capability increases from its rated short-circuit current $I = 19$ kA at rated maximum voltage $V_{max} = 72.5$ kV up to $I_{max} = KI = (1.21)(19) = 23$ kA at an operating voltage $V_{min} = V_{max}/K = 72.5/1.21 = 60$ kV. At operating voltages V between V_{min} and V_{max}, the symmetrical interrupting capability is $I \times V_{max}/V = 1378/V$ kA. At operating voltages below V_{min}, the symmetrical interrupting capability remains at $I_{max} = 23$ kA.

Breakers of the 115-kV class and higher have a voltage range factor $K = 1.0$; that is, their symmetrical interrupting current capability remains constant.

EXAMPLE 7.7   Circuit breaker selection

The calculated symmetrical fault current is 17 kA at a three-phase bus where the operating voltage is 64 kV. The X/R ratio at the bus is unknown. Select a circuit breaker from Table 7.10 for this bus.

SOLUTION   The 69-kV-class breaker has a symmetrical interrupting capability $I(V_{max}/V) = 19(72.5/64) = 21.5$ kA at the operating voltage $V = 64$ kV. The calculated symmetrical fault current, 17 kA, is less than 80% of this capability (less than $0.80 \times 21.5 = 17.2$ kA), which is a requirement when X/R is unknown. Therefore, we select the 69-kV-class breaker from Table 7.10. ∎

FUSES

Figure 7.9(a) shows a cutaway view of a fuse, which is one of the simplest overcurrent devices. The fuse consists of a metal "fusible" link or links encapsulated in a tube, packed in filler material, and connected to contact terminals. Silver is a typical link metal, and sand is a typical filler material.

During normal operation, when the fuse is operating below its continuous current rating, the electrical resistance of the link is so low that it simply acts as a conductor. If an overload current from one to about six times its continuous current rating occurs and persists for more than a short interval of time, the temperature of the link eventually reaches a level that causes a restricted segment of the link to melt. As shown in Figure 7.9(b), a gap is then formed and an electric arc is established. As the arc causes the link metal to burn back, the gap width increases. The resistance of the arc eventually reaches such a high level that the arc cannot be sustained and it is extinguished, as in Figure 7.9(c). The current flow within the fuse is then completely cut off.

FIGURE 7.9

Typical fuse

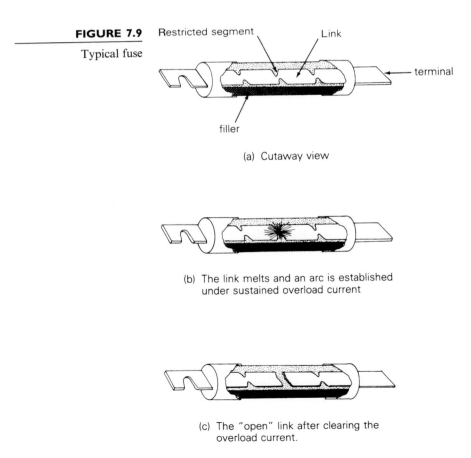

Restricted segment　　Link

terminal

filler

(a) Cutaway view

(b) The link melts and an arc is established under sustained overload current

(c) The "open" link after clearing the overload current.

FIGURE 7.10

Operation of a
current-limiting fuse

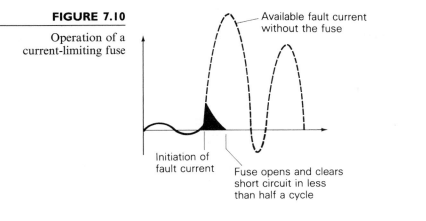

If the fuse is subjected to fault currents higher than about six times its continuous current rating, several restricted segments melt simultaneously, resulting in rapid arc suppression and fault clearing. Arc suppression is accelerated by the filler material in the fuse.

Many modern fuses are current limiting. As shown in Figure 7.10, a current-limiting fuse has such a high speed of response that it cuts off a high fault current in less than a half cycle—before it can build up to its full peak value. By limiting fault currents, these fuses permit the use of motors, transformers, conductors, and bus structures that could not otherwise withstand the destructive forces of high fault currents.

Fuse specification is normally based on the following four factors.

1. *Voltage rating.* This rms voltage determines the ability of a fuse to suppress the internal arc that occurs after the fuse link melts. A blown fuse should be able to withstand its voltage rating. Most low-voltage fuses have 250- or 600-V ratings. Ratings of medium-voltage fuses range from 2.4 to 34.5 kV.

2. *Continuous current rating.* The fuse should carry this rms current indefinitely, without melting and clearing.

3. *Interrupting current rating.* This is the largest rms asymmetrical current that the fuse can safely interrupt. Most modern, low-voltage current-limiting fuses have a 200-kA interrupting rating. Standard interrupting ratings for medium-voltage current-limiting fuses include 65, 80, and 100 kA.

4. *Time response.* The melting and clearing time of a fuse depends on the magnitude of the overcurrent or fault current, and is usually specified by a "time–current" curve. Figure 7.11 shows the time–current curve of a 15.5-kV, 100-A (continuous) current-limiting fuse. As shown, the fuse link melts within 2 s and clears within 5 s for a 500-A current. For a 5-kA current, the fuse link melts in less than 0.01 s and clears within 0.015 s.

FIGURE 7.11

Time–current curves for
a 15.5-kV, 100-A
current-limiting fuse

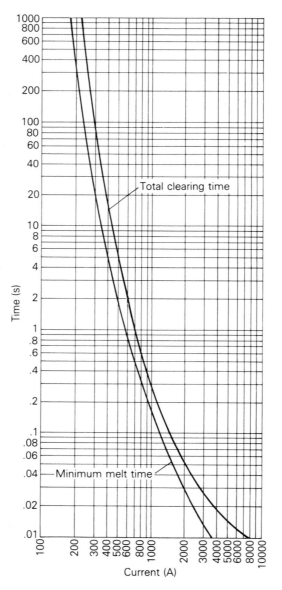

It is usually a simple matter to coordinate fuses in a power circuit such that
only the fuse closest to the fault opens the circuit. In a radial circuit, fuses
with larger continuous current ratings are located closer to the source, such
that the fuse closest to the fault clears before other, upstream fuses melt.

Fuses are inexpensive, fast operating, easily coordinated, and reliable,
and they do not require protective relays or instrument transformers. Their
chief disadvantage is that the fuse or the fuse link must be manually replaced
after it melts. They are basically one-shot devices that are, for example, inca-
pable of high-speed reclosing.

PROBLEMS

SECTION 7.1

7.1   In the circuit of Figure 7.1, $V = 220$ volts, $L = 3$ mH, $R = 0.5\ \Omega$, and $\omega = 2\pi 60$ rad/s. Determine (a) the rms symmetrical fault current; (b) the rms asymmetrical fault current at the instant the switch closes, assuming maximum dc offset; (c) the rms asymmetrical fault current 5 cycles after the switch closes, assuming maximum dc offset; (d) the dc offset as a function of time if the switch closes when the instantaneous source voltage is 244 volts.

7.2   Repeat Example 7.1 with $V = 4$ kV, $X = 3\ \Omega$, and $R = 1\ \Omega$.

7.3   In the circuit of Figure 7.1, let $R = 0.125\ \Omega$, $L = 10$ mH, and the source voltage is $e(t) = 151 \sin(377t + \alpha)$ V. Determine the current response after closing the switch for the following cases: (a) no dc offset; (b) maximum dc offset. Sketch the current waveform up to $t = 0.10$ s corresponding to case (a) and (b).

7.4   Consider the expression for $i(t)$ given by

$$i(t) = \sqrt{2} I_{\text{rms}}[\sin(\omega t - \theta_z) + \sin \theta_z \cdot \bar{e}^{(\omega R/X)t}]$$

where $\theta_z = \tan^{-1}(\omega L/R)$.
(a) For (X/R) equal to zero and infinity, plot $i(t)$ as a function of (ωt).
(b) Comment on the DC offset of the fault current waveforms.
(c) Find the asymmetrical current factor and the time of peak, t_p, in milliseconds, for (X/R) ratios of zero and infinity.

7.5   If the source impedance at a 13.2 kV distribution substation bus is $(0.5 + j1.5)\ \Omega$ per phase, compute the RMS and maximum peak instantaneous value of the fault current, for a balanced three-phase fault. For the system (X/R) ratio of 3.0, the asymmetrical factor is 1.9495 and the time of peak is 7.1 ms (see Problem 7.4). Comment on the withstanding peak current capability to which all substation electrical equipment need to be designed.

SECTION 7.2

7.6   A 1500-MVA 20-kV, 60-Hz three-phase generator is connected through a 1500-MVA 20-kV Δ/500-kV Y transformer to a 500-kV circuit breaker and a 500-kV transmission line. The generator reactances are $X_d'' = 0.17$, $X_d' = 0.30$, and $X_d = 1.5$ per unit, and its time constants are $T_d'' = 0.05$, $T_d' = 1.0$, and $T_A = 0.10$ s. The transformer series reactance is 0.10 per unit; transformer losses and exciting current are neglected. A three-phase short-circuit occurs on the line side of the circuit breaker when the generator is operated at rated terminal voltage and at no-load. The breaker interrupts the fault 3 cycles after fault inception. Determine (a) the subtransient current through the breaker in per-unit and in kA rms; and (b) the rms asymmetrical fault current the breaker interrupts, assuming maximum dc offset. Neglect the effect of the transformer on the time constants.

7.7   For Problem 7.6, determine (a) the instantaneous symmetrical fault current in kA in phase a of the generator as a function of time, assuming maximum dc offset occurs in this generator phase; and (b) the maximum dc offset current in kA as a function of time that can occur in any one generator phase.

7.8   A 300-MVA, 13.8-kV, three-phase, 60-Hz, Y-connected synchronous generator is adjusted to produce rated voltage on open circuit. A balanced three-phase fault is applied to the terminals at t = 0. After analyzing the raw data, the symmetrical transient current is obtained as

$$i_{ac}(t) = 10^4(1 + e^{-t/\tau_1} + 6e^{-t/\tau_2}) \quad A$$

where $\tau_1 = 200$ ms and $\tau_2 = 15$ ms. (a) Sketch $i_{ac}(t)$ as a function of time for $0 \leqslant t \leqslant 500$ ms. (b) Determine X_d'' and X_d in per-unit based on the machine ratings.

7.9   Two identical synchronous machines, each rated 60 MVA, 15 kV, with a subtransient reactance of 0.1 pu, are connected through a line of reactance 0.1 pu on the base of the machine rating. One machine is acting as a synchronous generator, while the other is working as a motor drawing 40 MW at 0.8 pf leading with a terminal voltage of 14.5 kV, when a symmetrical three-phase fault occurs at the motor terminals. Determine the subtransient currents in the generator, the motor, and the fault by using the internal voltages of the machines. Choose a base of 60 MVA, 15 kV in the generator circuit.

SECTION 7.3

7.10   Recalculate the subtransient current through the breaker in Problem 7.6 if the generator is initially delivering rated MVA at 0.80 p.f. lagging and at rated terminal voltage.

7.11   Solve Example 7.4, parts (a) and (c) without using the superposition principle. First calculate the internal machine voltages E_g'' and E_m'', using the prefault load current. Then determine the subtransient fault, generator, and motor currents directly from Figure 7.4(a). Compare your answers with those of Example 7.3.

7.12   Equipment ratings for the four-bus power system shown in Figure 7.12 are as follows:

  Generator G1:   500 MVA, 13.8 kV, $X'' = 0.20$ per unit
  Generator G2:   750 MVA, 18 kV, $X'' = 0.18$ per unit
  Generator G3:   1000 MVA, 20 kV, $X'' = 0.17$ per unit
  Transformer T1:   500 MVA, 13.8 Δ/500 Y kV, $X = 0.12$ per unit
  Transformer T2:   750 MVA, 18 Δ/500 Y kV, $X = 0.10$ per unit
  Transformer T3:   1000 MVA, 20 Δ/500 Y kV, $X = 0.10$ per unit
  Each 500-kV line: $X_1 = 50 \ \Omega$

A three-phase short circuit occurs at bus 1, where the prefault voltage is 525 kV. Prefault load current is neglected. Draw the positive-sequence reactance diagram in

FIGURE 7.12

Problems 7.12, 7.13, 7.19, 7.24, 7.25, 7.26

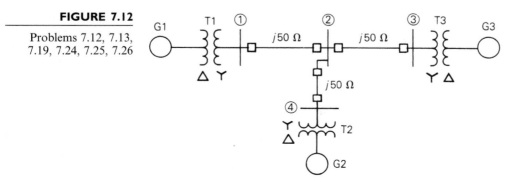

per-unit on a 1000-MVA, 20-kV base in the zone of generator G3. Determine (a) the Thévenin reactance in per-unit at the fault, (b) the subtransient fault current in per-unit and in kA rms, and (c) contributions to the fault current from generator G1 and from line 1–2.

7.13 For the power system given in Problem 7.12, a three-phase short circuit occurs at bus 2, where the prefault voltage is 525 kV. Prefault load current is neglected. Determine the (a) Thévenin equivalent at the fault, (b) subtransient fault current in per-unit and in kA rms, and (c) contributions to the fault from lines 1–2, 2–3, and 2–4.

7.14 Equipment ratings for the five-bus power system shown in Figure 7.13 are as follows:

Generator G1:	50 MVA, 12 kV, $X'' = 0.2$ per unit
Generator G2:	100 MVA, 15 kV, $X'' = 0.2$ per unit
Transformer T1:	50 MVA, 10 kV Y/138 kV Y, $X = 0.10$ per unit
Transformer T2:	100 MVA, 15 kV Δ/138 kV Y, $X = 0.10$ per unit
Each 138-kV line:	$X_1 = 40\ \Omega$

A three-phase short circuit occurs at bus 5, where the prefault voltage is 15 kV. Prefault load current is neglected. (a) Draw the positive-sequence reactance diagram in per-unit on a 100-MVA, 15-kV base in the zone of generator G2. Determine: (b) the Thévenin equivalent at the fault, (c) the subtransient fault current in per-unit and in kA rms, and (d) contributions to the fault from generator G2 and from transformer T2.

FIGURE 7.13

Problems 7.14, 7.15, 7.20

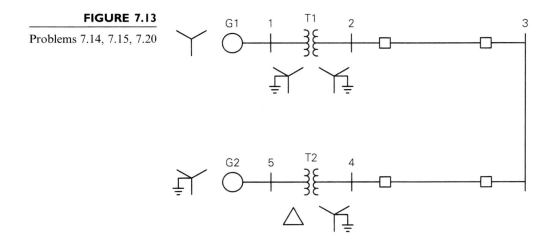

7.15 For the power system given in Problem 7.14, a three-phase short circuit occurs at bus 4, where the prefault voltage is 138 kV. Prefault load current is neglected. Determine (a) the Thévenin equivalent at the fault, (b) the subtransient fault current in per-unit and in kA rms, and (c) contributions to the fault from transformer T2 and from line 3–4.

7.16 In the system shown in Figure 7.14, a three-phase short circuit occurs at point F. Assume that prefault currents are zero and that the generators are operating at rated voltage. Determine the fault current.

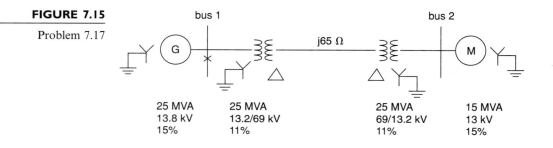

FIGURE 7.14
Problem 7.16

11 kV
20 MVA 15%
G1
11 kV
10 MVA 10%
G2
30 MVA 5%
1:3
(3 + j5) Ω
33 kV
F

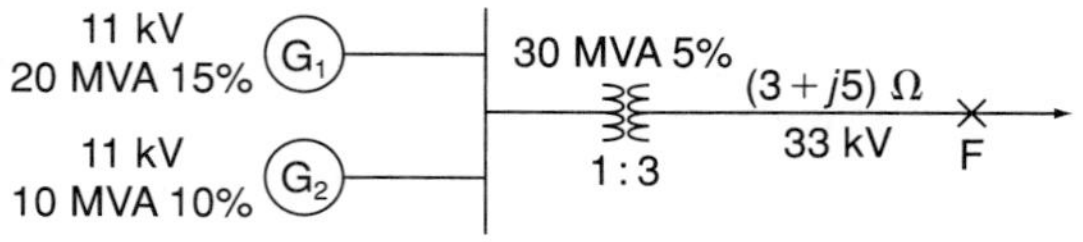

7.17 A three-phase short circuit occurs at the generator bus (bus 1) for the system shown in Figure 7.15. Neglecting prefault currents and assuming that the generator is operating at its rated voltage, determine the subtransient fault current using superposition.

FIGURE 7.15
Problem 7.17

bus 1
bus 2
G
j65 Ω
M
25 MVA
13.8 kV
15%
25 MVA
13.2/69 kV
11%
25 MVA
69/13.2 kV
11%
15 MVA
13 kV
15%

SECTION 7.4

7.18 The bus impedance matrix for a three-bus power system is

$$\boldsymbol{Z}_{\text{bus}} = j\begin{bmatrix} 0.12 & 0.08 & 0.04 \\ 0.08 & 0.12 & 0.06 \\ 0.04 & 0.06 & 0.08 \end{bmatrix} \quad \text{per unit}$$

where subtransient reactances were used to compute $\boldsymbol{Z}_{\text{bus}}$. Prefault voltage is 1.0 per unit and prefault current is neglected. (a) Draw the bus impedance matrix equivalent circuit (rake equivalent). Identify the per-unit self- and mutual impedances as well as the prefault voltage in the circuit. (b) A three-phase short circuit occurs at bus 2. Determine the subtransient fault current and the voltages at buses 1, 2, and 3 during the fault.

7.19 Determine $\boldsymbol{Y}_{\text{bus}}$ in per-unit for the circuit in Problem 7.12. Then invert $\boldsymbol{Y}_{\text{bus}}$ to obtain $\boldsymbol{Z}_{\text{bus}}$.

7.20 Determine $\boldsymbol{Y}_{\text{bus}}$ in per-unit for the circuit in Problem 7.14. Then invert $\boldsymbol{Y}_{\text{bus}}$ to obtain $\boldsymbol{Z}_{\text{bus}}$.

7.21 Figure 7.16 shows a system reactance diagram. (a) Draw the admittance diagram for the system by using source transformations. (b) Find the bus admittance matrix $\boldsymbol{Y}_{\text{bus}}$. (c) Find the bus impedance $\boldsymbol{Z}_{\text{bus}}$ matrix by inverting $\boldsymbol{Y}_{\text{bus}}$.

FIGURE 7.16

Problem 7.21

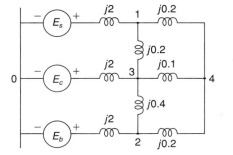

7.22 For the network shown in Figure 7.17, impedances labeled 1 through 6 are in per-unit. (a) Determine Y_{bus}. Preserve all buses. (b) Using MATLAB or a similar computer program, invert Y_{bus} to obtain Z_{bus}.

FIGURE 7.17

Problem 7.22

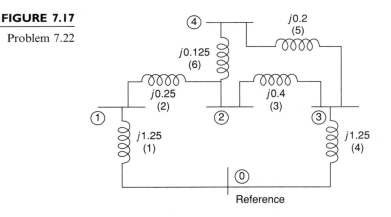

7.23 A single-line diagram of a four-bus system is shown in Figure 7.18, for which Z_{BUS} is given below:

$$Z_{BUS} = j \begin{bmatrix} 0.25 & 0.2 & 0.16 & 0.14 \\ 0.2 & 0.23 & 0.15 & 0.151 \\ 0.16 & 0.15 & 0.196 & 0.1 \\ 0.14 & 0.151 & 0.1 & 0.195 \end{bmatrix} \text{ per unit}$$

Let a three-phase fault occur at bus 2 of the network.
(a) Calculate the initial symmetrical RMS current in the fault.
(b) Determine the voltages during the fault at buses 1, 3, and 4.
(c) Compute the fault currents contributed to bus 2 by the adjacent unfaulted buses 1, 3, and 4.
(d) Find the current flow in the line from bus 3 to bus 1. Assume the prefault voltage V_f at bus 2 to be $1\underline{/0°}$ pu, and neglect all prefault currents.

FIGURE 7.18

Single-line diagram for Problem 7.21

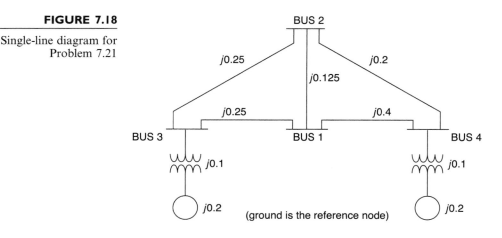

(ground is the reference node)

PW **7.24** PowerWorld Simulator case Problem 7.24 models the system shown in Figure 7.12 with all data on a 1000-MVA base. Using PowerWorld Simulator, determine the current supplied by each generator and the per-unit bus voltage magnitudes at each bus for a fault at bus 2.

PW **7.25** Repeat Problem 7.24, except place the fault at bus 6.

PW **7.26** Repeat Problem 7.24, except place the fault midway between buses 2 and 3. Determining the values for line faults requires that the line be split, with a fictitious bus added at the point of the fault. The original line's impedance is then allocated to the two new lines based on the fault location, 50% each for this problem. Fault calculations are then the same as for a bus fault. This is done automatically in PowerWorld Simulator by first right-clicking on a line, and then selecting "Fault..". The Fault dialog appears as before, except now the fault type is changed to "In-Line Fault." Set the location percentage field to 50% to model a fault midway between buses 2 and 3.

PW **7.27** One technique for limiting fault current is to place reactance in series with the generators. Such reactance can be modeled in Simulator by increasing the value of the generator's positive sequence internal impedance. For the Problem 7.24 case, how much per-unit reactance must be added to G3 to limit its maximum fault current to 3.0 per unit? Where is the location of the most severe bus fault?

PW **7.28** Using PowerWorld Simulator case Example 6.13, determine the per-unit current and actual current in amps supplied by each of the generators for a fault at the PETE69 bus. During the fault, what percentage of the system buses have voltage magnitudes below 0.75 per unit?

PW **7.29** Repeat Problem 7.28, except place the fault at the TIM69 bus.

PW **7.30** Redo Example 7.5, except first open the generator at bus 3.

SECTION 7.5

7.31 A three-phase circuit breaker has a 15.5-kV rated maximum voltage, 9.0-kA rated short-circuit current, and a 2.67-rated voltage range factor. (a) Determine the symmetrical interrupting capability at 10-kV and 5-kV operating voltages. (b) Can this breaker be safely installed at a three-phase bus where the symmetrical fault current is 10 kA, the operating voltage is 13.8 kV, and the (X/R) ratio is 12?

7.32 A 500-kV three-phase transmission line has a 2.2-kA continuous current rating and a 2.5-kA maximum short-time overload rating, with a 525-kV maximum operating voltage. Maximum symmetrical fault current on the line is 30 kA. Select a circuit breaker for this line from Table 7.10.

7.33 A 69-kV circuit breaker has a voltage range factor $K = 1.21$, a continuous current rating of 1200 A, and a rated short-circuit current of 19,000 A at the maximum rated voltage of 72.5 kV. Determine the maximum symmetrical interrupting capability of the breaker. Also, explain its significance at lower operating voltages.

7.34 As shown in Figure 7.18, a 25-MVA, 13.8-kV, 60-Hz synchronous generator with $X_d'' = 0.15$ per unit is connected through a transformer to a bus that supplies four identical motors. The rating of the three-phase transformer is 25 MVA, 13.8/6.9 kV, with a leakage reactance of 0.1 per unit. Each motor has a subtransient reactance $X_d'' = 0.2$ per unit on a base of 5 MVA and 6.9 kV. A three-phase fault occurs at point P, when the bus voltage at the motors is 6.9 kV. Determine: (a) the subtransient fault current, (b) the subtransient current through breaker A, (c) the symmetrical short-circuit interrupting current (as defined for circuit breaker applications) in the fault and in breaker A.

FIGURE 7.19

Problem 7.34

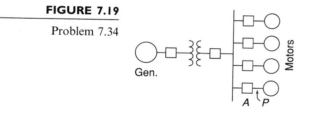

Gen.

CASE STUDY QUESTIONS

A. Why are arcing (high-impedance) faults more difficult to detect than low-impedance faults?

B. What methods are available to prevent the destructive effects of arcing faults from occurring?

DESIGN PROJECT 4 (*CONTINUED*): POWER FLOW/SHORT CIRCUITS

Additional time given: 3 weeks
Additional time required: 10 hours

This is a continuation of Design Project 4. Assignments 1 and 2 are given in Chapter 6.

Assignment 3: Symmetrical Short Circuits

For the single-line diagram that you have been assigned (Figure 6.13 or 6.14), convert the positive-sequence reactance data to per-unit using the given base quantities. For synchronous machines, use subtransient reactance. Then using PowerWorld Simulator, create the machine, transmission line, and transformer input data files. Next, run the program to compute subtransient fault currents for a bolted three-phase-to-ground fault at bus 1, then at bus 2, then at bus 3, and so on. Also compute bus voltages during the faults and the positive-sequence bus impedance matrix. Assume 1.0 per-unit prefault voltage. Neglect prefault load currents and all losses.

Your output for this assignment consists of three input data files and three output data (fault currents, bus voltages, and the bus impedance matrix) files.

This project continues in Chapter 9.

REFERENCES

1. Westinghouse Electric Corporation, *Electrical Transmission and Distribution Reference Book*, 4th ed. (East Pittsburgh, PA: 1964).

2. E. W. Kimbark, *Power System Stability, Synchronous Machines*, vol. 3 (New York: Wiley, 1956).

3. A. E. Fitzgerald, C. Kingsley, and S. Umans, *Electric Machinery*, 5th ed. (New York: McGraw-Hill, 1990).

4. M. S. Sarma, *Electric Machines* 2nd ed. (Boston: PWS Publishing, 1994).

5. J. R. Neuenswander, *Modern Power Systems* (New York: Intext Educational Publishers, 1971).

6. H. E. Brown, *Solution of Large Networks by Matrix Methods* (New York: Wiley, 1975).

7. G. N. Lester, "High Voltage Circuit Breaker Standards in the USA—Past, Present and Future," *IEEE Transactions PAS*, vol. PAS–93 (1974): pp. 590–600.

8. W. D. Stevenson, Jr., *Elements of Power System Analysis*, 4th ed. (New York: McGraw-Hill, 1982).

9. C. A. Gross, *Power System Analysis* (New York: Wiley, 1979).

10. *Application Guide for AC High-Voltage Circuit Breakers Rated on a Symmetrical Current Basis*, ANSI C 37.010 (New York: American National Standards Institute, 1972).

11. F. Shields, "The Problem of Arcing Faults in Low-Voltage Power Distribution Systems," *IEEE Transactions on Industry and General Applications*, vol. IGA-3, no. 1, (January/February 1967), pp. 15–25.

Generator stator showing completed windings for a 757-MVA, 3600-RPM, 60-Hz synchronous generator (Courtesy of General Electric)

8

SYMMETRICAL COMPONENTS

The method of symmetrical components, first developed by C. L. Fortescue in 1918, is a powerful technique for analyzing unbalanced three-phase systems. Fortescue defined a linear transformation from phase components to a new set of components called *symmetrical components*. The advantage of this transformation is that for balanced three-phase networks the equivalent circuits obtained for the symmetrical components, called *sequence networks*, are separated into three uncoupled networks. Furthermore, for unbalanced three-phase systems, the three sequence networks are connected only at points of unbalance. As a result, sequence networks for many cases of unbalanced three-phase systems are relatively easy to analyze.

The symmetrical component method is basically a modeling technique that permits systematic analysis and design of three-phase systems. Decoupling a detailed three-phase network into three simpler sequence networks reveals complicated phenomena in more simplistic terms. Sequence network results can then be superposed to obtain three-phase network results. As an example, the application of symmetrical components to unsymmetrical short-circuit studies (see Chapter 9) is indispensable.

The objective of this chapter is to introduce the concept of symmetrical components in order to lay a foundation and provide a framework for later chapters covering both equipment models as well as power system analysis and design methods. In Section 8.1, we define symmetrical components. In Sections 8.2–8.7, we present sequence networks of loads, series impedances, transmission lines, rotating machines, and transformers. We discuss complex power in sequence networks in Section 8.8. Although Fortescue's original work is valid for polyphase systems with n phases, we will consider only three-phase systems here.

CASE STUDY   The following article provides an overview of three energy storage technologies: (1) battery energy storage; (2) compressed air energy storage; and (3) flywheel energy storage. Energy storage can reduce the requirements for excess generation capacity by supplying power during peak load periods, a technique called *load leveling*. Energy storage can also reduce the requirements for online generator reserves (spinning reserves) by supplying power for short durations during generation capacity deficiencies until offline generators can be started and brought online [4].

Electrical Energy Storage—Challenges and New Market Opportunities

Electrical energy storage has been fundamental to the design of uninterruptible power supply (UPS) systems and various off-grid power supplies for many years. However, it is the new challenges posed by the ascendancy of distributed generation and renewables, the associated issues of intermittency, combined with the ever increasing pressures to maintain customer supply standards whilst optimising the utilisation of distribution assets that is now bringing storage far more centre-stage for power network applications. Whilst the UK has led the way with large-scale pumped hydro storage, the current emphasis is very much on smaller-scale packaged systems, which may be far more easily deployed throughout the power systems network. Much of

("Electrical Energy Storage—Challenges and New Market Opportunities" by John Baker, First published in Energy World magazine, Nov–Dec 2004, published by the UK Energy Institute, www.energyinst.org.uk.)

the activities in this latter area currently reside overseas, particularly in North America and Japan. However, there are important lessons to be learnt and still significant opportunities for their development, integration and deployment in the UK.

Storage systems span a considerable spectrum of technologies, ranging from short-term/high power technologies, such as superconducting magnetic energy storage (SMES), through to bulk energy storage technologies, which can include various flow cell, compressed air and pumped-hydro storage options.

The selection of the most appropriate technology for any given application is a function of the application's charge and discharge ratings, the actual energy storage required (eg. over seconds, minutes, hours or longer) and its daily operating cycle. Various other considerations also apply, including system acquisition and life costs, footprint, environmental tolerance, and overall developmental maturity.

BATTERY ENERGY STORAGE

Battery energy storage systems (BESS) represent perhaps the most well known form of electrical energy storage, albeit via their storage of chemical energy in the battery reagents and their reversible electrochemical conversion to and from electrical energy.

Lead-acid systems have been established as both larger and smaller-scale packaged systems. The core electrochemistry continues to evolve, particularly in terms of its development for partial-state-of-charge operation, a requirement that is common both to various renewables applications and also hybrid electric vehicles.

Some of the better known lead acid battery storage implementations include:

- the 8.5 MW (power)/8.5 MWh (energy) BEWAG plant in Berlin, constructed in 1986 when West Berlin was effectively an 'electrical island' in East Germany. The system provided a crucial spinning reserve and frequency control functionality; and
- the 3 MW/4.5 MWh Vernon plant at GNB's battery smelting facility in California. The system services a crucial security-of-supply requirement, to safeguard the smelter's environmental control systems in the event of loss of the utility supply, whilst also being used for peak shaving operations.

Alternatives to lead-acid include the nickel-cadmium and sodium-sulphur electrochemistries. The former offers significant advantages over lead-acid in terms of its chronological and cycle life expectancies, its short-term power rating and its low maintenance requirements. Although power utility applications to date have been limited, the technology has achieved significant prominence via its implementation by the Golden Valley Electric Association in Fairbanks, Alaska, as the 'world's largest battery' (pictured).

The system fulfils a critical spinning reserve application in what is essentially an electrical island and provides the Golden Valley utility with sufficient time to start up reserve generators, in the event

of individual units dropping off line. The system itself comprises four battery strings, each of 3,440 cells, with a string voltage of 5,200 V. It is rated at 27 MW for 15 minutes, or 40 MW for 7 minutes, up to a maximum transient limitation of 46 MVA, imposed by the power converter. The nickel-cadmium electrochemistry was chosen in view of Golden Valley's requirement for a 20 year life, with the system expected to perform 100 complete and 500 partial discharges during this period. The system was jointly implemented on a turnkey basis by ABB and SAFT, at a total project cost of $35 million.

The high temperature sodium-sulphur (NaS) system represents the third principal electrochemistry currently implemented in power systems networks, via the partnership agreement between NGK and the Tokyo Electric Power Company (TEPCO). Although the original developmental drivers for the battery system emanated from electric vehicle applications (including the Silent Power programme in the UK), it is uniquely through NGK's programme in Japan that the technology has been developed for the stationary applications sector.

The NGK system is offered on a modular basis in two basic variants. The PS module is rated at 50 kW/60 or 430 kWh capacity, and with the PQ module similarly rated at 40 kW/360 kWh, but with a short term 'pulse power' rating of up to five times its rated power. The system's principal advantages relative to lead-acid include its higher energy density, extended cycle and chronological lives, low O&M costs, short-term high power capability and insensitivity to external ambient temperature.

The 'world's most largest battery': Golden Valley, Alaska, Ni-Cd battery energy storage system. (Patrick Endres/Alaska&PhotoGraphics.com)

NGK announced its partnership agreement with TEPCO in October 2001, which has resulted in the latter implementing the system in its own power network, whilst also initiating direct sales to third party customers. In return, NGK has committed to build up production capacity to 65 MW per year, with the potential to increase this to 200 MW/year, in line with market demand. To date, in excess of 80 projects have been implemented, with some 500 MWh of storage capacity.

The largest installation to date, representing the world's highest capacity battery storage project, is the 8 MW/57.6 MWh system at a Hitachi plant in Japan. This installation is principally used for load levelling purposes, on a daily operational cycle.

REDOX FLOW CELLS

Redox flow cells are analogous to batteries in many respects, but with their chemical energy stored in electrolyte solutions, external to the flow cells (or modules) themselves, as shown in Figure 1. The electrolyte solutions are circulated through the flow cells, with electrochemical conversion taking place across an ion exchange membrane which separates the two electrolytes.

Power and energy then become independent variables, with system power rating being determined by the number of flow cells and their surface area, and energy capacity by the volume of the electrolyte solutions. Systems can therefore be designed to suit the requirements of particular applications, with the potential for the provision of medium to longer-term storage capacity via the in-

stallation of an increased quantity and/or capacity of electrolyte storage tanks.

Developmental and demonstration activities have centred around three principal electrochemistries to date, namely the polysulphide/bromide system, vanadium and zinc bromine.

The polysulphide/bromide system, better known as 'Regenesys' has previously been developed over the past twelve years by RWE Innogy and its predecessor companies (Innogy and National Power). The system has been marketed as a grid-connected utility scale storage system, for power ratings in excess of 5 MWe. Notwithstanding the significant scale-up of and commitment to Regenesys related activities, RWE Innogy announced in December 2003 that it would no longer be funding the technology's development and subsequent commercialisation. It has since announced (September 2004) the sale of an exclusive licence on the intellectual property and related physical assets to VRB Power Systems, for the sum of $1.3 million.

The vanadium redox battery (VRB) employs the V2/V3 and V4/V5 redox couples in sulphuric acid as the negative and positive electrolytes respectively. Vanadium redox batteries are potentially suitable for a wide range of energy storage applications, including power quality, uninterruptible power supplies, peak shaving, increased security of supply and integration with renewable energy systems.

The two principal developers and suppliers of vanadium redox systems are currently VRB Power Systems Inc and Sumitomo, with extensive cross linkages between the two. Further developmental programmes also being pursued by parties such as RE-fuel Technology, Magnam Technologies and the Cellennium Company. Systems installed to date by VRB Power and Sumitomo are summarised in Table 1.

The zinc-bromine battery was first developed by Exxon in the early 1970s and comprises a zinc cathode and a bromine anode separated by a microporous separator. Zinc-bromine batteries are suitable for a range of applications with discharge times ranging from seconds up to several hours. The primary focus of development and demonstration projects to date has been for grid connected utility

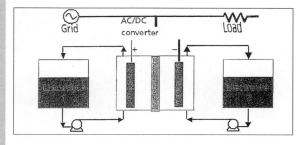

Figure 1
A redox flow cell energy storage system

TABLE 1   Energy storage systems supplied by VRB Power and Sumitomo

Customer	Basic specification	Application	Start date
Eskom	250 kW × 2 hours	Demonstration unit at Univ. of Stellenbosch	Sep 2001
Pacificorp	250 kW × 8 hours	Peak power capacity and end of line voltage support	Mar 2004
Hydro Tasmania	200 kW × 4 hours	Three way hybrid with wind turbines and diesel generator	Nov 2003
Institute of Applied Energy	AC 170 kW × 6 hours	Stabilisation of wind turbine output	Mar 2001
Tottori Sanyo Electric	AC 3 MW × 1.5 seconds	UPS	Apr 2001
	AC 1.5 MW × 1 hour	Peak shaving	
Obayashi Corporation	DC 30 kW × 8 hours	Hybrid with PV cells	Apr 2001
Kwansei Gakuin Univ.	AC 500 kW × 10 hours	Peak shaving	Nov 2001
Centro Elettrotecnico Sperimentale Italiano	AC 42 kW × 2 hours	Peak shaving	Nov 2001

TABLE 2   Zinc-bromine energy storage systems supplied by ZBB Energy

Customer	Basic specification	Application	Installation date
Detroit Edison	400 kWh	Peak shaving and voltage imbalance	June 2001
United Energy, Melbourne	200 kWh	Demonstration unit for network storage	November 2001
Australian Inland Energy	500 kWh	Hybrid with photovoltaic cells	June 2002
PowerLight Corporation	2 × 50 kWh	Hybrid with photovoltaic cells	November 2003
Pacific Gas and Electric Company	2 MWh	Peak power capacity (substation upgrade deferral)	2005

applications for load levelling and renewable energy system optimisation.

At the present time, the only company that is actively developing and supplying zinc-bromine batteries is ZBB Energy Corporation (ZBB). The company was established in 1982 and over the past 20 years has developed or acquired the intellectual property for the zinc-bromine battery. Its technology is now in the first stages of commercialisation, via the company's F2500 baseline turnkey product, a fully containerised 500 kWh (250 kW × 2 hours) grid-interactive storage system. In addition, it can supply individual 50 kWh modules for renewable energy applications. Key demonstration units installed by ZBB in recent years are summarised in Table 2.

LONGER-TERM STORAGE—COMPRESSED AIR

Compressed air energy storage (CAES) complements pumped hydro as a larger-scale (100 MW class), medium/longer-term (hours) storage option.

Input power, storage capacity and output power are independent variables, which provides for a great degree of design flexibility. A diagrammatic representation of a CAES plant is shown in Figure 2.

Only two CAES plant have been constructed and commissioned to date, namely the 290 MW Huntdorf plant in Germany (1978) and the 110 MW McIntosh plant in Alabama (1991). Operating experience on both plant is extremely favourable, with

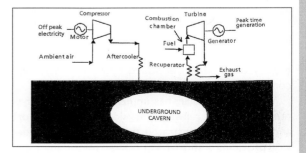

Figure 2
Compressed air energy storage system

the former having completed some 7,000 starts to date, with 90% availability and 99% start reliability.

Such large scale implementations rely on the availability of favourable geological conditions for their underground storage reservoirs. More recently, attention has focussed on the possibility of small scale CAES, utilising fabricated pressure vessel/piping storage and able to provide some 3 to 5 hours storage capacity, at ratings of 5 to 10 MW. Such small scale CAES systems are of particular interest in the US, in the context of buffering wind resources in several states.

SHORTER-TERM STORAGE—FLYWHEELS

Flywheel storage, more correctly referred to as kinetic energy storage, provides a high power rating storage medium, typically sized to discharge over some 10 to 100 seconds. Conventional steel rotor systems have been in place for many years and are often installed in combination with stand-by diesel generators, to provide extremely secure power supplies to such applications as primary broadcasting stations, financial processing centres and air traffic control hubs. The leading commercial suppliers include Piller, Active Power and Satcon.

Much of the current research and developmental effort in relation to kinetic energy storage is directed towards high speed machines, running at tens of thousands of RPM and utilising state-of-the-art composite materials technology. The high directional strength properties of such composites, in combination with their relatively low densities allows the designer considerable freedom in optimising the overall flywheel configuration and hence its specific energy and specific power. Units have already been supplied on a commercial basis by Urenco Power Technologies (UPT) and with further systems being developed by AFS-Trinity, Beacon Power, Piller and others.

UPT, in particular, has implemented various systems providing railway trackside voltage support and has also demonstrated the application of a device providing a short-term power smoothing capability in relation to wind turbine output. However, the company's future is now far from certain, following the decision by its Urenco parent, in May 2004, to cease funding the development of the technology.

MARKET APPLICATIONS

The power utilities sector is presently in a time of unprecedented change, with a shift away from large central generating resource in favour of smaller-scale, distributed resources. The implications of such a shift on power distribution networks are potentially massive, for the UK akin to 'rewiring Britain.' The electricity regulator, OFGEM, is therefore keen to promote innovative design solutions via such initiatives as the IFI (Innovation Funding Incentive) and RPZs (Registered Power Zones), which provide an opportunity to introduce such new technologies as storage.

The overall financial viability of any storage system is a function of its ability to extract value (revenue) from one or more value/revenue streams. A multi-faceted storage system will be able to extract value from multiple revenue streams and thereby enhance its overall financial viability. Value and revenue flows may be extracted from such functionalities as arbitrage and load levelling ('traditional' storage applications), spinning reserve, frequency regulation, network stability, voltage support, renewables integration, quality of supply, power quality and asset deferral. Such value/revenue flows are usually expressed in terms of £ per kW per annum; Table 3 summarises some illustrative values.

A summation of the total revenue flows may then be used to calculate a break-even capital cost

TABLE 3 Illustrative value/revenue flows

Application	Annualised benefit (£ per kW per year)
Arbitrage	25–60
Load levelling	150–200
Spinning reserve	50–120
Frequency regulation	50–130
Voltage support	20–50
Renewables integration	20–60
Power quality	50–500

for the storage system, based on its capital and O&M costs, and an assumed amortisation period and capital charge (discount rate). Provided the system is able to come in at under this break-even cost level, it is likely to be worthy of a more detailed feasibility study.

This article has demonstrated the considerable number of developments in hand in relation to energy storage technologies. The various systems available are able to span a full range of applications, from high power/short duration requirements to longer-term multiple-hour systems. The complementary developments in the wider power utilities sector present a whole range of new issues and challenges, including very specific opportunities for storage. The immediate challenge for storage systems is to demonstrate their technical and commercial viability, in early demonstration schemes.

8.1

DEFINITION OF SYMMETRICAL COMPONENTS

Assume that a set of three-phase voltages designated V_a, V_b, and V_c is given. In accordance with Fortescue, these phase voltages are resolved into the following three sets of sequence components:

1. *Zero-sequence* components, consisting of three phasors with equal magnitudes and with zero phase displacement, as shown in Figure 8.1(a)

2. *Positive-sequence* components, consisting of three phasors with equal magnitudes, $\pm 120°$ phase displacement, and positive sequence, as in Figure 8.1(b)

3. *Negative-sequence* components, consisting of three phasors with equal magnitudes, $\pm 120°$ phase displacement, and negative sequence, as in Figure 8.1(c)

FIGURE 8.1

Resolving phase voltages into three sets of sequence components

$V_{a0} \; V_{b0} \; V_{c0} = V_0$

(a) Zero-sequence components

V_{c1} $V_{a1} = V_1$ V_{b1}

(b) Positive-sequence components

V_{b2} $V_{a2} = V_2$ V_{c2}

(c) Negative-sequence components

Phase *a*

Phase *b*

Phase *c*

In this text we will work only with the zero-, positive-, and negative-sequence components of phase a, which are V_{a0}, V_{a1}, and V_{a2}, respectively. For simplicity, we drop the subscript a and denote these sequence components as V_0, V_1, and V_2. They are defined by the following transformation:

$$\begin{bmatrix} V_a \\ V_b \\ V_c \end{bmatrix} = \begin{bmatrix} 1 & 1 & 1 \\ 1 & a^2 & a \\ 1 & a & a^2 \end{bmatrix} \begin{bmatrix} V_0 \\ V_1 \\ V_2 \end{bmatrix} \tag{8.1.1}$$

where

$$a = 1\underline{/120^\circ} = \frac{-1}{2} + j\frac{\sqrt{3}}{2} \tag{8.1.2}$$

Writing (8.1.1) as three separate equations:

$$V_a = V_0 + V_1 + V_2 \tag{8.1.3}$$

$$V_b = V_0 + a^2 V_1 + a V_2 \tag{8.1.4}$$

$$V_c = V_0 + a V_1 + a^2 V_2 \tag{8.1.5}$$

In (8.1.2), a is a complex number with unit magnitude and a 120° phase angle. When any phasor is multiplied by a, that phasor rotates by 120° (counterclockwise). Similarly, when any phasor is multiplied by $a^2 = (1\underline{/120^\circ}) \cdot (1\underline{/120^\circ}) = 1\underline{/240^\circ}$, the phasor rotates by 240°. Table 8.1 lists some common identities involving a.

The complex number a is similar to the well-known complex number $j = \sqrt{-1} = 1\underline{/90^\circ}$. Thus the only difference between j and a is that the angle of j is 90°, and that of a is 120°.

Equation (8.1.1) can be rewritten more compactly using matrix notation. We define the following vectors V_p and V_s, and matrix A:

TABLE 8.1

Common identities involving $a = 1\underline{/120^\circ}$

$a^4 = a = 1\underline{/120^\circ}$
$a^2 = 1\underline{/240^\circ}$
$a^3 = 1\underline{/0^\circ}$
$1 + a + a^2 = 0$
$1 - a = \sqrt{3}\underline{/-30^\circ}$
$1 - a^2 = \sqrt{3}\underline{/+30^\circ}$
$a^2 - a = \sqrt{3}\underline{/270^\circ}$
$ja = 1\underline{/210^\circ}$
$1 + a = -a^2 = 1\underline{/60^\circ}$
$1 + a^2 = -a = 1\underline{/-60^\circ}$
$a + a^2 = -1 = 1\underline{/180^\circ}$

$$V_p = \begin{bmatrix} V_a \\ V_b \\ V_c \end{bmatrix} \tag{8.1.6}$$

$$V_s = \begin{bmatrix} V_0 \\ V_1 \\ V_2 \end{bmatrix} \tag{8.1.7}$$

$$A = \begin{bmatrix} 1 & 1 & 1 \\ 1 & a^2 & a \\ 1 & a & a^2 \end{bmatrix} \tag{8.1.8}$$

V_p is the column vector of phase voltages, V_s is the column vector of sequence voltages, and A is a 3×3 transformation matrix. Using these definitions, (8.1.1) becomes

$$V_p = AV_s \tag{8.1.9}$$

The inverse of the A matrix is

$$A^{-1} = \frac{1}{3} \begin{bmatrix} 1 & 1 & 1 \\ 1 & a & a^2 \\ 1 & a^2 & a \end{bmatrix} \tag{8.1.10}$$

Equation (8.1.10) can be verified by showing that the product AA^{-1} is the unit matrix. Also, premultiplying (8.1.9) by A^{-1} gives

$$V_s = A^{-1}V_p \tag{8.1.11}$$

Using (8.1.6), (8.1.7), and (8.1.10), then (8.1.11) becomes

$$\begin{bmatrix} V_0 \\ V_1 \\ V_2 \end{bmatrix} = \frac{1}{3} \begin{bmatrix} 1 & 1 & 1 \\ 1 & a & a^2 \\ 1 & a^2 & a \end{bmatrix} \begin{bmatrix} V_a \\ V_b \\ V_c \end{bmatrix} \tag{8.1.12}$$

Writing (8.1.12) as three separate equations,

$$V_0 = \tfrac{1}{3}(V_a + V_b + V_c) \tag{8.1.13}$$

$$V_1 = \tfrac{1}{3}(V_a + aV_b + a^2V_c) \tag{8.1.14}$$

$$V_2 = \tfrac{1}{3}(V_a + a^2V_b + aV_c) \tag{8.1.15}$$

Equation (8.1.13) shows that there is no zero-sequence voltage in a *balanced* three-phase system because the sum of three balanced phasors is zero. In an unbalanced three-phase system, line-to-neutral voltages may have a zero-sequence component. However, line-to-line voltages never have a zero-sequence component, since by KVL their sum is always zero.

The symmetrical component transformation can also be applied to currents, as follows. Let

$$I_p = AI_s \tag{8.1.16}$$

where I_p is a vector of phase currents,

$$I_p = \begin{bmatrix} I_a \\ I_b \\ I_c \end{bmatrix} \tag{8.1.17}$$

and I_s is a vector of sequence currents,

$$I_s = \begin{bmatrix} I_0 \\ I_1 \\ I_2 \end{bmatrix} \tag{8.1.18}$$

Also,

$$I_s = A^{-1}I_p \tag{8.1.19}$$

Equations (8.1.16) and (8.1.19) can be written as separate equations as follows. The phase currents are

$$I_a = I_0 + I_1 + I_2 \tag{8.1.20}$$

$$I_b = I_0 + a^2 I_1 + a I_2 \tag{8.1.21}$$

$$I_c = I_0 + a I_1 + a^2 I_2 \tag{8.1.22}$$

and the sequence currents are

$$I_0 = \tfrac{1}{3}(I_a + I_b + I_c) \tag{8.1.23}$$

$$I_1 = \tfrac{1}{3}(I_a + a I_b + a^2 I_c) \tag{8.1.24}$$

$$I_2 = \tfrac{1}{3}(I_a + a^2 I_b + a I_c) \tag{8.1.25}$$

In a three-phase Y-connected system, the neutral current I_n is the sum of the line currents:

$$I_n = I_a + I_b + I_c \tag{8.1.26}$$

Comparing (8.1.26) and (8.1.23),

$$I_n = 3I_0 \tag{8.1.27}$$

The neutral current equals three times the zero-sequence current. In a balanced Y-connected system, line currents have no zero-sequence component, since the neutral current is zero. Also, in any three-phase system with no neutral path, such as a Δ-connected system or a three-wire Y-connected system with an ungrounded neutral, line currents have no zero-sequence component.

The following three examples further illustrate symmetrical components.

EXAMPLE 8.1   **Sequence components: balanced line-to-neutral voltages**

Calculate the sequence components of the following balanced line-to-neutral voltages with *abc* sequence:

$$V_p = \begin{bmatrix} V_{an} \\ V_{bn} \\ V_{cn} \end{bmatrix} = \begin{bmatrix} 277\underline{/0^\circ} \\ 277\underline{/-120^\circ} \\ 277\underline{/+120^\circ} \end{bmatrix} \quad \text{volts}$$

SOLUTION   Using (8.1.13)–(8.1.15):

$$V_0 = \tfrac{1}{3}[277\underline{/0^\circ} + 277\underline{/-120^\circ} + 277\underline{/+120^\circ}] = 0$$

$$V_1 = \tfrac{1}{3}[277\underline{/0^\circ} + 277\underline{/(-120^\circ + 120^\circ)} + 277\underline{/(120^\circ + 240^\circ)}]$$

$$= 277\underline{/0^\circ} \quad \text{volts} = V_{an}$$

$$V_2 = \tfrac{1}{3}[277\underline{/0^\circ} + 277\underline{/(-120^\circ + 240^\circ)} + 277\underline{/(120^\circ + 120^\circ)}]$$

$$= \tfrac{1}{3}[277\underline{/0^\circ} + 277\underline{/120^\circ} + 277\underline{/240^\circ}] = 0$$

This example illustrates the fact that balanced three-phase systems with *abc* sequence (or positive sequence) have no zero-sequence or negative-sequence components. For this example, the positive-sequence voltage V_1 equals V_{an}, and the zero-sequence and negative-sequence voltages are both zero. ∎

EXAMPLE 8.2   Sequence components: balanced *acb* currents

A Y-connected load has balanced currents with *acb* sequence given by

$$\mathbf{I}_p = \begin{bmatrix} I_a \\ I_b \\ I_c \end{bmatrix} = \begin{bmatrix} 10\underline{/0^\circ} \\ 10\underline{/+120^\circ} \\ 10\underline{/-120^\circ} \end{bmatrix} \quad \text{A}$$

Calculate the sequence currents.

SOLUTION   Using (8.1.23)–(8.1.25):

$$I_0 = \tfrac{1}{3}[10\underline{/0^\circ} + 10\underline{/120^\circ} + 10\underline{/-120^\circ}] = 0$$

$$I_1 = \tfrac{1}{3}[10\underline{/0^\circ} + 10\underline{/(120^\circ + 120^\circ)} + 10\underline{/(-120^\circ + 240^\circ)}]$$

$$= \tfrac{1}{3}[10\underline{/0^\circ} + 10\underline{/240^\circ} + 10\underline{/120^\circ}] = 0$$

$$I_2 = \tfrac{1}{3}[10\underline{/0^\circ} + 10\underline{/(120^\circ + 240^\circ)} + 10\underline{/(-120^\circ + 120^\circ)}]$$

$$= 10\underline{/0^\circ} \text{ A} = I_a$$

This example illustrates the fact that balanced three-phase systems with *acb* sequence (or negative sequence) have no zero-sequence or positive-sequence components. For this example the negative-sequence current I_2 equals I_a, and the zero-sequence and positive-sequence currents are both zero. ∎

EXAMPLE 8.3   Sequence components: unbalanced currents

A three-phase line feeding a balanced-Y load has one of its phases (phase *b*) open. The load neutral is grounded, and the unbalanced line currents are

$$\mathbf{I}_p = \begin{bmatrix} I_a \\ I_b \\ I_c \end{bmatrix} = \begin{bmatrix} 10\underline{/0^\circ} \\ 0 \\ 10\underline{/120^\circ} \end{bmatrix} \quad \text{A}$$

Calculate the sequence currents and the neutral current.

FIGURE 8.2

Circuit for Example 8.3

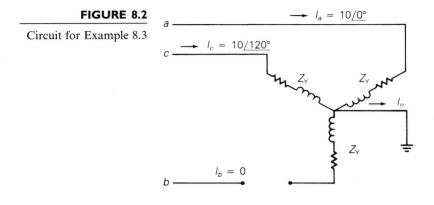

SOLUTION The circuit is shown in Figure 8.2. Using (8.1.23)–(8.1.25):

$$I_0 = \tfrac{1}{3}[10\underline{/0°} + 0 + 10\underline{/120°}]$$

$$= 3.333\underline{/60°} \quad \text{A}$$

$$I_1 = \tfrac{1}{3}[10\underline{/0°} + 0 + 10\underline{/(120° + 240°)}] = 6.667\underline{/0°} \quad \text{A}$$

$$I_2 = \tfrac{1}{3}[10\underline{/0°} + 0 + 10\underline{/(120° + 120°)}]$$

$$= 3.333\underline{/-60°} \quad \text{A}$$

Using (8.1.26) the neutral current is

$$I_n = (10\underline{/0°} + 0 + 10\underline{/120°})$$

$$= 10\underline{/60°} \; \text{A} = 3I_0$$

This example illustrates the fact that *unbalanced* three-phase systems may have nonzero values for all sequence components. Also, the neutral current equals three times the zero-sequence current, as given by (8.1.27). ∎

8.2

SEQUENCE NETWORKS OF IMPEDANCE LOADS

Figure 8.3 shows a balanced-Y impedance load. The impedance of each phase is designated Z_Y, and a neutral impedance Z_n is connected between the load neutral and ground. Note from Figure 8.3 that the line-to-ground voltage V_{ag} is

$$V_{ag} = Z_Y I_a + Z_n I_n$$

$$= Z_Y I_a + Z_n(I_a + I_b + I_c)$$

$$= (Z_Y + Z_n)I_a + Z_n I_b + Z_n I_c \qquad (8.2.1)$$

FIGURE 8.3

Balanced-Y impedance
load

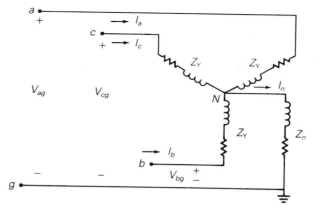

Similar equations can be written for V_{bg} and V_{cg}:

$$V_{bg} = Z_n I_a + (Z_Y + Z_n)I_b + Z_n I_c \tag{8.2.2}$$

$$V_{cg} = Z_n I_a + Z_n I_b + (Z_Y + Z_n)I_c \tag{8.2.3}$$

Equations (8.2.1)–(8.2.3) can be rewritten in matrix format:

$$\begin{bmatrix} V_{ag} \\ V_{bg} \\ V_{cg} \end{bmatrix} = \begin{bmatrix} (Z_Y + Z_n) & Z_n & Z_n \\ Z_n & (Z_Y + Z_n) & Z_n \\ Z_n & Z_n & (Z_Y + Z_n) \end{bmatrix} \begin{bmatrix} I_a \\ I_b \\ I_c \end{bmatrix} \tag{8.2.4}$$

Equation (8.2.4) is written more compactly as

$$V_p = Z_p I_p \tag{8.2.5}$$

where V_p is the vector of line-to-ground voltages (or phase voltages), I_p is the vector of line currents (or phase currents), and Z_p is the 3×3 phase impedance matrix shown in (8.2.4). Equations (8.1.9) and (8.1.16) can now be used in (8.2.5) to determine the relationship between the sequence voltages and currents, as follows:

$$AV_s = Z_p A I_s \tag{8.2.6}$$

Premultiplying both sides of (8.2.6) of A^{-1} gives

$$V_s = (A^{-1} Z_p A)I_s \tag{8.2.7}$$

or

$$V_s = Z_s I_s \tag{8.2.8}$$

where

$$Z_s = A^{-1} Z_p A \tag{8.2.9}$$

The impedance matrix Z_s defined by (8.2.9) is called the *sequence impedance matrix*. Using the definition of A, its inverse A^{-1}, and Z_p given by (8.1.8), (8.1.10), and (8.2.4), the sequence impedance matrix $\mathbf{Z}_s$ for the balanced-Y load is

$$
\mathbf{Z}_s = \frac{1}{3}
\begin{bmatrix}
1 & 1 & 1 \\
1 & a & a^2 \\
1 & a^2 & a
\end{bmatrix}
\begin{bmatrix}
(Z_Y + Z_n) & Z_n & Z_n \\
Z_n & (Z_Y + Z_n) & Z_n \\
Z_n & Z_n & (Z_Y + Z_n)
\end{bmatrix}
$$

$$
\times
\begin{bmatrix}
1 & 1 & 1 \\
1 & a^2 & a \\
1 & a & a^2
\end{bmatrix}
\tag{8.2.10}
$$

Performing the indicated matrix multiplications in (8.2.10), and using the identity $(1 + a + a^2) = 0$,

$$
\mathbf{Z}_s = \frac{1}{3}
\begin{bmatrix}
1 & 1 & 1 \\
1 & a & a^2 \\
1 & a^2 & a
\end{bmatrix}
\begin{bmatrix}
(Z_Y + 3Z_n) & Z_Y & Z_Y \\
(Z_Y + 3Z_n) & a^2 Z_Y & a Z_Y \\
(Z_Y + 3Z_n) & a Z_Y & a^2 Z_Y
\end{bmatrix}
$$

$$
=
\begin{bmatrix}
(Z_Y + 3Z_n) & 0 & 0 \\
0 & Z_Y & 0 \\
0 & 0 & Z_Y
\end{bmatrix}
\tag{8.2.11}
$$

As shown in (8.2.11), the sequence impedance matrix $\mathbf{Z}_s$ for the balanced-Y load of Figure 8.3 is a diagonal matrix. Since $\mathbf{Z}_s$ is diagonal, (8.2.8) can be written as three *uncoupled* equations. Using (8.1.7), (8.1.18), and (8.2.11) in (8.2.8),

$$
\begin{bmatrix}
V_0 \\
V_1 \\
V_2
\end{bmatrix}
=
\begin{bmatrix}
(Z_Y + 3Z_n) & 0 & 0 \\
0 & Z_Y & 0 \\
0 & 0 & Z_Y
\end{bmatrix}
\begin{bmatrix}
I_0 \\
I_1 \\
I_2
\end{bmatrix}
\tag{8.2.12}
$$

Rewriting (8.2.12) as three separate equations,

$$V_0 = (Z_Y + 3Z_n)I_0 = Z_0 I_0 \tag{8.2.13}$$

$$V_1 = Z_Y I_1 = Z_1 I_1 \tag{8.2.14}$$

$$V_2 = Z_Y I_2 = Z_2 I_2 \tag{8.2.15}$$

As shown in (8.2.13), the zero-sequence voltage V_0 depends only on the zero-sequence current I_0 and the impedance $(Z_Y + 3Z_n)$. This impedance is called the *zero-sequence impedance* and is designated Z_0. Also, the positive-sequence voltage V_1 depends only on the positive-sequence current I_1 and an impedance $Z_1 = Z_Y$ called the *positive-sequence impedance*. Similarly, V_2 depends only on I_2 and the *negative-sequence impedance* $Z_2 = Z_Y$.

Equations (8.2.13)–(8.2.15) can be represented by the three networks shown in Figure 8.4. These networks are called the *zero-sequence, positive-sequence,* and *negative-sequence networks*. As shown, each sequence network

FIGURE 8.4

Sequence networks of a
balanced-Y load

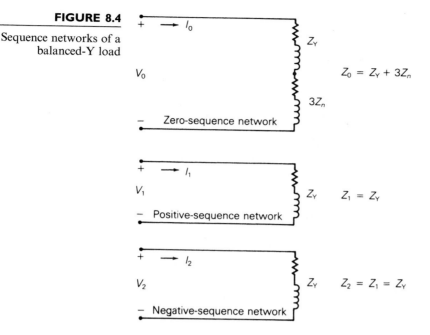

$$Z_0 = Z_Y + 3Z_n$$

$$Z_1 = Z_Y$$

$$Z_2 = Z_1 = Z_Y$$

is separate, uncoupled from the other two. The separation of these sequence networks is a consequence of the fact that Z_s is a diagonal matrix for a balanced-Y load. This separation underlies the advantage of symmetrical components.

Note that the neutral impedance does not appear in the positive- and negative-sequence networks of Figure 8.4. This illustrates the fact that positive- and negative-sequence currents do not flow in neutral impedances. However, the neutral impedance is multiplied by 3 and placed in the zero-sequence network of the figure. The voltage $I_0(3Z_n)$ across the impedance $3Z_n$ is the voltage drop $(I_n Z_n)$ across the neutral impedance Z_n in Figure 8.3, since $I_n = 3I_0$.

When the neutral of the Y load in Figure 8.3 has no return path, then the neutral impedance Z_n is infinite and the term $3Z_n$ in the zero-sequence network of Figure 8.4 becomes an open circuit. Under this condition of an open neutral, no zero-sequence current exists. However, when the neutral of the Y load is solidly grounded with a zero-ohm conductor, then the neutral impedance is zero and the term $3Z_n$ in the zero-sequence network becomes a short circuit. Under this condition of a solidly grounded neutral, zero-sequence current I_0 can exist when there is a zero-sequence voltage caused by unbalanced voltages applied to the load.

Figure 2.15 shows a balanced-Δ load and its equivalent balanced-Y load. Since the Δ load has no neutral connection, the equivalent Y load in Figure 2.15 has an open neutral. The sequence networks of the equivalent Y load corresponding to a balanced-Δ load are shown in Figure 8.5. As shown,

FIGURE 8.5

Sequence networks for
an equivalent Y
representation of a
balanced-Δ load

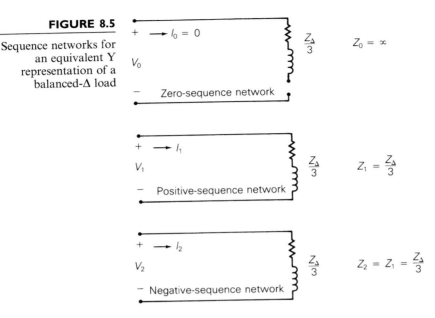

the equivalent Y impedance $Z_Y = Z_\Delta/3$ appears in each of the sequence networks. Also, the zero-sequence network has an open circuit, since $Z_n = \infty$ corresponds to an open neutral. No zero-sequence current occurs in the equivalent Y load.

The sequence networks of Figure 8.5 represent the balanced-Δ load as viewed from its terminals, but they do not represent the internal load characteristics. The currents I_0, I_1, and I_2 in Figure 8.5 are the sequence components of the line currents feeding the Δ load, not the load currents within the Δ. The Δ load currents, which are related to the line currents by (2.5.14), are not shown in Figure 8.5.

EXAMPLE 8.4   **Sequence networks: balanced-Y and balanced-Δ loads**

A balanced-Y load is in parallel with a balanced-Δ-connected capacitor bank. The Y load has an impedance $Z_Y = (3 + j4)$ Ω per phase, and its neutral is grounded through an inductive reactance $X_n = 2$ Ω. The capacitor bank has a reactance $X_c = 30$ Ω per phase. Draw the sequence networks for this load and calculate the load-sequence impedances.

SOLUTION   The sequence networks are shown in Figure 8.6. As shown, the Y-load impedance in the zero-sequence network is in series with three times the neutral impedance. Also, the Δ-load branch in the zero-sequence network is open, since no zero-sequence current flows into the Δ load. In the positive- and negative-sequence circuits, the Δ-load impedance is divided by 3 and placed in parallel with the Y-load impedance. The equivalent sequence impedances are

FIGURE 8.6

Sequence networks for
Example 8.4

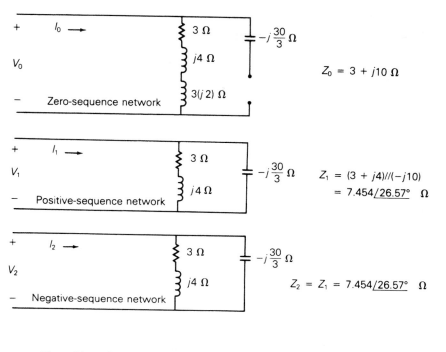

Zero-sequence network

$Z_0 = 3 + j10\ \Omega$

Positive-sequence network

$Z_1 = (3 + j4)//(-j10)$
$\quad = 7.454\underline{/26.57°}\ \ \Omega$

Negative-sequence network

$Z_2 = Z_1 = 7.454\underline{/26.57°}\ \ \Omega$

$$Z_0 = Z_Y + 3Z_n = 3 + j4 + 3(j2) = 3 + j10 \quad \Omega$$

$$Z_1 = Z_Y//(Z_\Delta/3) = \frac{(3 + j4)(-j30/3)}{3 + j4 - j(30/3)}$$

$$= \frac{(5\underline{/53.13°})(10\underline{/-90°})}{6.708\underline{/-63.43°}} = 7.454\underline{/26.57°} \quad \Omega$$

$$Z_2 = Z_1 = 7.454\underline{/26.57°} \quad \Omega \qquad\qquad \blacksquare$$

Figure 8.7 shows a general three-phase linear impedance load. The load could represent a balanced load such as the balanced-Y or balanced-Δ load, or an unbalanced impedance load. The general relationship between the line-to-ground voltages and line currents for this load can be written as

$$\begin{bmatrix} V_{ag} \\ V_{bg} \\ V_{cg} \end{bmatrix} = \begin{bmatrix} Z_{aa} & Z_{ab} & Z_{ac} \\ Z_{ab} & Z_{bb} & Z_{bc} \\ Z_{ac} & Z_{bc} & Z_{cc} \end{bmatrix} \begin{bmatrix} I_a \\ I_b \\ I_c \end{bmatrix} \tag{8.2.16}$$

or

$$V_p = Z_p I_p \tag{8.2.17}$$

where V_p is the vector of line-to-neutral (or phase) voltages, I_p is the vector of line (or phase) currents, and Z_p is a 3×3 phase impedance matrix. It is assumed here that the load is nonrotating, and that Z_p is a symmetric matrix, which corresponds to a bilateral network.

FIGURE 8.7

General three-phase
impedance load (linear,
bilateral network,
nonrotating equipment)

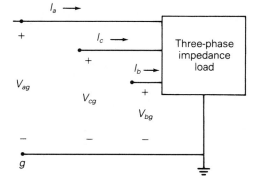

Since (8.2.17) has the same form as (8.2.5), the relationship between the sequence voltages and currents for the general three-phase load of Figure 8.6 is the same as that of (8.2.8) and (8.2.9), which are rewritten here:

$$\mathbf{V}_s = \mathbf{Z}_s \mathbf{I}_s \qquad (8.2.18)$$

$$\mathbf{Z}_s = \mathbf{A}^{-1} \mathbf{Z}_p \mathbf{A} \qquad (8.2.19)$$

The sequence impedance matrix $\mathbf{Z}_s$ given by (8.2.19) is a 3×3 matrix with nine sequence impedances, defined as follows:

$$\mathbf{Z}_s = \begin{bmatrix} Z_0 & Z_{01} & Z_{02} \\ Z_{10} & Z_1 & Z_{12} \\ Z_{20} & Z_{21} & Z_2 \end{bmatrix} \qquad (8.2.20)$$

The diagonal impedances Z_0, Z_1, and Z_2 in this matrix are the self-impedances of the zero-, positive-, and negative-sequence networks. The off-diagonal impedances are the mutual impedances between sequence networks. Using the definitions of $A, A^{-1}, \mathbf{Z}_p$, and $\mathbf{Z}_s$, (8.2.19) is

$$\begin{bmatrix} Z_0 & Z_{01} & Z_{02} \\ Z_{10} & Z_1 & Z_{12} \\ Z_{20} & Z_{21} & Z_2 \end{bmatrix} = \frac{1}{3} \begin{bmatrix} 1 & 1 & 1 \\ 1 & a & a^2 \\ 1 & a^2 & a \end{bmatrix} \begin{bmatrix} Z_{aa} & Z_{ab} & Z_{ac} \\ Z_{ab} & Z_{bb} & Z_{bc} \\ Z_{ac} & Z_{bc} & Z_{cc} \end{bmatrix} \begin{bmatrix} 1 & 1 & 1 \\ 1 & a^2 & a \\ 1 & a & a^2 \end{bmatrix}$$

$$(8.2.21)$$

Performing the indicated multiplications in (8.2.21), and using the identity $(1 + a + a^2) = 0$, the following separate equations can be obtained (see Problem 8.18):

Diagonal sequence impedances

$$Z_0 = \tfrac{1}{3}(Z_{aa} + Z_{bb} + Z_{cc} + 2Z_{ab} + 2Z_{ac} + 2Z_{bc}) \qquad (8.2.22)$$

$$Z_1 = Z_2 = \tfrac{1}{3}(Z_{aa} + Z_{bb} + Z_{cc} - Z_{ab} - Z_{ac} - Z_{bc}) \qquad (8.2.23)$$

Off-diagonal sequence impedances

$$Z_{01} = Z_{20} = \tfrac{1}{3}(Z_{aa} + a^2 Z_{bb} + a Z_{cc} - a Z_{ab} - a^2 Z_{ac} - Z_{bc}) \qquad (8.2.24)$$

$$Z_{02} = Z_{10} = \tfrac{1}{3}(Z_{aa} + a Z_{bb} + a^2 Z_{cc} - a^2 Z_{ab} - a Z_{ac} - Z_{bc}) \qquad (8.2.25)$$

$$Z_{12} = \tfrac{1}{3}(Z_{aa} + a^2 Z_{bb} + a Z_{cc} + 2a Z_{ab} + 2a^2 Z_{ac} + 2 Z_{bc}) \qquad (8.2.26)$$

$$Z_{21} = \tfrac{1}{3}(Z_{aa} + a Z_{bb} + a^2 Z_{cc} + 2a^2 Z_{ab} + 2a Z_{ac} + 2 Z_{bc}) \qquad (8.2.27)$$

A *symmetrical load* is defined as a load whose sequence impedance matrix is diagonal; that is, all the mutual impedances in (8.2.24)–(8.2.27) are zero. Equating these mutual impedances to zero and solving, the following conditions for a symmetrical load are determined. When both

$$Z_{aa} = Z_{bb} = Z_{cc} \qquad (8.2.28)$$

and $\qquad\qquad\qquad\qquad\qquad$ conditions for a symmetrical load

$$Z_{ab} = Z_{ac} = Z_{bc} \qquad (8.2.29)$$

then

$$Z_{01} = Z_{10} = Z_{02} = Z_{20} = Z_{12} = Z_{21} = 0 \qquad (8.2.30)$$

$$Z_0 = Z_{aa} + 2Z_{ab} \qquad (8.2.31)$$

$$Z_1 = Z_2 = Z_{aa} - Z_{ab} \qquad (8.2.32)$$

The conditions for a symmetrical load are that the diagonal phase impedances be equal and that the off-diagonal phase impedances be equal. These conditions can be verified by using (8.2.28) and (8.2.29) with the iden-

FIGURE 8.8

Sequence networks of a three-phase symmetrical impedance load (linear, bilateral network, nonrotating equipment)

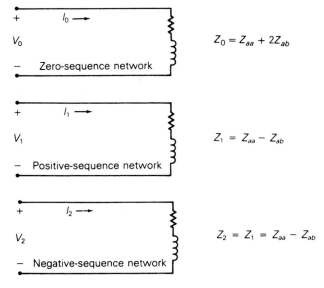

$Z_0 = Z_{aa} + 2Z_{ab}$

Zero-sequence network

$Z_1 = Z_{aa} - Z_{ab}$

Positive-sequence network

$Z_2 = Z_1 = Z_{aa} - Z_{ab}$

Negative-sequence network

tity $(1 + a + a^2) = 0$ in (8.2.24)–(8.2.27) to show that all the mutual sequence impedances are zero. Note that the positive- and negative-sequence impedances are equal for a symmetrical load, as shown by (8.2.32), and for a nonsymmetrical load, as shown by (8.2.23). This is always true for linear, symmetric impedances that represent nonrotating equipment such as transformers and transmission lines. However, the positive- and negative-sequence impedances of rotating equipment such as generators and motors are generally not equal. Note also that the zero-sequence impedance Z_0 is not equal to the positive- and negative-sequence impedances of a symmetrical load unless the mutual phase impedances $Z_{ab} = Z_{ac} = Z_{bc}$ are zero.

The sequence networks of a symmetrical impedance load are shown in Figure 8.8. Since the sequence impedance matrix Z_s is diagonal for a symmetrical load, the sequence networks are separate or uncoupled.

8.3

SEQUENCE NETWORKS OF SERIES IMPEDANCES

Figure 8.9 shows series impedances connected between two three-phase buses denoted abc and $a'b'c'$. Self-impedances of each phase are denoted Z_{aa}, Z_{bb}, and Z_{cc}. In general, the series network may also have mutual impedances between phases. The voltage drops across the series-phase impedances are given by

$$
\begin{bmatrix} V_{an} - V_{a'n} \\ V_{bn} - V_{b'n} \\ V_{cn} - V_{c'n} \end{bmatrix} = \begin{bmatrix} V_{aa'} \\ V_{bb'} \\ V_{cc'} \end{bmatrix} = \begin{bmatrix} Z_{aa} & Z_{ab} & Z_{ac} \\ Z_{ab} & Z_{bb} & Z_{bc} \\ Z_{ac} & Z_{cb} & Z_{cc} \end{bmatrix} \begin{bmatrix} I_a \\ I_b \\ I_c \end{bmatrix} \tag{8.3.1}
$$

Both self-impedances and mutual impedances are included in (8.3.1). It is assumed that the impedance matrix is symmetric, which corresponds to a bilateral network. It is also assumed that these impedances represent non-

FIGURE 8.9

Three-phase series impedances (linear, bilateral network, nonrotating equipment)

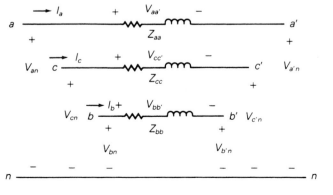

rotating equipment. Typical examples are series impedances of transmission lines and of transformers. Equation (8.3.1) has the following form:

$$V_p - V_{p'} = Z_p I_p \qquad (8.3.2)$$

where V_p is the vector of line-to-neutral voltages at bus abc, $V_{p'}$ is the vector of line-to-neutral voltages at bus $a'b'c'$, I_p is the vector of line currents, and Z_p is the 3×3 phase impedance matrix for the series network. Equation (8.3.2) is now transformed to the sequence domain in the same manner that the load-phase impedances were transformed in Section 8.2. Thus,

$$V_s - V_{s'} = Z_s I_s \qquad (8.3.3)$$

where

$$Z_s = A^{-1} Z_p A \qquad (8.3.4)$$

From the results of Section 8.2, this sequence impedance Z_s matrix is diagonal under the following conditions:

$$
\left.
\begin{aligned}
Z_{aa} &= Z_{bb} = Z_{cc} \\
\text{and} \\
Z_{ab} &= Z_{ac} = Z_{bc}
\end{aligned}
\right\}
\begin{aligned}
&\text{conditions for} \\
&\text{symmetrical} \\
&\text{series impedances}
\end{aligned}
\qquad (8.3.5)
$$

When the phase impedance matrix Z_p of (8.3.1) has both equal self-impedances and equal mutual impedances, then (8.3.4) becomes

$$
Z_s = \begin{bmatrix} Z_0 & 0 & 0 \\ 0 & Z_1 & 0 \\ 0 & 0 & Z_2 \end{bmatrix}
\qquad (8.3.6)
$$

where

$$Z_0 = Z_{aa} + 2Z_{ab} \qquad (8.3.7)$$

and

$$Z_1 = Z_2 = Z_{aa} - Z_{ab} \qquad (8.3.8)$$

and (8.3.3) becomes three uncoupled equations, written as follows:

$$V_0 - V_{0'} = Z_0 I_0 \qquad (8.3.9)$$

$$V_1 - V_{1'} = Z_1 I_1 \qquad (8.3.10)$$

$$V_2 - V_{2'} = Z_2 I_2 \qquad (8.3.11)$$

Equations (8.3.9)–(8.3.11) are represented by the three uncoupled sequence networks shown in Figure 8.10. From the figure it is apparent that for symmetrical series impedances, positive-sequence currents produce only positive-sequence voltage drops. Similarly, negative-sequence currents produce only negative-sequence voltage drops, and zero-sequence currents produce only zero-sequence voltage drops. However, if the series impedances

FIGURE 8.10

Sequence networks of
three-phase symmetrical
series impedances
(linear, bilateral
network, nonrotating
equipment)

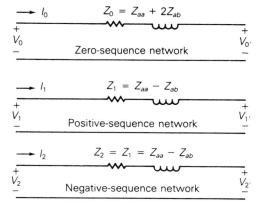

are not symmetrical, then $\boldsymbol{Z}_s$ is not diagonal, the sequence networks are coupled, and the voltage drop across any one sequence network depends on all three sequence currents.

8.4

SEQUENCE NETWORKS OF THREE-PHASE LINES

Section 4.7 develops equations suitable for computer calculation of the series phase impedances, including resistances and inductive reactances, of three-phase overhead transmission lines. The series phase impedance matrix $\boldsymbol{Z}_P$ for an untransposed line is given by Equation (4.7.19), and $\hat{\boldsymbol{Z}}_P$ for a completely transposed line is given by (4.7.21)–(4.7.23). Equation (4.7.19) can be transformed to the sequence domain to obtain

$$\boldsymbol{Z}_S = A^{-1}\boldsymbol{Z}_P A \tag{8.4.1}$$

$\boldsymbol{Z}_S$ is the 3×3 series sequence impedance matrix whose elements are

$$\boldsymbol{Z}_S = \begin{bmatrix} Z_0 & Z_{01} & Z_{02} \\ Z_{10} & Z_1 & Z_{12} \\ Z_{20} & Z_{21} & Z_2 \end{bmatrix} \ \ \Omega/\text{m} \tag{8.4.2}$$

In general $\boldsymbol{Z}_S$ is not diagonal. However, if the line is completely transposed,

$$\hat{\boldsymbol{Z}}_S = A^{-1}\hat{\boldsymbol{Z}}_P A = \begin{bmatrix} \hat{Z}_0 & 0 & 0 \\ 0 & \hat{Z}_1 & 0 \\ 0 & 0 & \hat{Z}_2 \end{bmatrix} \tag{8.4.3}$$

where, from (8.3.7) and (8.3.8),

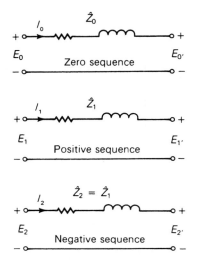

$$\hat{Z}_0 = \hat{Z}_{aaeq} + 2\hat{Z}_{abeq} \tag{8.4.4}$$

$$\hat{Z}_1 = \hat{Z}_2 = \hat{Z}_{aaeq} - \hat{Z}_{abeq} \tag{8.4.5}$$

A circuit representation of the series sequence impedances of a completely transposed three-phase line is shown in Figure 8.11.

Section 4.11 develops equations suitable for computer calculation of the shunt phase admittances of three-phase overhead transmission lines. The shunt admittance matrix Y_P for an untransposed line is given by Equation (4.11.16), and $\hat{Y}_P$ for a completely transposed three-phase line is given by (4.11.17).

Equation (4.11.16) can be transformed to the sequence domain to obtain

$$Y_S = A^{-1}Y_P A \tag{8.4.6}$$

where

$$Y_S = \mathbf{G}_S + j(2\pi f)\mathbf{C}_S \tag{8.4.7}$$

$$\mathbf{C}_S = \begin{bmatrix} C_0 & C_{01} & C_{02} \\ C_{10} & C_1 & C_{12} \\ C_{20} & C_{21} & C_2 \end{bmatrix} \text{ F/m} \tag{8.4.8}$$

In general, $\mathbf{C}_S$ is not diagonal. However, for the completely transposed line,

$$\hat{Y}_S = A^{-1}\hat{Y}_P A = \begin{bmatrix} \hat{y}_0 & 0 & 0 \\ 0 & \hat{y}_1 & 0 \\ 0 & 0 & \hat{y}_2 \end{bmatrix} = j(2\pi f)\begin{bmatrix} \hat{C}_0 & 0 & 0 \\ 0 & \hat{C}_1 & 0 \\ 0 & 0 & \hat{C}_2 \end{bmatrix} \tag{8.4.9}$$

where

FIGURE 8.12

Circuit representations
of the capacitances of a
completely transposed
three-phase line

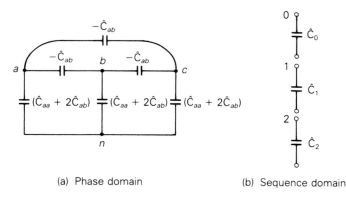

(a) Phase domain

(b) Sequence domain

$$\hat{C}_0 = \hat{C}_{aa} + 2\hat{C}_{ab} \quad \text{F/m} \tag{8.4.10}$$

$$\hat{C}_1 = \hat{C}_2 = \hat{C}_{aa} - \hat{C}_{ab} \quad \text{F/m} \tag{8.4.11}$$

Since $\hat{C}_{ab}$ is negative, the zero-sequence capacitance $\hat{C}_0$ is usually much less than the positive- or negative-sequence capacitance.

Circuit representations of the phase and sequence capacitances of a completely transposed three-phase line are shown in Figure 8.12.

8.5

SEQUENCE NETWORKS OF ROTATING MACHINES

A Y-connected synchronous generator grounded through a neutral impedance Z_n is shown in Figure 8.13. The internal generator voltages are designated E_a, E_b, and E_c, and the generator line currents are designated I_a, I_b, and I_c.

FIGURE 8.13

Y-connected
synchronous generator

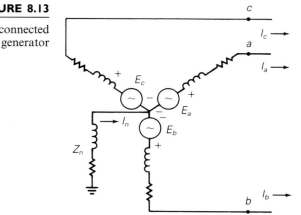

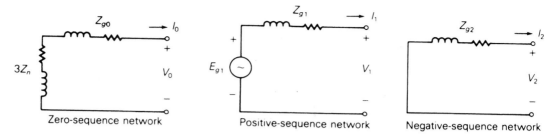

FIGURE 8.14   Sequence networks of a Y-connected synchronous generator

The sequence networks of the generator are shown in Figure 8.14. Since a three-phase synchronous generator is designed to produce balanced internal phase voltages E_a, E_b, E_c with only a positive-sequence component, a source voltage E_{g1} is included only in the positive-sequence network. The sequence components of the line-to-ground voltages at the generator terminals are denoted V_0, V_1, and V_2 in Figure 8.14.

The voltage drop in the generator neutral impedance is $Z_n I_n$, which can be written as $(3Z_n)I_0$, since, from (8.1.27), the neutral current is three times the zero-sequence current. Since this voltage drop is due only to zero-sequence current, an impedance $(3Z_n)$ is placed in the zero-sequence network of Figure 8.14 in series with the generator zero-sequence impedance Z_{g0}.

The sequence impedances of rotating machines are generally not equal. A detailed analysis of machine-sequence impedances is given in machine theory texts. We give only a brief explanation here.

When a synchronous generator stator has balanced three-phase positive-sequence currents under steady-state conditions, the net mmf produced by these positive-sequence currents rotates at the synchronous rotor speed in the same direction as that of the rotor. Under this condition, a high value of magnetic flux penetrates the rotor, and the positive-sequence impedance Z_{g1} has a high value. Under steady-state conditions, the positive-sequence generator impedance is called the *synchronous impedance*.

When a synchronous generator stator has balanced three-phase negative-sequence currents, the net mmf produced by these currents rotates at synchronous speed in the direction opposite to that of the rotor. With respect to the rotor, the net mmf is not stationary but rotates at twice synchronous speed. Under this condition, currents are induced in the rotor windings that prevent the magnetic flux from penetrating the rotor. As such, the negative-sequence impedance Z_{g2} is less than the positive-sequence synchronous impedance.

When a synchronous generator has only zero-sequence currents, which are line (or phase) currents with equal magnitude and phase, then the net mmf produced by these currents is theoretically zero. The generator zero-sequence impedance Z_{g0} is the smallest sequence impedance and is due to leakage flux, end turns, and harmonic flux from windings that do not produce a perfectly sinusoidal mmf.

Typical values of machine-sequence impedances are listed in Table A.1 in the Appendix. The positive-sequence machine impedance is synchronous, transient, or subtransient. *Synchronous* impedances are used for steady-state conditions, such as in power-flow studies, which are described in Chapter 6. *Transient* impedances are used for stability studies, which are described in Chapter 13, and *subtransient* impedances are used for short-circuit studies, which are described in Chapters 7 and 9. Unlike the positive-sequence impedances, a machine has only one negative-sequence impedance and only one zero-sequence impedance.

The sequence networks for three-phase synchronous motors and for three-phase induction motors are shown in Figure 8.15. Synchronous motors have the same sequence networks as synchronous generators, except that the sequence currents for synchronous motors are referenced *into* rather than out of the sequence networks. Also, induction motors have the same sequence networks as synchronous motors, except that the positive-sequence voltage

FIGURE 8.15

Sequence networks of three-phase motors

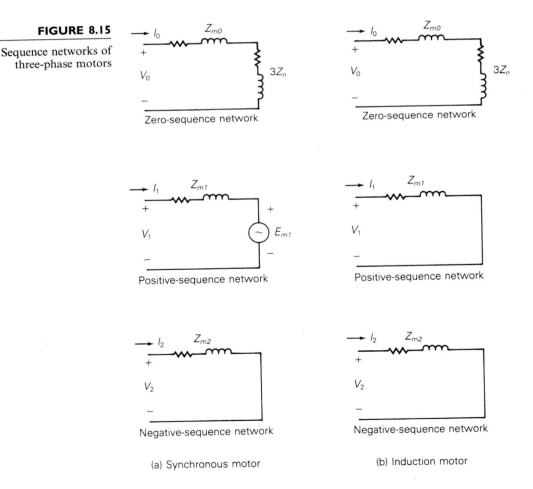

Zero-sequence network

Zero-sequence network

Positive-sequence network

Positive-sequence network

Negative-sequence network

Negative-sequence network

(a) Synchronous motor

(b) Induction motor

source E_{m1} is removed. Induction motors do not have a dc source of magnetic flux in their rotor circuits, and therefore E_{m1} is zero (or a short circuit).

The sequence networks shown in Figures 8.14 and 8.15 are simplified networks for rotating machines. The networks do not take into account such phenomena as machine saliency, saturation effects, and more complicated transient effects. These simplified networks, however, are in many cases accurate enough for power system studies.

EXAMPLE 8.5   Currents in sequence networks

Draw the sequence networks for the circuit of Example 2.5 and calculate the sequence components of the line current. Assume that the generator neutral is grounded through an impedance $Z_n = j10\ \Omega$, and that the generator sequence impedances are $Z_{g0} = j1\ \Omega$, $Z_{g1} = j15\ \Omega$, and $Z_{g2} = j3\ \Omega$.

SOLUTION   The sequence networks are shown in Figure 8.16. They are obtained by interconnecting the sequence networks for a balanced-Δ load, for

FIGURE 8.16

Sequence networks for Example 8.5

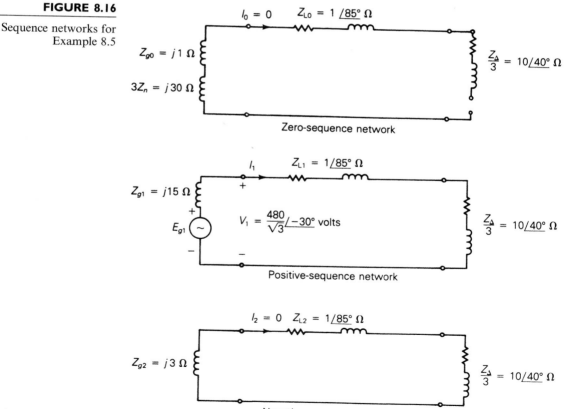

series-line impedances, and for a synchronous generator, which are given in Figures 8.5, 8.10, and 8.14.

It is clear from Figure 8.16 that $I_0 = I_2 = 0$ since there are no sources in the zero- and negative-sequence networks. Also, the positive-sequence generator terminal voltage V_1 equals the generator line-to-neutral terminal voltage. Therefore, from the positive-sequence network shown in the figure and from the results of Example 2.5,

$$I_1 = \frac{V_1}{(Z_{L1} + \frac{1}{3}Z_\Delta)} = 25.83\underline{/-73.78^\circ} \text{ A} = I_a$$

Note that from (8.1.20), I_1 equals the line current I_a, since $I_0 = I_2 = 0$. ∎

The following example illustrates the superiority of using symmetrical components for analyzing unbalanced systems.

EXAMPLE 8.6   Solving unbalanced three-phase networks using sequence components

A Y-connected voltage source with the following unbalanced voltage is applied to the balanced line and load of Example 2.5.

$$\begin{bmatrix} V_{ag} \\ V_{bg} \\ V_{cg} \end{bmatrix} = \begin{bmatrix} 277\underline{/0^\circ} \\ 260\underline{/-120^\circ} \\ 295\underline{/+115^\circ} \end{bmatrix} \text{ volts}$$

The source neutral is solidly grounded. Using the method of symmetrical components, calculate the source currents I_a, I_b, and I_c.

SOLUTION   Using (8.1.13)–(8.1.15), the sequence components of the source voltages are:

$$V_0 = \tfrac{1}{3}(277\underline{/0^\circ} + 260\underline{/-120^\circ} + 295\underline{/115^\circ})$$
$$= 7.4425 + j14.065 = 15.912\underline{/62.11^\circ} \text{ volts}$$

$$V_1 = \tfrac{1}{3}(227\underline{/0^\circ} + 260\underline{/-120^\circ + 120^\circ} + 295\underline{/115^\circ + 240^\circ})$$
$$= \tfrac{1}{3}(277\underline{/0^\circ} + 260\underline{/0^\circ} + 295\underline{/-5^\circ})$$
$$= 276.96 - j8.5703 = 277.1\underline{/-1.772^\circ} \text{ volts}$$

$$V_2 = \tfrac{1}{3}(277\underline{/0^\circ} + 260\underline{/-120^\circ + 240^\circ} + 295\underline{/115^\circ + 120^\circ})$$
$$= \tfrac{1}{3}(277\underline{/0^\circ} + 260\underline{/120^\circ} + 295\underline{/235^\circ})$$
$$= -7.4017 - j5.4944 = 9.218\underline{/216.59^\circ} \text{ volts}$$

These sequence voltages are applied to the sequence networks of the line and load, as shown in Figure 8.17. The sequence networks of this figure

FIGURE 8.17

Sequence networks for
Example 8.6

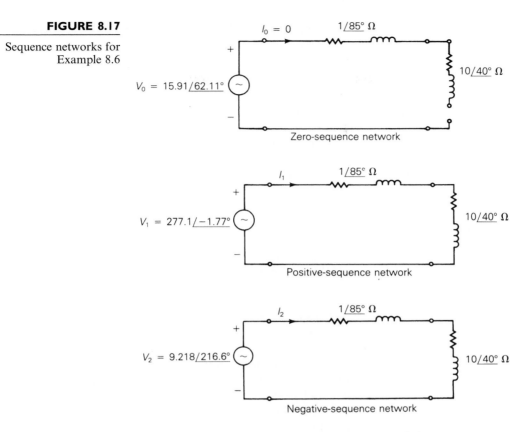

are uncoupled, and the sequence components of the source currents are easily calculated as follows:

$$I_0 = 0$$

$$I_1 = \frac{V_1}{Z_{L1} + \dfrac{Z_\Delta}{3}} = \frac{277.1\underline{/-1.772°}}{10.73\underline{/43.78°}} = 25.82\underline{/-45.55°} \quad \text{A}$$

$$I_2 = \frac{V_2}{Z_{L2} + \dfrac{Z_\Delta}{3}} = \frac{9.218\underline{/216.59°}}{10.73\underline{/43.78°}} = 0.8591\underline{/172.81°} \quad \text{A}$$

Using (8.1.20)–(8.1.22), the source currents are:

$$I_a = (0 + 25.82\underline{/-45.55°} + 0.8591\underline{/172.81°})$$

$$= 17.23 - j18.32 = 25.15\underline{/-46.76°} \quad \text{A}$$

$$I_b = (0 + 25.82\underline{/-45.55° + 240°} + 0.8591\underline{/172.81° + 120°})$$

$$= (25.82\underline{/194.45°} + 0.8591\underline{/292.81°})$$

$$= -24.67 - j7.235 = 25.71\underline{/196.34°} \quad \text{A}$$

$$I_c = (0 + 25.82\underline{/-45.55° + 120°} + 0.8591\underline{/172.81° + 240°})$$
$$= (25.82\underline{/74.45°} + 0.8591\underline{/52.81°})$$
$$= 7.441 + j25.56 = 26.62\underline{/73.77°} \quad \text{A}$$

You should calculate the line currents for this example without using symmetrical components, in order to verify this result and to compare the two solution methods (see Problem 8.33). Without symmetrical components, coupled KVL equations must be solved. With symmetrical components, the conversion from phase to sequence components decouples the networks as well as the resulting KVL equations, as shown above. ∎

8.6

PER-UNIT SEQUENCE MODELS OF THREE-PHASE TWO-WINDING TRANSFORMERS

Figure 8.18(a) is a schematic representation of an ideal Y–Y transformer grounded through neutral impedances Z_N and Z_n. Figures 8.18(b–d) show the per-unit sequence networks of this ideal transformer.

When balanced positive-sequence currents or balanced negative-sequence currents are applied to the transformer, the neutral currents are zero and there are no voltage drops across the neutral impedances. Therefore, the per-unit positive- and negative-sequence networks of the ideal Y–Y transformer, Figures 8.18(b) and (c), are the same as the per-unit single-phase ideal transformer, Figure 3.9(a).

Zero-sequence currents have equal magnitudes and equal phase angles. When per-unit sequence currents $I_{A0} = I_{B0} = I_{C0} = I_0$ are applied to the high-voltage windings of an ideal Y–Y transformer, the neutral current $I_N = 3I_0$ flows through the neutral impedance Z_N, with a voltage drop $(3Z_N)I_0$. Also, per-unit zero-sequence current I_0 flows in each low-voltage winding [from (3.3.9)], and therefore $3I_0$ flows through neutral impedance Z_n, with a voltage drop $(3I_0)Z_n$. The per-unit zero-sequence network, which includes the impedances $(3Z_N)$ and $(3Z_n)$, is shown in Figure 8.18(b).

Note that if either one of the neutrals of an ideal transformer is ungrounded, then no zero sequence can flow in either the high- or low-voltage windings. For example, if the high-voltage winding has an open neutral, then $I_N = 3I_0 = 0$, which in turn forces $I_0 = 0$ on the low-voltage side. This can be shown in the zero-sequence network of Figure 8.18(b) by making $Z_N = \infty$, which corresponds to an open circuit.

The per-unit sequence networks of a practical Y–Y transformer are shown in Figure 8.19(a). These networks are obtained by adding external impedances to the sequence networks of the ideal transformer, as follows. The leakage impedances of the high-voltage windings are series impedances like the series impedances shown in Figure 8.9, with no coupling between phases $(Z_{ab} = 0)$. If the phase a, b, and c windings have equal leakage impedances

FIGURE 8.18

Ideal Y–Y transformer

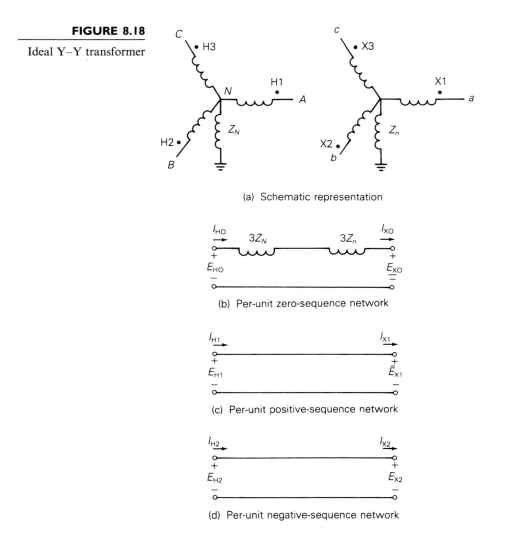

(a) Schematic representation

(b) Per-unit zero-sequence network

(c) Per-unit positive-sequence network

(d) Per-unit negative-sequence network

$Z_H = R_H + jX_H$, then the series impedances are *symmetrical* with sequence networks, as shown in Figure 8.10, where $Z_{H0} = Z_{H1} = Z_{H2} = Z_H$. Similarly, the leakage impedances of the low-voltage windings are symmetrical series impedances with $Z_{X0} = Z_{X1} = Z_{X2} = Z_X$. These series leakage impedances are shown in per-unit in the sequence networks of Figure 8.19(a).

The shunt branches of the practical Y–Y transformer, which represent exciting current, are equivalent to the Y load of Figure 8.3. Each phase in Figure 8.3 represents a core loss resistor in parallel with a magnetizing induc-tance. Assuming these are the same for each phase, then the Y load is *sym-metrical*, and the sequence networks are shown in Figure 8.4. These shunt branches are also shown in Figure 8.19(a). Note that $(3Z_N)$ and $(3Z_n)$ have already been included in the zero-sequence network.

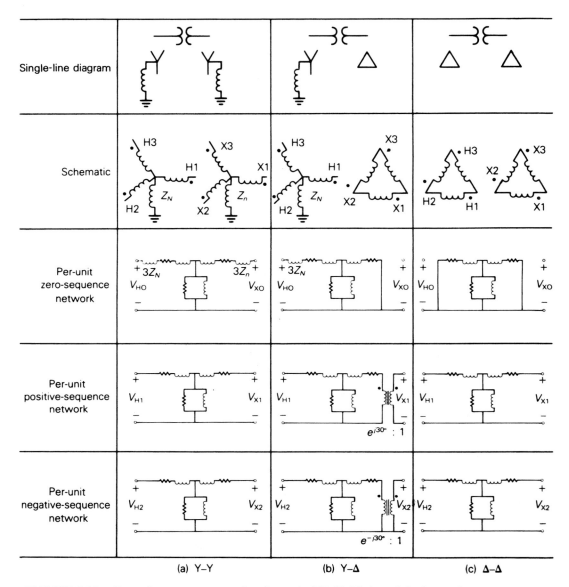

| | (a) Y–Y | (b) Y–Δ | (c) Δ–Δ |

FIGURE 8.19   Per-unit sequence networks of practical Y–Y, Y–Δ, and Δ–Δ transformers

The per-unit positive- and negative-sequence transformer impedances of the practical Y–Y transformer in Figure 8.19(a) are identical, which is always true for nonrotating equipment. The per-unit zero-sequence network, however, depends on the neutral impedances Z_N and Z_n.

The per-unit sequence networks of the Y–Δ transformer, shown in Figure 8.19(b), have the following features:

1. The per-unit impedances do not depend on the winding connections. That is, the per-unit impedances of a transformer that is connected Y–Y, Y–Δ, Δ–Y, or Δ–Δ are the same. However, the base voltages do depend on the winding connections.

2. A phase shift is included in the per-unit positive- and negative-sequence networks. For the American standard, the positive-sequence voltages and currents on the high-voltage side of the Y–Δ transformer lead the corresponding quantities on the low-voltage side by 30°. For negative sequence, the high-voltage quantities lag by 30°.

3. Zero-sequence currents can flow in the Y winding if there is a neutral connection, and corresponding zero-sequence currents flow within the Δ winding. However, no zero-sequence current enters or leaves the Δ winding.

The phase shifts in the positive- and negative-sequence networks of Figure 8.19(b) are represented by the phase-shifting transformer of Figure 3.4. Also, the zero-sequence network of Figure 8.19(b) provides a path on the Y side for zero-sequence current to flow, but no zero-sequence current can enter or leave the Δ side.

The per-unit sequence networks of the Δ–Δ transformer, shown in Figure 8.19(c), have the following features:

1. The positive- and negative-sequence networks, which are identical, are the same as those for the Y–Y transformer. It is assumed that the windings are labeled so there is no phase shift. Also, the per-unit impedances do not depend on the winding connections, but the base voltages do.

2. Zero-sequence currents *cannot* enter or leave either Δ winding, although they can circulate within the Δ windings.

EXAMPLE 8.7 **Solving unbalanced three-phase networks with transformers using per-unit sequence components**

A 75-kVA, 480-volt Δ/208-volt Y transformer with a solidly grounded neutral is connected between the source and line of Example 8.6. The transformer leakage reactance is $X_{eq} = 0.10$ per unit; winding resistances and exciting current are neglected. Using the transformer ratings as base quantities, draw the per-unit sequence networks and calculate the phase a source current I_a.

SOLUTION The base quantities are $S_{base1\phi} = 75/3 = 25$ kVA, $V_{baseHLN} = 480/\sqrt{3} = 277.1$ volts, $V_{baseXLN} = 208/\sqrt{3} = 120.1$ volts, and $Z_{baseX} = (120.1)^2/25{,}000 = 0.5770$ Ω. The sequence components of the actual source voltages are given in Figure 8.17. In per-unit, these voltages are

$$V_0 = \frac{15.91\underline{/62.11°}}{277.1} = 0.05742\underline{/62.11°} \quad \text{per unit}$$

$$V_1 = \frac{277.1\underline{/-1.772°}}{277.1} = 1.0\underline{/-1.772°} \quad \text{per unit}$$

$$V_2 = \frac{9.218\underline{/216.59°}}{277.1} = 0.03327\underline{/216.59°} \quad \text{per unit}$$

The per-unit line and load impedances, which are located on the low-voltage side of the transformer, are

$$Z_{L0} = Z_{L1} = Z_{L2} = \frac{1\underline{/85°}}{0.577} = 1.733\underline{/85°} \quad \text{per unit}$$

$$Z_{\text{load1}} = Z_{\text{load2}} = \frac{Z_\Delta}{3(0.577)} = \frac{10\underline{/40°}}{0.577} = 17.33\underline{/40°} \quad \text{per unit}$$

FIGURE 8.20

Per-unit sequence networks for Example 8.7

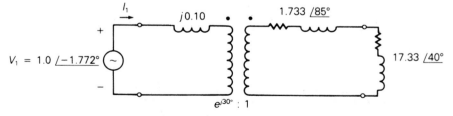

(a) Per-unit zero-sequence network

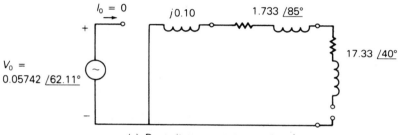

(b) Per-unit positive-sequence network

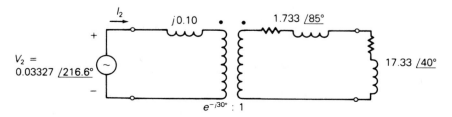

(c) Per-unit negative-sequence network

The per-unit sequence networks are shown in Figure 8.20. Note that the per-unit line and load impedances, when referred to the high-voltage side of the phase-shifting transformer, do not change (see (3.1.26)). Therefore, from Figure 8.20, the sequence components of the source currents are

$$I_0 = 0$$

$$I_1 = \frac{V_1}{jX_{eq} + Z_{L1} + Z_{load1}} = \frac{1.0\underline{/-1.772°}}{j0.10 + 1.733\underline{/85°} + 17.33\underline{/40°}}$$

$$= \frac{1.0\underline{/-1.772°}}{13.43 + j12.97} = \frac{1.0\underline{/-1.772°}}{18.67\underline{/44.0°}} = 0.05356\underline{/-45.77°} \quad \text{per unit}$$

$$I_2 = \frac{V_2}{jX_{eq} + Z_{L2} + Z_{load2}} = \frac{0.03327\underline{/216.59°}}{18.67\underline{/44.0°}}$$

$$= 0.001782\underline{/172.59°} \quad \text{per unit}$$

The phase a source current is then, using (8.1.20),

$$I_a = I_0 + I_1 + I_2$$

$$= 0 + 0.05356\underline{/-45.77°} + 0.001782\underline{/172.59°}$$

$$= 0.03511 - j0.03764 = 0.05216\underline{/-46.19°} \quad \text{per unit}$$

Using $I_{baseH} = \dfrac{75{,}000}{480\sqrt{3}} = 90.21 \text{ A},$

$$I_a = (0.05216)(90.21)\underline{/-46.19°} = 4.705\underline{/-46.19°} \quad \text{A} \qquad \blacksquare$$

8.7

PER-UNIT SEQUENCE MODELS OF THREE-PHASE THREE-WINDING TRANSFORMERS

Three identical single-phase three-winding transformers can be connected to form a three-phase bank. Figure 8.21 shows the general per-unit sequence networks of a three-phase three-winding transformer. Instead of labeling the windings 1, 2, and 3, as was done for the single-phase transformer, the letters H, M, and X are used to denote the high-, medium-, and low-voltage windings, respectively. By convention, a common S_{base} is selected for the H, M, and X terminals, and voltage bases V_{baseH}, V_{baseM}, and V_{baseX} are selected in proportion to the rated line-to-line voltages of the transformer.

For the general zero-sequence network, Figure 8.21(a), the connection between terminals H and H′ depends on how the high-voltage windings are connected, as follows:

1. Solidly grounded Y—Short H to H′.

2. Grounded Y through Z_N—Connect $(3Z_N)$ from H to H′.

FIGURE 8.21

Per-unit sequence networks of a three-phase three-winding transformer

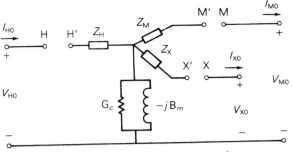

(a) Per-unit zero-sequence network

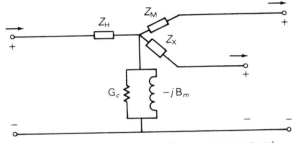

(b) Per-unit positive- or negative-sequence network
(phase shift not shown)

3. Ungrounded Y—Leave H–H′ open as shown.

4. Δ—Short H′ to the reference bus.

Terminals X–X′ and M–M′ are connected in a similar manner.

The impedances of the per-unit negative-sequence network are the same as those of the per-unit positive-sequence network, which is always true for nonrotating equipment. Phase-shifting transformers, not shown in Figure 8.21(b), can be included to model phase shift between Δ and Y windings.

EXAMPLE 8.8   **Three-winding three-phase transformer: per-unit sequence networks**

Three transformers, each identical to that described in Example 3.9, are connected as a three-phase bank in order to feed power from a 900-MVA, 13.8-kV generator to a 345-kV transmission line and to a 34.5-kV distribution line. The transformer windings are connected as follows:

> 13.8-kV windings (X):   Δ, to generator
>
> 199.2-kV windings (H):   solidly grounded Y, to 345-kV line
>
> 19.92-kV windings (M):   grounded Y through $Z_n = j0.10\ \Omega$, to 34.5-kV line

The positive-sequence voltages and currents of the high- and medium-voltage Y windings lead the corresponding quantities of the low-voltage Δ winding by 30°. Draw the per-unit sequence networks, using a three-phase base of 900 MVA and 13.8 kV for terminal X.

SOLUTION   The per-unit sequence networks are shown in Figure 8.22. Since $V_{baseX} = 13.8$ kV is the rated line-to-line voltage of terminal X, $V_{baseM} = \sqrt{3}(19.92) = 34.5$ kV, which is the rated line-to-line voltage of terminal M. The base impedance of the medium-voltage terminal is then

$$Z_{baseM} = \frac{(34.5)^2}{900} = 1.3225 \quad \Omega$$

Therefore, the per-unit neutral impedance is

$$Z_n = \frac{j0.10}{1.3225} = j0.07561 \quad \text{per unit}$$

FIGURE 8.22

Per-unit sequence networks for Example 8.8

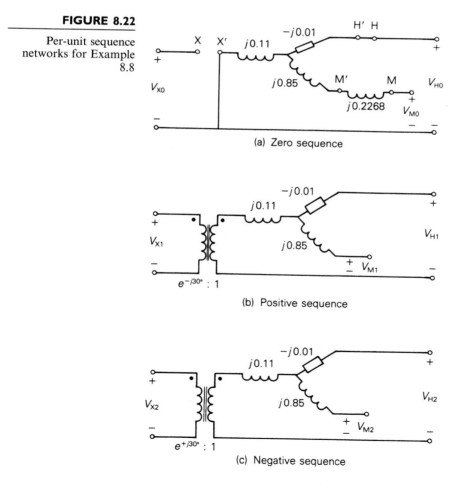

(a) Zero sequence

(b) Positive sequence

(c) Negative sequence

and $(3Z_n) = j0.2268$ is connected from terminal M to M' in the per-unit zero-sequence network. Since the high-voltage windings have a solidly grounded neutral, H to H' is shorted in the zero-sequence network. Also, phase-shifting transformers are included in the positive- and negative-sequence networks. ■

8.8

POWER IN SEQUENCE NETWORKS

The power delivered to a three-phase network can be determined from the power delivered to the sequence networks. Let S_p denote the total complex power delivered to the three-phase load of Figure 8.7, which can be calculated from

$$S_p = V_{ag}I_a^* + V_{bg}I_b^* + V_{cg}I_c^* \tag{8.8.1}$$

Equation (8.8.1) is also valid for the total complex power delivered by the three-phase generator of Figure 8.13, or for the complex power delivered to any three-phase bus. Rewriting (8.8.1) in matrix format,

$$S_p = [V_{ag}\,V_{bg}\,V_{cg}] \begin{bmatrix} I_a^* \\ I_b^* \\ I_c^* \end{bmatrix}$$

$$= V_p^{\mathrm{T}} I_p^* \tag{8.8.2}$$

where T denotes transpose and * denotes complex conjugate. Now, using (8.1.9) and (8.1.16),

$$S_p = (AV_s)^{\mathrm{T}}(AI_s)^*$$

$$= V_s^{\mathrm{T}}[A^{\mathrm{T}}A^*]I_s^* \tag{8.8.3}$$

Using the definition of A, which is (8.1.8), to calculate the term within the brackets of (8.8.3), and noting that a and a^2 are conjugates,

$$A^{\mathrm{T}}A^* = \begin{bmatrix} 1 & 1 & 1 \\ 1 & a^2 & a \\ 1 & a & a^2 \end{bmatrix}^{\mathrm{T}} \begin{bmatrix} 1 & 1 & 1 \\ 1 & a^2 & a \\ 1 & a & a^2 \end{bmatrix}^*$$

$$= \begin{bmatrix} 1 & 1 & 1 \\ 1 & a^2 & a \\ 1 & a & a^2 \end{bmatrix} \begin{bmatrix} 1 & 1 & 1 \\ 1 & a & a^2 \\ 1 & a^2 & a \end{bmatrix}$$

$$= \begin{bmatrix} 3 & 0 & 0 \\ 0 & 3 & 0 \\ 0 & 0 & 3 \end{bmatrix} = 3U \tag{8.8.4}$$

Equation (8.8.4) can now be used in (8.8.3) to obtain

$$S_p = 3 V_s^T I_s^*$$

$$= 3[V_0 + V_1 + V_2] \begin{bmatrix} I_0^* \\ I_1^* \\ I_2^* \end{bmatrix} \tag{8.8.5}$$

$$S_p = 3(V_0 I_0^* + V_1 I_1^* + V_2 I_2^*)$$

$$= 3 S_s \tag{8.8.6}$$

Thus, the total complex power S_p delivered to a three-phase network equals *three* times the total complex power S_s delivered to the sequence networks.

The factor of 3 occurs in (8.8.6) because $A^T A^* = 3U$, as shown by (8.8.4). It is possible to eliminate this factor of 3 by defining a new transformation matrix $A_1 = (1/\sqrt{3})A$ such that $A_1^T A_1^* = U$, which means that A_1 is a *unitary* matrix. Using A_1 instead of A, the total complex power delivered to three-phase networks would equal the total complex power delivered to the sequence networks. However, standard industry practice for symmetrical components is to use A, defined by (8.1.8).

EXAMPLE 8.9   Power in sequence networks

Calculate S_p and S_s delivered by the three-phase source in Example 8.6. Verify that $S_p = 3S_s$.

SOLUTION   Using (8.5.1),

$$S_p = (277\underline{/0°})(25.15\underline{/+46.76°}) + (260\underline{/-120°})(25.71\underline{/-196.34°})$$

$$+ (295\underline{/115°})(26.62\underline{/-73.77°})$$

$$= 6967\underline{/46.76°} + 6685\underline{/43.66°} + 7853\underline{/41.23°}$$

$$= 15{,}520 + j14{,}870 = 21{,}490\underline{/43.78°} \quad \text{VA}$$

In the sequence domain,

$$S_s = V_0 I_0^* + V_1 I_1^* + V_2 I_2^*$$

$$= 0 + (277.1\underline{/-1.77°})(25.82\underline{/45.55°})$$

$$+ (9.218\underline{/216.59°})(0.8591\underline{/-172.81°})$$

$$= 7155\underline{/43.78°} + 7.919\underline{/43.78°}$$

$$= 5172 + j4958 = 7163\underline{/43.78°} \quad \text{VA}$$

Also,

$$3S_s = 3(7163\underline{/43.78°}) = 21{,}490\underline{/43.78°} = S_p$$

∎

PROBLEMS

SECTION 8.1

8.1   Using the operator $a = 1/120°$, evaluate the following in polar form: (a) $(a+1)/(1+a-a^2)$, (b) $(a^2+a+j)/(ja-a^2)$, (c) $(1-a)(1+a^2)$, (d) $(a+a^2)(a^2+1)$.

8.2   Using $a = 1/120°$, evaluate the following in rectangular form:

    a. a^{10}

    b. $(ja)^{10}$

    c. $(1-a)^3$

    d. e^a

    Hint for (d): $e^{(x+jy)} = e^x e^{jy} = e^x/\underline{y}$, where y is in radians.

8.3   Determine the symmetrical components of the following line currents: (a) $I_a = 10/90°$, $I_b = 10/340°$, $I_c = 10/200°$ A; (b) $I_a = 100$, $I_b = j100$, $I_c = 0$ A.

8.4   Find the phase voltages V_{an}, V_{bn}, and V_{cn} whose sequence components are: $V_0 = 20/80°$, $V_1 = 100/0°$, $V_2 = 30/180°$ V.

8.5   For the unbalanced three-phase system described by

$$I_a = 12/0°\text{A}, \quad I_b = 6/-90°\text{A}, \quad I_C = 8/150°\text{A}$$

compute the symmetrical components I_0, I_1, I_2.

8.6   (a) Given the symmetrical components to be

$$V_0 = 10/0°\,V, \quad V_1 = 80/30°\,V, \quad V_2 = 40/-30°\,V$$

determine the unbalanced phase voltages V_a, V_b, and V_c.
(b) Using the results of part (a), calculate the line-to-line voltages V_{ab}, V_{bc}, and V_{ca}. Then determine the symmetrical components of these ling-to-line voltages, the symmetrical components of the corresponding phase voltages, and the phase voltages. Compare them with the result of part (a). Comment on why they are different, even though either set will result in the same line-to-line voltages.

8.7   One line of a three-phase generator is open circuited, while the other two are short-circuited to ground. The line currents are $I_a = 0$, $I_b = 1500/90°$, and $I_c = 1500/-30°$ A. Find the symmetrical components of these currents. Also find the current into the ground.

8.8   Let an unbalanced, three-phase, Wye-connected load (with phase impedances of Z_a, Z_b, and Z_c) be connected to a balanced three-phase supply, resulting in phase voltages of V_a, V_b, and V_c across the corresponding phase impedances. Choosing V_{ab} as the reference, show that

$$V_{ab,0} = 0; \quad V_{ab,1} = \sqrt{3}V_{a,1}e^{j30°}; \quad V_{ab,2} = \sqrt{3}V_{a,2}e^{-j30°}.$$

8.9   Reconsider Problem 8.8 and choosing V_{bc} as the reference, show that

$$V_{bc,0} = 0; \quad V_{bc,1} = -j\sqrt{3}V_{a,1}; \quad V_{bc,2} = j\sqrt{3}V_{a,2}.$$

8.10   Given the line-to-ground voltages $V_{ag} = 280/0°$, $V_{bg} = 290/-130°$, and $V_{cg} = 260/110°$ volts, calculate (a) the sequence components of the line-to-ground voltages, denoted V_{Lg0}, V_{Lg1}, and V_{Lg2}; (b) line-to-line voltages V_{ab}, V_{bc}, and V_{ca}; and (c) sequence components of the line-to-line voltages V_{LL0}, V_{LL1}, and V_{LL2}. Also, verify the following general relation: $V_{LL0} = 0$, $V_{LL1} = \sqrt{3}V_{Lg1}/+30°$, and $V_{LL2} = \sqrt{3}V_{Lg2}/-30°$ volts.

8.11   A balanced Δ-connected load is fed by a three-phase supply for which phase C is open and phase A is carrying a current of $10/0°$ A. Find the symmetrical components of the line currents. (Note that zero-sequence currents are not present for any three-wire system.)

8.12   A Y-connected load bank with a three-phase rating of 500 kVA and 2300 V consists of three identical resistors of 10.58 Ω. The load bank has the following applied voltages: $V_{ab} = 1840/82.8°$, $V_{bc} = 2760/-41.4°$, and $V_{ca} = 2300/180°$ V. Determine the symmetrical components of (a) the line-to-line voltages V_{ab0}, V_{ab1}, and V_{ab2}; (b) the line-to-neutral voltages V_{an0}, V_{an1}, and V_{an2}; (c) and the line currents I_{a0}, I_{a1}, and I_{a2}. (Note that the absence of a neutral connection means that zero-sequence currents are not present.)

SECTION 8.2

8.13   The currents in a Δ load are $I_{ab} = 10/0°$, $I_{bc} = 20/-90°$, and $I_{ca} = 15/90°$ A. Calculate (a) the sequence components of the Δ-load currents, denoted $I_{Δ0}$, $I_{Δ1}$, $I_{Δ2}$; (b) the line currents I_a, I_b, and I_c, which feed the Δ load; and (c) sequence components of the line currents I_{L0}, I_{L1}, and I_{L2}. Also, verify the following general relation: $I_{L0} = 0$, $I_{L1} = \sqrt{3}I_{Δ1}/-30°$, and $I_{L2} = \sqrt{3}I_{Δ2}/+30°$ A.

8.14   The voltages given in Problem 8.10 are applied to a balanced-Y load consisting of $(6 + j8)$ ohms per phase. The load neutral is solidly grounded. Draw the sequence networks and calculate I_0, I_1, and I_2, the sequence components of the line currents. Then calculate the line currents I_a, I_b, and I_c.

8.15   Repeat Problem 8.14 with the load neutral open.

8.16   Repeat Problem 8.14 for a balanced-Δ load consisting of $(12 + j16)$ ohms per phase.

8.17   Repeat Problem 8.14 for the load shown in Example 8.4 (Figure 8.6).

8.18   Perform the indicated matrix multiplications in (8.2.21) and verify the sequence impedances given by (8.2.22)–(8.2.27).

8.19   The following unbalanced line-to-ground voltages are applied to the balanced-Y load shown in Figure 3.3: $V_{ag} = 100/0°$, $V_{bg} = 75/180°$, and $V_{cg} = 50/90°$ volts. The Y load has $Z_Y = 3 + j4$ Ω per phase with neutral impedance $Z_n = j1$ Ω. (a) Calculate the line currents I_a, I_b, and I_c without using symmetrical components. (b) Calculate the line currents I_a, I_b, and I_c using symmetrical components. Which method is easier?

8.20   (a) Consider three equal impedances of $(j27)$ Ω connected in Δ. Obtain the sequence networks.
(b) Now, with a mutual impedance of $(j6)$ Ω between each pair of adjacent branches in the Δ-connected load of part (a), how would the sequence networks change?

8.21   The three-phase impedance load shown in Figure 8.7 has the following phase impedance matrix:

$$Z_p = \begin{bmatrix} (6 + j10) & 0 & 0 \\ 0 & (6 + j10) & 0 \\ 0 & 0 & (6 + j10) \end{bmatrix} \; \Omega$$

Determine the sequence impedance matrix Z_s for this load. Is the load symmetrical?

8.22   The three-phase impedance load shown in Figure 8.7 has the following sequence impedance matrix:

$$Z_S = \begin{bmatrix} (8 + j12) & 0 & 0 \\ 0 & 4 & 0 \\ 0 & 0 & 4 \end{bmatrix} \; \Omega$$

Determine the phase impedance matrix Z_p for this load. Is the load symmetrical?

8.23 Consider a three-phase balanced Y-connected load with self and mutual impedances as shown in Figure 8.23. Let the load neutral be grounded through an impedance Z_n. Using Kirchhoff's laws, develop the equations for line-to-neutral voltages, and then determine the elements of the phase impedance matrix. Also find the elements of the corresponding sequence impedance matrix.

FIGURE 8.23

Problem 8.23

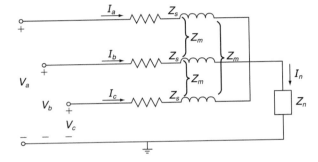

8.24 A three-phase balanced voltage source is applied to a balanced Y-connected load with ungrounded neutral. The Y-connected load consists of three mutually coupled re-actances, where the reactance of each phase is $j12 \; \Omega$ and the mutual coupling between any two phases is $j4 \; \Omega$. The line-to-line source voltage is $100 \sqrt{3}$ V. Determine the line currents (a) by mesh analysis without using symmetrical components, and (b) using symmetrical components.

8.25 A three-phase balanced Y-connected load with series impedances of $(8 + j24) \; \Omega$ per phase and mutual impedance between any two phases of $j4 \; \Omega$ is supplied by a three-phase unbalanced source with line-to-neutral voltages of $V_{an} = 200/\underline{25°}$, $V_{bn} = 100/\underline{-155°}$, $V_{cn} = 80/\underline{100°}$ V. The load and source neutrals are both solidly grounded. Determine: (a) the load sequence impedance matrix, (b) the symmetrical components of the line-to-neutral voltages, (c) the symmetrical components of the load currents, and (d) the load currents.

SECTION 8.3

8.26 Repeat Problem 8.14 but include balanced three-phase line impedances of $(3 + j4)$ ohms per phase between the source and load.

8.27 Consider the flow of unbalanced currents in the symmetrical three-phase line section with neutral conductor as shown in Figure 8.24. (a) Express the voltage drops across the line conductors given by $V_{aa'}$, $V_{bb'}$, and $V_{cc'}$ in terms of line currents, self-impedances defined by $Z_s = Z_{aa} + Z_{nn} - 2Z_{an}$, and mutual impedances defined by $Z_m = Z_{ab} + Z_{nn} - 2Z_{an}$. (b) Show that the sequence components of the voltage drops between the ends of the line section can be written as $V_{aa'0} = Z_0 I_{a0}$, $V_{aa'1} = Z_1 I_{a1}$, and $V_{aa'2} = Z_2 I_{a2}$, where $Z_0 = Z_s + 2Z_m = Z_{aa} + 2Z_{ab} + 3Z_{nn} - 6Z_{an}$ and $Z_1 = Z_2 = Z_s - Z_m = Z_{aa} - Z_{ab}$.

FIGURE 8.24

Problem 8.27

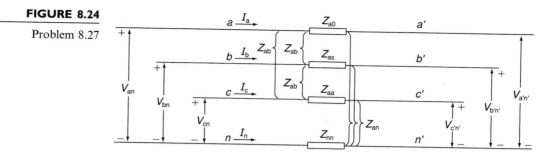

8.28 Let the terminal voltages at the two ends of the line section shown in Figure 8.24 be given by:

$$V_{an} = (182 + j70) \text{ kV} \qquad V_{an'} = (154 + j28) \text{ kV}$$
$$V_{bn} = (72.24 - j32.62) \text{ kV} \qquad V_{bn'} = (44.24 + j74.62) \text{ kV}$$
$$V_{cn} = (-170.24 + j88.62) \text{ kV} \quad V_{cn'} = (-198.24 + j46.62) \text{ kV}$$

The line impedances are given by:

$$Z_{aa} = j60 \ \Omega \qquad Z_{ab} = j20 \ \Omega \qquad Z_{nn} = j80 \ \Omega \qquad Z_{an} = 0$$

(a) Compute the line currents using symmetrical components. (*Hint:* See Problem 8.27.) (b) Compute the line currents without using symmetrical components.

8.29 A completely transposed three-phase transmission line of 200 km in length has the following symmetrical sequence impedances and sequence admittances:

$$Z_1 = Z_2 = j0.5 \ \Omega/\text{km}; \quad Z_0 = j2 \ \Omega/\text{km}$$
$$Y_1 = Y_2 = j3 \times 10^{-9} \ \text{s/m}; \quad Y_0 = j1 \times 10^{-9} \ \text{s/m}$$

Set up the nominal Π sequence circuits of this medium-length line.

SECTION 8.5

8.30 As shown in Figure 8.25, a balanced three-phase, positive-sequence source with $V_{AB} = 480\underline{/0°}$ volts is applied to an unbalanced Δ load. Note that one leg of the Δ is

FIGURE 8.25

Problem 8.30

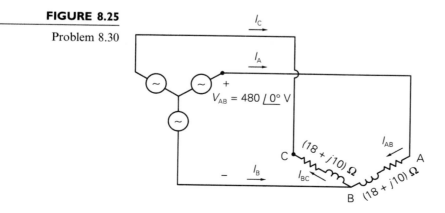

open. Determine: (a) the load currents I_{AB} and I_{BC}; (b) the line currents I_A, I_B, and I_C, which feed the Δ load; and (c) the zero-, positive-, and negative-sequence components of the line currents.

8.31   A balanced Y-connected generator with terminal voltage $V_{bc} = 480\underline{/90°}$ volts is connected to a balanced-Δ load whose impedance is $20\underline{/40°}$ ohms per phase. The line impedance between the source and load is $0.5\underline{/80°}$ ohm for each phase. The generator neutral is grounded through an impedance of $j5$ ohms. The generator sequence impedances are given by $Z_{g0} = j7$, $Z_{g1} = j15$, and $Z_{g2} = j10$ ohms. Draw the sequence networks for this system and determine the sequence components of the line currents.

8.32   In a three-phase system, a synchronous generator supplies power to a 208-volt synchronous motor through a line having an impedance of $0.5\underline{/80°}$ ohm per phase. The motor draws 10 kW at 0.8 p.f. leading and at rated voltage. The neutrals of both the generator and motor are grounded through impedances of $j5$ ohms. The sequence impedances of both machines are $Z_0 = j5$, $Z_1 = j15$, and $Z_2 = j10$ ohms. Draw the sequence networks for this system and find the line-to-line voltage at the generator terminals. Assume balanced three-phase operation.

8.33   Calculate the source currents in Example 8.6 without using symmetrical components. Compare your solution method with that of Example 8.6. Which method is easier?

8.34   A Y-connected synchronous generator rated 20 MVA at 13.8 kV has a positive-sequence reactance of $j2.38\ \Omega$, negative-sequence reactance of $j3.33\ \Omega$, and zero-sequence reactance of $j0.95\ \Omega$. The generator neutral is solidly grounded. With the generator operating unloaded at rated voltage, a so-called single line-to-ground fault occurs at the machine terminals. During this fault, the line-to-ground voltages at the generator terminals are $V_{ag} = 0$, $V_{bg} = 8.071\underline{/-102.25°}$, and $V_{cg} = 8.071\underline{/102.25°}$ kV. Determine the sequence components of the generator fault currents and the generator fault currents. Draw a phasor diagram of the pre-fault and post-fault generator terminal voltages. (*Note:* For this fault, the sequence components of the generator fault currents are all equal to each other.)

8.35   Figure 8.26 shows a single-line diagram of a three-phase, interconnected generator-reactor system, in which the given per-unit reactances are based on the ratings of the individual pieces of equipment. If a three-phase short-circuit occurs at fault point F, obtain the fault MVA and fault current in kA, if the pre-fault busbar line-to-line voltage is 13.2 kV. Choose 100 MVA as the base MVA for the system.

FIGURE 8.26

One-line diagram for
Problem 8.35

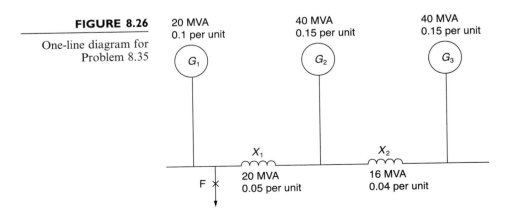

8.36 Consider Figures 8.13 and 8.14 of the text with reference to a Y-connected synchronous generator (grounded through a neutral impedance Z_n) operating at no load. For a line-to-ground fault occurring on phase a of the generator, list the constraints on the currents and voltages in the phase domain, transform those into the sequence domain, and then obtain a sequence-network representation. Also, find the expression for the fault current in phase a.

8.37 Reconsider the synchronous generator of Problem 8.36. Obtain sequence-network representations for the following fault conditions.
(a) A short-circuit between phases b and c.
(b) A double line-to-ground fault with phases b and c grounded.

SECTION 8.6

8.38 Three single-phase, two-winding transformers, each rated 450 MVA, 20 kV/288.7 kV, with leakage reactance $X_{eq} = 0.12$ per unit, are connected to form a three-phase bank. The high-voltage windings are connected in Y with a solidly grounded neutral. Draw the per-unit zero-, positive-, and negative-sequence networks if the low-voltage windings are connected: (a) in Δ with American standard phase shift, (b) in Y with an open neutral. Use the transformer ratings as base quantities. Winding resistances and exciting current are neglected.

8.39 The leakage reactance of a three-phase, 500-MVA, 345 Y/23 Δ-kV transformer is 0.09 per unit based on its own ratings. The Y winding has a solidly grounded neutral. Draw the sequence networks. Neglect the exciting admittance and assume American standard phase shift.

8.40 Choosing system bases to be 360/24 kV and 100 MVA, redraw the sequence networks for Problem 8.39.

8.41 Draw the zero-sequence reactance diagram for the power system shown in Figure 3.33. The zero-sequence reactance of each generator and of the synchronous motor is 0.05 per unit based on equipment ratings. Generator 2 is grounded through a neutral reactor of 0.06 per unit on a 100-MVA, 18-kV base. The zero-sequence reactance of each transmission line is assumed to be three times its positive-sequence reactance. Use the same base as in Problem 3.29.

8.42 Three identical Y-connected resistors of $1.0/0°$ per unit form a load bank, which is supplied from the low-voltage Y-side of a $Y - \Delta$ transformer. The neutral of the load is not connected to the neutral of the system. The positive- and negative-sequence currents flowing toward the resistive load are given by

$$I_{a,1} = 1/\underline{4.5°} \text{ per unit;} \quad I_{a,2} = 0.25/\underline{250°} \text{ per unit}$$

and the corresponding voltages on the low-voltage Y-side of the transformer are

$$V_{an,1} = 1/\underline{45°} \text{ per unit (Line-to-neutral voltage base)}$$

$$V_{an,2} = 0.25/\underline{250°} \text{ per unit (Line-to-neutral voltage base)}$$

Determine the line-to-line voltages and the line currents in per unit on the high-voltage side of the transformer. Account for the phase shift.

SECTION 8.7

8.43 Draw the positive-, negative-, and zero-sequence circuits for the transformers shown in Figure 3.34. Include ideal phase-shifting transformers showing phase shifts determined in Problem 3.32. Assume that all windings have the same kVA rating and that the equivalent leakage reactance of any two windings with the third winding open is 0.10 per unit. Neglect the exciting admittance.

8.44 A single-phase three-winding transformer has the following parameters: $Z_1 = Z_2 = Z_3 = 0 + j0.05$, $G_c = 0$, and $B_m = 0.2$ per unit. Three identical transformers, as described, are connected with their primaries in Y (solidly grounded neutral) and with their secondaries and tertiaries in Δ. Draw the per-unit sequence networks of this transformer bank.

SECTION 8.8

8.45 For Problem 8.14, calculate the real and reactive power delivered to the three-phase load.

8.46 A three-phase impedance load consists of a balanced-Δ load in parallel with a balanced-Y load. The impedance of each leg of the Δ load is $Z_\Delta = 6 + j6 \ \Omega$, and the impedance of each leg of the Y load is $Z_Y = 2 + j2 \ \Omega$. The Y load is grounded through a neutral impedance $Z_n = j1 \ \Omega$. Unbalanced line-to-ground source voltages V_{ag}, V_{bg}, and V_{cg} with sequence components $V_0 = 10\underline{/60°}$, $V_1 = 100\underline{/0°}$, and $V_2 = 15\underline{/200°}$ volts are applied to the load. (a) Draw the zero-, positive-, and negative-sequence networks. (b) Determine the complex power delivered to each sequence network. (c) Determine the total complex power delivered to the three-phase load.

8.47 For Problem 8.12, compute the power absorbed by the load using symmetrical components. Then verify the answer by computing directly without using symmetrical components.

8.48 For Problem 8.25, determine the complex power delivered to the load in terms of symmetrical components. Verify the answer by adding up the complex power of each of the three phases.

8.49 Using the voltages of Problem 8.6(a) and the currents of Problem 8.5, compute the complex power dissipated based on (a) phase components, and (b) symmetrical components.

CASE STUDY QUESTIONS

A. What are the advantages of energy storage in electric utility systems?

B. At what locations in a power system is battery energy storage most effective?

REFERENCES

1. Westinghouse Electric Corporation, *Applied Protective Relaying* (Newark, NJ: Westinghouse, 1976).

2. P. M. Anderson, *Analysis of Faulted Power Systems* (Ames, IA: Iowa State University Press, 1973).

3. W. D. Stevenson, Jr., *Elements of Power System Analysis*, 4th ed. (New York: McGraw-Hill, 1982).

4. J. Baker, "Electrical Energy Storage—Challenges and New Market Opportunities," *Energy World, 324* (November/December 2004), pp. 10–12.

9

UNSYMMETRICAL FAULTS

Short circuits occur in three-phase power systems as follows, in order of frequency of occurrence: single line-to-ground, line-to-line, double line-to-ground, and balanced three-phase faults. The path of the fault current may have either either zero impedance, which is called a *bolted* short circuit, or nonzero impedance. Other types of faults include one-conductor-open and two-conductors-open, which can occur when conductors break or when one or two phases of a circuit breaker inadvertently open.

Although the three-phase short circuit occurs the least, we considered it first, in Chapter 7, because of its simplicity. When a balanced three-phase fault occurs in a balanced three-phase system, there is only positive-sequence fault current; the zero-, positive-, and negative-sequence networks are completely uncoupled.

When an unsymmetrical fault occurs in an otherwise balanced system, the sequence networks are interconnected only at the fault location. As such, the computation of fault currents is greatly simplified by the use of sequence networks.

As in the case of balanced three-phase faults, unsymmetrical faults have two components of fault current: an ac or symmetrical component—including subtransient, transient, and steady-state currents—and a dc component. The simplified E/X method for breaker selection described in Section 7.5 is also applicable to unsymmetrical faults. The dc offset current need not be considered unless it is too large—for example, when the X/R ratio is too large.

We begin this chapter by using the per-unit zero-, positive-, and negative-sequence networks to represent a three-phase system. Also, we make certain assumptions to simplify fault-current calculations, and briefly review the balanced three-phase fault. We present single line-to-ground, line-to-line, and double line-to-ground faults in Sections 9.2, 9.3, and 9.4. The use of the positive-sequence bus impedance matrix for three-phase fault calculations in Section 7.4 is extended in Section 9.5 to unsymmetrical fault calculations by considering a bus impedance matrix for each sequence network. Examples using PowerWorld Simulator, which is based on the use of bus impedance matrices, are also included. The PowerWorld Simulator computes symmetrical fault currents for both three-phase and unsymmetrical faults. The Simulator may be used in power system design to select, set, and coordinate protective equipment.

CASE STUDY   When short circuits are not interrupted promptly, electrical fires and explosions can occur. To minimize the probability of electrical fire and explosion, the following are recommended:

> Careful design of electric power system layouts
> Quality equipment installation
> Power system protection that provides rapid detection and isolation of faults (see Chapter 10)
> Automatic fire-suppression systems
> Formal maintenance programs and inspection intervals
> Repair or retirement of damaged or decrepit equipment

The following article describes incidents at three U.S. utilities during the summer of 1990 [8].

Fires at U.S. Utilities

GLENN ZORPETTE

Electrical fires in substations were the cause of three major midsummer power outages in the

("Fires at U.S. Utilities" by Glenn Zorpette, © 1991 IEEE. Reprinted, with permission, from IEEE Spectrum, 28, 1 (Jan. 1991), p. 64.)

United States, two on Chicago's West Side and one in New York City's downtown financial district. In Chicago, the trouble began Saturday night, July 28, with a fire in switch house No. 1 at the Commonwealth Edison Co.'s Crawford substation, according to spokesman Gary Wald.

Some 40,000 residents of Chicago's West Side lost electricity. About 25,000 had service restored within a day or so and the rest, within three days. However, as part of the restoration, Commonwealth Edison installed a temporary line configuration around the Crawford substation. But when a second fire broke out on Aug. 5 in a different, nearby substation, some of the protective systems that would have isolated that fire were inoperable because of that configuration. Thus, what would have been a minor mishap resulted in a one-day loss of power to 25,000 customers—the same 25,000 whose electricity was restored first after the Crawford fire.

The New York outage began around midday on Aug. 13, after an electrical fire broke out in switching equipment at Consolidated Edison's Seaport substation, a point of entry into Manhattan for five 138-kilovolt transmission lines. To interrupt the flow of energy to the fire, Edison had to disconnect the five lines, which cut power to four networks in downtown Manhattan, according to Con Ed spokeswoman Martha Liipfert.

Power was restored to three of the networks within about five hours, but the fourth network, Fulton—which carried electricity to about 2400 separate residences and 815 businesses—was out until Aug. 21. Liipfert said much of the equipment in the Seaport substation will have to be replaced, at an estimated cost of about $25 million.

Mounting concern about underground electrical vaults in some areas was tragically validated by an explosion in Pasadena, Calif., that killed three city workers in a vault. Partly in response to the explosion, the California Public Utilities Commission adopted new regulations last Nov. 21 requiring that utilities in the state set up formal maintenance programs, inspection intervals, and guidelines for rejecting decrepit or inferior equipment. "They have to maintain a paper trail, and we as a commission will do inspections of underground vaults and review their records to make sure they're maintaining their vaults and equipment in good order," said Russ Copeland, head of the commission's utility safety branch.

9.1

SYSTEM REPRESENTATION

A three-phase power system is represented by its sequence networks in this chapter. The zero-, positive-, and negative-sequence networks of system components—generators, motors, transformers, and transmission lines—as developed in Chapter 8 can be used to construct system zero-, positive-, and negative-sequence networks. We make the following assumptions:

1. The power system operates under balanced steady-state conditions before the fault occurs. Thus the zero-, positive-, and negative-sequence networks are uncoupled before the fault occurs. During unsymmetrical faults they are interconnected only at the fault location.

2. Prefault load current is neglected. Because of this, the positive-sequence internal voltages of all machines are equal to the prefault voltage V_F. Therefore, the prefault voltage at each bus in the positive-sequence network equals V_F.

3. Transformer winding resistances and shunt admittances are neglected.

4. Transmission-line series resistances and shunt admittances are neglected.

5. Synchronous machine armature resistance, saliency, and saturation are neglected.

6. All nonrotating impedance loads are neglected.

7. Induction motors are either neglected (especially for motors rated 50 hp or less) or represented in the same manner as synchronous machines.

Note that these assumptions are made for simplicity in this text, and in practice should not be made for all cases. For example, in primary and secondary distribution systems, prefault currents may be in some cases comparable to short-circuit currents, and in other cases line resistances may significantly reduce fault currents.

Although fault currents as well as contributions to fault currents on the fault side of Δ–Y transformers are not affected by Δ–Y phase shifts, contributions to the fault from the other side of such transformers are affected by Δ–Y phase shifts for unsymmetrical faults. Therefore, we include Δ–Y phase-shift effects in this chapter.

We consider faults at the general three-phase bus shown in Figure 9.1. Terminals *abc*, denoted the *fault terminals*, are brought out in order to make external connections that represent faults. Before a fault occurs, the currents $I_a, I_b,$ and I_c are zero.

Figure 9.2(a) shows general sequence networks as viewed from the fault terminals. Since the prefault system is balanced, these zero-, positive-, and negative-sequence networks are uncoupled. Also, the sequence components of the fault currents, $I_0, I_1,$ and I_2, are zero before a fault occurs. The general sequence networks in Figure 9.2(a) are reduced to their Thévenin equivalents as viewed from the fault terminals in Figure 9.2(b). Each sequence network has a Thévenin equivalent impedance. Also, the positive-sequence network has a Thévenin equivalent voltage source, which equals the prefault voltage V_F.

FIGURE 9.1

General three-phase bus

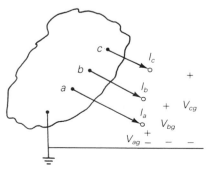

FIGURE 9.2

Sequence networks at a
general three-phase bus
in a balanced system

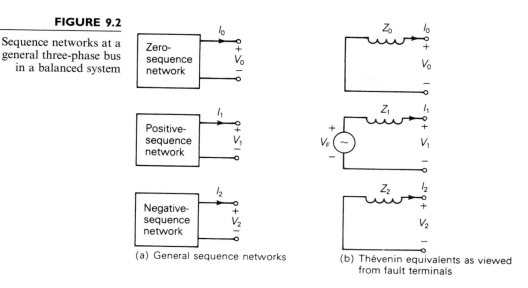

(a) General sequence networks

(b) Thévenin equivalents as viewed
from fault terminals

EXAMPLE 9.1   Power-system sequence networks and their Thévenin equivalents

A single-line diagram of the power system considered in Example 7.3 is
shown in Figure 9.3, where negative- and zero-sequence reactances are also
given. The neutrals of the generator and Δ–Y transformers are solidly
grounded. The motor neutral is grounded through a reactance $X_n = 0.05$ per
unit on the motor base. (a) Draw the per-unit zero-, positive-, and negative-
sequence networks on a 100-MVA, 13.8-kV base in the zone of the generator.
(b) Reduce the sequence networks to their Thévenin equivalents, as viewed
from bus 2. Prefault voltage is $V_F = 1.05\underline{/0^\circ}$ per unit. Prefault load current
and Δ–Y transformer phase shift are neglected.

FIGURE 9.3

Single-line diagram for
Example 9.1

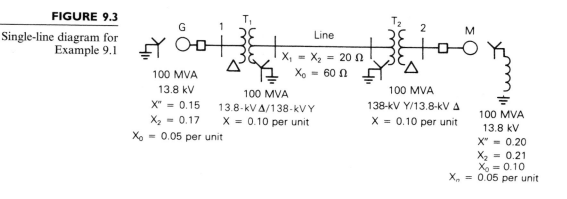

SOLUTION

a. The sequence networks are shown in Figure 9.4. The positive-sequence network is the same as that shown in Figure 7.4(a). The negative-sequence network is similar to the positive-sequence network, except that there are no sources, and negative-sequence machine reactances are shown. Δ–Y phase shifts are omitted from the positive- and negative-sequence networks for this example. In the zero-sequence network the zero-sequence generator, motor, and transmission-line reactances are shown. Since the motor neutral is grounded through a neutral reactance X_n, $3X_n$ is included in the zero-sequence motor circuit. Also, the zero-sequence Δ–Y transformer models are taken from Figure 8.19.

b. Figure 9.5 shows the sequence networks reduced to their Thévenin equivalents, as viewed from bus 2. For the positive-sequence equivalent, the Thévenin voltage source is the prefault voltage $V_F = 1.05\underline{/0°}$ per unit.

FIGURE 9.4

Sequence networks for Example 9.1

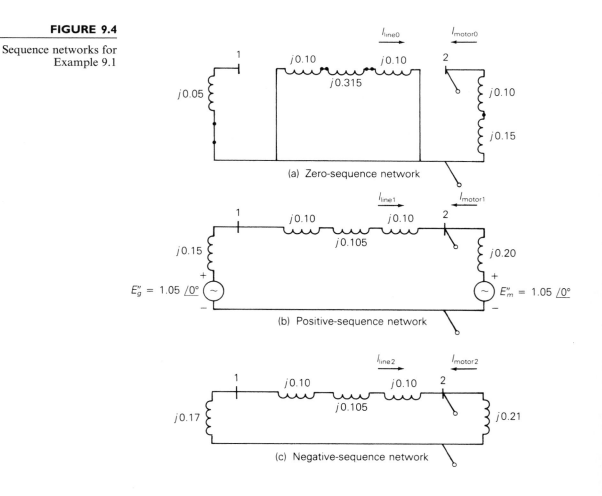

(a) Zero-sequence network

(b) Positive-sequence network

(c) Negative-sequence network

FIGURE 9.5

Thévenin equivalents of
sequence networks for
Example 9.1

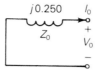

(a) Zero-sequence network

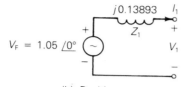

(b) Positive-sequence network

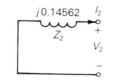

(c) Negative-sequence network

From Figure 9.4, the positive-sequence Thévenin impedance at bus 2 is the motor impedance $j0.20$, as seen to the right of bus 2, in parallel with $j(0.15 + 0.10 + 0.105 + 0.10) = j0.455$, as seen to the left; the parallel combination is $j0.20 // j0.455 = j0.13893$ per unit. Similarly, the negative-sequence Thévenin impedance is $j0.21 // j(0.17 + 0.10 + 0.105 + 0.10) = j0.21 // j0.475 = j0.14562$ per unit. In the zero-sequence network of Figure 9.4, the Thévenin impedance at bus 2 consists only of $j(0.10 + 0.15) = j0.25$ per unit, as seen to the right of bus 2; due to the Δ connection of transformer T_2, the zero-sequence network looking to the left of bus 2 is open. ∎

Recall that for three-phase faults, as considered in Chapter 7, the fault currents are balanced and have only a positive-sequence component. Therefore we work only with the positive-sequence network when calculating three-phase fault currents.

EXAMPLE 9.2   Three-phase short-circuit calculations using sequence networks

Calculate the per-unit subtransient fault currents in phases a, b, and c for a bolted three-phase-to-ground short circuit at bus 2 in Example 9.1.

SOLUTION   The terminals of the positive-sequence network in Figure 9.5(b) are shorted, as shown in Figure 9.6. The positive-sequence fault current is

FIGURE 9.6

Example 9.2: Bolted three-phase-to-ground fault at bus 2

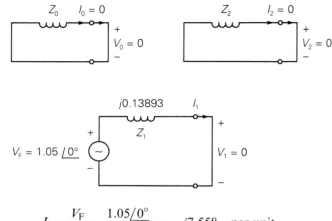

$$I_1 = \frac{V_F}{Z_1} = \frac{1.05\underline{/0°}}{j0.13893} = -j7.558 \quad \text{per unit}$$

which is the same result as obtained in part (c) of Example 7.4. Note that since subtransient machine reactances are used in Figures 9.4–9.6, the current calculated above is the positive-sequence subtransient fault current at bus 2. Also, the zero-sequence current I_0 and negative-sequence current I_2 are both zero. Therefore, the subtransient fault currents in each phase are, from (8.1.16),

$$\begin{bmatrix} I_a'' \\ I_b'' \\ I_c'' \end{bmatrix} = \begin{bmatrix} 1 & 1 & 1 \\ 1 & a^2 & a \\ 1 & a & a^2 \end{bmatrix} \begin{bmatrix} 0 \\ -j7.558 \\ 0 \end{bmatrix} = \begin{bmatrix} 7.558\underline{/-90°} \\ 7.558\underline{/150°} \\ 7.558\underline{/30°} \end{bmatrix} \quad \text{per unit} \qquad \blacksquare$$

The sequence components of the line-to-ground voltages at the fault terminals are, from Figure 9.2(b),

$$\begin{bmatrix} V_0 \\ V_1 \\ V_2 \end{bmatrix} = \begin{bmatrix} 0 \\ V_F \\ 0 \end{bmatrix} - \begin{bmatrix} Z_0 & 0 & 0 \\ 0 & Z_1 & 0 \\ 0 & 0 & Z_2 \end{bmatrix} \begin{bmatrix} I_0 \\ I_1 \\ I_2 \end{bmatrix} \qquad (9.1.1)$$

During a bolted three-phase fault, the sequence fault currents are $I_0 = I_2 = 0$ and $I_1 = V_F/Z_1$; therefore, from (9.1.1), the sequence fault voltages are $V_0 = V_1 = V_2 = 0$, which must be true since $V_{ag} = V_{bg} = V_{cg} = 0$. However, fault voltages need not be zero during unsymmetrical faults, which we consider next.

9.2

SINGLE LINE-TO-GROUND FAULT

Consider a single line-to-ground fault from phase a to ground at the general three-phase bus shown in Figure 9.7(a). For generality, we include a fault

FIGURE 9.7

Single line-to-ground
fault

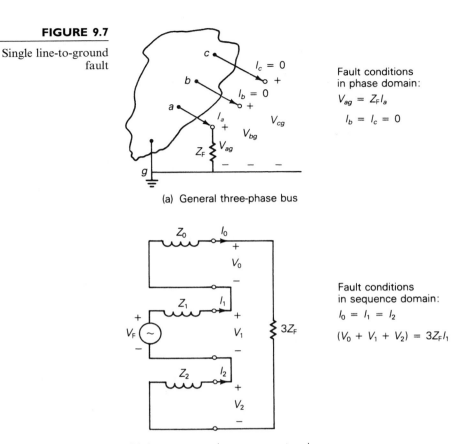

(a) General three-phase bus

Fault conditions
in phase domain:

$V_{ag} = Z_F I_a$

$I_b = I_c = 0$

(b) Interconnected sequence networks

Fault conditions
in sequence domain:

$I_0 = I_1 = I_2$

$(V_0 + V_1 + V_2) = 3Z_F I_1$

impedance Z_F. In the case of a bolted fault, $Z_F = 0$, whereas for an arcing fault, Z_F is the arc impedance. In the case of a transmission-line insulator flashover, Z_F includes the total fault impedance between the line and ground, including the impedances of the arc and the transmission tower, as well as the tower footing if there are no neutral wires.

The relations to be derived here apply only to a single line-to-ground fault on phase a. However, since any of the three phases can be arbitrarily labeled phase a, we do not consider single line-to-ground faults on other phases.

From Figure 9.7(a):

Fault conditions in phase domain ⎫ $I_b = I_c = 0$ (9.2.1)

Single line-to-ground fault   ⎭ $V_{ag} = Z_F I_a$ (9.2.2)

We now transform (9.2.1) and (9.2.2) to the sequence domain. Using (9.2.1) in (8.1.19),

$$\begin{bmatrix} I_0 \\ I_1 \\ I_2 \end{bmatrix} = \frac{1}{3} \begin{bmatrix} 1 & 1 & 1 \\ 1 & a & a^2 \\ 1 & a^2 & a \end{bmatrix} \begin{bmatrix} I_a \\ 0 \\ 0 \end{bmatrix} = \frac{1}{3} \begin{bmatrix} I_a \\ I_a \\ I_a \end{bmatrix} \qquad (9.2.3)$$

Also, using (8.1.3) and (8.1.20) in (9.2.2),

$$(V_0 + V_1 + V_2) = Z_F(I_0 + I_1 + I_2) \qquad (9.2.4)$$

From (9.2.3) and (9.2.4):

$$\left. \begin{array}{l} \text{Fault conditions in sequence domain} \\ \text{Single line-to-ground fault} \end{array} \right\} \begin{array}{ll} I_0 = I_1 = I_2 & (9.2.5) \\ (V_0 + V_1 + V_2) = (3Z_F)I_1 & \end{array}$$

$$(9.2.6)$$

Equations (9.2.5) and (9.2.6) can be satisfied by interconnecting the sequence networks in series at the fault terminals through the impedance $(3Z_F)$, as shown in Figure 9.7(b). From this figure, the sequence components of the fault currents are:

$$I_0 = I_1 = I_2 = \frac{V_F}{Z_0 + Z_1 + Z_2 + (3Z_F)} \qquad (9.2.7)$$

Transforming (9.2.7) to the phase domain via (8.1.20),

$$I_a = I_0 + I_1 + I_2 = 3I_1 = \frac{3V_F}{Z_0 + Z_1 + Z_2 + (3Z_F)} \qquad (9.2.8)$$

Note also from (8.1.21) and (8.1.22),

$$I_b = (I_0 + a^2 I_1 + a I_2) = (1 + a^2 + a)I_1 = 0 \qquad (9.2.9)$$

$$I_c = (I_0 + a I_1 + a^2 I_2) = (1 + a + a^2)I_1 = 0 \qquad (9.2.10)$$

These are obvious, since the single line-to-ground fault is on phase a, not phase b or c.

The sequence components of the line-to-ground voltages at the fault are determined from (9.1.1). The line-to-ground voltages at the fault can then be obtained by transforming the sequence voltages to the phase domain.

EXAMPLE 9.3   **Single line-to-ground short-circuit calculations using sequence networks**

Calculate the subtransient fault current in per-unit and in kA for a bolted single line-to-ground short circuit from phase a to ground at bus 2 in Example 9.1. Also calculate the per-unit line-to-ground voltages at faulted bus 2.

SOLUTION   The zero-, positive-, and negative-sequence networks in Figure 9.5 are connected in series at the fault terminals, as shown in Figure 9.8.

FIGURE 9.8

Example 9.3: Single line-to-ground fault at bus 2

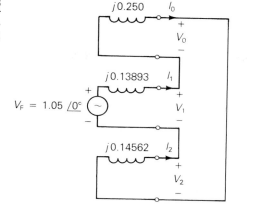

Since the short circuit is bolted, $Z_F = 0$. From (9.2.7), the sequence currents are:

$$I_0 = I_1 = I_2 = \frac{1.05/0°}{j(0.25 + 0.13893 + 0.14562)}$$

$$= \frac{1.05}{j0.53455} = -j1.96427 \quad \text{per unit}$$

From (9.2.8), the subtransient fault current is

$$I_a'' = 3(-j1.96427) = -j5.8928 \quad \text{per unit}$$

The base current at bus 2 is $100/(13.8\sqrt{3}) = 4.1837$ kA. Therefore,

$$I_a'' = (-j5.8928)(4.1837) = 24.65/\underline{-90°} \quad \text{kA}$$

From (9.1.1), the sequence components of the voltages at the fault are

$$\begin{bmatrix} V_0 \\ V_1 \\ V_2 \end{bmatrix} = \begin{bmatrix} 0 \\ 1.05/0° \\ 0 \end{bmatrix} - \begin{bmatrix} j0.25 & 0 & 0 \\ 0 & j0.13893 & 0 \\ 0 & 0 & j0.14562 \end{bmatrix} \begin{bmatrix} -j1.96427 \\ -j1.96427 \\ -j1.96427 \end{bmatrix}$$

$$= \begin{bmatrix} -0.49107 \\ 0.77710 \\ -0.28604 \end{bmatrix} \quad \text{per unit}$$

Transforming to the phase domain, the line-to-ground voltages at faulted bus 2 are

$$
\begin{bmatrix} V_{ag} \\ V_{bg} \\ V_{cg} \end{bmatrix} = \begin{bmatrix} 1 & 1 & 1 \\ 1 & a^2 & a \\ 1 & a & a^2 \end{bmatrix} \begin{bmatrix} -0.49107 \\ 0.77710 \\ -0.28604 \end{bmatrix} = \begin{bmatrix} 0 \\ 1.179\underline{/231.3°} \\ 1.179\underline{/128.7°} \end{bmatrix} \quad \text{per unit}
$$

Note that $V_{ag} = 0$, as specified by the fault conditions. Also $I_b'' = I_c'' = 0$.

Open PowerWorld Simulator case Example 9_3 to see an animated view of this example. The process for simulating an unsymmetrical fault is almost identical to that for a balanced fault. That is, from the one-line, first right-click on the bus symbol corresponding to the fault location. This displays the local menu. Select "Fault.." to display the Fault dialog. Verify that the correct bus is selected, and then set the Fault Type field to "Single Line-to-Ground." Finally, click on **Calculate** to determine the fault currents and voltages. The results are shown in the tables at the bottom of the dialog. Notice that with an unsymmetrical fault the phase magnitudes are no longer identical. The values can be animated on the one-line by changing the One-line Display field value.

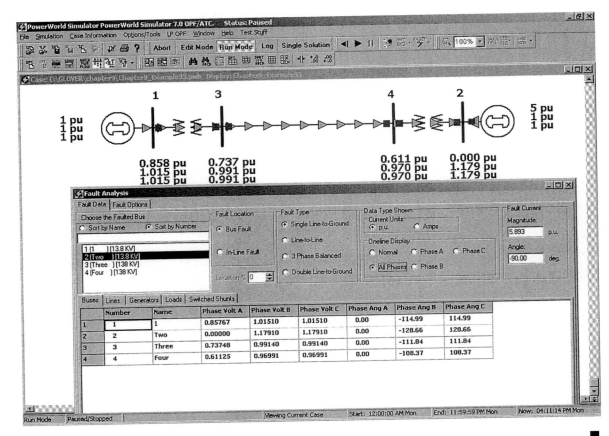

Screen for Example 9.3

9.3

LINE-TO-LINE FAULT

Consider a line-to-line fault from phase b to c, shown in Figure 9.9(a). Again, we include a fault impedance Z_F for generality. From Figure 9.9(a):

$$\left.\begin{array}{l}\text{Fault conditions in phase domain}\\\text{Line-to-line fault}\end{array}\right\} \quad \begin{array}{l} I_a = 0 \qquad\qquad (9.3.1)\\[6pt] I_c = -I_b \qquad\qquad (9.3.2)\\[6pt] V_{bg} - V_{cg} = Z_F I_b \qquad (9.3.3)\end{array}$$

We transform (9.3.1)–(9.3.3) to the sequence domain. Using (9.3.1) and (9.3.2) in (8.1.19),

FIGURE 9.9

Line-to-line fault

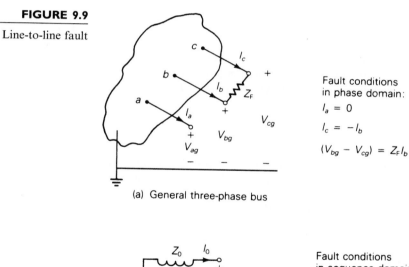

Fault conditions
in phase domain:

$I_a = 0$

$I_c = -I_b$

$(V_{bg} - V_{cg}) = Z_F I_b$

(a) General three-phase bus

Fault conditions
in sequence domain:

$I_0 = 0$

$I_2 = -I_1$

$(V_1 - V_2) = Z_F I_1$

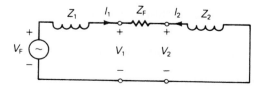

(b) Interconnected sequence networks

$$\begin{bmatrix} I_0 \\ I_1 \\ I_2 \end{bmatrix} = \frac{1}{3} \begin{bmatrix} 1 & 1 & 1 \\ 1 & a & a^2 \\ 1 & a^2 & a \end{bmatrix} \begin{bmatrix} 0 \\ I_b \\ -I_b \end{bmatrix} = \begin{bmatrix} 0 \\ \frac{1}{3}(a-a^2)I_b \\ \frac{1}{3}(a^2-a)I_b \end{bmatrix} \qquad (9.3.4)$$

Using (8.1.4), (8.1.5), and (8.1.21) in (9.3.3),

$$(V_0 + a^2 V_1 + a V_2) - (V_0 + a V_1 + a^2 V_2) = Z_F(I_0 + a^2 I_1 + a I_2) \quad (9.3.5)$$

Noting from (9.3.4) that $I_0 = 0$ and $I_2 = -I_1$, (9.3.5) simplifies to

$$(a^2 - a)V_1 - (a^2 - a)V_2 = Z_F(a^2 - a)I_1$$

or

$$V_1 - V_2 = Z_F I_1 \qquad (9.3.6)$$

Therefore, from (9.3.4) and (9.3.6):

Fault conditions in sequence domain $\Big\}$ $I_0 = 0$ (9.3.7)

Line-to-line fault $I_2 = -I_1$ (9.3.8)

$$V_1 - V_2 = Z_F I_1 \qquad (9.3.9)$$

Equations (9.3.7)–(9.3.9) are satisfied by connecting the positive- and negative-sequence networks in parallel at the fault terminals through the fault impedance Z_F, as shown in Figure 9.9(b). From this figure, the fault currents are:

$$I_1 = -I_2 = \frac{V_F}{(Z_1 + Z_2 + Z_F)} \qquad I_0 = 0 \qquad (9.3.10)$$

Transforming (9.3.10) to the phase domain and using the identity $(a^2 - a) = -j\sqrt{3}$, the fault current in phase b is

$$I_b = I_0 + a^2 I_1 + a I_2 = (a^2 - a)I_1$$

$$= -j\sqrt{3} I_1 = \frac{-j\sqrt{3} V_F}{(Z_1 + Z_2 + Z_F)} \qquad (9.3.11)$$

Note also from (8.1.20) and (8.1.22) that

$$I_a = I_0 + I_1 + I_2 = 0 \qquad (9.3.12)$$

and

$$I_c = I_0 + a I_1 + a^2 I_2 = (a - a^2)I_1 = -I_b \qquad (9.3.13)$$

which verify the fault conditions given by (9.3.1) and (9.3.2). The sequence components of the line-to-ground voltages at the fault are given by (9.1.1).

EXAMPLE 9.4   **Line-to-line short-circuit calculations using sequence networks**

Calculate the subtransient fault current in per-unit and in kA for a bolted line-to-line fault from phase b to c at bus 2 in Example 9.1.

FIGURE 9.10

Example 9.4: Line-to-line fault at bus 2

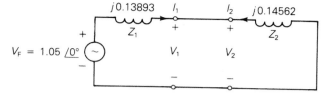

SOLUTION   The positive- and negative-sequence networks in Figure 9.5 are connected in parallel at the fault terminals, as shown in Figure 9.10. From (9.3.10) with $Z_F = 0$, the sequence fault currents are

$$I_1 = -I_2 = \frac{1.05\underline{/0°}}{j(0.13893 + 0.14562)} = 3.690\underline{/-90°}$$

$$I_0 = 0$$

From (9.3.11), the subtransient fault current in phase b is

$$I_b'' = (-j\sqrt{3})(3.690\underline{/-90°}) = -6.391 = 6.391\underline{/180°} \quad \text{per unit}$$

Using 4.1837 kA as the base current at bus 2,

$$I_b'' = (6.391\underline{/180°})(4.1837) = 26.74\underline{/180°} \quad \text{kA}$$

Also, from (9.3.12) and (9.3.13),

$$I_a'' = 0 \qquad I_c'' = 26.74\underline{/0°} \quad \text{kA}$$

The line-to-line fault results for this example can be shown in Power-World Simulator by repeating the Example 9.3 procedure, with the exception that the Fault Type field value should be "Line-to-Line." ∎

9.4

DOUBLE LINE-TO-GROUND FAULT

A double line-to-ground fault from phase b to phase c through fault impedance Z_F to ground is shown in Figure 9.11(a). From this figure:

$$\left.\begin{array}{l}\text{Fault conditions in the phase domain} \\ \text{Double line-to-ground fault}\end{array}\right\} \quad I_a = 0 \qquad (9.4.1)$$

$$V_{cg} = V_{bg} \qquad (9.4.2)$$

$$V_{bg} = Z_F(I_b + I_c) \qquad (9.4.3)$$

Transforming (9.4.1) to the sequence domain via (8.1.20),

$$I_0 + I_1 + I_2 = 0 \qquad (9.4.4)$$

Also, using (8.1.4) and (8.1.5) in (9.4.2),

$$(V_0 + aV_1 + a^2V_2) = (V_0 + a^2V_1 + aV_2)$$

FIGURE 9.11

Double line-to-ground
fault

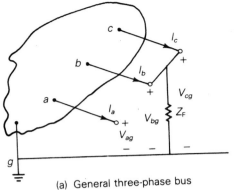

Fault conditions
in phase domain:

$I_a = 0$

$V_{bg} = V_{cg} = Z_F(I_b + I_c)$

(a) General three-phase bus

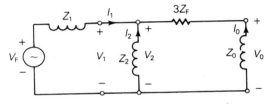

Fault conditions
in sequence domain:

$I_0 + I_1 + I_2 = 0$

$V_0 - V_1 = (3Z_F)I_0$

$V_1 = V_2$

(b) Interconnected sequence networks

Simplifying:

$$(a^2 - a)V_2 = (a^2 - a)V_1$$

or

$$V_2 = V_1 \tag{9.4.5}$$

Now, using (8.1.4), (8.1.21), and (8.1.22) in (9.4.3),

$$(V_0 + a^2 V_1 + a V_2) = Z_F(I_0 + a^2 I_1 + a I_2 + I_0 + a I_1 + a^2 I_2) \tag{9.4.6}$$

Using (9.4.5) and the identity $a^2 + a = -1$ in (9.4.6),

$$(V_0 - V_1) = Z_F(2I_0 - I_1 - I_2) \tag{9.4.7}$$

From (9.4.4), $I_0 = -(I_1 + I_2)$; therefore, (9.4.7) becomes

$$V_0 - V_1 = (3Z_F)I_0 \tag{9.4.8}$$

From (9.4.4), (9.4.5), and (9.4.8), we summarize:

Fault conditions in the sequence domain $\left.\right\}$ Double line-to-ground fault

$$I_0 + I_1 + I_2 = 0 \tag{9.4.9}$$

$$V_2 = V_1 \tag{9.4.10}$$

$$V_0 - V_1 = (3Z_F)I_0 \tag{9.4.11}$$

Equations (9.4.9)–(9.4.11) are satisfied by connecting the zero-, positive-, and negative-sequence networks in parallel at the fault terminal; additionally, $(3Z_F)$ is included in series with the zero-sequence network. This connection is shown in Figure 9.11(b). From this figure the positive-sequence fault current is

$$I_1 = \frac{V_F}{Z_1 + [Z_2 /\!/ (Z_0 + 3Z_F)]} = \frac{V_F}{Z_1 + \left[\dfrac{Z_2(Z_0 + 3Z_F)}{Z_2 + Z_0 + 3Z_F}\right]} \tag{9.4.12}$$

Using current division in Figure 9.11(b), the negative- and zero-sequence fault currents are

$$I_2 = (-I_1)\left(\frac{Z_0 + 3Z_F}{Z_0 + 3Z_F + Z_2}\right) \tag{9.4.13}$$

$$I_0 = (-I_1)\left(\frac{Z_2}{Z_0 + 3Z_F + Z_2}\right) \tag{9.4.14}$$

These sequence fault currents can be transformed to the phase domain via (8.1.16). Also, the sequence components of the line-to-ground voltages at the fault are given by (9.1.1).

EXAMPLE 9.5   Double line-to-ground short-circuit calculations using sequence networks

Calculate (a) the subtransient fault current in each phase, (b) neutral fault current, and (c) contributions to the fault current from the motor and from the transmission line, for a bolted double line-to-ground fault from phase b to c to ground at bus 2 in Example 9.1. Neglect the Δ–Y transformer phase shifts.

SOLUTION

a. The zero-, positive-, and negative-sequence networks in Figure 9.5 are connected in parallel at the fault terminals in Figure 9.12. From (9.4.12) with $Z_F = 0$,

FIGURE 9.12

Example 9.5: Double line-to-ground fault at bus 2

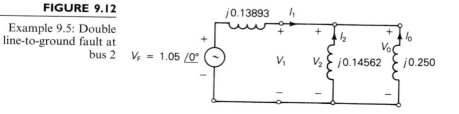

$$I_1 = \frac{1.05\underline{/0^\circ}}{j\left[0.13893 + \dfrac{(0.14562)(0.25)}{0.14562 + 0.25}\right]} = \frac{1.05\underline{/0^\circ}}{j0.23095}$$

$$= -j4.5464 \quad \text{per unit}$$

From (9.4.13) and (9.4.14),

$$I_2 = (+j4.5464)\left(\frac{0.25}{0.25 + 0.14562}\right) = j2.8730 \quad \text{per unit}$$

$$I_0 = (+j4.5464)\left(\frac{0.14562}{0.25 + 0.14562}\right) = j1.6734 \quad \text{per unit}$$

Transforming to the phase domain, the subtransient fault currents are:

$$\begin{bmatrix} I_a'' \\ I_b'' \\ I_c'' \end{bmatrix} = \begin{bmatrix} 1 & 1 & 1 \\ 1 & a^2 & a \\ 1 & a & a^2 \end{bmatrix} \begin{bmatrix} +j1.6734 \\ -j4.5464 \\ +j2.8730 \end{bmatrix} = \begin{bmatrix} 0 \\ 6.8983\underline{/158.66^\circ} \\ 6.8983\underline{/21.34^\circ} \end{bmatrix} \quad \text{per unit}$$

Using the base current of 4.1837 kA at bus 2,

$$\begin{bmatrix} I_a'' \\ I_b'' \\ I_c'' \end{bmatrix} = \begin{bmatrix} 0 \\ 6.8983\underline{/158.66^\circ} \\ 6.8983\underline{/21.34^\circ} \end{bmatrix} (4.1837) = \begin{bmatrix} 0 \\ 28.86\underline{/158.66^\circ} \\ 28.86\underline{/21.34^\circ} \end{bmatrix} \quad \text{kA}$$

b. The neutral fault current is

$$I_n = (I_b'' + I_c'') = 3I_0 = j5.0202 \quad \text{per unit}$$

$$= (j5.0202)(4.1837) = 21.00\underline{/90^\circ} \quad \text{kA}$$

c. Neglecting Δ–Y transformer phase shifts, the contributions to the fault current from the motor and transmission line can be obtained from Figure 9.4. From the zero-sequence network, Figure 9.4(a), the contribution to the zero-sequence fault current from the line is zero, due to the transformer connection. That is,

$$I_{\text{line } 0} = 0$$

$$I_{\text{motor } 0} = I_0 = j1.6734 \quad \text{per unit}$$

From the positive-sequence network, Figure 9.4(b), the positive terminals of the internal machine voltages can be connected, since $E_g'' = E_m''$. Then, by current division,

$$I_{\text{line } 1} = \frac{X_m''}{X_m'' + (X_g'' + X_{T1} + X_{\text{line } 1} + X_{T2})} I_1$$

$$= \frac{0.20}{0.20 + (0.455)}(-j4.5464) = -j1.3882 \quad \text{per unit}$$

$$I_{\text{motor } 1} = \frac{0.455}{0.20 + 0.455}(-j4.5464) = -j3.1582 \quad \text{per unit}$$

From the negative-sequence network, Figure 9.4(c), using current division,

$$I_{\text{line } 2} = \frac{0.21}{0.21 + 0.475}(j2.8730) = j0.8808 \quad \text{per unit}$$

$$I_{\text{motor } 2} = \frac{0.475}{0.21 + 0.475}(j2.8730) = j1.9922 \quad \text{per unit}$$

Transforming to the phase domain with base currents of 0.41837 kA for the line and 4.1837 kA for the motor,

$$\begin{bmatrix} I''_{\text{line } a} \\ I''_{\text{line } b} \\ I''_{\text{line } c} \end{bmatrix} = \begin{bmatrix} 1 & 1 & 1 \\ 1 & a^2 & a \\ 1 & a & a^2 \end{bmatrix} \begin{bmatrix} 0 \\ -j1.3882 \\ j0.8808 \end{bmatrix}$$

$$= \begin{bmatrix} 0.5074\underline{/-90^\circ} \\ 1.9813\underline{/172.643^\circ} \\ 1.9813\underline{/7.357^\circ} \end{bmatrix} \quad \text{per unit}$$

$$= \begin{bmatrix} 0.2123\underline{/-90^\circ} \\ 0.8289\underline{/172.643^\circ} \\ 0.8289\underline{/7.357^\circ} \end{bmatrix} \quad \text{kA}$$

$$\begin{bmatrix} I''_{\text{motor } a} \\ I''_{\text{motor } b} \\ I''_{\text{motor } c} \end{bmatrix} = \begin{bmatrix} 1 & 1 & 1 \\ 1 & a^2 & a \\ 1 & a & a^2 \end{bmatrix} \begin{bmatrix} j1.6734 \\ -j3.1582 \\ j1.9922 \end{bmatrix}$$

$$= \begin{bmatrix} 0.5074\underline{/90^\circ} \\ 4.9986\underline{/153.17^\circ} \\ 4.9986\underline{/26.83^\circ} \end{bmatrix} \quad \text{per unit}$$

$$= \begin{bmatrix} 2.123\underline{/90^\circ} \\ 20.91\underline{/153.17^\circ} \\ 20.91\underline{/26.83^\circ} \end{bmatrix} \quad \text{kA}$$

The double line-to-line fault results for this example can be shown in PowerWorld Simulator by repeating the Example 9.3 procedure, with the exception that the Fault Type field value should be "Double Line-to-Ground." ∎

EXAMPLE 9.6   Effect of Δ–Y transformer phase shift on fault currents

Rework Example 9.5, with the Δ–Y transformer phase shifts included. Assume American standard phase shift.

SOLUTION   The sequence networks of Figure 9.4 are redrawn in Figure 9.13 with ideal phase-shifting transformers representing Δ–Y phase shifts. In ac-

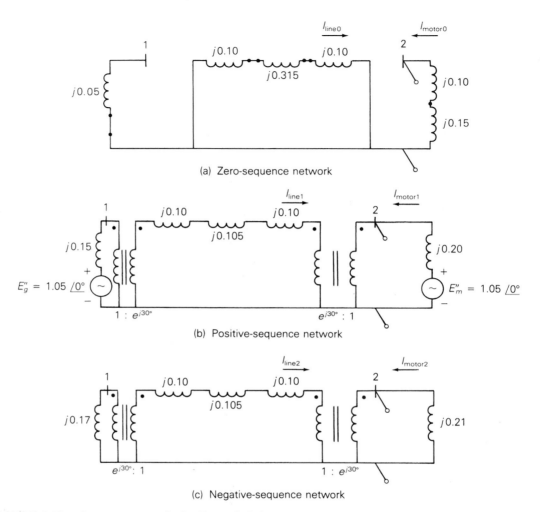

FIGURE 9.13   Sequence networks for Example 9.6

cordance with the American standard, positive-sequence quantities on the high-voltage side of the transformers lead their corresponding quantities on the low-voltage side by 30°. Also, the negative-sequence phase shifts are the reverse of the positive-sequence phase shifts.

a. Recall from Section 3.1 and (3.1.26) that per-unit impedance is unchanged when it is referred from one side of an ideal phase-shifting transformer to the other. Accordingly, the Thévenin equivalents of the sequence networks in Figure 9.13, as viewed from fault bus 2, are the same as those given in Figure 9.5. Therefore, the sequence components as well as the phase components of the fault currents are the same as those given in Example 9.5(a).

b. The neutral fault current is the same as that given in Example 9.5(b).

c. The zero-sequence network, Figure 9.13(a), is the same as that given in Figure 9.4(a). Therefore, the contributions to the zero-sequence fault current from the line and motor are the same as those given in Example 9.5(c).

$$I_{\text{line } 0} = 0 \qquad I_{\text{motor } 0} = I_0 = j1.6734 \quad \text{per unit}$$

The contribution to the positive-sequence fault current from the line in Figure 9.13(b) leads that in Figure 9.4(b) by 30°. That is,

$$I_{\text{line } 1} = (-j1.3882)(1\underline{/30°}) = 1.3882\underline{/-60°} \quad \text{per unit}$$

$$I_{\text{motor } 1} = -j3.1582 \quad \text{per unit}$$

Similarly, the contribution to the negative-sequence fault current from the line in Figure 9.13(c) lags that in Figure 9.4(c) by 30°. That is,

$$I_{\text{line } 2} = (j0.8808)(1\underline{/-30°}) = 0.8808\underline{/60°} \quad \text{per unit}$$

$$I_{\text{motor } 2} = j1.9922 \quad \text{per unit}$$

Thus, the sequence currents as well as the phase currents from the motor are the same as those given in Example 9.5(c). Also, the sequence currents from the line have the same magnitudes as those given in Example 9.5(c), but the positive- and negative-sequence line currents are shifted by +30° and −30°, respectively. Transforming the line currents to the phase domain:

$$
\begin{bmatrix} I''_{\text{line } a} \\ I''_{\text{line } b} \\ I''_{\text{line } c} \end{bmatrix} =
\begin{bmatrix} 1 & 1 & 1 \\ 1 & a^2 & a \\ 1 & a & a^2 \end{bmatrix}
\begin{bmatrix} 0 \\ 1.3882\underline{/-60°} \\ 0.8808\underline{/60°} \end{bmatrix}
$$

$$
= \begin{bmatrix} 1.2166\underline{/-21.17°} \\ 2.2690\underline{/180°} \\ 1.2166\underline{/21.17°} \end{bmatrix} \quad \text{per unit}
$$

$$
= \begin{bmatrix} 0.5090\underline{/-21.17°} \\ 0.9492\underline{/180°} \\ 0.5090\underline{/21.17°} \end{bmatrix} \quad \text{kA}
$$

In conclusion, Δ–Y transformer phase shifts have no effect on the fault currents and no effect on the contribution to the fault currents on the fault side of the Δ–Y transformers. However, on the other side of the Δ–Y transformers, the positive- and negative-sequence components of the contributions to the fault currents are shifted by ±30°, which affects both the magnitude as well as the angle of the phase components of these fault contributions for unsymmetrical faults. ∎

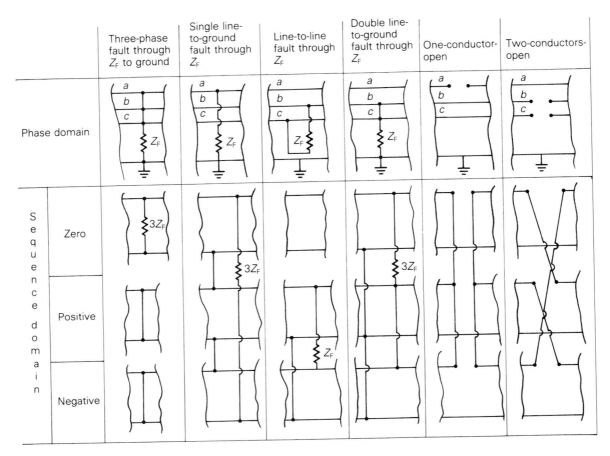

FIGURE 9.14 Summary of faults

Figure 9.14 summarizes the sequence network connections for both the balanced three-phase fault and the unsymmetrical faults that we have considered. Sequence network connections for two additional faults, one-conductor-open and two-conductors-open, are also shown in Figure 9.14 and are left as an exercise for you to verify (see Problems 9.26 and 9.27).

9.5

SEQUENCE BUS IMPEDANCE MATRICES

We use the positive-sequence bus impedance matrix in Section 7.4 for calculating currents and voltages during balanced three-phase faults. This method is extended here to unsymmetrical faults by representing each sequence network as a bus impedance equivalent circuit (or as a rake equivalent). A bus impedance matrix can be computed for each sequence network by inverting the corresponding bus admittance network. For simplicity, resistances, shunt admittances, nonrotating impedance loads, and prefault load currents are neglected.

Figure 9.15 shows the connection of sequence rake equivalents for both symmetrical and unsymmetrical faults at bus n of an N-bus three-phase power system. Each bus impedance element has an additional subscript, 0, 1, or 2, that identifies the sequence rake equivalent in which it is located. Mutual impedances are not shown in the figure. The prefault voltage V_F is in-

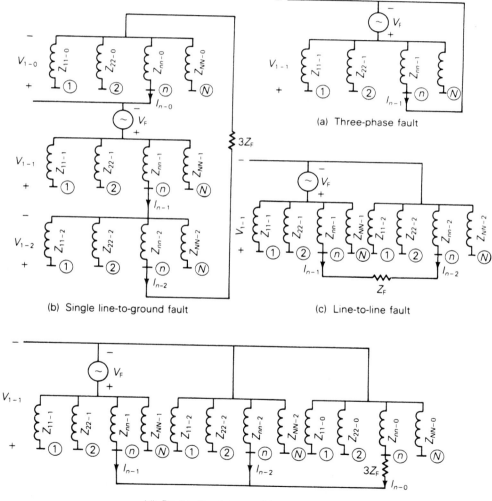

(a) Three-phase fault

(b) Single line-to-ground fault

(c) Line-to-line fault

(d) Double line-to-ground fault

FIGURE 9.15   Connection of rake equivalent sequence networks for three-phase system faults (mutual impedances not shown)

cluded in the positive-sequence rake equivalent. From the figure the sequence components of the fault current for each type of fault at bus n are as follows:

Balanced three-phase fault:

$$I_{n-1} = \frac{V_F}{Z_{nn-1}} \tag{9.5.1}$$

$$I_{n-0} = I_{n-2} = 0 \tag{9.5.2}$$

Single line-to-ground fault (phase a to ground):

$$I_{n-0} = I_{n-1} = I_{n-2} = \frac{V_F}{Z_{nn-0} + Z_{nn-1} + Z_{nn-2} + 3Z_F} \tag{9.5.3}$$

Line-to-line fault (phase b to c):

$$I_{n-1} = -I_{n-2} = \frac{V_F}{Z_{nn-1} + Z_{nn-2} + Z_F} \tag{9.5.4}$$

$$I_{n-0} = 0 \tag{9.5.5}$$

Double line-to-ground fault (phase b to c to ground):

$$I_{n-1} = \frac{V_F}{Z_{nn-1} + \left[\dfrac{Z_{nn-2}(Z_{nn-0} + 3Z_F)}{Z_{nn-2} + Z_{nn-0} + 3Z_F} \right]} \tag{9.5.6}$$

$$I_{n-2} = (-I_{n-1}) \left(\frac{Z_{nn-0} + 3Z_F}{Z_{nn-0} + 3Z_F + Z_{nn-2}} \right) \tag{9.5.7}$$

$$I_{n-0} = (-I_{n-1}) \left(\frac{Z_{nn-2}}{Z_{nn-0} + 3Z_F + Z_{nn-2}} \right) \tag{9.5.8}$$

Also from Figure 9.15, the sequence components of the line-to-ground voltages at any bus k during a fault at bus n are:

$$\begin{bmatrix} V_{k-0} \\ V_{k-1} \\ V_{k-2} \end{bmatrix} = \begin{bmatrix} 0 \\ V_F \\ 0 \end{bmatrix} - \begin{bmatrix} Z_{kn-0} & 0 & 0 \\ 0 & Z_{kn-1} & 0 \\ 0 & 0 & Z_{kn-2} \end{bmatrix} \begin{bmatrix} I_{n-0} \\ I_{n-1} \\ I_{n-2} \end{bmatrix} \tag{9.5.9}$$

If bus k is on the unfaulted side of a Δ–Y transformer, then the phase angles of V_{k-1} and V_{k-2} in (9.5.9) are modified to account for Δ–Y phase shifts. Also, the above sequence fault currents and sequence voltages can be transformed to the phase domain via (8.1.16) and (8.1.9).

EXAMPLE 9.7   **Single line-to-ground short-circuit calculations using $Z_{bus\ 0}$, $Z_{bus\ 1}$, and $Z_{bus\ 2}$**

Faults at buses 1 and 2 for the three-phase power system given in Example 9.1 are of interest. The prefault voltage is 1.05 per unit. Prefault load cur-

rent is neglected. (a) Determine the per-unit zero-, positive-, and negative-sequence bus impedance matrices. Find the subtransient fault current in per-unit for a bolted single line-to-ground fault current from phase a to ground (b) at bus 1 and (c) at bus 2. Find the per-unit line-to-ground voltages at (d) bus 1 and (e) bus 2 during the single line-to-ground fault at bus 1.

SOLUTION

a. Referring to Figure 9.4(a), the zero-sequence bus admittance matrix is

$$Y_{\text{bus }0} = -j \left[\begin{array}{c|c} 20 & 0 \\ \hline 0 & 4 \end{array} \right] \quad \text{per unit}$$

Inverting $Y_{\text{bus }0}$,

$$Z_{\text{bus }0} = j \left[\begin{array}{c|c} 0.05 & 0 \\ \hline 0 & 0.25 \end{array} \right] \quad \text{per unit}$$

Note that the transformer leakage reactances and the zero-sequence transmission-line reactance in Figure 9.4(a) have no effect on $Z_{\text{bus }0}$. The transformer Δ connections block the flow of zero-sequence current from the transformers to bus 1 and 2.

The positive-sequence bus admittance matrix, from Figure 9.4(b), is

$$Y_{\text{bus }1} = -j \left[\begin{array}{c|c} 9.9454 & -3.2787 \\ \hline -3.2787 & 8.2787 \end{array} \right] \quad \text{per unit}$$

Inverting $Y_{\text{bus }1}$,

$$Z_{\text{bus }1} = j \left[\begin{array}{c|c} 0.11565 & 0.04580 \\ \hline 0.04580 & 0.13893 \end{array} \right] \quad \text{per unit}$$

Similarly, from Figure 9.4(c)

$$Y_{\text{bus }2} = -j \left[\begin{array}{c|c} 9.1611 & -3.2787 \\ \hline -3.2787 & 8.0406 \end{array} \right]$$

Inverting $Y_{\text{bus }2}$,

$$Z_{\text{bus }2} = j \left[\begin{array}{c|c} 0.12781 & 0.05212 \\ \hline 0.05212 & 0.14562 \end{array} \right] \quad \text{per unit}$$

b. From (9.5.3), with $n = 1$ and $Z_F = 0$, the sequence fault currents are

$$I_{1-0} = I_{1-1} = I_{1-2} = \frac{V_F}{Z_{11-0} + Z_{11-1} + Z_{11-2}}$$

$$= \frac{1.05\underline{/0°}}{j(0.05 + 0.11565 + 0.12781)} = \frac{1.05}{j0.29346} = -j3.578 \quad \text{per unit}$$

The subtransient fault currents at bus 1 are, from (8.1.16),

$$\begin{bmatrix} I_{1a}'' \\ I_{1b}'' \\ I_{1c}'' \end{bmatrix} = \begin{bmatrix} 1 & 1 & 1 \\ 1 & a^2 & a \\ 1 & a & a^2 \end{bmatrix} \begin{bmatrix} -j3.578 \\ -j3.578 \\ -j3.578 \end{bmatrix} = \begin{bmatrix} -j10.73 \\ 0 \\ 0 \end{bmatrix} \quad \text{per unit}$$

c. Again from (9.5.3), with $n = 2$ and $Z_F = 0$,

$$I_{2-0} = I_{2-1} = I_{2-2} = \frac{V_F}{Z_{22-0} + Z_{22-1} + Z_{22-2}}$$

$$= \frac{1.05 \underline{/0^\circ}}{j(0.25 + 0.13893 + 0.14562)} = \frac{1.05}{j0.53455}$$

$$= -j1.96427 \quad \text{per unit}$$

and

$$\begin{bmatrix} I_{2a}'' \\ I_{2b}'' \\ I_{2c}'' \end{bmatrix} = \begin{bmatrix} 1 & 1 & 1 \\ 1 & a^2 & a \\ 1 & a & a^2 \end{bmatrix} \begin{bmatrix} -j1.96427 \\ -j1.96427 \\ -j1.96427 \end{bmatrix} = \begin{bmatrix} -j5.8928 \\ 0 \\ 0 \end{bmatrix} \quad \text{per unit}$$

This is the same result as obtained in Example 9.3.

d. The sequence components of the line-to-ground voltages at bus 1 during the fault at bus 1 are, from (9.5.9), with $k = 1$ and $n = 1$,

$$\begin{bmatrix} V_{1-0} \\ V_{1-1} \\ V_{1-2} \end{bmatrix} = \begin{bmatrix} 0 \\ 1.05 \underline{/0^\circ} \\ 0 \end{bmatrix} - \begin{bmatrix} j0.05 & 0 & 0 \\ 0 & j0.11565 & 0 \\ 0 & 0 & j0.12781 \end{bmatrix} \begin{bmatrix} -j3.578 \\ -j3.578 \\ -j3.578 \end{bmatrix}$$

$$= \begin{bmatrix} -0.1789 \\ 0.6362 \\ -0.4573 \end{bmatrix} \quad \text{per unit}$$

and the line-to-ground voltages at bus 1 during the fault at bus 1 are

$$\begin{bmatrix} V_{1-ag} \\ V_{1-bg} \\ V_{1-cg} \end{bmatrix} = \begin{bmatrix} 1 & 1 & 1 \\ 1 & a^2 & a \\ 1 & a & a^2 \end{bmatrix} \begin{bmatrix} -0.1789 \\ +0.6362 \\ -0.4573 \end{bmatrix}$$

$$= \begin{bmatrix} 0 \\ 0.9843 \underline{/254.2^\circ} \\ 0.9843 \underline{/105.8^\circ} \end{bmatrix} \quad \text{per unit}$$

e. The sequence components of the line-to-ground voltages at bus 2 during the fault at bus 1 are, from (9.5.9), with $k = 2$ and $n = 1$,

$$
\begin{bmatrix} V_{2-0} \\ V_{2-1} \\ V_{2-2} \end{bmatrix} = \begin{bmatrix} 0 \\ 1.05\underline{/0^\circ} \\ 0 \end{bmatrix} - \begin{bmatrix} 0 & 0 & 0 \\ 0 & j0.04580 & 0 \\ 0 & 0 & j0.05212 \end{bmatrix} \begin{bmatrix} -j3.578 \\ -j3.578 \\ -j3.578 \end{bmatrix}
$$

$$
= \begin{bmatrix} 0 \\ 0.8861 \\ -0.18649 \end{bmatrix} \text{ per unit}
$$

Note that since both bus 1 and 2 are on the low-voltage side of the Δ–Y transformers in Figure 9.3, there is no shift in the phase angles of these sequence voltages. From the above, the line-to-ground voltages at bus 2 during the fault at bus 1 are

$$
\begin{bmatrix} V_{2-ag} \\ V_{2-bg} \\ V_{2-cg} \end{bmatrix} = \begin{bmatrix} 1 & 1 & 1 \\ 1 & a^2 & a \\ 1 & a & a^2 \end{bmatrix} \begin{bmatrix} 0 \\ 0.8861 \\ -0.18649 \end{bmatrix}
$$

$$
= \begin{bmatrix} 0.70 \\ 0.9926\underline{/249.4^\circ} \\ 0.9926\underline{/110.6^\circ} \end{bmatrix} \text{ per unit} \qquad \blacksquare
$$

PowerWorld Simulator computes the symmetrical fault current for each of the following faults at any bus in an N-bus power system: balanced three-phase fault, single line-to-ground fault, line-to-line fault, or double line-to-ground fault. For each fault, the Simulator also computes bus voltages and contributions to the fault current from transmission lines and transformers connected to the fault bus.

Input data for the Simulator include machine, transmission-line, and transformer data, as illustrated in Tables 9.1, 9.2, and 9.3 as well as the prefault voltage V_F and fault impedance Z_F. When the machine positive-sequence reactance input data consist of direct axis subtransient reactances, the computed symmetrical fault currents are subtransient fault currents. Alternatively, transient or steady-state fault currents are computed when these

TABLE 9.1 Synchronous machine data for Example 9.8	Bus	X_0 per unit	$X_1 = X_d''$ per unit	X_2 per unit	Neutral Reactance X_n per unit
	1	0.0125	0.045	0.045	0
	3	0.005	0.0225	0.0225	0.0025

TABLE 9.2

Line data for Example 9.8

Bus-to-Bus	X_0 per unit	X_1 per unit
2–4	0.3	0.1
2–5	0.15	0.05
4–5	0.075	0.025

TABLE 9.3

Transformer data for Example 9.8

Low-Voltage (connection) bus	High-Voltage (connection) bus	Leakage Reactance per unit	Neutral Reactance per unit
1 (Δ)	5 (Y)	0.02	0
3 (Δ)	4 (Y)	0.01	0

$S_{base} = 100$ MVA

$V_{base} = \begin{cases} 15 \text{ kV at buses 1, 3} \\ 345 \text{ kV at buses 2, 4, 5} \end{cases}$

input data consist of direct axis transient or synchronous reactances. Transmission-line positive- and zero-sequence series reactances are those of the equivalent π circuits for long lines or of the nominal π circuit for medium or short lines. Also, recall that the negative-sequence transmission-line reactance equals the positive-sequence transmission-line reactance. All machine, line, and transformer reactances are given in per-unit on a common MVA base. Prefault load currents are neglected.

The Simulator computes (but does not show) the zero-, positive-, and negative-sequence bus impedance matrices $Z_{bus\,0}, Z_{bus\,1}$, and $Z_{bus\,2}$, by inverting the corresponding bus admittance matrices.

After $Z_{bus\,0}, Z_{bus\,1}$, and $Z_{bus\,2}$ are computed, (9.5.1)–(9.5.9) are used to compute the sequence fault currents and the sequence voltages at each bus during a fault at bus 1 for the fault type selected by the program user (for example, three-phase fault, or single line-to-ground fault, and so on). Contributions to the sequence fault currents from each line or transformer branch connected to the fault bus are computed by dividing the sequence voltage across the branch by the branch sequence impedance. The phase angles of positive- and negative-sequence voltages are also modified to account for Δ–Y transformer phase shifts. The sequence currents and sequence voltages are then transformed to the phase domain via (8.1.16) and (8.1.9). All these computations are then repeated for a fault at bus 2, then bus 3, and so on to bus N.

Output data for the fault type and fault impedance selected by the user consist of the fault current in each phase, contributions to the fault current from each branch connected to the fault bus for each phase, and the line-to-ground voltages at each bus—for a fault at bus 1, then bus 2, and so on to bus N.

EXAMPLE 9.8 **PowerWorld Simulator**

Consider the five-bus power system whose single-line diagram is shown in Figure 6.2. Machine, line, and transformer data are given in Tables 9.1, 9.2, and 9.3. Note that the neutrals of both transformers and generator 1 are solidly grounded, as indicated by a neutral reactance of zero for these equipments. However, a neutral reactance = 0.0025 per unit is connected to the generator 2 neutral. The prefault voltage is 1.05 per unit. Using PowerWorld Simulator, determine the fault currents and voltages for a bolted single line-to-ground fault at bus 1, then bus 2, and so on to bus 5.

SOLUTION Open PowerWorld Simulator case Example 9.8 to see an animated view of this example. Tables 9.4 and 9.5 summarize the PowerWorld Simulator results for each of the faults. Note that these fault currents are subtransient currents, since the machine positive-sequence reactance input consists of direct axis subtransient reactances.

TABLE 9.4

Fault currents for Example 9.8

Fault Bus	Single Line-to-Ground Fault Current (phase A) per unit/degrees	GEN LINE OR TRSF	Bus-to-Bus	Phase A	Phase B	Phase C
				per unit/degrees		
1	46.02/−90.00	G1	GRND−1	34.41/ −90.00	5.804/ −90.00	5.804/ −90.00
		T1	5−1	11.61/ −90.00	5.804/ 90.00	5.804/ 90.00
2	14.14/−90.00	L1	4−2	5.151/ −90.00	0.1124/ 90.00	0.1124/ 90.00
		L2	5−2	8.984/ −90.00	0.1124/ −90.00	0.1124/ −90.00
3	64.30/−90.00	G2	GRND−3	56.19/ −90.00	4.055/ −90.00	4.055/ −90.00
		T2	4−3	8.110/ −90.00	4.055/ 90.00	4.055/ 90.00
4	56.07/−90.00	L1	2−4	1.742/ −90.00	0.4464/ 90.00	0.4464/ 90.00
		L3	5−4	10.46/ −90.00	2.679/ 90.00	2.679/ 90.00
		T2	3−4	43.88/ −90.00	3.125/ −90.00	3.125/ −90.00
5	42.16/−90.00	L2	2−5	2.621/ −90.00	0.6716/ 90.00	0.6716/ 90.00
		L3	4−5	15.72/ −90.00	4.029/ 90.00	4.029/ 90.00
		T1	1−5	23.82/ −90.00	4.700/ −90.00	4.700/ −90.00

TABLE 9.5	$V_{prefault} = 1.05 \angle 0$		Bus Voltages during Fault		
Bus voltages for Example 9.8	Fault Bus	Bus	Phase A	Phase B	Phase C
	1	1	$0.0000 \angle 0.00$	$0.9537 \angle -107.55$	$0.9537 \angle 107.55$
		2	$0.5069 \angle 0.00$	$0.9440 \angle -105.57$	$0.9440 \angle 105.57$
		3	$0.7888 \angle 0.00$	$0.9912 \angle -113.45$	$0.9912 \angle 113.45$
		4	$0.6727 \angle 0.00$	$0.9695 \angle -110.30$	$0.9695 \angle 110.30$
		5	$0.4239 \angle 0.00$	$0.9337 \angle -103.12$	$0.9337 \angle 103.12$
	2	1	$0.8832 \angle 0.00$	$1.0109 \angle -115.90$	$1.0109 \angle 115.90$
		2	$0.0000 \angle 0.00$	$1.1915 \angle -130.26$	$1.1915 \angle 130.26$
		3	$0.9214 \angle 0.00$	$1.0194 \angle -116.87$	$1.0194 \angle 116.87$
		4	$0.8435 \angle 0.00$	$1.0158 \angle -116.47$	$1.0158 \angle 116.47$
		5	$0.7562 \angle 0.00$	$1.0179 \angle -116.70$	$1.0179 \angle 116.70$
	3	1	$0.6851 \angle 0.00$	$0.9717 \angle -110.64$	$0.9717 \angle 110.64$
		2	$0.4649 \angle 0.00$	$0.9386 \angle -104.34$	$0.9386 \angle 104.34$
		3	$0.0000 \angle 0.00$	$0.9942 \angle -113.84$	$0.9942 \angle 113.84$
		4	$0.3490 \angle 0.00$	$0.9259 \angle -100.86$	$0.9259 \angle 100.86$
		5	$0.5228 \angle 0.00$	$0.9462 \angle -106.04$	$0.9462 \angle 106.04$
	4	1	$0.5903 \angle 0.00$	$0.9560 \angle -107.98$	$0.9560 \angle 107.98$
		2	$0.2309 \angle 0.00$	$0.9401 \angle -104.70$	$0.9401 \angle 104.70$
		3	$0.4387 \angle 0.00$	$0.9354 \angle -103.56$	$0.9354 \angle 103.56$
		4	$0.0000 \angle 0.00$	$0.9432 \angle -105.41$	$0.9432 \angle 105.41$
		5	$0.3463 \angle 0.00$	$0.9386 \angle -104.35$	$0.9386 \angle 104.35$
	5	1	$0.4764 \angle 0.00$	$0.9400 \angle -104.68$	$0.9400 \angle 104.68$
		2	$0.1736 \angle 0.00$	$0.9651 \angle -109.57$	$0.9651 \angle 109.57$
		3	$0.7043 \angle 0.00$	$0.9751 \angle -111.17$	$0.9751 \angle 111.17$
		4	$0.5209 \angle 0.00$	$0.9592 \angle -108.55$	$0.9592 \angle 108.55$
		5	$0.0000 \angle 0.00$	$0.9681 \angle -110.07$	$0.9681 \angle 110.07$

■

PROBLEMS

SECTION 9.1

9.1 The single-line diagram of a three-phase power system is shown in Figure 9.16. Equipment ratings are given as follows:

Synchronous generators:

G1	1000 MVA	15 kV	$X_d'' = X_2 = 0.18, X_0 = 0.07$ per unit
G2	1000 MVA	15 kV	$X_d'' = X_2 = 0.20, X_0 = 0.10$ per unit
G3	500 MVA	13.8 kV	$X_d'' = X_2 = 0.15, X_0 = 0.05$ per unit
G4	750 MVA	13.8 kV	$X_d'' = 0.30, X_2 = 0.40, X_0 = 0.10$ per unit

FIGURE 9.16

Problem 9.1

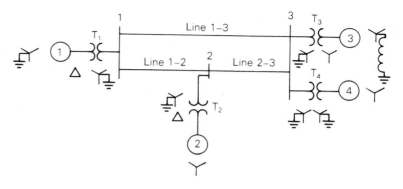

Transformers:

T1	1000 MVA	15 kV Δ/765 kV Y	X = 0.10 per unit
T2	1000 MVA	15 kV Δ/765 kV Y	X = 0.10 per unit
T3	500 MVA	15 kV Y/765 kV Y	X = 0.12 per unit
T4	750 MVA	15 kV Y/765 kV Y	X = 0.11 per unit

Transmission lines:

1–2	765 kV	$X_1 = 50\ \Omega$, $X_0 = 150\ \Omega$
1–3	765 kV	$X_1 = 40\ \Omega$, $X_0 = 100\ \Omega$
2–3	765 kV	$X_1 = 40\ \Omega$, $X_0 = 100\ \Omega$

The inductor connected to generator 3 neutral has a reactance of 0.05 per unit using generator 3 ratings as a base. Draw the zero-, positive-, and negative-sequence reactance diagrams using a 1000-MVA, 765-kV base in the zone of line 1–2. Neglect the Δ–Y transformer phase shifts.

9.2   Faults at bus n in Problem 9.1 are of interest (the instructor selects $n = 1, 2$, or 3). Determine the Thévenin equivalent of each sequence network as viewed from the fault bus. Prefault voltage is 1.0 per unit. Prefault load currents and Δ–Y transformer phase shifts are neglected. (*Hint*: Use the Y–Δ conversion in Figure 2.27.)

9.3   Determine the subtransient fault current in per-unit and in kA during a bolted three-phase fault at the fault bus selected in Problem 9.2.

9.4   Equipment ratings for the four-bus power system shown in Figure 7.10 are given as follows:

Generator G1:   500 MVA, 13.8 kV, $X_d'' = X_2 = 0.20$, $X_0 = 0.10$ per unit

Generator G2:   750 MVA, 18 kV, $X_d'' = X_2 = 0.18$, $X_0 = 0.09$ per unit

Generator G3:   1000 MVA, 20 kV, $X_d'' = 0.17$, $X_2 = 0.20$, $X_0 = 0.09$ per unit

Transformer T1:   500 MVA, 13.8 kV Δ/500 kV Y, X = 0.12 per unit

Transformer T2:   750 MVA, 18 kV Δ/500 kV Y, X = 0.10 per unit

Transformer T3:   1000 MVA, 20 kV Δ/500 kV Y, X = 0.10 per unit

Each line: $X_1 = 50$ ohms, $X_0 = 150$ ohms

The inductor connected to generator G3 neutral has a reactance of 0.028 Ω. Draw the zero-, positive-, and negative-sequence reactance diagrams using a 1000-MVA, 20-kV base in the zone of generator G3. Neglect Δ–Y transformer phase shifts.

9.5 Faults at bus n in Problem 9.4 are of interest (the instructor selects $n = 1, 2, 3$, or 4). Determine the Thévenin equivalent of each sequence network as viewed from the fault bus. Prefault voltage is 1.0 per unit. Prefault load currents and Δ–Y phase shifts are neglected.

9.6 Determine the subtransient fault current in per-unit and in kA during a bolted three-phase fault at the fault bus selected in Problem 9.5.

9.7 Equipment ratings for the five-bus power system shown in Figure 7.11 are given as follows:

Generator G1:   50 MVA, 12 kV, $X_d'' = X_2 = 0.20$, $X_0 = 0.10$ per unit

Generator G2:   100 MVA, 15 kV, $X_d'' = 0.2$, $X_2 = 0.23$, $X_0 = 0.1$ per unit

Transformer T1:   50 MVA, 10 kV Y/138 kV Y, $X = 0.10$ per unit

Transformer T2:   100 MVA, 15 kV Δ/138 kV Y, $X = 0.10$ per unit

Each 138-kV line: $X_1 = 40$ ohms, $X_0 = 100$ ohms

Draw the zero-, positive-, and negative-sequence reactance diagrams using a 100-MVA, 15-kV base in the zone of generator G2. Neglect Δ–Y transformer phase shifts.

9.8 Faults at bus n in Problem 9.7 are of interest (the instructor selects $n = 1, 2, 3, 4$, or 5). Determine the Thévenin equivalent of each sequence network as viewed from the fault bus. Prefault voltage is 1.0 per unit. Prefault load currents and Δ–Y phase shifts are neglected.

9.9 Determine the subtransient fault current in per-unit and in kA during a bolted three-phase fault at the fault bus selected in Problem 9.8.

9.10 Consider the system shown in Figure 9.17. (a) As viewed from the fault at F, determine the Thévenin equivalent of each sequence network. Neglect Δ–Y phase shifts. (b) Compute the fault currents for a balanced three-phase fault at fault point F through three fault impedances $Z_{FA} = Z_{FB} = Z_{FC} = j0.5$ per unit. Equipment data in per-unit on the same base are given as follows.

Synchronous generators:

G1 $X_1 = 0.2$ $X_2 = 0.12$ $X_0 = 0.06$

G2 $X_1 = 0.33$ $X_2 = 0.22$ $X_0 = 0.066$

Transformers:

T1 $X_1 = X_2 = X_0 = 0.2$

T2 $X_1 = X_2 = X_0 = 0.225$

T3 $X_1 = X_2 = X_0 = 0.27$

T4 $X_1 = X_2 = X_0 = 0.16$

Transmission lines:

L1 $X_1 = X_2 = 0.14$ $X_0 = 0.3$

L1 $X_1 = X_2 = 0.35$ $X_0 = 0.6$

FIGURE 9.17

Problem 9.10

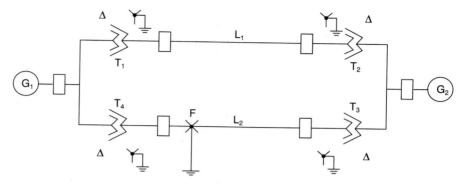

9.11   Equipment ratings and per-unit reactances for the system shown in Figure 9.18 are given as follows:

Synchronous generators:

G1   100 MVA   25 kV $X_1 = X_2 = 0.2$ $X_0 = 0.05$

G2   100 MVA   13.8 kV $X_1 = X_2 = 0.2$ $X_0 = 0.05$

Transformers:

T1   100 MVA   25/230 kV $X_1 = X_2 = X_0 = 0.05$

T2   100 MVA   13.8/230 kV $X_1 = X_2 = X_0 = 0.05$

Transmission lines:

TL12   100 MVA   230 kV $X_1 = X_2 = 0.1$ $X_0 = 0.3$

TL13   100 MVA   230 kV $X_1 = X_2 = 0.1$ $X_0 = 0.3$

TL23   100 MVA   230 kV $X_1 = X_2 = 0.1$ $X_0 = 0.3$

Using a 100-MVA, 230-kV base for the transmission lines, draw the per-unit sequence networks and reduce them to their Thévenin equivalents, "looking in" at bus 3. Neglect Δ–Y phase shifts. Compute the fault currents for a bolted three-phase fault at bus 3.

FIGURE 9.18

Problem 9.11

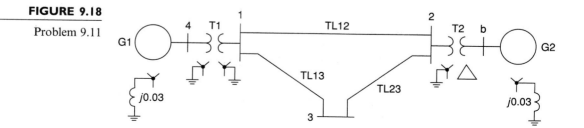

9.12 Consider the one-line diagram of a simple power system shown in Figure 9.19. System data in per-unit on a 100-MVA base are given as follows:

Synchronous generators:

G1 100 MVA 20 kV $X_1 = X_2 = 0.15$ $X_0 = 0.05$

G2 100 MVA 20 kV $X_1 = X_2 = 0.15$ $X_0 = 0.05$

Transformers:

T1 100 MVA 20/220 kV $X_1 = X_2 = X_0 = 0.1$

T2 100 MVA 20/220 kV $X_1 = X_2 = X_0 = 0.1$

Transmission lines:

L12 100 MVA 220 kV $X_1 = X_2 = 0.125$ $X_0 = 0.3$

L13 100 MVA 220 kV $X_1 = X_2 = 0.15$ $X_0 = 0.35$

L23 100 MVA 220 kV $X_1 = X_2 = 0.25$ $X_0 = 0.7125$

The neutral of each generator is grounded through a current-limiting reactor of 0.08333 per unit on a 100-MVA base. All transformer neutrals are solidly grounded. The generators are operating no-load at their rated voltages and rated frequency with their EMFs in phase. Determine the fault current for a balanced three-phase fault at bus 3 through a fault impedance $Z_F = 0.1$ per unit on a 100-MVA base. Neglect Δ–Y phase shifts.

FIGURE 9.19

Problem 9.12

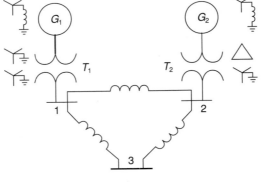

SECTIONS 9.2–9.4

9.13 Determine the subtransient fault current in per-unit and in kA, as well as the per-unit line-to-ground voltages at the fault bus for a bolted single line-to-ground fault at the fault bus selected in Problem 9.2.

9.14 Repeat Problem 9.13 for a single line-to-ground arcing fault with arc impedance $Z_F = 30 + j0 \ \Omega$.

9.15   Repeat Problem 9.13 for a bolted line-to-line fault.

9.16   Repeat Problem 9.13 for a bolted double line-to-ground fault.

9.17   Repeat Problems 9.1 and 9.13 including Δ–Y transformer phase shifts. Assume American standard phase shift. Also calculate the sequence components and phase components of the contribution to the fault current from generator n ($n = 1, 2,$ or 3 as specified by the instructor in Problem 9.2).

9.18   A 500-MVA, 13.8-kV synchronous generator with $X_d'' = X_2 = 0.20$ and $X_0 = 0.05$ per unit is connected to a 500-MVA, 13.8-kV Δ/500-kV Y transformer with 0.10 per-unit leakage reactance. The generator and transformer neutrals are solidly grounded. The generator is operated at no-load and rated voltage, and the high-voltage side of the transformer is disconnected from the power system. Compare the subtransient fault currents for the following bolted faults at the transformer high-voltage terminals: three-phase fault, single line-to-ground fault, line-to-line fault, and double line-to-ground fault.

9.19   Determine the subtransient fault current in per-unit and in kA, as well as contributions to the fault current from each line and transformer connected to the fault bus for a bolted single line-to-ground fault at the fault bus selected in Problem 9.5.

9.20   Repeat Problem 9.19 for a bolted line-to-line fault.

9.21   Repeat Problem 9.19 for a bolted double line-to-ground fault.

9.22   Determine the subtransient fault current in per-unit and in kA, as well as contributions to the fault current from each line, transformer, and generator connected to the fault bus for a bolted single line-to-ground fault at the fault bus selected in Problem 9.8.

9.23   Repeat Problem 9.22 for a single line-to-ground arcing fault with arc impedance $Z_F = 0.05 + j0$ per unit.

9.24   Repeat Problem 9.22 for a bolted line-to-line fault.

9.25   Repeat Problem 9.22 for a bolted double line-to-ground fault.

9.26   As shown in Figure 9.20(a), two three-phase buses abc and $a'b'c'$ are interconnected by short circuits between phases b and b' and between c and c', with an open circuit between phases a and a'. The fault conditions in the phase domain are $I_a = I_{a'} = 0$ and $V_{bb'} = V_{cc'} = 0$. Determine the fault conditions in the sequence domain and verify the interconnection of the sequence networks as shown in Figure 9.14 for this one-conductor-open fault.

FIGURE 9.20

Problems 9.26 and 9.27: open conductor faults

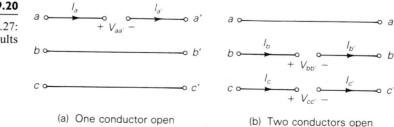

(a) One conductor open   (b) Two conductors open

9.27 Repeat Problem 9.26 for the two-conductors-open fault shown in Figure 9.20(b). The fault conditions in the phase domain are

$$I_b = I_{b'} = I_c = I_{c'} = 0 \quad \text{and} \quad V_{aa'} = 0$$

9.28 For the system of Problem 9.10, compute the fault current and voltages at the fault for the following faults at point F: (a) a bolted single line-to-ground fault; (b) a line-to-line fault through a fault impedance $Z_F = j0.05$ per unit; (c) a double line-to-ground fault from phase B to C to ground, where phase B has a fault impedance $Z_F = j0.05$ per unit, phase C also has a fault impedance $Z_F = j0.05$ per unit, and the common line-to-ground fault impedance is $Z_G = j0.033$ per unit.

9.29 For the system of Problem 9.11, compute the fault current and voltages at the fault for the following faults at bus 3: (a) a bolted single line-to-ground fault, (b) a bolted line-to-line fault, (c) a bolted double line-to-ground fault. Also, for the single line-to-ground fault at bus 3, determine the currents and voltages at the terminals of generators G1 and G2.

9.30 For the system of Problem 9.12, compute the fault current for the following faults at bus 3: (a) a single line-to-ground fault through a fault impedance $Z_F = j0.1$ per unit, (b) a line-to-line fault through a fault impedance $Z_F = j0.1$ per unit, (c) a double line-to-ground fault through a common fault impedance to ground $Z_F = j0.1$ per unit.

9.31 For the three-phase power system with single-line diagram shown in Figure 9.21, equipment ratings and per-unit reactances are given as follows:

Machines 1 and 2:	100 MVA	20 kV	$X_1 = X_2 = 0.2$
	$X_0 = 0.04$	$X_n = 0.04$	
Transformers 1 and 2:	100 MVA	$20\Delta/345Y$ kV	
	$X_1 = X_2 = X_0 = 0.08$		

Select a base of 100 MVA, 345 kV for the transmission line. On that base, the series reactances of the line are $X_1 = X_2 = 0.15$ and $X_0 = 0.5$ per unit. With a nominal system voltage of 345 kV at bus 3, machine 2 is operating as a motor drawing 50 MVA at 0.8 power factor lagging. Compute the change in voltage at bus 3 when the transmission line undergoes (a) a one-conductor-open fault, (b) a two-conductor-open fault along its span between buses 2 and 3.

FIGURE 9.21

Problem 9.31

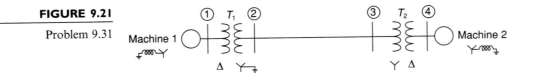

9.32 At the general three-phase bus shown in Figure 9.7 (a) of the text, consider a simultaneous single line-to-ground fault on phase a and line-to-line fault between phases b and c, with no fault impedances. Obtain the sequence-network interconnection satisfying the current and voltage constraints.

9.33   Thevenin equivalent sequence networks looking into the faulted bus of a power system are given with $Z_1 = j0.15$, $Z_2 = j0.15$, $Z_0 = j0.2$, and $E_1 = 1/\underline{0°}$ per unit. Compute the fault currents and voltages for the following faults occurring at the faulted bus:
(a) Balanced 3-phase fault
(b) Single line-to-ground fault
(c) Line-line fault
(d) Double line-to-ground fault
Which is the worst fault from the viewpoint of the fault current?

9.34   The single-line diagram of a simple power system is shown in Figure 9.22 with per unit values. Determine the fault current at bus 2 for a three-phase fault. Ignore the effect of phase shift.

FIGURE 9.22

For Problem 9.34

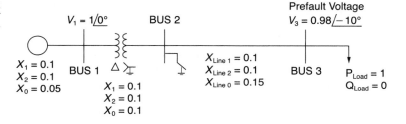

9.35   Consider a simple circuit configuration shown in Figure 9.23 to calculate the fault currents I_1, I_2, and I with the switch closed.
(a) Compute E_1 and E_2 prior to the fault based on the prefault voltage $V = 1/\underline{0°}$, and then, with the switch closed, determine I_1, I_2, and I.
(b) Start by ignoring prefault currents, with $E_1 = E_2 = 1/\underline{0°}$. Then superimpose the load currents, which are the prefault currents, $I_1 = -I_2 = 1/\underline{0°}$. Compare the results with those of part (a).

FIGURE 9.23

For Problem 9.35

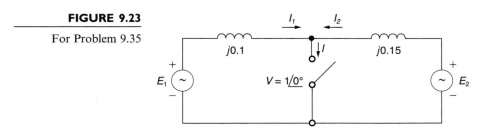

SECTION 9.5

9.36   The zero-, positive-, and negative-sequence bus impedance matrices for a three-bus three-phase power system are

$$\boldsymbol{Z}_{bus\,0} = j \begin{bmatrix} 0.10 & 0 & 0 \\ 0 & 0.20 & 0 \\ 0 & 0 & 0.10 \end{bmatrix} \quad per\ unit$$

$$\boldsymbol{Z}_{bus\,1} = \boldsymbol{Z}_{bus\,2} = j \begin{bmatrix} 0.12 & 0.08 & 0.04 \\ 0.08 & 0.12 & 0.06 \\ 0.04 & 0.06 & 0.08 \end{bmatrix}$$

Determine the per-unit fault current and per-unit voltage at bus 2 for a bolted three-phase fault at bus 1. The prefault voltage is 1.0 per unit.

9.37 Repeat Problem 9.36 for a bolted single line-to-ground fault at bus 1.

9.38 Repeat Problem 9.36 for a bolted line-to-line fault at bus 1.

9.39 Repeat Problem 9.36 for a bolted double line-to-ground fault at bus 1.

9.40 Compute the 3×3 per-unit zero-, positive-, and negative-sequence bus impedance matrices for the power system given in Problem 9.1. Use a base of 1000 MVA and 765 kV in the zone of line 1–2.

9.41 Using the bus impedance matrices determined in Problem 9.40, verify the fault currents for the faults given in Problems 9.3, 9.13, 9.14, 9.15, and 9.16.

9.42 Compute the 4×4 per-unit zero-, positive-, and negative-sequence bus impedance matrices for the power system given in Problem 9.4. Use a base of 1000 MVA and 20 kV in the zone of generator G3.

9.43 Using the bus impedance matrices determined in Problem 9.42, verify the fault currents for the faults given in Problems 9.6, 9.19, 9.20, and 9.21.

9.44 Compute the 5×5 per-unit zero-, positive-, and negative-sequence bus impedance matrices for the power system given in Problem 9.7. Use a base of 100 MVA and 15 kV in the zone of generator G2.

9.45 Using the bus impedance matrices determined in Problem 9.44, verify the fault currents for the faults given in Problems 9.9, 9.22, 9.23, 9.24, and 9.25.

9.46 The positive-sequence impedance diagram of a five-bus network with all values in per-unit on a 100-MVA base is shown in Figure 9.24. The generators at buses 1 and 3 are rated 270 and 225 MVA, respectively. Generator reactances include subtransient values plus reactances of the transformers connecting them to the buses. The turns ratios of the transformers are such that the voltage base in each generator circuit is equal to the voltage rating of the generator. (a) Develop the positive-sequence bus

FIGURE 9.24

Problems 9.46 and 9.47

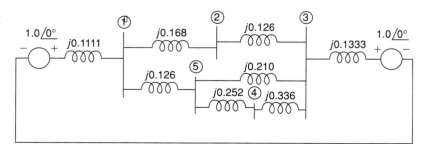

admittance matrix $Y_{bus\ 1}$. (b) Using MATLAB or another computer program, invert $Y_{bus\ 1}$ to obtain $Z_{bus\ 1}$. (c) Determine the subtransient current for a three-phase fault at bus 4 and the contributions to the fault current from each line. Neglect prefault currents and assume a prefault voltage of 1.0 per unit.

9.47 For the five-bus network shown in Figure 9.24, a bolted single-line-to-ground fault occurs at the bus 2 end of the transmission line between buses 1 and 2. The fault causes the circuit breaker at the bus 2 end of the line to open, but all other breakers remain closed. The fault is shown in Figure 9.25. Compute the subtransient fault current with the circuit breaker at the bus-2 end of the faulted line open. Neglect prefault current and assume a prefault voltage of 1.0 per unit.

FIGURE 9.25

Problem 9.47

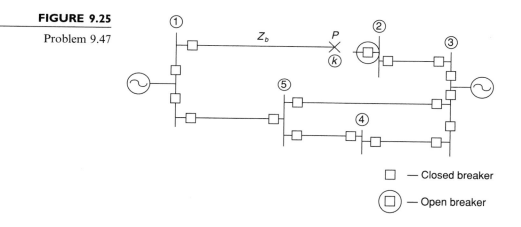

□ — Closed breaker

◻ — Open breaker

9.48 A single-line diagram of a four-bus system is shown in Figure 9.26. Equipment ratings and per-unit reactances are given as follows.

Machines 1 and 2:	100 MVA	20 kV	$X_1 = X_2 = 0.2$
	$X_0 = 0.04$	$X_n = 0.05$	
Transformers T_1 and T_2:	100 MVA	20Δ/345Y kV	
	$X_1 = X_2 = X_0 = 0.08$		

On a base of 100 MVA and 345 kV in the zone of the transmission line, the series reactances of the transmission line are $X_1 = X_2 = 0.15$ and $X_0 = 0.5$ per unit. (a) Draw each of the sequence networks and determine the bus impedance matrix for each of them. (b) Assume the system to be operating at nominal system voltage without prefault currents, when a bolted line-to-line fault occurs at bus 3. Compute the fault current, the line-to-line voltages at the faulted bus, and the line-to-line voltages at the

FIGURE 9.26

Problem 9.48

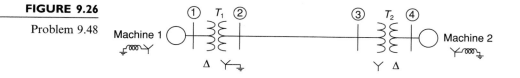

Machine 1 T_1 T_2 Machine 2

terminals of machine 2. (c) Assume the system to be operating at nominal system voltage without prefault currents, when a bolted double-line-to-ground fault occurs at the terminals of machine 2. Compute the fault current and the line-to-line voltages at the faulted bus.

9.49 The system shown in Figure 9.27 is the same as in Problem 9.48 except that the transformers are now Y–Y connected and solidly grounded on both sides. (a) Determine the bus impedance matrix for each of the three sequence networks. (b) Assume the system to be operating at nominal system voltage without prefault currents, when a bolted single-line-to-ground fault occurs on phase A at bus 3. Compute the fault current, the current out of phase C of machine 2 during the fault, and the line-to-ground voltages at the terminals of machine 2 during the fault.

FIGURE 9.27

Problem 9.49

9.50 The results in Table 9.5 show that during a phase "a" single line-to-ground fault the phase angle on phase "a" voltages is always zero. Explain why we would expect this result.

 9.51 The results in Table 9.5 show that during the single line-to-ground fault at bus 2 the "b" and "c" phase voltage magnitudes at bus 2 actually rise above the pre-fault voltage of 1.05 per unit. Use PowerWorld Simulator with case Example 9_8 to determine the type of bus 2 fault that gives the highest per-unit voltage magnitude.

PW **9.52** Using PowerWorld Simulator case Example 9_8, plot the variation in the phase "a," "b," and "c" voltage magnitudes during a single line-to-ground fault as the fault reactance is varied from 0 to 2.0 per unit in 0.25 per-unit steps (the fault impedance is specified on the Fault Options page of the Fault Analysis dialog).

PW **9.53** Redo Example 9.8, except with a line-to-line fault at each of the buses. Compare the fault currents with the values given in Table 9.4.

PW **9.54** Redo Example 9.8, except with a bolted double line-to-ground fault at each of the buses. Compare the fault currents with the values given in Table 9.4.

PW **9.55** Redo Example 9.8, except with a new line installed between buses 2 and 4. The parameters for this new line should be identical to those of the existing line between buses 2 and 4. The new line is not mutually coupled to any other line. Are the fault currents larger or smaller than the Example 9.8 values?

PW **9.56** Redo Example 9.8, except with a second generator added at bus 3. The parameters for the new generator should be identical to those of the existing generator at bus 3. Are the fault currents larger or smaller than the Example 9.8 values?

PW **9.57** Using PowerWorld Simulator Design case 6, calculate the per-unit fault current and the current supplied by each of the generators for a single line-to-ground fault at the PETE69 bus. During the fault, what percentage of buses have voltage magnitude below 0.75 per unit?

PW **9.58** Repeat Problem 9.57, except place the fault at the TIM69 bus.

DESIGN PROJECT 4 (CONTINUED): POWER FLOW/SHORT CIRCUITS

Additional time given: 3 weeks
Additional time required: 10 hours

This is a continuation of Design Project 4. Assignments 1 and 2 are given in Chapter 6. Assignment 3 is given in Chapter 7.

Assignment 4: Short Circuits—Breaker/Fuse Selection

For the single-line diagram that you have been assigned (Figure 6.13 or 6.14), convert the zero-, positive-, and negative-sequence reactance data to per-unit using the given system base quantities. Use subtransient machine reactances. Then using PowerWorld Simulator, create the generator, transmission line, and transformer input data files. Next run the Simulator to compute subtransient fault currents for (1) single-line-to-ground, (2) line-to-line, and (3) double-line-to-ground bolted faults at each bus. Also compute the zero-, positive-, and negative-sequence bus impedance matrices. Assume 1.0 per-unit prefault voltage. Also, neglect prefault load currents and all losses.

For students assigned to Figure 6.13: Select a suitable circuit breaker from Table 7.10 for *each* location shown on your single-line diagram. Each breaker that you select should: (1) have a rated voltage larger than the maximum system operating voltage, (2) have a rated continuous current at least 30% larger than normal load current (normal load currents are computed in Assignment 2), and (3) have a rated short-circuit current larger than the maximum fault current for any type of fault at the bus where the breaker is located (fault currents are computed in Assignments 3 and 4). This conservative practice of selecting a breaker to interrupt the entire fault current, not just the contribution to the fault through the breaker, allows for future increases in fault currents. *Note:* Assume that the (X/R) ratio at each bus is less than 15, such that the breakers are capable of interrupting the dc-offset in addition to the subtransient fault current. Circuit breaker cost should also be a factor in your selection. Do *not* select a breaker that interrupts 63 kA if a 40-kA or a 31.5-kA breaker will do the job.

For students assigned to Figure 6.14: Enclosed [9, 10] are "melting time" and "total clearing time" curves for K rated fuses with continuous current ratings from 15 to 200 A. Select suitable branch and tap fuses from these curves for each of the following three locations on your single-line diagram: bus 2, bus 4, and bus 7. Each fuse you select should have a continuous current rating that is at least 15% higher but not more than 50% higher than the normal load current at that bus (normal load currents are computed in Assignment 2). Assume that cables to the load can withstand 50% continuous overload currents. Also, branch fuses should be coordinated with tap fuses; that is, for every fault current, the tap fuse should clear before the branch fuse melts. For each of the three buses, assume a reasonable X/R ratio and determine the asymmetrical fault current for a three-phase bolted fault (subtransient current is computed in Assignment 3). Then for the fuses that you select from [9, 10], determine the clearing time CT of tap fuses and the melting time MT of branch fuses. The ratio MT/CT should be less than 0.75 for good coordination.

DESIGN PROJECT 6

Time given: 3 weeks
Approximate time required: 10 hours

As a protection engineer for Metropolis Light and Power (MLP) your job is to ensure that the transmission line and transformer circuit breaker ratings are sufficient to interrupt the fault current associated with any type of fault (balanced three phase, single line-to-ground, line-to-line, and double line-to-ground). The MLP power system is modeled in case Chapter9_Design. This case models the positive, negative and zero sequence values for each system device. Note that the 69/138 kV transformers are grounded wye on the low side and delta on the high side; the 138 kV/345 kV transformers grounded wye on both sides. In this design problem your job is to evaluate the circuit breaker ratings for the three 345 kV transmission lines and the six 345/138 kV transformers. You need not consider the 138 or 69 kV transmission lines, or the 138/69 kV transformers.

Design Procedure

1. Load Chapter9_Design into PowerWorld Simulator. Perform an initial power flow solution to get the base case system operating point.

2. Apply each of the four fault types to each of the 345 kV buses and to the 138 kV buses attached to 345/138 kV transformers to determine the maximum fault current that each of the 345 kV lines and 345/138 kV transformers will experience.

3. For each device select a suitable circuit breaker from Table 7.10. Each breaker that you select should a) have a rated voltage larger than the maximum system operating voltage, b) have a rated continuous current at least 30% larger than the normal rated current for the line, c) have a rated short circuit current larger than the maximum fault current for any type of fault at the bus where the breaker is located. This conservative practice of selecting a breaker to interrupt the entire fault current, not just the contribution to the fault current through the breaker allows for future increases in fault currents. Since higher rated circuit breakers cost more, you should select the circuit breaker with the lowest rating that satisfies the design constraints.

Simplifying Assumptions

1. You need only consider the base case conditions given in the Chapter9_Design case.

2. You may assume that the X/R ratios at each bus is sufficiently small (less than 15) so that the dc offset has decayed to a sufficiently low value (see Section 7.7 for details).

3. As is common with commercial software, including PowerWorld Simulator, the Δ-Y transformer phase shifts are neglected.

CASE STUDY QUESTIONS

A. Are safety hazards associated with generation, transmission, and distribution of electric power by the electric utility industry greater than or less than safety hazards associated with the transportation industry? The chemical products industry? The medical services industry? The agriculture industry?

B. What is the public's perception of the electric utility industry's safety record?

REFERENCES

1. Westinghouse Electric Corporation, *Electrical Transmission and Distribution Reference Book*, 4th ed. (East Pittsburgh, PA, 1964).

2. Westinghouse Electric Corporation, *Applied Protective Relaying* (Newark, NJ, 1976).

3. P. M. Anderson, *Analysis of Faulted Power Systems* (Ames: Iowa State University Press, 1973).

4. J. R. Neuenswander, *Modern Power Systems* (New York: Intext Educational Publishers, 1971).

5. H. E. Brown, *Solution of Large Networks by Matrix Methods* (New York: Wiley, 1975).

6. W. D. Stevenson, Jr., *Elements of Power System Analysis*, 4th ed. (New York: McGraw-Hill, 1982).

7. C. A. Gross, *Power System Analysis* (New York: Wiley, 1979).

8. Glenn Zorpette, "Fires at U.S. Utilities," *IEEE Spectrum, 28*, 1 (January, 1991), p. 64.

9. McGraw Edison Company, *Fuse Catalog*, R240-91-1 (Canonsburg, PA: Mcgraw Edison, April 1985).

10. Westinghouse Electric Corporation, *Electric Utility Engineering Reference Book: Distribution Systems* (Pittsburgh, PA: Westinghouse, 1959).

10

SYSTEM PROTECTION

Short circuits occur in power systems when equipment insulation fails, due to system overvoltages caused by lightning or switching surges, to insulation contamination, or to other mechanical and natural causes. Careful design, operation, and maintenance can minimize the occurrence of short circuits but cannot eliminate them. We discussed methods for calculating short-circuit currents for balanced and unbalanced faults in Chapters 7 and 9. Such currents can be several orders of magnitude larger than normal operating currents and, if allowed to persist, may cause insulation damage, conductor melting, fire, and explosion. Windings and busbars may also suffer mechanical damage due to high magnetic forces during faults. Clearly, faults must be quickly removed from a power system. Standard EHV protective equipment is designed to clear faults within 3 cycles, whereas lower-voltage protective equipment typically operates within 5–20 cycles.

This chapter provides an introduction to power system protection. Blackburn defines protection as "the science, skill, and art of applying and setting relays and/or fuses to provide maximum sensitivity to faults and undesirable conditions, but to avoid their operation on all permissible or tolerable conditions" [1]. The basic idea is to define the undesirable conditions and look for differences between the undesirable and permissible conditions that relays or fuses can sense. It is also important to remove only the faulted equipment from the system while maintaining as much of the unfaulted system as possible in service, in order to continue to supply as much of the load as possible.

Although fuses and reclosers (circuit breakers with built-in instrument transformers and relays) are widely used to protect primary distribution systems (with voltages in the 2.4–46 kV range), we focus primarily in this chapter on circuit breakers and relays, which are used to protect HV (115–230 kV) and EHV (345–765 kV) power systems. The IEEE defines a relay as "a device whose function is to detect defective lines or apparatus or other power system conditions of an abnormal or dangerous nature and to initiate appropriate control action" [1]. In practice, a relay is a device that closes or opens a contact when energized. Relays are also used in low-voltage (600-V and below) power systems and almost anywhere that electricity is used. They are used in heating, air conditioning, stoves, clothes washers and dryers, refrigerators, dishwashers, telephone networks, traffic controls, airplane and other transportation systems, and robotics, as well as many other applications.

Problems with the protection equipment itself can occur. A second line of defense, called *backup* relays, may be used to protect the first line of defense, called *primary* relays. In HV and EHV systems, separate current- or voltage-measuring devices, separate trip coils on the circuit breakers, and separate batteries for the trip coils may be used. Also, the various protective devices must be properly coordinated such that primary relays assigned to protect equipment in a particular zone operate first. If the primary relays fail, then backup relays should operate after a specified time delay.

This chapter begins with a discussion of the basic system-protection components.

CASE STUDY For a given power system configuration, the performance of protection, monitoring, and control equipment determines how the system responds to contingencies and catastrophic events. The following article provides an overview of relaying practices and describes relay misoperations, modes of relay operations under abnormal system conditions, and advances in computer relays and adaptive relaying [14].

Blackouts and Relaying Considerations

S. H. HOROWITZ AND A. G. PHADKE

In recent years, we have seen several catastrophic failures of power systems throughout the world. The usual scenario of such events is that the power system is in a stressed state, followed by faults on critical facilities, followed by unanticipated tripping of other facilities, finally leading to system breakup and blackouts. It is well recognized that, in most of these events, protection systems—relays—play an important role in that they operate in an unexpected manner and are often a contributing factor to the cascading phenomena. In many postmortem analyses of such events, relays have been unfairly singled out as the main culprits, and often draconian measures have been suggested to correct the perceived faults in the protection system.

This article addresses the broad question of relaying philosophies, modes of operation of relays under abnormal system conditions, the categories of their failure modes, and advances made possible by modern technology in computer relays, adaptive relaying, and the use of wide-area measurements for improved protection and the ability to mitigate wide-area outages. It is hoped that this discussion will lead to a dialogue among system planners, relay engineers, and operating personnel to explore the realities of relay system performance under stressed system conditions and to adapt solutions made possible by new technological innovations.

AN OVERVIEW OF RELAYING PRACTICES

The number of relays on a modern high-voltage transmission system is very high. A very rough estimate of relays in service in the North American grid is on the order of 5 million. The primary task of relays is to trip associated circuit breakers (i.e., transmission lines, buses, generators, transformers, etc.) in response to faults or other conditions for

("Blackouts and Relaying Considerations" by S. H. Horowitz and A. G. Phadke, IEEE Power and Energy (Sept/Oct 2006), pg. 60–67.)

which the protection system is designed. All protection systems usually consist of several relays with well-defined conditions under which they should trip. In most high-voltage networks, there are high-speed relays that operate in less than three cycles of the fundamental power system frequency. Their main application is in protecting equipment against damage due to faults. These relays are autonomous, utilizing locally measured signals, often supplemented by remote terminal data over pilot channels. These relays, referred to as primary protection systems, may be duplicated for reliability. In addition, backup relays are provided, which, in general, operate more slowly and disconnect a larger portion of the power system.

Traditionally, the relaying philosophy has been biased towards dependability, meaning that if a fault should occur, the primary relay must operate to clear the fault. This bias is a result of decades of experience with operating ac power systems where the main threat to the system, aside from damage to equipment, is a loss of synchronism among generators due to slowly cleared faults. High-speed fault clearing is considered essential on modern power systems since operation of backup relaying would increase the probability of system instability and extensive outages when primary protection fails to clear a fault.

The bias towards high dependability inevitably leads to a reduction in security of the protection system. In other words, by making the protection system highly dependable, the protection system is also made more prone to false trips when no trips are warranted. For most traditional power systems, this increased probability of an occasional false trip was considered an acceptable risk to ensure a dependable high-speed clearing of all faults coupled with the fact that the system itself must withstand the loss of a single system element. However, as the power systems undergo fundamental changes, such as those brought on by open access and deregulation, one must reexamine this traditional protection philosophy, especially when the power system is stressed.

WHAT ARE THE TYPES OF RELAY MISOPERATIONS?

In how many ways can a relay misoperate? Of course, there is a possibility of the failure to trip for a fault. One could rule out this type of misoperation in light of the multiple tripping relays employed in most modern power systems. This leaves the mode of failure that could be generically labeled an unanticipated trip. To further understand this phenomenon, consider the following ways in which the protection system may operate.

Correct and appropriate

Correct and appropriate is the operation we all desire. The protection system operates as designed and its operation is appropriate for the prevailing system condition.

Correct and inappropriate

In the case of *correct and inappropriate*, the protection system has operated as designed and set, but its operation was not appropriate for the prevailing system condition. This type of misoperation is quite common and can occur for many reasons. For example, when the relays were set, the network conditions for which the settings should be designed were not properly identified. In other words, the prevailing network condition when the relay operated was not anticipated in the specification stage. Depending on the utility practices, this may be an oversight on the part of the planning department or engineering department. We are seeing more and more such oversights as the number of trained engineering staff in most electric utility companies is severely depleted. Since loading patterns on the networks have changed due to open access and deregulation, the engineers who specify conditions under which the relays must operate have to revisit their network contingencies.

Another reason for this type of operation could be a wrong setting made when the relay was installed or calibrated. The result is that the relay operated as it was designed or set to do, but the result was not what was intended.

Incorrect and appropriate

Although, for completeness, we include *incorrect and appropriate* in this category, we do not believe this is a common occurrence. One could include those operations in which a relay operated incorrectly, but in so doing prevented another, more serious trip and avoided a major cascading failure.

Incorrect and inappropriate

In the case of *incorrect and inappropriate*, the relays have produced an incorrect operation, which has contributed to a cascading outage. This phenomenon has been investigated from the point of view of hidden failures in protection systems. Given the very large number of protective devices in service, it is inevitable that some of these will have failed in a manner that does not exhibit itself when the power system is in a normal state. However, when another triggering event (such as a fault) occurs, the hidden failure causes the offending relay to misoperate. Recent research results on these phenomena, and possible countermeasures against them, have been published in the technical literature.

DISTANCE RELAYS

A distance relay is a protective relay in which the response to the local voltage and current is primarily a function of the impedance between the relay location and the point of fault. It is independent of a communication link and is the most commonly used backup protective system. This is the most ubiquitous relay found on any power system. In addition to transmission line protection, it is also used for out-of-step relaying and loss-of-field relaying and is often a component of many remedial action schemes.

In transmission line protection, which is the principal application of distance relays, several zones are employed in a step-distance configuration, in which each zone is separated by a difference in operating times. Figure 1 shows a typical step-distance relay timing arrangement. Zone 1 provides instantaneous protection for 80–90% of the line between its terminals. Zone 2 is applied to protect the remaining 10–20% of the line but with time delay. Zone 2 also provides some level of backup to the adjacent

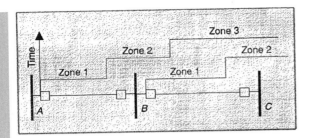

Figure 1
Step distance protection of a transmission system; zone coverage and time delay coordination.

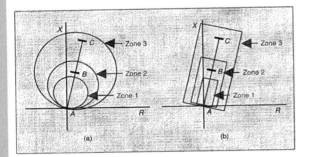

Figure 2
Step distance relay characteristics: (a) typical electromechanical relay characteristic and (b) representative electronic or computer relay characteristics.

transmission line. Zone 3 is applied as a remote backup to Zones 1 and 2 of all adjacent lines in the event that the protective systems of those lines fail to clear a fault. Figure 2 shows the operating characteristics of the separate zones of the step-distance relay protection on an R-X diagram. Circular characteristics shown in Figure 2(a) are common in electromechanical relays and certain computer relays. More common in modern electronic and computer relays are the quadrilateral characteristics shown in Figure 2(b).

THE LIABILITY OF LOADABILITY

Loadability of a relay is the maximum load that may be carried on the protected line such that a load

greater than the loadability limit will be indistinguishable from a fault to the relay and will lead to a trip of the line. The loadability of a relay is affected by the prevailing voltage at the line terminal and also by the load power factor. When the power system is stressed, the reactive power flows may become unusual. When the power system is undergoing electromechanical swings, the voltages near the electrical center of the system may get depressed significantly. When system contingencies are specified to confirm the appropriateness of relay characteristic settings, it is imperative that these effects on loadability be taken into account.

Loadability problems of overreaching zones of protection have been recognized since the early days of protection. Where a third zone as a remote backup zone is deemed to be necessary, certain technical innovations are available to alter its loadability limit. These innovations are more readily implemented in modern computer relays. However, even the electromechanical relays do have limited ability to improve their loadability.

If the voltage and current phasors at the line terminal are such that the loadability limit is reached, the apparent impedance seen by a distance relay must be on the relay characteristic. For simplicity of analysis, consider a rectangular characteristic of a line relay (such as a Zone 3 setting since it has the lowest loadability limit), where the line has negligible resistance. If the line is exporting both active and reactive power at a power factor angle ϕ, the apparent impedance at this loadability lies in the first quadrant of the R-X diagram as shown in Figure 3. For the rectangular characteristic considered here, let the intercept on the R-axis be R_0. The loadability limit S_0 is given by (per unit)

$$S_0 = (P_0 + jQ_0)$$
$$= (E^2/R_0)(\cos^2 \phi + j \sin \phi \cos \phi).$$

It is clear that the loadability limit is a function of both the voltage and the power factor angle. If the loadability limit at unity power factor and with the line voltage of 1.0 per unit is S_{max}, the loadability limit at a voltage of E per unit and power factor angle ϕ is given by

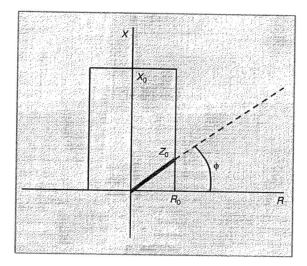

Figure 3
Loadability limit for a simple rectangular impedance characteristic. The line is exporting active and reactive power.

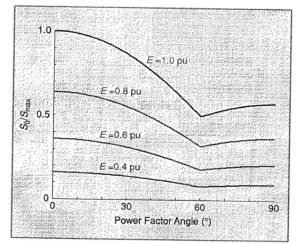

Figure 4
Loadability limit of a rectangular zone shape. Note the dependence on power factor angle and the line terminal voltage.

$$S_0 = S_{max}E^2 \cos \phi.$$

The dependence of loadability on line voltage and power factor is shown on Figure 4. The break in the curves is due to the corner of the rectangle, which is assumed to occur at a power factor angle of 60°. It should be clear that the loadability dependence on voltage and power factor angle is strongly influenced by the shape of the relay characteristic and must be evaluated for the characteristic actually in use. Note also that the ordinate of Figure 4 shows the mega-voltampere loadability; the megawatt loadability is obtained by multiplying the MVA loadability by the appropriate power factor.

WHEN IS A THIRD ZONE NEEDED?

Zone 3 of a distance relay is used to provide remote backup protection if the primary protection system should fail. Although in common usage a protection system may mean only the relays, the actual protection system consists of many other subsystems that contribute to the detection and removal of faults. Each of these subsystems may fail, making the primary protection inoperative. Figure 5 shows the

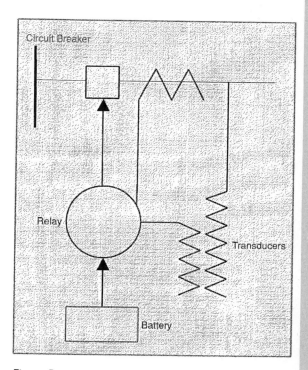

Figure 5
Elements of a protection system.

basic elements of the protection system. In the following discussion, we examine all of the protective elements in a substation with regard to the application of Zone 3 to provide backup.

Batteries

If only one battery is available at the substation, a system control, alarm, and data acquisition system (SCADA) alerts the engineering or operating department to take corrective action if the battery becomes defective. At the higher transmission voltage levels, it is not uncommon to provide two batteries, in which case providing backup protection for battery failure may not be necessary. However, if only one battery is available, even with a SCADA warning it may be advisable to add Zone 3 at the remote stations(s) if the failed battery is at a location that is not easily reached and if maintenance personnel may not have the time to correct the problem quickly.

Relays

To cover any single relay failure, it is common practice to use multiple relays covering the phase and ground faults. At the higher voltages or at more critical stations, there would be two sets of relays including pilot protection. One may therefore conclude that remote backup protection may be unnecessary, but care must be taken to be sure that no common-mode failures exist within the circuitry of multiple relay sets.

Transducers

At the lower voltage levels, the transducers, current transformers (CTs) and potential transformers (PTs) are not normally duplicated, and a failure of the potential or current transformers could go unnoticed and result in a failure to trip. In this instance, a Zone 3 remote backup would be desirable. At the higher voltage levels, the current transformer secondary windings are duplicated, each serving a separate set of relays. The potential transformers or devices are also duplicated or fused separately to maintain voltage integrity to each set of relays.

Circuit breakers

Circuit breakers are not duplicated, and failure of a circuit breaker to clear a fault must be considered. Circuit breaker failure tripping schemes are sensitive to system and station configuration. In some cases, it is sufficient to open all local breakers that can contribute to the fault upon detecting a breaker failure. This may not be sufficient to clear a fault, and a transfer trip scheme is required. This involves expenditures for communication equipment, which may not be justified, and a remote Zone 3 would be preferable. It is not our purpose here to catalog all possible bus arrangements, but some examples are instructive.

As extra-high voltage (EHV) transmission systems matured, local backup replaced remote backup, and breaker failure schemes started to evolve as a subset of local backup protection. At first, a separate set of relays was used to initiate the required tripping, but this quickly was replaced by a special isolated circuit that employed all of the protective relays that simultaneously initiated tripping the appropriate circuit breaker(s) and started a timer. When the timer timed out, a tripping relay opened all of the breakers that could see the fault for which the protective relays had operated. The timer was set just beyond the normal clearing time of the circuit breaker, usually on the order of 7–10 cycles.

Substations are designed for reliability of service and flexibility of operation and to allow for equipment maintenance with a minimum interruption of service consistent with an economic evaluation of the costs involved versus the benefits derived. Bus arrangements range from a single-bus, single-breaker scheme common to low-voltage distribution stations to the breaker-and-a half configuration that is almost universal for EHV stations. In situations where breaker failure schemes trip all of the breakers that can sense a fault, including transfer trip, a Zone 3 application may not have any advantage. However, in the absence of a communication channel for technical or economic reasons, a Zone 3 setting may be of some advantage. Nonetheless, the Zone 3 setting must encompass all possible infeeds. In the case of tapped loads, it is not always possible to set Zone 2 to see the apparent impedance to the "far end" fault and not overreach other Zone 2 relays. In such cases, a possible solution is to set a remote Zone 3 to see the apparent impedance to the remote location.

As each component failure is identified, a backup scheme capable of covering that failure must be designed. In general, in the absence of a communication channel, duplicate transducers or batteries, and for certain bus configurations, remote Zone 3 provides the most effective backup protection. Of course, catastrophic physical failures such as earthquakes or storms that destroy the building, panels, or multiple primary equipment will also require a remote backup system such as Zone 3. We may also classify as a catastrophic failure certain human errors such as incorrect settings or equipment outages during maintenance.

COUNTERMEASURES FOR IMPROVING ZONE 3 LOADABILITY

One thing to note is that load excursions are balanced phenomena, so that the presence of unbalanced currents (negative sequence) would indicate a fault. Thus, loadability should not be an issue when negative sequence currents are present. The third zone characteristic for three-phase faults is the only one where load conditions could be confused with remote faults. It is also unlikely that three-phase faults would have a significant fault resistance, so the third zone shape for such faults could be considerably modified to reduce the resistive reach of the relay, thus increasing the loadability of the relay. These and other improvements are generally available only in computer relays.

Electromechanical relays
The traditional electromechanical relays have fixed zone boundaries, and usually they are made up of circular shapes or straight lines. Directional relays can be applied as blinders restricting the admittance (mho) circular characteristics of Zone 3 to high X/R values as opposed to normal load power factor angles. As indicated above, the presence of negative sequence current is a good indicator that a fault, not a balanced load condition, exists. Some innovations in electromechanical relays have addressed the loadability problem by proposing the use of a figure-eight shape (a forward offset characteristic) to increase the loadability limits of relays. This offset mho characteristic is easily attainable with electromechanical

relays and may offer one of the best and most inexpensive solutions to the loadablity problem.

Electronic relays
Solid-state relays can shape their characteristics or introduce load encroachment elements that are also attractive solutions. Quadrilateral characteristics, which shape the protection zones to any desired shape, are a definite improvement.

Computer relays
Computer relays have the potential for solving many of the problems that could not be addressed by conventional relays. Many computer relays now offer multiple groups of settings, so that Zone 3 settings could be adapted to changing system configurations and the accompanying in-feed variations. It is also possible to include a load-encroachment characteristic which outlines an area within the relay characteristic that prevents operation for a defined load impedance. Computer relays could also be made to block trip if a balanced condition exists and the power factor of the line current is characteristic of system loading conditions. This approach is indeed reasonable and should provide security in the case of a heavy load. However, one should take note of the fact that often under unusual system conditions where significant amounts of vars are being transmitted, the power factor may not always be a sure indicator that a load, rather than a fault, exists on the line. Under normal system conditions, vars and watts are tightly coupled through the generator operating parameters. Under system stress or fault conditions, vars and watts may be decoupled and the 30° power factor criteria may not apply (see Figure 3 for the effect of power factor and voltage on loadability).

If one postulates communication between relays at a station or with relays at remote stations, it may be possible to design more effective logic for differentiating between a load and a fault. Trending observed changes in line currents, correlating changes seen by different relays, and using information received from control centers indicative of the state of the power system could all be integrated in computer relays with a very effective check on the operation of backup protection systems.

Adaptive relaying

The intelligent supervision of Zone 3 is an excellent example of adaptive relaying, with its ability to adjust its performance to match the prevailing power system conditions. The problem of the sensitivity of the Zone 3 characteristic to emergency load or power system instability can be solved with the use of adaptive relaying principles. Several solutions using electromechanical or electronic relays have already been suggested. Each of these solutions is more readily applied with computer relays. Using computer logic instead of complex wiring immediately improves the reliability of the relay. Prefault load could be taken into account, and prevailing stability margins could be introduced into the relaying algorithms. The overwhelming thrust of the North American Electric Reliability Council (NERC) rules and other instructions regarding the application of Zone 3 elements has been to prevent their operation during emergency conditions. Although this is a desirable goal, it should be recognized that even with all the intelligence available to modern computer relays, the problem of distinguishing a fault from a heavy load in a relatively short time and using only the current and voltage signals available to the relay cannot be solved in every single imaginable (and some unimaginable) power system scenario. If Zone 3 is the protection system of last resort, and if it is incorrectly prevented from operating when it should do so, the consequences to the power system may be much worse than those resulting from an unwanted Zone 3 trip.

Hidden failures

A vexing problem for protection engineers is that of hidden failures in the protection system. These are failures that go undetected when the power system is in a normal state and that contribute unnecessary trips when the system is disturbed by faults or severe dynamic conditions. Although increased maintenance, calibration, and frequent reviews of protective settings would reduce the likelihood of hidden failures, this is not always the case. Known cases abound where the very act of maintenance has induced a hidden failure mode that remains unexposed until cascading failures begin to unfold.

Given the large number of protective devices present on a power system, it is not possible to rule out the possibility of hidden failures in some relays. An approach for providing countermeasures against false trips due to hidden failures is to identify locations in the network where hidden failures would be particularly damaging to the power system and then provide a computer-based relay as a supervisor of the relay operation. In addition, a voting scheme among all the protection systems can be implemented at such critical sites.

INTEGRATING PROTECTION AND CONTROL

Our concern up to this point has been the load-ability of backup protection schemes. In any realistic scenario, the total protection, operation, and control package must be considered as an integrated system. It is being recognized that the advent of new technologies has provided two of the most valuable elements of this total package: system integrity protective schemes (SIPSs) and wide-area-based protection systems (WAPSs).

SIPS

SIPS is an increasingly valuable tool to initiate system corrective actions as opposed to equipment protection. Among the many system stress scenarios that a SIPS may act upon are transmission congestion, transient instability, voltage and frequency degradation, and thermal overloading. Many corrective actions are available to respond to these potential system stresses. An SIPS program can initiate any one or combination of the following: load or generator rejection or fast valving, load shedding due to underfreqency or undervoltage, out-of-step blocking and tripping, system separation, etc. These measures are normally inactive and are armed automatically when some system stress condition occurs; alternatively, they may be armed manually from a control center. The remedial action schemes (RAS) and system protection schemes (SPS) that have become popular in recent years are SIPS systems. To create more intelligent SIPSs, one must rely on the newly available wide-area measurement systems being deployed in many power systems.

Wide-area protection systems (WAPSs)

To implement effective, intelligent SIPS appropriate for the prevailing power system condition, it is essential that a real-time, fast-acting, system-wide data collection system be available. Wide-area measurement systems (WAMSs) are rapidly becoming an important element in system operation and control. The use of real-time measurements to determine the state of the power system goes back to the late 1960s. State estimation, contingency evaluations, and optimization of the operating state have been practiced by most modern power systems. However, the constraining contingencies for power systems are usually rooted in the dynamic phenomena. Thus, there has been a disconnect between what is actually needed in real time and what could be achieved by the prevailing technology. All this is changing with the introduction of the global positioning system (GPS), which can be used to provide synchronized phasor measurements for the entire power system, thus providing high-quality system data at a very rapid rate (once every few cycles of the power system frequency). High-speed broadband communication has also become available so that these measurements could be transferred over great distances with latencies on the order of 50 ms, and progress is being made in developing analytical techniques that can effectively use these measurements in real time.

Brittle and ductile systems

Modern power systems deliver power to the local load centers efficiently and economically. However, they are brittle in the sense that they tend to break up through the loss of synchronism when faced with a catastrophic disturbance. The break up can be compared to the shattering of a brittle structure upon being struck with a heavy blow. A ductile structure, on the other hand, would deform around the disturbance and prevent the disturbance from cascading into a total system collapse. This is illustrated conceptually in Figure 6. It is possible that the developing stress levels in modern power systems will compel engineers to bring some measure of ductility to the power systems by judicious use of high-power electronic devices.

CHANGING NATURE OF POWER SYSTEMS

It is clear that the power systems of today are not what they were before open access and deregulation were introduced in many countries. We are witnessing unusual generation and power flow pat-

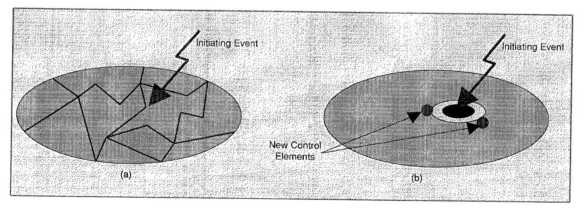

Figure 6
(a) Brittle power system breaks up into synchronous islands when a strong disturbance occurs. (b) By using high-power electronic devices at optimum locations, the damage due to the initiating event can be confined to a small region, rendering the power system ductile.

terns, unplanned congestion, and the introduction of renewable generation with its associated uncertainties regarding protection and control. The emphasis on stockholder good versus public good has transformed the industry culture from reliability to economy. This change has resulted in a lack of incentives for infrastructure improvements, restrictions on information exchange, and, most disturbingly, the lack of signals for technical manpower needs, both present and future. To meet these challenges, the industry must embrace the latest technologies, recruit the best and brightest of our graduates, and increase the exchange of technical concepts and innovation though industry conferences and literature.

FOR FURTHER READING

S. H. Horowitz and A. G. Phadke, *Power System Relaying*, 2nd ed. Somerset, England: Research Studies Press, 1995.

S. H. Horowitz and A. G. Phadke, "Third zone revisited," *IEEE Trans. Power Syst. Delivery*, pp. 23–29, Jan. 2006.

S. H. Horowitz, A. G. Phadke, and J. S. Thorp, "Adaptive transmission system relaying," in *Proc. IEEE Summer Meeting*, *1987 Trans. PWRD*, Oct. 1988, pp. 1436–1445.

S. H. Horowitz and A. G. Phadke, "Boosting immunity to blackouts," *IEEE Power Energy Mag.*, vol. 1, pp. 47–53, Sep./Oct. 2003.

BIOGRAPHIES

S. H. Horowitz received a bachelor's degree in electrical engineering from the City College of New York. He is a consultant, author, and lecturer. He is a Life Fellow of the IEEE and a member of the National Academy of Engineering.

A. G. Phadke received a B.Sc. degree from Agra University, a B.Tech. degree from the Indian Institute of Technology, an M.S. degree from the Illinois Institute of Technology, and a Ph.D. from the University of Wisconsin. He is a University Distinguished Professor (emeritus) at Virginia Tech in Blacksburg, Virginia. He is a Life Fellow of the IEEE and a member of the National Academy of Engineering.

10.1

SYSTEM PROTECTION COMPONENTS

Protection systems have three basic components:

1. Instrument transformers

2. Relays

3. Circuit breakers

Figure 10.1 shows a simple overcurrent protection schematic with: (1) one type of instrument transformer—the current transformer (CT), (2) an overcurrent relay (OC), and (3) a circuit breaker (CB) for a single-phase line. The function of the CT is to reproduce in its secondary winding a current I' that is proportional to the primary current I. The CT converts primary currents in the kiloamp range to secondary currents in the 0–5 ampere range for convenience of measurement, with the following advantages.

Safety: Instrument transformers provide electrical isolation from the power system so that personnel working with relays will work in a safer environment.

FIGURE 10.1

Overcurrent protection schematic

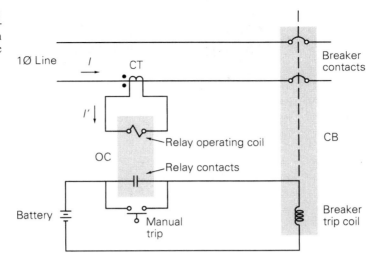

Economy: Lower-level relay inputs enable relays to be smaller, simpler, and less expensive.

Accuracy: Instrument transformers accurately reproduce power system currents and voltages over wide operating ranges.

The function of the relay is to discriminate between normal operation and fault conditions. The OC relay in Figure 10.1 has an operating coil, which is connected to the CT secondary winding, and a set of contacts. When $|I'|$ exceeds a specified "pickup" value, the operating coil causes the normally open contacts to close. When the relay contacts close, the trip coil of the circuit breaker is energized, which then causes the circuit breaker to open.

Note that the circuit breaker does not open until its operating coil is energized, either manually or by relay operation. Based on information from instrument transformers, a decision is made and "relayed" to the trip coil of the breaker, which actually opens the power circuit—hence the name *relay*.

System-protection components have the following design criteria [2]:

Reliability: Operate dependably when fault conditions occur, even after remaining idle for months or years. Failure to do so may result in costly damages.

Selectivity: Avoid unnecessary, false trips.

Speed: Operate rapidly to minimize fault duration and equipment damage. Any intentional time delays should be precise.

Economy: Provide maximum protection at minimum cost.

Simplicity: Minimize protection equipment and circuitry.

Since it is impossible to satisfy all these criteria simultaneously, compromises must be made in system protection.

10.2

INSTRUMENT TRANSFORMERS

There are two basic types of instrument transformers: voltage transformers (VTs), formerly called potential transformers (PTs), and current transformers (CTs). Figure 10.2 shows a schematic representation for the VT and CT. The transformer primary is connected to or into the power system and is insulated for the power system voltage. The VT reduces the primary voltage and the CT reduces the primary current to much lower, standardized levels suitable for operation of relays. Photos of VTs and CTs are shown in Figures 10.3–10.6.

For system-protection purposes, VTs are generally considered to be sufficiently accurate. Therefore, the VT is usually modeled as an ideal transformer, where

$$V' = (1/\mathrm{n})V \tag{10.2.1}$$

V' is a scaled-down representation of V and is in phase with V. A standard VT secondary voltage rating is 115 V (line-to-line). Standard VT ratios are given in Table 10.1.

Ideally, the VT secondary is connected to a voltage-sensing device with infinite impedance, such that the entire VT secondary voltage is across the sensing device. In practice, the secondary voltage divides across the high-impedance sensing device and the VT series leakage impedances. VT leakage impedances are kept low in order to minimize voltage drops and phase-angle differences from primary to secondary.

The primary winding of a current transformer usually consists of a single turn, obtained by running the power system's primary conductor through the CT core. The normal current rating of CT secondaries is standardized at 5 A in the United States, whereas 1 A is standard in Europe and

FIGURE 10.2

VT and CT schematic

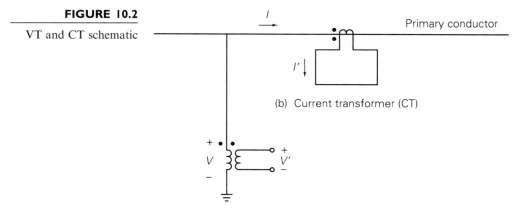

(b) Current transformer (CT)

(a) Voltage transformer (VT)

FIGURE 10.3

Three 34.5-kV voltage transformers with 34.5 kV : 115/67 volt VT ratios, at Lisle substation, Lisle, Illinois (Courtesy of Commonwealth Edison, an Exelon company)

FIGURE 10.4

Three 500-kV coupling capacitor voltage transformers with 303.1 kV : 115/67 volt VT ratios, Westwing 500-kV Switching Substation (Courtesy of Arizona Public Service)

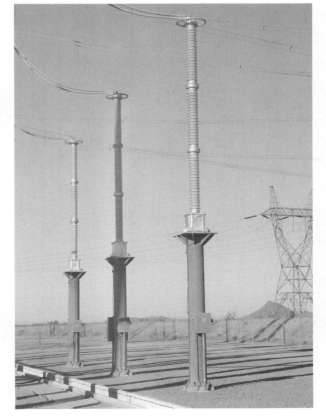

FIGURE 10.5

Three 25 kV class
current transformers–
window design
(Courtesy of Kuhlman
Electric Corporation)

FIGURE 10.6

500-kV class current
transformers with
2000 : 5 CT ratios in
front of 500-kV SF6
circuit breakers,
Westwing 500-kV
Switching Substation
(Courtesy of Arizona
Public Service
Company)

TABLE 10.1	Voltage Ratios						
Standard VT ratios	1:1	2:1	2.5:1	4:1	5:1	20:1	40:1
	60:1	100:1	200:1	300:1	400:1	600:1	800:1
	1000:1	2000:1	3000:1	4500:1			

TABLE 10.2	Current Ratios						
Standard CT ratios	50:5	100:5	150:5	200:5	250:5	300:5	400:5
	450:5	500:5	600:5	800:5	900:5	1000:5	1200:5
	1500:5	1600:5	2000:5	2400:5	2500:5	3000:5	3200:5
	4000:5	5000:5	6000:5				

some other regions. Currents of 10 to 20 times (or greater) normal rating often occur in CT windings for a few cycles during short circuits. Standard CT ratios are given in Table 10.2.

Ideally, the CT secondary is connected to a current-sensing device with zero impedance, such that the entire CT secondary current flows through the sensing device. In practice, the secondary current divides, with most flowing through the low-impedance sensing device and some flowing through the CT shunt excitation impedance. CT excitation impedance is kept high in order to minimize excitation current.

An approximate equivalent circuit of a CT is shown in Figure 10.7, where

Z' = CT secondary leakage impedance

X_e = (Saturable) CT excitation reactance

Z_B = Impedance of terminating device (relay, including leads)

The total impedance Z_B of the terminating device is called the *burden* and is typically expressed in values of less than an ohm. The burden on a CT may also be expressed as volt-amperes at a specified current.

Associated with the CT equivalent circuit is an excitation curve that determines the relationship between the CT secondary voltage E' and excitation current I_e. Excitation curves for a multiratio bushing CT with ANSI classification C100 are shown in Figure 10.8.

FIGURE 10.7

CT equivalent circuit

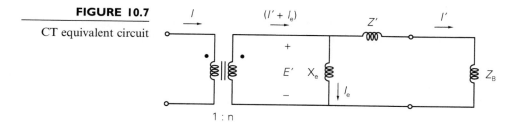

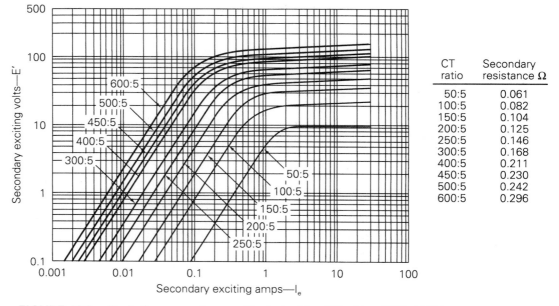

CT ratio	Secondary resistance Ω
50:5	0.061
100:5	0.082
150:5	0.104
200:5	0.125
250:5	0.146
300:5	0.168
400:5	0.211
450:5	0.230
500:5	0.242
600:5	0.296

FIGURE 10.8   Excitation curves for a multiratio bushing CT with a C100 ANSI accuracy classification [3] (Westinghouse Relay Manual, A New Silent Sentinels Publication (Newark, NJ: Westinghouse Electric Corporation, 1972))

Current transformer performance is based on the ability to deliver a secondary output current I' that accurately reproduces the primary current I. Performance is determined by the highest current that can be reproduced without saturation to cause large errors. Using the CT equivalent circuit and excitation curves, the following procedure can be used to determine CT performance.

STEP 1   Assume a CT secondary output current I'.

STEP 2   Compute $E' = (Z' + Z_B)I'$.

STEP 3   Using E', find I_e from the excitation curve.

STEP 4   Compute $I = n(I' + I_e)$.

STEP 5   Repeat Steps 1–4 for different values of I', then plot I' versus I.

For simplicity, approximate computations are made with magnitudes rather than with phasors. Also, the CT error is the percentage difference between $(I' + I_e)$ and I', given by:

$$\text{CT error} = \frac{I_e}{I' + I_e} \times 100\% \qquad (10.2.2)$$

The following examples illustrate the procedure.

EXAMPLE 10.1   Current transformer (CT) performance

Evaluate the performance of the multiratio CT in Figure 10.8 with a 100:5 CT ratio, for the following secondary output currents and burdens: (a) $I' = 5$ A and $Z_B = 0.5$ Ω; (b) $I' = 8$ A and $Z_B = 0.8$ Ω; and (c) $I' = 15$ A and $Z_B = 1.5$ Ω. Also, compute the CT error for each output current.

SOLUTION From Figure 10.8, the CT with a 100:5 CT ratio has a secondary resistance $Z' = 0.082$ Ω. Completing the above steps:

a. **STEP 1** $I' = 5$ A

  STEP 2   From Figure 10.7,

$$E' = (Z' + Z_B)I' = (0.082 + 0.5)(5) = 2.91 \text{ V}$$

  STEP 3   From Figure 10.8, $I_e = 0.25$ A

  STEP 4   From Figure 10.7, $I = (100/5)(5 + 0.25) = 105$ A

  CT error $= \dfrac{0.25}{5.25} \times 100 = 4.8\%$

b. **STEP 1** $I' = 8$ A

  STEP 2   From Figure 10.7,

$$E' = (Z' + Z_B)I' = (0.082 + 0.8)(8) = 7.06 \text{ V}$$

  STEP 3   From Figure 10.8, $I_e = 0.4$ A

  STEP 4   From Figure 10.7, $I = (100/5)(8 + 0.4) = 168$ A

  CT error $= \dfrac{0.4}{8.4} \times 100 = 4.8\%$

c. **STEP 1** $I' = 15$ A

  STEP 2   From Figure 10.7,

$$E' = (Z' + Z_B)I' = (0.082 + 1.5)(15) = 23.73 \text{ V}$$

  STEP 3   From Figure 10.8, $I_e = 20$ A

  STEP 4   From Figure 10.7, $I = (100/5)(15 + 20) = 700$ A

  CT error $= \dfrac{20}{35} \times 100 = 57.1\%$

Note that for the 15-A secondary current in (c), high CT saturation causes a large CT error of 57.1%. Standard practice is to select a CT ratio to give a little less than 5-A secondary output current at maximum normal load. From (a), the 100:5 CT ratio and 0.5 Ω burden are suitable for a maximum primary load current of about 100 A. This example is extended in Problem 10.2 to obtain a plot of I' versus I. ∎

EXAMPLE 10.2   **Relay operation versus fault current and CT burden**

An overcurrent relay set to operate at 8 A is connected to the multiratio CT in Figure 10.8 with a 100:5 CT ratio. Will the relay detect a 200-A primary fault current if the burden Z_B is (a) 0.8 Ω, (b) 3.0 Ω?

SOLUTION   Note that if an ideal CT is assumed, $(100/5) \times 8 = 160$-A primary current would cause the relay to operate.

a. From Example 10.1(b), a 168-A primary current with $Z_B = 0.8\ \Omega$ produces a secondary output current of 8 A, which would cause the relay to operate. Therefore, the higher 200-A fault current will also cause the relay to operate.

b. STEP 1 $I' = 8$ A

 STEP 2   From Figure 10.7,

$$E' = (Z' + Z_B)I' = (0.05 + 3.0)(8) = 24.4 \quad V$$

 STEP 3   From Figure 10.8, $I_e = 30$ A

 STEP 4   From Figure 10.7, $I = (100/5)(8 + 30) = 760$ A

With a 3.0-Ω burden, 760 A is the lowest primary current that causes the relay to operate. Therefore, the relay will not operate for the 200-A fault current.   ∎

10.3

OVERCURRENT RELAYS

As shown in Figure 10.1, the CT secondary current I' is the input to the overcurrent relay operating coil. Instantaneous overcurrent relays respond to the magnitude of their input current, as shown by the trip and block regions in Figure 10.9. If the current magnitude $I' = |I'|$ exceeds a specified adjustable current magnitude I_p, called the *pickup* current, then the relay contacts close "instantaneously" to energize the circuit breaker trip coil. If I' is less than the pickup current I_p, then the relay contacts remain open, blocking the trip coil.

Time-delay overcurrent relays also respond to the magnitude of their input current, but with an intentional time delay. As shown in Figure 10.10, the time delay depends on the magnitude of the relay input current. If I' is a large multiple of the pickup current I_p, then the relay operates (or trips) after a small time delay. For smaller multiples of pickup, the relay trips after a longer time delay. And if $I' < I_p$, the relay remains in the blocking position.

FIGURE 10.9

FIGURE 10.9

Instantaneous
overcurrent relay block
and trip regions

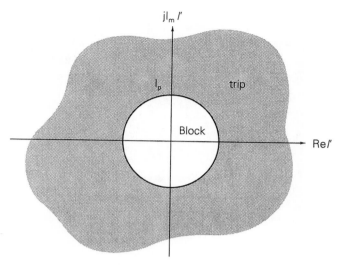

FIGURE 10.10

Time-delay overcurrent
relay block and trip
regions

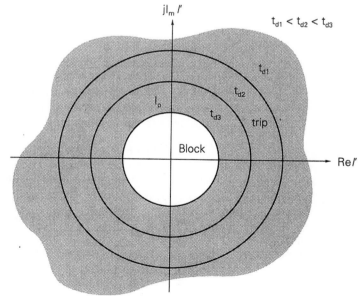

FIGURE 10.11

Electromechanical
time-delay overcurrent
relay (Courtesy of
ABB-Westinghouse)

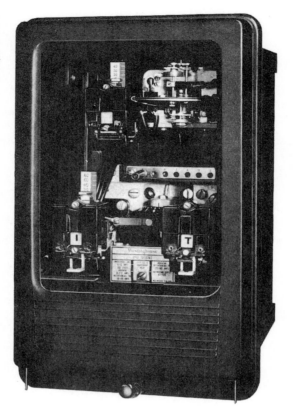

The ABB-Westinghouse series of CO relays, one of which is shown in Figure 10.11, is a typical time-delay overcurrent relay product line. Characteristic curves of the CO-8 relay are shown in Figure 10.12. These relays have two settings:

Current tap setting: The pickup current in amperes.

Time-dial setting: The adjustable amount of time delay.

The characteristic curves are usually shown with operating time in seconds versus relay input current as a multiple of the pickup current. The curves are asymptotic to the vertical axis and decrease with some inverse power of current magnitude for values exceeding the pickup current. This inverse time characteristic can be shifted up or down by adjustment of the time-dial setting. Although discrete time-dial settings are shown in Figure 10.12, intermediate values can be obtained by interpolating between the discrete curves.

FIGURE 10.12

CO-8 time-delay overcurrent relay characteristics (Courtesy of Westinghouse Electric Corporation)

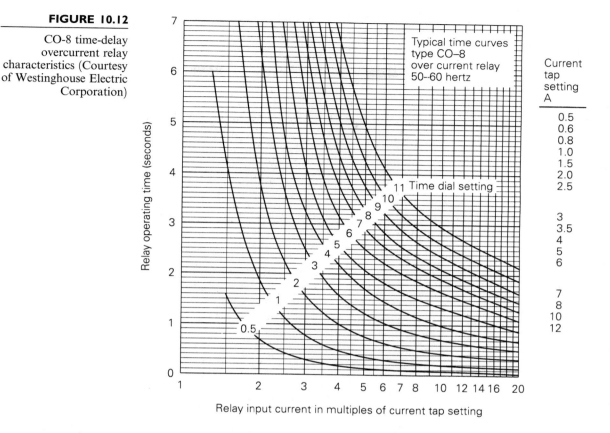

Relay operating time (seconds)

Typical time curves type CO–8 over current relay 50–60 hertz

11 Time dial setting

Relay input current in multiples of current tap setting

Current tap setting A
0.5
0.6
0.8
1.0
1.5
2.0
2.5
3
3.5
4
5
6
7
8
10
12

EXAMPLE 10.3   **Operating time for a CO-8 time-delay overcurrent relay**

The CO-8 relay with a current tap setting of 6 amperes and a time-dial setting of 1 is used with the $100:5$ CT in Example 10.1. Determine the relay operating time for each case.

SOLUTION

a. From Example 10.1(a)

$$I' = 5 \text{ A} \qquad \frac{I'}{I_p} = \frac{5}{6} = 0.83$$

The relay does not operate. It remains in the blocking position.

b.
$$I' = 8 \text{ A} \qquad \frac{I'}{I_p} = \frac{8}{6} = 1.33$$

Using curve 1 in Figure 10.12, $t_{operating} = 6$ seconds.

c.
$$I' = 15 \text{ A} \qquad \frac{I'}{I_p} = \frac{15}{6} = 2.5$$

From curve 1, $t_{operating} = 1.2$ seconds.

FIGURE 10.13

Comparison of CO relay
characteristics (Courtesy
of ABB-Westinghouse)

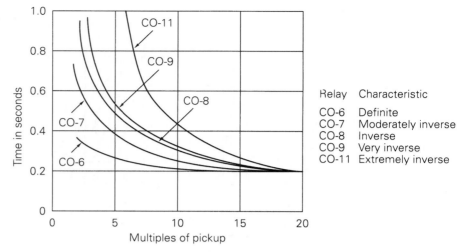

Relay	Characteristic
CO-6	Definite
CO-7	Moderately inverse
CO-8	Inverse
CO-9	Very inverse
CO-11	Extremely inverse

Figure 10.13 shows the time-current characteristics of five CO time-delay overcurrent relays used in transmission and distribution lines. The time-dial settings are selected in the figure so that all relays operate in 0.2 seconds at 20 times the pickup current. The choice of relay time-current characteristic depends on the sources, lines, and loads. The definite (CO-6) and moderately inverse (CO-7) relays maintain a relatively constant operating time above 10 times pickup. The inverse (CO-8), very inverse (CO-9), and extremely inverse (CO-11) relays operate respectively faster on higher fault currents.

Figure 10.14 illustrates the operating principle of an electromechanical time-delay overcurrent relay. The ac input current to the relay operating coil sets up a magnetic field that is perpendicular to a conducting aluminum disc. The disc can rotate and is restrained by a spiral spring. Current is induced in the disc, interacts with the magnetic field, and produces a torque. If the input current exceeds the pickup current, the disc rotates through an angle θ to close the relay contacts. The larger the input current, the larger is the torque and the faster the contact closing. After the current is removed or reduced below the pickup, the spring provides reset of the contacts.

A solid state relay panel between older-style electromechanical relays is shown in Figure 10.15.

FIGURE 10.14

Electromechanical time-delay overcurrent
relay—induction disc
type

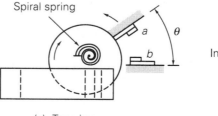

(a) Top view

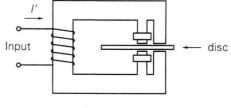

(b) Side view

FIGURE 10.15

Solid-state relay panel
(center) for a 345-kV
transmission line, with
electromechanical relays
on each side, at Electric
Junction Substation,
Naperville, Illinois
(Courtesy of
Commonwealth Edison,
an Exelon Company)

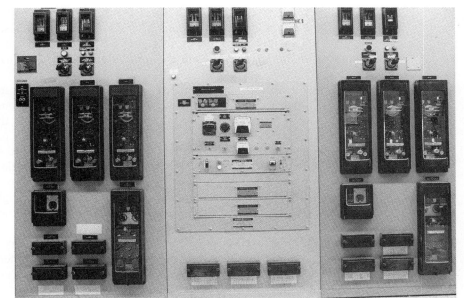

10.4

RADIAL SYSTEM PROTECTION

Many radial systems are protected by time-delay overcurrent relays. Adjustable time delays can be selected such that the breaker closest to the fault opens, while other upstream breakers with larger time delays remain closed. That is, the relays can be coordinated to operate in sequence so as to interrupt minimum load during faults. Successful relay coordination is obtained when fault currents are much larger than normal load currents. Also, coordination of overcurrent relays usually limits the maximum number of breakers in a radial system to five or less, otherwise the relay closest to the source may have an excessive time delay.

Consider a fault at P_1 to the right of breaker B3 for the radial system of Figure 10.16. For this fault we want breaker B3 to open while B2 (and B1)

FIGURE 10.16

Single-line diagram of a
34.5-kV radial system

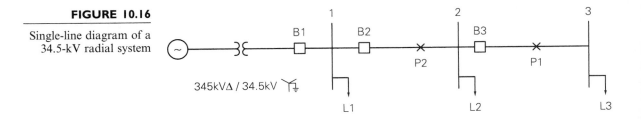

remains closed. Under these conditions, only load L3 is interrupted. We could select a longer time delay for the relay at B2, so that B3 operates first. Thus, for any fault to the right of B3, B3 provides primary protection. Only if B3 fails to open will B2 open, after time delay, thus providing backup protection.

Similarly, consider a fault at P_2 between B2 and B3. We want B2 to open while B1 remains closed. Under these conditions, loads L2 and L3 are interrupted. Since the fault is closer to the source, the fault current will be larger than for the previous fault considered. B2, set to open for the previous, smaller fault current after time delay, will open more rapidly for this fault. We also select the B1 relay with a longer time delay than B2, so that B2 opens first. Thus, B2 provides primary protection for faults between B2 and B3, as well as backup protection for faults to the right of B3. Similarly, B1 provides primary protection for faults between B1 and B2, as well as backup protection for further downstream faults.

The *coordination time interval* is the time interval between the primary and remote backup protective devices. It is the difference between the time that the backup relaying operates and the time that circuit breakers clear the fault under primary relaying. Precise determination of relay operating times is complicated by several factors, including CT error, dc offset component of fault current, and relay overtravel. Therefore, typical coordination time intervals from 0.2 to 0.5 seconds are selected to account for these factors in most practical applications.

EXAMPLE 10.4   Coordinating time-delay overcurrent relays in a radial system

Data for the 60-Hz radial system of Figure 10.16 are given in Tables 10.3, 10.4, and 10.5. Select current tap settings (TSs) and time-dial settings (TDSs) to protect the system from faults. Assume three CO-8 relays for each breaker,

---

TABLE 10.3

Maximum loads— Example 10.4

Bus	S MVA	Lagging p.f.
1	11.0	0.95
2	4.0	0.95
3	6.0	0.95

TABLE 10.4

Symmetrical fault currents—Example 10.4

Bus	Maximum Fault Current (Bolted Three-Phase) A	Minimum Fault Current (L–G or L–L) A
1	3000	2200
2	2000	1500
3	1000	700

Breaker	Breaker Operating Time	CT Ratio	Relay
B1	5 cycles	400:5	CO-8
B2	5 cycles	200:5	CO-8
B3	5 cycles	200:5	CO-8

TABLE 10.5

Breaker, CT, and relay data—Example 10.4

FIGURE 10.17

Relay connections to trip all three phases

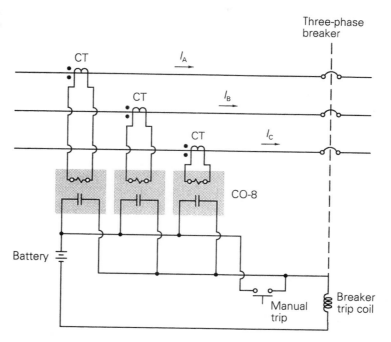

one for each phase, with a 0.3-second coordination time interval. The relays for each breaker are connected as shown in Figure 10.17, so that all three phases of the breaker open when a fault is detected on any one phase. Assume a 34.5-kV (line-to-line) voltage at all buses during normal operation. Also, future load growth is included in Table 10.3, such that maximum loads over the operating life of the radial system are given in this table.

SOLUTION First, select TSs such that the relays do not operate for maximum load currents. Starting at B3, the primary and secondary CT currents for maximum load L3 are

$$I_{L3} = \frac{S_{L3}}{V_3\sqrt{3}} = \frac{6 \times 10^6}{(34.5 \times 10^3)\sqrt{3}} = 100.4 \text{ A}$$

$$I'_{L3} = \frac{100.4}{(200/5)} = 2.51 \text{ A}$$

From Figure 10.12, we select for the B3 relay a 3-A TS, which is the lowest TS above 2.51 A.

Note that $|S_{L2} + S_{L3}| = |S_{L2}| + |S_{L3}|$ because the load power factors are identical. Thus, at B2, the primary and secondary CT currents for maximum load are

$$I_{L2} = \frac{S_{L2} + S_{L3}}{V_2\sqrt{3}} = \frac{(4+6) \times 10^6}{(34.5 \times 10^3)\sqrt{3}} = 167.3 \text{ A}$$

$$I'_{L2} = \frac{167.3}{(200/5)} = 4.18 \text{ A}$$

From Figure 10.12, select for the B2 relay a 5-A TS, the lowest TS above 4.18 A. At B1,

$$I_{L1} = \frac{S_{L1} + S_{L2} + S_{L3}}{V_1\sqrt{3}} = \frac{(11+4+6) \times 10^6}{(34.5 \times 10^3)\sqrt{3}} = 351.4 \text{ A}$$

$$I'_{L1} = \frac{351.4}{(400/5)} = 4.39 \text{ A}$$

Select a 5-A TS for the B1 relay.

Next select the TDSs. We first coordinate for the maximum fault currents in Table 10.4, checking coordination for minimum fault currents later. Starting at B3, the largest fault current through B3 is 2000 A, which occurs for the three-phase fault at bus 2 (just to the right of B3). Neglecting CT saturation, the fault-to-pickup current ratio at B3 for this fault is

$$\frac{I'_{3\text{Fault}}}{\text{TS3}} = \frac{2000/(200/5)}{3} = 16.7$$

Since we want to clear faults as rapidly as possible, select a 1/2 TDS for the B3 relay. Then, from the 1/2 TDS curve in Figure 10.12, the relay operating time is T3 = 0.05 seconds. Adding the breaker operating time (5 cycles = 0.083 s), primary protection clears this fault in $T3 + T_{\text{breaker}} = 0.05 + 0.083 = 0.133$ seconds.

For this same fault, the fault-to-pickup current ratio at B2 is

$$\frac{I'_{2\text{Fault}}}{\text{TS2}} = \frac{2000/(200/5)}{5} = 10.0$$

Adding the B3 relay operating time (T3 = 0.05 s), breaker operating time (0.083 s), and 0.3 s coordination time interval, we want a B2 relay operating time

$$T2 = T3 + T_{\text{breaker}} + T_{\text{coordination}} = 0.05 + 0.083 + 0.3 \approx 0.43 \text{ s}$$

From Figure 10.12, select TDS2 = 2.

Next select the TDS at B1. The largest fault current through B2 is 3000 A, for a three-phase fault at bus 1 (just to the right of B2). The fault-to-pickup current ratio at B2 for this fault is

Breaker	Relay	TS	TDS
B1	CO-8	5	3
B2	CO-8	5	2
B3	CO-8	3	1/2

TABLE 10.6

Solution—Example 10.4

$$\frac{I'_{2\text{Fault}}}{\text{TS2}} = \frac{3000/(200/5)}{5} = 15.0$$

From the 2 TDS curve in Figure 10.12, T2 = 0.38 s. For this same fault,

$$\frac{I'_{1\text{Fault}}}{\text{TS1}} = \frac{3000/(400/5)}{5} = 7.5$$

$$\text{T1} = \text{T2} + \text{T}_{\text{breaker}} + \text{T}_{\text{coordination}} = 0.38 + 0.083 + 0.3 \approx 0.76 \text{ s}$$

From Figure 10.12, select TDS1 = 3. The relay settings are shown in Table 10.6. Note that for reliable relay operation the fault-to-pickup current ratios with minimum fault currents should be greater than 2. Coordination for minimum fault currents listed in Table 10.4 is evaluated in Problem 10.11. ∎

Note that separate relays are used for each phase in Example 10.4, and therefore these relays will operate for three-phase as well as line-to-line, single line-to-ground, and double line-to-ground faults. However, in many cases single line-to-ground fault currents are much lower than three-phase fault currents, especially for distribution feeders with high zero-sequence impedances. In these cases a separate ground relay with a lower current tap setting than the phase relays is used. The ground relay is connected to operate on zero-sequence current from three of the phase CTs connected in parallel or from a CT in the grounded neutral.

10.5

RECLOSERS AND FUSES

Automatic circuit reclosers are commonly used for distribution circuit protection. A *recloser* is a self-controlled device for automatically interrupting and reclosing an ac circuit with a preset sequence of openings and reclosures. Unlike circuit breakers, which have separate relays to control breaker opening and reclosing, reclosers have built-in controls. More than 80% of faults on overhead distribution circuits are temporary, caused by tree limb contact, by animal interference, by wind bringing bare conductors in contact, or by lightning. The automatic tripping-reclosing sequence of reclosers clears these temporary faults and restores service with only momentary outages, thereby significantly improving customer service. A disadvantage of reclosers is the increased hazard when a circuit is physically contacted by people—for exam-

FIGURE 10.18

Single-line diagram of a
13.8-kV radial
distribution feeder with
fuse/recloser/relay
protection

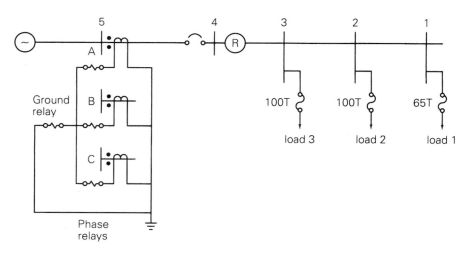

ple, in the case of a broken conductor at ground level that remains energized. Also, reclosing should be locked out during live-line maintenance by utility personnel.

Figure 10.18 shows a common protection scheme for radial distribution circuits utilizing fuses, reclosers, and time-delay overcurrent relays. Data for the 13.8-kV feeder in this figure is given in Table 10.7. There are three load taps protected by fuses. The recloser ahead of the fuses is set to open and reclose for faults up to and beyond the fuses. For temporary faults the recloser can be set for one or more instantaneous or time-delayed trips and reclosures in order to clear the faults and restore service. If faults persist, the fuses operate for faults to their right (downstream), or the recloser opens after time delay and locks out for faults between the recloser and fuses. Separate time-delay overcurrent phase and ground relays open the substation breaker after multiple reclosures of the recloser.

Coordination of the fuses, recloser, and time-delay overcurrent relays is shown via the time-current curves in Figure 10.19. Type T (slow) fuses are selected because their time-current characteristics coordinate well with reclosers. The fuses are selected on the basis of maximum loads served from the taps. A 65 T fuse is selected for the bus 1 tap, which has a 60-A maximum

TABLE 10.7

Data for Figure 10.18

Bus	Maximum Load Current A	3φ Fault Current A	IL-G Fault Current A
1	60	1000	850
2	95	1500	1300
3	95	2000	1700
4	250	3000	2600
5	250	4000	4050

FIGURE 10.19

Time-current curves for the radial distribution circuit of Figure 10.18

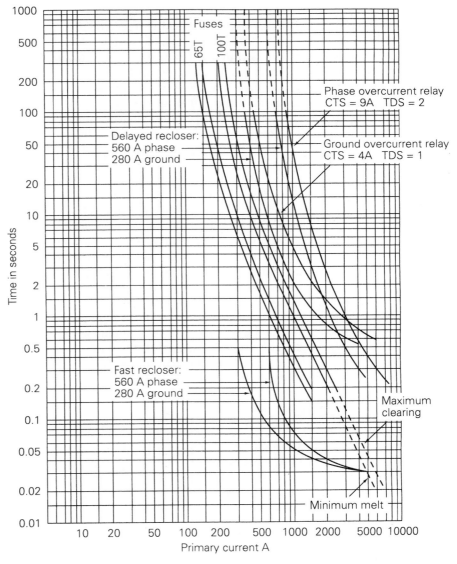

load current, and 100 T is selected for the bus 2 and 3 taps, which have 95-A maximum load currents. The fuses should also have a rated voltage larger than the maximum bus voltage and an interrupting current rating larger than the maximum asymmetrical fault current at the fuse location. Type T fuses with voltage ratings of 15 kV and interrupting current ratings of 10 kA and higher are standard.

Standard reclosers have minimum trip ratings of 50, 70, 100, 140, 200, 280, 400, 560, 800, 1120, and 1600 A, with voltage ratings up to 38 kV and maximum interrupting currents up to 16 kA. A minimum trip rating of 200–250% of maximum load current is typically selected for the phases, in order to override cold load pickup with a safety factor. The minimum trip rating of the ground unit is typically set at maximum load and should be higher than the maximum allowable load unbalance. For the recloser in Figure 10.18, which carries a 250-A maximum load, minimum trip ratings of 560 A for each phase and 280 A for the ground unit are selected.

A popular operation sequence for reclosers is two fast operations, without intentional time delay, followed by two delayed operations. The fast operations allow temporary faults to self-clear, whereas the delayed operations allow downstream fuses to clear permanent faults. Note that the time-current curves of the fast recloser lie below the fuse curves in Figure 10.19, such that the recloser opens before the fuses melt. The fuse curves lie below the delayed recloser curves, such that the fuses clear before the recloser opens. The recloser is typically programmed to reclose $\frac{1}{2}$ s after the first fast trip, 2 s after the second fast trip, and 5–10 s after a delayed trip.

Time-delay overcurrent relays with an extremely inverse characteristic coordinate with both reclosers and type T fuses. A 300:5 CT ratio is selected to give a secondary current of $250 \times (5/300) = 4.17$ A at maximum load. Relay settings are selected to allow the recloser to operate effectively to clear faults before relay operation. A current tap setting of 9 A is selected for the CO-11 phase relays so that minimum pickup exceeds twice the maximum load. A time-dial setting of 2 is selected so that the delayed recloser trips at least 0.2 s before the relay. The ground relay is set with a current tap setting of 4 A and a time-dial setting of 1.

EXAMPLE 10.5   **Fuse/recloser coordination**

For the system of Figure 10.18, describe the operating sequence of the protective devices for the following faults: (a) a self-clearing, temporary, three-phase fault on the load side of tap 2; and (b) a permanent three-phase fault on the load side of tap 2.

SOLUTION

a. From Table 10.7, the three-phase fault current at bus 2 is 1500 A. From Figure 10.19, the 560-A fast recloser opens 0.05 s after the 1500-A fault current occurs, and then recloses $\frac{1}{2}$ s later. Assuming the fault has self-cleared, normal service is restored. During the 0.05-s fault duration, the 100 T fuse does not melt.

b. For a permanent fault the fast recloser opens after 0.05 s, recloses $\frac{1}{2}$ s later into the permanent fault, opens again after 0.05 s, and recloses into the fault a second time after a 2-s delay. Then the 560-A delayed recloser opens 3 seconds later. During this interval the 100 T fuse clears the fault. The delayed recloser then recloses 5 to 10 s later, restoring service to loads 1 and 3. ∎

10.6

DIRECTIONAL RELAYS

Directional relays are designed to operate for fault currents in only one direction. Consider the directional relay D in Figure 10.20, which is required to operate only for faults to the right of the CT. Since the line impedance is mostly reactive, a fault at P_1 to the right of the CT will have a fault current I from bus 1 to bus 2 that lags the bus voltage V by an angle of almost 90°. This fault current is said to be in the forward direction. On the other hand, a fault at P_2, to the left of the CT, will have a fault current I that leads V by almost 90°. This fault current is said to be in the reverse direction.

The directional relay has two inputs: the reference voltage $V = \text{V}\underline{/0°}$, and current $I = \text{I}\underline{/\phi}$. The relay trip and block regions, shown in Figure 10.21, can be described by

$$-180° < (\phi - \phi_1) < 0° \quad \text{(Trip)}$$

$$\text{Otherwise} \quad \text{(Block)} \qquad (10.6.1)$$

where ϕ is the angle of the current with respect to the voltage and ϕ_1, typically 2° to 8°, defines the boundary between the trip and block regions.

The contacts of the overcurrent relay OC and the directional relay D are connected in series in Figure 10.20, so that the breaker trip coil is ener-

FIGURE 10.20

Directional relay in series with overcurrent relay (only phase A is shown)

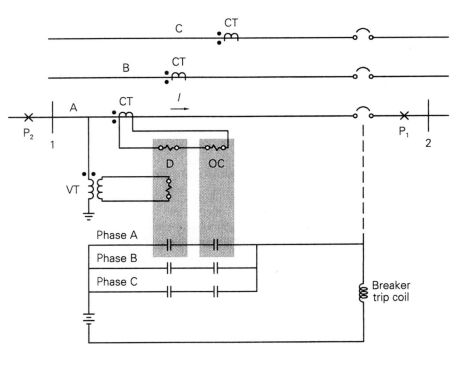

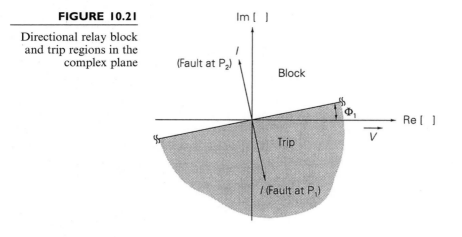

gized only when the CT secondary current (1) exceeds the OC relay pickup value, and (2) is in the forward tripping direction.

Although construction details differ, the operating principle of an electromechanical directional relay is similar to that of a watt-hour meter. There are two input coils, a voltage coil and a current coil, both located on a stator, and there is a rotating disc element. Suppose that the reference voltage is passed through a phase-shifting element to obtain $V_1 = V\underline{/\phi_1 - 90°}$. If V_1 and $I = I\underline{/\phi}$ are applied to a watt-hour meter, the torque on the rotating element is

$$\text{T} = \text{kVI}\cos(\phi_1 - \phi - 90°) = \text{kVI}\sin(\phi_1 - \phi) \qquad (10.6.2)$$

Note that for faults in the forward direction the current lags the voltage, and the angle $(\phi_1 - \phi)$ in (10.6.2) is close to 90°. This results in maximum positive torque on the rotating disc, which would cause the relay contacts to close. On the other hand, for faults in the reverse direction the current leads the voltage, and $(\phi_1 - \phi)$ is close to $-90°$. This results in maximum negative torque tending to rotate the disc element in the backward direction. Backward motion can be restrained by mechanical stops.

10.7

PROTECTION OF TWO-SOURCE SYSTEM WITH DIRECTIONAL RELAYS

It becomes difficult and in some cases impossible to coordinate overcurrent relays when there are two or more sources at different locations. Consider the system with two sources shown in Figure 10.22. Suppose there is a fault at P_1. We want B23 and B32 to clear the fault so that service to the three loads

FIGURE 10.22

System with two sources

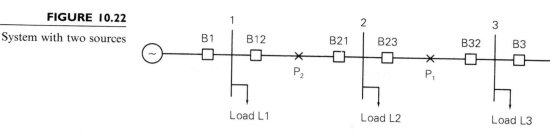

continues without interruption. Using time-delay overcurrent relays, we could set B23 faster than B21. Now consider a fault at P_2 instead. Breaker B23 will open faster than B21, and load L2 will be disconnected. When a fault can be fed from both the left and right, overcurrent relays cannot be coordinated. However, directional relays can be used to overcome this problem.

EXAMPLE 10.6   **Two-source system protection with directional and time-delay overcurrent relays**

Explain how directional and time-delay overcurrent relays can be used to protect the system in Figure 10.22. Which relays should be coordinated for a fault (a) at P_1, (b) at P_2? (c) Is the system also protected against bus faults?

SOLUTION   Breakers B12, B21, B23, and B32 should respond only to faults on their "forward" or "line" sides. Directional overcurrent relays connected as shown in Figure 10.20 can be used for these breakers. Overcurrent relays alone can be used for breakers B1 and B3, which do not need to be directional.

a. For a fault at P_1, the B21 relay would not operate; B12 should coordinate with B23 so that B23 trips before B12 (and B1). Also, B3 should coordinate with B32.

b. For a fault at P_2, B23 would not operate; B32 should coordinate with B21 so that B21 trips before B32 (and B3). Also, B1 should coordinate with B12.

c. Yes, the directional overcurrent relays also protect the system against bus faults. If the fault is at bus 2, relays at B21 and B23 will not operate, but B12 and B32 will operate to clear the fault. B1 and B21 will operate to clear a fault at bus 1. B3 and B23 will clear a fault at bus 3.   ∎

10.8

ZONES OF PROTECTION

Protection of simple systems has been discussed so far. For more general power system configurations, a fundamental concept is the division of a sys-

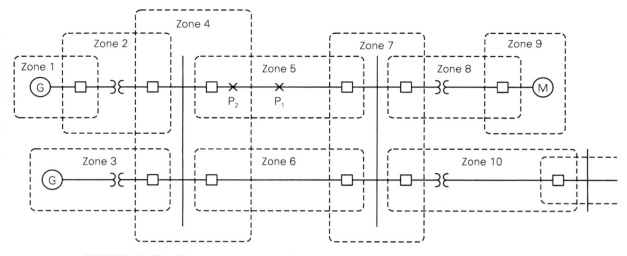

FIGURE 10.23   Power system protective zones

tem into protective zones [1]. If a fault occurs anywhere within a zone, action will be taken to isolate that zone from the rest of the system. Zones are defined for:

generators,

transformers,

buses,

transmission and distribution lines, and

motors.

Figure 10.23 illustrates the protective zone concept. Each zone is defined by a closed, dashed line. Zone 1, for example, contains a generator and connecting leads to a transformer. In some cases a zone may contain more than one component. For example, zone 3 contains a generator-transformer unit and connecting leads to a bus, and zone 10 contains a transformer and a line. Protective zones have the following characteristics:

Zones are overlapped.

Circuit breakers are located in the overlap regions.

For a fault anywhere in a zone, all circuit breakers in that zone open to isolate the fault.

FIGURE 10.24

Overlapping protection around a circuit breaker

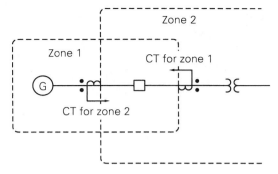

Neighboring zones are overlapped to avoid the possibility of unprotected areas. Without overlap the small area between two neighboring zones would not be located in any zone and thus would not be protected.

Since isolation during faults is done by circuit breakers, they should be inserted between equipment in a zone and each connection to the system. That is, breakers should be inserted in each overlap region. As such, they identify the boundaries of protective zones. For example, zone 5 in Figure 10.23 is connected to zones 4 and 7. Therefore, a circuit breaker is located in the overlap region between zones 5 and 4, as well as between zones 5 and 7.

If a fault occurs anywhere within a zone, action is taken to open all breakers in that zone. For example, if a fault occurs at P_1 on the line in zone 5, then the two breakers in zone 5 should open. If a fault occurs at P_2 within the overlap region of zones 4 and 5, then all five breakers in zones 4 and 5 should open. Clearly, if a fault occurs within an overlap region, two zones will be isolated and a larger part of the system will be lost from service. To minimize this possibility, overlap regions are kept as small as possible.

Overlap is accomplished by having two sets of instrument transformers and relays for each circuit breaker. For example, the breaker in Figure 10.24 shows two CTs, one for zone 1 and one for zone 2. Overlap is achieved by the order of the arrangement: first the equipment in the zone, second the breaker, and then the CT for that zone.

EXAMPLE 10.7   Zones of protection

Draw the protective zones for the power system shown in Figure 10.25. Which circuit breakers should open for a fault at P_1? at P_2?

SOLUTION   Noting that circuit breakers identify zone boundaries, protective zones are drawn with dashed lines as shown in Figure 10.26. For a fault at P_1, located in zone 5, breakers B24 and B42 should open. For a fault at P_2, located in the overlap region of zones 4 and 5, breakers B24, B42, B21, and B23 should open.

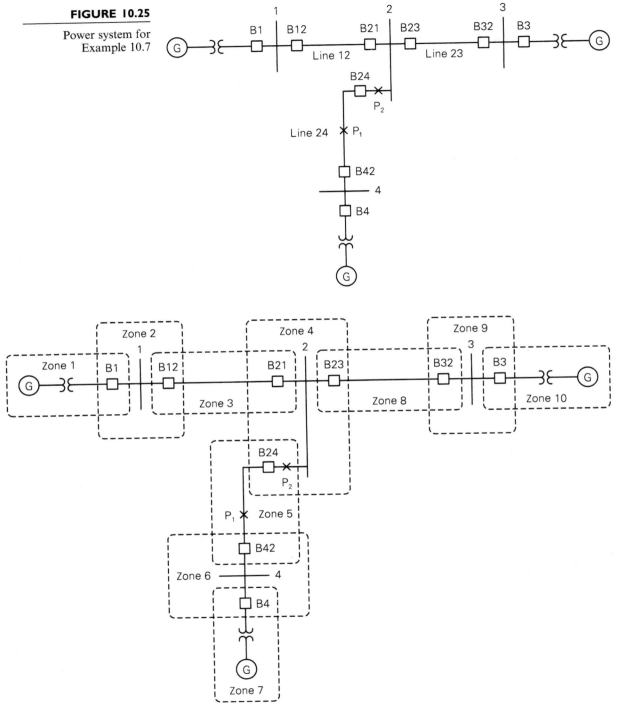

FIGURE 10.25

Power system for
Example 10.7

FIGURE 10.26 Protective zones for Example 10.7

10.9

LINE PROTECTION WITH IMPEDANCE (DISTANCE) RELAYS

Coordinating time-delay overcurrent relays can also be difficult for some radial systems. If there are too many radial lines and buses, the time delay for the breaker closest to the source becomes excessive.

Also, directional overcurrent relays are difficult to coordinate in transmission loops with multiple sources. Consider the use of these relays for the transmission loop shown in Figure 10.27. For a fault at P_1, we want the B21 relay to operate faster than the B32 relay. For a fault at P_2, we want B32 faster than B13. And for a fault at P_3, we want B13 faster than B21. Proper coordination, which depends on the magnitudes of the fault currents, becomes a tedious process. Furthermore, when consideration is given to various lines or sources out of service, coordination becomes extremely difficult.

To overcome these problems, relays that respond to a voltage-to-current ratio can be used. Note that during a three-phase fault, current increases while bus voltages close to the fault decrease. If, for example, current increases by a factor of 5 while voltage decreases by a factor of 2, then the voltage-to-current ratio decreases by a factor of 10. That is, the voltage-to-

FIGURE 10.27

345-kV transmission loop

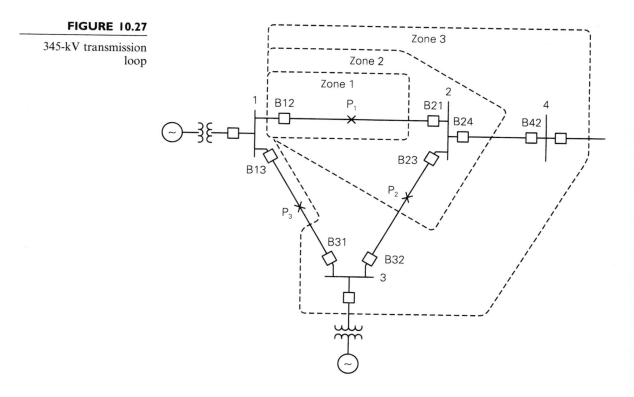

FIGURE 10.28

Impedance relay block
and trip regions

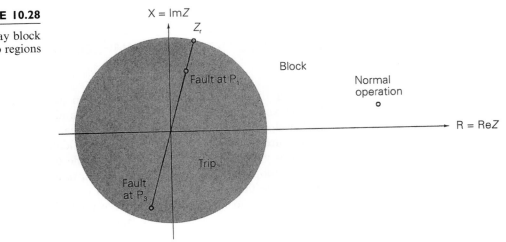

current ratio is more sensitive to faults than current alone. A relay that operates on the basis of voltage-to-current ratio is called an *impedance* relay. It is also called a *distance* relay or a *ratio* relay.

Impedance relay block and trip regions are shown in Figure 10.28, where the impedance Z is defined as the voltage-to-current ratio at the relay location. The relay trips for $|Z| < |Z_r|$, where Z_r is an adjustable relay setting. The impedance circle that defines the border between the block and trip regions passes through Z_r.

A straight line called the *line impedance locus* is shown for the impedance relay in Figure 10.28. This locus is a plot of positive sequence line impedances, predominantly reactive, as viewed between the relay location and various points along the line. The relay setting Z_r is a point in the R-X plane through which the impedance circle that defines the trip-block boundary must pass.

Consider an impedance relay for breaker B12 in Figure 10.27, for which $Z = V_1/I_{12}$. During normal operation, load currents are usually much smaller than fault currents, and the ratio Z has a large magnitude (and some arbitrary phase angle). Therefore, Z will lie outside the circle of Figure 10.28, and the relay will not trip during normal operation.

During a three-phase fault at P_1, however, Z appears to relay B12 to be the line impedance from the B12 relay to the fault. If $|Z_r|$ in Figure 10.28 is set to be larger than the magnitude of this impedance, then the B12 relay will trip. Also, during a three-phase fault at P_3, Z appears to relay B12 to be the negative of the line impedance from the relay to the fault. If $|Z_r|$ is larger than the magnitude of this impedance, the B12 relay will trip. Thus, the impedance relay of Figure 10.28 is not directional; a fault to the left or right of the relay can cause a trip.

Two ways to include directional capability with an impedance relay are shown in Figure 10.29. In Figure 10.29(a), an impedance relay with directional restraint is obtained by including a directional relay in series with an

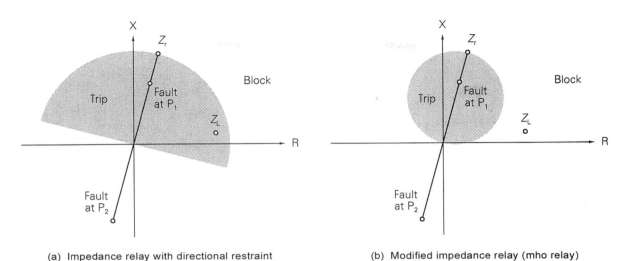

(a) Impedance relay with directional restraint   (b) Modified impedance relay (mho relay)

FIGURE 10.29   Impedance relays with directional capability

impedance relay, just as was done previously with an overcurrent relay. In Figure 10.29(b), a modified impedance relay is obtained by offsetting the center of the impedance circle from the origin. This modified impedance relay is sometimes called a *Mho* relay. If either of these relays is used at B12 in Figure 10.27, a fault at P_1 will result in a trip decision, but a fault at P_3 will result in a block decision.

Note that the radius of the impedance circle for the modified impedance relay is half of the corresponding radius for the impedance relay with directional restraint. The modified impedance relay has the advantage of better selectivity for high power factor loads. For example, the high power factor load Z_L lies outside the trip region of Figure 10.29(b) but inside the trip region of Figure 10.29(a).

The *reach* of an impedance relay denotes how far down the line the relay detects faults. For example, an 80% reach means that the relay will detect any (solid three-phase) fault between the relay and 80% of the line length. This explains the term *distance* relay.

It is common practice to use three directional impedance relays per phase, with increasing reaches and longer time delays. For example, Figure 10.27 shows three protection zones for B12. The zone 1 relay is typically set for an 80% reach and instantaneous operation, in order to provide primary protection for line 1–2. The zone 2 relay is set for about 120% reach, extending beyond bus 2, with a typical time delay of 0.2 to 0.3 seconds. The zone 2 relay provides backup protection for faults on line 1–2 as well as remote backup for faults on line 2–3 or 2–4 in zone 2.

Note that in the case of a fault on line 2–3 we want the B23 relay to trip, not the B12 relay. Since the impedance seen by B12 for faults near bus 2, either on line 1–2 or line 2–3, is essentially the same, we cannot set the B12 zone 1 relay for 100% reach. Instead, an 80% reach is selected to avoid in-

FIGURE 10.30

Three-zone, directional
impedance relay

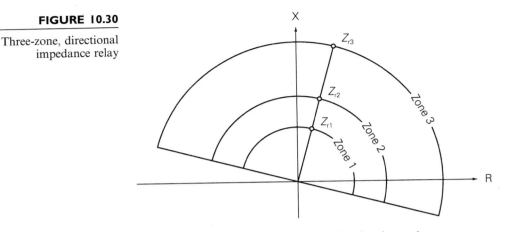

(a) Impedance relay with directional restraint

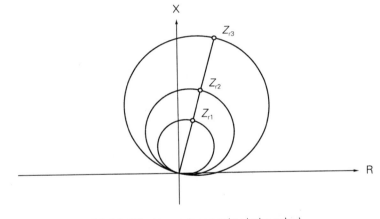

(b) Modified impedance relay (mho relay)

stantaneous operation of B12 for a fault on line 2–3 near bus 2. For example, if there is a fault at P_2 on line 2–3, B23 should trip instantaneously; if it fails, B12 will trip after time delay. Other faults at or near bus 2 also cause tripping of the B12 zone 2 relay after time delay.

Reach for the zone 3 B12 relay is typically set to extend beyond buses 3 and 4 in Figure 10.27, in order to provide remote backup for neighboring lines. As such, the zone 3 reach is set for 100% of line 1–2 plus 120% of either line 2–3 or 2–4, whichever is longer, with an even larger time delay, typically one second.

Typical block and trip regions are shown in Figure 10.30 for both types of three-zone, directional impedance relays. Relay connections for a three-zone impedance relay with directional restraint are shown in Figure 10.31.

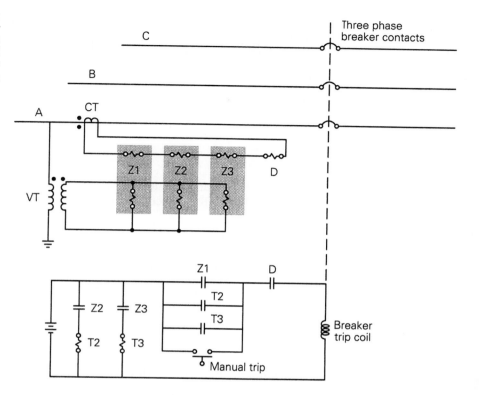

T2 : zone 2 timing relay

T3 : zone 3 timing relay

EXAMPLE 10.8 **Three-zone impedance relay settings**

Table 10.8 gives positive-sequence line impedances as well as CT and VT ratios at B12 for the 345-kV system shown in Figure 10.27. (a) Determine the settings Z_{r1}, Z_{r2}, and Z_{r3} for the B12 three-zone, directional impedance relays connected as shown in Figure 10.31. Consider only solid, three-phase faults.

TABLE 10.8

Data for Example 10.8

Line	Positive-Sequence Impedance Ω
1–2	8 + j50
2–3	8 + j50
2–4	5.3 + j33
1–3	4.3 + j27

Breaker	CT Ratio	VT Ratio
B12	1500 : 5	3000 : 1

(b) Maximum current for line 1–2 during emergency loading conditions is 1500 A at a power factor of 0.95 lagging. Verify that **B12** does not trip during normal and emergency loadings.

SOLUTION

a. Denoting V_{LN} as the line-to-neutral voltage at bus 1 and I_L as the line current through **B12**, the primary impedance Z viewed at **B12** is

$$Z = \frac{V_{LN}}{I_L} \quad \Omega$$

Using the CT and VT ratios given in Table 10.8, the secondary impedance viewed by the **B12** impedance relays is

$$Z' = \frac{V_{LN} \left/ \left(\frac{3000}{1}\right)\right.}{I_L \left/ \left(\frac{1500}{5}\right)\right.} = \frac{Z}{10}$$

We set the **B12** zone 1 relay for 80% reach, that is, 80% of the line 1–2 (secondary) impedance:

$$Z_{r1} = 0.80(8 + j50)/10 = 0.64 + j4 = 4.05\underline{/80.9°}\ \Omega \quad \text{secondary}$$

Setting the **B12** zone 2 relay for 120% reach:

$$Z_{r2} = 1.2(8 + j50)/10 = 0.96 + j6 = 6.08\underline{/80.9°}\ \Omega \quad \text{secondary}$$

From Table 10.8, line 2–4 has a larger impedance than line 2–3. Therefore, we set the **B12** zone 3 relay for 100% reach of line 1–2 plus 120% reach of line 2–4.

$$Z_{r3} = 1.0(8 + j50)/10 + 1.2(5.3 + j33)/10$$

$$= 1.44 + j8.96 = 9.07\underline{/80.9°}\ \Omega \quad \text{secondary}$$

b. The secondary impedance viewed by **B12** during emergency loading, using $V_{LN} = 345/\sqrt{3}\underline{/0°} = 199.2\underline{/0°}$ kV and $I_L = 1500\underline{/-\cos^{-1}(0.95)} = 1500\underline{/-18.19°}$ A, is

$$Z' = Z/10 = \left(\frac{199.2 \times 10^3}{1500\underline{/-18.19°}}\right)\bigg/10 = 13.28\underline{/18.19°}\ \Omega \quad \text{secondary}$$

Since this impedance exceeds the zone 3 setting of $9.07\underline{/80.9°}\ \Omega$, the impedance during emergency loading lies outside the trip regions of the three-zone, directional impedance relay. Also, lower line loadings during normal operation will result in even larger impedances farther away from the trip regions. **B12** will trip during faults but not during normal and emergency loadings. ∎

Remote backup protection of adjacent lines using zone 3 of an impedance relay may be ineffective. In practice, buses have multiple lines of different lengths with sources at their remote ends. Contributions to fault currents from the multiple lines may cause the zone 3 relay to underreach. This "infeed effect" is illustrated in Problem 10.21.

The impedance relays considered so far use line-to-neutral voltages and line currents and are called *ground fault relays*. They respond to three-phase, single line-to-ground, and double line-to-ground faults very effectively. The impedance seen by the relay during unbalanced faults will generally not be the same as seen during three-phase faults and will not be truly proportional to the distance to the fault location. However, the relay can be accurately set for any fault location after computing impedance to the fault using fault currents and voltages. For other fault locations farther away (or closer), the impedance to the fault will increase (or decrease).

Ground fault relays are relatively insensitive to line-to-line faults. Impedance relays that use line-to-line voltages V_{ab}, V_{bc}, V_{ca} and line-current differences $I_a - I_b$, $I_b - I_c$, $I_c - I_a$ are called *phase relays*. Phase relays respond effectively to line-to-line faults and double line-to-ground faults but are relatively insensitive to single line-to-ground faults. Therefore, both phase and ground fault relays need to be used.

10.10

DIFFERENTIAL RELAYS

Differential relays are commonly used to protect generators, buses, and transformers. Figure 10.32 illustrates the basic method of differential relaying

FIGURE 10.32

Differential relaying for
generator protection
(protection for one
phase shown)

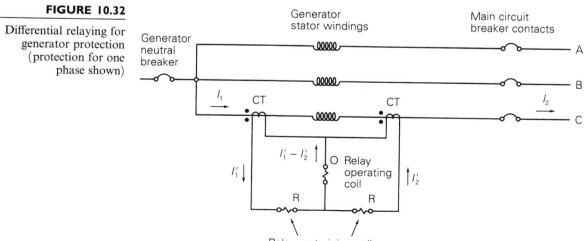

for generator protection. The protection of only one phase is shown. The method is repeated for the other two phases. When the relay in any one phase operates, all three phases of the main circuit breaker will open, as well as the generator neutral and field breakers (not shown).

For the case of no internal fault within the generator windings, $I_1 = I_2$, and, assuming identical CTs, $I_1' = I_2'$. For this case the current in the relay operating coil is zero, and the relay does not operate. On the other hand, for an internal fault such as a phase-to-ground or phase-to-phase short within the generator winding, $I_1 \neq I_2$, and $I_1' \neq I_2'$. Therefore, a difference current $I_1' - I_2'$ flows in the relay operating coil, which may cause the relay to operate. Since this relay operation depends on a *difference* current, it is called a *differential* relay.

An electromechanical differential relay called a *balance beam* relay is shown in Figure 10.33. The relay contacts close if the downward force on the right side exceeds the downward force on the left side. The electromagnetic force on the right, operating coil is proportional to the square of the operating coil mmf—that is, to $[N_0(I_1' - I_2')]^2$. Similarly, the electromagnetic force on the left, restraining coil is proportional to $[N_r(I_1' + I_2')/2]^2$. The condition for relay operation is then

$$[N_0(I_1' - I_2')]^2 > [N_r(I_1' + I_2')/2]^2 \tag{10.10.1}$$

Taking the square root:

$$|I_1' - I_2'| > k|(I_1' + I_2')/2| \tag{10.10.2}$$

where

$$k = N_r/N_0 \tag{10.10.3}$$

Assuming I_1' and I_2' are in phase, (10.10.2) is solved to obtain

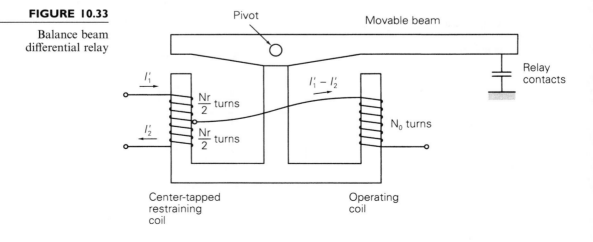

FIGURE 10.33

Balance beam
differential relay

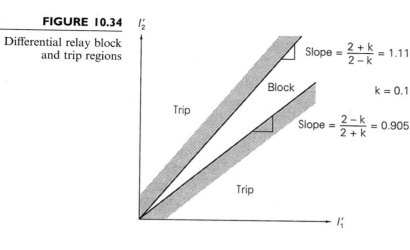

FIGURE 10.34

Differential relay block and trip regions

$$I_2' > \frac{2 + k}{2 - k} I_1' \quad \text{for } I_2' > I_1'$$

$$I_2' < \frac{2 - k}{2 + k} I_1' \quad \text{for } I_2' < I_1' \tag{10.10.4}$$

Equation (10.10.4) is plotted in Figure 10.34 to obtain the block and trip regions of the differential relay for $k = 0.1$. Note that as k increases, the block region becomes larger; that is, the relay becomes less sensitive. In practice, no two CTs are identical, and the differential relay current $I_1' - I_2'$ can become appreciable during external faults, even though $I_1 = I_2$. The balanced beam relay solves this problem without sacrificing sensitivity during normal currents, since the block region increases as the currents increase, as shown in Figure 10.34. Also, the relay can be easily modified to enlarge the block region for very small currents near the origin, in order to avoid false trips at low currents.

Note that differential relaying provides primary zone protection without backup. Coordination with protection in adjacent zones is eliminated, which permits high speed tripping. Precise relay settings are unnecessary. Also, the need to calculate system fault currents and voltages is avoided.

10.11

BUS PROTECTION WITH DIFFERENTIAL RELAYS

Differential bus protection is illustrated by the single-line diagram of Figure 10.35. In practice, three differential relays are required, one for each phase. Operation of any one relay would cause all of the three-phase circuit breakers connected to the bus to open, thereby isolating the three-phase bus from service.

FIGURE 10.35

Single-line diagram of
differential bus
protection

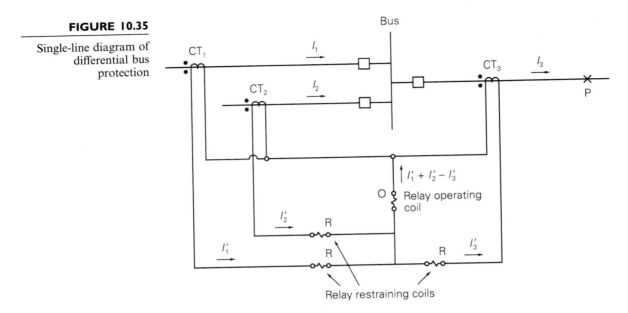

For the case of no internal fault between the CTs—that is, no bus fault—$I_1 + I_2 = I_3$. Assuming identical CTs, the differential relay current $I_1' + I_2' - I_3'$ equals zero, and the relay does not operate. However, if there is a bus fault, the differential current $I_1' + I_2' - I_3'$, which is not zero, flows in the operating coil to operate the relay. Use of the restraining coils overcomes the problem of nonidentical CTs.

A problem with differential bus protection can result from different levels of fault currents and varying amounts of CT saturation. For example, consider an external fault at point P in Figure 10.35. Each of the CT_1 and CT_2 primaries carries part of the fault current, but the CT_3 primary carries the sum $I_3 = I_1 + I_2$. CT_3, energized at a higher level, will have more saturation, such that $I_3' \neq I_1' + I_2'$. If the saturation is too high, the differential current in the relay operating coil could result in a false trip. This problem becomes more difficult when there are large numbers of circuits connected to the bus. Various schemes have been developed to overcome this problem [1].

10.12

TRANSFORMER PROTECTION WITH DIFFERENTIAL RELAYS

The protection method used for power transformers depends on the transformer MVA rating. Fuses are often used to protect transformers with small MVA ratings, whereas differential relays are commonly used to protect transformers with ratings larger than 10 MVA.

FIGURE 10.36

Differential protection of a single-phase, two-winding transformer

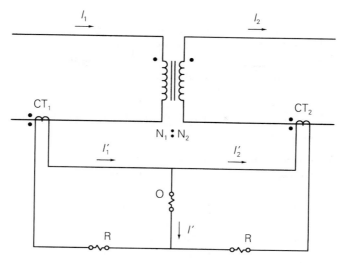

The differential protection method is illustrated in Figure 10.36 for a single-phase, two-winding transformer. Denoting the turns ratio of the primary and secondary CTs by $1/n_1$ and $1/n_2$, respectively (a CT with 1 primary turn and n secondary turns has a turns ratio a $= 1/n$), the CT secondary currents are

$$I_1' = \frac{I_1}{n_1} \qquad I_2' = \frac{I_2}{n_2} \qquad\qquad (10.12.1)$$

and the current in the relay operating coil is

$$I' = I_1' - I_2' = \frac{I_1}{n_1} - \frac{I_2}{n_2} \qquad\qquad (10.12.2)$$

For the case of no fault between the CTs—that is, no internal transformer fault—the primary and secondary currents for an ideal transformer are related by

$$I_2 = \frac{N_1 I_1}{N_2} \qquad\qquad (10.12.3)$$

Using (10.12.3) in (10.12.2),

$$I' = \frac{I_1}{n_1}\left(1 - \frac{N_1/N_2}{n_2/n_1}\right) \qquad\qquad (10.12.4)$$

To prevent the relay from tripping for the case of no internal transformer fault, where (10.12.3) and (10.12.4) are satisfied, the differential relay current I' must be zero. Therefore, from (10.12.4), we select

$$\frac{n_2}{n_1} = \frac{N_1}{N_2} \qquad\qquad (10.12.5)$$

If an internal transformer fault between the CTs does occur, (10.12.3) is not satisfied and the differential relay current $I' = I_1' - I_2'$ is not zero. The relay will trip if the operating condition given by (10.10.4) is satisfied. Also, the value of k in (10.10.4) can be selected to control the size of the block region shown in Figure 10.34, thereby controlling relay sensitivity.

EXAMPLE 10.9   **Differential relay protection for a single-phase transformer**

A single-phase two-winding, 10-MVA, 80 kV/20 kV transformer has differential relay protection. Select suitable CT ratios. Also, select k such that the relay blocks for up to 25% mismatch between I_1' and I_2'.

SOLUTION   The transformer-rated primary current is

$$I_{1rated} = \frac{10 \times 10^6}{80 \times 10^3} = 125 \text{ A}$$

From Table 10.2, select a 150:5 primary CT ratio to give $I_1' = 125(5/150) = 4.17$ A at rated conditions. Similarly, $I_{2rated} = 500$ A. Select a 600:5 secondary CT ratio to give $I_2' = 500(5/600) = 4.17$ A and a differential current $I' = I_1' - I_2' = 0$ (neglecting magnetizing current) at rated conditions. Also, for a 25% mismatch between I_1' and I_2', select a 1.25 upper slope in Figure 10.34. That is,

$$\frac{2+k}{2-k} = 1.25 \qquad k = 0.2222$$ ∎

A common problem in differential transformer protection is the mismatch of relay currents that occurs when standard CT ratios are used. If the primary winding in Example 10.9 has a 138-kV instead of 80-kV rating, then $I_{1rated} = 10 \times 10^6/138 \times 10^3 = 72.46$ A, and a 100:5 primary CT would give $I_1' = 72.46(5/100) = 3.62$ A at rated conditions. This current does not balance $I_2' = 4.17$ A using a 5:600 secondary CT, nor $I_2' = 3.13$ A using a 5:800 secondary CT. The mismatch is about 15%.

One solution to this problem is to use auxiliary CTs, which provide a wide range of turns ratios. A 5:5.76 auxiliary CT connected to the 5:600 secondary CT in the above example would reduce I_2' to $4.17(5/5.76) = 3.62$ A, which does balance I_1'. Unfortunately, auxiliary CTs add their own burden to the main CTs and also increase transformation errors. A better solution is to use tap settings on the relays themselves, which have the same effect as auxiliary CTs. Most transformer differential relays have taps that provide for differences in restraining windings in the order of 2 or 3 to 1.

When a transformer is initially energized, it can draw a large "inrush" current, a transient current that flows in the shunt magnetizing branch and decays after a few cycles to a small steady-state value. Inrush current appears as a differential current since it flows only in the primary winding. If a large inrush current does occur upon transformer energization, a differential relay will see a large differential current and trip out the transformer unless the protection method is modified to detect inrush current.

One method to prevent tripping during transformer inrush is based on the fact that inrush current is nonsinusoidal with a large second-harmonic component. A filter can be used to pass fundamental and block harmonic components of the differential current I' to the relay operating coil. Another method is based on the fact that inrush current has a large dc component, which can be used to desensitize the relay. Time-delay relays may also be used to temporarily desensitize the differential relay until the inrush current has decayed to a low value.

Figure 10.37 illustrates differential protection of a three-phase Y–Δ

FIGURE 10.37

Differential protection of a three-phase, Y–Δ, two-winding transformer

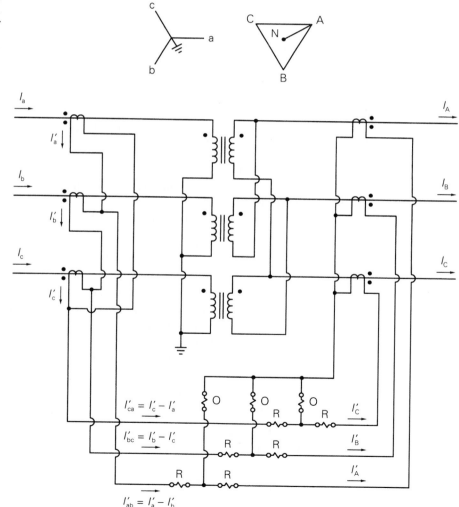

two-winding transformer. Note that a Y–Δ transformer produces 30° phase shifts in the line currents. The CTs must be connected to compensate for the 30° phase shifts, such that the CT secondary currents as seen by the relays are in phase. The correct phase-angle relationship is obtained by connecting CTs on the Y side of the transformer in Δ, and CTs on the Δ side in Y.

EXAMPLE 10.10   **Differential relay protection for a three-phase transformer**

A 30-MVA, 34.5 kV Y/138 kV Δ transformer is protected by differential relays with taps. Select CT ratios, CT connections, and relay tap settings. Also determine currents in the transformer and in the CTs at rated conditions. Assume that the available relay tap settings are 5:5, 5:5.5, 5:6.6, 5:7.3, 5:8, 5:9, and 5:10, giving relay tap ratios of 1.00, 1.10, 1.32, 1.46, 1.60, 1.80, and 2.00.

SOLUTION   As shown in Figure 10.37, CTs are connected in Δ on the (34.5-kV) Y side of the transformer, and CTs are connected in Y on the (138-kV) Δ side, in order to obtain the correct phasing of the relay currents.

Rated current on the 138-kV side of the transformer is

$$I_{A\,rated} = \frac{30 \times 10^6}{\sqrt{3}(138 \times 10^3)} = 125.51 \text{ A}$$

Select a 150:5 CT on the 138-kV side to give $I'_A = 125.51(5/150) = 4.184$ A in the 138-kV CT secondaries and in the righthand restraining windings of Figure 10.37.

Next, rated current on the 34.5-kV side of the transformer is

$$I_{a\,rated} = \frac{30 \times 10^6}{\sqrt{3}(34.5 \times 10^3)} = 502.04 \text{ A}$$

Select a 500:5 CT on the 34.5-kV side to give $I'_a = 502.0(5/500) = 5.02$ A in the 34.5-kV CT secondaries and $I'_{ab} = 5.02\sqrt{3} = 8.696$ A in the lefthand restraining windings of Figure 10.37.

Finally, select relay taps to balance the currents in the restraining windings. The ratio of the currents in the left- to righthand restraining windings is

$$\frac{I'_{ab}}{I'_A} = \frac{8.696}{4.184} = 2.078$$

The closest relay tap ratio is $T'_{AB}/T'_A = 2.0$, corresponding to a relay tap setting of $T'_A : T'_{ab} = 5 : 10$. The percentage mismatch for this tap setting is

$$\left| \frac{(I'_A/T'_A) - (I'_{ab}/T'_{ab})}{(I'_{ab}/T'_{ab})} \right| \times 100 = \left| \frac{(4.184/5) - (8.696/10)}{(8.696/10)} \right| \times 100 = 3.77\%$$

This is a good mismatch; since transformer differential relays typically have their block regions adjusted between 20% and 60% (by adjusting k in Figure 10.34), a 3.77% mismatch gives an ample safety margin in the event of CT and relay differences.   ∎

For three-phase transformers (Y–Y, Y–Δ, Δ–Y, Δ–Δ), the general rule is to connect CTs on the Y side in Δ and CTs on the Δ side in Y. This arrangement compensates for the 30° phase shifts in Y–Δ or Δ–Y banks. Note also that zero-sequence current cannot enter a Δ side of a transformer or the CTs on that side, and zero-sequence current on a grounded Y side cannot enter the Δ-connected CTs on that side. Therefore, this arrangement also blocks zero-sequence currents in the differential relays during external ground faults. For internal ground faults, however, the relays can operate from the positive- and negative-sequence currents involved in these faults.

Differential protection methods have been modified to handle multi-winding transformers, voltage-regulating transformers, phase-angle regulating transformers, power-rectifier transformers, transformers with special connections (such as zig-zag), and other, special-purpose transformers. Also, other types of relays such as gas-pressure detectors for liquid-filled transformers are used.

10.13

PILOT RELAYING

Pilot relaying refers to a type of differential protection that compares the quantities at the terminals via a communication channel rather than by a direct wire interconnection of the relays. Differential protection of generators, buses, and transformers considered in previous sections does not require pilot relaying because each of these devices is at one geographical location where CTs and relays can be directly interconnected. However, differential relaying of transmission lines requires pilot relaying because the terminals are widely separated (often by many kilometers). In actual practice, pilot relaying is typically applied to short transmission lines (up to 80 km) with 69 to 115 kV ratings.

Four types of communication channels are used for pilot relaying:

1. *Pilot wires:* Separate electrical circuits operating at dc, 50 to 60 Hz, or audio frequencies. These could be owned by the power company or leased from the telephone company.

2. *Power-line carrier:* The transmission line itself is used as the communication circuit, with frequencies between 30 and 300 kHz being transmitted. The communication signals are applied to all three phases using an L–C voltage divider and are confined to the line under protection by blocking filters called *line traps* at each end.

3. *Microwave:* A 2 to 12 GHz signal transmitted by line-of-sight paths between terminals using dish antennas.

4. *Fiber optic cable:* Signals transmitted by light modulation through electrically nonconducting cable. This cable eliminates problems due to electrical insulation, inductive coupling from other circuits, and atmospheric disturbances.

Two common fault detection methods are *directional comparison*, where the power flows at the line terminals are compared, and *phase comparison*, where the relative phase angles of the currents at the terminals are compared. Also, the communication channel can either be required for trip operations, which is known as a *transfer trip system*, or not be required for trip operations, known as a *blocking system*. A particular pilot-relaying method is usually identified by specifying the fault-detection method and the channel use. The four basic combinations are directional comparison blocking, directional comparison transfer trip, phase comparison blocking, and phase comparison transfer trip.

Like differential relays, pilot relays provide primary zone protection without backup. Thus, coordination with protection in adjacent zones is eliminated, resulting in high-speed tripping. Precise relay settings are unnecessary. Also, the need to calculate system fault currents and voltages is eliminated.

10.14

DIGITAL RELAYING

In previous sections we described the operating principle of relays built with electromechanical components, including the induction disc time-delay overcurrent relay, Figure 10.14; the directional relay, similar in operation to a watt-hour meter; and the balance-beam differential relay, Figure 10.33. These electromechanical relays, introduced in the early 1900s, have performed well over the years and continue in relatively maintenance-free operation today. Solid-state relays using analog circuits and logic gates, with block-trip regions similar to those of electromechanical relays and with newer types of block/trip regions, have been available since the late 1950s. Such relays, widely used in HV and EHV systems, offer the reliability and ruggedness of their electromechanical counterparts at a competitive price. Beyond solid-state analog relays, a new generation of relays based on digital computer technology has been under development since the 1980s.

Benefits of digital relays include accuracy, improved sensitivity to faults, better selectivity, flexibility, user-friendliness, easy testing, and relay event monitoring/recording capabilities. Digital relaying also has the advantage that modifications to tripping characteristics, either changes in conventional settings or shaping of entirely new block/trip regions, could be made by updating software from a remote computer terminal. For example, the relay engineer could reprogram tripping characteristics of field-installed, in-service relays without leaving the engineering office. Alternatively, relay software could be updated in real time, based on operating conditions, from a central computer.

An important feature of power system protection is the decentralized, local nature of relays. Except for pilot relaying, each relay receives information from nearby local CTs and VTs and trips only local breakers. Interest in digital relaying is not directed at replacing local relays by a central computer. Instead, each electromechanical or solid-state analog relay would be replaced by a dedicated, local digital relay with a similar operating principle, such as time-delay overcurrent, impedance, or differential relaying. The central computer would interact with local digital relays in a supervisory role.

PROBLEMS

SECTION 10.2

10.1 The primary conductor in Figure 10.2 is one phase of a three-phase transmission line operating at 345 kV, 600 MVA, 0.95 power factor lagging. The CT ratio is 1200:5 and the VT ratio is 3000:1. Determine the CT secondary current I' and the VT secondary voltage V'. Assume zero CT error.

10.2 A CO-8 relay with a current tap setting of 5 amperes is used with the 100:5 CT in Example 10.1. The CT secondary current I' is the input to the relay operating coil. The CO-8 relay burden is shown in the following table for various relay input currents.

CO-8 relay input current I', A	5	8	10	13	15
CO-8 relay burden Z_B, Ω	0.5	0.8	1.0	1.3	1.5

Primary current and CT error are computed in Example 10.1 for the 5-, 8-, and 15-A relay input currents. Compute the primary current and CT error for (a) $I' = 10$ A and $Z_B = 1.0\ \Omega$, and for (b) $I' = 13$ A and $Z_B = 1.3\ \Omega$. (c) Plot I' versus I for the above five values of I'. (d) For reliable relay operation, the fault-to-pickup current ratio with minimum fault current should be greater than two. Determine the minimum fault current for application of this CT and relay with 5-A tap setting.

10.3 An overcurrent relay set to operate at 10 A is connected to the CT in Figure 10.8 with a 200:5 CT ratio. Determine the minimum primary fault current that the relay will detect if the burden Z_B is (a) 1.0 Ω, (b) 4.0 Ω, and (c) 5.0 Ω.

10.4 Given the open-delta VT connection shown in Figure 10.38, both VTs having a voltage rating of 240 kV : 120 V, the voltages are specified as $V_{AB} = 230\underline{/0°}$, $V_{BC} = 230\underline{/-120°}$, and $V_{CA} = 230\underline{/120°}$ kV. Determine V_{ab}, V_{bc}, and V_{ca} for the following cases: (a) The dots are shown in Figure 10.38. (b) The dot near c is moved to b in Figure 10.38.

FIGURE 10.38

Problem 10.4

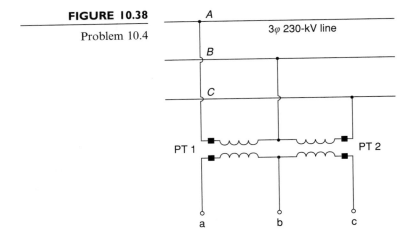

10.5 A CT with an excitation curve given in Figure 10.39 has a rated current ratio of 500 : 5 A and a secondary leakage impedance of $0.1 + j0.5\ \Omega$. Calculate the CT secondary output current and the CT error for the following cases: (a) The impedance of the terminating device is $4.9 + j0.5\ \Omega$ and the primary CT load current is 400 A. (b) The impedance of the terminating device is $4.9 + j0.5\ \Omega$ and the primary CT fault current is 1200 A. (c) The impedance of the terminating device is $14.9 + j1.5\ \Omega$ and the primary CT load current is 400 A. (d) The impedance of the terminating device is $14.9 + j1.5\ \Omega$ and the primary CT fault current is 1200 A.

FIGURE 10.39

Problem 10.5

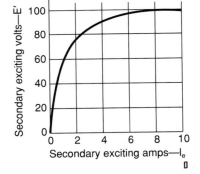

10.6 The CT of Problem 10.5 is utilized in conjunction with a current-sensitive device that will operate at current levels of 8 A or above. Check whether the device will detect the 1200-A fault current for cases (b) and (d) in Problem 10.5.

SECTION 10.3

10.7 The input current to a CO-8 relay is 10 A. Determine the relay operating time for the following current tap settings (TS) and time dial settings (TDS): (a) TS = 1.0, TDS = 1/2; (b) TS = 2.0, TDS = 1.5; (c) TS = 2.0, TDS = 7; (d) TS = 3.0, TDS = 7; and (e) TS = 12.0, TDS = 1.

10.8 The relay in Problem 10.2 has a time-dial setting of 4. Determine the relay operating time if the primary fault current is 500 A.

10.9   An RC circuit used to produce time delay is shown in Figure 10.40. For a step input voltage $v_i(t) = 2u(t)$ and $C = 10\ \mu F$, determine T_{delay} for the following cases: (a) $R = 100\ k\Omega$; and (b) $R = 1\ M\Omega$. Sketch the output $v_o(t)$ versus time for cases (a) and (b).

FIGURE 10.40

Problem 10.9

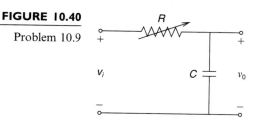

10.10   Reconsider case (b) of Problem 10.5. Let the load impedance $4.9 + j0.5\ \Omega$ be the input impedance to a CO-7 induction disc time-delay overcurrent relay. The CO-7 relay characteristic is shown in Figure 10.41. For a tap setting of 5 A and a time dial setting of 2, determine the relay operating time.

FIGURE 10.41

Problems 10.10 and
10.14

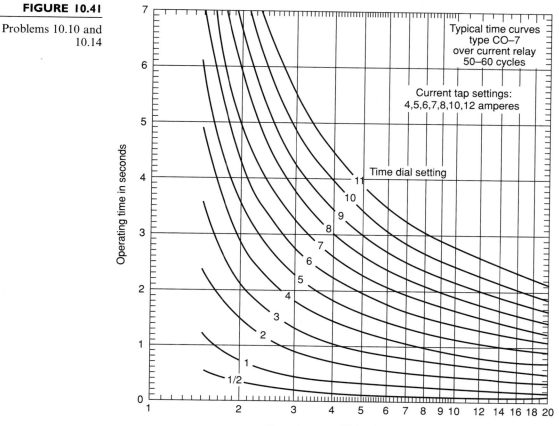

SECTION 10.4

10.11 Evaluate relay coordination for the minimum fault currents in Example 10.4. For the selected current tap settings and time dial settings, (a) determine the operating time of relays at B2 and B3 for the 700-A fault current. (b) Determine the operating time of relays at B1 and B2 for the 1500-A fault current. Are the fault-to-pickup current ratios $\geqslant 2.0$ (a requirement for reliable relay operation) in all cases? Are the coordination time intervals $\geqslant 0.3$ seconds in all cases?

10.12 Repeat Example 10.4 for the following system data. Coordinate the relays for the maximum fault currents.

Bus	Maximum Load		Symmetrical Fault Current	
	MVA	Lagging p.f.	Maximum A	Minimum A
1	9.0	0.95	5000	3750
2	9.0	0.95	3000	2250
3	9.0	0.95	2000	1500

Breaker	Breaker Operating Time	CT Ratio	Relay
B1	5 cycles	600:5	CO-8
B2	5 cycles	400:5	CO-8
B3	5 cycles	200:5	CO-8

10.13 Using the current tap settings and time dial settings that you have selected in Problem 10.12, evaluate relay coordination for the minimum fault currents. Are the fault-to-pickup current ratios $\geqslant 2.0$, and are the coordination time delays $\geqslant 0.3$ seconds in all cases?

10.14 An 11-kV radial system is shown in Figure 10.42. Assuming a CO-7 relay with relay characteristic given in Figure 10.41 and the same power factor for all loads, select relay settings to protect the system.

FIGURE 10.42
Problem 10.14

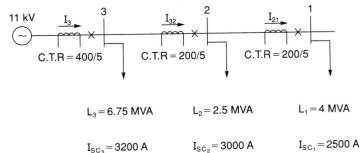

SECTION 10.5

10.15 Rework Example 10.5 for the following faults: (a) a three-phase, permanent fault on the load side of tap 3; (b) a single line-to-ground, permanent fault at bus 4 on the load side of the recloser; and (c) a three-phase, permanent fault at bus 4 on the source side of the recloser.

10.16 A three-phase 34.5-kV feeder supplying a 4-MVA load is protected by 80E power fuses in each phase, in series with a recloser. The time-current characteristic of the 80E fuse is shown in Figure 10.43. Analysis yields maximum and minimum fault currents of 1000 and 500 A, respectively. (a) To have the recloser clear the fault, find the maximum clearing time necessary for recloser operation. (b) To have the fuses clear the fault, find the minimum recloser clearing time. Assume that the recloser operating time is independent of fault current magnitude.

FIGURE 10.43

Problem 10.16

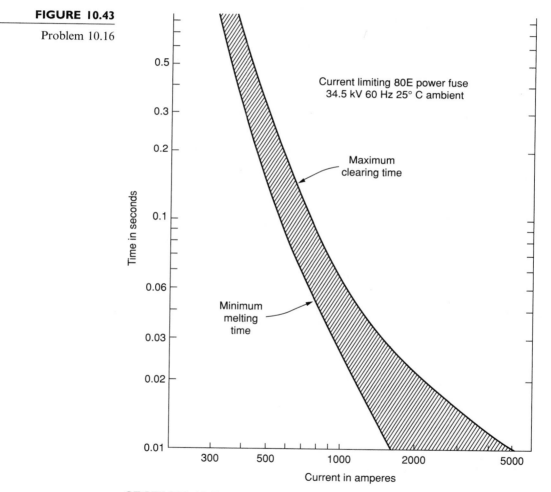

SECTION 10.7

10.17 For the system shown in Figure 10.44, directional overcurrent relays are used at breakers B12, B21, B23, B32, B34, and B43. Overcurrent relays alone are used at B1 and B4. (a) For a fault at P_1, which breakers do not operate? Which breakers should be coordinated? Repeat (a) for a fault at (b) P_2, (c) P_3. (d) Explain how the system is protected against bus faults.

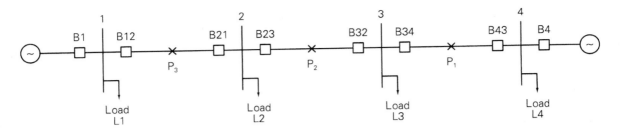

FIGURE 10.44 Problem 10.17

SECTION 10.8

10.18 (a) Draw the protective zones for the power system shown in Figure 10.45. Which circuit breakers should open for a fault at (a) P_1, (b) P_2, and (c) P_3?

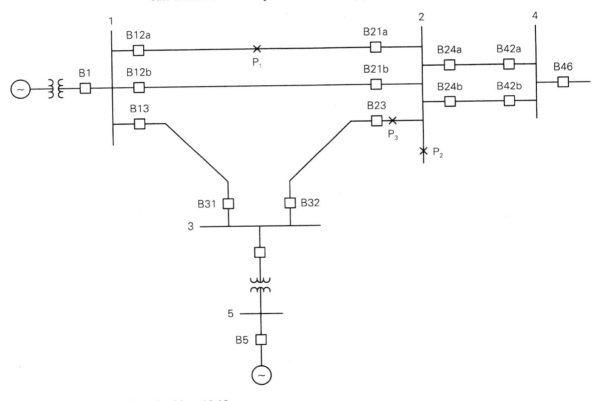

FIGURE 10.45 Problem 10.18

10.19 Figure 10.46 shows three typical bus arrangements. Although the number of lines connected to each arrangement varies widely in practice, four lines are shown for convenience and comparison. Note that the required number of circuit breakers per line is 1 for the ring bus, $1\frac{1}{2}$ for the breaker-and-a-half double-bus, and 2 for the double-breaker double-bus arrangement. For each arrangement: (a) Draw the protective zones. (b) Identify the breakers that open under primary protection for a fault on line

FIGURE 10.46

Problem 10.19—typical bus arrangements

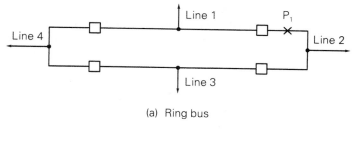

(a) Ring bus

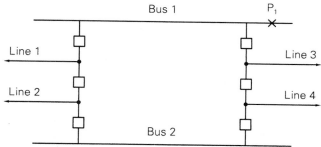

(b) Breaker-and-a-half double bus

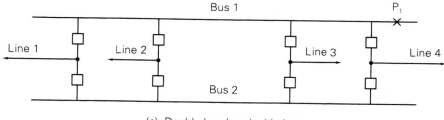

(c) Double-breaker double bus

1. (c) Identify the lines that are removed from service under primary protection during a bus fault at P_1. (d) Identify the breakers that open under backup protection in the event a breaker fails to clear a fault on line 1 (that is, a stuck breaker during a fault on line 1).

SECTION 10.9

10.20   Three-zone mho relays are used for transmission line protection of the power system shown in Figure 10.25. Positive-sequence line impedances are given as follows.

Line	Positive-Sequence Impedance, Ω
1–2	$6 + j60$
2–3	$4 + j40$
2–4	$5 + j50$

Rated voltage for the high-voltage buses is 500 kV. Assume a 1500:5 CT ratio and a 4500:1 VT ratio at B12. (a) Determine the settings Z_{t1}, Z_{t2}, and Z_{t3} for the mho relay at B12. (b) Maximum current for line 1–2 under emergency loading conditions is 1400 A at 0.90 power factor lagging. Verify that B12 does not trip during emergency loading conditions.

10.21   Line impedances for the power system shown in Figure 10.47 are $Z_{12} = Z_{23} = 3.0 + j40.0\ \Omega$, and $Z_{24} = 6.0 + j80.0\ \Omega$. Reach for the zone 3 B12 impedance relays is set for 100% of line 1–2 plus 120% of line 2–4. (a) For a bolted three-phase fault at bus 4, show that the apparent primary impedance "seen" by the B12 relays is

$$Z_{\text{apparent}} = Z_{12} + Z_{24} + (I_{32}/I_{12})Z_{24}$$

where (I_{32}/I_{12}) is the line 2–3 to line 1–2 fault current ratio. (b) If $|I_{32}/I_{12}| > 0.20$, does the B12 relay see the fault at bus 4?

   Note: This problem illustrates the "infeed effect." Fault currents from line 2–3 can cause the zone 3 B12 relay to underreach. As such, remote backup of line 2–4 at B12 is ineffective.

FIGURE 10.47

Problem 10.21

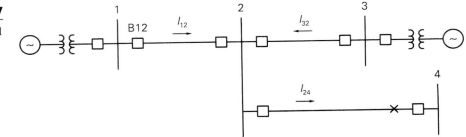

10.22   Consider the transmission line shown in Figure 10.48 with series impedance Z_L, negligible shunt admittance, and a load impedance Z_R at the receiving end. (a) Determine Z_R for the given conditions of $V_R = 1.0$ per unit and $S_R = 2 + j0.8$ per unit. (b) Construct the impedance diagram in the R-X plane for $Z_L = 0.1 + j0.3$ per unit. (c) Find Z_S for this condition and the angle δ between Z_S and Z_R.

FIGURE 10.48

Problem 10.22

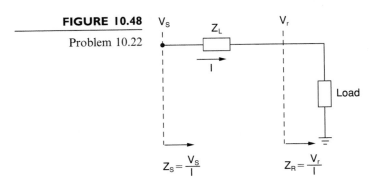

10.23   A simple system with circuit breaker-relay locations is shown in Figure 10.49. The six transmission-line circuit breakers are controlled by zone distance and directional relays, as shown in Figure 10.50. The three transmission lines have the same positive-

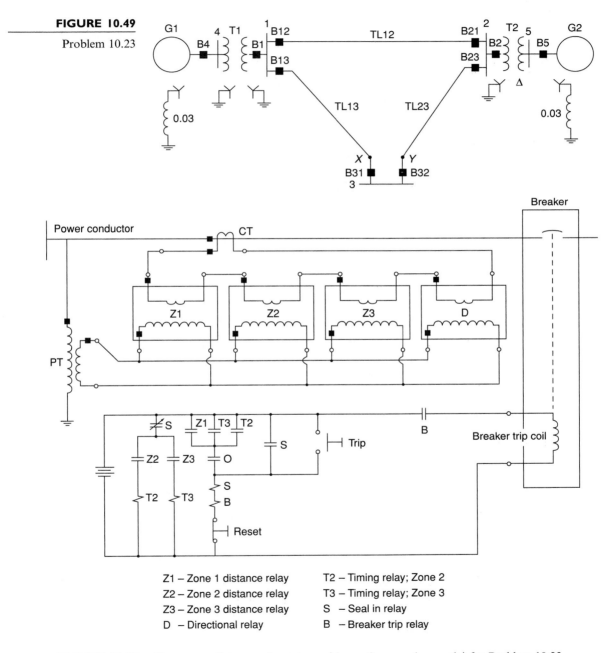

FIGURE 10.49

Problem 10.23

FIGURE 10.50   Three-zone distance-relay scheme (shown for one phase only) for Problem 10.23

Z1 – Zone 1 distance relay   T2 – Timing relay; Zone 2
Z2 – Zone 2 distance relay   T3 – Timing relay; Zone 3
Z3 – Zone 3 distance relay   S – Seal in relay
D – Directional relay   B – Breaker trip relay

sequence impedance of j0.1 per unit. The reaches for zones 1, 2, and 3 are 80, 120, and 250%, respectively. Consider only three-phase faults. (a) Find the settings Z_r in per unit for all distance relays. (b) Convert the settings in Ω if the VTs are rated 133 kV : 115 V and the CTs are rated 400 : 5 A. (c) For a fault at location X, which is 10% down line TL31 from bus 3, discuss relay operations.

SECTION 10.10

10.24   Select k such that the differential relay characteristic shown in Figure 10.34 blocks for up to 20% mismatch between I_1' and I_2'.

SECTION 10.11

10.25   Consider a protected bus that terminates four lines, as shown in Figure 10.51. Assume that the linear couplers have the standard $X_m = 5$ mΩ and a three-phase fault externally located on line 3 causes the fault currents shown in Figure 10.51. Note that the infeed current on line 3 to the fault is $-j10$ kA. (a) Determine V_o. (b) Let the fault be moved to an internal location on the protected bus between lines 3 and 4. Find V_o and discuss what happens. (c) By moving the external fault from line 3 to a corresponding

FIGURE 10.51

Problem 10.25—Bus differential protection using linear couplers

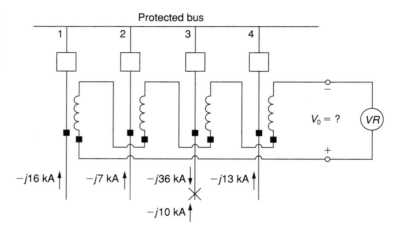

point on (i) line 2 and (ii) line 4, determine V_o in each case.

SECTION 10.12

10.26   A single-phase, 5-MVA, 20/8.66-kV transformer is protected by a differential relay with taps. Available relay tap settings are 5 : 5, 5 : 5.5, 5 : 6.6, 5 : 7.3, 5 : 8, 5 : 9, and 5 : 10, giving tap ratios of 1.00, 1.10, 1.32, 1.46, 1.60, 1.80, and 2.00. Select CT ratios and relay tap settings. Also, determine the percentage mismatch for the selected tap setting.

10.27   A three-phase, 500-MVA, 345 kV Δ/500 kV Y transformer is protected by differential relays with taps. Select CT ratios, CT connections, and relay tap settings. Determine the currents in the transformer and in the CTs at rated conditions. Also determine the percentage mismatch for the selected relay tap settings. Available relay tap settings are given in Problem 10.26.

10.28   For a Δ–Y connected, 15-MVA, 33 : 11 kV transformer with differential relay protection and CT ratios shown in Figure 10.52, determine the relay currents at full load and calculate the minimum relay current setting to allow 125% overload.

FIGURE 10.52

Problem 10.28

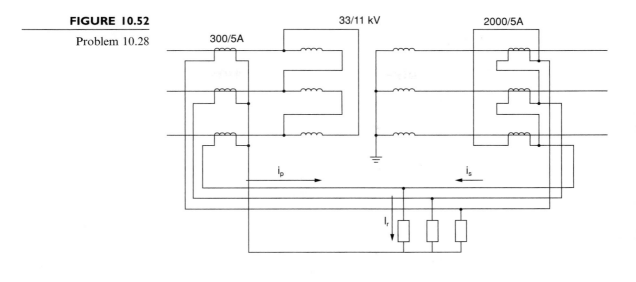

10.29   Consider a three-phase Δ–Y connected, 30-MVA, 33:11 kV transformer with differential relay protection. If the CT ratios are 500:5 A on the primary side and 2000:5 A on the secondary side, compute the relay current setting for faults drawing up to 200% of rated transformer current.

10.30   Determine the CT ratios for differential protection of a three-phase, Δ–Y connected, 15-MVA, 33:11 kV transformer, such that the circulating current in the transformer Δ does not exceed 5 A.

CASE STUDY QUESTIONS

A.   What are the differences among computer relays, solid-state relays, and electro-mechanical relays?

B.   What are the benefits and risks of computer relays?

C.   What is adaptive relaying?

REFERENCES

I.   J. L. Blackburn, *Protective Relaying* (New York: Dekker, 1997).

2.   J. L. Blackburn et al., *Applied Protective Relaying* (Newark, NJ: Westinghouse Electric Corporation, 1976).

3.   *Westinghouse Relay Manual, A New Silent Sentinels Publication* (Newark, NJ: Westinghouse Electric Corporation, 1972).

4.   J. W. Ingleson et al., "Bibliography of Relay Literature. 1986–1987. IEEE Committee Report," *IEEE Transactions on Power Delivery, 4*, 3, pp. 1649–1658 (July 1989).

5. *IEEE Recommended Practice for Protection and Coordination of Industrial and Commercial Power Systems*—IEEE Buff Book, IEEE Standard 242-2001 (www.ieee.org, January 2001).

6. *Distribution Manual* (New York: Ebasco/Electrical World, 1990).

7. C. Russel Mason, *The Art and Science of Protective Relaying* (New York: Wiley, 1956).

8. C. A. Gross, *Power System Analysis* (New York: Wiley, 1979).

9. W. D. Stevenson, Jr., *Elements of Power System Analysis*, 4th ed. (New York: McGraw-Hill, 1982).

10. A. R. Bergen, *Power System Analysis* (Englewood Cliffs, NJ: Prentice-Hall, 1986).

11. S. H. Horowitz and A. G. Phadke, *Power System Relaying* (New York: Research Studies Press, 1992).

12. A. G. Phadke and J. S. Thorpe, *Computer Relaying for Power Systems* (New York: Wiley, 1988).

13. C. F. Henville, "Digital Relay Reports Verify Power System Models," *IEEE Computer Applications in Power*, 13, 1 (January 2000), pp. 26–32.

14. S. H. Horowitz and A. G. Phadke, "Blackouts and Relaying Considerations," *IEEE Power and Energy*, 4, 3 (September/October 2006), pp. 60–67.

15. D. Reimert, *Protective Relaying for Power Generation Systems* (Boca Raton, FL: CRC Press, 2005).

11

POWER SYSTEM CONTROLS

Automatic control systems are used extensively in power systems. Local controls are employed at turbine-generator units and at selected voltage-controlled buses. Central controls are employed at area control centers.

Figure 11.1 shows two basic controls of a steam turbine-generator: the voltage regulator and turbine-governor. The voltage regulator adjusts the power output of the generator exciter in order to control the magnitude of generator terminal voltage V_t. When a reference voltage V_{ref} is raised (or lowered), the output voltage V_r of the regulator increases (or decreases) the exciter voltage E_{fd} applied to the generator field winding, which in turn acts to increase (or decrease) V_t. Also a voltage transformer and rectifier monitor V_t, which is used as a feedback signal in the voltage regulator. If V_t decreases,

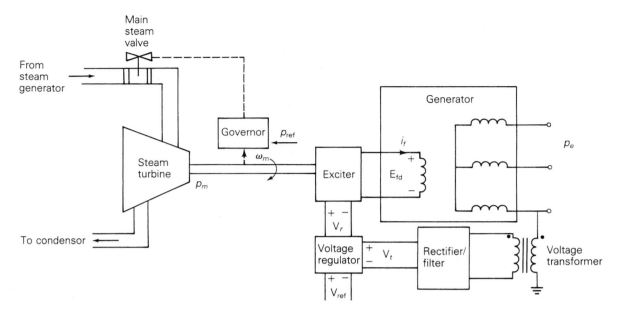

FIGURE II.I Voltage regulator and turbine-governor controls for a steam-turbine generator

the voltage regulator increases V_r to increase E_{fd}, which in turn acts to increase V_t.

The turbine-governor shown in Figure 11.1 adjusts the steam valve position to control the mechanical power output p_m of the turbine. When a reference power level p_{ref} is raised (or lowered), the governor moves the steam valve in the open (or close) direction to increase (or decrease) p_m. The governor also monitors rotor speed ω_m, which is used as a feedback signal to control the balance between p_m and the electrical power output p_e of the generator. Neglecting losses, if p_m is greater than p_e, ω_m increases, and the governor moves the steam valve in the close direction to reduce p_m. Similarly, if p_m is less than p_e, ω_m decreases, and the governor moves the valve in the open direction.

In addition to voltage regulators at generator buses, equipment is used to control voltage magnitudes at other selected buses. Tap-changing transformers, switched capacitor banks, and static var systems can be automatically regulated for rapid voltage control.

Central controls also play an important role in modern power systems. Today's systems are composed of interconnected areas, where each area has its own control center. There are many advantages to interconnections. For example, interconnected areas can share their reserve power to handle anticipated load peaks and unanticipated generator outages. Interconnected areas can also tolerate larger load changes with smaller frequency deviations than an isolated area.

FIGURE 11.2

Daily load cycle

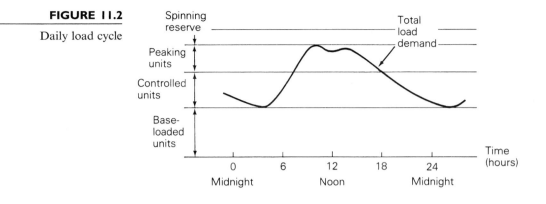

Figure 11.2 shows how a typical area meets its daily load cycle. The base load is carried by base-loaded generators running at 100% of their rating for 24 hours. Nuclear units and large fossil-fuel units are typically base-loaded. The variable part of the load is carried by units that are controlled from the central control center. Medium-sized fossil-fuel units and hydro units are used for control. During peak load hours, smaller, less efficient units such as gas-turbine or diesel-generating units are employed. In addition, generators operating at partial output (with *spinning reserve*) and standby generators provide a reserve margin.

The central control center monitors information including area frequency, generating unit outputs, and tie-line power flows to interconnected areas. This information is used by automatic *load-frequency control* (LFC) in order to maintain area frequency at its scheduled value (60 Hz) and net tie-line power flow out of the area at its scheduled value. Raise and lower reference power signals are dispatched to the turbine-governors of controlled units.

Operating costs vary widely among controlled units. Larger units tend to be more efficient, but the varying cost of different fuels such as coal, oil, and gas is an important factor. *Economic dispatch* determines the megawatt outputs of the controlled units that minimize the total operating cost for a given load demand. Economic dispatch is coordinated with LFC such that reference power signals dispatched to controlled units move the units toward their economic loadings and satisfy LFC objectives. *Optimal power flow* combines economic dispatch with power flow so as to optimize generation without exceeding limits on transmission line loadability.

In this chapter we investigate automatic controls employed in power systems under normal operation. Sections 11.1 and 11.2 describe the operation of the two generator controls: voltage regulator and turbine-governor. We discuss load-frequency control in Section 11.3, economic dispatch in Section 11.4, and optimal power flow in Section 11.5.

CASE   STUDY   Restructuring of the electric utility industry throughout the world has had a significant impact on the controls, operations, and planning of interconnected power systems. The following article compares the transmission planning process of a regulated utility industry with the planning process during the transition to a de-regulated industry. Also, the present structure of transmission planning in the United States is compared with the present structure in Australia [12].

Transmission System Planning—The Old World Meets the New

ROBERT J. THOMAS, JAMES T. WHITEHEAD, HUGH OUTHRED, AND TIMOTHY D. MOUNT

INVITED PAPER

Transmission systems in the United States and around the world have become more difficult to plan under deregulation. Uncertainties in generation type, location, and pricing coupled with market mechanisms that encourage transfers not designed for in the legacy transmission system have made the transmission planner's life difficult. This paper describes the planning process under regulation, some of the issues encountered during the present transition to a deregulated industry, and some of the open questions that need to be addressed in the United States.

Keywords—Planning, transmission system.

Manuscript received October 1, 2004; revised June 1, 2005. This work was supported in part by the National Science Foundation Power Systems Engineering Research Center (PSERC).

R. J. Thomas is with the School of Electrical and Computer Engineering, Cornell University, Ithaca, NY 14853 USA (e-mail: rjt1@cornell.edu).

J. T. Whitehead is with the Tennessee Valley Authority, Chattanooga, TN 37402-2801 USA (e-mail: jtwhitehead@tva.gov).

H. Outhred is with the Centre for Energy and Environmental Markets, University of New South Wales, Sydney, N.S.W. 2052, Australia (e-mail: h.outhred@unsw.edu.au).

T. D. Mount is with the Department of Applied Economics and Management, Cornell University Ithaca, NY 14853 USA (e-mail: tdm2@cornell.edu).

Digital Object Identifier 10.1109/JPROC.2005.857489

("Transmission System Planning—The Old World Meets the New" by Robert J. Thomas, James T. Whitehead, Hugh Outhred and Timothy D. Mount, Proceedings of the IEEE (Nov. 2005), pg. 2026–2035.)

I.   INTRODUCTION

The function of an electricity supply system is to maintain voltage waveforms of appropriate quality at the points of connection of end-use equipment (loads) and thus provide a continuous flow of electrical energy to meet end users' requirements. The transmission network is a shared resource in an electricity industry that makes an essential contribution to this capability by:

- providing connectivity between all large generators and all load centers and thus compensating for differences between the geographical distributions of generation and electricity demand;
- improving supply availability by automatically exploiting the diversity between the stochastic processes of generator availability and electricity demand;
- improving supply quality by contributing to the management of voltage magnitude, phase balance, and waveform purity, particularly when contingencies occur.

Planning is a process that regulated electric utilities traditionally went through in order to demonstrate due diligence with respect to their "obligation to serve" load. The results of the planning process usually produced several plans for expansion of either generation or transmission, or both. Transmission planning is the process of designing future network configurations that meet predicted future needs. It is inherently a cooperative process because the transmission network is a resource that is shared by all network users (generators and loads). Under regulation, most utilities were vertically integrated regional monopolies that were responsible for all generation, transmission, and distribution in one or more contiguous regions. These utilities could make planning decisions centrally for their own service territories. In the United States, there was additional coordination provided by power pools, federal and state regulators, and industry oversight organizations, such as the North American Electric Reliability Council (NERC). An acceptable plan for expanding transmission carried with it an obligation by the regulators to permit the utilities to earn an allowed rate on and return of all capital costs that were "used and useful." Under this form of regulation, it was possible to maintain transmission and generation adequacy, and, some would argue, the supply system was "overbuilt."

A regulated utility was obligated to meet all reasonable requests by end users for future supply needs, usually with little expectation that they would provide advance notice of their requirements in terms of either timing or location. In return for accepting this broad obligation, the regulated utility was left with considerable planning autonomy and discretion to exercise engineering judgement in designing and implementing a supply-side solution. When the load doubled every ten years, as it did up to the 1970s, network planning was usually subordinated to generation planning. In addition, there were relatively few constraints from the public on acquiring easements for new transmission. After the oil embargo in 1973, there was more public opposition to expansion when load growth slowed down and the capital costs of nuclear power plants escalated (Nelson and Peck [1]). However, the basic responsibilities for providing a reliable supply system did not change substantially.

Since restructuring of the electric utility industry began in the 1990s in the United States, the planning process has become more complicated. A major reason is that many investment decisions, particularly for generators, are now determined by market forces rather than by a centralized decision process.

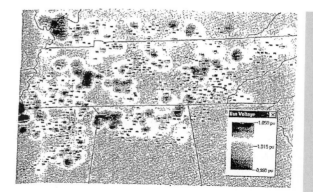

Figure 1
Normal peak load voltage profile on the TVA system (July 1999).

The financial risk of this investment falls on the investors and is no longer backed by the customers as it was under regulation. Consequently, there is no guarantee that market forces will meet all legitimate investment needs, even for maintaining generation adequacy. Although the financing of most transmission is still regulated, it is no longer clear cut how to assign the financial responsibilities for serving load, particularly in terms of maintaining the reliability of supply. For example, there has been a substantial increase in the quantity of power transferred over long distances through the TVA territory (see, for example, Figs. 1 and 2). Both thermal and voltage

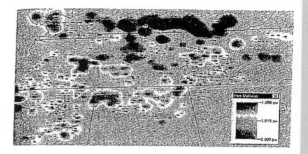

Figure 2
The voltage profile with 8000 MW of loop-flow (August 1999).

constraints on transmission have been experienced in locations that previously were rarely congested.

One solution to the problem of congestion caused by power transfers is to expand the capacity of the transmission network. However, it is not a simple task to divide the cost of this expansion between customers in the service territory and the many generators and loads in other service territories that benefit from the power transfers. When commercial power transfers through a network contribute to congestion, it is more difficult for an individual transmission owner to predict how market forces will affect future congestion (possibly in new locations) on an expanded network. Since the transfers generally depend on decisions made in other service territories, will the transfers continue in the future, or do they represent a short-term arbitrage opportunity caused by temporary regional differences in fuel prices?

Finally, even if the transfers are temporary and transmission expansion is not required, it is still difficult to allocate the true system costs of transfers to individual transactions. Congestion due to transfers may result in higher nodal prices within a service territory, but there is no guarantee that the revenue collected for transfers from wheeling charges, for example, will end up compensating customers for the higher energy prices. Since there is no global optimization of the dispatch in the eastern or the western interconnections, the financial payouts for some transfers may reflect anomalies due to price inconsistencies across the "seams" between control areas rather than true reductions in the cost of meeting load. In other words, completing a transaction that is commercially viable under the current conditions does not guarantee that there will be positive net benefits to the system.

In addition to the new challenges associated with deregulation, the current structure of the utility industry in the United States is inherently complicated due to different combinations of public and private ownership, state and federal regulation, and merchant and regulated companies. This complicated structure suggests that the path to deregulation should be cautious. In particular, the reliability of supply is a shared responsibility for all users of a transmission network. It is important to determine what markets can and cannot do, particularly for the transmission network, if high standards of reliability are to be maintained in the future.

The purpose of this paper is to describe the current state of transmission planning in the United States (Section II) and to compare it with the current structure in Australia (Section IV). The Australian electricity market provides a good example of successful deregulation. It is a relatively small system, in terms of load, that is now centrally dispatched and no longer has to deal with seams problems. More importantly, the Australian regulators have generally embraced market forces by tolerating high price volatility for energy in the spot market, with a few price spikes reaching \$8000/MWh (\$10,000 A/MWh) (see [11]). However, average spot prices in Australia have remained low, sufficient new investment has been made to maintain generation adequacy, and there has been no need to establish a capacity auction to ensure there is enough generation capacity available to meet the load. Generators simply cannot afford to miss the opportunity of getting paid \$8000/MWh. In contrast, the high price volatility tolerated in the Australian market is not politically acceptable in the United States at present. For this and other reasons, concerns about generation adequacy in the future have arisen in the United States. Deregulated power pools in the northeastern states are currently proposing to maintain generation adequacy by supplementing payments to generators through capacity auctions.

In spite of the reliance on market forces to maintain generation adequacy in Australia, the transmission system in Australia is still fully regulated. Merchant transmission is restricted to dc interties. The dividing line between regulation and market forces is quite clear in Australia compared to the United States. The main thesis of this paper is that the overall reliability of the supply system in the United States is being jeopardized by the uncertainty that currently exists over who is responsible for maintaining generation and transmission adequacy. This is a particularly relevant question for transmission planning. Should planning decisions for transmission be centralized, or can these decisions be left to decentralized market forces? We argue that it is better to follow the Australian approach of caution over transmission planning.

Standard engineering principles imply that a new design should be tested and verified first before imposing it on the existing supply system. The interdependent characteristics of a transmission network make it inappropriate to use a piecemeal approach to transmission planning. In spite of this, the initial strategy followed by the Federal Energy Regulatory Commission (FERC) has been to allow a wide variety of approaches to deregulation in different regions. This strategy may work for generation, but it is not appropriate for transmission. The primary reason why more coordination is needed for the transmission network is that it plays a central role in maintaining the reliability of supply for customers. Unlike real energy, the benefits of reliability are widely shared by all users of the network. Consequently, it is difficult to divide up the responsibilities for maintaining reliability among individual users of the network.

Section II of the paper describes transmission planning in the United States and specific issues relating to reliability are discussed in Section III. The main features of the Australian supply system are presented in Section IV, challenges to transmission planning are discussed in Section V, and the conclusions are summarized in Section VI.

II. PLANNING TRANSMISSION IN THE UNITED STATES

Planning the expansion of a transmission system requires an investigation of several plans for interconnecting generation and load centers. In effect, this effort produces a number of feasible transmission expansion plans and, with the help of various tools, identifies one plan that best meets forecasted needs. This is a complex task, since the constraints imposed on transmission lines severely limit options. Not only do transmission systems have to satisfy rigid technical and financial constraints, they must also comply with a host of major environmental constraints that may change during the planning process. The criterion that any transmission system must satisfy is that it must be able to transmit electric energy economically and reliably from generation to all load centers at an acceptable voltage level and in the quantity desired through environmentally acceptable corridors using environmentally acceptable lines and towers over the range of load expected during the life of the line.

Prior to the 1950s, utilities in the United States would try, to the extent possible, to locate generation near load centers. For a variety of reasons, including economies of scale in generation and increased real estate prices, it became increasingly necessary to locate generation far from load centers. Around 1950, U.S. utilities began to experience difficulty in obtaining transmission rights-of-way [2]. As a result, the economics of transmission-line loading became an important aspect of transmission expansion planning. The economics of loading fall into two categories: costs associated with the transmission of power and those that are not. Those not directly associated with the transmission of power are: 1) cost of rights-of-way; 2) the transmission line itself; 3) circuit breakers and steel structures; and 4) direct and indirect overhead costs. Those directly associated with transmission of power are: 5) transformer costs; 6) cost of line losses; and 7) the cost of reactive compensation/generation.

Curve (a) in Fig. 3 represents items 1–4 summed and divided by megawatts to be transmitted. Clearly the most economic loading occurs for infinite megawatts transmitted if only costs 1–4 were taken into account. Curve (b) represents the sum of costs 5–7 divided by megawatt loading. Here the most economical loading is 0 MW. The two curves intersect, or, when combined, result in curve (c). Curve (c) shows that when both costs are considered, there is a minimum cost loading point. Unfortunately, transmission lines cannot always be loaded to their optimum as dictated by costs alone. While the economics have not changed, the ability to load lines optimally is more difficult in systems where market forces determine generation patterns.

For many years prior to the 1970s, the business of transmission planning was relatively straightfor-

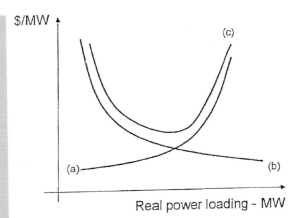

Figure 3
A generic portrayal of economic costs associated with transmission. Curve (a) represents operating costs and curve (b) construction costs. Curve (c) is the composite.

ward. Most of the problems encountered were about the load exceeding the thermal capability of transmission equipment. While there were some voltage problems on the system, they were mostly local and could easily be identified and solved effectively using a power flow program, such as the Philadelphia Electric Company program. Creating a base case was relatively easy because the primary inputs of connectivity, load, and generation were well known. Those were the days when a utility knew that it had to serve its native load with its own generation, when load forecasts were fairly accurate, and when the plan for generation expansion and operation was solid. Load forecasting techniques were based simply on historical load growth trends and knowledge of new developments that affected load growth. Generation plans were based on the load forecasts, and companies developed long-range plans to add generation and transmission to serve its native load.

During those years, the transmission system was conservatively planned and facilities were relatively unloaded. The power flow program was quite adequate for defining what were mostly steady-state problems. Transient stability analysis was just beginning to be a consideration, and few transient problems were known to exist. Utilities knew that they had to depend on their own generation plan in order to serve their future native loads. Any errors in the load forecasts resulted in moving the plan a year or two forward or backward in time depending on actual load growth. Power transfers between utilities could be predicted by identifying who held the long-term contracts, and these contracts changed little. Thus, a transmission expansion plan could be determined for the next several years with a high degree of confidence.

It was fairly certain that any reasonable plan could be implemented. Costs were much lower than they are today, and budget impacts were not too restrictive. Public and environmental acceptance of new construction was not a big problem. Therefore, once a plan was proposed, it had a good chance of actually being implemented on time. Gradually over the years, costs for new facilities, budget pressures, and concerns over environmental impacts have increased, while public acceptance of new transmission has decreased. Even without the changes brought on by deregulation, transmission planning has become more difficult because of increased project lead times caused by public and environmental issues that have led to difficulties in implementing any plan on time.

When FERC Orders 888 and 889 were issued in 1996, the transmission world began to change substantially, and the major impacts on transmission planning were: 1) the implementation of market-based generation and 2) the separation of transmission planning from generation planning. The first sign of this change was the rapid proliferation of independent power producers (IPPs) during the latter part of the 1990s. The construction of numerous generating plants by IPPs often exceeded the amount of capacity built by the existing utility in a region. This led to substantial uncertainties about future power flows because it was not known how much the IPPs would generate, or whether they would replace native generation or sell their output to neighboring systems and, if so, to which ones. At the same time, there was growing uncertainty about generation expansion because the implementation of the standards of conduct required by FERC

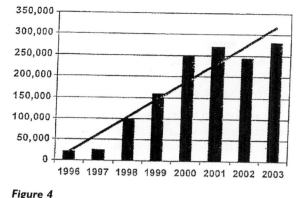

Figure 4
Growth in TVA transactions (Source: TVA).

Order 889 required the physical and informational separation of utility generation planning and transmission planning. This separation had the effect of creating further uncertainty about future patterns of generation in power flow models and eliminated the traditional form of integrated transmission and generation planning.

In the TVA region, an important manifestation of the increased generation by IPPs was the appearance of periods of extremely high and unanticipated transfers across interconnections and through the utility's transmission systems (see Fig. 4). These flows were much higher than the levels forecasted in existing transmission plans, and they created problems of high line-loading and low voltages across the network. This was the first evidence that the deregulated power markets had changed the pattern of generation and that power flows would not be consistent with prior plans. Even today, power flows across interconnections are changing rapidly as market conditions continue to change. A transmission planner's ability to predict flows on a transmission system accurately for future years has been adversely affected by these changes. As a result, the only practical posture for a transmission planner is to treat transfers through the network as temporary, plan for increases of native load, and leave the impacts of the transfers to the transmission security experts in real-time operations. In this way, transfers can be curtailed if reliability standards are threatened. For example, the number of transmission loading reliefs (TLRs) has increased dramatically in the TVA region. The overall result is that enhancements in transfer capacities on a network will tend to occur only when transmission improvements are implemented to meet the needs of native load. Consequently, investment in transmission will be insufficient to support the growth of transfers even when this increased capacity would lead to higher net benefits for the supply system as a whole.

There are growing indications that there has been insufficient investment in transmission since deregulation began. In a recent report on long-term reliability assessment by the NERC [3], the section on "Transmission Issues" raises some concerns about the future adequacy of the transmission grid. The report states [3, p. 34]:

> Over the past decade, the increased demands placed on the transmission system in response to industry restructuring and market-related needs are causing the grid to be operated closer to its reliability limits more of the time. The demand for electricity continued to grow in the 1980s and 1990s, but transmission additions have not kept pace. The uncertainty associated with transmission financing and cost recovery and the impediments to siting and building new transmission facilities have resulted in a general slowdown in construction of new transmission. In some areas of North America, increases in generating capability have surpassed the capability of the transmission system to simultaneously move all of the electricity capable of being produced. In addition, market-based electricity transactions flowing across the grid have increased, as has the incidence of grid congestion. The result is increased loading on existing transmission systems and tighter transmission operating margins.

In the United States, there are about 150,000 mi of transmission ranging in voltage levels from 138 to 765 kV. While projected growth in load was about 18% from 1996 to 2007, the projected growth in transmission was less than 4%. TVA is a typical company where about 64% of transmission structures are 25 years or older, 73% of transformers are 25 years or older, 53% of breakers are 25 years or older, and 31% are 35 years or older. It is clear that the transmission infrastructure is aging substantially at a time when there is an increased dependence on it to facilitate economic transactions while attempting to maintain the reliability of the past. For example, in TVA alone transactions have grown from 20,000 annually in 1996 to 250,000 annually in 2001. This growth has been just as prolific in areas such as New York, New England, PJM, California, and Texas that have introduced deregulated markets.

III. MAINTAINING SYSTEM RELIABILITY IN A DEREGULATED MARKET ENVIRONMENT

The U.S. Department of Energy (DOE) National Transmission Grid Study [4] determined that congestion in the U.S. electricity transmission system places daily constraints on electricity trade, increases electricity costs to consumers, and threatens reliable operations. To address the problems of transmission congestion, the Secretary of Energy chartered an Electricity Advisory Board that established a Transmission Grid Solution Subcommittee. The report from this subcommittee defines transmission congestion or "bottlenecks" as follows: "Bottlenecks occur when the system is constrained such that it cannot accommodate the flow of electricity and systematically inhibits transactions. Thus, a bottleneck has economic and/or reliability impacts." National interest transmission "bottlenecks" are power flow constraints on regional transmission lines that:

- create congestion that significantly decreases reliability;
- restrict competition;
- enhance opportunities for suppliers to exploit market power;
- increase prices to consumers;
- increase infrastructure vulnerabilities;
- increase the risk of blackouts.

The report cites the following sources and amounts as indicators of the true costs of congestion:

- FERC study of 16 constraints: $700 million for six summer months;
- DOE study of four regions: $447 million;
- ISO New England: $125 million–$600 million;
- California Path 15: $222 million for the 16-month period before December 2000.

In addition to the economic cost caused by inadequate transmission, such a system will impair security, create inefficient markets by limiting market transactions, and reduce overall system reliability. Real power flow may be restricted on certain lines under normal operating conditions or as a result of equipment failures and system disturbance conditions. Bottlenecks are persistent flow limits that may be major inhibitors to economic efficiency and system reliability. Specifically, they undermine the physical security of the supply system exposing vulnerable infrastructure elements or critical facilities whose loss or impairment would substantially reduce the transmission of power into or out of key load or resource centers.

Transmission bottlenecks clearly affect system reliability. Criteria established by the North American Electric Reliability Council (NERC), the relevant Regional Reliability Council, or by authorities with local jurisdiction such as state service commissions or Independent System Operators (ISOs) provide normal limits to transmission capability due to:

- stability limits—anticipated power flows after a contingency in the list of specified contingencies would exceed stability limits, resulting in an unstable power system;
- thermal limits—anticipated power flows after a contingency would exceed the thermal limit of a line or a component of the network (e.g., a transformer);
- voltage collapse limits—anticipated power flows after a contingency would create a reactive demand that would exceed the local reactive resources, resulting in rapid voltage decay;
- loop flow—unscheduled power flows on lines or facilities that result in a violation of reliability criteria;
- resource deficiency—installed capacity levels are inadequate to support the load.

Since relieving one reliability limit will simply cause the next most limiting element to appear, these limits should be viewed as "system" limits needing "system" solutions. However, since there has been little growth in the transmission system, the current regulatory focus is on relieving specific bottlenecks rather than on a more comprehensive plan to build new lines to maintain reliability for future growth. Neither the technical definition of the problem nor the tools for planning under these uncertainties have been fully developed. Even if a suitable plan for transmission can be developed, the discussion in the previous section explains why the new deregulated markets pose difficult questions about how to implement the plan. Although understanding the effects of deregulation on investment decisions is particularly important for planning purposes, current knowledge and information about investment behavior is limited.

Since all users share the performance and the level of reliability provided by a network, reliability is primarily a "public" good. (All customers benefit from their level of reliability without "consuming" it. In contrast, real energy is a "private" good because the real energy used by one customer is no longer available to other customers.) Markets can work well for private goods but tend to undersupply public goods like reliability (and oversupply public bads like pollution). The reason is that customers are generally unwilling to pay their fair share of a pubic good because it is possible to rely on others to provide it (i.e. they are "free riders"). Generally, some form of regulatory intervention is needed to make a market for a public good socially efficient.

Even if the desired level of reliability on a network is specified by a regulatory agency, such as FERC, or by an industry oversight organization, such as NERC, it is difficult to partition the benefits of reliability to individual transmission companies (TransCos). As a result, it is also difficult to rely on decentralized decisions by different TransCos to reach economically efficient ways of meeting a required level of reliability. All proposed additions to a network should be evaluated together to form a single comprehensive plan for providing transmission services. If the transmission network happens to be owned by a single TransCo, it is possible, in theory, for this TransCo, like a vertically integrated utility under regulation, to make efficient decisions about how to meet a given reliability standard. In reality, the three transmission networks in the United States (the eastern and western interconnections and Texas) have many owners. As Section II showed, even the large public supply systems, such as TVA, have difficulties making efficient investment decisions for transmission due to power transfers by IPPs through their networks.

One of the potential problems arising from the current regulatory focus on bottlenecks is that it leads to "pipeline" thinking. This type of thinking considers that the main role of transmission is to transfer power from low-cost regions to high-cost regions. The existence of large price differences between regions in a market is an indication that there is congestion limiting the amount of power that can be transferred. Increasing the transfer capacity allows more low cost power into the high-cost region, and the price difference times the flow can be used to compensate the investor who builds the new transmission. For example, this is the standard way of financing a merchant dc intertie. There are many examples of this type of economic transfer, such as the transfers between the Pacific Northwest and California, and transfers from Hydro Quebec to the northeastern states. However, a network also helps to provide a reliable supply system that is able to respond to contingencies, and the main economic challenge for transmission is how to pay for reliability.

A continuing manifestation of pipeline thinking is the regulatory effort expended to support physical bilateral contracts. An important reason for the prevalence of these contracts is that they formed the basis for long-term trades between different utilities when the industry was fully regulated. Many of these contracts still exist or have been renewed. In deregulated markets, FERC has allowed the parties in short-run bilateral trades to identify a transmission path between a generator and a load, and to pay the regulated wheeling charges to individual TransCos along this path. As long as a transfer can be supported on the network without jeopardizing reliability, it should be allowed. This is true even though the specified contract path is only an accounting mechanism, and it represents a way to bypass the optimum pattern of dispatch determined by a market or by a traditional regulated supply system. The actual flows of real energy may be very different, and they will change over time in response to changes in the patterns of dispatch and load.

The differences between the contract path and the actual flows of a transfer are likely to diverge substantially when parts of the network are congested. More importantly, it is also likely that the true costs of accommodating a bilateral contract on a congested network are substantially higher than the regulated wheeling charges. The inevitable differences that exist between the true costs and the accounting costs mean that commercially attractive contracts can be found when there is no real gain in net benefits for the whole system, and vice versa. In other words, finding and exploiting arbitrage opportunities for transfers does not necessarily lead to a more efficient market outcome using this type of accounting.

One way of dealing with physical bilateral contracts on a congested network is to identify congested transmission links, or "flowgates," and to require the parties in a contract to purchase a share of the appropriate flowgates [5]. It is, however, still difficult to use the concept of flowgates effectively on a dense network, like the eastern interconnection, because typically there are multiple and variable pathways associated with a particular contract. The basic problem with a physical contract is still there. The realized accounting costs may be very different from the true costs of a transfer, and therefore, there is no guarantee that contracts will be economically efficient.

Hogan [6] has shown that a financial transfer right (FTR) can be used instead of a flowgate right to hedge the price difference between two regions on a congested network. As a result, physical bilateral contracts can be replaced by a portfolio of financial contracts to provide the same reduction of financial risk in a forward contract for real energy between two regions. The advantage of an FTR is that the differences in nodal prices do reflect the true cost differences between locations (unless there are artificial price differences on seams between control areas). Even though the volume of trading in forward financial instruments is growing, a large proportion of the supply of electricity is still covered by physical bilateral contracts. Furthermore, there has been little leadership from regulators to encourage traders to switch from physical to financial instruments for hedging transfers, and this issue was not a major focus in the draft for a standard market design (FERC 2000).

The success of deregulation depends in part on having a viable institutional structure that supports financial hedging of spot prices throughout the nation and provides incentives for all generators and loads to become full participants in the market. This type of spatial forward market would form a solid basis for making efficient investment decisions for generation and for transmission if it was solely for transporting power. The problem of paying for reliability remains unresolved, and the discussion turns now to how these important issues are handled in Australia.

IV. TRANSMISSION PLANNING IN THE AUSTRALIAN NATIONAL ELECTRICITY MARKET

As an example of a transmission planning process in a different restructuring context we discuss the one in process in Australia. The Australian National Electricity Market (NEM) is implemented on an interconnected, multistate power system that extends over 4000 km. The geographical extent of this power system and its relatively small demand (approximately 30 GW) mean that network issues are important in power system operation and planning.

The NEM is an energy-only market and its core feature is a set of spot energy and ancillary markets that solve a constrained economic dispatch problem every 5 min based on 5-min demand forecasts, with commercial energy trading based on 30-min average spot prices and associated derivative instruments.

The power system, including the transmission network, is represented in the NEM by a hub-and-spoke approximation to nodal pricing, which has the following features.

- Major flow constraints between market regions due to thermal limits or power system security considerations are represented by a set of linear flow constraints, updated as necessary to reflect evolving power system operating conditions.
- Within market regions, network losses are represented but flow constraints are not, except to the extent that they impact on flow constraints between market regions. Otherwise they are managed by targeted market rules and contractual arrangements.
- The market rules provide for market regions to be redefined from year to year to adequately reflect evolving patterns of major flow constraints, although this concept has not been followed in practice. A report on the criteria for setting region boundaries and representing flow constraints between regions has recently been released [7].
- Settlement residue auctions allow market participants to bid for access to the revenue streams associated with the "profits" made by regulated interconnections between market regions. This may be regarded as a type of financial transmission right.

The reasoning behind this market design choice is discussed in [8]. A key feature is that this design internalizes network losses and security constraints in the market solution. Transmission loading relief procedures are not required.

The market rules support the concept of a market network service provider (MNSP) in the guise of a two-terminal link that is independently controllable and has its terminals in different market regions. An MNSP offers into the NEM in a similar fashion to a generator, except that it offers to transfer electrical energy from one market region to another if the price differential exceeds the nominated amount.

Two MNSP dc links have been built, one of which later applied for and was granted transfer to regulated status. One MNSP link is currently under construction, a nominal 400-MW dc link that will connect the island of Tasmania to the mainland, and

which is scheduled to enter service toward the end of 2005. However, most transmission in the NEM is regulated, and this is likely to continue to be the case. This is partly because market prices do not fully reflect the value of transmission services and partly because of government reluctance to trust the market to provide adequate transmission services.

A. Transmission planning in the restructured Australian electricity industry

A parallel process for managing near-term power system security and assessing future supply reliability (planning) supplements the NEM, described above. The National Electricity Market Management Company (NEMMCO) is market and power system operator and planning manager. This enhances compatibility and consistency between the planning functions and market and system operation. The key planning functions are as follows.

- A group of supply and demand industry representatives and government officials (the reliability panel) sets transmission-level reliability targets for the multistate power system that underlies the NEM.
- NEMMCO operates the power system with a power system security envelope that it deems to be consistent with the reliability targets. This is done via the spot market design (which implements a security-constrained dispatch according to security constraints that are continually updated in an on-line fashion) and by NEMMCO acquiring appropriate levels of ancillary services.
- NEMMCO produces a set of projections that look forward up to ten years. These are:
  — a predispatch process that projects spot prices and dispatch levels for the next 24 h, which is updated every 5 min;
  — a short-term projection of assessed system adequacy (STPASA), which projects operating reserve margins for the next week, which is updated every 2 h;
  — a medium-term projection of assessed system adequacy (MTPASA), which projects

operating reserve margins for the next two years and is updated weekly;
— a statement of opportunities (SOO), which includes demand forecasts and committed generation construction projects (those for which contracts have been signed), forecasts operating reserve margins for the next ten years and is updated annually;
— an annual national transmission statement (ANTS), which identifies key transmission constraints and is updated annually in conjunction with the SOO.

The SOO and the ANTS, in conjunction with more detailed annual planning reports produced by each regulated transmission service providers, provide a basis of transmission network planning. For example, the SOO contains a list of committed generation projects, and the ANTS identifies transmission constraints as seen from a system-wide perspective.

The regulated transmission network service providers undertake transmission planning within their franchise service territories and jointly consider interconnectors between their service territories, according to a "regulatory test" promulgated by the federal-level regulator, the Australian Competition and Consumer Commission (ACCC).

The regulatory test contains two mechanisms that transmission and distribution network service providers can use to justify network investment:

• a cost–benefit test, which must identify the best option, with a positive net benefit under most scenarios of future market development, considering of distributed resource options as well as network investment options;
• a reliability test, which must show that the proposed investment is required to meet the specified reliability criterion.

There have been practical difficulties in applying the cost–benefit test (e.g., [9]) and most transmission investment is currently undertaken under the reliability test. The Council of Australian Governments (COAG), in which the federal and state governments work together, is currently reviewing the design and implementation of electricity industry restructuring in Australia. This review aims to identify incremental improvements rather than make radical changes, and part of its work program relates to the representation of transmission networks in the NEM and planning procedures for regulated transmission services. Reference [10] provides a discussion of the objectives and early outcomes of this review.

B. Lessons from the Australian experience

The Australian experience underlies the challenges in successfully dealing with transmission issues in a restructured electricity industry. Transmission services are not readily incorporated into wholesale electricity markets, nor are they easy to regulate in a restructured industry.

The current Australian model implements a mix of competitive and regulated approaches to network services that is not entirely satisfactory. Resolving this dilemma appears to depend on more active end-user participation in the market processes, which would permit more satisfactory commercial substitutes to emerge for the traditional "obligation to serve."

However, it is not yet clear how far this substitution can be taken, as there are both political and technical barriers. The latter derive from the fundamentally shared nature of network services, which makes it very difficult to establish unambiguous, separable, physical property rights.

V. CHALLENGES TO FUTURE TRANSMISSION SYSTEM PLANNING

The concept of obligation to serve, which is a "prime directive" in regulated utilities, must be translated into clear contractual requirements in a restructured electricity industry in which the ownership of the supply industry has been fragmented. This is difficult to achieve because the cooperative, "best intentions" nature of planning runs counter to the competitive nature of the restructured industry with its associated reliance on confidentiality and the use of misinformation as a competitive device.

Another problem in a restructured electricity industry is that the services provided by transmission networks are difficult to value because they are not readily detectable by end users except when quality of supply is unsatisfactory. Thus the value of transmission services is measured more by the rare failures to maintain quality of supply than by their usual satisfactory state, and to value these, spot market prices in supply-constrained parts of the network must be allowed to rise to demand-rationing levels (which may be very high) rather than be cost based. Also, much of the value of transmission services derives from availability and quality of supply reasons that are difficult to price.

Moreover, those rare rationing events are more likely to be due to complex, multifactor causes than clearly identifiable transmission network phenomena. Thus, the best way to relax a particular constraint may be both difficult to identify and unique to that event, characteristics that do not support an objective investment decision making process.

A further complicating factor is that distributed resources (embedded generation and demand side options) can partly substitute for investment in transmission services. However, this equivalence is only partial, is difficult to quantify and may not hold in particular outage scenarios. Therefore, it is also difficult to choose between investment options in transmission and distributed resources in an objective manner.

Rapid changes in market conditions have also brought to the surface the question of which company is going to provide power to which customer. Previously, transmission planners knew which customers their transmission systems were going to serve for many years into the future. Consequently, an adequate and reliable transmission system consistent with the accuracy of the load forecasts could be planned. Today utilities are beginning to see customers shift to other suppliers, sometimes with very short notice. Obviously, long-range transmission planning in this new environment is much harder to implement in the most efficient way. As a result, there is a growing reluctance on the part of transmission owners to make large investments in transmission improvements when there is evidence that some customers may not be there to pay for their share of the improvement.

A development which affects the accuracy of load forecasts for the future is the trend toward implementation of local distributed generation. While these technologies offer the transmission planner an ability to address load growth problems without the construction of major facilities, they are also a large source of planning uncertainty, since customers could elect to serve themselves.

These new and difficult problems notwithstanding, planning must be done. In systems that have restructured the planning process is dictated by the RTOs. Figs. 5–7 show flow diagrams for the planning processes of the NYISO, the NERTO, and PJMRTO.

The challenges of transmission planning in a modern, restructured industry environment may thus be summarized as follows.

- The traditional obligation to serve has usually not been adequately replaced in a restructured industry context and is now typically shared in an ambiguous manner between industry participants, regulators, and government.

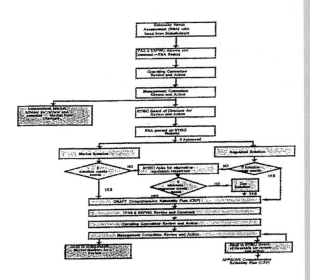

Figure 5
The NYISO planning process. (This figure is the original work of Thomas Gentile, National Grid Company (Thomas.gentile@ngrid.com), who generously provided them for use in this paper.)

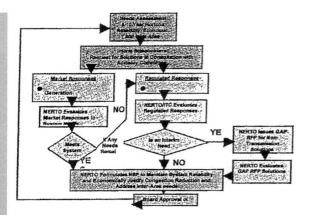

Figure 6
The NERTO regional planning process. (This figure is the original work of Thomas Gentile, National Grid Company (Thomas.gentile@ngrid.com) who generously provided them for use in this paper.)

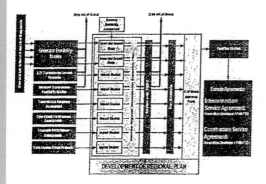

Figure 7
The PJM regional planning process (available at PJM Website—PJM 101 The Basics).

- Supply-rationing events (which are still regarded as infringements of the obligation to serve rather than a characteristic of a competitive industry) are often interpreted as evidence of inadequate transmission capacity, even though they are often the result of more complex phenomena that involve generators, loads, and pricing policies. Consequently, there should be shared accountability among all participants as well as with regulators and government policy makers.

- Supply-rationing events and the risks of supply rationing events are difficult to price effectively, and thus it is difficult to design market processes to manage them.

- Transmission planning has been made more difficult because of the characteristics of competition, including confidentiality, misinformation, and time-varying competitive strategies that may lead to radically differing patterns of network flows, including operating conditions that threaten power system security.

- Transmission planning has also been made more difficult by increased uncertainty about load growth, as well as greater difficulties in obtaining transmission-line easements.

Indeed, the transmission planner's "crystal ball" has become extremely "cloudy" because of greater unknowns in future load, generation, and transfers—three most important inputs to an accurate planning model. There are two other factors that are "clouding the crystal ball." First, the higher cost of constructing new facilities has come at the same time when most transmission owners are facing hard economic times with restricted budgets. This often leads to delays in completing a transmission improvement beyond the time when it was originally planned and needed. The second factor is the increasing lead time needed to initiate a transmission improvement due to environmental issues and decreased public acceptance. Again, transmission improvements are being delayed, and planners have to lengthen their planning horizons to account for these delays. Lengthening lead times and delays in construction occurring at a time when uncertainties about future needs have grown substantially are leading to the inability of transmission planners to ensure that the transmission system meets adequacy standards in the future.

VI. CONCLUSION

All of the economic and engineering factors described above coupled with the underinvestment in new transmission facilities in the United States over the last decade have combined to produce a transmission system that is loaded to a much higher degree than in the past with little hope of improvement in the foreseeable future. The transmission planner's world has changed from the simpler times of mainly worrying about load growth versus generation adequacy to one that is much more complex. Now, the planner must worry about voltage problems and transient and voltage stability in order to plan a reliable supply system that can operate closer to the edge.

The dual roles of the transmission network in maintaining reliability and enabling interregional transfers of real energy are highly interdependent. Given the growing uncertainties that now exist about which organizations have the responsibility for meeting load and maintaining the reliability of the supply system, transmission owners should still be eligible to receive a regulated rate of return for all transmission services. Better performance-based incentives and planning procedures are needed to direct investment into the best places to maintain the traditional high reliability of the network. The final responsibility for maintaining the reliability of a transmission network rests with the regulators, and it cannot be left to decentralized decisions by transmission owners and load-serving entities.

Federal regulators should be more proactive in encouraging the conversion of physical bilateral contracts to financial contracts for inter-regional transfers. Facilitating the establishment of viable forward markets for electricity throughout the country would be an important step in the right direction. There is no economic justification for continuing to support the accounting procedures used to legitimize short-term transfers based on physical bilateral contracts. Some of these contracts may increase net social benefits for all users of a network, but there is no guarantee that a financially attractive contract produces a net gain for all. This is particularly true when the transmission system is congested.

The increased loadings of transmission capacity have led to concerns about reactive power utilization because reactive losses and concerns about voltage stability greatly increase when congestion occurs. While the power flow and transient stability programs are still the primary tools used for transmission planning, new tools have been made available to improve planning, such as optimal power flow (OPF) models. Nevertheless, new tools are still needed to study many more scenarios more efficiently as planners consider more uncertainties. Data such as machine data, exciter and governor models, power factors, and generator reactive capabilities have become more important and new tools must be made available to verify these data and models. These are the responsibilities undertaken through a traditional planning process, and there is no evidence to justify a greater dependence on decentralized decision making as a replacement for transmission planning at this time. In addition, today's transmission planners must communicate closely with real-time operations to better understand how the system is being operated and how the new challenges posed by markets are being managed in today's changing world. Although many improvements in the regulation of transmission can be made, it is important to determine what deregulated markets really can do before transferring the responsibility for system reliability to market forces.

REFERENCES

[1] C. R. Nelson and S. C. Peck, "The NERC fan: a retrospective analysis of the NERC summary forecasts," *J. Business Econom. Stat.*, vol. 3, pp. 179–187, Jul. 1985.

[2] D. K. Blake, R. M. Butler, and R. A. Schmidt Jr., "Problems mount in transmission system planning," *Elec. World*, 1955.

[3] "Long-term reliability assessment" Sep. 2004 [Online]. Available: http://www.nerc.com/~filez/rasreports.html, NERC Report

[4] U.S. Dept. Energy, "National transmission grid study" May 2002 [Online]. Available: http://www.eh.doe.gov/ntgs/reports.html

[5] H.-P. Chao and S. Peck, "A market mechanism for electric power transmission," *J. Regulat. Econom.*, vol. 10, no. 1, pp. 25–59, 1996.

[6] W. W. Hogan, "Financial transmission right formulations" Mar. 31, 2002 [Online]. Available: http://ksghome.harvard. edu/~whogan/

[7] Charles River Assoc., "NEM—Transmission region boundary structure" Sep. 2004 [Online]. Available: http://www.mce.gov.au

[8] H. R. Outhred and R. J. Kaye, "Incorporating network effects in a competitive electricity industry: an Australian perspective," in *Issues in Transmission Pricing and Technology*, M. Einhorn and R. Siddiqui, Eds. Norwell, MA: Kluwer, 1996, ch. 9, pp. 207–228.

[9] S. Littlechild, "Transmission regulation, merchant investment, and the experience of SNI and Murraylink in the Australian electricity market" 2003 [Online]. Available: www.ksg.harvard.edu/hepg

[10] H. R. Outhred, "A review of experience with commercializing and regulating network services in the Australian electricity industry," *Int. Energy J.*, vol. 6, no. 1, pt. 4, pp. 4-99–4-108, Jun. 2004.

[11] NEMMCO, "An introduction to Australia's national electricity market, 6th ed.," Jun. 2005 [Online]. Available: http://www.nemmco.com.au/nemgeneral/NEM_general.htm#Publications

Robert J. Thomas (Fellow, IEEE) received the Ph.D. degree from Wayne State University, Detroit, MI, in 1973.

He is currently Professor of Electrical and Computer Engineering at Cornell University, Ithaca, NY. He has published in the areas of transient control and voltage collapse problems as well as technical, economic, and institutional impacts of restructuring. His technical background is broadly in the areas of systems analysis and control of large-scale electric power systems.

Dr. Thomas has been a member of the IEEE-USA Energy Policy Committee since 1991 and was the committee's Chair from 1997 to 1998. He has served as the IEEE-USA Vice President for Technology Policy.

James T. Whitehead (Senior Member, IEEE) received the B.S. degree in electrical engineering from Tennessee Technological University, Cookeville.

He has worked for the Tennessee Valley Authority in Chattanooga for about 34 years. He is currently Manager of Transmission Planning. He is author of several technical papers.

Mr. Whitehead is a Registered Professional Engineer and is known as an expert in lightning having served as past chairman of the IEEE Working Group on Estimating the Lightning Performance of Transmission Lines.

Hugh Outhred received the Ph.D. degree in electrical engineering from the University of Sydney, Sydney, N.S.W., Australia.

He is Joint Director (Engineering) and Presiding Director of the Centre for Energy and Environmental Markets at the University of New South Wales, Sydney. His 30-year career includes research on power system analysis, electricity industry restructuring, energy and sustainability policy, and renewable energy technology and its interaction with power systems. He has contributed to the theory of electricity industry restructuring since 1979 and to its practical implementation in Australia from its inception. In 1985 and 1986, he was seconded to the government of New South Wales as an advisor on electricity restructuring and sustainability. In 1995 and 1996, he led a project for the National Grid Management Council to undertake electricity-trading experiments to trial the proposed National Electricity Market trading rules prior to their formal implementation. In 1997, he was appointed as an inaugural member of the NSW Licence Compliance Advisory Board, and in 1998 he was appointed as an inaugural member of the National Electricity Tribunal.

Dr. Outhred is a Fellow of the Institution of Engineers, Australia.

Timothy D. Mount received the Ph.D. degree from the University of California, Berkeley, in 1970.

He is a professor of applied economics and management at Cornell University, Ithaca, NY. His research interests include econometric modeling and policy analysis relating to the use of fuels and electricity, and to their environmental consequences (acid rain, smog, and global warming). He is currently conducting research on the restructuring of markets for electricity and the implications for price behavior in auctions for electricity, the rates charged to customers, and the environment.

CASE STUDY   An important, but often overlooked aspect of power system operations is the restoration of the system following a large blackout. During restoration the operating condition of the power system is usually quite different from that seen during normal operation. The following article presents an overview of some of the unique issues that need to be considered during system restoration [15].

Overcoming Restoration Challenges Associated with Major Power System Disturbances

M. M. ADIBI AND L. H. FINK

Recognizing that power system blackouts are likely to occur, it is prudent to consider the necessary measures that reduce their extent, intensity, and duration. Immediately after a major disturbance, the power system's frequency rise and decay are arrested automatically by load rejection, load shedding, controlled separation, and isolation mechanisms. The success rate of these automatic restoration mechanisms has been about 50%! The challenge is to coordinate the control and protective mechanisms with the operation of the generating plants and the electrical system. During the subsequent restoration, plant operators, in coordination with system operators, attempt manually to maintain a balance between load and generation. The duration of these manual procedures has invariably been much longer than equipment limitations can accommodate. Especially in light of the industry's reconfiguration, there is a danger that the operation of power plants and the power system may not maintain the necessary coordination resulting in greater impacts.

Records of major disturbances indicate that the initial system faults have been cleared in milliseconds, and systems have separated into unbalanced load and generation subsystems several seconds later. Blackouts have taken place several minutes after the separations, and the power systems have been re-

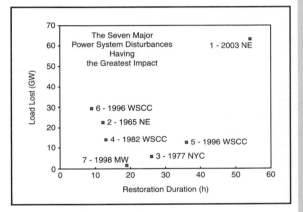

Figure 1
The seven disturbances with the greatest impact.

stored several hours after the blackouts. Most of these power outages have been of extended duration. For instance, as found in a review covering 24 recent power failures, seven have lasted more than 6 h. The U.S.-Canada report on the 14 August 2003 blackout cites seven major power disturbances with the greatest impact, lasting between 10–50 h, as shown in Figure 1. These failures clearly indicate a need for renewed emphasis on developing restoration methodologies and implementation plans.

RESTORATION ISSUES

A study of annual system disturbances reported by the North American Electric Reliability Council (NERC) over a ten-year period shows 117 power

("Overcoming Restoration Challenges Associated with Major Power System Disturbances" by M. M. Adibi and L. H. Fink from IEEE Power and Energy Magazine, (Sept/Oct 2006), pg. 68–77.)

system disturbances that have had one or more restoration problem(s) belonging to a number of functional groups.

In 23 cases, problems were due to reactive power unbalance, involving generator underexcitations, sustained overvoltage, and switched capacitors/reactors. In 11 cases, problems were due to load and generation unbalance, including responses to sudden increases in load, and underfrequency load shedding. In 29 cases, problems were due to inadequate load and generation coordination, including lack of black-start capability, problems with switching operations, line overloads, and control center coordination. In 56 cases, problems were due to monitoring and control inadequacies, including communication, supervisory control and data acquisition (SCADA) system capabilities, computer overloading, display capabilities, simulation tools, and system status determination. In 15 cases, problems were due to protective systems, including interlocking schemes, synchronization and synchrocheck, standing phase angles, and problems with other types of relays as described later. In 20 cases, problems were due to depletion of energy storage, including low-pressure compressed air/gas and discharged batteries. In 41 cases, problems were due to system restoration plan inadequacies, including lack of planned procedure, outdated procedure, procedure not being followed, inadequate training, and, incredibly, lack of standard communication vocabulary.

Certainly, this summary does not include all the restoration problems encountered. The more common and significant problems are briefly described in this article.

RESTORATION PLANNING

Most operating companies maintain restoration plans based on their restoration objectives, operating philosophies and practices, and familiarity with the characteristics of their power plant restart capabilities and power system reintegration peculiarities. While these plans have successfully restored power systems in the past, they can be improved significantly by simulating steady-state, transient, and dynamic behavior of the power system under various restoration operating conditions and by employing engineering and operating judgment reflecting many factors not readily modeled.

Most power systems have certain characteristics in common and behave in a similar manner during the restoration process. It is therefore possible to establish a general procedure and guidelines to enhance rapid restoration. A detailed plan, however, must be developed specifically to meet the particular requirements of an individual power system. Once a plan has been developed and tested (by simulation and training drills), an online restoration guidance program capable of guiding the operator in making decisions on what steps to take and when to take them goes a long way toward minimizing the duration of blackout and, consequently, the impact of the blackout.

Figure 2 shows a general procedure comprising three temporal restoration stages. The basic distinction between the first stage and the succeeding stages is that, during the first stage, time is critical and many urgent actions must be taken quickly. The basic distinction between the two initial stages and the third stage is that, in the initial stages, blocks of load are control means to maintain stability, whereas in the third stage, the restoration of load is the primary objective.

In the first stage, the postdisturbance system status is evaluated, a "target system" for restoration

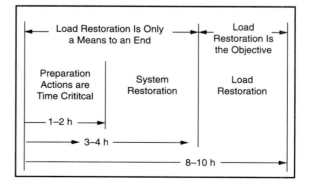

Figure 2
Typical restoration stages.

TABLE I   Initial sources of power and critical loads.

	Minutes	Success Probability
Availability of Initial Sources		
Run-of-the-river hydro	5–10	High
Pump-storage hydro	5–10	High
Combustion turbine (CT)	5–15	I in 2 or 3 CTs
Full or partial load rejection	Short	G T 50%
Low-frequency isolation scheme	Short	G T 50%
Controlled islanding	Short	Special cases
Tie-line with adjacent systems	Short	Not relied on*
Critical Loads		**Priorities**
Cranking drum-type units		High
Pipe-type cables pumping system		High
Transmission stations		Medium
Distribution stations		Medium
Industrial loads		Low**

*Policy: Provide remote cranking power
**Used in the initial stage to an advantage

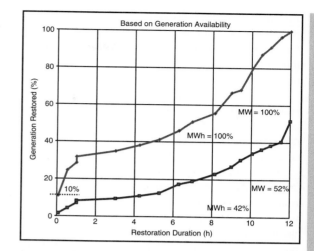

Figure 3
Significance of initial source.

is defined, a strategy for rebuilding the transmission network is selected, the system is sectionalized into a few subsystems, and steps are taken to supply the critical loads with the initial sources of power that are available in each subsystem. The postrestoration "target system" will be more or less like the predisturbance system, depending on the severity of the disturbance, but it is important that it be clearly defined in advance to avoid missteps causing the system to go "back to square one." Table I lists the types of initial sources of power that may be available and the critical loads. The effect of having an initial source of power, both on the duration of the restoration and on the minimization of the unserved load, is illustrated in Figure 3. In this particular case, the choices for the initial source of power were installing combustion turbines, providing a low-frequency isolation mechanism, or equipping the base-loaded unit with full-load rejection capability. The full-load rejection alternative was selected as providing the best balance between cost and reliability.

In the second stage, the overall goal is reintegration of the bulk power network, as a means of achieving the goals of load restoration in the third stage. To this end, skeleton transmission paths are energized, subsystems defined in the first stage are resynchronized, and sufficient load is restored to stabilize generation and voltage. Larger, base-load units are prepared for restart. Such tasks as energizing higher voltage lines and synchronizing subsystems require either reliable guidelines that have been prepared in advance or tools for analysis prior to critical switching actions.

In the third stage, the primary objective of restoration is to minimize the unserved load, and the scheduling of load pickup will be based on response rate capabilities of available generators. The effective system response rate and the responsive reserve increase with the increase in the number of generators and load restoration can be accomplished in increasingly larger steps.

FREQUENCY CONTROL

Sustained operation of power systems is impossible unless generator frequencies and bus voltages are kept within strict limits. During normal operation, these requirements are met by automatic control loops under operator supervision. During restoration, when individual generators are being brought up to speed and large blocks of load are being reconnected, perturbations outside the range of automatic controls are inevitable; hence, hands-on control by system operators is necessary.

"System" frequency is the mean frequency of all the machines that are online, and deviations by individual machines must be strictly minimized to avoid mechanical damage to the generator and disruption of the entire system. This is generally accomplished by picking up loads in increments that can be accommodated by the inertia and response of the restored and synchronized generators.

Smaller radial loads should be restored prior to larger loads while maintaining a reasonably constant real-to-reactive power ratio. Feeders equipped with underfrequency relays are picked up at the subsequent phases of restoration when system frequency has stabilized. Common practice in the initial stages of restart and reintegration is to rely on black-start combustion turbines (CT units), low-head short-conduit hydro units (hydro units), and drum-type boiler-turbine units (steam units). Figure 4 shows typical frequency response of these units to a 10% sudden load increase.

During restoration, operators must consider a prime mover's frequency response to a sudden increase in load. Such sudden load increases occur when picking up large network loads or when one of the online generators trips off. Load pickup in small increments tends to prolong the restoration duration, but in picking up large increments, there is

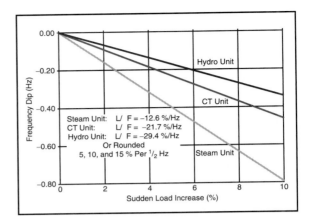
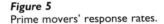

Figure 5
Prime movers' response rates.

always the risk of triggering a frequency decline and causing a recurrence of the system outage. The allowable size of load pickup depends on the rate of response of prime movers already online, which are likely to be under manual control at this point. Typical response rates are 5, 10, and 15% load for a frequency dip of about 0.5 Hz for steam, CT, and hydro units, respectively, as shown in Figure 5.

A set of guidelines for controlling system frequency has been established. These include: 1) black-starting combustion turbines in the automatic mode and at their maximum ramp rate; 2) placing these units under manual mode soon after they are paralleled (for black-start CT units with no automatic mode, these steps become 1) adjusting the governor speed droop to about 2%, and 2) returning the governor speed droop back to 5% soon after it is paralleled); 3) firming generation to meet the largest contingency, i.e., loss of the largest unit; 4) distributing reserve according to the online generators' dynamic response rates; and 5) ensuring that the size of load to be picked up is less than online generators' response rates.

The above guidelines permit the largest load increment that would keep the frequency within acceptable limits, the effective generation reserve that would meet the largest contingency, and the gover-

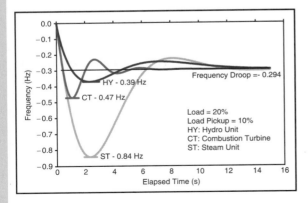

Figure 4
Prime movers' frequency responses.

nor speed droop that would improve prime movers' frequency response.

VOLTAGE CONTROL

During the early stages of restoring high-voltage overhead and underground transmission lines, there are concerns with three related overvoltage areas: sustained power frequency overvoltages, switching transients, and harmonic resonance.

Overhead transmission system

Sustained power frequency overvoltages are caused by charging currents of lightly loaded transmission lines. If excessive, these currents may cause generator underexcitation or even self-excitation and instability. Sustained overvoltages also overexcite transformers, generate harmonic distortions, and cause transformer overheating.

Switching transients are caused by energizing large segments of the transmission system or by switching capacitive elements. Such transients are usually highly damped and of short duration. However, in conjunction with sustained overvoltages, they may result in arrester failures. They are not usually a significant factor at transmission voltages below 100 kV. At higher voltages, however, they may become significant because arrester operating voltages are relatively close to normal system voltage, and high-voltage lines are usually long so that energy stored on the lines may be large. In most cases, though, with no sustained traveling wave transients, surge arresters have sufficient energy-absorbing capability to clamp harmful overvoltages to safe levels without sustaining damage. The likely effects of transient overvoltages are determined by the study of special system conditions. Computer-aided analysis has proven to be a valuable tool in understanding switching surge overvoltages.

Harmonic resonance voltages are oscillatory undamped or only weakly damped temporary overvoltages of long duration. They originate from switching operations and nonlinear equipment, reflecting several factors that are characteristic of the networks during restoration. First, the natural frequency of the series-resonant circuit formed by the source inductance and line-charging capacitance may, under normal operating conditions, be a low multiple of 60 Hz. Next, "magnetizing in-rush" caused by energizing a transformer produces many harmonics. Finally, during early stages of restoration, the lines are lightly loaded; resonance therefore is lightly damped, which in turn means the resulting resonance voltages may be very high. If transformers become overexcited due to power frequency overvoltage, harmonic resonance voltages will be sustained or even escalate.

Power transformers, surge arresters, and circuit breakers are the equipment first affected by overvoltages. For transformers, concerns and constraints are with exceeding basic insulation levels (BILs), overexcitation, harmonic generation, and excessive heating. For circuit breakers, concerns are with higher transient recovery voltages, restriking, flashover, and lowering of the interruption capability. For surge arresters, overvoltages cause operation, prevent resealing, and damage the arresters.

Thus, for control of voltages during system restoration, factors to be considered are the length of line to be energized, the size of underlying loads, and the adequacy of online generation (minimum source impedance). In general, it is desirable to energize as large a section of a line as the resulting sustained and transient overvoltages will allow. Energizing small sections tends to prolong the restoration process, but energizing a large section involves a risk of damaging equipment insulation. Energizing lines with inadequate source impedance could result in higher sustained and transient voltages than equipment can withstand. The startup of more out-of-sequence generators, however, would use critical time and delay the overall restoration process. Underlying loads at the receiving or sending end of lines tend to reduce the sustained and transient voltages. In the case of switching transients, operators need to know the minimum load that would avoid transient overvoltages.

In developing restoration guidelines, the above concerns can be addressed by simple analysis and simulation, as shown in Figures 6 and 7. Sending- and receiving-end sustained and transient voltages

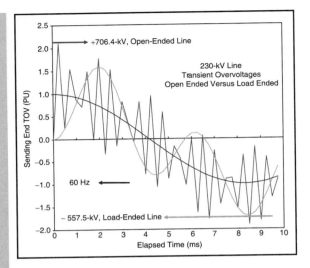

Figure 6
Sending-end transient overvoltages (TOV) of a 230-kV line.

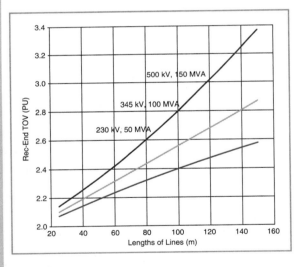

Figure 7
Sending-end transient overvoltages (TOV) of open-ended lines.

can be determined for energizing lines of different voltage levels and lengths, energized by different sizes of generators, and with different sizes of line-end cold loads. These results can be used to provide qualitative guidelines to assist operators in energizing high (230 kV) and extrahigh (345 and 500 kV) voltage lines, as shown in Figure 7.

Underground transmission systems

Over 90% of underground transmission lines are of high-pressure oil-filled (HPOF) pipe-type cable, with voltages ranging from 69–500 kV. The primary concern with such cables is the integrity of the insulation. After a blackout, power supply to the pumping plants that maintain the oil pressure is lost. As the cable cools, dissolved gases are liberated, forming gas pockets in the insulation. Reenergization of the cable could then result in immediate failure of pothead terminators. Hence, the pumping plant is a critical load of very high priority.

Another concern during cable reenergization is the ability of the energizing system to absorb the cable's charging current. It should be noted that: 1) cables are loaded at well below their surge impedance loading, 2) the MVAr charging currents per mile of a cable is about ten times that of an overhead line of the same voltage class, and 3) about 2 MW of load is picked up per one MVAr of charging.

GENERATOR REACTIVE CAPABILITY

Transformer tap selection

The generator reactive capability (GRC) curves furnished by manufacturers and used in operation planning typically have a greater range than can be realized during actual operation. Generally, these manufacturers' GRC curves are strictly a function of the synchronous machine design parameters and do not consider plant and system operating conditions as limiting factors. Concern over GRC is warranted by the need for reactive power to provide voltage support for large blocks of power transfer.

Figure 8 shows the rated and actual reactive power capability limits for a 460-MVA generator at 237-kV system bus voltage. The rated limits represent, respectively, the overexcitation limit due to rotor overheating, the underexcitation limit due to the stator core-end overheating (and the minimum excitation limiter relay settings), and the overload limit due to the stator overheating. However, more restrictive operating limits are imposed by the plant

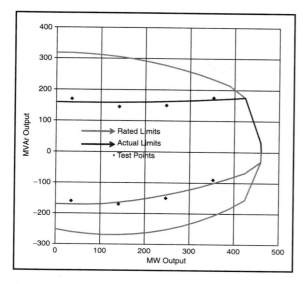

Figure 8
Rated and actual generator MVAr capability.

auxiliary bus voltage limits (typically $\pm$5%), the generator terminal voltage limits ($\pm$5%), and the system bus minimum and maximum voltages during peak and light load conditions.

The high- and low-voltage limits for the auxiliary bus, generator terminal, and system bus are interrelated by the tap positions on the generator step-up (GSU) and auxiliary (AUX) transformers. It should be noted that, in general (particularly in the United States), the GSU and AUX transformers are not equipped with underload tap changers, and therefore these tap positions are very infrequently changed after installation. Consequently, as power system operating conditions change, it is necessary to check these transformer tap positions and ascertain that adequate over- and underexcitation reactive power is available to meet the needs of the power system under both peak and light load conditions.

Remote black-start
Black-start combustion turbines are often considered for remote cranking of steam electric stations under partial or complete power system collapse. Typically, this type of combustion turbine can be

started within 5–15 min, which is well within the 30–45 min critical time interval allowed for the hot restart of drum-type boiler-turbine-generators (B-T-Gs).

In planning for black-start of a steam plant, a number of constraints must be considered. Among the more important ones are the "sustained" voltage drop (up to 20%, lasting over 10 s) at the steam plant's large auxiliary motor terminals, the over- and underexcitation limits of the combustion turbine, and the settings of the protective relays installed in the system between the black-start source and the steam plant.

In black-start operation, there are many limiting factors that impose severe demands on the reactive capability of the black-start source. One extreme condition is the initial absorption of the charging currents of the high-voltage (HV) and extra-high-voltage (EHV) lines to the steam plant. Another is supplying the reactive power demand for starting the largest auxiliary motor in the steam plant.

As shown in Figure 9, the 42-MVA CT must be capable of absorbing about 15 MVAr when energiz-

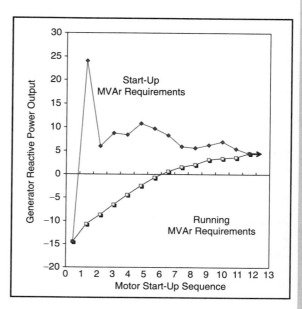

Figure 9
Motor start-up sequence.

TABLE 2 Cumulative starting and running reactive loads of a 75-MW steam plant.

Sequence of Starting	Horsepower	Starting MVAr*
0–Local load		4.5
1–Induced draft fan	6,000	34.9
2–Forced draft fan	3,000	23.5
3–Circulating water pump	3,000	25.3
4–Primary air fan	2,500	24.6
5–Startup BF pump	2,500	26.2
6–Boiler circulating pump	2,000	25.2
7–Condensate pump	1,500	23.9
8–River water pump	900	21.8
9–Auxilary cooling water pump	700	21.4
10–Coal mill	700	21.8
11–Air compressor	700	22.2
12–Closed cooling water	350	20.9

*Motor's starting reactive power plus previous motors' running reactive powers.

ing the 230- and 345-kV path between the CT and the steam plant and of supplying about 25 MVAr when stating the 6,000 hp induced draft fan in the 75-MW drum-type boiler-turbine-generator (BTG). These limiting conditions can be met by optimum selections of the GSU and AUX transformer taps and by adjusting the voltage set-point at the cranking source. Table 2 lists, and Figure 9 shows, the cumulative starting and running reactive power requirements of the auxiliary motors in the steam plant.

To determine the set-point and the optimum tap positions for all the transformers in the path between the cranking source and the steam unit's auxiliary bus, a number of analytical tools, including optimal power flows, are used to arrive at an approximate solution.

Nuclear plant requirements

A high restoration priority for utilities having nuclear plants is to provide two independent off-site sources of power within 4 h after an outage to enable controlled reactor shutdown and subsequent restoration to service within 24 h after the scram. Otherwise, the reactor must go through a cooling-down cycle which renders it unavailable for two

or more days. Therefore, in areas with significant nuclear power, full power restoration may not be achievable for several days after a blackout.

Nuclear units are typically large—over 600 MW—and usually remotely located. They cannot be provided with the required off-site power sources until the EHV transmission lines are restored. There are a variety of factors, such as adequate reactive absorbing capability, the minimum source requirement, and negative sequence currents, that must be considered before EHV transmission lines are energized. In general, these requirements cannot be met until the third stage of restoration.

PROTECTION SYSTEM ISSUES

Distance, differential, and excitation relays

The performance of protective systems may be measured by the relative percentages of 1) correct and desired relay operations, 2) correct and undesired operations, 3) wrong tripping, and 4) failure to trip. The primary reason for the second and fourth categories is change in the power system topology. During restoration, the power system undergoes continual changes and, therefore, is subject to correct and undesired operations and failure to trip.

It is important that the performance of relays and relay schemes be evaluated under restoration conditions. The foreknowledge that certain system operating conditions could cause correct and undesired operations or failure to trip makes it possible to avoid such operating conditions during development and execution of the restoration plan.

The protective relays that could affect the restoration procedure include:

- distance relays without potential restraints
- out-of-step relays
- synchro-check relays
- negative sequence voltage relays
- differential relays lacking harmonic restraints
- V/Hz relays
- generator underexcitation relays
- loss-of-field relays
- underfrequency switched reactor/capacitor relays.

Standing phase angle reduction

The presence of excessive standing phase angle (SPA) differences across open circuit breakers causes significant delays in restoration. The SPA may occur across a tie-line between two connected systems or between two connected subsystems. It must be brought to a safe limit before an attempt is made to close the breaker.

To determine a safe SPA value, the impact on the T-G shaft torque of closing a breaker should be evaluated. The T-G shaft torque is the sum of its constant-load torque that exists before and the transient mechanical torque immediately after closing the breaker. The constant-load torque can readily be reduced by lowering the generator's real power output, but the transient torque can be reduced only by reducing the SPA, a feat not readily accomplished.

There has been a need for an efficient methodology to serve as a guideline for reducing excessive SPA without resorting to the raising and lowering of various generation levels on a trial and error basis.

Asymmetry issues during restoration

Many existing EHV lines have asymmetrical (horizontal) conductor spacing without being transposed. These characteristics generate unacceptable negative sequence currents (NSCs). Under light load conditions and during restoration, NSC has caused cascade tripping of a number of generators (resulting in wide-area blackouts), has prevented synchronization of incoming generators, and has blocked remote black-start of large thermal units.

As shown in Figure 10, typically the generator NSC relays are set at 4% for alarms and 10% for tripping. It is important that when attempting to provide an off-site source to a nuclear or other large thermal power plant (e.g., in a remote black-start), the extent of NSC be determined and, if unacceptable, either the operation be deferred to after the initial restoration phase or the appropriate "underlying load" be determined for connection to the receiving end of the EHV lines.

ESTIMATING RESTORATION DURATION

Typically, restoration duration is estimated based on the availability of various prime movers. In these estimates, it is assumed that load can be picked up as soon as generation becomes available and that the time required for switching operations to energize transformers, lines, start-up of large motors, etc. is much less than the time required for generation availability. Case studies supported by field tests have shown that the above assumptions are not necessarily correct, and the restoration duration should be estimated using both the generation availability and the switching operations.

For example, in using the 20-MW combustion turbine of Table 3 for remote cranking of the 275-MW drum-type B-T-G, if the time estimate for cranking operation is well within 30–45 min, the hot restart of the B-T-G could be planned; if not, the B-T-G should be scheduled for a cold start-up, which requires an elapsed time of 3–4 h after the system collapse.

The critical path method (CPM) is a technique that can be used in restoration planning, scheduling, and evaluation. Its strength lies in the fact that pre-

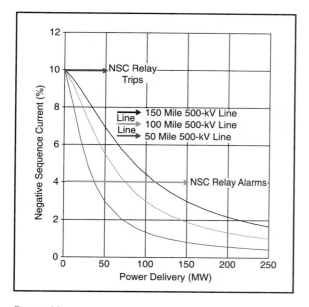

Figure 10
Current imbalance in 500-kV lines.

TABLE 3 Typical prime movers start-up timings.

General Data	CT	Drum	SCOT*
Type	CT	Drum	SCOT*
Unit size, MW	20	275	800
Fuel	Gas	Oil	Coal
Duty	Peaking	Cycle	Base
Min. load, MW	5	30	420**
Hot Restart (h)			
Max. elapsed time	0.0	$\frac{1}{2}-\frac{3}{4}$	3
Light-off to synch.	0.1	$1\frac{1}{2}$	4
Synch to min. load	0.1	$\frac{3}{4}$	$1\frac{1}{2}$
Min. load to full load	0.1	1	$1\frac{1}{2}$
Cold Start-Up (h)			
Min. elapsed time	0.0	**3–4**	8
Light-off to synch.	0.1	7	16
Synch to min. load	0.0	$1\frac{3}{4}$	3
Min. load to full load	0.1	1	$1\frac{1}{2}$

*SCOT: Super-critical once-through.
**Gen. is manually loaded to this load level.

cise time estimates do not have to be made for each action. Using CPM, the restoration plan is broken down to levels, tasks, and basic operating actions. Operating actions include opening/closing breakers, raising/lowering transformer taps, adjusting voltage and frequency set-points, and starting auxiliary motors. Then, based on operator experience, the optimistic time, the pessimistic time, and the most likely time for each action can be determined. Estimation of duration of various tasks may dictate revision of the overall restoration plan. In any case, one can estimate the duration of the restoration with some degree of probability and not base the duration estimate merely on the timing of the prime movers.

RESTORATION TRAINING

During restoration, system operators are faced with a state of their system that is quite different from that to which they are accustomed in day-to-day operation and for which the EMS application programs at their disposal were not designed and are not well adopted.

There are distinct differences between the normal state and the restorative state in the type of models relevant to each, the objective pursued, and the information available at the control center to support the models. In the normal state, the primary objective is to minimize the cost of producing power, subject to observing certain security constraints. EMS application programs that help attain this goal incorporate models that represent the simplified, primarily steady-state behavior of power systems and incorporate data that is primarily obtained automatically from the system.

During the restoration process, by contrast, the objective is quite different. The objective here is to minimize the restoration time and the amount of unserved kilowatthours of energy, with security as a subsidiary objective, and without regard to production cost. At the same time, many dynamic phenomena neglected (or unimportant) during normal operation play a critical role during restoration and must be taken into account.

Available generic simulators can provide procedural training in the early stages of operator training. However, for exercising and preparing operators to cope with system-specific and time-critical emergency operations such as restoration, high-fidelity system-referenced simulators are needed. It is important to note that during the restart and reintegration phases of restoration, a power system often consists of one or more islands, most of the automatic controls have tripped or are deactivated, and the system is primarily under manual control. During these early phases of restoration, wider voltage and frequency ranges are tolerated. Under these large perturbations of long duration, the models and simulations that have been developed for small perturbations will not accurately represent the behavior of the power system and its components.

As shown in Figure 11, drills provide an excellent testing ground for the restoration plan and training of personnel. If conducted realistically, they will uncover potential problems with the existing plan. A key to good training and problem solving lies in the extent of the exercise. It should involve as many of the people and events that would be involved in an actual bulk power system restoration as possible. The exercises must be run frequently, and conditions must be varied so that operators will be trained in the handling of unpredictable events.

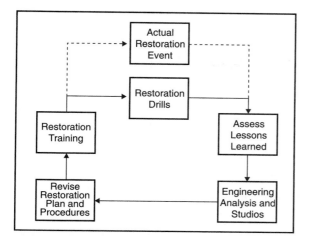

Figure 11
System restoration cycle.

ACKNOWLEDGMENTS

This article has been compiled from the IEEE PES Power System Restoration Working Group papers. These papers are included in the book referenced under "For Further Reading." The contributing members included M. M. Adibi (chair); R. W. Alexander, PP&L; J. N. Borkoski, BG&E; L. H. Fink, consultant; R. J. Kafka, PEPCO; D. P. Milanicz, BG&E; T. L. Volkmann, NSP; and J. N. Wrubel, PSE&G.

FOR FURTHER READING

M. M. Adibi, *Power System Restoration—Methodologies and Implementation Strategies.* New York: Wiley, 2000.

BIOGRAPHIES

M. M. Adibi, as a program manager at IBM, conducted a 1967 investigation of the 1965 Northeast blackout for the Department of Public Service of the State of New York. In 1969, following the PJM blackout of 1967, he investigated bulk power security assessment for Edison Electric Institute. Since 1979 and in the aftermath of the 1977 New York blackout, he has developed restoration plans for over a dozen international utilities and has chaired the IEEE Power System Restoration Working Group. He is a Fellow of the IEEE.

L. H. Fink has had nearly 50 years of experience in electric power utility systems engineering and research with the Philadelphia Electric Company, the U.S. Department of Energy (where he developed and for five years managed the national research program Systems Engineering for Power), and subsequent consulting. His numerous publications deal with high-voltage underground cable systems, power plant and power system control, voltage dynamic phenomena, and system security analysis. He is a fellow of the IEEE.

11.1

GENERATOR-VOLTAGE CONTROL

The *exciter* delivers dc power to the field winding on the rotor of a synchronous generator. For older generators, the exciter consists of a dc generator driven by the rotor. The dc power is transferred to the rotor via slip rings and brushes. For newer generators, *static* or *brushless* exciters are often employed.

For static exciters, ac power is obtained directly from the generator terminals or a nearby station service bus. The ac power is then rectified via thyristors and transferred to the rotor of the synchronous generator via slip rings and brushes.

For brushless exciters, ac power is obtained from an "inverted" synchronous generator whose three-phase armature windings are located on the main generator rotor and whose field winding is located on the stator.

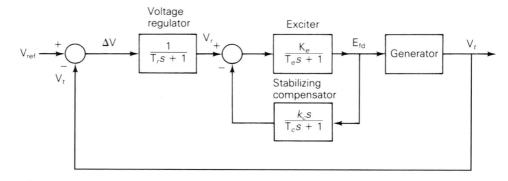

FIGURE 11.3   Simplified block diagram—generator-voltage control

The ac power from the armature windings is rectified via diodes mounted on the rotor and is transferred directly to the field winding. For this design, slip rings and brushes are eliminated.

Block diagrams of several standard types of generator-voltage control systems have been developed by the IEEE Working Group on Exciters [1]. A simplified block diagram of generator-voltage control, similar to those given in [1], is shown in Figure 11.3. Nonlinearities due to exciter saturation and limits on exciter output are not shown in this figure.

The generator terminal voltage V_t in Figure 11.3 is compared with a voltage reference V_{ref} to obtain a voltage error signal ΔV, which in turn is applied to the voltage regulator. The $1/(T_r s + 1)$ block accounts for voltage-regulator time delay, where s is the Laplace operator and T_r is the voltage-regulator time constant. Note that if a unit step is applied to a $1/(T_r s + 1)$ block, the output rises exponentially to unity with time constant T_r.

Neglecting the stabilizing compensator in Figure 11.3, the output V_r of the voltage regulator is applied to the exciter, which is represented by a $K_e/(T_e s + 1)$ block. The output of this exciter block is the field voltage E_{fd}, which is applied to the generator field winding and acts to adjust the generator terminal voltage. The generator block, which relates the effect of changes in E_{fd} to V_t, can be developed from synchronous-machine equations [2].

The stabilizing compensator shown in Figure 11.3 is used to improve the dynamic response of the exciter by reducing excessive overshoot. The compensator is represented by a $(K_c s)/(T_c s + 1)$ block, which provides a filtered first derivative. The input to this block is the exciter voltage E_{fd}, and the output is a stabilizing feedback signal that is subtracted from the regulator voltage V_r.

Block diagrams such as those shown in Figure 11.3 are used for computer representation of generator-voltage control in transient stability computer programs (see Chapter 13). In practice, high-gain, fast-responding exciters provide large, rapid increases in field voltage E_{fd} during short circuits at the generator terminals in order to improve transient stability after fault clearing. Equations represented in the block diagram can be used to compute the transient response of generator-voltage control.

11.2

TURBINE-GOVERNOR CONTROL

Turbine-generator units operating in a power system contain stored kinetic energy due to their rotating masses. If the system load suddenly increases, stored kinetic energy is released to initially supply the load increase. Also, the electrical torque T_e of each turbine-generating unit increases to supply the load increase, while the mechanical torque T_m of the turbine initially remains constant. From Newton's second law, $J\alpha = T_m - T_e$, the acceleration α is therefore negative. That is, each turbine-generator decelerates and the rotor speed drops as kinetic energy is released to supply the load increase. The electrical frequency of each generator, which is proportional to rotor speed for synchronous machines, also drops.

From this, we conclude that either rotor speed or generator frequency indicates a balance or imbalance of generator electrical torque T_e and turbine mechanical torque T_m. If speed or frequency is decreasing, then T_e is greater than T_m (neglecting generator losses). Similarly, if speed or frequency is increasing, T_e is less than T_m. Accordingly, generator frequency is an appropriate control signal for governing the mechanical output power of the turbine.

The steady-state frequency–power relation for turbine-governor control is

$$\Delta p_m = \Delta p_{\text{ref}} - \frac{1}{R}\Delta f \qquad (11.2.1)$$

where Δf is the change in frequency, Δp_m is the change in turbine mechanical power output, and Δp_{ref} is the change in a reference power setting. R is called

FIGURE 11.4

Steady-state frequency–power relation for a turbine-governor

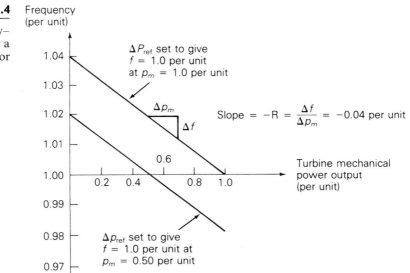

the *regulation constant*. The equation is plotted in Figure 11.4 as a family of curves, with Δp_{ref} as a parameter. Note that when Δp_{ref} is fixed, Δp_m is directly proportional to the drop in frequency.

Figure 11.4 illustrates a steady-state frequency–power relation. When an electrical load change occurs, the turbine-generator rotor accelerates or decelerates, and frequency undergoes a transient disturbance. Under normal operating conditions, the rotor acceleration eventually becomes zero, and the frequency reaches a new steady-state, shown in the figure.

The regulation constant R in (11.2.1) is the negative of the slope of the Δf versus Δp_m curves shown in Figure 11.4. The units of R are Hz/MW when Δf is in Hz and Δp_m is in MW. When Δf and Δp_m are given in per-unit, however, R is also in per-unit.

EXAMPLE 11.1   **Turbine-governor response to frequency change at a generating unit**

A 500-MVA, 60-Hz turbine-generator has a regulation constant R = 0.05 per unit based on its own rating. If the generator frequency increases by 0.01 Hz in steady-state, what is the decrease in turbine mechanical power output? Assume a fixed reference power setting.

SOLUTION   The per-unit change in frequency is

$$\Delta f_{\text{p.u.}} = \frac{\Delta f}{f_{\text{base}}} = \frac{0.01}{60} = 1.6667 \times 10^{-4} \quad \text{per unit}$$

Then, from (11.2.1), with $\Delta p_{\text{ref}} = 0$,

$$\Delta p_{m\text{p.u.}} = \left(\frac{-1}{0.05}\right)(1.6667 \times 10^{-4}) = -3.3333 \times 10^{-4} \quad \text{per unit}$$

$$\Delta p_m = (\Delta p_{m\text{p.u.}})S_{\text{base}} = (-3.3333 \times 10^{-4})(500) = -1.6667 \quad \text{MW}$$

The turbine mechanical power output decreases by 1.67 MW.   ∎

The steady-state frequency–power relation for one area of an interconnected power system can be determined by summing (11.2.1) for each turbine-generating unit in the area. Noting that Δf is the same for each unit,

$$\Delta p_m = \Delta p_{m1} + \Delta p_{m2} + \Delta p_{m3} + \cdots$$

$$= (\Delta p_{\text{ref1}} + \Delta p_{\text{ref2}} + \cdots) - \left(\frac{1}{R_1} + \frac{1}{R_2} + \cdots\right)\Delta f$$

$$= \Delta p_{\text{ref}} - \left(\frac{1}{R_1} + \frac{1}{R_2} + \cdots\right)\Delta f \qquad (11.2.2)$$

where Δp_m is the total change in turbine mechanical powers and Δp_{ref} is the total change in reference power settings within the area. We define the *area frequency response characteristic β* as

$$\beta = \left(\frac{1}{R_1} + \frac{1}{R_2} \cdots \right) \tag{11.2.3}$$

Using (11.2.3) in (11.2.2),

$$\Delta p_m = \Delta p_{\text{ref}} - \beta \Delta f \tag{11.2.4}$$

Equation (11.2.4) is the area steady-state frequency–power relation. The units of β are MW/Hz when Δf is in Hz and Δp_m is in MW. β can also be given in per-unit. In practice, β is somewhat higher than that given by (11.2.3) due to system losses and the frequency dependence of loads.

A standard figure for the regulation constant is R = 0.05 per unit. When all turbine-generating units have the same per-unit value of R based on their own ratings, then each unit shares total power changes in proportion to its own ratings. This desirable feature is illustrated by the following example.

EXAMPLE 11.2   **Response of turbine-governors to a load change in an interconnected power system**

An interconnected 60-Hz power system consists of one area with three turbine-generator units rated 1000, 750, and 500 MVA, respectively. The regulation constant of each unit is R = 0.05 per unit based on its own rating. Each unit is initially operating at one-half of its own rating, when the system load suddenly increases by 200 MW. Determine (a) the per-unit area frequency response characteristic β on a 1000 MVA system base, (b) the steady-state drop in area frequency, and (c) the increase in turbine mechanical power output of each unit. Assume that the reference power setting of each turbine-generator remains constant. Neglect losses and the dependence of load on frequency.

SOLUTION

a. The regulation constants are converted to per-unit on the system base using

$$R_{\text{p.u.new}} = R_{\text{p.u.old}} \frac{S_{\text{base(new)}}}{S_{\text{base(old)}}}$$

We obtain

$$R_{1\text{p.u.new}} = R_{1\text{p.u.old}} = 0.05$$

$$R_{2\text{p.u.new}} = (0.05)\left(\frac{1000}{750}\right) = 0.06667$$

$$R_{3\text{p.u.new}} = (0.05)\left(\frac{1000}{550}\right) = 0.10 \quad \text{per unit}$$

Using (11.2.3),

$$\beta = \frac{1}{R_1} + \frac{1}{R_2} + \frac{1}{R_3} = \frac{1}{0.05} + \frac{1}{0.06667} + \frac{1}{0.10} = 45.0 \quad \text{per unit}$$

b. Neglecting losses and dependence of load on frequency, the steady-state increase in total turbine mechanical power equals the load increase, 200 MW or 0.20 per unit. Using (11.2.4) with $\Delta p_{\text{ref}} = 0$,

$$\Delta f = \left(\frac{-1}{\beta}\right)\Delta p_m = \left(\frac{-1}{45}\right)(0.20) = -4.444 \times 10^{-3} \quad \text{per unit}$$

$$= (-4.444 \times 10^{-3})(60) = -0.2667 \quad \text{Hz}$$

The steady-state frequency drop is 0.2667 Hz.

c. From (11.2.1), using $\Delta f = -4.444 \times 10^{-3}$ per unit,

$$\Delta p_{m1} = \left(\frac{-1}{0.05}\right)(-4.444 \times 10^{-3}) = 0.08888 \quad \text{per unit}$$

$$= 88.88 \quad \text{MW}$$

$$\Delta p_{m2} = \left(\frac{-1}{0.06667}\right)(-4.444 \times 10^{-3}) = 0.06666 \quad \text{per unit}$$

$$= 66.66 \quad \text{MW}$$

$$\Delta p_{m3} = \left(\frac{-1}{0.10}\right)(-4.444 \times 10^{-3}) = 0.04444 \quad \text{per unit}$$

$$= 44.44 \quad \text{MW}$$

Note that unit 1, whose MVA rating is $33\frac{1}{3}\%$ larger than that of unit 2 and 100% larger than that of unit 3, picks up $33\frac{1}{3}\%$ more load than unit 2 and 100% more load than unit 3. That is, each unit shares the total load change in proportion to its own rating. ∎

Figure 11.5 shows a block diagram of a nonreheat steam turbine-governor, which includes nonlinearities and time delays that were not included in (11.2.1). The deadband block in this figure accounts for the fact that speed governors do not respond to changes in frequency or to reference power settings that are smaller than a specified value. The limiter block ac-

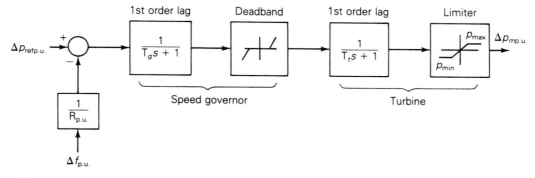

FIGURE 11.5   Turbine-governor block diagram

counts for the fact that turbines have minimum and maximum outputs. The $1/(Ts + 1)$ blocks account for time delays, where s is the Laplace operator and T is a time constant. Typical values are $T_g = 0.10$ and $T_t = 1.0$ seconds. Block diagrams for steam turbine-governors with reheat and hydro turbine-governors are also available [3].

11.3

LOAD-FREQUENCY CONTROL

As shown in Section 11.2, turbine-governor control eliminates rotor accelerations and decelerations following load changes during normal operation. However, there is a steady-state frequency error Δf when the change in turbine-governor reference setting Δp_{ref} is zero. One of the objectives of load-frequency control (LFC), therefore, is to return Δf to zero.

In a power system consisting of interconnected areas, each area agrees to export or import a scheduled amount of power through transmission-line interconnections, or tie-lines, to its neighboring areas. Thus, a second LFC objective is to have each area absorb its own load changes during normal operation. This objective is achieved by maintaining the net tie-line power flow out of each area at its scheduled value.

The following summarizes the two basic LFC objectives for an interconnected power system:

1. Following a load change, each area should assist in returning the steady-state frequency error Δf to zero.

2. Each area should maintain the net tie-line power flow out of the area at its scheduled value, in order for the area to absorb its own load changes.

The following control strategy developed by N. Cohn [4] meets these LFC objectives. We first define the *area control error* (ACE) as follows:

$$
\begin{aligned}
\text{ACE} &= (p_{tie} - p_{tie, sched}) + B_f(f - 60) \\
&= \Delta p_{tie} + B_f \Delta f
\end{aligned}
\tag{11.3.1}
$$

where Δp_{tie} is the deviation in net tie-line power flow out of the area from its scheduled value $p_{tie, sched}$, and Δf is the deviation of area frequency from its scheduled value (60 Hz). Thus, the ACE for each area consists of a linear combination of tie-line error Δp_{tie} and frequency error Δf. The constant B_f is called a *frequency bias constant*.

The change in reference power setting Δp_{refi} of each turbine-governor operating under LFC is proportional to the integral of the area control error. That is,

$$
\Delta p_{refi} = -K_i \int \text{ACE} \, dt
\tag{11.3.2}
$$

Each area monitors its own tie-line power flows and frequency at the area control center. The ACE given by (11.3.1) is computed and a percentage of the ACE is allocated to each controlled turbine-generator unit. Raise or lower commands are dispatched to the turbine-governors at discrete time intervals of two or more seconds in order to adjust the reference power settings. As the commands accumulate, the integral action in (11.3.2) is achieved.

The constant K_i in (11.3.2) is an integrator gain. The minus sign in (11.3.2) indicates that if either the net tie-line power flow out of the area or the area frequency is low—that is, if the ACE is negative—then the area should increase its generation.

When a load change occurs in any area, a new steady-state operation can be obtained only after the power output of every turbine-generating unit in the interconnected system reaches a constant value. This occurs only when all reference power settings are zero, which in turn occurs only when the ACE of every area is zero. Furthermore, the ACE is zero in every area only when both Δp_{tie} and Δf are zero. Therefore, in steady-state, both LFC objectives are satisfied.

EXAMPLE 11.3 **Response of LFC to a load change in an interconnected power system**

As shown in Figure 11.6, a 60-Hz power system consists of two interconnected areas. Area 1 has 2000 MW of total generation and an area frequency response characteristic $\beta_1 = 700$ MW/Hz. Area 2 has 4000 MW of total generation and $\beta_2 = 1400$ MW/Hz. Each area is initially generating one-half of its total generation, at $\Delta p_{\text{tie1}} = \Delta p_{\text{tie2}} = 0$ and at 60 Hz when the load in area 1 suddenly increases by 100 MW. Determine the steady-state frequency error Δf and the steady-state tie-line error Δp_{tie} of each area for the following two cases: (a) without LFC, and (b) with LFC given by (11.3.1) and (11.3.2). Neglect losses and the dependence of load on frequency.

SOLUTION

a. Since the two areas are interconnected, the steady-state frequency error Δf is the same for both areas. Adding (11.2.4) for each area,

$$(\Delta p_{m1} + \Delta p_{m2}) = (\Delta p_{\text{ref1}} + \Delta p_{\text{ref2}}) - (\beta_1 + \beta_2)\Delta f$$

FIGURE 11.6

Example 11.3

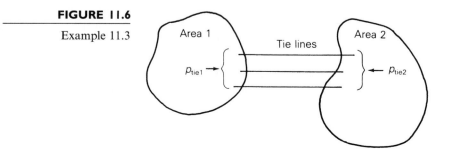

Neglecting losses and the dependence of load on frequency, the steady-state increase in total mechanical power of both areas equals the load increase, 100 MW. Also, without LFC, Δp_{ref1} and Δp_{ref2} are both zero. The above equation then becomes

$$100 = -(\beta_1 + \beta_2)\Delta f = -(700 + 1400)\Delta f$$

$$\Delta f = -100/2100 = -0.0476 \quad \text{Hz}$$

Next, using (11.2.4) for each area, with $\Delta p_{ref} = 0$,

$$\Delta p_{m1} = -\beta_1\Delta f = -(700)(-0.0476) = 33.33 \quad \text{MW}$$

$$\Delta p_{m2} = -\beta_2\Delta f = -(1400)(-0.0476) = 66.67 \quad \text{MW}$$

In response to the 100-MW load increase in area 1, area 1 picks up 33.33 MW and area 2 picks up 66.67 MW of generation. The 66.67-MW increase in area 2 generation is transferred to area 1 through the tie-lines. Therefore, the change in net tie-line power flow out of each area is

$$\Delta p_{tie2} = +66.67 \quad \text{MW}$$

$$\Delta p_{tie1} = -66.67 \quad \text{MW}$$

b. From (11.3.1), the area control error for each area is

$$ACE_1 = \Delta p_{tie1} + B_1\Delta f_1$$

$$ACE_2 = \Delta p_{tie2} + B_2\Delta f_2$$

Neglecting losses, the sum of the net tie-line flows must be zero; that is, $\Delta p_{tie1} + \Delta p_{tie2} = 0$ or $\Delta p_{tie2} = -\Delta p_{tie1}$. Also, in steady-state $\Delta f_1 = \Delta f_2 = \Delta f$. Using these relations in the above equations,

$$ACE_1 = \Delta p_{tie1} + B_1\Delta f$$

$$ACE_2 = -\Delta p_{tie1} + B_2\Delta f$$

In steady-state, $ACE_1 = ACE_2 = 0$; otherwise, the LFC given by (11.3.2) would be changing the reference power settings of turbine-governors on LFC. Adding the above two equations,

$$ACE_1 + ACE_2 = 0 = (B_1 + B_2)\Delta f$$

Therefore, $\Delta f = 0$ and $\Delta p_{tie1} = \Delta p_{tie2} = 0$. That is, in steady-state the frequency error is returned to zero, area 1 picks up its own 100-MW load increase, and area 2 returns to its original operating condition—that is, the condition before the load increase occurred.

We note that turbine-governor controls act almost instantaneously, subject only to the time delays shown in Figure 11.5. However, LFC acts more slowly. LFC raise and lower signals are dispatched from the area control center to turbine-governors at discrete-time intervals of 2 or more seconds. Also, it takes time for the raise or lower signals to accumulate. Thus, case (a) represents the first action. Turbine-governors in both areas rapidly respond to the load increase in area 1 in order to stabilize the frequency drop. Case (b) represents the second action. As LFC signals are dispatched to turbine-governors, Δf and Δp_{tie} are slowly returned to zero. ∎

The choice of the B_f and K_i constants in (11.3.1) and (11.3.2) affects the transient response to load changes—for example, the speed and stability of the response. The frequency bias B_f should be high enough such that each area adequately contributes to frequency control. Cohn [4] has shown that choosing B_f equal to the area frequency response characteristic, $B_f = \beta$, gives satisfactory performance of the interconnected system. The integrator gain K_i should not be too high; otherwise, instability may result. Also, the time interval at which LFC signals are dispatched, 2 or more seconds, should be long enough so that LFC does not attempt to follow random or spurious load changes. A detailed investigation of the effect of B_f, K_i and LFC time interval on the transient response of LFC and turbine-governor controls is beyond the scope of this text.

Two additional LFC objectives are to return the integral of frequency error and the integral of net tie-line error to zero in steady-state. By meeting these objectives, LFC controls both the time of clocks that are driven by 60-Hz motors and energy transfers out of each area. These two objectives are achieved by making temporary changes in the frequency schedule and tie-line schedule in (11.3.1).

Finally, note that LFC maintains control during normal changes in load and frequency—that is, changes that are not too large. During emergencies, when large imbalances between generation and load occur, LFC is bypassed and other, emergency controls are applied.

11.4

ECONOMIC DISPATCH

Section 11.3 describes how LFC adjusts the reference power settings of turbine-governors in an area to control frequency and net tie-line power flow out of the area. This section describes how the real power output of each controlled generating unit in an area is selected to meet a given load and to minimize the total operating costs in the area. This is the *economic dispatch* problem [5].

We begin this section by considering an area with only fossil-fuel generating units, with no constraints on maximum and minimum generator outputs, and with no transmission losses. The economic dispatch problem is first solved for this idealized case. Then we include inequality constraints on generator outputs; then we include transmission losses. Next we discuss the coordination of economic dispatch with LFC. Finally, we briefly discuss the dispatch of other types of units including nuclear, pumped-storage hydro, and hydro units.

FOSSIL-FUEL UNITS, NO INEQUALITY CONSTRAINTS, NO TRANSMISSION LOSSES

Figure 11.7 shows the operating cost C_i of a fossil-fuel generating unit versus real power output P_i. Fuel cost is the major portion of the variable cost of operation, although other variable costs, such as maintenance, could have been included in the figure. Fixed costs, such as the capital cost of installing the unit, are not included. Only those costs that are a function of unit power output—that is, those costs that can be controlled by operating strategy—enter into the economic dispatch formulation.

In practice, C_i is constructed of piecewise continuous functions valid for ranges of output P_i, based on empirical data. The discontinuities in Figure 11.7 may be due to the firing of equipment such as additional boilers or condensers as power output is increased. It is often convenient to express C_i in terms of BTU/hr, which is relatively constant over the lifetime of the unit, rather than \$/hr, which can change monthly or daily. C_i can be converted to \$/hr by multiplying the fuel input in BTU/hr by the cost of fuel in \$/BTU.

Figure 11.8 shows the unit incremental operating cost dC_i/dP_i versus unit output P_i, which is the slope or derivative of the C_i versus P_i curve in Figure 11.7. When C_i consists of only fuel costs, dC_i/dP_i is the ratio of the incremental fuel energy input in BTU to incremental energy output in kWhr, which is called incremental *heat rate*. Note that the reciprocal of the heat rate, which is the ratio of output energy to input energy, gives a measure of fuel efficiency for the unit. For the unit shown in Figure 11.7, maximum efficiency occurs at $P_i = 600$ MW, where the heat rate is $C/P = 5.4 \times 10^9/600 \times 10^3 = 9000$ BTU/kWhr. The efficiency at this output is

$$\text{percentage efficiency} = \left(\frac{1}{9000} \frac{\text{kWhr}}{\text{BTU}} \right) \left(3413 \frac{\text{BTU}}{\text{kWhr}} \right) \times 100 = 37.92\%$$

FIGURE 11.7

Unit operating cost versus real power output—fossil-fuel generating unit

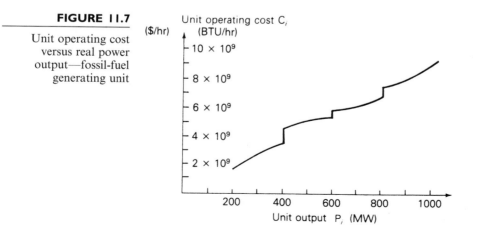

FIGURE 11.8

Unit incremental operating cost $\dfrac{dC_i}{dP_i}$

Unit incremental operating cost versus real power output— fossil-fuel generating unit

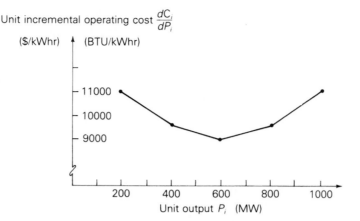

The dC_i/dP_i curve in Figure 11.8 is also represented by piecewise continuous functions valid for ranges of output P_i. For analytical work, the actual curves are often approximated by straight lines. The ratio dC_i/dP_i can also be converted to $/kWhr by multiplying the incremental heat rate in BTU/kWhr by the cost of fuel in $/BTU.

For the area of an interconnected power system consisting of N units operating on economic dispatch, the total variable cost C_T of operating these units is

$$C_T = \sum_{i=i}^{N} C_i$$

$$= C_1(P_1) + C_2(P_2) + \cdots + C_N(P_N) \quad \$/hr \tag{11.4.1}$$

where C_i, expressed in $/hr, includes fuel cost as well as any other variable costs of unit i. Let P_T equal the total load demand in the area. Neglecting transmission losses,

$$P_1 + P_2 + \cdots + P_N = P_T \tag{11.4.2}$$

Due to relatively slow changes in load demand, P_T may be considered constant for periods of 2 to 10 minutes. The economic dispatch problem can be stated as follows:

Find the values of unit outputs P_1, $P_2, \ldots, P_N$ that minimize C_T given by (11.4.1), subject to the equality constraint given by (11.4.2).

A criterion for the solution to this problem is: All units on economic dispatch should operate at equal incremental operating cost. That is,

$$\frac{dC_1}{dP_1} = \frac{dC_2}{dP_2} = \cdots = \frac{dC_N}{dP_N} \tag{11.4.3}$$

An intuitive explanation of this criterion is the following. Suppose one unit is operating at a higher incremental operating cost than the other units. If the output power of that unit is reduced and transferred to units with lower incremental operating costs, then the total operating cost C_T decreases. That is, reducing the output of the unit with the *higher* incremental cost results in a *greater cost decrease* than the cost increase of adding that same output reduction to units with lower incremental costs. Therefore, all units must operate at the same incremental operating cost (the economic dispatch criterion).

A mathematical solution to the economic dispatch problem can also be given. The minimum value of C_T occurs when the total differential dC_T is zero. That is,

$$dC_T = \frac{\partial C_T}{\partial P_1} dP_1 + \frac{\partial C_T}{\partial P_2} dP_2 + \cdots + \frac{\partial C_T}{\partial P_N} dP_N = 0 \tag{11.4.4}$$

Using (11.4.1), (11.4.4) becomes

$$dC_T = \frac{dC_1}{dP_1} dP_1 + \frac{dC_2}{dP_2} dP_2 + \cdots + \frac{dC_N}{dP_N} dP_N = 0 \tag{11.4.5}$$

Also, assuming P_T is constant, the differential of (11.4.2) is

$$dP_1 + dP_2 + \cdots + dP_N = 0 \tag{11.4.6}$$

Multiplying (11.4.6) by λ and subtracting the resulting equation from (11.4.5),

$$\left(\frac{dC_1}{dP_1} - \lambda\right) dP_1 + \left(\frac{dC_2}{dP_2} - \lambda\right) dP_2 + \cdots + \left(\frac{dC_N}{dP_N} - \lambda\right) dP_N = 0 \tag{11.4.7}$$

Equation (11.4.7) is satisfied when each term in parentheses equals zero. That is,

$$\frac{dC_1}{dP_1} = \frac{dC_2}{dP_2} = \cdots = \frac{dC_N}{dP_N} = \lambda \tag{11.4.8}$$

Therefore, all units have the same incremental operating cost, denoted here by λ, in order to minimize the total operating cost C_T.

EXAMPLE 11.4   **Economic dispatch solution neglecting generator limits and line losses**

An area of an interconnected power system has two fossil-fuel units operating on economic dispatch. The variable operating costs of these units are given by

$$C_1 = 10P_1 + 8 \times 10^{-3}P_1^2 \quad \$/hr$$
$$C_2 = 8P_2 + 9 \times 10^{-3}P_2^2 \quad \$/hr$$

where P_1 and P_2 are in megawatts. Determine the power output of each unit, the incremental operating cost, and the total operating cost C_T that minimizes C_T as the total load demand P_T varies from 500 to 1500 MW. Generating unit inequality constraints and transmission losses are neglected.

SOLUTION   The incremental operating costs of the units are

$$\frac{dC_1}{dP_1} = 10 + 16 \times 10^{-3}P_1 \quad \$/\text{MWhr}$$

$$\frac{dC_2}{dP_2} = 8 + 18 \times 10^{-3}P_2 \quad \$/\text{MWhr}$$

Using (11.4.8), the minimum total operating cost occurs when

$$\frac{dC_1}{dP_1} = 10 + 16 \times 10^{-3}P_1 = \frac{dC_2}{dP_2} = 8 + 18 \times 10^{-3}P_2$$

Using $P_2 = P_T - P_1$, the preceding equation becomes

$$10 + 16 \times 10^{-3}P_1 = 8 + 18 \times 10^{-3}(P_T - P_1)$$

Solving for P_1,

$$P_1 = \frac{18 \times 10^{-3}P_T - 2}{34 \times 10^{-3}} = 0.5294P_T - 58.82 \quad \text{MW}$$

Also, the incremental operating cost when C_T is minimized is

$$\frac{dC_2}{dP_2} = \frac{dC_1}{dP_1} = 10 + 16 \times 10^{-3}P_1 = 10 + 16 \times 10^{-3}(0.5294P_T - 58.82)$$

$$= 9.0589 + 8.4704 \times 10^{-3}P_T \quad \$/\text{MWhr}$$

and the minimum total operating cost is

$$C_T = C_1 + C_2 = (10P_1 + 8 \times 10^{-3}P_1^2) + (8P_2 + 9 \times 10^{-3}P_2^2) \quad \$/\text{hr}$$

The economic dispatch solution is shown in Table 11.1 for values of P_T from 500 to 1500 MW.

TABLE 11.1 Economic dispatch solution for Example 11.4	P_T MW	P_1 MW	P_2 MW	dC_1/dP_1 $\$/\text{MWhr}$	C_T $\$/\text{hr}$
	500	206	294	13.29	5529
	600	259	341	14.14	6901
	700	312	388	14.99	8358
	800	365	435	15.84	9899
	900	418	482	16.68	11525
	1000	471	529	17.53	13235
	1100	524	576	18.38	15030
	1200	576	624	19.22	16910
	1300	629	671	20.07	18875
	1400	682	718	20.92	20924
	1500	735	765	21.76	23058

∎

EFFECT OF INEQUALITY CONSTRAINTS

Each generating unit must not operate above its rating or below some minimum value. That is,

$$P_{i\min} < P_i < P_{i\max} \qquad i = 1, 2, \ldots, N \tag{11.4.9}$$

Other inequality constraints may also be included in the economic dispatch problem. For example, some unit outputs may be restricted so that certain transmission lines or other equipments are not overloaded. Also, under adverse weather conditions, generation at some units may be limited to reduce emissions.

When inequality constraints are included, we modify the economic dispatch solution as follows. If one or more units reach their limit values, then these units are held at their limits, and the remaining units operate at equal incremental operating cost λ. The incremental operating cost of the area equals the common λ for the units that are not at their limits.

EXAMPLE 11.5   **Economic dispatch solution including generator limits**

Rework Example 11.4 if the units are subject to the following inequality constraints:

$$100 \leq P_1 \leq 600 \quad \text{MW}$$

$$400 \leq P_2 \leq 1000 \quad \text{MW}$$

SOLUTION   At light loads, unit 2 operates at its lower limit of 400 MW, where its incremental operating cost is $dC_2/dP_2 = 15.2$ \$/MWhr. Additional load comes from unit 1 until $dC_1/dP_1 = 15.2$ \$/MWhr, or

$$\frac{dC_1}{dP_1} = 10 + 16 \times 10^{-3}P_1 = 15.2$$

$$P_1 = 325 \quad \text{MW}$$

For P_T less than 725 MW, where P_1 is less than 325 MW, the incremental operating cost of the area is determined by unit 1 alone.

At heavy loads, unit 1 operates at its upper limit of 600 MW, where its incremental operating cost is $dC_1/dP_1 = 19.60$ \$/MWhr. Additional load comes from unit 2 for all values of dC_2/dP_2 greater than 19.60 \$/MWhr. At $dC_2/dP_2 = 19.60$ \$/MWhr,

$$\frac{dC_2}{dP_2} = 8 + 18 \times 10^{-3}P_2 = 19.60$$

$$P_2 = 644 \quad \text{MW}$$

For P_T greater than 1244 MW, where P_2 is greater than 644 MW, the incremental operating cost of the area is determined by unit 2 alone.

For $725 < P_T < 1244$ MW, neither unit has reached a limit value, and the economic dispatch solution is the same as that given in Table 11.1.

The solution to this example is summarized in Table 11.2 for values of P_T from 500 to 1500 MW.

	P_T MW	P_1 MW	P_2 MW	dC/dP $/MWhr	C_T $/hr
TABLE 11.2 Economic dispatch solution for Example 11.5	500	100	400		5720
	600	200	400	$\frac{dC_1}{dP_1}$ 13.20	6960
	700	300	400	14.80	8360
	725	325	400	15.20	8735
	800	365	435	15.84	9899
	900	418	482	16.68	11525
	1000	471	529	17.53	13235
	1100	524	576	18.38	15030
	1200	576	624	19.22	16910
	1244	600	644	19.60	17765
	1300	600	700	$\frac{dC_2}{dP_2}$ 20.60	18890
	1400	600	800	22.40	21040
	1500	600	900	24.20	23370

The dC/dP column groups: $\frac{dC_1}{dP_1}$ values 11.60, 13.20, 14.80, 15.20 for the first rows, and $\frac{dC_2}{dP_2}$ values 19.60, 20.60, 22.40, 24.20 for the last rows.

■

EXAMPLE 11.6 PowerWorld Simulator—economic dispatch, including generator limits

PowerWorld Simulator case Example 11_6 uses a five-bus, three-generator lossless case to show the interaction between economic dispatch and the transmission system. The variable operating costs for each of the units are given by

$$C_1 = 10P_1 + 0.016P_1^2 \ \$/hr$$

$$C_2 = 8P_2 + 0.018P_2^2 \ \$/hr$$

$$C_4 = 12P_4 + 0.018P_4^2 \ \$/hr$$

where P_1, P_2, and P_4 are the generator outputs in megawatts. Each generator has minimum/maximum limits of

$$100 \le P_1 \le 400 \ \text{MW}$$

$$150 \le P_2 \le 500 \ \text{MW}$$

$$50 \le P_4 \le 300 \ \text{MW}$$

In addition to solving the power-flow equations, PowerWorld Simulator can simultaneously solve the economic dispatch problem to optimally allocate the generation in an area. To turn on this option, select **Case Information, Areas...** to view a list of each of the control areas in a case (just one in this example). Then toggle the AGC Status field to ED. Now anytime the power-flow equations are solved, the generator outputs are also changed using the economic dispatch.

Initially the case has a total load of 392 MW, with an economic dispatch of $P_1 = 141$ MW, $P_2 = 181$, and $P_4 = 70$, with an incremental operating cost, λ, of 14.52 \$/MWh. To view a graph showing the incremental cost curves for all of the area generators, right-click on any generator to display the generator's local menu, and then select "All Area Gen IC Curves" (right-click on the graph's axes to change their scaling).

To see how changing load impacts the economic dispatch and power-flow solutions, first select **Simulation, Solve and Animate** to begin the simulation. Then, on the one-line, click on the up/down arrows next to the Load Scalar field. This field is used to scale the load at each bus in the system. Notice that the change in the Total Hourly Cost field is well approximated by the change in the load multiplied by the incremental operating cost.

Determine the maximum amount of load this system can supply without overloading any transmission line with the generators dispatched using economic dispatch.

SOLUTION   The maximum system economic loading is determined numerically to be 655 MW (which occurs with a Load Scalar of 1.67), with the line from bus 2 to bus 5 being the critical element.

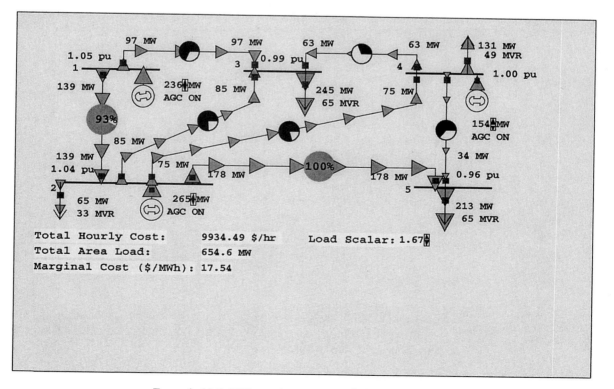

Example 11.6: With maximum economic loading

EFFECT OF TRANSMISSION LOSSES

Although one unit may be very efficient with a low incremental operating cost, it may also be located far from the load center. The transmission losses associated with this unit may be so high that the economic dispatch solution requires the unit to decrease its output, while other units with higher incremental operating costs but lower transmission losses increase their outputs.

When transmission losses are included in the economic dispatch problem, (11.4.2) becomes

$$P_1 + P_2 + \cdots + P_N - P_L = P_T \tag{11.4.10}$$

where P_T is the total load demand and P_L is the total transmission loss in the area. In general, P_L is not constant, but depends on the unit outputs $P_1, P_2, \ldots, P_N$. The total differential of (11.4.10) is

$$(dP_1 + dP_2 + \cdots + dP_N) - \left(\frac{\partial P_L}{\partial P_1} dP_1 + \frac{\partial P_L}{\partial P_2} dP_2 + \cdots + \frac{\partial P_L}{\partial P_N} dP_N \right) = 0 \tag{11.4.11}$$

Multiplying (11.4.11) by λ and subtracting the resulting equation from (11.4.5),

$$\left(\frac{dC_1}{dP_1} + \lambda \frac{\partial P_L}{\partial P_1} - \lambda \right) dP_1 + \left(\frac{dC_2}{dP_2} + \lambda \frac{\partial P_L}{\partial P_2} - \lambda \right) dP_2$$

$$+ \cdots + \left(\frac{dC_N}{dP_N} + \lambda \frac{\partial P_L}{\partial P_N} - \lambda \right) dP_N = 0 \tag{11.4.12}$$

Equation (11.4.12) is satisfied when each term in parentheses equals zero. That is,

$$\frac{dC_i}{dP_i} + \lambda \frac{\partial P_L}{\partial P_i} - \lambda = 0$$

or

$$\lambda = \frac{dC_i}{dP_i} (L_i) = \frac{dC_i}{dP_i} \left(\frac{1}{1 - \dfrac{\partial P_L}{\partial P_i}} \right) \qquad i = 1, 2, \ldots, N \tag{11.4.13}$$

Equation (11.4.13) gives the economic dispatch criterion, including transmission losses. Each unit that is not at a limit value operates such that its incremental operating cost dC_i/dP_i multiplied by the *penalty factor* L_i is the same. Note that when transmission losses are negligible, $\partial P_L/\partial P_i = 0$, $L_i = 1$, and (11.4.13) reduces to (11.4.8).

EXAMPLE 11.7   **Economic dispatch solution including generator limits and line losses**

Total transmission losses for the power system area given in Example 11.5 are given by

$$P_L = 1.5 \times 10^{-4}P_1^2 + 2 \times 10^{-5}P_1P_2 + 3 \times 10^{-5}P_2^2 \quad MW$$

where P_1 and P_2 are given in megawatts. Determine the output of each unit, total transmission losses, total load demand, and total operating cost C_T when the area $\lambda = 16.00$ \$/MWhr.

SOLUTION   Using the incremental operating costs from Example 11.4 in (11.4.13),

$$\frac{dC_1}{dP_1}\left(\frac{1}{1 - \dfrac{\partial P_L}{\partial P_1}}\right) = \frac{10 + 16 \times 10^{-3}P_1}{1 - (3 \times 10^{-4}P_1 + 2 \times 10^{-5}P_2)} = 16.00$$

$$\frac{dC_2}{dP_2}\left(\frac{1}{1 - \dfrac{\partial P_L}{\partial P_2}}\right) = \frac{8 + 18 \times 10^{-3}P_2}{1 - (6 \times 10^{-5}P_2 + 2 \times 10^{-5}P_1)} = 16.00$$

Rearranging the above two equations,

$$20.8 \times 10^{-3}P_1 + 32 \times 10^{-5}P_2 = 6.00$$

$$32 \times 10^{-5}P_1 + 18.96 \times 10^{-3}P_2 = 8.00$$

Solving,

$$P_1 = 282 \quad MW \qquad P_2 = 417 \quad MW$$

Using the equation for total transmission losses,

$$P_L = 1.5 \times 10^{-4}(282)^2 + 2 \times 10^{-5}(282)(417) + 3 \times 10^{-5}(417)^2$$

$$= 19.5 \quad MW$$

From (11.4.10), the total load demand is

$$P_T = P_1 + P_2 - P_L = 282 + 417 - 19.5 = 679.5 \quad MW$$

Also, using the cost formulas given in Example 11.4, the total operating cost is

$$C_T = C_1 + C_2 = 10(282) + 8 \times 10^{-3}(282)^2 + 8(417) + 9 \times 10^{-3}(417)^2$$

$$= 8357 \quad \$/hr$$

Note that when transmission losses are included, λ given by (11.4.13) is no longer the incremental operating cost of the area. Instead, λ is the unit incremental operating cost dC_i/dP_i multiplied by the unit penalty factor L_i.   ∎

EXAMPLE 11.8 **PowerWorld Simulator—economic dispatch, including generator limits and line losses**

Example 11.8 repeats the Example 11.6 power system, except that now losses are included, with each transmission line modeled with an R/X ratio of 1/3. The current value of each generator's loss sensitivity, $\partial P_L / \partial P_G$, is shown immediately below the generator's MW output field. Calculate the penalty factors L_i, and verify that the economic dispatch shown in the figure is optimal. Assume a Load Scalar of 1.0.

SOLUTION From (11.4.13) the condition for optimal dispatch is

$$\lambda = dC_i/dP_i(1/(1 - \partial P_L/\partial P_i) = dC_i/dP_i L_i \qquad i = 1, 2, \ldots, N$$

with

$$L_i = 1/(1 - \partial P_L/\partial P_i)$$

Therefore, $L_1 = 1.0$, $L_2 = 0.9733$, and $L_4 = 0.9238$.

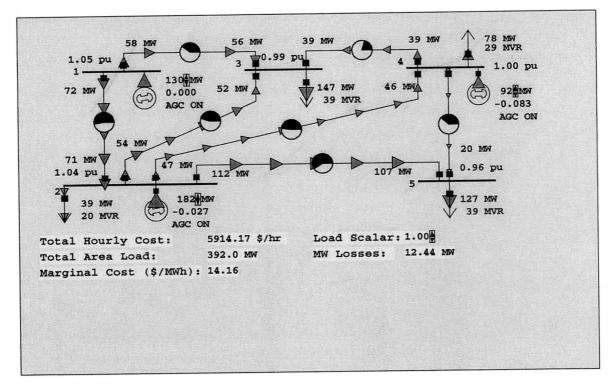

Example 11.8: Five-bus case with transmission line losses

With $P_1 = 130.1$ MW, $dC_1/dP_1 * L_1 = (10 + 0.032 * 130.1) * 1.0$
$$= 14.16 \quad \$/MWh$$

With $P_2 = 181.8$ MW, $dC_2/dP_2 * L_2 = (8 + 0.036 * 181.8) * 0.9733$
$$= 14.16 \quad \$/MWh$$

With $P_4 = 92.4$ MW, $dC_4/dP_4 * L_4 = (12 + 0.036 * 92.4) * 0.9238$
$$= 14.16 \quad \$/MWh \quad \blacksquare$$

In Example 11.7, total transmission losses are expressed as a quadratic function of unit output powers. For an area with N units, this formula generalizes to

$$P_L = \sum_{i=1}^{N} \sum_{j=1}^{N} P_i B_{ij} P_j \tag{11.4.14}$$

where the B_{ij} terms are called *loss coefficients* or B *coefficients*. The B coefficients are not truly constant, but vary with unit loadings. However, the B coefficients are often assumed constant in practice since the calculation of $\partial P_L/\partial P_i$ is thereby simplified. Using (11.4.14),

$$\frac{\partial P_L}{\partial P_i} = 2 \sum_{j=1}^{N} B_{ij} P_j \tag{11.4.15}$$

This equation can be used to compute the penalty factor L_i in (11.4.13).

Various methods of evaluating B coefficients from power-flow studies are available [6]. In practice, more than one set of B coefficients may be used during the daily load cycle.

When the unit incremental cost curves are linear, an analytic solution to the economic dispatch problem is possible, as illustrated by Examples 11.4–11.6. However, in practice, the incremental cost curves are nonlinear and contain discontinuities. In this case, an iterative solution by digital computer can be obtained. Given the load demand P_T, the unit incremental cost curves, generator limits, and B coefficients, such an iterative solution can be obtained by the following nine steps. Assume that the incremental cost curves are stored in tabular form, such that a unique value of P_i can be read for each dC_i/dP_i.

STEP 1   Set iteration index $m = 1$.

STEP 2   Estimate mth value of λ.

STEP 3   Skip this step for all $m > 1$. Determine initial unit outputs P_i ($i = 1, 2, \ldots, N$). Use $dC_i/dP_i = \lambda$ and read P_i from each incremental operating cost table. Transmission losses are neglected here.

STEP 4   Compute $\partial P_L/\partial P_i$ from (11.4.15) ($i = 1, 2, \ldots, N$).

STEP 5   Compute dC_i/dP_i from (11.4.13) ($i = 1, 2, \ldots, N$).

STEP 6   Determine updated values of unit output P_i $(i = 1, 2, \ldots, N)$. Read P_i from each incremental operating cost table. If P_i exceeds a limit value, set P_i to the limit value.

STEP 7   Compare P_i determined in Step 6 with the previous value $(i = 1, 2, \ldots, N)$. If the change in each unit output is less than a specified tolerance ε_1, go to Step 8. Otherwise, return to Step 4.

STEP 8   Compute P_L from (11.4.14).

STEP 9   If $\left| \left(\sum_{i=1}^{N} P_i \right) - P_L - P_T \right|$ is less than a specified tolerance ε_2, stop. Otherwise, set $m = m + 1$ and return to Step 2.

Instead of having their values stored in tabular form for this procedure, the incremental cost curves could instead be represented by nonlinear functions such as polynomials. Then, in Step 3 and Step 5 each unit output P_i would be computed from the nonlinear functions instead of being read from a table. Note that this procedure assumes that the total load demand P_T is constant. In practice, this economic dispatch program is executed every few minutes with updated values of P_T.

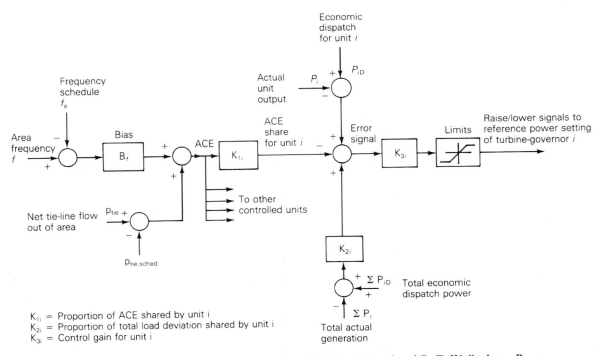

K_{1i} = Proportion of ACE shared by unit i
K_{2i} = Proportion of total load deviation shared by unit i
K_{3i} = Control gain for unit i

FIGURE 11.9   Automatic generation control [11] (A. J. Wood and B. F. Wollenberg, Power Generation, Operation, and Control (New York: Wiley, 1989))

COORDINATION OF ECONOMIC DISPATCH WITH LFC

Both the load-frequency control (LFC) and economic dispatch objectives are achieved by adjusting the reference power settings of turbine-governors on control. Figure 11.9 shows an *automatic generation control* strategy for achieving both objectives in a coordinated manner. As shown, the area control error (ACE) is first computed, and a share K_{1i} ACE is allocated to each unit. Second, the deviation of total actual generation from total desired generation is computed, and a share $K_{2i} \sum (P_{iD} - P_i)$ is allocated to unit i. Third, the deviation of actual generation from desired generation of unit i is computed, and $(P_{iD} - P_i)$ is allocated to unit i. An error signal formed from these three components and multiplied by a control gain K_{3i} determines the raise or lower signals that are sent to the turbine-governor of each unit i on control.

In practice, raise or lower signals are dispatched to the units at discrete time intervals of 2 to 10 seconds. The desired outputs P_{iD} of units on control, determined from the economic dispatch program, are updated at slower intervals, typically every 2 to 10 minutes.

OTHER TYPES OF UNITS

The economic dispatch criterion has been derived for a power system area consisting of fossil-fuel generating units. In practice, however, an area has a mix of different types of units including fossil-fuel, nuclear, pumped-storage hydro, hydro, wind, and other types.

Although the fixed costs of a nuclear unit may be high, their operating costs are low due to inexpensive nuclear fuel. As such, nuclear units are normally base-loaded at their rated outputs. That is, the reference power settings of turbine-governors for nuclear units are held constant at rated output; therefore, these units do not participate in LFC or economic dispatch.

Pumped-storage hydro is a form of energy storage. During off-peak hours these units are operated as synchronous motors to pump water to a higher elevation. Then during peak-load hours the water is released and the units are operated as synchronous generators to supply power. As such, pumped-storage hydro units are used for light-load build-up and peak-load shaving. Economic operation of the area is improved by pumping during off-peak hours when the area λ is low, and by generating during peak-load hours when λ is high. Techniques are available for incorporating pumped-storage hydro units into economic dispatch of fossil-fuel units [7].

In an area consisting of hydro plants located along a river, the objective is to maximize the energy generated over the yearly water cycle rather than to minimize total operating costs. Reservoirs are used to store water during high-water or light-load periods, although some water may have to be released through spillways. Also, there are constraints on water levels due to river transportation, irrigation, or fishing requirements. Optimal strategies are available for coordinating outputs of plants along a river [8]. Economic dispatch strategies for mixed fossil-fuel/hydro systems are also available [9, 10, 11].

Techniques are also available for including reactive power flows in the economic dispatch formulation, whereby both active and reactive powers are selected to minimize total operating costs. In particular, reactive injections from generators, switched capacitor banks, and static var systems, along with transformer tap settings, can be selected to minimize transmission-line losses [11]. However, electric utility companies usually control reactive power locally. That is, the reactive power output of each generator is selected to control the generator terminal voltage, and the reactive power output of each capacitor bank or static var system located at a power system bus is selected to control the voltage magnitude at that bus. In this way, the reactive power flows on transmission lines are low, and the need for central dispatch of reactive power is eliminated.

11.5

OPTIMAL POWER FLOW

Economic dispatch has one significant shortcoming—it ignores the limits imposed by the devices in the transmission system. Each transmission line and transformer has a limit on the amount of power that can be transmitted through it, with the limits arising because of thermal, voltage, or stability considerations (Section 5.6). Traditionally, the transmission system was designed so that when the generation was dispatched economically there would be no limit violations. Hence, just solving economic dispatch was usually sufficient. However, with the worldwide trend toward deregulation of the electric utility industry, the transmission system is becoming increasingly constrained. For example, in the PJM power market in the eastern United States the costs associated with active transmission line and transformer limit violations increased from $65 million in 1999 to almost $2.1 billion in 2005 [14].

The solution to the problem of optimizing the generation while enforcing the transmission lines is to combine economic dispatch with the power flow. The result is known as the optimal power flow (OPF). There are several methods for solving the OPF, with the linear programming (LP) approach the most common [13] (this is the technique used with Power-World Simulator). The LP OPF solution algorithm iterates between solving the power flow to determine the flow of power in the system devices and solving an LP to economically dispatch the generation (and possibility other controls) subject to the transmission system limits. In the absence of system elements loaded to their limits, the OPF generation dispatch will be identical to the economic dispatch solution, and the marginal cost of energy at each bus will be identical to the system λ. However, when one or more elements are loaded to their limits the economic dispatch becomes constrained, and the bus marginal energy prices are no longer identical. In some electricity markets these marginal prices are known as the Locational Marginal Prices (LMPs) and are used to determine the wholesale price of electricity at various locations in the system. For example, the real-time LMPs for the Midwest ISO are available online at www.midwestmarket.org.

EXAMPLE 11.9 PowerWorld Simulator—optimal power flow

PowerWorld Simulator case Example 11_9 duplicates the five-bus case from Example 11.6, except that the case will be solved using PowerWorld Simulator's LP OPF algorithm. To turn on the OPF option, first select **Case Information, Areas...**, and toggle the AGC Status field to OPF. Finally, rather than solving the case with the "Single Solution" button, select **LP OPF, Primal LP** to solve using the LP OPF. Initially the OPF solution matches the ED solution from Example 11.6 since there are no overloaded lines. The green-colored fields immediately to the right of the buses show the marginal cost of supplying electricity to each bus in the system (i.e., the bus

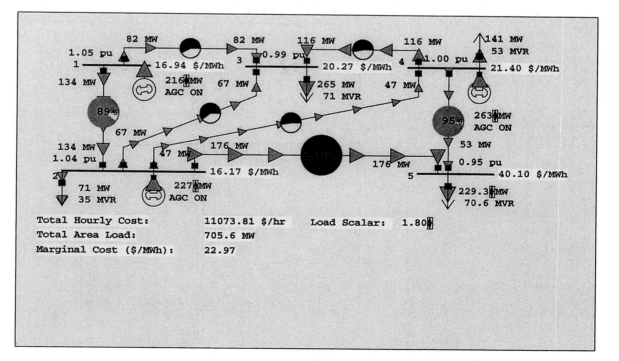

Example 11.9: Optimal Power-Flow solution with load multiplier = 1.80

LMPs). With the system initially unconstrained, the bus marginal prices are all identical at \$14.53/MWh, with a Load Scalar of 1.0.

Now increase the Load Scalar field from 1.00 to the maximum economic loading value, determined to be 1.67 in Example 11.6, and again select **LP OPF, Primal LP**. The bus marginal prices are still all identical, now at a value of \$17.52/MWh, with the line from bus 2 to 5 just reaching its maximum value. For load scalar values above 1.67, the line from bus 2 to bus 5 becomes constrained, with a result that the bus marginal prices on the constrained side of the line become higher than those on the unconstrained side.

With the load scalar equal to 1.80, numerically verify that the price of power at bus 5 is approximately \$40.08/MWh.

SOLUTION   The easiest way to numerically verify the bus 5 price is to increase the load at bus 5 by a small amount and compare the change in total system operating cost. With a load scalar of 1.80, the bus 5 MW load is 229.3 MW with a case hourly cost of \$11,074. Increasing the bus 5 load by 1.8 MW and resolving the LP OPF gives a new cost of \$11,147, a change of about \$40.5/MWh (note that this increase in load also increases the bus 5 price to over \$42/MWh). Because of the constraint, the price of power at bus 5 is actually more than double the incremental cost of the most expensive generator! ∎

PROBLEMS

SECTION 11.1

11.1   The block-diagram representation of a closed-loop automatic regulating system, in which generator voltage control is accomplished by controlling the exciter voltage, is shown in Figure 11.10. T_a, T_e, T_f are the time constants associated with the amplifier, exciter, and generator field circuit, respectively. (a) Find the open-loop transfer function $G(s)$. (b) Evaluate the minimum open-loop gain such that the steady-state error Δe_{ss} does not exceed 1%. (c) Discuss the nature of the dynamic response of the system to a step change in the reference input voltage.

FIGURE 11.10

Problem 11.1

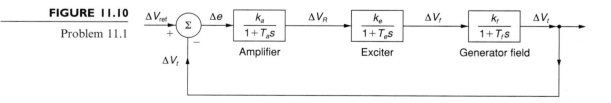

11.2   The Automatic Voltage Regulator (AVR) system of a generator is represented by the simplified block diagram shown in Figure 11.11, in which the sensor is modeled by a simple first-order transfer function. The voltage is sensed through a voltage transformer and then rectified through a bridge rectifier. Parameters of the AVR system are given as follows.

	Gain	Time Constant (seconds)
Amplifier	K_A	$\tau_A = 0.1$
Exciter	$K_E = 1$	$\tau_E = 0.4$
Generator	$K_G = 1$	$\tau_G = 1.0$
Sensor	$K_R = 1$	$\tau_R = 0.05$

(a) Determine the open-loop transfer function of the block diagram and the closed-loop transfer function relating the generator terminal voltage $V_t(s)$ to the reference voltage $V_{ref}(s)$. (b) For the range of K_A from 0 to 12.16, comment on the stability of the system. (c) For $K_A = 10$, evaluate the steady-state step response and steady-state error.

FIGURE 11.11

Problem 11.2

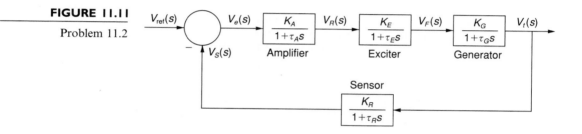

11.3   Let a rate feedback stabilizer be added to the AVR system of Problem 11.2. The block diagram of the compensated AVR system is shown in Figure 11.12. Let the stabilizer time constant be $\tau_F = 0.04$ s and the derivative gain be $K_F = 2$. (a) Obtain the closed-loop transfer function $V_t(s)/V_{ref}(s)$. (b) Using MATLAB, evaluate the step response and find the steady-state step response.

FIGURE 11.12

Problem 11.3

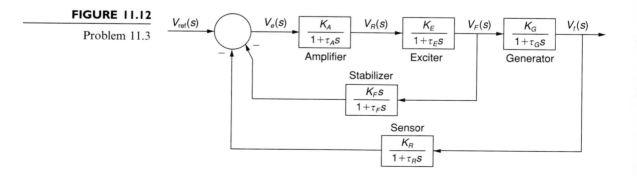

11.4   The Proportional-Integral-Derivative (PID) controller may be used to improve the dynamic response as well as reduce or even eliminate the steady-state error for the AVR system of Problem 11.2. Let the transfer function of the PID controller be given as follows $G_C(s) = K_P + K_I/s + K_D s$. Draw the block diagram of the AVR compensated with the PID controller and comment on the functions of the derivative as well as the integral component of the controller.

SECTION 11.2

11.5   An area of an interconnected 60-Hz power system has three turbine-generator units rated 200, 300, and 500 MVA. The regulation constants of the units are 0.03, 0.04, and 0.06 per unit, respectively, based on their ratings. Each unit is initially operating at one-half its own rating when the load suddenly decreases by 100 MW. Determine (a) the unit area frequency response characteristic β on a 100-MVA base, (b) the steady-state increase in area frequency, and (c) the MW decrease in mechanical power output of each turbine. Assume that the reference power setting of each turbine-governor remains constant. Neglect losses and the dependence of load on frequency.

11.6   Each unit in Problem 11.5 is initially operating at one-half its own rating when the load suddenly increases by 75 MW. Determine (a) the steady-state decrease in area frequency, and (b) the MW increase in mechanical power output of each turbine. Assume that the reference power setting of each turbine-generator remains constant. Neglect losses and the dependence of load on frequency.

11.7   Each unit in Problem 11.5 is initially operating at one-half its own rating when the frequency increases by 0.003 per unit. Determine the MW decrease of each unit. The reference power setting of each turbine-governor is fixed. Neglect losses and the dependence of load on frequency.

11.8   Repeat Problem 11.7 if the frequency decreases by 0.005 per unit. Determine the MW increase of each unit.

11.9   Based on a generator rating of 100 MVA, a turbine-generator unit operating in a 60-Hz interconnected power system has a regulation constant $R = 0.06$ per unit. Following a system disturbance, the system adjusts to steady-state operation with a frequency decrease of 0.025 Hz. Find the increase in turbine output power for this turbine-generator unit.

11.10   An interconnected 60-Hz power system consisting of one area has two turbine-generator units, rated 500 and 750 MVA, with regulation constants of 0.04 and 0.05 per unit, respectively, based on their respective ratings. When each unit carries a 300-MVA steady-state load, let the area load suddenly increase by 250 MVA. (a) Compute the area frequency response characteristic β on a 1000-MVA base. (b) Calculate Δf in per-unit on a 60-Hz base and in Hz.

11.11   A single area consists of two generating units rated (unit 1) 750 MVA with speed regulation $R_1 = 0.07$ per unit on its MVA base and (unit 2) 500 MVA with $R_2 = 0.04$ per unit on its MVA base. The units are operating in parallel at 60 Hz with unit 1 supplying 600 MW and unit 2 supplying 300 MW. Let the load be increased by 90 MW. (a) Assuming that there is no frequency-dependent component of the load, find the steady-state frequency deviation and the corresponding new generation of each unit. (b) Let the load vary linearly versus frequency with a 1.5% increase (decrease) in load for every 1% increase (decrease) in frequency. That is, $\Delta P_L = \Delta P_{LC} + D\Delta f$, where ΔP_L is the total load change, $\Delta P_{LC} = 90$ MW is the component of the load change that is independent of frequency, $D\Delta f$ is the frequency-dependent component of load change, and $D = 1.5$ with all quantities expressed in per unit. Calculate the steady-state frequency deviation and the corresponding new generation of each unit.

SECTION 11.3

11.12 A 60-Hz power system consists of two interconnected areas. Area 1 has 1200 MW of generation and an area frequency response characteristic $\beta_1 = 600$ MW/Hz. Area 2 has 1800 MW of generation and $\beta_2 = 800$ MW/Hz. Each area is initially operating at one-half its total generation, at $\Delta p_{tie1} = \Delta p_{tie2} = 0$ and at 60 Hz, when the load in area 1 suddenly increases by 400 MW. Determine the steady-state frequency error and the steady-state tie-line error Δp_{tie} of each area. Assume that the reference power settings of all turbine-governors are fixed. That is, LFC is not employed in any area. Neglect losses and the dependence of load on frequency.

11.13 Repeat Problem 11.12 if LFC is employed in area 2 alone. The area 2 frequency bias coefficient is set at $B_{f2} = \beta_2 = 800$ MW/Hz. Assume that LFC in area 1 is inoperative due to a computer failure.

11.14 Repeat Problem 11.12 if LFC is employed in both areas. The frequency bias co-efficients are $B_{f1} = \beta_1 = 600$ MW/Hz and $B_{f2} = \beta_2 = 800$ MW/Hz.

11.15 Repeat Problem 11.13 if there is a third area, with 3000 MW of generation and $\beta_3 = 1500$ MW/Hz. The load in area 1 increases by 400 MW. LFC is employed in area 2 alone. All three areas are interconnected.

11.16 Rework Problems 11.13 through 11.15 when the load in area 2 suddenly decreases by 400 MW. The load in area 1 does not change.

11.17 A 60-Hz interconnected power system has two areas denoted area 1 and area 2, where the area frequency response characteristics are given by $\beta_1 = 400$ MW/Hz and $\beta_2 = 300$ MW/Hz. The total power generated in each of these areas is 1000 and 750 MW, respectively. Each area is initially generating power in steady state with $\Delta P_{tie1} = \Delta P_{tie2} = 0$, when the load in area 1 suddenly increases by 60 MW. Compute the resulting steady-state change in frequency Δf as well as the steady-state changes in tie-line flows ΔP_{tie1} and ΔP_{tie2} (a) without LFC, and (b) with LFC. Neglect all losses and the dependence of load on frequency.

11.18 On a 1000-MVA common base, a two-area system interconnected by a tie line has the following parameters:

Area	1	2
Area Frequency Response Characteristic	$\beta_1 = 0.05$ per unit	$\beta_2 = 0.0625$ per unit
Frequency-Dependent Load Coefficient	$D_1 = 0.6$ per unit	$D_2 = 0.9$ per unit
Base Power	1000 MVA	1000 MVA
Governor Time Constant	$\tau_{g1} = 0.25$ s	$\tau_{g2} = 0.3$ s
Turbine Time Constant	$\tau_{t1} = 0.5$ s	$\tau_{t2} = 0.6$ s

The two areas are operating in parallel at the nominal frequency of 60 Hz. The areas are initially operating in steady state with each area supplying 1000 MW when a sudden load change of 187.5 MW occurs in area 1. Compute the new steady-state frequency and change in tie-line power flow (a) without LFC, and (b) with LFC.

SECTION 11.4

11.19 An area of an interconnected power system has two fossil-fuel units operating on economic dispatch. The variable operating costs of these units are given by

$$C_1 = \begin{cases} 4P_1 + 0.02P_1^2 & \text{for } 0 < P_1 \leqslant 100 \quad \text{MW} \\ 8P_1 \dfrac{\$}{\text{hr}} & \text{for } P_1 > 100 \quad \text{MW} \end{cases}$$

$$C_2 = 0.03P_2^2 \dfrac{\$}{\text{hr}}$$

where P_1 and P_2 are in megawatts. Determine the power output of each unit, the incremental operating cost, and the total cost C_T that minimizes C_T as the load demand P_T varies from 200 to 700 MW. Generating-unit inequalities and transmission losses are neglected.

11.20 Rework Problem 11.19 if the units are subject to the following inequality constraints:

$$100 \leqslant P_1 \leqslant 500 \quad \text{MW}$$
$$50 \leqslant P_2 \leqslant 300 \quad \text{MW}$$

11.21 Rework Problem 11.19 if total transmission losses for the power system are given by

$$P_L = 2 \times 10^{-4} P_1^2 + 1 \times 10^{-4} P_2^2 \quad \text{MW}$$

11.22 Rework Problems 11.19 through 11.21 if the operating cost of unit 2 is changed to

$$C_2 = 0.05P_2^2 \quad \$/\text{hr}$$

Compare the results with those of Problems 11.19 through 11.21.

11.23 Expand the summations in (11.4.14) for $N = 2$, and verify the formula for $\partial P_L / \partial P_i$ given by (11.4.15). Assume $B_{ij} = B_{ji}$.

11.24 Given two generating units with their respective variable operating costs as:

$$C_1 = 0.01P^2{}_{G1} + 2P_{G1} + 100 \quad \$/\text{hr} \qquad \text{for } 25 \leq P_{G1} \leq 150 \text{ MW}$$
$$C_2 = 0.004P^2{}_{G2} + 2.6P_{G2} + 80 \quad \$/\text{hr} \qquad \text{for } 30 \leq P_{G2} \leq 200 \text{ MW}$$

Determine the economically optimum division of generation for $55 \leq P_L \leq 350$ MW. In particular, for $P_L = 282$ MW, compute P_{G1} and P_{G2}. Neglect transmission losses.

11.25 The system of Problem 11.24 is represented by a single-line diagram shown in Figure 11.13. For convenience this system is also modeled in PowerWorld case Problem 11.25. A base case power-flow study on the system yields the following data in per-unit on a 100-MVA base:

FIGURE 11.13

Problem 11.25

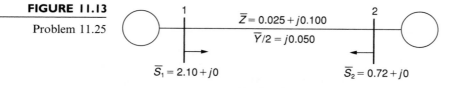

$\bar{Z} = 0.025 + j0.100$

$\bar{Y}/2 = j0.050$

$\bar{S}_1 = 2.10 + j0$

$\bar{S}_2 = 0.72 + j0$

All values in per-unit

$$V_1 = 1\underline{/0°} \text{ per unit} \qquad P_{G1} = 1.0313 \text{ per unit} \qquad P_{L1} = 2.1000 \text{ per unit}$$

$$V_2 = 1\underline{/6.616°} \text{ per unit} \qquad P_{G2} = 1.8200 \text{ per unit} \qquad P_{L2} = 0.7200 \text{ per unit}$$

Transmission losses $P_{TL} = 0.0313$ per unit.

(a) Increasing the load at each bus by 10%, compute the A coefficients defined by:

$$A_{kj} = \partial\delta_k/\partial P_{Gj} \approx \Delta\delta_k/\Delta P_{Gj} \qquad k = 1, 2, \ldots, b \text{ buses}$$
$$j = 1, 2, \ldots, n \text{ generators}$$

(b) Determine the B coefficients defined by:

$$B_{ij} = \frac{1}{2}(\partial^2 P_{TL}/\partial P_{Gj}\partial P_{Gj}) = \frac{1}{2}\left\{ \sum_{m=1}^{b}\sum_{k=1}^{b}(\partial^2 P_{TL}/\partial\delta_m\partial\delta_k)A_{mi}A_{kj} \right\}$$

By using the relations:

$$\partial^2 P_{TL}/\partial\delta_m\partial\delta_k = 2V_m V_k g_{mk} \cos(\delta_m - \delta_k), \qquad m \neq k$$

$$= -2\sum_{\substack{i=1 \\ i \neq m}}^{b} V_i V_m g_{im} \cos(\delta_i - \delta_m), \qquad m = k$$

in which conductances (g's) can be obtained from the bus admittance matrix **Y**.

(c) Solve for the economically optimum division of load with consideration of losses.

Hint: The condition for economically optimum generation requires the weighted incremental cost functions for all units on economic dispatch to be equal, and the weighting factor or penalty factor of generator i is given by

$$Li = \frac{1}{1 - (\partial P_{TL}/\partial P_{Gi})}$$

PW **11.26** Resolve Example 11.6, except with the generation at bus 2 set to a fixed value (i.e., modeled as off of AGC). Plot the variation in the total hourly cost as the generation at bus 2 is varied between 1000 and 200 MW in 5-MW steps, resolving the economic dispatch at each step. What is the relationship between bus 2 generation at the minimum point on this plot and the value from economic dispatch in Example 11.6? Assume a Load Scalar of 1.0.

PW **11.27** Using PowerWorld case Example 11_8, with the Load Scalar equal to 1.0, determine the generation dispatch that minimizes system losses (*Hint:* Manually vary the generation at buses 2 and 4 until their loss sensitivity values are zero). Compare the operating cost between this solution and the Example 11.8 economic dispatch result. Which is better?

PW **11.28** Repeat Problem 11.27, except with the Load Scalar equal to 1.3.

SECTION 11.5

PW **11.29** Using LP OPF with PowerWorld Simulator case Example 11_9, plot the variation in the bus 5 marginal price as the Load Scalar is increased from 1.0 in steps of 0.02. What is the maximum possible load scalar without overloading any transmission line? Why is it impossible to operate without violations above this value?

PW   **11.30**   Load PowerWorld Simulator case Problem 11_30. This case models a slightly modified version of the 37 bus case from Example 6.13 with generator cost information, but also with two of the three lines between buses BLT69 and UIUC69 open. When the case is loaded the "Total Cost" field shows the economic dispatch solution, which results in an overload on the remaining line between buses BLT69 and UIUC69. Before solving the case, select **LP OPF, OFP Buses** to view the bus LMPs, noting that they are all identical. Then Select **LP OPF, Primal LP** to solve the case using the OPF, and again view the bus LMPs. Verify the LMP at the UIUC69 bus by manually changing the load at the bus by one MW, and then noting the change in the Total Cost field. Repeat for the DEMAR69 bus. Note, because of solution convergence tolerances the manually calculated results may not exactly match the OFP calculated bus LMPs.

CASE STUDY QUESTIONS

1.   What is meant by generator black-start capability, why is it needed, and what types of generators are best at providing black-start capability?

2.   Why are overvoltages a concern during the restoration of the transmission system?

3.   Research a recent large power system outage, describing some of the unique aspects associated with the restoration of the power system.

A.   How has the ongoing deregulation of the electric utility industry affected planning of transmission systems?

B.   What are the differences between transmission planning in the United States and transmission planning in Australia?

C.   Has deregulation made power systems more reliable or less reliable?

REFERENCES

1.   IEEE Committee Report, "Computer Representation of Excitation Systems," *IEEE Transactions PAS*, vol. PAS-87 (June 1968), pp. 1460–1464.

2.   M. S. Sarma, *Electric Machines* 2d ed. (Boston, PWS Publishing 1994).

3.   IEEE Committee Report, "Dynamic Models for Steam and Hydro Turbines in Power System Studies," *IEEE Transactions PAS*, vol. PAS-92, no. 6 (November/December 1973), pp. 1904–1915.

4.   N. Cohn, *Control of Generation and Power Flow on Interconnected Systems* (New York: Wiley, 1971).

5.   L. K. Kirchmayer, *Economic Operation of Power Systems* (New York: Wiley, 1958).

6.   L. K. Kirchmayer and G. W. Stagg, "Evaluation of Methods of Coordinating Incremental Fuel Costs and Incremental Transmission Losses," *Transactions AIEE*, vol. 71, part III (1952), pp. 513–520.

7.   G. H. McDaniel and A. F. Gabrielle, "Dispatching Pumped Storage Hydro," *IEEE Transmission PAS*, vol. PAS-85 (May 1966), pp. 465–471.

8. E. B. Dahlin and E. Kindingstad, "Adaptive Digital River Flow Predictor for Power Dispatch," *IEEE Transactions PAS*, vol. PAS-83 (April 1964), pp. 320–327.

9. L. K. Kirchmayer, *Economic Control of Interconnected Systems* (New York: Wiley, 1959).

10. J. H. Drake et al., "Optimum Operation of a Hydrothermal System," *Transactions AIEE* (*Power Apparatus and Systems*), vol. 62 (August 1962), pp. 242–250.

11. A. J. Wood and B. F. Wollenberg, *Power Generation, Operation, and Control* (New York: Wiley, 1989).

12. R. J. Thomas et al., "Transmission System Planning—The Old World Meets The New," *Proceedings of The IEEE*, 93, 11 (November 2005), pp. 2026–2035.

13. B. Stott and J. L. Marinho, "Linear Programming for Power System Network Security Applications," *IEEE Trans. on Power Apparatus and Systems*, vol. PAS-98, (May/June 1979), pp. 837–848.

14. 2005 PJM State of the Market Report, available online at http://www.pjm.com/markets/market-monitor/som.html.

15. M. M. Adibi and L. H. Fink, "Overcoming Restoration Challenges Associated with Major Power System Disturbances", *IEEE Power and Energy Magazine*, 4, 5 (September/October 2006), pp. 68–77.

Three-phase 600 MVA, 22-kV Generator Step-up Transformer at the Jim Bridger Power Plant, Rock Springs, WY, USA (Courtesy of PacifiCorp)

12

TRANSMISSION LINES: TRANSIENT OPERATION

Transient overvoltages caused by lightning strikes to transmission lines and by switching operations are of fundamental importance in selecting equipment insulation levels and surge-protection devices. We must, therefore, understand the nature of transmission-line transients.

During our study of the steady-state performance of transmission lines in Chapter 5, the line constants R, L, G, and C were recognized as distributed rather than lumped constants. When a line with distributed constants is subjected to a disturbance such as a lightning strike or a switching operation, voltage and current waves arise and travel along the line at a velocity near the speed of light. When these waves arrive at the line terminals, reflected voltage and current waves arise and travel back down the line, superimposed on the initial waves.

Because of line losses, traveling waves are attenuated and essentially die out after a few reflections. Also, the series inductances of transformer windings effectively block the disturbances, thereby preventing them from entering generator windings. However, due to the reinforcing action of several reflected waves, it is possible for voltage to build up to a level that could cause transformer insulation or line insulation to arc over and suffer damage.

Circuit breakers, which can operate within 50 ms, are too slow to protect against lightning or switching surges. Lightning surges can rise to peak levels within a few microseconds and switching surges within a few hundred microseconds—fast enough to destroy insulation before a circuit breaker could open. However, protective devices are available. Called surge arresters, these can be used to protect equipment insulation against transient overvoltages. These devices limit voltage to a ceiling level and absorb the energy from lightning and switching surges.

We begin this chapter with a discussion of traveling waves on single-phase lossless lines (Section 12.1). We present boundary conditions in Section 12.2 and the Bewley lattice diagram for organizing reflections in Section 12.3. We derive discrete-time models of single-phase lines and of lumped RLC elements in Section 12.4, and discuss the effects of line losses and multi-conductor lines in Sections 12.5 and 12.6. In Section 12.7 we discuss power system overvoltages including lightning surges, switching surges, and power-frequency overvoltages, followed by an introduction to insulation coordination in Section 12.8.

CASE STUDY   Two case-study reports are presented here. The first describes metal oxide varistor (MOV) arresters used by electric utilities to protect power and substation equipment against transient overvoltages in power systems with rated voltages through 345 kV [22]. The second, in preparation for Chapter 13 (Transient Stability) describes a wide-area stability and voltage control system being designed and implemented by Bonneville Power Administration in the northwestern United States [23].

VariSTAR® Type AZE Surge Arresters for Systems through 345 kV ANSI/IEEE C62.11 Certified Station Class Arresters*

GENERAL

The VariSTAR AZE Surge Arrester offers the latest in metal oxide varistor (MOV) technology for the economical protection of power and substation equipment. This arrester is gapless and constructed of a single series column of MOV disks. The arrester is designed and tested to the requirements of ANSI/IEEE Standard C62.11, and is available in ratings suitable for the transient overvoltage protection of electrical equipment on systems through 345 kV.

Cooper Power Systems assures the design integrity of the AZE arrester through a rigorous testing

TABLE I AZE Series S (AZES) Ratings and Characteristics

Arrester Characteristic	Rating
System Application Voltages	3–345 kV
Arrester Voltage Ratings	3–360 kV
Rated Discharge Energy, (kJ/kV of MCOV)	
Arrester Ratings: 3–108 kV	3.4
120–240 kV	5.6
258–360 kV	8.9
System Frequency	50/60 Hz
Impulse Classifying Current	10 kA
High Current Withstand	100 kA
Pressure Relief Rating, kA rms sym	
Metal-Top Designs	65 kA
Cubicle-Mount Designs	40 kA
Cantilever Strength (in-lbs)*	
Metal-Top Designs:	
3–240 kV	90,000
258–360 kV	120,000

*Maximum working load should not exceed 40% of this value.
(August 1997. New Issue. © Cooper Industries, Inc.)

program conducted at our Thomas A. Edison Technical Center and at the factory in Olean, NY. The availability of complete "in-house" testing facilities assures that as continuous process improvements are made, they are professionally validated to high technical standards.

Table 1, shown above, contains information on some of the specific ratings and characteristics of AZE Series S (AZES) surge arresters.

CONSTRUCTION

External

The Type AZE station class arrester is available in two design configurations—a metal-top design in ratings 3–360 kV and a cubicle-mount design in ratings 3–48 kV. Cubicle-mount designs are ideally suited for confined spaces where clearances between live parts are limited.

The wet-process porcelain housing features an alternating shed design (ratings > 48 kV) that provides excellent resistance to the effects of atmospheric housing contamination. AZE arresters are available with optional extra creepage porcelains for use in areas with extreme natural atmospheric and man-made pollution.

The dielectric properties of the porcelain are coordinated with the electrical protective characteristics of the arrester. The unit end castings are of a corrosion-resistant aluminum alloy configured for interchangeable mounting with other manufacturers' arresters for ease in upgrading to the

Figure I
120 kV rated VariSTAR Type AZE surge arrester.
(August 1997. New Issue. © Cooper Industries, Inc.)

TABLE 2 Discharge Voltages—Maximum Guaranteed Protective Characteristics for AZES Surge Arresters

Arrester Rating (kV, rms)	Arrester MCOV (kV, rms)	Front-of-Wave Protective Level (kV)* 10 kA	Lightning Impulse Discharge Voltages (8/20 μsec, kV)						Switching Impulse Discharge Voltages (kV)**		
			1.5 kA	3 kA	5 kA	10 kA	20 kA	40 kA	500 A	1000 A	2000 A
3	2.55	9.7	7.4	7.8	8.1	8.6	9.8	12.2	6.8		
6	5.10	19.2	14.8	15.5	16.1	17.0	19.1	23.2	13.5		
9	7.65	28.8	22.1	23.3	24.1	25.5	28.5	34.1	20.2		
10	8.40	31.5	24.3	25.6	26.5	27.9	31.2	37.3	22.2		
12	10.2	38.3	29.5	31.0	32.1	33.9	37.8	45.0	27.0		
15	12.7	47.6	36.7	38.6	39.9	42.1	47.0	55.8	33.6		
18	15.3	57.3	44.2	46.5	48.1	50.7	56.5	66.9	40.4		
21	17.0	63.6	49.1	51.7	53.4	56.3	62.7	74.2	44.9		
24	19.5	73.0	56.3	59.3	61.3	64.6	71.9	84.9	51.5		
27	22.0	81.4	63.6	66.9	69.1	72.8	81.0	95.7	58.1		
30	24.4	91.2	70.5	74.1	76.6	80.7	89.8	106	64.4		
33	27.5	103	79.4	83.6	86.3	91.0	101	119	72.6		
36	29.0	108	83.8	88.1	91.0	95.9	107	126	76.6		
39	31.5	118	91.0	95.7	98.9	104	116	136	83.2		
42	34.0	127	98.2	103	107	112	125	147	89.8		
45	36.5	136	105	111	115	120	134	158	96.4		
48	39.0	146	113	118	122	129	143	169	103		
54	42.0	157	121	128	132	139	154	181	111		
60	48.0	179	139	146	151	158	176	207	127		
66	53.0	198	153	161	166	175	195	229	140		
72	57.0	212	165	173	179	188	209	246	151		
78	62.0	232	179	188	194	205	228	267	164		
84	68.0	253	196	207	213	224	250	293	180		
90	70.0	261	202	213	220	231	257	302	185		
96	76.0	284	219	231	238	251	279	327	201		
108	84.0	313	243	255	263	277	308	362	222		
120	98.0	337	267	277	283	298	326	379	241	250	
132	106	365	288	300	306	323	354	411	261	271	
138	111	382	302	314	321	338	370	430	273	284	
144	115	396	313	325	332	350	383	446	283	294	
162	130	447	354	368	376	396	433	504	320	332	
168	131	451	356	371	379	399	437	508	323	335	
172	140	481	381	396	405	426	467	542	345	358	
180	144	495	392	407	416	438	480	558	355	368	
192	152	523	414	430	439	463	506	589	374	388	
198	160	550	435	453	462	487	533	620	394	409	
204	165	567	449	467	477	502	550	639	406	421	
216	174	598	473	492	503	529	580	674	428	444	
228	182	626	495	515	526	554	606	705	448	465	
240	190	653	517	537	549	578	633	736	468	485	
258	209	684	547	568	580	605	666	771	502	526	535
264	212	693	555	576	588	613	675	782	509	533	543
276	220	720	575	598	611	637	701	811	528	553	563
288	230	751	602	625	639	665	732	848	552	578	589
294	235	767	615	639	652	679	748	866	564	591	602
300	239	781	625	650	663	691	761	881	574	601	612
312	245	801	630	655	669	709	780	903	578	606	617
330	267	872	698	726	741	772	850	985	641	671	683
336	269	879	704	731	747	778	856	991	645	676	689
360	289	945	756	785	802	836	920	1064	693	727	740

*Based on a current impulse that results in a discharge voltage cresting in 0.5 μs.
**45–60 μs rise time current surge.
(August 1997. New Issue. © Cooper Industries, Inc.)

VariSTAR arrester technology. This three-footed mounting is provided on a 8.75 to 10 inch diameter pattern for customer supplied 0.5 inch diameter hardware.

High cantilever strength assures mechanical integrity (Table I lists the cantilever strength of metal-top AZES arresters). Cooper Power Systems recommends that a load limit of 100 pounds not be exceeded on the line terminal of cubicle mount designs. Loads exceeding this limit could cause a shortening of arrester life. Housings are available in standard grey or optional brown glaze color.

Standard line and ground terminal connectors accommodate up to a 0.75 inch diameter conductor. Insulating bases and discharge counters are optionally available for in-service monitoring of arrester discharge activity.

The end fittings and porcelain housing of each arrester unit are sealed and tested by means of a sensitive helium mass spectrometer; this assures that the quality and insulation protection provided by the arrester is never compromised over its lifetime by the entrance of moisture. A corrosion-resistant nameplate is provided and contains all information required by Standards. In addition, stacking arrangement information is provided for multi-unit arresters. Voltage grading rings are included for arresters rated 172 kV and above.

WACS—Wide-Area Stability and Voltage Control System: R&D and Online Demonstration

CARSON W. TAYLOR, DENNIS C. ERICKSON, KENNETH E. MARTIN, ROBERT E. WILSON, AND VAITHIANATHAN VENKATASUBRAMANIAN

INVITED PAPER

As background, we describe frequently used feed-forward wide-area discontinuous power system stability controls. Then we describe online demonstration of a new response-based (feedback) Wide-Area stability and voltage Control System (WACS). The control system uses powerful discontinuous actions for power system stabilization. The control system comprises phasor measurements at many substations, fiber-optic communications, real-time deterministic computers, and transfer trip output signals to circuit breakers at many other substations and power plants. Finally, we describe future development of WACS. WACS is developed as a flexible platform to prevent blackouts and facilitate electrical commerce.

Keywords—Blackout prevention, emergency control, phasor measurements, power system stability, unstable limit cycle, voltage stability, wide-area measurements and control.

Manuscript received May 3, 2002; revised October 25, 2003. The work at Washington State University was supported in part by the Power System Electric Engineering Research Center (PSERC) and in part by the Consortium for Electric Reliability Technology Solutions (CERTS).

C. W. Taylor and K. E. Martin are with the Bonneville Power Administration, Vancouver, WA 98666 USA (e-mail: cwtaylor@bpa.gov; kemartin@bpa.gov).

D. C. Erickson is with Ciber, Inc., Beaverton, OR 97007 USA (e-mail: derickson@ciber.com).

R. E. Wilson is with Western Area Power Administration, Lakewood, CO 80228 USA (e-mail: rewilson@wapa.gov).

V. Venkatasubramanian is with the Department of Electrical and Computer Science, Washington State University, Pullman, WA 99165 USA (e-mail: mani@eecs.wsu.edu).

Digital Object Identifier 10.1109/JPROC.2005.846338

*WACS—Wide-Area Stability and Voltage Control System ..."
by Carson W. Taylor, Dennis C. Erickson, Kenneth E. Martin,
Robert E. Wilson and Vaithianathan Venkatasubramanian,
Proceedings of the IEEE (Nov. 2005), pg. 892–905.)*

I.   INTRODUCTION

The Bonneville Power Administration (BPA), Portland, OR; Ciber Inc., Beaverton, OR; and Wash-

CASE STUDY **613**

ington State University (WSU), Pullman, are designing and implementing a Wide-Area stability and voltage Control System termed *WACS*. WACS provides a flexible platform for rapid implementation of generator tripping and reactive power compensation switching for transient stability and voltage support of a large power system. Features include synchronized positive sequence phasor measurements, digital fiber-optic communications from 500-kV substations, a real-time control computer programmed in the G language, and output communications for generator tripping and 500-kV capacitor/reactor bank switching. The WACS software runs two algorithms in parallel.

As background, we describe widely used emergency controls termed Special Protection Systems (SPS). SPS is based on *direct detection* of predefined outages, with high-speed binary (transfer trip) signals to control centers for logic decisions, and then to power plants and substations for generator tripping and capacitor/reactor bank switching Disadvantages of SPS include control for only predefined events, complexity, and relatively high cost.

In contrast with SPS, WACS employs strategically placed sensors to react to the power system *response* to arbitrary disturbances. WACS provides single discontinuous stabilizing actions or true feedback control. As true feedback control, the need for discontinuous action is determined and commanded, the power system response is observed, and further discontinuous action such as generator tripping or capacitor bank switching is taken as necessary. The WACS platform may also be used for wide-area modulation control of generators and transmission-level power electronic devices and for control center operator alarms and monitoring.

We describe WACS benefits and describe large-scale simulations showing the interarea stabilization of large disturbances by WACS.

We also describe WACS design (measurement, communications, control), and initial online implementation. Online testing results include statistics of communications delay from global positioning system (GPS) time-tagged substation measurements to GPS-timed receipt by the control computer.

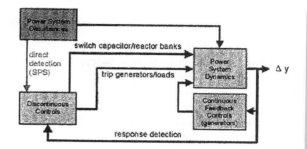

Figure 1
Local and wide-area, continuous and discontinuous power system stability controls. Adapted from drawings by Dr. J. F. Hauer.

II. POWER SYSTEM STABILITY CONTROLS

Power system stability controls are described in several books and reports [1]–[5]. Reference [6] is an early, but still valuable, paper describing discontinuous controls.

Fig. 1 shows a block diagram of the power system stability control environment. Power system stability encompasses electromechanical (rotor angle) stability among groups of synchronous generators, and voltage stability involving load response to disturbances [7]. RMS-type sensors are generally used—electromechanical oscillations and slow voltage variations amplitude-modulate the 50- or 60-Hz power frequency waveforms. Electromagnetic transients are not of primary interest except for sensor filtering considerations.

Most stability controls are continuous feedback controls at power plants: automatic voltage regulator and power system stabilizer for generator excitation control, and prime mover control (speed governor). Controls are largely the single-input–single-output type, designed via classical feedback control methods [1]. Additional local continuous stability controls are at transmission-level power electronic devices such as static var compensators. There are also local discontinuous controls for reactive power compensation (capacitor/reactor banks) switching and load shedding.

Installed wide-area stability controls are mainly based on direct detection of selected outages. These emergency controls are termed SPS or remedial action schemes. We describe these controls in Section IV.

Advanced wide-area stability controls measure power system response to disturbances. There are very few implementations at present. We describe present-day projects in Sections V and VI.

Discontinuous controls supplement the basic continuous controls by relieving stress for very large disturbances, providing a region of attraction and a secure postdisturbance operating or equilibrium point. Continuous controls then operate effectively over a smaller nonlinear range.

Wide-area controls—feedforward and feedback, continuous and discontinuous—obviously have potential for improved observability and controllability.

III. WESTERN NORTH AMERICAN POWER SYSTEM

Fig. 8 in Section VI shows the western North American interconnected power system. Connections to the eastern North American synchronous interconnection are by small back-to-back HVdc converter stations.

Long-distance interarea transmission lines characterize the western interconnection. Major lines are 500-, 345-, and 230-kV. There are two ±500-kV dc links: the 3100-MW 1360-km Pacific HVdc intertie from the Columbia River to Los Angeles and the 1920-MW 787-km Intermountain Power Project link from Utah to southern California. Hydro power predominates in the Pacific Northwest (PNW—British Columbia, Washington State, and Oregon). Large coal-based power plants predominate in the eastern and southern portions of the interconnection. Most generation in California is natural gas or oil based.

In spring and summer, with good hydro generation conditions, the dominant interarea power flows are from the PNW to California and also from coal-based generation in the eastern areas (Wyoming, Utah, Arizona, and Nevada) to California. The northern California portion of the Pacific ac intertie comprises three series capacitor compensated 500-kV lines with nonsimultaneous rating of 4800 MW for the three Oregon to California lines.

Large-scale power flow and transient stability simulations of potential disturbances are necessary to determine transfer limits for many defined transmission paths. Power flow (steady-state) simulations model over 10,000 buses (nodes representing generation and load injection stations, and transformer substations), requiring solving over 20,000 nonlinear algebraic equations. Transient stability simulation adds thousands of nonlinear differential equations. In defining simultaneous power transfer limits, major sensitivities are shown on nomograms. Portions of nomogram boundaries are limited by either first swing transient stability, transient damping of oscillations, or postdisturbance voltage support criterion.

Controls described in this paper are oriented toward high-power transfers from the PNW and British Columbia to California, but are adaptable to other applications.

IV. FEEDFORWARD WIDE-AREA STABILITY CONTROLS

The widely used SPS provide control for potential single and multiple-related outages identified in the power system planning process. Compared to the financial and permitting difficulties of transmission line construction, SPS are low cost and easy to install. A large increase in power transfer capability is realized.

As generation and load increase without corresponding increase in transmission lines, SPS controls proliferate [8]. At BPA, there are many schemes for a myriad of operating and disturbance conditions. Tens of millions of dollars have been invested over many years. Additional schemes were added following the cascading power failures in summer 1996 [9], [10].

Control actions are mainly for detection of transmission outages, but also for some generation outages. The most complex scheme involves the Pacific ac intertie where high-speed outage detection of around fifty 500-kV lines is installed (detection at both line ends). Fault tolerant programmable logic controllers are at BPA's two control centers: one near Portland and the other in Spokane, WA.

The most important control action is tripping of PNW hydroelectric generators. There are few difficulties with tripping hydro generators and they can be rapidly returned to service. The generators are at the sending end of the PNW to California power transfer path, with the generator tripping braking remaining Northwest generators that are accelerating relative to southwest generators. For outages of either the Pacific ac or HVdc intertie, up to 2700 MW of generation may be tripped.

Other control actions are energizing 500-kV series and shunt capacitor banks, and disconnecting shunt reactors. BPA 500-kV shunt capacitor banks are in the 200–380 MVAr range.

Control actions take place as fast as 150 ms after the outage. The delay time includes detection time, communications to central logic, logic computer processing time, communications to power plants and substations, and power circuit breaker operating time. Communication of SPS activation signals use the same high-speed "transfer trip" used for isolation of transmission line short circuits. At BPA these are primarily frequency shift key audio tones over analog microwave. Newer systems use digital messages over digital microwave or fiber optics (SONET).

The consequences of SPS failure can be large-scale blackouts. Controls are clearly not as robust as additional transmission lines and must be highly reliable by design. High redundancy in detection, communication, and logic computers is required.

The complexity of the SPS is ever increasing. BPA has a full-time operator devoted to prearming (enabling) and monitoring the many schemes.

Besides complexity, a shortcoming of preplanned event driven control is that other disturbances may occur that have not been considered in planning. These may originate in other parts of the interconnected power system.

V. FEEDBACK WIDE-AREA STABILITY CONTROLS

Feedback controls measure power system variables and can respond to arbitrary disturbances. Control can be continuous or discontinuous.

Stability control using remote signals are not new and are a simplified form of wide-area control. In 1976, BPA implemented modulation of the Pacific HVdc intertie using active power and later current magnitude signals from a remote substation on the parallel Pacific ac intertie [11], [12]. The continuous control damped electromechanical oscillations between groups of PNW generators and groups of Pacific Southwest generators; the oscillation period was around 3 s. Analog microwave communications and analog controls were used. The modulation was single input–four outputs, the outputs being active and reactive power at the northern HVdc terminal (rectifier operation) in Oregon and active and reactive power at the southern terminal (inverter operation) near Los Angeles. Modulation was discontinued after a major expansion of the HVdc intertie from two terminals to four terminals in 1989.

A. Phasor measurements

Although various types of rms sensors may be used, digital positive sequence, GPS-synchronized phasor measurements [4], [5], [13], [14] are most often considered for wide-area control. "Positive sequence" refers to transformation of an unbalanced set of three-phase voltages or currents into a set of positive, negative, and zero sequence "symmetrical components," where positive sequence is a set of three-phase voltages or currents with equal magnitudes, 120° phase difference, and normal phase rotation [5, Ch. 8]. In normal operation without short circuits or individual phase outages, the phase voltages and currents are nearly equal to the positive sequence voltages and currents.

Several manufacturers offer phasor measurement sensors. Typically, channels for multiple three-phase voltage and current measurements are provided. The positive sequence voltage and current phasors are computed and GPS time tagged once every two cycles, or in newer equipment, once every cycle of the power frequency (30- or 60-Hz data rate for 60-Hz power frequency). Power system frequency deviation from nominal is also computed, with GPS providing a precise time and frequency reference. There are tradeoffs between response speed and filtering. The phasor measurements are grouped,

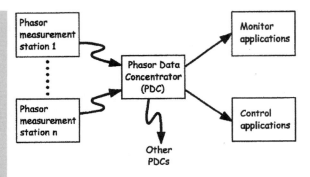

Figure 2
Control center PDC.

and data packets are transmitted to a central site where packets from several measurement locations (substations) are organized by time stamp [15], [16]. Outputs from a "phasor data concentrator" (PDC) are networked to monitoring and control applications (Fig. 2).

From the voltage and current phasors, applications may compute active and reactive power.

In coming years, phasor measurements will become more common as part of IT advances such as substation automation. The phasor measurements can be made available at small cost as part of other substation measurements, for example, protective relaying [17].

Networked phasor measurements are a key part of a BPA/U.S. Department of Energy/Electric Power Research Institute (EPRI)/Western Area Power Administration program for wide-area measurement systems (WAMS) [5, Ch. 11.8], [18]. WAMS is valuable for power system identification, power system monitoring, control center state estimation, and power system dynamic performance analysis following disturbances—including large blackouts.

B.   Continuous wide-area controls
Continuous wide-area controls offer observability and controllability benefits where conventional local continuous controls have shortcomings. Possibilities include "wide-area power system stabilizers" [19] and controls for powerful transmission-level power electronic devices such as HVdc, thyristor-controlled series capacitors, and static var compensators.

Wide-area controls are especially attractive for unusual system structures. Remote signals may augment control using local measurements.

Because of increased control leverage and continuous exposure to adverse interactions, caution compared to local control is required. Communications latency is one concern. Dynamics mimicking electromechanical oscillations are another [20], [21]. These may include sensor processing artifacts such as aliasing of network resonances or harmonics, or generator shaft torsional dynamics. Hydro plant water column oscillations may appear to be electromechanical oscillations. Extra monitoring and supervision of control is desirable.

C.   Discontinuous wide-area controls
Compared to continuous control, discontinuous control lends to be safer—action is only taken when necessary. Discontinuous control has similarities with biological systems where stimuli must be above an activation threshold.

Similar to feedforward controls (SPS), feedback discontinuous controls initiate a large stabilizing action that improves first swing transient stability, reduces stress to improve oscillation damping, and provides a larger region of attraction for a more secure postdisturbance operating point.

We next describe a specific wide-area discontinuous feedback control (WACS) in development and demonstration.

VI.   WACS

A.   Overview
Fig. 3 shows a pictorial block diagram of WACS. Selected existing phasor measurements are used for inputs, and existing SPS transfer trip circuits are available for outputs. The new development is the real-time controller.

Based partly on the 10 August 1996 cascading failure [10], the original BPA concept was to combine voltage magnitude measurements with generator reactive power measurements using fuzzy logic. The premise is that generator reactive power measurements can be a more sensitive indicator of insecurity than voltage magnitude—voltages can be near normal but generator reactive power outputs near limits indicate insecurity. R&D at WSU

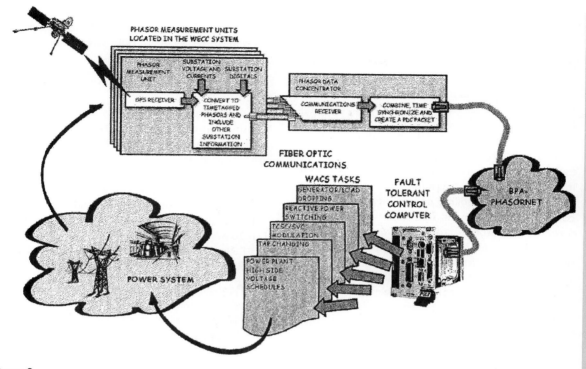

Figure 3
WACS block diagram.

showed, however, that a voltage magnitude based control is faster and simpler for transient stability [22]. Both methods are now used. Recent field experience has actually shown that the two methods, Vmag and VmagQ algorithms, have similar speed. This is partly due to recent replacement of slow rotating generator field winding excitation equipment with modern thyristor exciters at two large power plants.

References [23]–[25] describe recent WACS research and development.

Twelve voltage magnitude measurements from seven 500-kV stations are used. Two stations are near the Oregon–California border (Malin and Captain Jack), one is in central Oregon (Summer Lake), and three are near the Columbia River in northern Oregon or southern Washington (John Day, Slatt, Ashe, and McNary). Fifteen generator reactive power measurements at five power plant switching stations near the Columbia River are used

(Big Eddy, John Day, Slatt, Ashe, and McNary). The hydro power plants feeding into Big Eddy, John Day, and McNary comprise 18, 16, and 12 generators, respectively; two to four generators are connected to a transmission line from the power plant to the switching station where phasor measurements are made.

We designed WACS so that loss of measurements from a single location or even multiple locations will only slightly degrade control. Measurements at widely spaced locations (hundreds of kilometers) provide spatial averaging or filtering against the aliasing effects discussed above. Spatial filtering along with discontinuous control action biases the phasor measurement requirements toward fast response rather than secure filtering.

B. Allowable time for control actions
For first swing transient stability, control action must be taken prior to the peak of the forward in-

terarea angle swing—the sooner the better. For a simple second-order undamped dynamic system with natural frequency of 1/3 Hz (3-s period), the step response peak is at 1.5 s. The impulse response peak is at 0.75 s. Most disturbances are closer to a step response than an impulse (the rare three-phase short circuit approaches an impulse, but opening the faulted line provides a step response effect). Nowadays the frequency of the Pacific intertie mode is around 0.25 Hz (4-s period), allowing more time for control action. The oscillation frequency is even lower for high-stress operation.

For transient stability, control action should be completed within around 1 s—especially for the less powerful capacitor/reactor bank switching.

The delay time for phasor measurement, fiber-optic communications, PDC throughput including wait time for slowly arriving packets, transfer trip, and circuit breaker tripping (of generators or shunt reactors) or closing (shunt capacitor bank insertion) are approximately 3, 2, 2, 1, and 2–5 60-Hz cycles respectively, or around 10 cycles for tripping and 13 cycles for closing (167 and 217 ms). Time for several control execution loops, intentional time delay, and throughput delay will be 67 ms or longer. Thus, it appears that control action can be taken within 0.3 s after sufficient power system electromechanical response to the disturbance. For capacitor/reactor bank switching, the local supervising voltage measurement sensors will be responding during the same time as the WACS measurements and processing.

The more sensitive $VmagQ$ algorithm may operate following longer time frame dynamics. In fact, need for the existing SPS action is determined by power flow simulation of a point in time several minutes after the disturbance. The SPS helps ensure postdisturbance voltage support for angle stability following generator overexcitation limiting, tap changing, and other slower actions. Many seconds are available for taking sequential feedback actions as necessary.

The conditions existing in northeastern Ohio preceding the 14 August 2003 blackout were exactly what the $VmagQ$ algorithm caters to. Voltage magnitudes were mildly depressed, but Cleveland-area generators were at or near their reactive power limits [26]. Automatic load shedding by control similar to WACS could have prevented the blackout.

C. Phasor measurement communications and PDC

BPA legacy communications is analog microwave. Transmission of phasor measurement packets using modems has high latency (60–100 ms) and relatively high dropout rates. Thus, BPA-owned fiber-optic communications (SONET) are used for WACS. BPA has an extensive fiber-optic network, with links and terminal equipment still being added.

Fig. 4 shows tests of fiber-optic latency of less than 26 ms for a link from the Slatt switching station to a BPA control center. The link uses direct digital transfer into SONET.

D. Real-time hardware and software

We selected National Instruments' LabVIEW Real-Time (RT) hardware and software [27]. The software is a true dataflow language that prevents race conditions and allows for parallel tasking (multitasking and multithreading are supported). It has many needed programming features and library components, including data acquisition/processing/output, TCP/UDP, signal processing, filtering, math

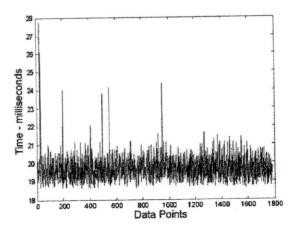

Figure 4
Fiber-optic communications latency over 1 min, Slatt PMU to PDC.

Figure 5
WACS real-time hardware.

operations, execution logic and state machine, execution tracing and timing, display graphics, and fuzzy logic tools with graphical editors. The graphical code is largely self-documenting. Modular architecture aids the testing and certification of critical modules by designing virtual simulated "real-world" conditions around them.

Controller development is done on a PC, and code is then downloaded to real-time deterministic hardware and software [27]. Fig. 5 shows the rack-mounted WACS hardware. A full-featured host PC that connects via Ethernet to the WACS RT engine is available to monitor, test, develop, and upload software and to run model studies for tuning. This is important, as the RT processor (engine) itself is always operated "headless" (no monitor, keyboard, or mouse). We do use a small text display status monitor and a sequence of events recorder.

WACS parameters are managed using an .ini file. This allows easy changes to the application for tuning the algorithms, defining station names, scale factors, etc. This is important in avoiding the pitfalls of "hardwiring."

1) *Control Execution Rate:* The rate at present is 30 control executions per second (33.3-ms intervals), which is the same rate as the phasor measurement packets. The time for a control execution cycle is around 8 ms, allowing addition of other features and moving to a future 60 packets per second data rate. For security against inadvertent actions during faults, measurement noise, and other dis-

turbances, output must be above a threshold for two or more control executions.

2) *Input Processing:* Tasks include reading and decoding the data from the UDP Ethernet connection to the PDC and data sanity checks for transmission errors, nonvalid data, missing packets, and extra long latency. Close coordination with PDC data processing is required, and a customized message of only the needed measurements is transmitted to WACS. For missing or corrupt data, there are two options: one is to block the data and not use it in the algorithm, and the second is to pad the data with the last valid packet for a few control executions. The former is used currently. Both fatal and nonfatal errors are managed and reports issued. Also, the data is weighted and limit tested before being passed to the algorithm subroutines. The WACS software can also run in a library mode using simulation data or archived data from the PDC as inputs to the algorithms. This validates performance, facilitating certification.

3) *Output Processing:* Following the algorithm computations, a pattern consisting of up to 32 isolated outputs is sent to a relay control stage, where several masking operations are done before passing to the transfer trip communications. Any fatal or nonfatal error will mask outputs, with a user-defined mask available to disable one or more outputs either temporarily or long term or to disable the results of either or both the fast or slow algorithms.

E. The Vmag algorithm

The voltage magnitude based Vmag algorithm provides first swing transient stability stabilization and relieves stress to improve transient damping. For growing oscillations, it will operate at some point for stabilization.

The algorithm is fairly simple, based on 12 voltage magnitude measurements at seven 500-kV stations. A weighted average voltage is computed from the 12 measurements, with highest weight for measurements close to the Oregon–California border where the voltage swings are usually greatest (Malin, Captain Jack, Summer Lake). A nonlinear accumulator (integrator) computes volt-seconds below a threshold setting that is currently 525 kV for capacitor/reactor bank switching and 520 kV for generator tripping (normal voltage is around 540 kV). Accumulation is blocked for voltage recovery. Control action results when the volt-second accumulation reaches a setpoint; also, the weighted voltage must be below 490 kV for generator tripping. The algorithm thus has semblance to proportional–integral (PI) control. Beneficially, faster operation results for more severe disturbances.

1) *Critical Disturbance:* A critical disturbance is near simultaneous outage of two nuclear generating units at a nuclear power plant in Arizona or California. Such events have occurred several times, and Western Electricity Coordinating Council (WECC) rules specify that cascading failure not result from these outages. Multigenerator outages at large coal-based power plants have also occurred.

 The largest disturbance is outage of two nuclear generators at the Palo Verde nuclear plant near Phoenix, AZ. The combined power loss is around 2700 MW. The lost power is made up by inertia power from rotors of all other generators, and by response to decaying frequency by the speed governors of other generators. About half of the response comes from hydro generators in the northern half of the interconnected power system, resulting in a large increase in the Pacific ac intertie north to south loading. Instability (loss of synchronism between northern and southern generators) results if the initial intertie loading is high. There is no SPS for this outage.

2) *Simulation of Two Palo Verde Generator Outages Without and With WACS:* The Appendix describes simulation methodology. Fig. 6 shows 500-kV voltage responses for the two Palo Verde generator outages with existing controls. The initial Pacific ac intertie loading is 4700 MW. The largest voltage swing is at

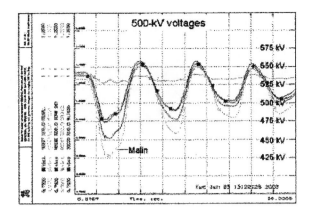

Figure 6
BPA 500-kV voltages for outage of two Palo Verde generators with existing controls, with Pacific ac intertie loading of 4700 MW.

the Malin station near the Oregon–California border, and the performance is considered marginally acceptable.

 Fig. 7 shows similar response with WACS at an initial intertie loading of 5000 MW—a 300-MW gain. The main WACS action is 916 MW of generator tripping at two hydro power plants in British Columbia 1.4 s after

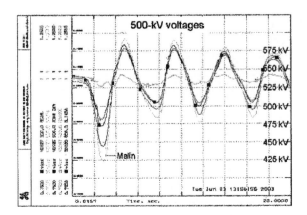

Figure 7
BPA 500-kV voltages for outage of two Palo Verde generators with WACS and with Pacific ac intertie loading of 5000 MW.

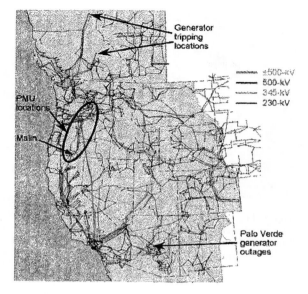

Figure 8
Western North American interconnected power system showing disturbance location, and WACS input measurement locations and output action locations.

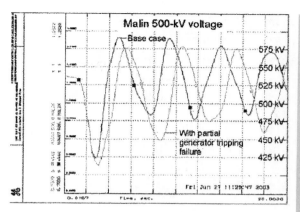

Figure 9
Malin voltage for outage of two Palo Verde generators with partial failure of WACS generator tripping.

the outage. WACS also inserted two 500-kV shunt capacitor banks 1.2 s after the outage. The transfer trip circuits from the BPA control center to the British Columbia power plants and capacitor bank locations exist as part of BPA's SPS. The higher oscillation frequency of Fig. 7 tends to indicate improved stability.

Fig. 8 shows the wide-area nature: an outage in Arizona, measurements in Oregon and southern Washington, and control actions in British Columbia.

3) *Partial Failure of Measurements:* Using the 5000-MW intertie loading with WACS case as reference, we simulate failure of the most important phasor measurement device. The failure is two voltage measurements at Captain Jack near Malin on the Oregon–California border. Simulation results are nearly identical to the reference case, so are not shown.

4) *Partial Failure of Generator Tripping:* Fig. 9 compares the reference case to simulation of

failure of generator tripping at one of the two British Columbia power plants. Stability is maintained, but performance is notably degraded. The oscillation frequency decreases.

Other sensitivity cases lead to improved tuning of the algorithm. We simulate other outages to verify expected performance.

F. The VmagQ algorithm

The *VmagQ* algorithm combines voltage magnitude measurements and generator reactive power measurements using fuzzy logic. Similar to the Vmag algorithm, we compute a weighted average 500-kV voltage magnitude from 12 phasor measurements at seven locations. More complicated is computation of weighted average reactive power from 15 transmission lines emanating from six large power plants. We compute active and reactive powers for these lines from the voltage and current phasors.

First, we estimate the number of connected generators—up to four generators per line may be connected on lines from hydro plants. While this information may be available at a slower data rate within a control center, we avoid interface and dependence on other data networks. The number of generators is estimated based on the normal loading range of individual generators. An error with estimation on one of the many lines is not serious.

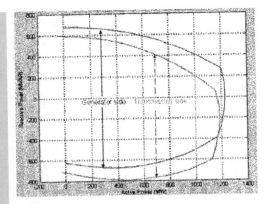

Figure 10
Active/reactive power capability curve for a nuclear plant generator.

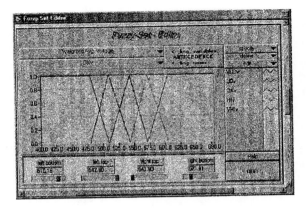

Figure 11
Weighted average voltage fuzzy set input. Linguistic variables are VLOv (very low voltage), LOv (low voltage), OKv (OK voltage), HIv (high voltage), and VHIv (very high voltage).

A normalization procedure follows, based on generator active/reactive power capability curves. We map the capability curves from the generator terminals to the transmission side where the phasor measurement sensors are located, accounting for station service load, and generator step-up transformer impedance and tap ratio. Fig. 10 shows this one-time mapping for a large nuclear plant generator.

Normalization results in reactive power output on a scale of approximately ± 1 (generator controls allow large temporary and small steady-state operation outside limits). The transmission-side reactive power limits corresponding to the active power are noted: Q_{max} and Q_{min}. We compute the normalized value from transmission side values as follows:

$$Q_{half} = \frac{(Q_{max} - Q_{min})}{2}$$

$$Q_{zero} = Q_{min} + Q_{half}$$

$$Q_{norm} = \frac{(Q - Q_{zero})}{Q_{half}}.$$

For example, for Fig. 10 with transmission side $P = 1000$ MW and $Q = 100$ MVAr, $Q_{norm} = 0.77$.

A weighted average of the normalized reactive powers is now computed. The individual weights are the product of the generator or generator group MVA rating and a factor. The factor is based on location and voltage support sensitivity. We have higher value to generators providing more sensitive control of transmission voltage by automatic voltage regulator line drop compensation [28] or by automatic high side voltage control.

We next combine the weighted average voltage magnitude and the weighted average generator reactive power are using basic fuzzy logic. To date, linguistic variable tuning is based on rules of thumb such as overlap cross points at 0.5. Figs. 11–13 show the input and output fuzzy sets. Fig. 14 shows the rule editor, and Fig. 15 shows the output control surface as a function of weighted average voltage for a weighted average generator reactive power input value of zero. The crisp output range is approximately ± 1, but elimination of unrealistic rule combinations result in a practical 0–1 range.

A crisp output value above a threshold enters an accumulator. With accumulator setpoint reached, capacitor/reactor bank switching or generator tripping is commanded. Capacitor/reactor bank switching occurs first, with generator tripping only in more severe situations where capacitor/reactor bank switching is not available or not sufficient.

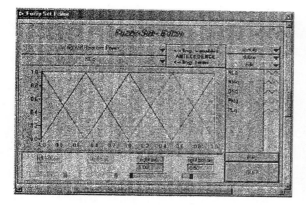

Figure 12
Weighted average generator reactive power fuzzy set input. Linguistic variables are NLq (negative large reactive power), NMq (negative medium), Okq, PMq (positive medium), and PLq (positive large).

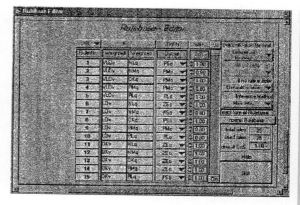

Figure 14
Rule base. Output default value of zero if no rule is active. Center of gravity defuzzification method is selected. The MAX–MIN inference method is used, which is explained in the Fig. 15 caption for rule 8.

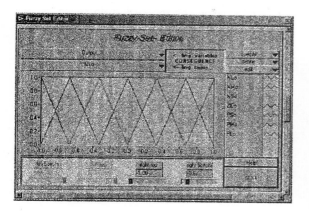

Figure 13
Output (consequence) fuzzy set. Linguistic variables are NLo (negative large output), NMo (negative medium output), NSo (negative small output), ZEo (zero output), PSo (positive small output), etc.

The main tuning is of the input measurement weights and the thresholds for output action—rather than the fuzzy sets.

As a power flow program steady-state simulation example, we simulated a simultaneous transfer nomogram limit point. Table 1 shows the fuzzy logic inputs and crisp outputs for a predisturbance case,

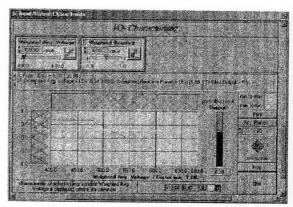

Figure 15
Fuzzy controller voltage versus output characteristic for reactive power input of zero. At the cursor point, the voltage input is 512 kV, the output is 0.30, and rules 8 and 13 are active. As shown, Rule 8 fires with value of 0.94, which is MIN of tile LOv and Okq linguistic variable values for 512-kV voltage and zero reactive power inputs (Figs. 11 and 12). For the output variable PSo, the area below 0.94 contributes to the center of gravity output (Fig. 13). If another rule activates output variable PSo, the MAX value of the two rules is used in the center of gravity computation.

TABLE I Power Flow Simulation Results, 4800 MW Intertie Power

	V_{Malin} kV	Q_{JohnD} MVAr	Vavg FL input	Qavg FL input	FL output
Pre outage	541.0	−158	540.7	0.273	0.10
2 Palo Verde outage	528.5	114	529.2	0.883	0.48
2 Palo Verde w/WACS[1]	530.5	89	533.7	0.826	0.45
2 Palo Verde w/WACS[2]	534.5	25	534.5	0.685	0.38

[1] 500-kV shunt capacitor/reactor bank switching at the 3 available locations.
[2] Capacitor/reactor bank switching plus 500 MW generator tripping.

the double Palo Verde outage case with existing controls, the outage case with WACS-initiated 500-kV capacitor/reactor bank switching at three locations, and the outage case with the capacitor/reactor bank switching and 500 MW of generator tripping. Also shown are key individual inputs to the fuzzy logic: voltage magnitude at Malin and reactive power from John Day power plant line 1 (John Day reactive power has the high weighting because of location and excellent voltage control performance).

The crisp fuzzy logic output, perhaps rescaled to 0–100, could also be used for a thermometer-type voltage security index for control center operators.

G. Tuning, testing, and monitoring

Since large disturbances are rare, algorithm verification and tuning is normally via offline large-scale simulation (Appendix I). The real-time controller is tested by inputting simulation results, and also by inputting archived phasor measurements from actual events.

H. WACS response for a large real event

At 07:40:56 on Monday 14 June 2004, a short circuit occurred near the Palo Verde Nuclear Plant west of Phoenix. The fault was not completely cleared for almost 39 s! Approximately 4589 MW of generation tripped, at and near Palo Verde in the southern part of the western North American interconnection. All three Palo Verde units tripped.

Pacific intertie stability was threatened, but maintained. With one line between Oregon and California out of service, the intertie limit prior to the event was 3200 MW. North to south intertie flow swung from the initial 2750 MW to 5500 MW and settled at 4500 MW several minutes later. Malin and Captain Jack voltages near the Oregon–California border swung from the initial 548 kV to 443 kV at 07:41:21.6. Operators and an existing "response-based" scheme switched BPA series capacitors and shunt capacitor/reactor banks during the swings and the subsequent intertie power increase. The power increase is from governor action at PNW hydro plants, which carry large amounts of spinning reserve.

On 14 June, WACS was in a monitor mode at a laboratory installation 5 km from the control center and PDC. Adjusting for communications and PDC features that were not yet in service, WACS operated correctly on the forward angle swing, before voltage swing minimum, as recorded by a sequence of events recorder [25].

To further validate the WACS algorithms, we played archived data from the 14 June event into the WACS code on an offline personal computer. One parameter was retuned to increase operating speed.

Fig. 16 shows voltages that are WACS inputs and includes the weighted average voltage computed by WACS (same weights used for both algorithms). For the Vmag algorithm, the accumulator thresholds

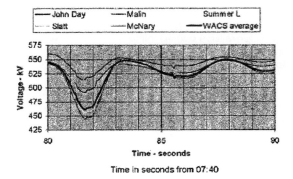

Figure 16
Northwest voltages for first three swings.

for capacitor/reactor bank switching are 525 kV and the thresholds for generator tripping are 520 kV. Accumulator setting for capacitor/reactor bank switching is 2 kV-s and accumulator setting for generator tripping is 4 kV-s. For generator tripping, the weighted average voltage must also be below 490 kV.

For first swing stabilization, discontinuous control action should occur before the voltage minimum at around 81.6 s after 07:40. This time is estimated for the real signal without measurement delay (PMU timetags are at the last sample of the phasor computation window; the PMU uses a four-cycle moving average filter with phasors calculated over one cycle).

Using the archived data as input to the real-time code, the WACS *Vmag* algorithm output for capacitor/reactor bank switching occurs at 81.033 s after 07:40. Adding 170 ms for communications, PDC, and circuit breaker delay, switching would be at 81.203 s, or around 0.4 s before the real signal voltage minimum.

WACS *Vmag* algorithm output for generator tripping occurs at 81.233 s after 07:40. Adding 170 ms the delays, tripping would be at 81.403 s, or around 0.2 s before the real signal voltage minimum.

For the *VmagQ* algorithm, Fig. 17 shows the weighted average voltage and weighted average reactive power that are combined using fuzzy logic.

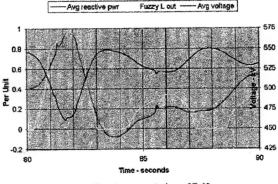

Figure 17
VmagQ fuzzy logic inputs and outputs.

The crisp (center of gravity) fuzzy logic output is also shown. High fuzzy logic output correctly occurs for the combination of low voltage and high reactive power output. Fuzzy logic output above a threshold is accumulated. For capacitor/reactor bank switching the thresholds are 0.40 per unit (p.u.). For generator tripping, the thresholds are 0.45 p.u. Accumulator settings are 0.05 p.u.-s for capacitor/reactor bank switching and 0.2 p.u.-s for generator tripping.

The *VmagQ* algorithm WACS output for capacitor/reactor bank switching occurs 81.033 s after 07:40 (same time as the voltage magnitude based algorithm). Adding 170 ms for communications, PDC, and circuit breaker delay, switching would be at 81.203 s, or around 0.4 s before the real signal voltage minimum.

WACS *VmagQ* algorithm output for generator tripping occurs at 81.533 s after 07:40 (300 ms later than the voltage magnitude based algorithm). Adding 170 ms for the delays, tripping would be at 81.703 s, or around 0.1 s after the real signal voltage minimum. While this will cause a larger backswing, the generator tripping reduces intertie loading and system stress.

The 14 June massive loss of generation was well beyond planning and operating reliability criteria. While unusual, power systems *are* continuously exposed to unusual events.

If stress (Pacific intertie loading) was somewhat higher on 14 June, instability would have occurred and caused controlled islanding with massive generation/load imbalance in the importing southern island that suffered the initial 4600-MW loss. Either massive underfrequency load shedding or a widespread blackout in California, Nevada, Arizona, and New Mexico would have resulted. Stress would have been higher later in the day as temperatures and load increased.

Events like the 14 June event are exactly what WACS can protect against. Reference [25] provides more details on this event.

VII. WACS STATUS

Since March 2003, WACS has been installed in a laboratory with real-time phasor measurement inputs from a PDC and recording of contact outputs. Based on the following, we have demonstrated "proof of concept," and tuned and validated the real-time controller hardware and software:

- large-scale simulations including playback of simulation results into real-time code;
- monitoring of real system performance over a two-year period;
- playback of archived data into real-time code, particularly for the 14 June 2003 massive generation outage event.

With successful R&D, BPA is considering a capital budget installation for commercial use as described below.

VIII. FURTHER DEVELOPMENT OF WACS

In parallel with the demonstration project, design requirements for permanent commercial implementation are being developed. Similar to SPS, very high reliability is required if WACS is used to increase power transfer limits. Dedicated, latest generation phasor measurement sensors likely will be required. A 60 packets per second data rate and a 60 times per second control execution rate will probably be used. Control computers will likely be at both BPA control centers, and all communications will be self-healing (geographically separated fiber-optic or digital microwave rings). The PDC function may be incorporated in the WACS computers.

After experience with single discontinuous control actions, bang–bang switching of capacitor banks can be considered for oscillation damping enhancement. Other control actions may be added, such as raising transmission voltage schedules at power plants during emergency situations; see Fig. 3. Nonlinear modulation of a static var compensator near Seattle, WA, using a voltage magnitude signal from Malin has been found effective [29] and could be implemented using WACS technology.

Many voltage and current phasor measurements plus system frequency measurements are available. Available measurements increase year by year. Binary signals from substations can be added to the phasor measurement packets. From this data,

WACS can provide monitoring and alarms for control center operators. Possibilities include event detection and monitoring of oscillation activity [30], [31, Ch. 6]. The software platform has extensive signal processing tools, and poor damping following severe disturbances is relatively easy to detect. A challenge is to minimize false alarms—logic or artificial intelligence (AI) methods applied to the large measurement base might be developed for this purpose.

Integration with other control center networks and application may offer synergies. For example, online security assessment simulations perhaps could be used for automatic tuning/learning/adaptation of WACS.

WACS is currently oriented toward improving stability of a specific interarea power transfer path. Application in other networks may be straightforward. For example, in a load area, either wide-area voltage measurements or combined voltage measurements and generator reactive power measurements could be used for reactive power compensation switching or load tripping. Such control in northern Ohio could have prevented the 14 August 2003 blackout.

Research is ongoing at WSU on further generalizing of wide-area controls to meshed networks where control strategy (measurements, disturbance classification, and generator or load tripping actions) is more difficult. WSU is also researching theoretical aspects, and use of voltage phase angles as control input. Appendices II and III describe this research.

IX. CONCLUSION

Automatic control experts state: "A modern view of control sees feedback as a tool for uncertainty management" [32, p. 1].

Given the many changes in the electric power industry with increasing complexity and reduced investment in transmission lines, the possibility and actual occurrences of large-scale blackouts are a worldwide concern. Clearly, new means to improve power system reliability and robustness are desirable. The addition of wide-area feedback control to frequently used wide-area feedforward control is an effective additional layer of defense against blackouts [33], as well as facilitator of electrical commerce.

WACS exploits advances in digital/optical communications and computation. Specific advantages include the following.

- Control for outages and conditions not covered by feedforward controls (SPS).
- Potentially simplifies operations for changing system conditions—currently, operators are required to reduce power transfers when unstudied conditions are encountered.
- Improved observability and controllability compared to local control. Discontinuous control reduces exposure to adverse interactions.
- Flexible, high reliability "open system" platform for rapid, low-cost control and monitoring additions, including wide-area continuous control.
- Provides a combination of reliability increase and power transfer capability increase.
- Caters to uncertainty in simulation results used to determine operating rules and limits.
- Future potential with cost reductions and further IT advances. Potential for application in meshed grid as well as intertie corridors. Control inputs and outputs may be extended over a larger geographical area such as the entire western North American power system.

Moving from WAMS to WACS (wide-area measurements to wide-area stability control) is a challenge in the new century.

ACKNOWLEDGMENT

Many BPA engineers have assisted in the development of WACS. Dr. Y. Chen contributed to the algorithms as a Ph.D. candidate at Washington State University, Pullman.

REFERENCES

[1] P. Kundur, *Power System Stability and Control*. New York: McGraw-Hill, 1994.

[2] C. W. Taylor, *Power System Voltage Stability*. New York: McGraw-Hill, 1994.

[3] T. Van Cutsem and C. Vournas, *Voltage Stability of Electric Power Systems*. Norwell, MA: Kluwer, 1998.

[4] "Advanced angle stability controls," CIGRE, Task Force 17, Advisory Group 02, Study Committee 38, Paris, France, Brochure 155, Apr. 2000.

[5] L. L. Grigsby, Ed., *The Electric Power Engineering Handbook*: CRC Press/IEEE Press, 2001, ch. 11, Power System Dynamics and Stability.

[6] IEEE Committee Report, "A description of discrete supplementary controls for stability," *IEEE Trans. Power App. Syst.*, vol. PAS-97, no. 1, pp. 149–165, Jan./Feb. 1978.

[7] IEEE/CIGRE Joint Task Force on Stability Terms and Definitions, "Definition and classification of power system stability," *IEEE Trans. Power Syst.*, vol. 19, no. 3, pp. 1387–1401, Aug. 2004.

[8] P. M. Anderson and B. K. LeReverend, "Industry experience with special protection schemes," *IEEE Trans. Power Syst.*, vol. 11, no. 3, pp. 1166–1179, Aug. 1996.

[9] C. W. Taylor and D. C. Erickson, "Recording and analyzing the July 2 cascading outage," *IEEE Comput. Appl. Power*, vol. 10, no. 1, pp. 26–30, Jan. 1997.

[10] D. M. Kosterev, C. W. Taylor, and W. A. Mittelstadt, "Model validation for the August 10, 1996 WSCC system outage," *IEEE Trans. Power Syst.*, vol. 14, no. 3, pp. 967–979, Aug. 1999.

[11] R. L. Cresap, D. N. Scott, W. A. Mittelstadt, and C. W. Taylor, "Operating experience with modulation of the Pacific HVDC intertie," *IEEE Trans. Power App. Syst.*, vol. PAS-98, pp. 1053–1059, Jul./Aug. 1978.

[12] CIGRE, Paris, France, CIGRE 14-05, 1978.

[13] A. G. Phadke, J. S. Thorp, and M. G. Adamiak, "New measurement technique for tracking voltage phasors, local system frequency, and rate of change of frequency," *IEEE Trans. Power App. Syst.*, vol. PAS-102, no. 5, pp. 1025–1038, May 1982.

[14] R. E. Wilson, "Uses of precise time and frequency in power systems," *Proc. IEEE (Special Issue on Time and Frequency)*, vol. 79, no. 7, pp. 1009–1018, Jul. 1991.

[15] IEEE Power Syst. Relaying Comm. (WG H-7, A. G. Phadke, Chairman), "Synchronized sampling and phasor measurements for relaying and control," *IEEE Trans. Power Del.*, vol. 9, no. 1, pp. 442–452, Jan. 1994.

[16] IEEE Power Syst. Relaying Comm. (WG H-8, K. E. Martin, Chairman), "IEEE standard for synchrophasors for power systems," *IEEE Trans. Power Del.*, vol. 13, no. 1, pp. 73–77, Jan. 1998.

[17] G. Benmouyal, E. E. Schweitzer, and A. Guzman, "Synchronized phasor measurement in protective relays for protection, control, and analysis of electric power systems," presented at the 29th Annu. Western Protective Relay Conf., Spokane, WA, 2002.

[18] J. Hauer, D. Trudnowski, G. Rogers, B. Mittelstadt, W. Litzenberger, and J. Johnson, "Keeping an eye on power system dynamics," *IEEE Comput. Appl. Power*, vol. 10, no. 4, pp. 50–54, Oct. 1997.

[19] I. Kamwa, R. Grondin, and Y. Hebert, "Wide-area measurement based stabilizing control of large power systems—a decentralized/hierarchical approach," *IEEE Trans. Power Syst.*, vol. 16, no. 1, pp. 136–153, Feb. 2001.

[20] J. F. Hauer and C. W. Taylor, "Information, reliability, and control in the new power system," in *Proc. 1998 American Control Conf.*, vol. 5, pp. 2986–2991.

[21] J. Lambert, A. G. Phadke, and D. McNabb, "Accurate voltage phasor measurement in a series-compensated network," *IEEE Trans. Power Del.*, vol. 9, no. 1, pp. 501–509, Jan. 1994.

[22] C. W. Taylor, V. Venkatasubramanian, and Y. Chen, "Wide-area stability and voltage control," presented at the Seventh Symp. Specialists in Electric Operational and Expansion Planning (VII SEPOPE), Curitiba, Brazil, 2000.

[23] C. W. Taylor and R. E. Wilson, "BPA's Wide-Area stability and voltage Control System (WACS): overview and large-scale simulations," presented at the Sixth Symp. Specialists in Electric Operational and Expansion Planning (IX SEPOPE), Rio de Janeiro, Brazil, 2004.

[24] R. E. Wilson and C. W. Taylor, "Using dynamic simulations to design the wide-area stability and voltage control system (WACS)," presented at the IEEE Power Engineering Soc. Power System Conf. Exposition, New York, 2004.

[25] C. W. Taylor, D. C. Erickson, and R. E. Wilson, "Reducing blackout risk by a wide-area control system (WACS): Adding a new layer of defense," in *Proc. 2005 Power System Computation Conf.*, submitted for publication.

[26] U.S.-Canada Power System Outage Task Force, *Final Report on the August 14, 2003 Blackout in the United States and Canada: Causes and Recommendations*, Apr. 2004.

[27] National Instruments [Online]. Available: http://www.ni.com

[28] D. Kosterev, "Design, installation, and initial operating experience with line drop compensation at John Day powerhouse," *IEEE Trans. Power Syst.*, vol. 16, no. 2, pp. 261–265, May 2001.

[29] V. Venkatasubramanian, K. W. Schneider, and C. W. Taylor, "Improving Pacific inter-tie stability using existing static VAR compensators and thyristor controlled series compensation," in *Proc. Bulk Power System Dynamics and Control IV—Restructuring*, 1998, pp. 647–650.

[30] J. F. Hauer and F. Vakili, "An oscillation trigger for power system monitoring," *IEEE Trans. Power Syst.*, vol. 5, no. 1, pp. 74–79, Feb. 1990.

[31] J. F. Hauer et al., "A Dynamic Information Manager for Networked Monitoring of Large Power Systems," EPRI, Palo Alto, CA, Final Rep. TR-112031, 1999.

[32] R. M. Murray. (2002, Jun.) Control in an information rich world. [Online]. Available: http://www.cds.caltech.edu/~murray/cdspanel/report/cdspanel-15aug02.pdf

[33] C. W. Taylor, "Improving grid behavior," *IEEE Spectrum*, vol. 36, no. 6, pp. 40–45, Jun. 1999.

[34] V. Venkatasubramanian, H. Schattler, and J. Zaborszky, "A taxonomy of the dynamics of large differential-algebraic systems," *Proc. IEEE*, vol. 83, no. 11, pp. 1530–1561, Nov. 1995.

[35] V. Vittal and A. A. Fouad, *Power System Transient Stability Analysis Using the Transient Energy Function Method.* Englewood Cliffs, NJ: Prentice-Hall, 1991.

[36] J. Li and V. Venkatasubramanian, "Study of unstable limit cycles in power system models," *IEEE Trans. Power Syst.*, submitted for publication.

[37] V. Venkatasubramanian and Y. Li, "Analysis of 1996 western American electric black-outs," presented at the Bulk Power System Dynamics and Control VI Conf., Cortina, Italy, 2004.

[38] Y. Li and V. Venkatasubramanian, "Coordination of transmission path transfers," *IEEE Trans. Power Syst.*, vol. 19, no. 3, pp. 1607–1615, Aug. 2004.

12.1

TRAVELING WAVES ON SINGLE-PHASE LOSSLESS LINES

We first consider a single-phase two-wire lossless transmission line. Figure 12.1 shows a line section of length Δx meters. If the line has a loop inductance L H/m and a line-to-line capacitance C F/m, then the line section has a series inductance L Δx H and shunt capacitance C Δx F, as shown. In Chapter 5, the direction of line position x was selected to be from the receiving end $(x = 0)$ to the sending end $(x = l)$; this selection was unimportant, since the variable x was subsequently eliminated when relating the steady-state sending-end quantities V_s and I_s to the receiving-end quantities V_R and I_R. Here, however, we are interested in voltages and current waveforms traveling along the line. Therefore, we select the direction of increasing x as being from the sending end $(x = 0)$ toward the receiving end $(x = l)$.

Writing a KVL and KCL equation for the circuit in Figure 12.1,

$$v(x + \Delta x, t) - v(x, t) = -L\Delta x \frac{\partial i(x, t)}{\partial t} \qquad (12.1.1)$$

$$i(x + \Delta x, t) - i(x, t) = -C\Delta x \frac{\partial v(x, t)}{\partial t} \qquad (12.1.2)$$

Dividing (12.1.1) and (12.1.2) by Δx and taking the limit as $\Delta x \to 0$, we obtain

$$\frac{\partial v(x, t)}{\partial x} = -L \frac{\partial i(x, t)}{\partial t} \qquad (12.1.3)$$

$$\frac{\partial i(x, t)}{\partial x} = -C \frac{\partial v(x, t)}{\partial t} \qquad (12.1.4)$$

We use partial derivatives here because $v(x, t)$ and $i(x, t)$ are differentiated with respect to both position x and time t. Also, the negative signs in (12.1.3) and (12.1.4) are due to the reference direction for x. For example, with a positive value of $\partial i/\partial t$ in Figure 12.1, $v(x, t)$ decreases as x increases.

Taking the Laplace transform of (12.1.3) and (12.1.4),

$$\frac{dV(x, s)}{dx} = -sLI(x, s) \qquad (12.1.5)$$

FIGURE 12.1

Single-phase two-wire lossless line section of length Δx

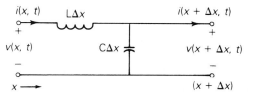

$$\frac{d\mathrm{I}(x,s)}{dx} = -s\mathrm{C}\mathrm{V}(x,s) \tag{12.1.6}$$

where zero initial conditions are assumed. $\mathrm{V}(x,s)$ and $\mathrm{I}(x,s)$ are the Laplace transforms of $v(x,t)$ and $i(x,t)$. Also, ordinary rather than partial derivatives are used since the derivatives are now with respect to only one variable, x.

Next we differentiate (12.1.5) with respect to x and use (12.1.6), in order to eliminate $\mathrm{I}(x,s)$:

$$\frac{d^2\mathrm{V}(x,s)}{dx^2} = -s\mathrm{L}\frac{d\mathrm{I}(x,s)}{dx} = s^2\mathrm{L}\mathrm{C}\mathrm{V}(x,s)$$

or

$$\frac{d^2\mathrm{V}(x,s)}{dx^2} - s^2\mathrm{L}\mathrm{C}\mathrm{V}(x,s) = 0 \tag{12.1.7}$$

Similarly, (12.1.6) can be differentiated in order to obtain

$$\frac{d^2\mathrm{I}(x,s)}{dx^2} - s^2\mathrm{L}\mathrm{C}\mathrm{I}(x,s) = 0 \tag{12.1.8}$$

Equation (12.1.7) is a linear, second-order homogeneous differential equation. By inspection, its solution is

$$\mathrm{V}(x,s) = \mathrm{V}^+(s)e^{-sx/v} + \mathrm{V}^-(s)e^{+sx/v} \tag{12.1.9}$$

where

$$v = \frac{1}{\sqrt{\mathrm{L}\mathrm{C}}} \quad \mathrm{m/s} \tag{12.1.10}$$

Similarly, the solution to (12.1.8) is

$$\mathrm{I}(x,s) = \mathrm{I}^+(s)e^{-sx/v} + \mathrm{I}^-(s)e^{+sx/v} \tag{12.1.11}$$

You can quickly verify that these solutions satisfy (12.1.7) and (12.1.8). The "constants" $\mathrm{V}^+(s), \mathrm{V}^-(s), \mathrm{I}^+(s)$, and $\mathrm{I}^-(s)$, which in general are functions of s but are independent of x, can be evaluated from the boundary conditions at the sending and receiving ends of the line. The superscripts $+$ and $-$ refer to waves traveling in the positive x and negative x directions, soon to be explained.

Taking the inverse Laplace transform of (12.1.9) and (12.1.11), and recalling the time shift properly, $\mathscr{L}[f(t-\tau)] = \mathrm{F}(s)e^{-s\tau}$, we obtain

$$v(x,t) = v^+\left(t - \frac{x}{v}\right) + v^-\left(t + \frac{x}{v}\right) \tag{12.1.12}$$

$$i(x,t) = i^+\left(t - \frac{x}{v}\right) + i^-\left(t + \frac{x}{v}\right) \tag{12.1.13}$$

where the functions $v^+(\ ), v^-(\ ), i^+(\ )$, and $i^-(\ )$, can be evaluated from the boundary conditions.

FIGURE 12.2

The function $f^+(u)$,

where $u = \left(t - \dfrac{x}{v}\right)$

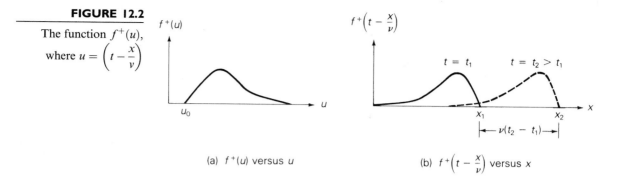

(a) $f^+(u)$ versus u (b) $f^+\left(t - \dfrac{x}{v}\right)$ versus x

We now show that $v^+(t - x/v)$ represents a voltage wave traveling in the positive x direction with velocity $v = 1/\sqrt{LC}$ m/s. Consider any wave $f^+(u)$, where $u = t - x/v$. Suppose that this wave begins at $u = u_0$, as shown in Figure 12.2(a). At time $t = t_1$, the wavefront is at $u_0 = (t_1 - x_1/v)$, or at $x_1 = v(t_1 - u_0)$. At a later time, t_2, the wavefront is at $u_0 = (t_2 - x_2/v)$ or at $x_2 = v(t_2 - u_0)$. As shown in Figure 12.2(b), the wavefront has moved in the positive x direction a distance $(x_2 - x_1) = v(t_2 - t_1)$ during time $(t_2 - t_1)$. The velocity is therefore $(x_2 - x_1)/(t_2 - t_1) = v$.

Similarly, $i^+(t - x/v)$ represents a current wave traveling in the positive x direction with velocity v. We call $v^+(t - x/v)$ and $i^+(t - x/v)$ the *forward* traveling voltage and current waves. It can be shown analogously that $v^-(t + x/v)$ and $i^-(t + x/v)$ travel in the negative x direction with velocity v. We call $v^-(t + x/v)$ and $i^-(t + x/v)$ the *backward* traveling voltage and current waves.

Recall from (5.4.16) that for a lossless line $f\lambda = 1/\sqrt{LC}$. It is now evident that the term $1/\sqrt{LC}$ in this equation is v, the velocity of propagation of voltage and current waves along the lossless line. Also, recall from Chapter 4 that L is proportional to μ and C is proportional to ε. For overhead lines, $v = 1/\sqrt{LC}$ is approximately equal to $1/\sqrt{\mu\varepsilon} = 1/\sqrt{\mu_0\varepsilon_0} = 3 \times 10^8$ m/s, the speed of light in free space. For cables, the relative permitivity $\varepsilon/\varepsilon_0$ may be 3 to 5 or even higher, resulting in a value of v lower than that for overhead lines.

We next evaluate the terms $I^+(s)$ and $I^-(s)$. Using (12.1.9) and (12.1.10) in (12.1.6),

$$\frac{s}{v}[-I^+(s)e^{-sx/v} + I^-(s)e^{+sx/v}] = -sC[V^+(s)e^{-sx/v} + V^-(s)e^{+sx/v}]$$

Equating the coefficients of $e^{-sx/v}$ on both sides of this equation,

$$I^+(s) = (vC)V^+(s) = \frac{V^+(s)}{\sqrt{\dfrac{L}{C}}} = \frac{V^+(s)}{Z_c} \qquad (12.1.14)$$

where

$$Z_c = \sqrt{\frac{L}{C}} \quad \Omega \tag{12.1.15}$$

Similarly, equating the coefficients of $e^{+sx/v}$,

$$I^-(s) = \frac{-V^-(s)}{Z_c} \tag{12.1.16}$$

Thus, we can rewrite (12.1.11) and (12.1.13) as

$$I(x,s) = \frac{1}{Z_c}[V^+(s)e^{-sx/v} - V^-(s)e^{+sx/v}] \tag{12.1.17}$$

$$i(x,t) = \frac{1}{Z_c}\left[v^+\left(t - \frac{x}{v}\right) - v^-\left(t + \frac{x}{v}\right)\right] \tag{12.1.18}$$

Recall from (5.4.3) that $Z_c = \sqrt{L/C}$ is the characteristic impedance (also called surge impedance) of a lossless line.

12.2

BOUNDARY CONDITIONS FOR SINGLE-PHASE LOSSLESS LINES

Figure 12.3 shows a single-phase two-wire lossless line terminated by an impedance $Z_R(s)$ at the receiving end and a source with Thévenin voltage $E_G(s)$ and with Thévenin impedance $Z_G(s)$ at the sending end. $V(x,s)$ and $I(x,s)$ are the Laplace transforms of the voltage and current at position x. The line has length l, surge impedance $Z_c = \sqrt{L/C}$, and velocity $v = 1/\sqrt{LC}$. We assume that the line is initially unenergized.

From Figure 12.3, the boundary condition at the receiving end is

$$V(l,s) = Z_R(s)I(l,s) \tag{12.2.1}$$

Using (12.1.9) and (12.1.17) in (12.2.1),

$$V^+(s)e^{-sl/v} + V^-(s)e^{+sl/v} = \frac{Z_R(s)}{Z_c}[V^+(s)e^{-sl/v} - V^-(s)e^{+sl/v}]$$

Solving for $V^-(l,s)$

FIGURE 12.3

Single-phase two-wire lossless line with source and load terminations

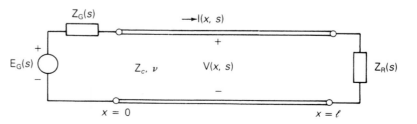

$$\mathbf{V}^-(l, s) = \Gamma_R(s)\mathbf{V}^+(s)e^{-2s\tau} \tag{12.2.2}$$

where

$$\Gamma_R(s) = \frac{\dfrac{Z_R(s)}{Z_c} - 1}{\dfrac{Z_R(s)}{Z_c} + 1} \quad \text{per unit} \tag{12.2.3}$$

$$\tau = \frac{l}{v} \quad \text{seconds} \tag{12.2.4}$$

$\Gamma_R(s)$ is called the *receiving-end voltage reflection coefficient*. Also, τ, called the *transit time* of the line, is the time it takes a wave to travel the length of the line.

Using (12.2.2) in (12.1.9) and (12.1.17),

$$\mathbf{V}(x, s) = \mathbf{V}^+(s)[e^{-sx/v} + \Gamma_R(s)e^{s[(x/v)-2\tau]}] \tag{12.2.5}$$

$$\mathbf{I}(x, s) = \frac{\mathbf{V}^+(s)}{Z_c}[e^{-sx/v} - \Gamma_R(s)e^{s[(x/v)-2\tau]}] \tag{12.2.6}$$

From Figure 12.3 the boundary condition at the sending end is

$$\mathbf{V}(0, s) = \mathbf{E}_G(s) - Z_G(s)\mathbf{I}(0, s) \tag{12.2.7}$$

Using (12.2.5) and (12.2.6) in (12.2.7),

$$\mathbf{V}^+(s)[1 + \Gamma_R(s)e^{-2s\tau}] = \mathbf{E}_G(s) - \left[\frac{Z_G(s)}{Z_c}\right]\mathbf{V}^+(s)[1 - \Gamma_R(s)e^{-2s\tau}]$$

Solving for $\mathbf{V}^+(s)$,

$$\mathbf{V}^+(s)\left\{\left[\frac{Z_G(s)}{Z_c} + 1\right] - \Gamma_R(s)e^{-2s\tau}\left[\frac{Z_G(s)}{Z_c} - 1\right]\right\} = \mathbf{E}_G(s)$$

$$\mathbf{V}^+(s)\left[\frac{Z_G(s)}{Z_c} + 1\right]\{1 - \Gamma_R(s)\Gamma_S(s)e^{-2s\tau}\} = \mathbf{E}_G(s)$$

or

$$\mathbf{V}^+(s) = \mathbf{E}_G(s)\left[\frac{Z_c}{Z_G(s) + Z_c}\right]\left[\frac{1}{1 - \Gamma_R(s)\Gamma_S(s)e^{-2s\tau}}\right] \tag{12.2.8}$$

where

$$\Gamma_S(s) = \frac{\dfrac{Z_G(s)}{Z_c} - 1}{\dfrac{Z_G(s)}{Z_c} + 1} \tag{12.2.9}$$

$\Gamma_S(s)$ is called the *sending-end voltage reflection coefficient*. Using (12.2.9) in (12.2.5) and (12.2.6), the complete solution is

$$V(x,s) = E_G(s) \left[\frac{Z_c}{Z_G(s) + Z_c} \right] \left[\frac{e^{-sx/v} + \Gamma_R(s)e^{s[(x/v)-2\tau]}}{1 - \Gamma_R(s)\Gamma_S(s)e^{-2s\tau}} \right] \qquad (12.2.10)$$

$$I(x,s) = \left[\frac{E_G(s)}{Z_G(s) + Z_c} \right] \left[\frac{e^{-sx/v} - \Gamma_R(s)e^{s[(x/v)-2\tau]}}{1 - \Gamma_R(s)\Gamma_S(s)e^{-2s\tau}} \right] \qquad (12.2.11)$$

where

$$\Gamma_R(s) = \frac{\dfrac{Z_R(s)}{Z_c} - 1}{\dfrac{Z_R(s)}{Z_c} + 1} \qquad \text{per unit}$$

$$\Gamma_S(s) = \frac{\dfrac{Z_G(s)}{Z_c} - 1}{\dfrac{Z_G(s)}{Z_c} + 1} \qquad \text{per unit} \qquad (12.2.12)$$

$$Z_c = \sqrt{\frac{L}{C}} \ \ \Omega \qquad v = \frac{1}{\sqrt{LC}} \ \ \text{m/s} \qquad \tau = \frac{l}{v} \ \ \text{s} \qquad (12.2.13)$$

The following four examples illustrate this general solution. All four examples refer to the line shown in Figure 12.3, which has length l, velocity v, characteristic impedance Z_c, and is initially unenergized.

EXAMPLE 12.1   **Single-phase lossless-line transients: step-voltage source at sending end, matched load at receiving end**

Let $Z_R = Z_c$ and $Z_G = 0$. The source voltage is a step, $e_G(t) = Eu_{-1}(t)$. (a) Determine $v(x,t)$ and $i(x,t)$. Plot the voltage and current versus time t at the center of the line and at the receiving end.

SOLUTION

a. From (12.2.12) with $Z_R = Z_c$ and $Z_G = 0$,

$$\Gamma_R(s) = \frac{1-1}{1+1} = 0 \qquad \Gamma_S(s) = \frac{0-1}{0+1} = -1$$

The Laplace transform of the source voltage is $E_G(s) = E/s$. Then, from (12.2.10) and (12.2.11),

$$V(x,s) = \left(\frac{E}{s}\right)(1)(e^{-sx/v}) = \frac{Ee^{-sx/v}}{s}$$

$$I(x,s) = \frac{(E/Z_c)}{s}e^{-sx/v}$$

FIGURE 12.4

Voltage and current
waveforms for Example
12.1

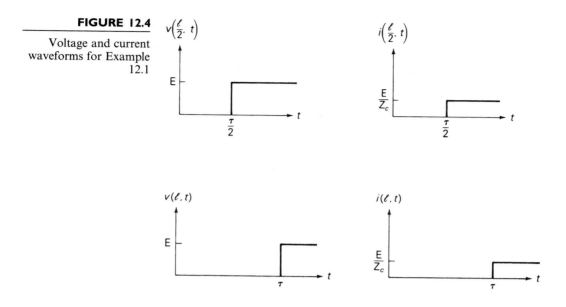

Taking the inverse Laplace transform,

$$v(x, t) = E u_{-1}\left(t - \frac{x}{v}\right)$$

$$i(x, t) = \frac{E}{Z_c} u_{-1}\left(t - \frac{x}{v}\right)$$

b. At the center of the line, where $x = l/2$,

$$v\left(\frac{l}{2}, t\right) = E u_{-1}\left(t - \frac{\tau}{2}\right) \qquad i\left(\frac{l}{2}, t\right) = \frac{E}{Z_c} u_{-1}\left(t - \frac{\tau}{2}\right)$$

At the receiving end, where $x = l$,

$$v(l, t) = E u_{-1}(t - \tau) \qquad i(l, t) = \frac{E}{Z_c} u_{-1}(t - \tau)$$

These waves, plotted in Figure 12.4, can be explained as follows. At $t = 0$
the ideal step voltage of E volts, applied to the sending end, encounters Z_c,
the characteristic impedance of the line. Therefore, a forward traveling
step voltage wave of E volts is initiated at the sending end. Also, since the
ratio of the forward traveling voltage to current is Z_c, a forward traveling
step current wave of (E/Z_c) amperes is initiated. These waves travel in
the positive x direction, arriving at the center of the line at $t = \tau/2$, and at
the end of the line at $t = \tau$. The receiving-end load is *matched* to the line;
that is, $Z_R = Z_c$. For a matched load, $\Gamma_R = 0$, and therefore no backward
traveling waves are initiated. In steady-state, the line with matched load is
energized at E volts with current E/Z_c amperes. ∎

EXAMPLE 12.2   Single-phase lossless-line transients: step-voltage source matched at sending end, open receiving end

The receiving end is open. The source voltage at the sending end is a step $e_G(t) = Eu_{-1}(t)$, with $Z_G(s) = Z_c$. (a) Determine $v(x,t)$ and $i(x,t)$. (b) Plot the voltage and current versus time t at the center of the line.

SOLUTION

a. From (12.2.12),

$$\Gamma_R(s) = \lim_{Z_R \to \infty} \frac{\dfrac{Z_R}{Z_c} - 1}{\dfrac{Z_R}{Z_c} + 1} = 1 \qquad \Gamma_S(s) = \frac{1-1}{1+1} = 0$$

The Laplace transform of the source voltage is $E_G(s) = E/s$. Then, from (12.2.10) and (12.2.11),

$$V(x,s) = \frac{E}{s}\left(\frac{1}{2}\right)\left[e^{-sx/v} + e^{s[(x/v)-2\tau]}\right]$$

$$I(x,s) = \frac{E}{s}\left(\frac{1}{2Z_c}\right)\left[e^{-sx/v} - e^{s[(x/v)-2\tau]}\right]$$

Taking the inverse Laplace transform,

$$v(x,t) = \frac{E}{2}u_{-1}\left(t - \frac{x}{v}\right) + \frac{E}{2}u_{-1}\left(t + \frac{x}{v} - 2\tau\right)$$

$$i(x,t) = \frac{E}{2Z_c}u_{-1}\left(t - \frac{x}{v}\right) - \frac{E}{2Z_c}u_{-1}\left(t + \frac{x}{v} - 2\tau\right)$$

b. At the center of the line, where $x = l/2$,

$$v\left(\frac{l}{2},t\right) = \frac{E}{2}u_{-1}\left(t - \frac{\tau}{2}\right) + \frac{E}{2}u_{-1}\left(t - \frac{3\tau}{2}\right)$$

$$i\left(\frac{l}{2},t\right) = \frac{E}{2Z_c}u_{-1}\left(t - \frac{\tau}{2}\right) - \frac{E}{2Z_c}u_{-1}\left(t - \frac{3\tau}{2}\right)$$

These waves are plotted in Figure 12.5. At $t = 0$ the step voltage source of E volts encounters the source impedance $Z_G = Z_c$ in series with the characteristic impedance of the line, Z_c. Using voltage division, the sending-end voltage at $t = 0$ is E/2. Therefore, a forward traveling step voltage wave of E/2 volts and a forward traveling step current wave of $E/(2Z_c)$ amperes are initiated at the sending end. These waves arrive at the center of the line at $t = \tau/2$. Also, with $\Gamma_R = 1$, the backward traveling voltage wave equals the forward traveling voltage wave, and the backward traveling current wave is the negative of the forward traveling current wave. These backward traveling waves, which are initiated at the receiving end at $t = \tau$ when the forward traveling waves

FIGURE 12.5

Voltage and current
waveforms for Example
12.2

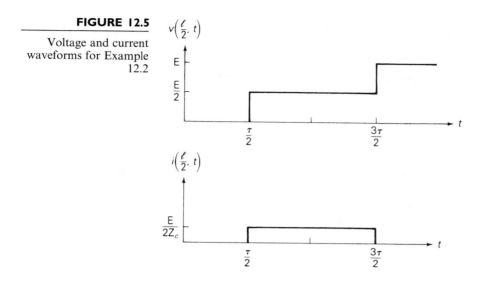

arrive there, arrive at the center of the line at $t = 3\tau/2$ and are superimposed on the forward traveling waves. No additional forward or backward traveling waves are initiated because the source impedance is matched to the line; that is, $\Gamma_S(s) = 0$. In steady-state the line, which is open at the receiving end, is energized at E volts with zero current. ∎

EXAMPLE 12.3 Single-phase lossless-line transients: step-voltage source matched at sending end, capacitive load at receiving end

The receiving end is terminated by a capacitor with C_R farads, which is initially unenergized. The source voltage at the sending end is a unit step $e_G(t) = Eu_{-1}(t)$, with $Z_G = Z_c$. Determine and plot $v(x, t)$ versus time t at the sending end of the line.

SOLUTION   From (12.2.12) with $Z_R = \dfrac{1}{sC_R}$ and $Z_G = Z_c$,

$$\Gamma_R(s) = \frac{\dfrac{1}{sC_R Z_c} - 1}{\dfrac{1}{sC_R Z_c} + 1} = \frac{-s + \dfrac{1}{Z_c C_R}}{s + \dfrac{1}{Z_c C_R}}$$

$$\Gamma_S(s) = \frac{1 - 1}{1 + 1} = 0$$

Then, from (12.2.10), with $E_G(s) = E/s$,

$$V(x,s) = \frac{E}{s}\left(\frac{1}{2}\right)\left[e^{-sx/v} + \left(\frac{-s + \dfrac{1}{Z_cC_R}}{s + \dfrac{1}{Z_cC_R}}\right)e^{s[(x/v)-2\tau]}\right]$$

$$= \frac{E}{2}\left[\frac{e^{-sx/v}}{s} + \frac{1}{s}\left(\frac{-s + \dfrac{1}{Z_cC_R}}{s + \dfrac{1}{Z_cC_R}}\right)e^{s[(x/v)-2\tau]}\right]$$

Using partial fraction expansion of the second term above,

$$V(x,s) = \frac{E}{2}\left[\frac{e^{-sx/v}}{s} + \left(\frac{1}{s} - \frac{2}{s + \dfrac{1}{Z_cC_R}}\right)e^{s[(x/v)-2\tau]}\right]$$

The inverse Laplace transform is

$$v(x,t) = \frac{E}{2}u_{-1}\left(t - \frac{x}{v}\right) + \frac{E}{2}[1 - 2e^{(-1/Z_cC_R)(t+x/v-2\tau)}]u_{-1}\left(t + \frac{x}{v} - 2\tau\right)$$

At the sending end, where $x = 0$,

$$v(0,t) = \frac{E}{2}u_{-1}(t) + \frac{E}{2}[1 - 2e^{(-1/Z_cC_R)(t-2\tau)}]u_{-1}(t - 2\tau)$$

$v(0,t)$ is plotted in Figure 12.6. As in Example 12.2, a forward traveling step voltage wave of E/2 volts is initiated at the sending end at $t = 0$. At $t = \tau$, when the forward traveling wave arrives at the receiving end, a backward traveling wave is initiated. The backward traveling voltage wave, an exponential with initial value $-E/2$, steady-state value $+E/2$, and time constant Z_cC_R, arrives at the sending end at $t = 2\tau$, where it is superimposed on the forward traveling wave. No additional waves are initiated, since the source impedance is matched to the line. In steady-state, the line and the capacitor at the receiving end are energized at E volts with zero current.

FIGURE 12.6

Voltage waveform for Example 12.3

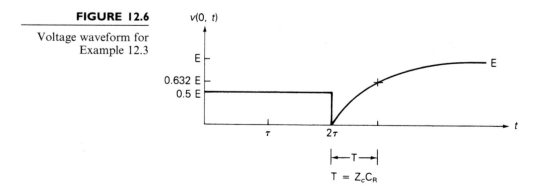

The capacitor at the receiving end can also be viewed as a short circuit at the instant $t = \tau$, when the forward traveling wave arrives at the receiving end. For a short circuit at the receiving end, $\Gamma_R = -1$, and therefore the backward traveling voltage wavefront is $-E/2$, the negative of the forward traveling wave. However, in steady-state the capacitor is an open circuit, for which $\Gamma_R = +1$, and the steady-state backward traveling voltage wave equals the forward traveling voltage wave.   ∎

EXAMPLE 12.4   **Single-phase lossless-line transients: step-voltage source with unmatched source resistance at sending end, unmatched resistive load at receiving end**

At the receiving end, $Z_R = Z_c/3$. At the sending end, $e_G(t) = E u_{-1}(t)$ and $Z_G = 2Z_c$. Determine and plot the voltage versus time at the center of the line.

SOLUTION   From (12.2.12),

$$\Gamma_R = \frac{\frac{1}{3} - 1}{\frac{1}{3} + 1} = -\frac{1}{2} \qquad \Gamma_S = \frac{2 - 1}{2 + 1} = \frac{1}{3}$$

From (12.2.10), with $E_G(s) = E/s$,

$$V(x,s) = \frac{E}{s}\left(\frac{1}{3}\right) \frac{[e^{-sx/v} - \frac{1}{2}e^{s[(x/v)-2\tau]}]}{1 + (\frac{1}{6}e^{-2s\tau})}$$

The preceding equation can be rewritten using the following geometric series:

$$\frac{1}{1 + y} = 1 - y + y^2 - y^3 + y^4 - \cdots$$

with $y = \frac{1}{6}e^{-2s\tau}$,

$$V(x,s) = \frac{E}{3s}\left[e^{-sx/v} - \frac{1}{2}e^{s[(x/v)-2\tau]}\right]$$

$$\times \left[1 - \frac{1}{6}e^{-2s\tau} + \frac{1}{36}e^{-4s\tau} - \frac{1}{216}e^{-6s\tau} + \cdots\right]$$

Multiplying the terms within the brackets,

$$V(x,s) = \frac{E}{3s}\left[e^{-sx/v} - \frac{1}{2}e^{s[(x/v)-2\tau]} - \frac{1}{6}e^{-s[(x/v)+2\tau]} + \frac{1}{12}e^{s[(x/v)-4\tau]}\right.$$

$$\left. + \frac{1}{36}e^{-s[(x/v)+4\tau]} - \frac{1}{72}e^{s[(x/v)-6\tau]} + \cdots\right]$$

Taking the inverse Laplace transform,

$$v(x,t) = \frac{E}{3}\left[u_{-1}\left(t - \frac{x}{v}\right) - \frac{1}{2}u_{-1}\left(t + \frac{x}{v} - 2\tau\right) - \frac{1}{6}u_{-1}\left(t - \frac{x}{v} - 2\tau\right)\right.$$

$$+ \frac{1}{12}u_{-1}\left(t + \frac{x}{v} - 4\tau\right) + \frac{1}{36}u_{-1}\left(t - \frac{x}{v} - 4\tau\right)$$

$$\left. - \frac{1}{72}u_{-1}\left(t + \frac{x}{v} - 6\tau\right)\cdots\right]$$

At the center of the line, where $x = l/2$,

$$v\left(\frac{l}{2}, t\right) = \frac{E}{3}\left[u_{-1}\left(t - \frac{\tau}{2}\right) - \frac{1}{2}u_{-1}\left(t - \frac{3\tau}{2}\right) - \frac{1}{6}u_{-1}\left(t - \frac{5\tau}{2}\right)\right.$$

$$\left. + \frac{1}{12}u_{-1}\left(t - \frac{7\tau}{2}\right) + \frac{1}{36}u_{-1}\left(t - \frac{9\tau}{2}\right) - \frac{1}{72}u_{-1}\left(t - \frac{11\tau}{2}\right)\cdots\right]$$

$v(l/2, t)$ is plotted in Figure 12.7(a). Since neither the source nor the load is matched to the line, the voltage at any point along the line consists of an infinite series of forward and backward traveling waves. At the center of the line, the first forward traveling wave arrives at $t = \tau/2$; then a backward traveling wave arrives at $3\tau/2$, another forward traveling wave arrives at $5\tau/2$, another backward traveling wave at $7\tau/2$, and so on.

The steady-state voltage can be evaluated from the final value theorem. That is,

$$v_{ss}(x) = \lim_{t \to \infty} v(x, t) = \lim_{s \to 0} sV(x, s)$$

$$= \lim_{s \to 0}\left\{ s\left(\frac{E}{s}\right)\left(\frac{1}{3}\right)\frac{[e^{-sx/v} - \frac{1}{2}e^{s[(x/v) - 2\tau]}]}{1 + \frac{1}{6}e^{-2s\tau}}\right\}$$

$$= E\left(\frac{1}{3}\right)\left(\frac{1 - \frac{1}{2}}{1 + \frac{1}{6}}\right) = \frac{E}{7}$$

The steady-state solution can also be evaluated from the circuit in Figure 12.7(b). Since there is no steady-state voltage drop across the lossless line when a dc source is applied, the line can be eliminated, leaving only the

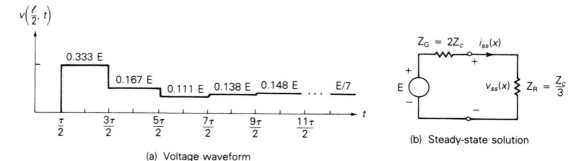

(a) Voltage waveform

(b) Steady-state solution

FIGURE 12.7   Example 12.4

source and load. The steady-state voltage is then, by voltage division,

$$v_{ss}(x) = E\left(\frac{Z_R}{Z_R + Z_G}\right) = E\left(\frac{\frac{1}{3}}{\frac{1}{3} + 2}\right) = \frac{E}{7}$$ ∎

12.3

BEWLEY LATTICE DIAGRAM

A lattice diagram developed by L. V. Bewley [2] conveniently organizes the reflections that occur during transmission-line transients. For the Bewley lattice diagram, shown in Figure 12.8, the vertical scale represents time and is scaled in units of τ, the transient time of the line. The horizontal scale represents line position x, and the diagonal lines represent traveling waves. Each reflection is determined by multiplying the incident wave arriving at an end by the reflection coefficient at that end. The voltage $v(x, t)$ at any point x and t on the diagram is determined by adding all the terms directly above that point.

FIGURE 12.8

Bewley lattice diagram
for Example 12.5

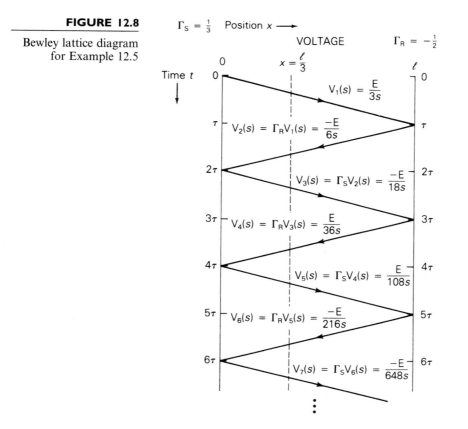

EXAMPLE 12.5   **Lattice diagram: single-phase lossless line**

For the line and terminations given in Example 12.4, draw the lattice diagram and plot $v(l/3, t)$ versus time t.

SOLUTION   The lattice diagram is shown in Figure 12.8. At $t = 0$, the source voltage encounters the source impedance and the line characteristic impedance, and the first forward traveling wave is determined by voltage division:

$$V_1(s) = E_G(s)\left[\frac{Z_c}{Z_c + Z_G}\right] = \frac{E}{s}\left[\frac{1}{1+2}\right] = \frac{E}{3s}$$

which is a step with magnitude $(E/3)$ volts. The next traveling wave, a backward one, is $V_2(s) = \Gamma_R(s)V_1(s) = (-\frac{1}{2})V_1(s) = -E/(6s)$, and the next wave, a forward one, is $V_3(s) = \Gamma_s(s)V_2(s) = (\frac{1}{3})V_2(s) = -E/(18s)$. Subsequent waves are calculated in a similar manner.

The voltage at $x = l/3$ is determined by drawing a vertical line at $x = l/3$ on the lattice diagram, shown dashed in Figure 12.8. Starting at the top of the dashed line, where $t = 0$, and moving down, each voltage wave is added at the time it intersects the dashed line. The first wave v_1 arrives at $t = \tau/3$, the second v_2 arrives at $5\tau/3$, v_3 at $7\tau/3$, and so on. $v(l/3, t)$ is plotted in Figure 12.9.

FIGURE 12.9

Voltage waveform for Example 12.5

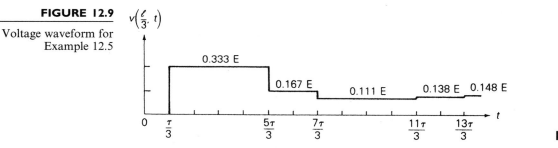

Figure 12.10 shows a forward traveling voltage wave V_A^+ arriving at the junction of two lossless lines A and B with characteristic impedances Z_A and Z_B, respectively. This could be, for example, the junction of an overhead line and a cable. When V_A^+ arrives at the junction, both a reflection V_A^- on line A and a refraction V_B^+ on line B will occur. Writing a KVL and KCL equation at the junction,

FIGURE 12.10

Junction of two single-phase lossless lines

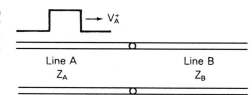

$$V_A^+ + V_A^- = V_B^+ \tag{12.3.1}$$

$$I_A^+ + I_A^- = I_B^+ \tag{12.3.2}$$

Recall that $I_A^+ = V_A^+/Z_A$, $I_A^- = -V_A^-/Z_A$, and $I_B^+ = V_B^+/Z_B$. Using these relations in (12.3.2),

$$\frac{V_A^+}{Z_A} - \frac{V_A^-}{Z_A} = \frac{V_B^+}{Z_B} \tag{12.3.3}$$

Solving (12.3.1) and (12.3.3) for V_A^- and V_B^+ in terms of V_A^+ yields

$$V_A^- = \Gamma_{AA} V_A^+ \tag{12.3.4}$$

where

$$\Gamma_{AA} = \frac{\dfrac{Z_B}{Z_A} - 1}{\dfrac{Z_B}{Z_A} + 1} \tag{12.3.5}$$

and

$$V_B^+ = \Gamma_{BA} V_A \tag{12.3.6}$$

where

$$\Gamma_{BA} = \frac{2\left(\dfrac{Z_B}{Z_A}\right)}{\dfrac{Z_B}{Z_A} + 1} \tag{12.3.7}$$

Note that Γ_{AA}, given by (12.3.5), is similar to Γ_R, given by (12.2.12), except that Z_B replaces Z_R. Thus, for waves arriving at the junction from line A, the "load" at the receiving end of line A is the characteristic impedance of line B.

EXAMPLE 12.6   **Lattice diagram: overhead line connected to a cable, single-phase lossless lines**

As shown in Figure 12.10, a single-phase lossless overhead line with $Z_A = 400\ \Omega$, $v_A = 3 \times 10^8$ m/s, and $l_A = 30$ km is connected to a single-phase lossless cable with $Z_B = 100\ \Omega$, $v_B = 2 \times 10^8$ m/s, and $l_B = 20$ km. At the sending end of line A, $e_g(t) = Eu_{-1}(t)$ and $Z_G = Z_A$. At the receiving end of line B, $Z_R = 2Z_B = 200\ \Omega$. Draw the lattice diagram for $0 \le t \le 0.6$ ms and plot the voltage at the junction versus time. The line and cable are initially unenergized.

SOLUTION   From (12.2.13),

$$\tau_A = \frac{30 \times 10^3}{3 \times 10^8} = 0.1 \times 10^{-3}\ \text{s} \qquad \tau_B = \frac{20 \times 10^3}{2 \times 10^8} = 0.1 \times 10^{-3}\ \text{s}$$

From (12.2.12), with $Z_G = Z_A$ and $Z_R = 2Z_B$,

$$\Gamma_S = \frac{1-1}{1+1} = 0 \qquad \Gamma_R = \frac{2-1}{2+1} = \frac{1}{3}$$

From (12.3.5) and (12.3.6), the reflection and refraction coefficients for waves arriving at the junction from line A are

$$\left. \Gamma_{AA} = \frac{\dfrac{100}{400} - 1}{\dfrac{100}{400} + 1} = \frac{-3}{5} \qquad \Gamma_{BA} = \frac{2\dfrac{100}{400}}{\dfrac{100}{400} + 1} = \frac{2}{5} \right\} \text{ from line A}$$

Reversing A and B, the reflection and refraction coefficients for waves returning to the junction from line B are

$$\left. \Gamma_{BB} = \frac{\dfrac{400}{100} - 1}{\dfrac{400}{100} + 1} = \frac{3}{5} \qquad \Gamma_{AB} = \frac{2\dfrac{400}{100}}{\dfrac{400}{100} + 1} = \frac{8}{5} \right\} \text{ from line B}$$

The lattice diagram is shown in Figure 12.11. Using voltage division, the first forward traveling voltage wave is

FIGURE 12.11

Lattice diagram for
Example 12.6

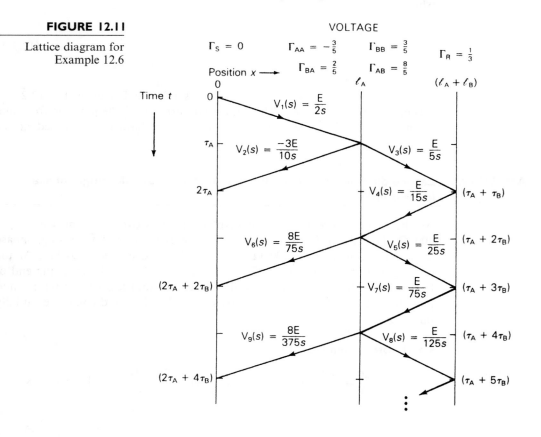

VOLTAGE

FIGURE 12.12

Junction voltage for
Example 12.6

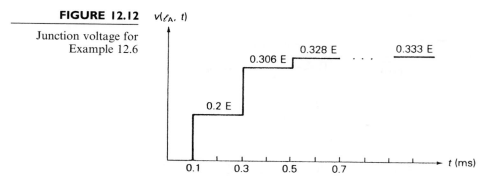

$$V_1(s) = E_G(s)\left(\frac{Z_A}{Z_A + Z_G}\right) = \frac{E}{s}\left(\frac{1}{2}\right) = \frac{E}{2s}$$

When v_1 arrives at the junction, a reflected wave v_2 and refracted wave v_3 are initiated. Using the reflection and refraction coefficients for line A,

$$V_2(s) = \Gamma_{AA}V_1(s) = \left(\frac{-3}{5}\right)\left(\frac{E}{2s}\right) = \frac{-3E}{10s}$$

$$V_3(s) = \Gamma_{BA}V_1(s) = \left(\frac{2}{5}\right)\left(\frac{E}{2s}\right) = \frac{E}{5s}$$

When v_2 arrives at the receiving end of line B, a reflected wave $V_4(s) = \Gamma_R V_3(s) = \frac{1}{3}(E/5s) = (E/15s)$ is initiated. When v_4 arrives at the junction, reflected wave v_5 and refracted wave v_6 are initiated. Using the reflection and refraction coefficients for line B,

$$V_5(s) = \Gamma_{BB}V_4(s) = \left(\frac{3}{5}\right)\left(\frac{E}{15s}\right) = \frac{E}{25s}$$

$$V_6(s) = \Gamma_{AB}V_4(s) = \left(\frac{8}{5}\right)\left(\frac{E}{15s}\right) = \frac{8E}{75s}$$

Subsequent reflections and refractions are calculated in a similar manner.

The voltage at the junction is determined by starting at $x = l_A$ at the top of the lattice diagram, where $t = 0$. Then, moving down the lattice diagram, voltage waves either just to the left or just to the right of the junction are added when they occur. For example, looking just to the right of the junction at $x = l_A^+$, the voltage wave v_3, a step of magnitude E/5 volts occurs at $t = \tau_A$. Then at $t = (\tau_A + 2\tau_B)$, two waves v_4 and v_5, which are steps of magnitude E/15 and E/25, are added to v_3. $v(l_A, t)$ is plotted in Figure 12.12.

The steady-state voltage is determined by removing the lossless lines and calculating the steady-state voltage across the receiving-end load:

$$v_{ss}(x) = E\left(\frac{Z_R}{Z_R + Z_G}\right) = E\left(\frac{200}{200 + 400}\right) = \frac{E}{3}$$ ∎

FIGURE 12.13

Junction of lossless lines A, B, C, D, and so on

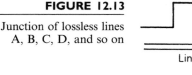
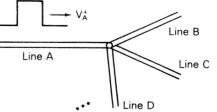

The preceding analysis can be extended to the junction of more than two lossless lines, as shown in Figure 12.13. Writing a KVL and KCL equation at the junction for a voltage V_A^+ arriving at the junction from line A,

$$V_A^+ + V_A^- = V_B^+ = V_C^+ = V_D^+ = \cdots \tag{12.3.8}$$

$$I_A^+ + I_A^- = I_B^+ + I_C^+ + I_D^+ + \cdots \tag{12.3.9}$$

Using $I_A^+ = V_A^+/Z_A$, $I_A^- = -V_A^-/Z_A$, $I_B^+ = V_B^+/Z_B$, and so on in (12.3.9),

$$\frac{V_A^+}{Z_A} - \frac{V_A^-}{Z_A} = \frac{V_B^+}{Z_B} + \frac{V_C^+}{Z_C} + \frac{V_D^+}{Z_D} + \cdots \tag{12.3.10}$$

Equations (12.3.8) and (12.3.10) can be solved for V_A^+, V_B^+, V_C^+, V_D^+, and so on in terms of V_A^+. (See Problem 12.14.)

12.4

DISCRETE-TIME MODELS OF SINGLE-PHASE LOSSLESS LINES AND LUMPED RLC ELEMENTS

Our objective in this section is to develop discrete-time models of single-phase lossless lines and of lumped RLC elements suitable for computer calculation of transmission-line transients at discrete-time intervals $t = \Delta t$, $2\Delta t$, $3\Delta t$, and so on. The discrete-time models are presented as equivalent circuits consisting of lumped resistors and current sources. The current sources in the models represent the past history of the circuit—that is, the history at times $t - \Delta t$, $t - 2\Delta t$, and so on. After interconnecting the equivalent circuits of all the components in any given circuit, nodal equations can then be written for each discrete time. Discrete-time models, first developed by L. Bergeron [3], are presented first.

SINGLE-PHASE LOSSLESS LINE

From the general solution of a single-phase lossless line, given by (12.1.12) and (12.1.18), we obtain

FIGURE 12.14

Single-phase two-wire
lossless line

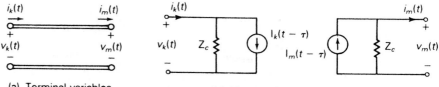

(a) Terminal variables   (b) Discrete-time equivalent circuit

$$v(x, t) + Z_c i(x, t) = 2v^+\left(t - \frac{x}{v}\right) \tag{12.4.1}$$

$$v(x, t) - Z_c i(x, t) = 2v^-\left(t + \frac{x}{v}\right) \tag{12.4.2}$$

In (12.4.1), the left side $(v + Z_c i)$ remains constant when the argument $(t - x/v)$ is constant. Therefore, to a fictitious observer traveling at velocity v in the positive x direction along the line, $(v + Z_c i)$ remains constant. If τ is the transit time from terminal k to terminal m of the line, the value of $(v + Z_c i)$ observed at time $(t - \tau)$ at terminal k must equal the value at time t at terminal m. That is,

$$v_k(t - \tau) + Z_c i_k(t - \tau) = v_m(t) + Z_c i_m(t) \tag{12.4.3}$$

where k and m denote terminals k and m, as shown in Figure 12.14(a).

Similarly, $(v - Z_c i)$ in (12.4.2) remains constant when $(t + x/v)$ is constant. To a fictitious observer traveling at velocity v in the negative x direction, $(v - Z_c i)$ remains constant. Therefore, the value of $(v - Z_c i)$ at time $(t - \tau)$ at terminal m must equal the value at time t at terminal k. That is,

$$v_m(t - \tau) - Z_c i_m(t - \tau) = v_k(t) - Z_c i_k(t) \tag{12.4.4}$$

Equation (12.4.3) is rewritten as

$$i_m(t) = I_m(t - \tau) - \frac{1}{Z_c} v_m(t) \tag{12.4.5}$$

where

$$I_m(t - \tau) = i_k(t - \tau) + \frac{1}{Z_c} v_k(t - \tau) \tag{12.4.6}$$

Similarly, (12.4.4) is rewritten as

$$i_k(t) = I_k(t - \tau) + \frac{1}{Z_c} v_k(t) \tag{12.4.7}$$

where

$$I_k(t - \tau) = i_m(t - \tau) - \frac{1}{Z_c} v_m(t - \tau) \tag{12.4.8}$$

Also, using (12.4.7) in (12.4.6),

$$I_m(t - \tau) = I_k(t - 2\tau) + \frac{2}{Z_c} v_k(t - \tau) \tag{12.4.9}$$

and using (12.4.5) in (12.4.8),

$$I_k(t - \tau) = I_m(t - 2\tau) - \frac{2}{Z_c} v_m(t - \tau) \tag{12.4.10}$$

Equations (12.4.5) and (12.4.7) are represented by the circuit shown in Figure 12.14(b). The current sources $I_m(t - \tau)$ and $I_k(t - \tau)$ shown in this figure, which are given by (12.4.9) and (12.4.10), represent the past history of the transmission line.

Note that in Figure 12.14(b) terminals k and m are not directly connected. The conditions at one terminal are "felt" indirectly at the other terminal after a delay of τ seconds.

LUMPED INDUCTANCE

As shown in Figure 12.15(a) for a constant lumped inductance L,

$$v(t) = L \frac{di(t)}{dt} \tag{12.4.11}$$

Integrating this equation from time $(t - \Delta t)$ to t,

$$\int_{t-\Delta t}^{t} di(t) = \frac{1}{L} \int_{t-\Delta t}^{t} v(t)\, dt \tag{12.4.12}$$

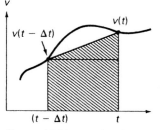

Trapezoidal Integration Rule

FIGURE 12.15

Lumped inductance

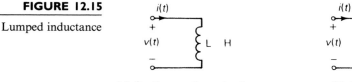

(a) Continuous time circuit   (b) Discrete-time circuit

Using the trapezoidal rule of integration,

$$i(t) - i(t - \Delta t) = \left(\frac{1}{L}\right)\left(\frac{\Delta t}{2}\right)[v(t) + v(t - \Delta t)]$$

Rearranging gives

$$i(t) = \frac{v(t)}{(2L/\Delta t)} + \left[i(t - \Delta t) + \frac{v(t - \Delta t)}{(2L/\Delta t)}\right]$$

or

$$i(t) = \frac{v(t)}{(2L/\Delta t)} + I_L(t - \Delta t) \qquad (12.4.13)$$

where

$$I_L(t - \Delta t) = i(t - \Delta t) + \frac{v(t - \Delta t)}{(2L/\Delta t)} = I_L(t - 2\Delta t) + \frac{v(t - \Delta t)}{(L/\Delta t)} \qquad (12.4.14)$$

Equations (12.4.13) and (12.4.14) are represented by the circuit shown in Figure 12.15(b). As shown, the inductor is replaced by a resistor with resistance $(2L/\Delta t)$ Ω. A current source $I_L(t - \Delta t)$ given by (12.4.14) is also included. $I_L(t - \Delta t)$ represents the past history of the inductor. Note that the trapezoidal rule introduces an error of the order $(\Delta t)^3$.

LUMPED CAPACITANCE

As shown in Figure 12.16(a) for a constant lumped capacitance C,

$$i(t) = C\frac{dv(t)}{dt} \qquad (12.4.15)$$

Integrating from time $(t - \Delta t)$ to t,

$$\int_{t-\Delta t}^{t} dv(t) = \frac{1}{C}\int_{t-\Delta t}^{t} i(t)\, dt \qquad (12.4.16)$$

Using the trapezoidal rule of integration,

$$v(t) - v(t - \Delta t) = \frac{1}{C}\left(\frac{\Delta t}{2}\right)[i(t) + i(t - \Delta t)]$$

FIGURE 12.16

Lumped capacitance

(a) Continuous time circuit (b) Discrete-time circuit

Rearranging gives

$$i(t) = \frac{v(t)}{(\Delta t/2C)} - I_C(t - \Delta t) \qquad (12.4.17)$$

where

$$I_C(t - \Delta t) = i(t - \Delta t) + \frac{v(t - \Delta t)}{(\Delta t/2C)} = -I_C(t - 2\Delta t) + \frac{v(t - \Delta t)}{(\Delta t/4C)} \quad (12.4.18)$$

Equations (12.4.17) and (12.4.18) are represented by the circuit in Figure 12.16(b). The capacitor is replaced by a resistor with resistance $(\Delta t/2C)$ Ω. A current source $I_C(t - \Delta t)$, which represents the capacitor's past history, is also included.

FIGURE 12.17

Lumped resistance

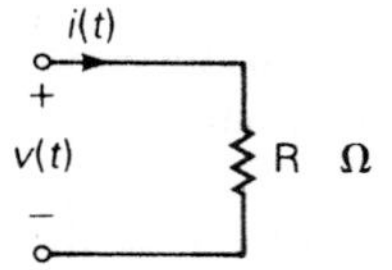

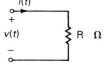

$i(t)$

$+$

$v(t)$ ⟩R Ω

$-$

(a) Continuous time circuit

LUMPED RESISTANCE

The discrete model of a constant lumped resistance R, as shown in Figure 12.17, is the same as the continuous model. That is,

$$v(t) = Ri(t) \qquad (12.4.19)$$

NODAL EQUATIONS

$i(t)$

$+$

$v(t)$ ⟩R Ω

$-$

(b) Discrete-time circuit

A circuit consisting of single-phase lossless transmission lines and constant lumped RLC elements can be replaced by the equivalent circuits given in Figures 12.14(b), 12.15(b), 12.16(b), and 12.17(b). Then, writing nodal equations, the result is a set of linear algebraic equations that determine the bus voltages at each instant t.

EXAMPLE 12.7   **Discrete-time equivalent circuit, single-phase lossless line transients, computer solution**

For the circuit given in Example 12.3, replace the circuit elements by their discrete-time equivalent circuits and write the nodal equations that determine the sending-end and receiving-end voltages. Then, using a digital computer, compute the sending-end and receiving-end voltages for $0 \le t \le 9$ ms. For numerical calculations, assume $E = 100$ V, $Z_c = 400$ Ω, $C_R = 5$ μF, $\tau = 1.0$ ms, and $\Delta t = 0.1$ ms.

SOLUTION   The discrete model is shown in Figure 12.18, where $v_k(t)$ represents the sending-end voltage $v(0, t)$ and $v_m(t)$ represents the receiving-end voltage $v(l, t)$. Also, the sending-end voltage source $e_G(t)$ in series with Z_G is converted to an equivalent current source in parallel with Z_G. Writing nodal equations for this circuit,

FIGURE 12.18

Discrete-time equivalent circuit for Example 12.7

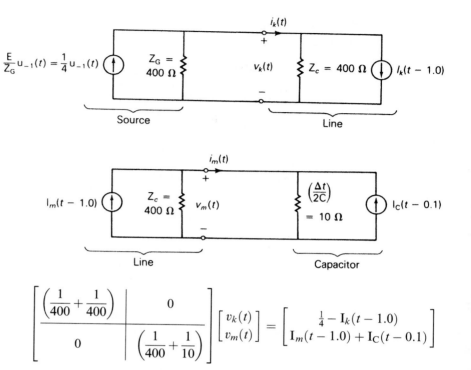

Solving,

$$v_k(t) = 200\left[\tfrac{1}{4} - I_k(t - 1.0)\right] \tag{a}$$

$$v_m(t) = 9.75610[I_m(t - 1.0) + I_C(t - 0.1)] \tag{b}$$

The current sources in these equations are, from (12.4.9), (12.4.10), and (12.4.18), with the argument $(t - \tau)$ replaced by t,

$$I_m(t) = I_k(t - 1.0) + \frac{2}{400}v_k(t) \tag{c}$$

$$I_k(t) = I_m(t - 1.0) - \frac{2}{400}v_m(t) \tag{d}$$

$$I_C(t) = -I_C(t - 0.1) + \frac{1}{5}v_m(t) \tag{e}$$

Equations (a) through (e) above are in a form suitable for digital computer solution. A scheme for iteratively computing v_k and v_m is as follows, starting at $t = 0$:

1. Compute $v_k(t)$ and $v_m(t)$ from equations (a) and (b).

2. Compute $I_m(t)$, $I_k(t)$, and $I_C(t)$ from equations (c), (d), and (e). Store $I_m(t)$ and $I_k(t)$.

3. Change t to $(t + \Delta t) = (t + 0.1)$ and return to (1) above.

FIGURE 12.19

Example 12.7

Output			
Time ms	VK Volts	VM Volts	Computer Program Listing
0.00	50.00	0.00	10 REM EXAMPLE 12.7
0.20	50.00	0.00	20 LPRINT "TIME VK VM"
0.40	50.00	0.00	30 LPRINT "ms Volts Volts"
0.60	50.00	0.00	40 IC = 0
0.80	50.00	0.00	50 T = 0
1.00	50.00	2.44	60 KPRINT = 2
1.20	50.00	11.73	65 REM T IS TIME. KPRINT DETERMINES THE PRINTOUT INTERVAL.
1.40	50.00	20.13	70 REM LINES 110 to 210 COMPUTE EQS(a)–(e) FOR THE FIRST
1.60	50.00	27.73	80 REM TEN TIME STEPS (A TOTAL OF ONE ms) DURING WHICH
1.80	50.00	34.61	90 REM THE CURRENT SOURCES ON THE RIGHT HAND SIDE
2.00	2.44	40.83	100 REM OF THE EQUATIONS ARE ZERO. TEN VALUES OF
2.20	11.73	46.46	105 REM CURRENT SOURCES IK(J) AND IM(J) ARE STORED.
2.40	20.13	51.56	110 FOR J = 1 TO 10
2.60	27.73	56.17	120 VK = 200/4
2.80	34.61	60.34	130 VM = 9.7561*IC
3.00	40.83	64.12	140 IM(J) = (2/400)*VK
3.20	46.46	67.53	150 IK(J) = (−2/400)*VM
3.40	51.56	70.62	160 IC = −IC + (1/5)*VM
3.60	56.17	73.42	170 Z = (J − 1)/KPRINT
3.80	60.34	75.95	180 M = INT(Z)
4.00	64.12	78.24	190 IF M = Z THEN LPRINT USING "*** **"; T,VK,VM
4.20	67.53	80.31	200 T = T + .1
4.40	70.62	82.18	210 NEXT J
4.60	73.42	83.88	220 REM LINES 250 to 420 COMPUTE EQS(a)–(e) FOR TIME T
4.80	75.95	85.41	230 REM EQUAL TO AND GREATER THAN 1.0 ms. THE PAST TEN
5.00	78.24	86.80	240 REM VALUES OF IK(J) AND IM(J) ARE STORED
5.20	80.31	88.06	250 FOR J = 1 TO 10
5.40	82.18	89.20	260 REM LINE 270 IS EQ(a).
5.60	83.88	90.22	270 VK = 200*((1/4) − IK(J))
5.80	85.41	91.15	280 REM LINE 290 IS EQ(b).
6.00	86.80	92.00	290 VM = 9.7561*(IM(J) + IC)
6.20	88.06	92.76	300 REM LINE 310 IS EQ(e).
6.40	89.20	93.45	310 IC = −IC + (1/5)*VM
6.60	90.22	94.07	320 REM LINES 330–360 ARE EQS (c) and (d).
6.80	91.15	94.64	330 C1 = IK(J) + (2/400)*VK
7.00	92.00	95.15	340 C2 = IM(J) − (2/400)*VM
7.20	92.76	95.61	350 IM(J) = C1
7.40	93.45	96.03	360 IK(J) = C2
7.60	94.07	96.40	370 Z = (J − 1)/KPRINT
7.80	94.64	96.75	380 M = INT(Z)
8.00	95.15	97.06	390 IF M = Z THEN LPRINT USING "*** **"; T,VK,VM
8.20	95.61	97.34	400 T = T + .1
8.40	96.03	97.59	410 NEXT J
8.60	96.40	97.82	420 IF T < 9.0 THEN GOTO 250
8.80	96.75	98.03	430 STOP
9.00	97.06	98.22	

FIGURE 12.19

(*continued*)

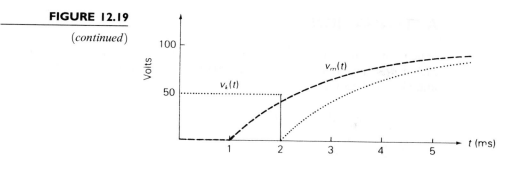

Note that since the transmission line and capacitor are unenergized for time t less than zero, the current sources $I_m(\ )$, $I_k(\ )$, and $I_C(\ )$ are zero whenever their arguments $(\ )$ are negative. Note also from equations (a) through (e) that it is necessary to store the past ten values of $I_m(\ )$ and $I_k(\ )$.

A personal computer program written in BASIC that executes the above scheme and the computational results are shown in Figure 12.19. The plotted sending-end voltage $v_k(t)$ can be compared with the results of Example 12.3.   ∎

Example 12.7 can be generalized to compute bus voltages at discrete-time intervals for an arbitrary number of buses, single-phase lossless lines, and lumped RLC elements. When current sources instead of voltage sources are employed, the unknowns are all bus voltages, for which nodal equations $\mathbf{YV} = \mathbf{I}$ can be written at each discrete-time instant. Also, the dependent current sources in $\mathbf{I}$ are written in terms of bus voltages and current sources at prior times. For computational convenience, the time interval Δt can be chosen constant so that the bus admittance matrix $\mathbf{Y}$ is a constant real symmetric matrix as long as the RLC elements are constant.

12.5

LOSSY LINES

Transmission-line series resistance or shunt conductance causes the following:

1. Attenuation

2. Distortion

3. Power losses

We briefly discuss these effects as follows.

ATTENUATION

When constant series resistance R Ω/m and shunt conductance G S/m are included in the circuit of Figure 12.1 for a single-phase two-wire line, (12.1.3) and (12.1.4) become

$$\frac{\partial v(x,t)}{\partial x} = -Ri(x,t) - L\frac{\partial i(x,t)}{\partial t} \tag{12.5.1}$$

$$\frac{\partial i(x,t)}{\partial x} = -Gv(x,t) - C\frac{\partial v(x,t)}{\partial t} \tag{12.5.2}$$

Taking the Laplace transform of these, equations analogous to (12.1.7) and (12.1.8) are

$$\frac{d^2V(x,s)}{dx^2} - \gamma^2(s)V(x,s) = 0 \tag{12.5.3}$$

$$\frac{d^2I(x,s)}{dx^2} - \gamma^2(s)I(x,s) = 0 \tag{12.5.4}$$

where

$$\gamma(s) = \sqrt{(R+sL)(G+sC)} \tag{12.5.5}$$

The solution to these equations is

$$V(x,s) = V^+(s)e^{-\gamma(s)x} + V^-(s)e^{+\gamma(s)x} \tag{12.5.6}$$

$$I(x,s) = I^+(s)e^{-\gamma(s)x} + I^-(s)e^{+\gamma(s)x} \tag{12.5.7}$$

In general, it is impossible to obtain a closed form expression for $v(x,t)$ and $i(x,t)$, which are the inverse Laplace transforms of these equations. However, for the special case of a *distortionless* line, which has the property R/L = G/C, the inverse Laplace transform can be obtained as follows. Rewrite (12.5.5) as

$$\gamma(s) = \sqrt{LC[(s+\delta)^2 - \sigma^2]} \tag{12.5.8}$$

where

$$\delta = \frac{1}{2}\left(\frac{R}{L} + \frac{G}{C}\right) \tag{12.5.9}$$

$$\sigma = \frac{1}{2}\left(\frac{R}{L} - \frac{G}{C}\right) \tag{12.5.10}$$

For a distortionless line, $\sigma = 0$, $\delta = $ R/L, and (12.5.6) and (12.5.7) become

$$V(x,s) = V^+(s)e^{-\sqrt{LC}[s+(R/L)]x} + V^-(s)e^{+\sqrt{LC}[s+(R/L)]x} \tag{12.5.11}$$

$$I(x,s) = I^+(s)e^{-\sqrt{LC}[s+(R/L)]x} + I^-(s)e^{+\sqrt{LC}[s+(R/L)]x} \tag{12.5.12}$$

Using $v = 1/\sqrt{LC}$ and $\sqrt{LC}(R/L) = \sqrt{RG} = \alpha$ for the distortionless line, the inverse transform of these equations is

$$v(x, t) = e^{-\alpha x}v^{+}\left(t - \frac{x}{v}\right) + e^{+\alpha x}v^{-}\left(t + \frac{x}{v}\right) \tag{12.5.13}$$

$$i(x, t) = e^{-\alpha x}i^{+}\left(t - \frac{x}{v}\right) + e^{\alpha x}i^{-}\left(t + \frac{x}{v}\right) \tag{12.5.14}$$

These voltage and current waves consist of forward and backward traveling waves similar to (12.1.12) and (12.1.13) for a lossless line. However, for the lossy distortionless line, the waves are attenuated versus x due to the $e^{\pm \alpha x}$ terms. Note that the attenuation term $\alpha = \sqrt{RG}$ is constant. Also, the attenuated waves travel at constant velocity $v = 1/\sqrt{LC}$. Therefore, waves traveling along the distortionless line do not change their shape; only their magnitudes are attenuated.

DISTORTION

For sinusoidal steady-state waves, the propagation constant $\gamma(j\omega)$ is, from (12.5.5), with $s = j\omega$

$$\gamma(j\omega) = \sqrt{(R + j\omega L)(G + j\omega C)} = \alpha + j\beta \tag{12.5.15}$$

For a lossless line, $R = G = 0$; therefore, $\alpha = 0$, $\beta = \omega\sqrt{LC}$, and the phase velocity $v = \omega/\beta = 1/\sqrt{LC}$ is constant. Thus, sinusoidal waves of all frequencies travel at constant velocity v without attenuation along a lossless line.

For a distortionless line $(R/L) = (G/C)$, and $\gamma(j\omega)$ can be rewritten, using (12.5.8)–(12.5.10), as

$$\gamma(j\omega) = \sqrt{LC\left(j\omega + \frac{R}{L}\right)^{2}} = \sqrt{LC}\left(j\omega + \frac{R}{L}\right)$$

$$= \sqrt{RG} + j\frac{\omega}{v} = \alpha + j\beta \tag{12.5.16}$$

Since $\alpha = \sqrt{RG}$ and $v = 1/\sqrt{LC}$ are constant, sinusoidal waves of all frequencies travel along the distortionless line at constant velocity with constant attenuation—that is, without distortion.

It can also be shown that for frequencies above 1 MHz, practical transmission lines with typical constants R, L, G, and C tend to be distortionless. Above 1 MHz, α and β can be approximated by

$$\alpha \simeq \frac{R}{2}\sqrt{\frac{C}{L}} + \frac{G}{2}\sqrt{\frac{L}{C}} \tag{12.5.17}$$

$$\beta \simeq \omega\sqrt{LC} = \frac{\omega}{v} \tag{12.5.18}$$

Therefore, sinusoidal waves with frequencies above 1 MHz travel along a practical line undistorted at constant velocity $v = 1/\sqrt{LC}$, with attenuation α given by (12.5.17).

FIGURE 12.20

Distortion and attenuation of surges on a 132-kV overhead line [4] (H. M. Lacey, "The Lightning Protection of High-Voltage Overhead Transmission and Distribution Systems," Proceedings of the IEEE, 96 (1949), p. 287)

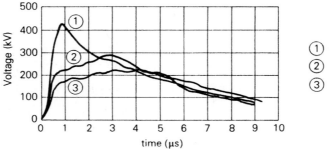

① Start (0. miles)

② Tower 150 (1.449 miles)

③ Tower 130 (4.97 miles)

At frequencies below 1 MHz these approximations do not hold, and lines are generally not distortionless. For typical transmission and distribution lines, (R/L) is much greater than (G/C) by a factor of 1000 or so. Therefore, the condition (R/L) = (G/C) for a distortionless line does not hold.

Figure 12.20 shows the effect of distortion and attenuation of voltage surges based on experiments on a 132-kV overhead transmission line [4]. The shapes of the surges at three points along the line are shown. Note how distortion reduces the front of the wave and builds up the tail as it travels along the line.

POWER LOSSES

Power losses are associated with series resistance R and shunt conductance G. When a current I flows along a line, I^2R losses occur, and when a voltage V appears across the conductors, V^2G losses occur. V^2G losses are primarily due to insulator leakage and corona for overhead lines, and to dielectric losses for cables. For practical lines operating at rated voltage and rated current, I^2R losses are much greater than V^2G losses.

As discussed above, the analysis of transients on single-phase two-wire lossy lines with constant parameters R, L, G, and C is complicated. The analysis becomes more complicated when skin effect is included, which means that R is not constant but frequency-dependent. Additional complications arise for a single-phase line consisting of one conductor with earth return, where Carson [5] has shown that both series resistance and inductance are frequency-dependent.

In view of these complications, the solution of transients on lossy lines is best handled via digital computation techniques. A single-phase line of length l can be approximated by a lossless line with half the total resistance $(Rl/2)$ Ω lumped in series with the line at both ends. For improved accuracy, the line can be divided into various line sections, and each section can be approximated by a lossless line section, with a series resistance lumped at both ends. Simulations have shown that accuracy does not significantly improve with more than two line sections.

FIGURE 12.21

Approximate model of a
lossy line segment

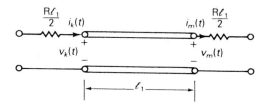

(a) Lossless line segment of length ℓ_1 with lumped line resistance

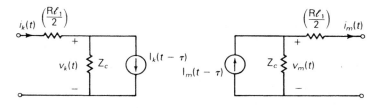

(b) Discrete-time model

Discrete-time equivalent circuits of a single-phase lossless line, Figure
12.14, together with a constant lumped resistance, Figure 12.17, can be used
to approximate a lossy line section, as shown in Figure 12.21. Also, digital
techniques for modeling frequency-dependent line parameters [6, 7] are available but we do not discuss them here.

12.6

MULTICONDUCTOR LINES

Up to now we have considered transients on single-phase two-wire lines. For
a transmission line with n conductors above a ground plane, waves travel in n
"modes," where each mode has its own wave velocity and its own surge impedance. In this section we illustrate "model analysis" for a relatively simple
three-phase line [8].

Given a three-phase, lossless, completely transposed line consisting of
three conductors above a perfectly conducting ground plane, the transmission-
line equations are

$$\frac{d\mathbf{V}(x,s)}{dx} = -s\mathbf{L}\mathbf{I}(x,s) \tag{12.6.1}$$

$$\frac{d\mathbf{I}(x,s)}{dx} = -s\mathbf{C}\mathbf{V}(x,s) \tag{12.6.2}$$

where

$$\mathbf{V}(x,s) = \begin{bmatrix} V_{ag}(x,s) \\ V_{bg}(x,s) \\ V_{cg}(x,s) \end{bmatrix} \qquad \mathbf{I}(x,s) = \begin{bmatrix} I_a(x,s) \\ I_b(x,s) \\ I_c(x,s) \end{bmatrix} \tag{12.6.3}$$

Equations (12.6.1) and (12.6.2) are identical to (12.1.5) and (12.1.6) except that scalar quantities are replaced by vector quantities. $\mathbf{V}(x,s)$ is the vector of line-to-ground voltages and $\mathbf{I}(x,s)$ is the vector of line currents. For a completely transposed line, the 3×3 inductance matrix $\mathbf{L}$ and capacitance matrix $\mathbf{C}$ are given by

$$\mathbf{L} = \begin{bmatrix} L_s & L_m & L_m \\ L_m & L_s & L_m \\ L_m & L_m & L_s \end{bmatrix} \quad \text{H/m} \tag{12.6.4}$$

$$\mathbf{C} = \begin{bmatrix} C_s & C_m & C_m \\ C_m & C_s & C_m \\ C_m & C_m & C_s \end{bmatrix} \quad \text{F/m} \tag{12.6.5}$$

For any given line configuration, $\mathbf{L}$ and $\mathbf{C}$ can be computed from the equations given in Sections 4.7 and 4.11. Note that L_s, L_m, and C_s are positive, whereas C_m is negative.

We now transform the phase quantities to modal quantities. First, we define

$$\begin{bmatrix} V_{ag}(x,s) \\ V_{bg}(x,s) \\ V_{cg}(x,s) \end{bmatrix} = \mathbf{T_V} \begin{bmatrix} V^0(x,s) \\ V^+(x,s) \\ V^-(x,s) \end{bmatrix} \tag{12.6.6}$$

$$\begin{bmatrix} I_a(x,s) \\ I_b(x,s) \\ I_c(x,s) \end{bmatrix} = \mathbf{T_I} \begin{bmatrix} I^0(x,s) \\ I^+(x,s) \\ I^-(x,s) \end{bmatrix} \tag{12.6.7}$$

$V^0(x,s)$, $V^+(x,s)$, and $V^-(x,s)$ are denoted *zero-mode*, *positive-mode*, and *negative-mode* voltages, respectively. Similarly, $I^0(x,s)$, $I^+(x,s)$, and $I^-(x,s)$ are *zero-*, *positive-*, and *negative-mode* currents. $\mathbf{T_V}$ and $\mathbf{T_I}$ are 3×3 constant transformation matrices, soon to be specified. Denoting $\mathbf{V}_m(x,s)$ and $\mathbf{I}_m(x,s)$ as the modal voltage and modal current vectors,

$$\mathbf{V}(x,s) = \mathbf{T_V} \mathbf{V}_m(x,s) \tag{12.6.8}$$

$$\mathbf{I}(x,s) = \mathbf{T_I} \mathbf{I}_m(x,s) \tag{12.6.9}$$

Using (12.6.8) and (12.6.9) in (12.6.1),

$$\mathbf{T_V} \frac{d\mathbf{V}_m(x,s)}{dx} = -s\mathbf{L}\mathbf{T_I}\mathbf{I}_m(x,s)$$

or

$$\frac{d\mathbf{V}_m(x,s)}{dx} = -s(\mathbf{T}_v^{-1}\mathbf{L}\mathbf{T}_I)\mathbf{I}_m(x,s) \tag{12.6.10}$$

Similarly, using (12.6.8) and (12.6.9) in (12.6.2),

$$\frac{d\mathbf{I}_m(x,s)}{dx} = -s(\mathbf{T}_I^{-1}\mathbf{C}\mathbf{T}_v)\mathbf{V}_m(x,s) \tag{12.6.11}$$

The objective of the modal transformation is to diagonalize the matrix products within the parentheses of (12.6.10) and (12.6.11), thereby decoupling these vector equations. For a three-phase completely transposed line, $\mathbf{T}_V$ and $\mathbf{T}_I$ are given by

$$\mathbf{T}_V = \mathbf{T}_I = \begin{bmatrix} 1 & 1 & 1 \\ 1 & -2 & 1 \\ 1 & 1 & -2 \end{bmatrix} \tag{12.6.12}$$

Also, the inverse transformation matrices are

$$\mathbf{T}_V^{-1} = \mathbf{T}_I^{-1} = \frac{1}{3}\begin{bmatrix} 1 & 1 & 1 \\ 1 & -1 & 0 \\ 1 & 0 & -1 \end{bmatrix} \tag{12.6.13}$$

Substituting (12.6.12), (12.6.13), (12.6.4), and (12.6.5) into (12.6.10) and (12.6.11) yields

$$\frac{d}{dx}\begin{bmatrix} \mathbf{V}^0(x,s) \\ \mathbf{V}^+(x,s) \\ \mathbf{V}^-(x,s) \end{bmatrix} = \begin{bmatrix} -s(L_s + 2L_m) & 0 & 0 \\ 0 & -s(L_s - L_m) & 0 \\ 0 & 0 & -s(L_s - L_m) \end{bmatrix}$$

$$\times \begin{bmatrix} \mathbf{I}^0(x,s) \\ \mathbf{I}^+(x,s) \\ \mathbf{I}^-(x,s) \end{bmatrix} \tag{12.6.14}$$

$$\frac{d}{dx}\begin{bmatrix} \mathbf{I}^0(x,s) \\ \mathbf{I}^+(x,s) \\ \mathbf{I}^-(x,s) \end{bmatrix} = \begin{bmatrix} -s(C_s + 2C_m) & 0 & 0 \\ 0 & -s(C_s - C_m) & 0 \\ 0 & 0 & -s(C_s - C_m) \end{bmatrix}$$

$$\times \begin{bmatrix} \mathbf{V}^0(x,s) \\ \mathbf{V}^+(x,s) \\ \mathbf{V}^-(x,s) \end{bmatrix} \tag{12.6.15}$$

From (12.6.14) and (12.6.15), the zero-mode equations are

$$\frac{d\mathbf{V}^0(x,s)}{dx} = -s(L_s + 2L_m)\mathbf{I}^0(x,s) \tag{12.6.16}$$

$$\frac{d\mathbf{I}^0(x,s)}{dx} = -s(C_s + 2C_m)\mathbf{V}^0(x,s) \tag{12.6.17}$$

These equations are identical in form to those of a two-wire lossless line, (12.1.5) and (12.1.6). By analogy, the zero-mode waves travel at velocity

$$v^0 = \frac{1}{\sqrt{(L_s + 2L_m)(C_s + 2C_m)}} \quad \text{m/s} \tag{12.6.18}$$

and the zero-mode surge impedance is

$$Z_c^0 = \sqrt{\frac{L_s + 2L_m}{C_s + 2C_m}} \quad \Omega \tag{12.6.19}$$

Similarly, the positive- and negative-mode velocities and surge impedances are

$$v^+ = v^- = \frac{1}{\sqrt{(L_s - L_m)(C_s - C_m)}} \quad \text{m/s} \tag{12.6.20}$$

$$Z_c^+ = Z_c^- = \sqrt{\frac{L_s - L_m}{C_s - C_m}} \quad \Omega \tag{12.6.21}$$

These equations can be extended to more than three conductors—for example, to a three-phase line with shield wires or to a double-circuit three-phase line. Although the details are more complicated, the modal transformation is straightforward. There is one mode for each conductor, and each mode has its own wave velocity and its own surge impedance.

The solution of transients on multiconductor lines is best handled via digital computer methods, and such programs are available [9, 10]. Digital techniques are also available to model the following effects:

1. Nonlinear and time-varying RLC elements [8]

2. Lossy lines with frequency-dependent line parameters [6, 7, 12]

12.7

POWER SYSTEM OVERVOLTAGES

Overvoltages encountered by power system equipment are of three types:

1. Lightning surges

2. Switching surges

3. Power frequency (50 or 60 Hz) overvoltages

LIGHTNING

Cloud-to-ground (CG) lightning is the greatest single cause of overhead transmission and distribution line outages. Data obtained over a 14-year period from electric utility companies in the United States and Canada and covering 25,000 miles of transmission show that CG lightning accounted for about 26% of outages on 230-kV circuits and about 65% of outages on 345-kV circuits [13]. A similar study in Britain, also over a 14-year period, covering 50,000 faults on distribution lines shows that CG lightning accounted for 47% of outages on circuits up to and including 33 kV [14].

The electrical phenomena that occur within clouds leading to a lightning strike are complex and not totally understood. Several theories [15, 16, 17] generally agree, however, that charge separation occurs within clouds. Wilson [15] postulates that falling raindrops attract negative charges and therefore leave behind masses of positively charged air. The falling raindrops bring the negative charge to the bottom of the cloud, and upward air drafts carry the positively charged air and ice crystals to the top of the cloud, as shown in Figure 12.22. Negative charges at the bottom of the cloud induce a positively charged region, or "shadow," on the earth directly below the cloud. The electric field lines shown in Figure 12.22 originate from the positive charges and terminate at the negative charges.

When voltage gradients reach the breakdown strength of the humid air within the cloud, typically 5 to 15 kV/cm, an ionized path or downward *leader* moves from the cloud toward the earth. The leader progresses somewhat randomly along an irregular path, in steps. These leader steps, about 50 m long, move at a velocity of about 10^5 m/s. As a result of the opposite charge distribution under the cloud, another upward leader may rise to meet the downward leader. When the two leaders meet, a lightning discharge occurs, which neutralizes the charges.

The current involved in a CG lightning stroke typically rises to a peak value within 1 to 10 μs, and then diminishes to one-half the peak within 20 to 100 μs. The distribution of peak currents is shown in Figure 12.23 [20]. This curve represents the percentage of strokes that exceed a given peak current.

FIGURE 12.22

Postulation of charge separation within clouds [16] (G. B. Simpson and F. J. Scrase, "The Distribution of Electricity in Thunderclouds," Proc. Royal Soc., Series A, 161 (1937), p. 309)

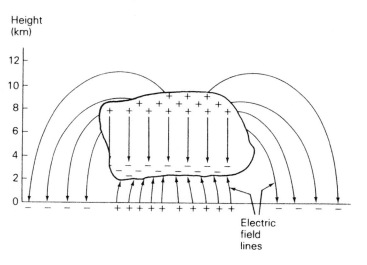

FIGURE 12.23

Frequency of occurrence of lightning currents that exceed a given peak value [20] (IEEE Guide for the Application of Metal-Oxide Surge Arresters for Alternating-Current Systems, IEEE std. C62.22-1997 (New York: The Institute of Electrical and Electronics Engineers, http://standards.ieee.org/1998))

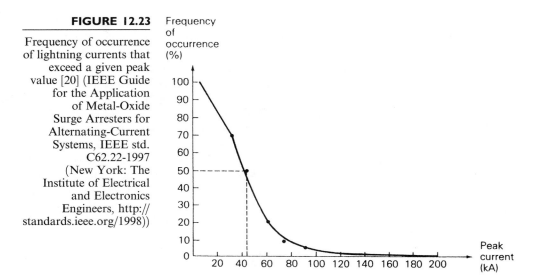

For example, 50% of all strokes have a peak current greater than 45 kA. In extreme cases, the peak current can exceed 200 kA. Also, test results indicate that approximately 90% of all strokes are negative.

It has also been shown that what appears to the eye as a single flash of lightning is often the cumulative effect of many strokes. A typical flash consists of typically 3 to 5, and occasionally as many as 40, strokes, at intervals of 50 ms.

The U.S. National Lightning Detection Network® (NLDN), owned and operated by Global Atmospherics, Inc., is a system that senses the electromagnetic fields radiated by individual return strokes in CG flashes. As of 2001, the NLDN employed more than 100 ground-based sensors geographically distributed throughout the 48 contiguous United States. The sensors transmit lightning data to a network control center in Tucson, Arizona via a satellite communication system. Data from the remote sensors are recorded and processed in real time at the network control center to provide the time, location, polarity, and an estimate of the peak current in each return stroke. The real-time data are then sent back through the communications network for satellite broadcast dissemination to real-time users, all within 30–40 seconds of each CG lightning flash. Recorded data are also reprocessed off-line within a few days of acquisition and stored in a permanent database for access by users who do not require real-time information. NLDN's archive data library contains over 160 million flashes dating from 1989 [25, www.LightningStorm.com].

Figure 12.24 shows a lightning flash density contour map providing a representation of measured annual CG flash density detected by the NLDN from 1989 to 1998. As shown, average annual CG lightning flash densities

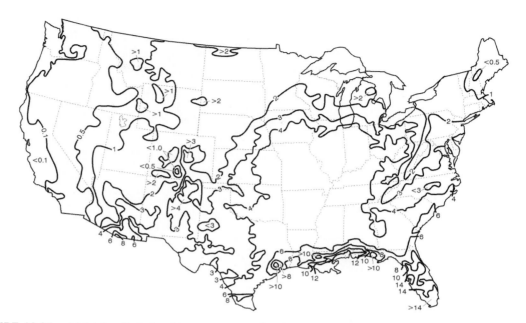

FIGURE 12.24 Lightning flash density contour map showing annual CG flash densities ($\#$flashes/km^2/year) in the contiguous United States, as detected by the NLDN from 1989 to 1998. (Courtesy of Valsala, Inc.)

range from about 0.1 flashes/km^2/year near the West Coast to more than 14 flashes/km^2/year in portions of the Florida peninsula. Figure 12.25 shows a high-resolution, 2 mi $\times$ 2 mi, CG flash density map in grid format.

Figure 12.26 shows an "asset exposure map" of all CG strikes in 1995 in a region that contains a 15-mile 69-kV transmission line. This map provides an indication of the level of exposure to lightning within an exposure area that surrounds the transmission line. By combining this data with estimates of peak stroke currents and transmission line fault records, the lightning performance of the line and individual line segments can be quantified. Improvements in line design and line protection can also be evaluated.

Electric utilities use real-time lightning maps to monitor the approach of lightning storms, estimate their severity, and then either position repair crews in advance of storms, call them out, or hold them over as required. Utilities also use real-time lightning data together with on-line monitoring of circuit breakers, relays, and/or substation alarms to improve operations, minimize or avert damage, and speed up the restoration of their systems.

A typical transmission-line design goal is to have an average of less than 0.50 lightning outages per year per 100 miles of transmission. For a given overhead line with a specified voltage rating, the following factors affect this design goal:

FIGURE 12.25

High-resolution CG flash density map in grid format. (Courtesy of Valsala, Inc.)

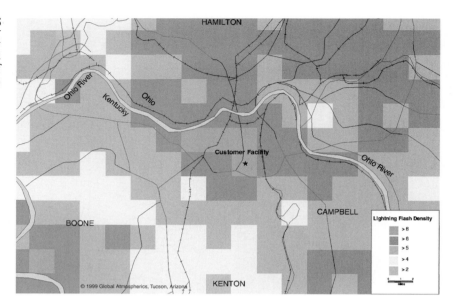

FIGURE 12.26

Asset exposure map showing CG strikes in 1995 in a region that contains a 15-mile 69-kV transmission line. (Courtesy of Valsala, Inc.)

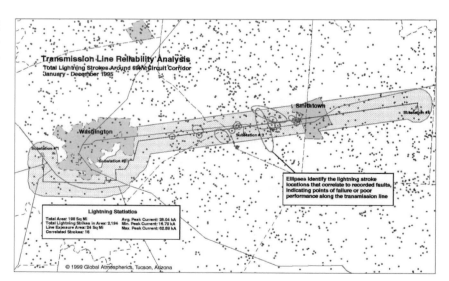

1. Tower height
2. Number and location of shield wires
3. Number of standard insulator discs per phase wire
4. Tower impedance and tower-to-ground impedance

FIGURE 12.27

Effect of shield wires

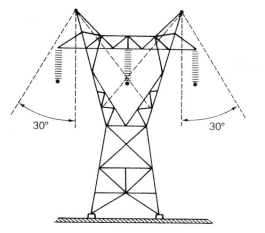

It is well known that lightning strikes tall objects. Thus, shorter, H-frame structures are less susceptible to lightning strokes than taller, lattice towers. Also, shorter span lengths with more towers per kilometer can reduce the number of strikes.

Shield wires installed above phase conductors can effectively shield the phase conductors from direct lightning strokes. Figure 12.27 illustrates the effect of shield wires. Experience has shown that the chance of a direct hit to phase conductors located within $\pm 30°$ arcs beneath the shield wires is reduced by a factor of 1000 [18]. Some lightning strokes are, therefore, expected to hit these overhead shield wires. When this occurs, traveling voltage and current waves propagate in both directions along the shield wire that is hit. When a wave arrives at a tower, a reflected wave returns toward the point where the lightning hit, and two refracted waves occur. One refracted wave moves along the shield wire into the next span. Since the shield wire is electrically connected to the tower, the other refracted wave moves down the tower, its energy being harmlessly diverted to ground.

However, if the tower impedance or tower-to-ground impedance is too high, IZ voltages that are produced could exceed the breakdown strength of the insulator discs that hold the phase wires. The number of insulator discs per string (see Table 4.1) is selected to avoid insulator flashover. Also, tower impedances and tower footing resistances are designed to be as low as possible. If the inherent tower construction does not give a naturally low resistance to ground, driven ground rods can be employed. Sometimes buried conductors running under the line (called *counterpoise*) are employed.

SWITCHING SURGES

The magnitudes of overvoltages due to lightning surges are not significantly affected by the power system voltage. On the other hand, overvoltages due to switching surges are directly proportional to system voltage. Consequently,

FIGURE 12.28

Energizing an open-
circuited line

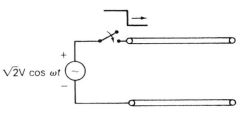

lightning surges are less important for EHV transmission above 345 kV and for UHV transmission, which has improved insulation. Switching surges become the limiting factor in insulation coordination for system voltages above 345 kV.

One of the simplest and largest overvoltages can occur when an open-circuited line is energized, as shown in Figure 12.28. Assume that the circuit breaker closes at the instant the sinusoidal source voltage has a peak value $\sqrt{2}$ V. Assuming zero source impedance, a forward traveling voltage wave of magnitude $\sqrt{2}$ V occurs. When this wave arrives at the open-circuited receiving end, where $\Gamma_R = +1$, the reflected voltage wave superimposed on the forward wave results in a maximum voltage of $2\sqrt{2}$ V $= 2.83$ V. Even higher voltages can occur when a line is reclosed after momentary interruption.

In order to reduce overvoltages due to line energizing or reclosing, resistors are almost always preinserted in circuit breakers at 345 kV and above. Resistors ranging from 200 to 800 Ω are preinserted when EHV circuit breakers are closed, and subsequently bypassed. When a circuit breaker closes, the source voltage divides across the preinserted resistors and the line, thereby reducing the initial line voltage. When the resistors are shorted out, a new transient is initiated, but the maximum line voltage can be substantially reduced by careful design.

Dangerous overvoltages can also occur during a single line-to-ground fault on one phase of a transmission line. When such a fault occurs, a voltage equal and opposite to that on the faulted phase occurs at the instant of fault inception. Traveling waves are initiated on both the faulted phase and, due to capacitive coupling, the unfaulted phases. At the line ends, reflections are produced and are superimposed on the normal operating voltages of the unfaulted phases. Kimbark and Legate [19] show that a line-to-ground fault can create an overvoltage on an unfaulted phase as high as 2.1 times the peak line-to-neutral voltage of the three-phase line.

POWER FREQUENCY OVERVOLTAGES

Sustained overvoltages at the fundamental power frequency (60 Hz in the United States) or at higher harmonic frequencies (such as 120 Hz, 180 Hz, and so on) occur due to load rejection, to ferroresonance, or to permanent faults. These overvoltages are normally of long duration, seconds to minutes, and are weakly damped.

12.8

INSULATION COORDINATION

Insulation coordination is the process of correlating electric equipment insulation strength with protective device characteristics so that the equipment is protected against expected overvoltages. The selection of equipment insulation strength and the protected voltage level provided by protective devices depends on engineering judgment and cost.

As shown by the top curve in Figure 12.29, equipment insulation strength is a function of time. Equipment insulation can generally withstand high transient overvoltages only if they are of sufficiently short duration. However, determination of insulation strength is somewhat complicated. During repeated tests with identical voltage waveforms under identical conditions, equipment insulation may fail one test and withstand another.

For purposes of insulation testing, a standard impulse voltage wave, as shown in Figure 12.30, is defined. The impulse wave shape is specified by giving the time T_1 in microseconds for the voltage to reach its peak value and the time T_2 for the voltage to decay to one-half its peak. One standard wave

FIGURE 12.29

Equipment insulation strength

FIGURE 12.30

Standard impulse voltage waveform

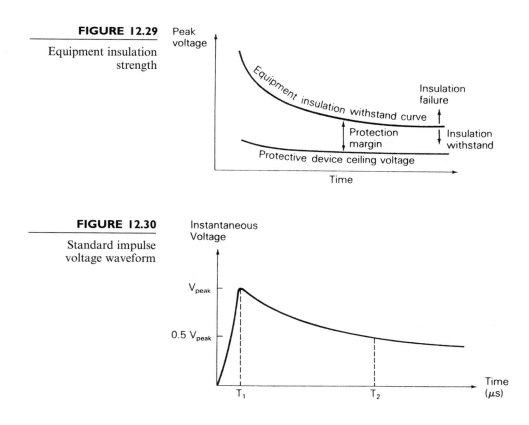

is a 1.2×50 wave, which rises to a peak value at $T_1 = 1.2$ μs and decays to one-half its peak at $T_2 = 50$ μs.

Basic insulation level or BIL is defined as the peak value of the standard impulse voltage wave in Figure 12.30. Standard BILs adopted by the IEEE are shown in Table 12.1. Equipment conforming to these BILs must be capable of withstanding repeated applications of the standard waveform of positive or negative polarity without insulation failure. Also, these standard BILs apply to equipment regardless of how it is grounded. For nominal system voltages 115 kV and above, solidly grounded equipment with the reduced BILs shown in the table have been used.

BILs are often expressed in per-unit, where the base voltage is the maximum value of nominal line-to-ground system voltage. Consider for example a 345-kV system, for which the maximum value of nominal line-to-ground voltage is $\sqrt{2}(345/\sqrt{3}) = 281.7$ kV. The 1550-kV standard BIL shown in Table 12.1 is then $(1550/281.7) = 5.5$ per unit.

Note that overhead-transmission-line insulation, which is external insulation, is usually self-restoring. When a transmission-line insulator string flashes over, a short circuit occurs. After circuit breakers open to deenergize the line, the insulation of the string usually recovers, and the line can be rapidly reenergized. However, transformer insulation, which is internal, is not

TABLE 12.1

Standard and reduced basic insulation levels [18]

Nominal System Voltage kVrms	Standard BIL kV	Reduced BIL* kV
1.2	45	
2.5	60	
5.0	75	
8.7	95	
15	110	
23	150	
34.5	200	
46	250	
69	350	
92	450	
115	550	450
138	650	550
161	750	650
196	900	750
230	1050	825–900
287	1300	1000–1100
345	1550	1175–1300
500		1300–1800
765		1675–2300

*For solidly grounded systems
These BILs are based on 1.2×50 μs voltage waveforms. They apply to internal (or non-self-restoring) insulation such as transformer insulation, as well as to external (or self-restoring) insulation, such as transmission-line insulation, on a statistical basis.
(Westinghouse Electric Corporation, Electrical Transmission and Distribution Reference Book, 4th ed. (East Pittsburgh, PA: 1964).))

FIGURE 12.31

Single-line diagram of
equipment and
protective device

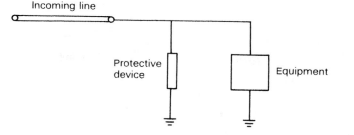

self-restoring. When transformer insulation fails, the transformer must be removed for repair or replaced.

To protect equipment such as a transformer against overvoltages higher than its BIL, a protective device, such as that shown in Figure 12.31, is employed. Such protective devices are generally connected in parallel with the equipment from each phase to ground. As shown in Figure 12.29, the function of the protective device is to maintain its voltage at a ceiling voltage below the BIL of the equipment it protects. The difference between the equipment breakdown voltage and the protective device ceiling voltage is the *protection margin*.

Protective devices should satisfy the following four criteria:

1. Provide a high or infinite impedance during normal system voltages, to minimize steady-state losses.

2. Provide a low impedance during surges, to limit voltage.

3. Dissipate or store the energy in the surge without damage to itself.

4. Return to open-circuit conditions after the passage of a surge.

One of the simplest protective devices is the rod gap, two metal rods with a fixed air gap, which is designed to spark over at specified overvoltages. Although it satisfies the first two protective device criteria, it dissipates very little energy and it cannot clear itself after arcing over.

A surge arrester, consisting of an air gap in series with a nonlinear silicon carbide resistor, satisfies all four criteria. The gap eliminates losses at normal voltages and arcs over during overvoltages. The resistor has the property that its resistance decreases sharply as the current through it increases, thereby limiting the voltage across the resistor to a specified ceiling. The resistor also dissipates the energy in the surge. Finally, following the passage of a surge, various forms of arc control quench the arc within the gap and restore the surge arrester to normal open-circuit conditions.

The "gapless" surge arrester, consisting of a nonlinear metal oxide resistor with no air gap, also satisfies all four criteria. At normal voltages the resistance is extremely high, limiting steady-state currents to microamperes and steady-state losses to a few watts. During surges, the resistance sharply decreases, thereby limiting overvoltage while dissipating surge energy. After

TABLE 12.2 Typical characteristics of station- and intermediate-class metal-oxide surge arresters [20]

	Steady-State Operation: System Voltage and Arrester Ratings			Protective Levels: Range of Industry Maxima per Unit of MCOV			Durability Characteristics: IEEE Std C62.11-1993		
Max System Voltage L-L kV-rms[a]	Max System Voltage L-G kV-rms[a]	Min MCOV Rating kV-rms	Duty Cycle Ratings kV-rms	0.5 μs FOW Protective Level[b]	8/20 μs Protective Level[b]	Switching Surge Protective Level[c]	High Current Withstand Crest Amperes	Trans. Line Discharge Miles	Pressure Relief kA rms (symmetrical)[d]
						Station Class			
4.37	2.52	2.55	3	2.32–2.48	2.10–2.20	1.70–1.85	65 000	150	40–80
8.73	5.04	5.1	6–9	2.33–2.48	1.97–2.23	1.70–1.85	65 000	150	40–80
13.1	7.56	7.65	9–12	2.33–2.48	1.97–2.23	1.70–1.85	65 000	150	40–80
13.9	8.00	8.4	10–15	2.33–2.48	1.97–2.23	1.70–1.85	65 000	150	40–80
14.5	8.37	8.4	10–15	2.33–2.48	1.97–2.23	1.70–1.85	65 000	150	40–80
26.2	15.1	15.3	18–27	2.33–2.48	1.97–2.23	1.70–1.85	65 000	150	40–80
36.2	20.9	22	27–36	2.43–2.48	1.97–2.23	1.70–1.85	65 000	150	40–80
48.3	27.8	29	36–48	2.43–2.48	1.97–2.23	1.70–1.85	65 000	150	40–80
72.5	41.8	42	54–72	2.19–2.40	1.97–2.23	1.70–1.85	65 000	150	40–80
121	69.8	70	90–120	2.19–2.40	1.97–2.18	1.64–1.84	65 000	150	40–80
145	83.7	84	108–144	2.19–2.39	1.97–2.17	1.64–1.84	65 000	150	40–80
169	97.5	98	120–172	2.19–2.39	1.97–2.17	1.64–1.84	65 000	150	40–80
242	139	140	172–240	2.19–2.36	1.97–2.15	1.64–1.84	65 000	175	40–80
362	209	209	258–312	2.19–2.36	1.97–2.15	1.64–1.84	65 000	175	40–80
550	317	318	396–564	2.01–2.47	2.01–2.25	1.71–1.85	65 000	200	40–80
800	461	462	576–612	2.01–2.47	2.01–2.25	1.71–1.85	65 000	200	40–80
						Intermediate class			
4.37–145	2.52–83.72	2.8–84	3–144	2.38–2.85	2.28–2.55	1.71–1.85	65 000	100	16.1[d]

[a] Voltage range A, ANSI C84.1-1989

[b] Equivalent front-of-wave protective level producing a voltage wave cresting in 0.5 μs. Protective level is maximum discharge voltage (DV) for 10 kA impulse current wave on arrester duty cycle rating through 312 kV, 15 kA for duty cycle ratings 396–564 kV and 20 kA for duty cycle ratings 576–612 kV, per IEEE Std C62.11-1993.

[c] Switching surge characteristics based on maximum switching surge classifying current (based on an impulse current wave with a time to actual crest of 45 μs to 60 μs) of 500 A on arrester duty cycle ratings 3–108 kV, 1000 A on duty cycle ratings 120–240 kV, and 2000 A on duty cycle ratings above 240 kV, per IEEE Std C62.11-1993.

[d] Test values for arresters with porcelain tops have not been standardized. Pressure relief classification is in 5 kA steps.

(IEEE Guide for the Application of Metal-Oxide Surge Arresters for Alternating-Current Systems, IEEE std. C62.22-1997 (New York: The Institute of Electrical and Electronics Engineers, http://standards.ieee.org/1998).)

the surge passes, the resistance naturally returns to its original high value. One advantage of the gapless arrester is that its ceiling voltage is closer to its normal operating voltage than is the conventional arrester, thus permitting reduced BILs and potential savings in the capital cost of equipment insulation.

There are four classes of surge arresters: station, intermediate, distribution, and secondary. Station arresters, which have the heaviest construction, are designed for the greatest range of ratings and have the best protective characteristics. Intermediate arresters, which have moderate construction, are designed for systems with nominal voltages 138 kV and below. Distribution arresters are employed with lower-voltage transformers and lines, where there is a need for economy. Secondary arresters are used for nominal system voltages below 1000 V. A summary of the protective characteristics of station- and intermediate-class metal-oxide surge arresters is given in Table 12.2 [20]. This summary is based on manufacturers' catalog information.

Note that arrester currents due to lightning surges are generally less than the lightning currents shown in Figure 12.23. In the case of direct strokes to transmission-line phase conductors, traveling waves are set up in two directions from the point of the stroke. Flashover of line insulation diverts part of the lightning current from the arrester. Only in the case of a direct stroke to a phase conductor very near an arrester, where no line flashover occurs, does the arrester discharge the full lightning current. The probability of this occurrence can be significantly reduced by using overhead shield wires to shield transmission lines and substations. Recommended practice for substations with unshielded lines is to select an arrester discharge current of at least 20 kA (even higher if the isokeraunic level is above 40 thunderstorm days per year). For substations with shielded lines, lower arrester discharge currents, from 5 to 20 kA, have been found satisfactory in most situations [20].

EXAMPLE 12.8   Metal-oxide surge arrester selection

Consider the selection of a station-class metal-oxide surge arrester for a 345-kV system in which the maximum 60-Hz voltage under normal system conditions is 1.04 per unit. (a) Select a station-class arrester from Table 12.2 with a maximum continuous operating voltage (MCOV) rating that exceeds the 1.04 per-unit maximum 60-Hz voltage of the system under normal system conditions. (b) For the selected arrester, determine the protective margin for equipment in the system with a 1300-kV BIL, based on a 10-kA impulse current wave cresting in 0.5 µs.

SOLUTION (a) The maximum 60 Hz line-to-neutral voltage under normal system conditions is $1.04(345/\sqrt{3}) = 207$ kV. From Table 12.2, select a station-class surge arrester with a 209-kV MCOV. This is the lowest MCOV rating that exceeds the 207 kV providing the greatest protective margin as well as economy. (b) From Table 12.2 for the selected surge arrester, the maximum discharge voltage for a 10-kA impulse current wave cresting in 0.5 μs ranges from 2.19 to 2.36 in per-unit of MCOV, or 457–493 kV, depending on arrester manufacturer. Therefore, the protective margin ranges from $(1300 - 493) = 807$ kV to $(1300 - 457) = 843$ kV.   ∎

When selecting a metal-oxide surge arrester, it is important that the arrester MCOV exceeds the maximum 60-Hz system voltage (line-to-neutral) under normal conditions. In addition to considerations affecting the selection of arrester MCOV, metal-oxide surge arresters should also be selected to withstand temporary overvoltages in the system at the arrester location— for example, the voltage rise on unfaulted phases during line-to-ground faults. That is, the temporary overvoltage (TOV) capability of metal-oxide surge arresters should not be exceeded. Additional considerations in the selection of metal-oxide surge arresters are discussed in reference [22] (see www.cooperpower.com).

PROBLEMS

SECTION 12.2

12.1 From the results of Example 12.2, plot the voltage and current profiles along the line at times $\tau/2$, τ, and 2τ. That is, plot $v(x, \tau/2)$ and $i(x, \tau/2)$ versus x for $0 \leqslant x \leqslant l$; then plot $v(x, \tau)$, $i(x, \tau)$, $v(x, 2\tau)$, and $i(x, 2\tau)$ versus x.

12.2 Rework Example 12.2 if the source voltage at the sending end is a ramp, $e_G(t) = Eu_{-2}(t) = Etu_{-1}(t)$, with $Z_G = Z_c$.

12.3 Referring to the single-phase two-wire lossless line shown in Figure 12.3, the receiving end is terminated by an inductor with L_R henries. The source voltage at the sending end is a step, $e_G(t) = Eu_{-1}(t)$ with $Z_G = Z_c$. Both the line and inductor are initially unenergized. Determine and plot the voltage at the center of the line $v(l/2, t)$ versus time t.

12.4 Rework Problem 12.3 if $Z_R = Z_c$ at the receiving end and the source voltage at the sending end is $e_G(t) = Eu_{-1}(t)$, with an inductive source impedance $Z_G(s) = sL_G$. Both the line and source inductor are initially unenergized.

12.5 Rework Example 12.4 with $Z_R = 4Z_c$ and $Z_G = Z_c/3$.

12.6 The single-phase, two-wire lossless line in Figure 12.3 has a series inductance $L = (1/3) \times 10^{-6}$ H/m, a shunt capacitance $C = (1/3) \times 10^{-10}$ F/m, and a 30-km line length. The source voltage at the sending end is a step $e_G(t) = 100u_{-1}(t)$ kV with $Z_G(s) = 100$ Ω. The receiving-end load consists of a 100-Ω resistor in parallel with a 2-mH inductor. The line and load are initially unenergized. Determine (a) the characteristic impedance in ohms, the wave velocity in m/s, and the transit time in ms for this line; (b) the sending- and receiving-end voltage reflection coefficients in per-unit; (c) the Laplace transform of the receiving-end current, $I_R(s)$; and (d) the receiving-end current $i_R(t)$ as a function of time.

12.7 The single-phase, two-wire lossless line in Figure 12.3 has a series inductance $L = 2 \times 10^{-6}$ H/m, a shunt capacitance $C = 1.25 \times 10^{-11}$ F/m, and a 100-km line length. The source voltage at the sending end is a step $e_G(t) = 100u_{-1}(t)$ kV with a source impedance equal to the characteristic impedance of the line. The receiving-end load consists of a 100-mH inductor in series with a 1-μF capacitor. The line and load are initially unenergized. Determine (a) the characteristic impedance in Ω, the wave velocity in m/s, and the transit time in ms for this line; (b) the sending- and receiving-end voltage reflection coefficients in per-unit; (c) the receiving-end voltage $v_R(t)$ as a function of time; and (d) the steady-state receiving-end voltage.

12.8 The single-phase, two-wire lossless line in Figure 12.3 has a series inductance $L = 0.999 \times 10^{-6}$ H/m, a shunt capacitance $C = 1.112 \times 10^{-11}$ F/m, and a 60-km line length. The source voltage at the sending end is a ramp $e_G(t) = Etu_{-1}(t) = Eu_{-2}(t)$ kV with a source impedance equal to the characteristic impedance of the line. The receiving-end load consists of a 150-Ω resistor in parallel with a 1-μF capacitor. The line and load are initially unenergized. Determine (a) the characteristic impedance in Ω, the wave velocity in m/s, and the transit time in ms for this line; (b) the sending- and receiving-end voltage reflection coefficients in per-unit; (c) the Laplace transform of the sending-end voltage, $V_S(s)$; and (d) the sending-end voltage $v_S(t)$ as a function of time.

SECTION 12.3

12.9 Draw the Bewley lattice diagram for Problem 12.5, and plot $v(l/3, t)$ versus time t for $0 \leqslant t \leqslant 5\tau$. Also plot $v(x, 3\tau)$ versus x for $0 \leqslant x \leqslant l$.

12.10 Rework Problem 12.9 if the source voltage is a pulse of magnitude E and duration $\tau/10$; that is, $e_G(t) = E[u_{-1}(t) - u_{-1}(t - \tau/10)]$. $Z_R = 4Z_c$ and $Z_G = Z_c/3$ are the same as in Problem 12.9.

12.11 Rework Example 12.6 if the source impedance at the sending end of line A is $Z_G = Z_A/4 = 100 \ \Omega$, and the receiving end of line B is short-circuited, $Z_R = 0$.

12.12 Rework Example 12.6 if the overhead line and cable are interchanged. That is, $Z_A = 100 \ \Omega$, $v_A = 2 \times 10^8$ m/s, $l_A = 20$ km, $Z_B = 400 \ \Omega$, $v_B = 3 \times 10^8$ m/s, and $l_B = 30$ km. The step voltage source $e_G(t) = Eu_{-1}(t)$ is applied to the sending end of line A with $Z_G = Z_A = 100 \ \Omega$, and $Z_R = 2Z_B = 800 \ \Omega$ at the receiving end. Draw the lattice diagram for $0 \leqslant t \leqslant 0.6$ ms and plot the junction voltage versus time t.

12.13 As shown in Figure 12.32, a single-phase two-wire lossless line with $Z_c = 400 \ \Omega$, $v = 3 \times 10^8$ m/s, and $l = 100$ km has a 400-Ω resistor, denoted R_J, installed across the center of the line, thereby dividing the line into two sections, A and B. The source voltage at the sending end is a pulse of magnitude 100 V and duration 0.1 ms. The source impedance is $Z_G = Z_c = 400 \ \Omega$, and the receiving end of the line is short-

FIGURE 12.32

Circuit for Problem 12.13

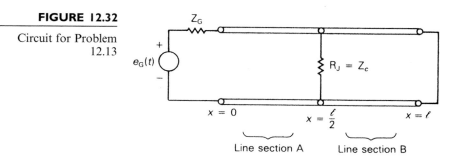

circuited. (a) Show that for an incident voltage wave arriving at the center of the line from either line section, the voltage reflection and refraction coefficients are given by

$$\Gamma_{BB} = \Gamma_{AA} = \frac{\left(\dfrac{Z_{eq}}{Z_c}\right) - 1}{\left(\dfrac{Z_{eq}}{Z_c}\right) + 1} \qquad \Gamma_{AB} = \Gamma_{BA} = \frac{2\left(\dfrac{Z_{eq}}{Z_c}\right)}{\left(\dfrac{Z_{eq}}{Z_c}\right) + 1}$$

where

$$Z_{eq} = \frac{R_J Z_c}{R_J + Z_c}$$

(b) Draw the Bewley lattice diagram for $0 \leqslant t \leqslant 6\tau$. (c) Plot $v(l/2, t)$ versus time t for $0 \leqslant t \leqslant 6\tau$ and plot $v(x, 6\tau)$ versus x for $0 \leqslant x \leqslant l$.

12.14 The junction of four single-phase two-wire lossless lines, denoted A, B, C, and D, is shown in Figure 12.13. Consider a voltage wave v_A^+ arriving at the junction from line A. Using (12.3.8) and (12.3.9), determine the voltage reflection coefficient Γ_{AA} and the voltage refraction coefficients Γ_{BA}, Γ_{CA}, and Γ_{DA}.

12.15 Referring to Figure 12.3, the source voltage at the sending end is a step $e_G(t) = Eu_{-1}(t)$ with an inductive source impedance $Z_G(s) = sL_G$, where $L_G/Z_c = \tau/3$. At the receiving end, $Z_R = Z_c/4$. The line and source inductance are initially unenergized. (a) Draw the Bewley lattice diagram for $0 \leqslant t \leqslant 5\tau$. (b) Plot $v(l, t)$ versus time t for $0 \leqslant t \leqslant 5\tau$.

12.16 As shown in Figure 12.33, two identical, single-phase, two-wire, lossless lines are connected in parallel at both the sending and receiving ends. Each line has a 400-Ω characteristic impedance, 3×10^8 m/s velocity of propagation, and 100-km line length. The source voltage at the sending end is a 100-kV step with source impedance $Z_G = 100\ \Omega$. The receiving end is shorted ($Z_R = 0$). Both lines are initially unenergized. (a) Determine the first forward traveling voltage waves that start at time $t = 0$ and travel on each line toward the receiving end. (b) Determine the sending- and receiving-end voltage reflection coefficients in per-unit. (c) Draw the Bewley lattice diagram for $0 < t < 2.0$ ms. (d) Plot the voltage at the center of one line versus time t for $0 < t < 2.0$ ms.

FIGURE 12.33

Circuit for Problem 12.16

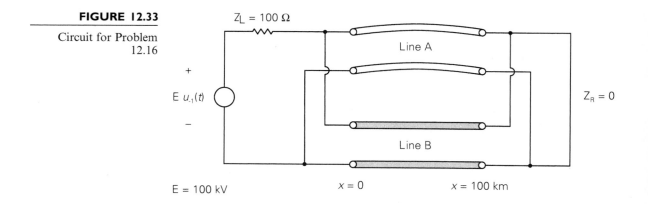

12.17 As shown in Figure 12.34, an ideal current source consisting of a 10-kA pulse with 50-μs duration is applied to the junction of a single-phase, lossless cable and a single-phase, lossless overhead line. The cable has a 200-Ω characteristic impedance, 2×10^8 m/s velocity of propagation, and 20-km length. The overhead line has a 300-Ω characteristic impedance, 3×10^8 m/s velocity of propagation, and 60-km length. The sending end of the cable is terminated by a 400-Ω resistor, and the receiving end of the overhead line is terminated by a 100-Ω resistor. Both the line and cable are initially unenergized. (a) Determine the voltage reflection coefficients Γ_S, Γ_R, Γ_{AA}, Γ_{AB}, Γ_{BA}, and Γ_{BB}. (b) Draw the Bewley lattice diagram for $0 < t < 0.8$ ms. (c) Determine and plot the voltage $v(0, t)$ at $x = 0$ versus time t for $0 < t < 0.8$ ms.

FIGURE 12.34

Circuit for Problem 12.17

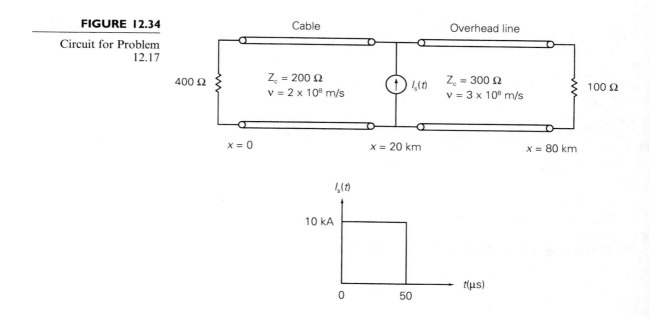

SECTION 12.4

12.18 For the circuit given in Problem 12.3, replace the circuit elements by their discrete-time equivalent circuits and write nodal equations in a form suitable for computer solution of the sending-end and receiving-end voltages. Give equations for all dependent sources. Assume $E = 1000$ V, $L_R = 10$ mH, $Z_c = 100$ Ω, $v = 2 \times 10^8$ m/s, $l = 40$ km, and $\Delta t = 0.02$ ms.

12.19 Repeat Problem 12.18 for the circuit given in Problem 12.13. Assume $\Delta t = 0.03333$ ms.

12.20 For the circuit given in Problem 12.7, replace the circuit elements by their discrete-time equivalent circuits. Use $\Delta t = 100$ μs $= 1 \times 10^{-4}$ s. Determine and show all resistance values on the discrete-time circuit. Write nodal equations for the discrete-time circuit, giving equations for all dependent sources. Then solve the nodal equations and determine the sending- and receiving-end voltages at the following times: $t = 100, 200, 300, 400, 500,$ and 600 μs.

12.21 For the circuit given in Problem 12.8, replace the circuit elements by their discrete-time equivalent circuits. Use $\Delta t = 50$ μs $= 5 \times 10^{-5}$ s and E $= 100$ kV. Determine and show all resistance values on the discrete-time circuit. Write nodal equations for the discrete-time circuit, giving equations for all dependent sources. Then solve the nodal equations and determine the sending- and receiving-end voltages at the following times: $t = 50, 100, 150, 200, 250,$ and 300 μs.

SECTION 12.5

12.22 Rework Problem 12.18 for a lossy line with a constant series resistance R $= 0.3$ Ω/km. Lump half of the total resistance at each end of the line.

SECTION 12.8

12.23 Repeat Example 12.8 for a 115-kV system with a 1.08 per-unit maximum 60-Hz voltage under normal operating conditions and with a 450-kV BIL.

12.24 Select a station-class metal-oxide surge arrester from Table 12.2 for the high-voltage side of a three-phase 400 MVA, 345-kV Y/13.8-kV Δ transformer. The maximum 60-Hz operating voltage of the transformer under normal operating conditions is 1.10 per unit. The high-voltage windings of the transformer have a BIL of 1300 kV and a solidly grounded neutral. A minimum protective margin of 1.4 per unit based on a 10-kA impulse current wave cresting in 0.5 μs is required. (*Note*: Additional considerations for the selection of metal-oxide surge arresters are given in reference [22] (www.cooperpower.com).

CASE STUDY QUESTIONS

A. Why are circuit breakers and fuses ineffective in protecting against transient overvoltages due to lightning and switching surges?

B. Where are surge arresters located in power systems?

C. How does one select a surge arrester to protect specific equipment?

REFERENCES

1. A. Greenwood, *Electrical Transients in Power Systems*, 2d ed. (New York: Wiley Interscience, 1991).

2. L. V. Bewley, *Travelling Waves on Transmission Systems*, 2d ed. (New York: Wiley, 1951).

3. L. Bergeron, *Water Hammer in Hydraulics and Wave Surges in Electricity* (New York: Wiley, 1961).

4. H. M. Lacey, "The Lightning Protection of High-Voltage Overhead Transmission and Distribution Systems," *Proc. IEE*, 96 (1949), p. 287.

5. J. R. Carson, "Wave Propagation in Overhead Wires with Ground Return," *Bell System Technical Journal 5* (1926), pp. 539–554.

6. W. S. Meyer and H. W. Dommel, "Numerical Modelling of Frequency-Dependent Transmission Line Parameters in an Electromagnetic Transients Program," *IEEE Transactions PAS*, vol. PAS-99 (September/October 1974), pp. 1401–1409.

7. A. Budner, "Introduction of Frequency-Dependent Line Parameters into an Electromagnetics Transients Program," *IEEE Transactions PAS*, vol. PAS-89 (January 1970), pp. 88–97.

8. D. E. Hedman, "Propagation on Overhead Transmission Lines: I—Theory of Modal Analysis and II—Earth Conduction Effects and Practical Results," *IEEE Transactions PAS* (March 1965), pp. 200–211.

9. H. W. Dommel, "A Method for Solving Transient Phenomena in Multiphase Systems," *Proceedings 2nd Power Systems Computation Conference*, Stockholm, 1966.

10. H. W. Dommel, "Digital Computer Solution of Electromagnetic Transients in Single- and Multiphase Networks," *IEEE Transactions PAS*, vol. PAS-88 (1969), pp. 388–399.

11. H. W. Dommel, "Nonlinear and Time-Varying Elements in Digital Simulation of Electromagnetic Transients," *IEEE Transactions PAS*, vol. PAS-90 (November/December 1971), pp. 2561–2567.

12. S. R. Naidu, *Transitorios Electromagnéticos em Sistemas de Poténcia*, Eletrobras/UFPb, Brazil, 1985.

13. "Report of Joint IEEE-EEI Committee on EHV Line Outages," *IEEE Transactions PAS*, 86 (1967), p. 547.

14. R. A. W. Connor and R. A. Parkins, "Operations Statistics in the Management of Large Distribution System," *Proc. IEEE, 113* (1966), p. 1823.

15. C. T. R. Wilson, "Investigations on Lightning Discharges and on the Electrical Field of Thunderstorms," *Phil. Trans. Royal Soc.*, Series A, *221* (1920), p. 73.

16. G. B. Simpson and F. J. Scrase, "The Distribution of Electricity in Thunderclouds," *Proc. Royal Soc.*, Series A, *161* (1937), p. 309.

17. B. F. J. Schonland and H. Collens, "Progressive Lightning," *Proc. Royal Soc.*, Series A, *143* (1934), p. 654.

18. Westinghouse Electric Corporation, *Electrical Transmission and Distribution Reference Book*, 4th ed. (East Pittsburgh, PA: 1964).

19. E. W. Kimbark and A. C. Legate, "Fault Surge Versus Switching Surge, A Study of Transient Voltages Caused by Line to Ground Faults," *IEEE Transactions PAS, 87* (1968), p. 1762.

20. *IEEE Guide for the Application of Metal-Oxide Surge Arresters for Alternating-Current Systems*, IEEE std. C62.22-1997 (New York: The Institute of Electrical and Electronics Engineers, http://standards.ieee.org/1998).

21. C. Concordia, "The Transient Network Analyzer for Electric Power System Problems," Supplement to *CIGRE Committee No. 13 Report*, 1956.

22. *Varistar Type AZE Surge Arresters for Systems through 345 kV*, Cooper Power Systems Catalog 235-87 (Waukesha, WI: Cooper Power Systems, http://www.cooperpower.com, August 1997).

23.   C. W. Taylor et al., "WACS—Wide-Area Stability and Voltage Control System: R&D and On-Line Demonstration," *Proceedings of The IEEE*, 93, 5 (May 2005), pp. 892–905.

24.   W. R. Newcott, "Lightning, Nature's High-Voltage Spectacle," *National Geographic*, 184, 1 (July 1993), pp. 83–103.

25.   K. L. Cummins, E. P. Krider, and M. D. Malone, "The U.S. National Lightning Detection Network TM and Applications of Cloud-to-Ground Lightning Data by Electric Power Utilities," *IEEE Transactions on Electromagnetic Compatibility*, 40, 4 (November 1998), pp. 465–480.

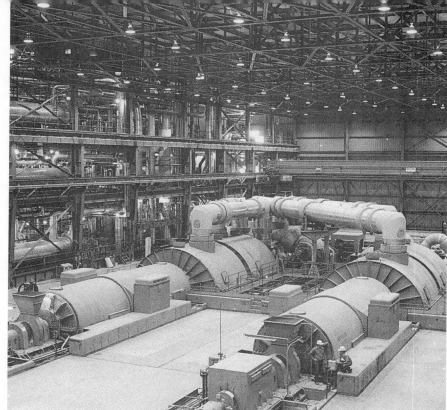

13

TRANSIENT STABILITY

Power system stability refers to the ability of synchronous machines to move from one steady-state operating point following a disturbance to another steady-state operating point, without losing synchronism [1]. There are three types of power system stability: steady-state, transient, and dynamic.

Steady-state stability, discussed in Chapter 5, involves slow or gradual changes in operating points. Steady-state stability studies, which are usually performed with a power-flow computer program (Chapter 6), ensure that phase angles across transmission lines are not too large, that bus voltages are close to nominal values, and that generators, transmission lines, transformers, and other equipment are not overloaded.

Transient stability, the main focus of this chapter, involves major disturbances such as loss of generation, line-switching operations, faults, and sudden load changes. Following a disturbance, synchronous machine frequencies undergo transient deviations from synchronous frequency (60 Hz), and machine power angles change. The objective of a transient stability study is to determine whether or not the machines will return to synchronous frequency with new steady-state power angles. Changes in power flows and bus voltages are also of concern.

Elgerd [2] gives an interesting mechanical analogy to the power system transient stability program. As shown in Figure 13.1, a number of masses representing synchronous machines are interconnected by a network of elastic strings representing transmission lines. Assume that this network is initially at rest in steady-state, with the net force on each string below its break point, when one of the strings is cut, representing the loss of a transmission line. As a result, the masses undergo transient oscillations and the forces on the strings fluctuate. The system will then either settle down to a new steady-state operating point with a new set of string forces, or additional strings will break, resulting in an even weaker network and eventual system collapse. That is, for a given disturbance, the system is either transiently stable or unstable.

In today's large-scale power systems with many synchronous machines interconnected by complicated transmission networks, transient stability studies are best performed with a digital computer program. For a specified disturbance, the program alternately solves, step by step, algebraic power-flow equations representing a network and nonlinear differential equations representing synchronous machines. Both predisturbance, disturbance, and postdisturbance computations are performed. The program output includes power angles and frequencies of synchronous machines, bus voltages, and power flows versus time.

In many cases, transient stability is determined during the first swing of machine power angles following a disturbance. During the first swing, which typically lasts about 1 second, the mechanical output power and the internal voltage of a generating unit are often assumed constant. However, where multiswings lasting several seconds are of concern, models of turbine-governors and excitation systems (for example, see Figures 11.3 and 11.5) as well as more detailed machine models can be employed to obtain accurate transient stability results over the longer time period.

Dynamic stability involves an even longer time period, typically several minutes. It is possible for controls to affect dynamic stability even though transient stability is maintained. The action of turbine-governors, excitation systems, tap-changing transformers, and controls from a power system dis-

FIGURE 13.1

Mechanical analog of power system transient stability [2] (O. I. Elgerd, Electric Energy Systems Theory, 2d. Ed. Page 457 Figure 12.6 (New York: McGraw-Hill, 1982))

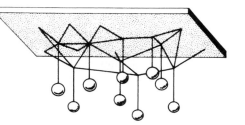

patch center can interact to stabilize or destabilize a power system several minutes after a disturbance has occurred.

To simplify transient stability studies, the following assumptions are made:

1. Only balanced three-phase systems and balanced disturbances are considered. Therefore, only positive-sequence networks are employed.

2. Deviations of machine frequencies from synchronous frequency (60 Hz) are small, and dc offset currents and harmonics are neglected. Therefore, the network of transmission lines, transformers, and impedance loads is essentially in steady-state; and voltages, currents, and powers can be computed from algebraic power-flow equations.

In Section 13.1 we introduce the swing equation, which determines synchronous machine rotor dynamics. In Section 13.2 we give a simplified model of a synchronous machine and a Thévenin equivalent of a system consisting of lines, transformers, loads, and other machines. Then in Section 13.3 we present the equal-area criterion; this gives a direct method for determining the transient stability of one machine connected to a system equivalent. We discuss numerical integration techniques for solving swing equations step by step in Section 13.4 and use them in Section 13.5 to determine multimachine stability. Finally, Section 13.6 discusses design methods for improving power system transient stability.

CASE STUDY   One of the challenges to the operation of a large, interconnected power system is to insure that the generators will remain in synchronism with one another following a large system disturbance such as the loss of a large generator or transmission line. Traditionally, such a stability assessment has been done by engineers performing lots of off-line studies using a variety of assumed system operating conditions. But the actual system operating point never exactly matches the assumed system conditions. The following article discusses a newer method for doing such an assessment in near real-time using the actual system operating conditions [13].

Real-Time Dynamic Security Assessment

ROBERT SCHAINKER, PETER MILLER, WADAD DUBBELDAY, PETER HIRSCH, AND GUORUI ZHANG

The electrical grid changes constantly with generation plants coming online or off-line as required to

("Real-Time Dynamic Security Assessment" by R. Schainker, P. Miller, W. Dubbelday, P. Hirsch, and G. Zhang, IEEE Power and Energy, (March/April 2006), pg. 51–58.)

meet diurnal electrical demand, and with transmission lines coming online or off-line due to transmission outage events or according to maintenance schedules. In state-of-the-art electric utility control centers (illustrative example shown in Figure 1), grid operators use energy management systems (EMSs)

Figure 1
Illustrative example of a state-of-the-art electric grid
energy management system.

to perform network and load monitoring, perform necessary grid control actions, and to manage grid power flows within its territory or region of responsibility.

Limits to flows and voltages on the transmission system are assigned on the basis of transmission line thermal limits and/or off-line studies of voltage and transient dynamic stability. Power flow limits for each transmission line determined in these off-line studies are, by design, conservative, since system operators must always maintain the security and economic operation of their power system over a wide range of operating conditions. Also, the assumption that the grid power flows settle down to steady-state condition is reexamined in real time as the transmission grid conditions change in real time.

Dynamic security assessment (DSA) software analyses allow for the study of the transient and dynamic responses to a large number of potential system disturbances (contingencies) in a transient time frame, which is normally up to about 10 s after a disturbance/outage. Currently, these analyses are performed off line since the simulation process takes hours of computer time to complete for a typically large grid network, which must be simulated for each condition of a large set of all possible outage conditions that could occur. The current, long simulation time makes DSA calculations impractical for use in a real-time application, wherein an operator would need to perform real-world control actions within tens of minutes after a real-world outage to be sure that the grid will not go into an unstable voltage instability and/or a cascading blackout condition.

Therefore, if the DSA calculation could be completed in less than about 10 min, operators who control the grid during emergency conditions (terrorist induced or "nature" induced) can indeed have sufficient time to take appropriate corrective or preventive control actions to handle the identified critial events, which may cause grid instability, or cause cascading outages that would severely impact their utility grid region or their neighboring utility regions, which would potentially avert billion dollar expenses associated with regional blackouts.

The work that lead up to this article was motivated by the attempt to dramatically reduce the time for DSA calculations so that DSA analyses can be converted from off-line studies to routine, online use in order to aid grid operators in their real-time controller analyses. The large amount of time for DSA calculations occurs because grid transients for a large grid network must be calculated over about a 10 s time interval and properly represent a large interconnected power system network system, which must properly represent detailed static and/or dynamic models of power system components, such as transmission network solid-state flexible ac transmission system devices, all types of generators, power system stabilizers, various types of relays/protection systems, load models, and various types of faults or disturbances.

This article describes the methods and successful results obtained in developing a real-time version of the DSA tool. The material below is organized by first providing a description of the DSA software package generally used by the U.S. electric utility industry. Then discussed is a way to dramatically reduce the computation time to perform DSA calculations, which, among other useful techniques, uses a new distributed computational architecture. The results from applying this new version of DSA are then presented using a large utility system as an example. The results clearly show that, indeed, using the new DSA approach, calculations for a large power system can be performed fast enough for the real-time application to EMSs that operate today's grid systems. The article then ends with some insights and concluding remarks.

DYNAMIC SECURITY ASSESSMENT (DSA)

DSA software performs simulations of the impact of potential electric grid fault conditions for a preset time frame after a potential grid disturbance, usually over a time interval of 5–10 s after an outage contingency condition occurs. Contingency conditions studied include "normal" transmission line and/or power plant outages caused by acts of nature or equipment (e.g., outages due to lightning and/or generator "trips"), wear and tear (for example, equipment age failures), and outage conditions caused by human error and/or potential terrorist-induced equipment failures.

Recent efforts by the authors of this article have focused on improving the performance of the DSA calculation process with the eventual goal of implementing the DSA evaluation process in an online utility energy management system (EMS). Past DSA research projects have resulted in significant achievements in determining which outage contingency conditions are significant and not significant by rapidly separating the outage contingencies into "definitely safe" and "potentially harmful" groups. The "potentially harmful" group must be studied in more detail to accurately determine whether a "potentially harmful" contingency is in fact harmful.

DYNAMIC SECURITY ASSESSMENT MODELS

The DSA program uses a complete representation of all the generators (for example, fossil, nuclear, gas, oil, hydro, and wind generators) including their exciters, governors and stabilizers, transmission lines and many other linear and nonlinear components. For example, nonlinear devices embedded into DSA software include such items as:

1) synchronous machines
2) induction motors
3) static VAR compensators
4) thyristor-controlled series compensations
5) thyristor-controlled tap changers and/or phase regulators
6) thyristor-controlled braking resistors or braking capacitors
7) static load models (nonlinear loads)
8) high-voltage dc link
9) user-defined models, as appropriate.

In addition, DSA software models different types of electric grid protection relays:

1) load shedding relay
2) underfrequency load shedding relay
3) voltage difference load dropping relay
4) underfrequency generation rejection relay
5) underfrequency line tripping relay
6) impedance/default distance relay
7) series capacitor gap relay
8) rate of change of power relay.

Also modeled within the DSA software are static nonlinear load models, which are different from constant impedance load models.

DSA also models the following four types of static nonlinear loads:

1) constant current load
2) constant mega-voltage-ampere load
3) general exponential voltage and frequency-dependent load
4) thermostatically controlled load.

Additionally, DSA models each transmission line as a network impedance model with capacitance, inductance, and resistance. Each line also has thermal line rating limits. In addition, tap- and phase-shifting transformers are modeled.

Contingencies for DSA are defined in terms of the fault type, location, duration, and sequence of events making up a contingency scenario. Typical short-circuit faults are three-phase faults, single-line-to-ground and/or double-line-to-ground short-circuit faults. Automatic switching actions taken into account in the computation simulation are line removal or line closure into the grid network. The location of the short-circuit fault can be at the electrical bus, line end, or line section.

DSA ALGORITHM

The solution to all these devices operating in an electric grid requires solving a large set of differential equations. For a 5,000-node network with 300 generators, over 14,000 nonlinear differential equations must be simultaneously solved. DSA uses a numerical analysis method to solve these nonlinear differential equations. The numerical method uses a small time step of about 0.01 s, and at each time step, the method linearizes the equations to calculate the future time response. A classic Newton-Rahpson iteration approach is incorporated into the numerical method, and for a 10-s simulation, 1,000 such time steps are used.

The solution for a conventional transient stability program can take considerable time to solve for one contingency and even longer for multiple contingencies. Typically, for a 5,000-node network in which 300 contingencies are investigated, a 30-s transient stability simulation may take over two hours of calculation time, dependent on the type of computer used in the calculation.

One would need over 100-fold improvement in DSA simulation time performance to be able to do this calculation in about 10 min or less.

Based on these requirements, some of the significant ways for improving the DSA performance deployed by the authors herein are described below.

1) An improved stopping (called *early termination*) criteria when evaluating each contingency is used to reduce the overall time each contingency is simulated. That is, if the program simulation is for 10 s and after a short time duration, say less than 2 s of simulation time, it can be determined that the contingency case being investigated is unstable or stable, then the DSA program evaluating that contingency is stopped, and a flag is set to unstable or stable for the contingency case being investigated. If no stable/unstable determination can be made, then the DSA program for that contingency case runs the full 10-s simulation time period specified. Using this technique, the DSA program does not have to be run to completion for every contingency.

It is run to completion only for those contingencies that are moderately stable or moderately unstable.

2) A novel distributed computing architecture (see Figure 2) was also used to improve the time it takes to perform the numerous contingency cases investigated. In general, there are two ways of performing distributed computation, and both were investigated. One is to parallelize the DSA algorithm and its calculation approach, using central processing units (CPUs) in parallel to perform the calculations. This will improve the performance somewhat, but due to the sparse nature of the differential equation matrices involved, this improvement has been found to be not very useful. A better technique is to run the full DSA software application on each of *n* computers (set up to communicate with each other) and distribute the contingencies (so each computer runs a different set of contingencies). The master computer distributes the contingencies to each of the slave/server computers as needed. Of course, this will work as long as the number of contingencies is equal, or exceeds the number of computers, which is certainly the case. Full distributed computation is thus achieved and the only slow down is due to the use of one master computer to orchestrate/distribute the contingencies to the other computers and receive/catalogue the solution results from the other computers as the results become available.

Using the above methods, and others, the authors developed a new DSA computation architecture and approach, which did improve the computation time by a factor of about 100+, based on the following improvement components: an improvement factor of about 2, due to not having to move data among computers and hard disk storage locations, an improvement factor of about 3 by using the "stopping" criteria discussed above, an improvement factor of about 4 by using five computers in the distributed computer architecture discussed above, and an im-

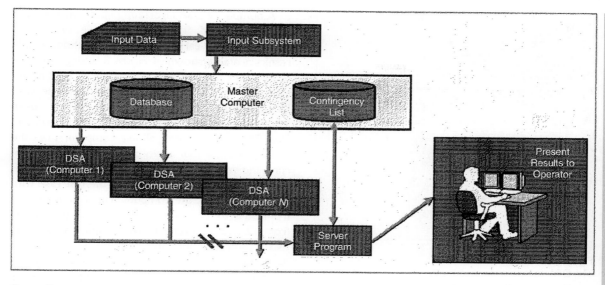

Figure 2
A schematic of the distributed computer architecture used to improve the DSA computation time.

provement factor of about 6 due to faster CPUs used to perform the calculations, as compared to those used circa 2000.

DSA INPUT DATA

The DSA input data consist of three sets of data:

- the power flow data, which contain all the transmission line configurations, tap-changing transformers, phaseshifting transformers, load representation, electric breaker information, relay information, and the type and location of the generation plants
- The dynamic data, which contain various types of generator models, including the generator exciter models, governor models, power system stabilizers, the exciter models, governor models, power system stabilizers, basic generator parameters (along with their limits and time constants), load models, and protection relay models
- The contingency data, which include the various types of faults, including the type, location,

duration, and clearance of the faults and the switching actions after faults are cleared.

DSA OUTPUT DATA

The DSA program produces output results for each contingency and for each generator. The results are data and information on items such as the relative generator angle, the speed of the generator, and the voltage at each generator. This output is temporarily saved on the computer running the contingency and is then transferred to the master computer at the end of the contingency run. On the master computer, time-dependent plots for each contingency of the top three worst grid node response cases are made available to the user.

Figure 3 was produced using the DSA improvement methods described above. On the vertical axis of this figure is the computer run time needed to perform a DSA calculation for a utility grid system that has 5,839 electric buses, 11,680 transmission lines, and 779 generators. The computation time needed to run this large, representative utility test case with only the master computer and then,

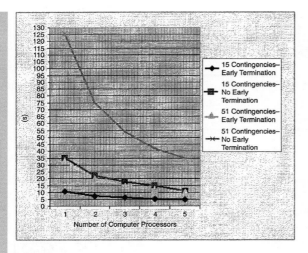

Figure 3
The DSA computation time performance, with and without early termination method.

sequentially, with one, two, three, four, and the available portion of the master computer used as a "fifth" slave computer. Each point on the plots in the figure show the time required to do all the DSA calculations. Comparison data were plotted for cases where the number of contingencies was 15 and 51. Also, for comparison purposes, data were plotted for cases where the early termination logic was used for each contingency case computed and for when no early termination logic was used for each contingency case computed. The results were impressive showing a significant improvement in computing time. For the test case with 51 contingencies, the computing time ranged from 125 s (using only one computer) down to 35 s using all five computers (i.e., the master and the four slave computers). This set of runs showed an improvement factor of about 3.6 in computing time. For the test case with 15 contingencies, the computer run time ranged from 35 s (using only one computer) down to 10.3 s using all five computers (i.e., the master and the four slave computers). This set of runs showed an improvement factor of about 3.4 in computer run time. These results clearly show the power of the master-slave computer architecture developed and successfully investigated and tested.

A number of DSA algorithm improvements were also investigated. The most effective one investigated was the "early termination" method. Sample results are also shown in Figure 3. Using the early termination method and comparing it to the "no early termination" method, for the test case with 51 contingencies, the computer run time improved from 125 s to 33 s (using only one computer) and from 35 s to 10 s (using all five computers, i.e., the master and the four slave computers). This set of runs showed about an improvement factor of 3.8 to 3.5 in computer run time. For the test case with 15 contingencies, the computer run time improved from 35 s to 11 s (using only one computer) and from 12 s to 5 s (using all five computers—i.e., the master and the four slave computers). This set of runs showed an improvement factor of about 3.2 to 2.4 in computer run time. These results also clearly show the power of the master-slave computer architecture system developed and successfully tested.

DSA GRAPHICAL OUTPUT DISPLAYS

The DSA program provides several graphical output displays to show the following types of output results (some of which are illustrated in Figures 4–7):

- largest generator speed angles, for both stable and unstable contingency cases
- highest frequencies, for both stable and unstable contingency cases.

CONCLUSIONS

Using the distributed computer architecture for DSA calculations, grid operators can now quickly analyze a large number of system contingency outage events. Thus, they can evaluate the appropriate preventive or corrective control actions to effectively handle various severe system disturbances or even mitigate costly cascading blackouts, events that are either initiated by nature or terrorist induced.

Online dynamic security analysis (DSA) requires extensive computer resources, particularly for large electric power systems. With the recent advances in computer technology and the intra- and interenterprise communication networking, it now becomes

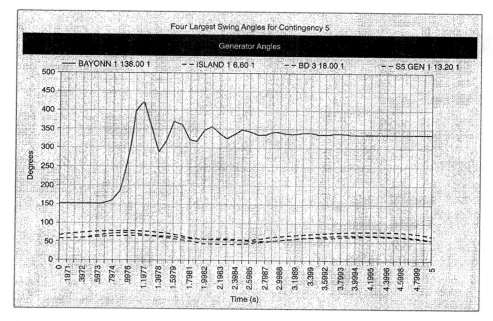

Figure 4
The DSA output plot for the largest generator swing angle for a stable contingency case.

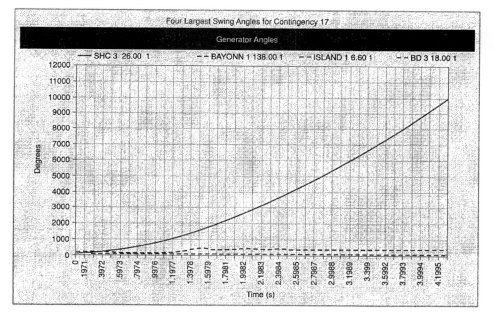

Figure 5
The DSA output plot for the largest generator swing angle for an unstable contingency case.

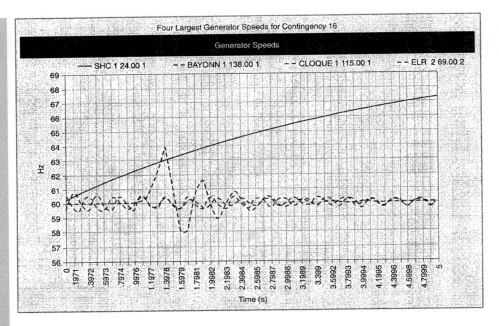

Figure 6
DSA output plot for largest generator speeds for an unstable case.

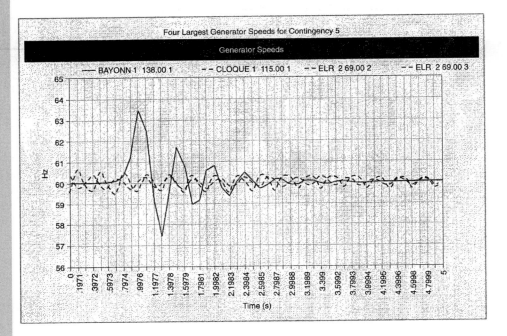

Figure 7
DSA output plot for largest generator speed display for stable case.

cost-effective and possible to apply distributed computing to online DSA in order to meet real-time performance requirements needed in the electric utility industry.

Thus, the major conclusions of the work presented herein are:

- The distributed computing architecture to perform the dynamic security assessment (DSA) analysis of a large interconnected power system with a large number of contingencies has been demonstrated to be extremely fast. As such, this computerized approach should be implemented for real-time decision-making conditions, which are faced by utility and grid operators when any unplanned outage condition occurs that might lead to system instability or even cascading blackout conditions.
- The distributed computer approach developed was tested successfully with five computers in a master-slave arrangement that is scalable to any number of extra slave computers.
- The dynamic security analysis (DSA) using distributed computing can be fully integrated with utility operator EMSs using real-time operating conditions and grid State Estimation estimators.
- The dynamic security analysis (DSA) using distributed computing can also be used for performing operational planning studies for large power systems.
- The dynamic security analysis (DSA) using the distributed computing technology presented herein used the Oracle 9i relational database and its related software. This enables flexible software integration with a wide variety of IT infrastructure systems currently used by many electric utilities and/or grid operators.
- The proposed approach can be used to better utilize existing computer resources and communication networks of electric utilities. This will significantly improve the performance of DSA computations electric utilities perform routinely.
- The performance of the DSA approach presented herein is also fast enough for the real-time calculation of the interface transfer limits using real-time operating conditions for large interconnected power systems.

ACKNOWLEDGMENTS

The material presented in this paper was sponsored by the Department of Homeland through a Space and Naval Warfare Systems Center, San Diego, Contract N66001-04-C-0076.

FOR FURTHER READING

"Analytical methods for contingency selection and ranking for dynamic security analysis," EPRI, Palo Alto, CA, TR-104352, Project 3103-03 Final Rep., Sep. 1994.

"Simulation program for on-line dynamic security Assessment," EPRI, Palo Alto, CA, TR-109751, Jan. 1998.

"Standard test cases for dynamic security assessment," EPRI, Palo Alto, CA, TR-105885, Dec. 1995.

A. A. Fouad and V. Vittal, *Power System Transient Stability Analysis Using the Transient Energy Function Method.* Englewood Cliffs, NJ: Prentice Hall 1992.

C. K. Tang, C. E. Graham, M. El-Kady, and R. T. H. Alden, "Transient stability index from conventional time domain simulation," *IEEE Trans. Power Syst.,* vol. 9, no. 3, Aug. 1993.

G. D. Irisarri, G. C. Ejebe, W. F. Tinney, and J. G. Waight, "Efficient computation of equilibrium points for transient energy analysis," *IEEE Trans. Power Syst.,* vol. 9, no. 2, May 1994.

BIOGRAPHIES

Robert Schainker is a technical executive and manager of the Security Program Department at the Electric Power Research Institute. He received his D.Sc. in applied mathematics and control systems, his M.S. in electrical engineering, and his B.S. in mechanical engineering at Washington University in St. Louis, Missouri.

Peter Miller is program manager, Homeland Security Advance Research Project Agency, Mission Support Office, Science and Technology, U.S. Department of Homeland Security. He holds a B.S. in mathematics from the City University of New

York, where he received the Borden Prize, and he holds an S.M. in computer science and electrical engineering from the Massachusetts Institute of Technology.

Wadad Dubbelday is an engineer at the Navy at the Space and Naval Warfare Systems Command (SPAWAR) office in San Diego, California. She holds a B.S. in physics from the Florida Institute of Technology and holds M.S. and Ph.D. degrees in electrical engineering-applied physics from the University of California at San Diego.

Peter Hirsch is a project manager in the Power Systems Assets Planning and Operations Department of the Electric Power Research Institute. He

also is the manager of the software quality group within EPRI's Power Delivery and Markets Sector. He holds a B.S. in applied mathematics and engineering physics and M.S. and Ph.D. degrees in mathematics from the University of Wisconsin. He is a Senior Member of the IEEE.

Guorui Zhang is principle engineer at EPRI-Solutions, a subsidiary of the Electric Power Research Institute. He received his B.S. in computer software engineering at Singh University, China; he received his Ph.D. in electrical engineering at the University of Manchester Institute of Science and Technology, Manchester, England. He is a Senior Member of the IEEE.

CASE STUDY   In the United States, electric utilities grew first as isolated systems. Gradually, however, neighboring utilities began to interconnect, which allowed utility companies to draw on each others' generation reserves during time of need and to schedule power transfers that take advantage of energy-cost differences. Although overall system reliability and economy have improved dramatically through interconnection, there is a remote possibility that an initial disturbance may lead to instability and a regional blackout. The following article reviews the August 14, 2003 blackout that occurred in the northeastern United States and Canada [9, 10].

Causes of the 14 August Blackout

On 14 August 2003, large portions of the mid-west and northeast United States and Ontario, Canada, experienced an electric power blackout affecting an estimated 50 million people in eight US states and the Canadian province of Ontario. Power was not restored for two days in some parts of the US, while parts of Ontario suffered rolling blackouts for more than a week. The joint US-Canada Power System Outage Task Force was established to investigate the causes of the blackout and ways to reduce the possibility of future outages.

The resulting document (of which this is an edited version) provides an interim report, presenting the facts found by the bi-national investigation.

("Causes of the 14 August Blackout", IEEE Power and Energy Magazine, (June 2004), pg. 30–37.)

THE SYSTEM AND ITS RELIABILITY ORGANIZATIONS

The North American power grid represents more than US$1 trillion in asset value, more than 200,000 miles of transmission lines operating at 230,000 volts and greater, 950,000 megawatts of generating capability, and nearly 3500 utility organizations serving well over 100 million customers and 283 million people.

While the power system in North America is commonly referred to as 'the grid', there are actually three distinct power grids or 'interconnections'. Within each interconnection, electricity is produced the instant it is used and flows over virtually all transmission lines from generators to loads. Only the northeastern portion of the Eastern Interconnection was affected by the 14 August blackout.

PLANNING AND RELIABLE OPERATION

Reliable operation of the power grid is complex and demanding for two fundamental reasons. First, electricity flows at the speed of light and is not economically storable in large quantities. Second, the

flow of alternating current (AC) electricity cannot be controlled like a liquid or gas by opening or closing a valve in a pipe, or switched like calls over a long-distance telephone network. Electricity flows freely along all available paths from the generators to the loads in accordance with the laws of physics.

Maintaining reliability is a complex enterprise that requires trained and skilled operators, sophisticated computers and communications, and careful planning and design. The North American Electric Reliability Council (NERC) and its 10 Regional Reliability Councils have developed system operating and planning standards for ensuring the reliability of a transmission grid, based on seven key concepts.

Balance power generation and demand continuously

To enable customers to use as much electricity as they wish at any moment, production by the generators must be scheduled or 'dispatched' to meet constantly changing demands. Demand is somewhat predictable, appearing as a daily demand curve, but failure to match generation to demand causes the frequency of an AC power system to increase (when generation exceeds demand) or decrease (when generation is less than demand). Large deviations in frequency can cause the rotational speed of generators to fluctuate, leading to vibrations that can damage generator turbine blades and other equipment. Extreme low frequencies can trigger automatic under-frequency 'load shedding', which takes blocks of customers off-line in order to prevent a total collapse of the electric system.

Balance reactive power supply and demand to maintain scheduled voltages

Reactive power sources, such as capacitor banks and generators, must be adjusted during the day to maintain voltages within a secure range pertaining to all system electrical equipment (stations, transmission lines, and customer equipment). Most generators have automatic voltage regulators. Low voltage can cause electric system instability or collapse and, at distribution, can cause damage to motors and the failure of electronic equipment. High voltages can exceed the insulation capabilities of equipment and cause dangerous electric arcs ('flashovers').

Monitor flows over transmission lines and other facilities to ensure that thermal (heating) limits are not exceeded

All lines, transformers and other equipment carrying electricity are heated by the flow of electricity, so frequency flow must be limited to avoid overheating and subsequent damage. In the case of overhead power lines, heating also causes the metal conductor to stretch or expand and sag closer to ground level. Conductor heating is also affected by ambient temperature, wind and other factors. Flow on overhead lines must be limited to ensure the line

doesn't sag into obstructions below or violate the minimum safety clearances between the energized lines and other objects (a flashover can then occur). All current-carrying devices are monitored continuously to ensure that they do not become overloaded or violate other operating constraints, typically using multiple ratings.

Keep the system in a stable condition

Because the electric system is interconnected and dynamic, electrical stability limits must be observed. The main concern is to ensure that generation dispatch and the resulting power flows and voltages are such that the system is stable at all times. There are two types of stability limits: voltage stability limits are set to ensure that the loss of a line or generator won't cause voltages to fall to dangerously low levels. Power (angle) stability limits are set to ensure that a short circuit or a loss of a line, transformer or generator won't cause the remaining generators and loads being served to lose synchronism with one another.

Operate the system so that it remains in a reliable condition even if a contingency occurs, such as the loss of a key generator or transmission facility

Because a generator or line trip can occur at any time from random failure, the power system must be operated in a preventive mode so that the loss of the most important generator or transmission facility does not jeopardize the remaining facilities in the system by causing them to exceed their emergency ratings or stability limits, which could lead to a cascading outage. Further, when a contingency does occur, the operators are required to identify and assess immediately the new worst contingencies, and make any adjustments to ensure the system would still remain operational and safe.

NERC operating policy requires that the system be restored within no more than 30 minutes to compliance with normal limits, and to a condition where it can once again withstand the next-worst single contingency without violating thermal, voltage or stability limits. A few areas of the grid are operated to withstand the concurrent loss of two or more facilities (i.e. 'N-2').

PLAN, DESIGN, AND MAINTAIN THE SYSTEM TO OPERATE RELIABLY

A utility that serves retail customers must estimate future loads and, in some cases, arrange for adequate sources of supplies and plan adequate transmission and distribution infrastructure. NERC planning standards identify a range of possible contingencies and set corresponding expectations for system performance under several categories of possible events.

Prepare for emergencies

Even after taking the above steps, emergencies can still occur because of external factors (severe weather, operator error, equipment failures, etc.) that exceed planning, design or operating criteria. Operators must have emergency procedures covering a range of scenarios. To deal with a system emergency that results in a blackout, there must be procedures and capabilities to use 'black start' generators (capable of restarting with no external power source) and to coordinate operations in order to restore the system as quickly as possible to a normal and reliable condition.

THE CAUSES OF THE BLACKOUT

The 14 August 2003 blackout was initiated by coinciding deficiencies in specific practices, equipment and human decisions. There were three groups of causes:

Inadequate situational awareness at FirstEnergy Corporation (FE)
For example:

- It failed to ensure the security of its transmission system after significant unforeseen contingencies by not routinely using an effective contingency analysis capability.
- It lacked procedures to ensure that its operators were continually aware of the functional state of their critical monitoring tools.

FE failed to adequately manage tree growth in its transmission rights-of-way. This failure was the common cause of the outage of three FE 345 kV transmission lines.

Failure of the interconnected grid's reliability organizations to provide effective diagnostic support
For example:

- MISO's reliability coordinators were using non-real-time data to support real-time 'flowgate' monitoring. This prevented it from detecting an N-1 security violation in FE's system and from assisting FE in relief actions.
- MISO lacked an effective means of identifying the location and significance of transmission line breaker operations reported by their Energy Management System (EMS).

WHY DOES A BLACKOUT CASCADE?

Major blackouts are rare, and no two scenarios are the same with varying initiating events (human actions or inactions, system topology, load/generation balances, etc.) and other factors (distance between generating stations and load centers, voltage profiles, etc.). Most wide-area blackouts start with short circuits (faults) on several transmission lines in short succession, sometimes resulting from natural causes or, as on 14 August, from inadequate tree management in right-of-way areas. A fault causes a high current and low voltage on the line containing the fault. This is detected by a protective relay, which quickly trips the circuit breakers to isolate that line from the rest of the system. Cascade occurs when there is a sequential tripping of numerous transmission lines and generators in a widening geographic area. A cascade can be triggered by just a few initiating events, as was seen on 14 August.

Power swings and voltage fluctuations caused by these initial events can cause other lines to detect high currents and low voltages that appear to be faults, even when faults do not actually exist on those other lines. Generators are tripped off during a cascade to protect them from severe power and voltage swings. Relay protection systems work well to protect lines and generators from damage and to isolate them from the system under normal, steady conditions. However, when power system operating and design criteria are violated as a result of several concurrent outages, most common protective relays cannot distinguish between the currents and voltages seen in a system cascade from those caused by a fault. This leads to more and more lines and generators being tripped, widening the blackout area.

HOW DID THE CASCADE EVOLVE ON 14 AUGUST?

At 16:05:57 EDT, the trip and lock-out of FE's Sammis-Star 345 kV line set off a cascade of interruptions on the high voltage system, causing electrical fluctuations and facility trips. Within seven minutes, the blackout rippled across much of the northeast US and Canada. By 16:13 EDT, more than 263 power plants (531 individual generating units) had been lost, and 10 s of millions of people in the US and Canada were without electricity.

The collapse of FE's transmission system induced unplanned power surges across the region. Shortly before the collapse, large electricity flows were moving across FE's system from the south to northern Ohio, eastern Michigan, and Ontario. This pathway in northeastern Ohio became unavailable with the collapse of FE's transmission system and the electricity took alternative paths to the load centers on the shore of Lake Erie. Power surged in from western Ohio and Indiana on one side and from Pennsylvania through New York and Ontario around the northern side of Lake Erie. Some transmission lines in these areas, already heavily loaded with normal flows, began to trip.

WHY THE BLACKOUT STOPPED WHERE IT DID

Extreme system conditions can damage equipment in several ways, from melting aluminum conductors (excessive currents) to breaking turbine blades on a generator (frequency excursions). The power system is designed to ensure that if conditions on the grid threaten the safe operation of the transmission lines, transformers or power plants, the threatened equipment automatically separates from the network to protect itself. Relays are the devices that effect this protection. Generators are usually the most expensive units on an electrical system, so system protection schemes are designed to drop a power plant off the system as a self-protective measure if grid conditions become unacceptable. When unstable power swings develop between a group of generators that are losing synchronization (matching frequency) with the rest of the system, to stop the flows, and thus the oscillations, all interconnections or ties between the unstable generators and the system must be separated. The most common method is for the transmission system to detect the power swings and trip at those locations—ideally before the swing reaches and harms the generator.

On 14 August, the cascade became a race between the power surges and the relays. The lines that tripped first were generally the longer lines, because the relay settings required to protect these lines use a longer apparent impedance tripping zone, which a power swing enters sooner than on shorter, networked lines.

VOLTAGE COLLAPSE

Although the blackout of 14 August has been labeled as a voltage collapse, this was not the case. Voltage collapse typically occurs on power systems that are heavily loaded, faulted (reducing the number of available paths for power to flow to loads), or have reactive power shortages. A classic voltage collapse occurs when an electricity system experiences a disturbance that causes a progressive and uncontrollable decline in voltage.

On 14 August, the northern Ohio electricity system did not experience a classic voltage collapse because low voltage never became the primary cause of line and generator tripping. Although voltage was a factor in some of the events that led to the ultimate cascading of the system in Ohio and beyond, the event was not a classic reactive power-driven voltage collapse.

COMMON OR SIMILAR FACTORS AMONG MAJOR OUTAGES

Among those common or similar to major outages and to the 14 August blackout are:

- Conductor contact with trees
- Underestimation of dynamic reactive output of system generators

- Inability to visualize events on the entire system
- Failure to ensure that system operation was within safe limits
- Lack of coordination on system protection
- Ineffective communication
- Lack of 'safety nets'
- Inadequate training of operating personnel

CONDUCTOR CONTACT WITH TREES

This factor was an initiating trigger in several of the outages and a contributing factor in the severity of several more. Unlike lightning strikes, for which system operators have fair storm-tracking tools, system operators generally do not have direct knowledge that a line has contacted a tree and faulted. They will sometimes test the line by trying to restore it to service, if that is deemed to be a safe operation. Even if it does go back into service, the line may fault and trip out again as load heats it up. Lines usually sag into right-of-way obstructions when the need to retain transmission interconnection is significant. High inductive load composition, such as air conditioning or irrigation pumping, accompanies hot weather and places higher burdens on transmission lines.

DYNAMIC REACTIVE OUTPUT OF GENERATORS

Reactive supply is an important ingredient in maintaining healthy power system voltages and facilitating power transfers. Inadequate reactive supply was a factor in most of the events. Shunt capacitors and generating resources are the most significant suppliers of reactive power. Operators perform contingency analysis based on how power system elements will perform under various power system conditions. They determine and set transfer limits based on these analyses. Modeling the dynamic reactive output of generators under stressed system conditions has proven to be more challenging. If the model is incorrect, estimating transfer limits will also be incorrect. In most of the events, the assumed contribution of dynamic reactive output of system generators was greater than the generators actually produced, resulting in more significant voltage problems.

SYSTEM VISIBILITY PROCEDURES AND OPERATOR TOOLS

Each control area operates as part of a single synchronous interconnection. However, the parties with various geographic or functional responsibilities for reliable operation of the grid do not have visibility of the entire system. Events in neighboring systems may not be visible to an operator or reliability coordinator, or power system data may be available in a control center but not be presented to operators or coordinators as information they can use in making appropriate operating decisions.

SYSTEM OPERATION WITHIN SAFE LIMITS

In several of the events, operators were unaware of the vulnerability of the system to the next contingency. The reasons were varied: inaccurate modeling for simulation, no visibility of the loss of key transmission elements, no operator monitoring of stability measures (reactive reserve monitor, power transfer angle), and no reassessment of system conditions following the loss of an element and readjustment of safe limits.

EFFECTIVENESS OF COMMUNICATIONS

Under normal conditions, parties with reliability responsibility need to communicate important and prioritized information to each other in a timely way. This is especially important in emergencies, during which operators should be relieved of duties unrelated to preserving the grid. A common factor in several of the outage events was that information from one system was not provided to neighboring ones.

NEED FOR SAFETY NETS

A safety net is a protective scheme that activates automatically if a pre-specified, significant contingency occurs. Such schemes can prevent some disturbances from getting out of control and involve actions such as shedding load, dropping generation or islanding. If a safety net had not been taken out of service in the west in August 1996, it would have lessened the severity of the disturbance from 28,000 MW of load lost to less than 7200 MW. However, safety nets should not be relied upon to establish transfer limits.

13.1

THE SWING EQUATION

Consider a generating unit consisting of a three-phase synchronous generator and its prime mover. The rotor motion is determined by Newton's second law, given by

$$J\alpha_m(t) = T_m(t) - T_e(t) = T_a(t) \qquad (13.1.1)$$

where J = total moment of inertia of the rotating masses, kgm^2

α_m = rotor angular acceleration, rad/s^2

T_m = mechanical torque supplied by the prime mover minus the retarding torque due to mechanical losses, Nm

T_e = electrical torque that accounts for the total three-phase electrical power output of the generator, plus electrical losses, Nm

T_a = net accelerating torque, Nm

Also, the rotor angular acceleration is given by

$$\alpha_m(t) = \frac{d\omega_m(t)}{dt} = \frac{d^2\theta_m(t)}{dt^2} \qquad (13.1.2)$$

$$\omega_m(t) = \frac{d\theta_m(t)}{dt} \qquad (13.1.3)$$

where ω_m = rotor angular velocity, rad/s

θ_m = rotor angular position with respect to a stationary axis, rad

T_m and T_e are positive for generator operation. In steady-state T_m equals T_e, the accelerating torque T_a is zero, and, from (13.1.1), the rotor acceleration α_m is zero, resulting in a constant rotor velocity called *synchronous speed*. When T_m is greater than T_e, T_a is positive and α_m is therefore positive, resulting in increasing rotor speed. Similarly, when T_m is less than T_e, the rotor speed is decreasing.

It is convenient to measure the rotor angular position with respect to a synchronously rotating reference axis instead of a stationary axis. Accordingly, we define

$$\theta_m(t) = \omega_{msyn}t + \delta_m(t) \qquad (13.1.4)$$

where ω_{msyn} = synchronous angular velocity of the rotor, rad/s

δ_m = rotor angular position with respect to a synchronously rotating reference, rad

Using (13.1.2) and (13.1.4), (13.1.1) becomes

$$J\frac{d^2\theta_m(t)}{dt^2} = J\frac{d^2\delta_m(t)}{dt^2} = T_m(t) - T_e(t) = T_a(t) \qquad (13.1.5)$$

It is also convenient to work with power rather than torque, and to work in per-unit rather than in actual units. Accordingly, we multiply (13.1.5) by $\omega_m(t)$ and divide by S_{rated}, the three-phase voltampere rating of the generator:

$$\frac{J\omega_m(t)}{S_{rated}}\frac{d^2\delta_m(t)}{dt^2} = \frac{\omega_m(t)T_m(t) - \omega_m(t)T_e(t)}{S_{rated}}$$

$$= \frac{p_m(t) - p_e(t)}{S_{rated}} = p_{mp.u.}(t) - p_{ep.u.}(t) = p_{ap.u.}(t) \quad (13.1.6)$$

where $p_{mp.u.}$ = mechanical power supplied by the prime mover minus mechanical losses, per unit

$p_{ep.u.}$ = electrical power output of the generator plus electrical losses, per unit

Finally, it is convenient to work with a normalized inertia constant, called the H constant, which is defined as

$$H = \frac{\text{stored kinetic energy at synchronous speed}}{\text{generator voltampere rating}}$$

$$= \frac{\frac{1}{2}J\omega_{msyn}^2}{S_{rated}} \quad \text{joules/VA or per unit-seconds} \quad (13.1.7)$$

The H constant has the advantage that it falls within a fairly narrow range, normally between 1 and 10 p.u.-s, whereas J varies widely, depending on generating unit size and type. Solving (13.1.7) for J and using in (13.1.6),

$$2H\frac{\omega_m(t)}{\omega_{msyn}^2}\frac{d^2\delta_m(t)}{dt^2} = p_{mp.u.}(t) - p_{ep.u.}(t) = p_{ap.u.}(t) \quad (13.1.8)$$

Defining per-unit rotor angular velocity,

$$\omega_{p.u.}(t) = \frac{\omega_m(t)}{\omega_{msyn}} \quad (13.1.9)$$

Equation (13.1.8) becomes

$$\frac{2H}{\omega_{msyn}}\omega_{p.u.}(t)\frac{d^2\delta_m(t)}{dt^2} = p_{mp.u.}(t) - p_{ep.u.}(t) = p_{ap.u.}(t) \quad (13.1.10)$$

For a synchronous generator with P poles, the electrical angular acceleration α, electrical radian frequency ω, and power angle δ are

$$\alpha(t) = \frac{P}{2}\alpha_m(t) \quad (13.1.11)$$

$$\omega(t) = \frac{P}{2}\omega_m(t) \quad (13.1.12)$$

$$\delta(t) = \frac{P}{2}\delta_m(t) \quad (13.1.13)$$

Similarly, the synchronous electrical radian frequency is

$$\omega_{syn} = \frac{P}{2}\omega_{msyn} \tag{13.1.14}$$

The per-unit electrical frequency is

$$\omega_{p.u.}(t) = \frac{\omega(t)}{\omega_{syn}} = \frac{\frac{2}{P}\omega(t)}{\frac{2}{P}\omega_{syn}} = \frac{\omega_m(t)}{\omega_{msyn}} \tag{13.1.15}$$

Therefore, using (13.1.13–13.1.15), (13.1.10) can be written as

$$\frac{2H}{\omega_{syn}}\omega_{p.u.}(t)\frac{d^2\delta(t)}{dt^2} = p_{mp.u.}(t) - p_{ep.u.}(t) = p_{ap.u.}(t) \tag{13.1.16}$$

Frequently (13.1.16) is modified to also include a term that represents a damping torque anytime the generator deviates from its synchronous speed, with its value proportional to the speed deviation

$$2H/\omega_{syn}w_{p.u.}(t)(d^2\delta(t)/(dt^2))$$
$$= p_{mp.u.}(t) - p_{ep.u.}(t) - D/\omega_{syn}(d\delta(t)/(dt))$$
$$= p_{ap.u.}(t) \tag{13.1.17}$$

where D is either zero or a relatively small positive number with typical values between 0 and 2. The units of D are per unit power divided by per unit speed deviation.

Equation (13.1.17), called the per-unit *swing equation*, is the fundamental equation that determines rotor dynamics in transient stability studies. Note that it is nonlinear due to $p_{ep.u.}(t)$, which is shown in Section 13.2 to be a nonlinear function of δ. Equation (13.1.17) is also nonlinear due to the $\omega_{p.u.}(t)$ term. However, in practice the rotor speed does not vary significantly from synchronous speed during transients. That is, $\omega_{p.u.}(t) \simeq 1.0$, which is often assumed in (13.1.17) for hand calculations.

Equation (13.1.17) is a second-order differential equation that can be rewritten as two first-order differential equations. Differentiating (13.1.4), and then using (13.1.3) and (13.1.12)–(13.1.14), we obtain

$$\frac{d\delta(t)}{dt} = \omega(t) - \omega_{syn} \tag{13.1.18}$$

Using (13.1.18) in (13.1.17),

$$\frac{2H}{\omega_{syn}}\omega_{p.u.}(t)\frac{d\omega(t)}{dt} = p_{mp.u.}(t) - p_{ep.u.}(t) - D/\omega_{syn}\frac{d\delta(t)}{dt} = p_{ap.u.}(t) \tag{13.1.19}$$

Equations (13.1.18) and (13.1.19) are two first-order differential equations.

EXAMPLE 13.1   **Generator per-unit swing equation and power angle during a short circuit**

A three-phase, 60-Hz, 500-MVA, 15-kV, 32-pole hydroelectric generating unit has an H constant of 2.0 p.u.-s and $D = 0$. (a) Determine ω_{syn} and ω_{msyn}. (b) Give the per-unit swing equation for this unit. (c) The unit is initially operating at $p_{mp.u.} = p_{ep.u.} = 1.0$, $\omega = \omega_{syn}$, and $\delta = 10°$ when a three-phase-to-ground bolted short circuit at the generator terminals causes $p_{ep.u.}$ to drop to zero for $t \geqslant 0$. Determine the power angle 3 cycles after the short circuit commences. Assume $p_{mp.u.}$ remains constant at 1.0 per unit. Also assume $\omega_{p.u.}(t) = 1.0$ in the swing equation.

SOLUTION

a. For a 60-Hz generator,

$$\omega_{syn} = 2\pi60 = 377 \text{ rad/s}$$

and, from (13.1.14), with P = 32 poles,

$$\omega_{msyn} = \frac{2}{P}\omega_{syn} = \left(\frac{2}{32}\right)377 = 23.56 \quad \text{rad/s}$$

b. From (13.1.16), with H = 2.0 p.u.-s,

$$\frac{4}{2\pi60}\omega_{p.u.}(t)\frac{d^2\delta(t)}{dt^2} = p_{mp.u.}(t) - p_{ep.u.}(t)$$

c. The initial power angle is

$$\delta(0) = 10° = 0.1745 \quad \text{radian}$$

Also, from (13.1.17), at $t = 0$,

$$\frac{d\delta(0)}{dt} = 0$$

Using $p_{mp.u.}(t) = 1.0$, $p_{ep.u.} = 0$, and $\omega_{p.u.}(t) = 1.0$, the swing equation from (**b**) is

$$\left(\frac{4}{2\pi60}\right)\frac{d^2\delta(t)}{dt^2} = 1.0 \quad t \geqslant 0$$

Integrating twice and using the above initial conditions,

$$\frac{d\delta(t)}{dt} = \left(\frac{2\pi60}{4}\right)t + 0$$

$$\delta(t) = \left(\frac{2\pi60}{8}\right)t^2 + 0.1745$$

At $t = 3$ cycles $= \dfrac{3 \text{ cycles}}{60 \text{ cycles/second}} = 0.05$ second,

$$\delta(0.05) = \left(\frac{2\pi60}{8}\right)(0.05)^2 + 0.1745$$

$$= 0.2923 \text{ radian} = 16.75°$$ ∎

EXAMPLE 13.2   Equivalent swing equation: two generating units

A power plant has two three-phase, 60-Hz generating units with the following ratings:

> *Unit 1:* 500 MVA, 15 kV, 0.85 power factor, 32 poles, $H_1 = 2.0$ p.u.-s, $D = 0$

> *Unit 2:* 300 MVA, 15 kV, 0.90 power factor, 16 poles, $H_2 = 2.5$ p.u.-s, $D = 0$

(a) Give the per-unit swing equation of each unit on a 100-MVA system base. (b) If the units are assumed to "swing together," that is, $\delta_1(t) = \delta_2(t)$, combine the two swing equations into one equivalent swing equation.

SOLUTION

a. If the per-unit powers on the right-hand side of the swing equation are converted to the system base, then the H constant on the left-hand side must also be converted. That is,

$$H_{\text{new}} = H_{\text{old}} \frac{S_{\text{old}}}{S_{\text{new}}} \quad \text{per unit}$$

Converting H_1 from its 500-MVA rating to the 100-MVA system base,

$$H_{1\text{new}} = H_{1\text{old}} \frac{S_{\text{old}}}{S_{\text{new}}} = (2.0)\left(\frac{500}{100}\right) = 10 \quad \text{p.u.-s}$$

Similarly, converting H_2,

$$H_{2\text{new}} = (2.5)\left(\frac{300}{100}\right) = 7.5 \quad \text{p.u.-s}$$

The per-unit swing equations on the system base are then

$$\frac{2H_{1\text{new}}}{\omega_{\text{syn}}} \omega_{1\text{p.u.}}(t) \frac{d^2\delta_1(t)}{dt^2} = \frac{20.0}{2\pi60} \omega_{1\text{p.u.}}(t) \frac{d^2\delta_1(t)}{dt^2}$$

$$= p_{m1\text{p.u.}}(t) - p_{e1\text{p.u.}}(t)$$

$$\frac{2H_{2\text{new}}}{\omega_{\text{syn}}} \omega_{2\text{p.u.}}(t) \frac{d^2\delta_2(t)}{dt^2} = \frac{15.0}{2\pi60} \omega_{2\text{p.u.}}(t) \frac{d^2\delta_2(t)}{dt^2} = p_{m2\text{p.u.}}(t) - p_{e2\text{p.u.}}$$

b. Letting:

$$\delta(t) = \delta_1(t) = \delta_2(t)$$

$$\omega_{\text{p.u.}}(t) = \omega_{1\text{p.u.}}(t) = \omega_{2\text{p.u.}}(t)$$

$$p_{mp.u.}(t) = p_{m1p.u.}(t) + p_{m2p.u.}(t)$$

$$p_{ep.u.}(t) = p_{e1p.u.}(t) + p_{e2p.u.}(t)$$

and adding the above swing equations

$$\frac{2(H_{1\text{new}} + H_{2\text{new}})}{\omega_{\text{syn}}}\omega_{\text{p.u.}}(t)\frac{d^2\delta(t)}{dt^2}$$

$$= \frac{35.0}{2\pi60}\omega_{\text{p.u.}}(t)\frac{d^2\delta(t)}{dt^2} = p_{mp.u.}(t) - p_{ep.u.}(t)$$

When transient stability studies involving large-scale power systems with many generating units are performed with a digital computer, computation time can be reduce by combining the swing equations of those units that swing together. Such units, which are called *coherent machines*, usually are connected to the same bus or are electrically close, and they are usually remote from network disturbances under study. ∎

13.2

SIMPLIFIED SYNCHRONOUS MACHINE MODEL AND SYSTEM EQUIVALENTS

Figure 13.2 shows a simplified model of a synchronous machine, called the classical model, that can be used in transient stability studies. As shown, the synchronous machine is represented by a constant internal voltage E' behind its direct axis transient reactance X'_d. This model is based on the following assumptions:

1. The machine is operating under balanced three-phase positive-sequence conditions.

2. Machine excitation is constant.

3. Machine losses, saturation, and saliency are neglected.

FIGURE 13.2

Simplified synchronous machine model for transient stability studies

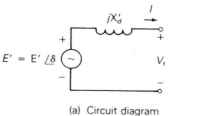

$E' = E' \underline{/\delta}$

(a) Circuit diagram

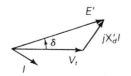

(b) Phasor diagram

FIGURE 13.3

Synchronous generator
connected to a system
equivalent

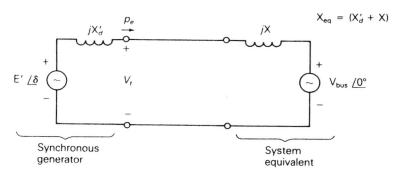

FIGURE 13.3

Synchronous generator connected to a system equivalent

In transient stability programs, more detailed models can be used to represent exciters, losses, saturation, and saliency. However, the simplified model reduces model complexity while maintaining reasonable accuracy in stability calculations.

Each generator in the model is connected to a system consisting of transmission lines, transformers, loads, and other machines. To a first approximation the system can be represented by an "infinite bus" behind a system reactance. An infinite bus is an ideal voltage source that maintains constant voltage magnitude, constant phase, and constant frequency.

Figure 13.3 shows a synchronous generator connected to a system equivalent. The voltage magnitude V_{bus} and $0°$ phase of the infinite bus are constant. The phase angle δ of the internal machine voltage is the machine power angle with respect to the infinite bus.

The equivalent reactance between the machine internal voltage and the infinite bus is $X_{eq} = (X_d' + X)$. From (6.7.3), the real power delivered by the synchronous generator to the infinite bus is

$$p_e = \frac{E'V_{bus}}{X_{eq}} \sin \delta \qquad (13.2.1)$$

During transient disturbances both E' and V_{bus} are considered constant in (13.2.1). Thus p_e is a sinusoidal function of the machine power angle δ.

EXAMPLE 13.3   Generator internal voltage and real power output versus power angle

Figure 13.4 shows a single-line diagram of a three-phase, 60-Hz synchronous generator, connected through a transformer and parallel transmission lines to an infinite bus. All reactances are given in per-unit on a common system base. If the infinite bus receives 1.0 per unit real power at 0.95 p.f. lagging, determine (a) the internal voltage of the generator and (b) the equation for the electrical power delivered by the generator versus its power angle δ.

SOLUTION

a. The equivalent circuit is shown in Figure 13.5, from which the equivalent reactance between the machine internal voltage and infinite bus is

FIGURE 13.4

Single-line diagram for
Example 13.3

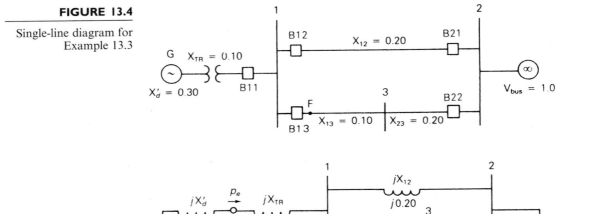

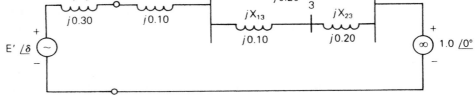

FIGURE 13.5 Equivalent circuit for Example 13.3

$$X_{eq} = X'_d + X_{TR} + X_{12}\|(X_{13} + X_{23})$$
$$= 0.30 + 0.10 + 0.20\|(0.10 + 0.20)$$
$$= 0.520 \quad \text{per unit}$$

The current into the infinite bus is

$$I = \frac{P}{V_{bus}(\text{p.f.})}\angle{-\cos^{-1}(\text{p.f.})} = \frac{(1.0)}{(1.0)(0.95)}\angle{-\cos^{-1} 0.95}$$
$$= 1.05263\angle{-18.195°} \quad \text{per unit}$$

and the machine internal voltage is

$$E' = E'\angle\delta = V_{bus} + jX_{eq}I$$
$$= 1.0\angle{0°} + (j0.520)(1.05263\angle{-18.195°})$$
$$= 1.0\angle{0°} + 0.54737\angle{71.805°}$$
$$= 1.1709 + j0.5200$$
$$= 1.2812\angle{23.946°} \quad \text{per unit}$$

b. From (13.2.1),

$$p_e = \frac{(1.2812)(1.0)}{0.520} \sin\delta = 2.4638 \sin\delta \quad \text{per unit} \qquad ■$$

13.3

THE EQUAL-AREA CRITERION

Consider a synchronous generating unit connected through a reactance to an infinite bus. Plots of electrical power p_e and mechanical power p_m versus power angle δ are shown in Figure 13.6. p_e is a sinusoidal function of δ, as given by (13.2.1).

Suppose the unit is initially operating in steady-state at $p_e = p_m = p_{m0}$ and $\delta = \delta_0$, when a step change in p_m from p_{m0} to p_{m1} occurs at $t = 0$. Due to rotor inertia, the rotor position cannot change instantaneously. That is, $\delta_m(0^+) = \delta_m(0^-)$; therefore, $\delta(0^+) = \delta(0^-) = \delta_0$ and $p_e(0^+) = p_e(0^-)$. Since $p_m(0^+) = p_{m1}$ is greater than $p_e(0^+)$, the acceleration power $p_a(0^+)$ is positive and, from (13.1.16), $(d^2\delta)/(dt^2)(0^+)$ is positive. The rotor accelerates and δ increases. When δ reaches δ_1, $p_e = p_{m1}$ and $(d^2\delta)/(dt^2)$ becomes zero. However, $d\delta/dt$ is still positive and δ continues to increase, overshooting its final steady-state operating point. When δ is greater than δ_1, p_m is less than p_e, p_a is negative, and the rotor decelerates. Eventually, δ reaches a maximum value δ_2 and then swings back toward δ_1. Using (13.1.16), which has no damping, δ would continually oscillate around δ_1. However, damping due to mechanical and electrical losses causes δ to stabilize at its final steady-state operating point δ_1. Note that if the power angle exceeded δ_3, then p_m would exceed p_e and the rotor would accelerate again, causing a further increase in δ and loss of stability.

One method for determining stability and maximum power angle is to solve the nonlinear swing equation via numerical integration techniques using a digital computer. We describe this method, which is applicable to multi-machine systems, in Section 13.4. However, there is also a direct method for

FIGURE 13.6

p_e and p_m versus δ

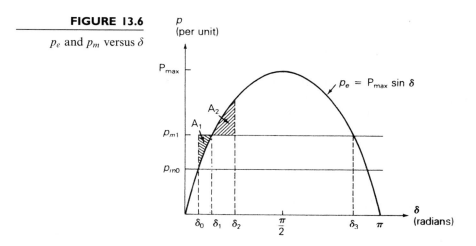

determining stability that does not involve solving the swing equation; this method is applicable for one machine connected to an infinite bus or for two machines. We describe the method, called the *equal-area criterion*, in this section.

In Figure 13.6, p_m is greater than p_e during the interval $\delta_0 < \delta < \delta_1$, and the rotor is accelerating. The shaded area A_1 between the p_m and p_e curves is called the accelerating area. During the interval $\delta_1 < \delta < \delta_2$, p_m is less than p_e, the rotor is decelerating, and the shaded area A_2 is the decelerating area. At both the initial value $\delta = \delta_0$ and the maximum value $\delta = \delta_2$, $d\delta/dt = 0$. The equal-area criterion states that $A_1 = A_2$.

To derive the equal-area criterion for one machine connected to an infinite bus, assume $\omega_{p.u.}(t) = 1$ in (13.1.16), giving

$$\frac{2H}{\omega_{syn}} \frac{d^2\delta}{dt^2} = p_{mp.u.} - p_{ep.u.} \tag{13.3.1}$$

Multiplying by $d\delta/dt$ and using

$$\frac{d}{dt}\left[\frac{d\delta}{dt}\right]^2 = 2\left(\frac{d\delta}{dt}\right)\left(\frac{d^2\delta}{dt^2}\right)$$

(13.3.1) becomes

$$\frac{2H}{\omega_{syn}}\left(\frac{d^2\delta}{dt^2}\right)\left(\frac{d\delta}{dt}\right) = \frac{H}{\omega_{syn}} \frac{d}{dt}\left[\frac{d\delta}{dt}\right]^2 = (p_{mp.u.} - p_{ep.u.})\frac{d\delta}{dt} \tag{13.3.2}$$

Multiplying (13.3.2) by dt and integrating from δ_0 to δ,

$$\frac{H}{\omega_{syn}}\int_{\delta_0}^{\delta} d\left[\frac{d\delta}{dt}\right]^2 = \int_{\delta_0}^{\delta}(p_{mp.u.} - p_{ep.u.})\,d\delta$$

or

$$\frac{H}{\omega_{syn}}\left[\frac{d\delta}{dt}\right]^2\bigg|_{\delta_0}^{\delta} = \int_{\delta_0}^{\delta}(p_{mp.u.} - p_{ep.u.})\,d\delta \tag{13.3.3}$$

The above integration begins at δ_0 where $d\delta/dt = 0$, and continues to an arbitrary δ. When δ reaches its maximum value, denoted δ_2, $d\delta/dt$ again equals zero. Therefore, the left-hand side of (13.3.3) equals zero for $\delta = \delta_2$ and

$$\int_{\delta_0}^{\delta_2}(p_{mp.u.} - p_{ep.u.})\,d\delta = 0 \tag{13.3.4}$$

Separating this integral into positive (accelerating) and negative (decelerating) areas, we arrive at the equal-area criterion

$$\int_{\delta_0}^{\delta_1} (p_{mp.u.} - p_{ep.u.}) \, d\delta + \int_{\delta_1}^{\delta_2} (p_{mp.u.} - p_{ep.u.}) \, d\delta = 0$$

or

$$\int_{\delta_0}^{\delta_1} \underbrace{(p_{mp.u.} - p_{ep.u.}) \, d\delta}_{A_1} = \int_{\delta_1}^{\delta_2} \underbrace{(p_{ep.u.} - p_{mp.u.}) \, d\delta}_{A_2} \qquad (13.3.5)$$

In practice, sudden changes in mechanical power usually do not occur, since the time constants associated with prime mover dynamics are on the order of seconds. However, stability phenomena similar to that described above can also occur from sudden changes in electrical power, due to system faults and line switching. The following three examples are illustrative.

EXAMPLE 13.4   **Equal-area criterion: transient stability during a three-phase fault**

The synchronous generator shown in Figure 13.4 is initially operating in the steady-state condition given in Example 13.3, when a temporary three-phase-to-ground bolted short circuit occurs on line 1–3 at bus 1, shown as point F in Figure 13.4. Three cycles later the fault extinguishes by itself. Due to a relay misoperation, all circuit breakers remain closed. Determine whether stability is or is not maintained and determine the maximum power angle. The inertia constant of the generating unit is 3.0 per unit-seconds on the system base. Assume p_m remains constant throughout the disturbance. Also assume $\omega_{p.u.}(t) = 1.0$ in the swing equation.

SOLUTION   Plots of p_e and p_m versus δ are shown in Figure 13.7. From Example 13.3 the initial operating point is $p_e(0^-) = p_m = 1.0$ per unit and $\delta(0^+) = \delta(0^-) = \delta_0 = 23.95° = 0.4179$ radian. At $t = 0$, when the short circuit occurs, p_e instantaneously drops to zero and remains at zero during the

---

FIGURE 13.7

p–δ plot for Example 13.4

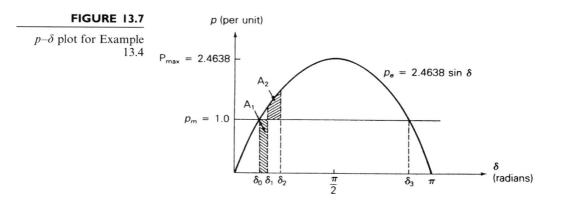

fault since power cannot be transferred past faulted bus 1. From (13.1.16), with $\omega_{\text{p.u.}}(t) = 1.0$,

$$\frac{2\text{H}}{\omega_{\text{syn}}} \frac{d^2\delta(t)}{dt^2} = p_{mp.u.} \qquad 0 \leqslant t \leqslant 0.05 \quad \text{s}$$

Integrating twice with initial condition $\delta(0) = \delta_0$ and $\dfrac{d\delta(0)}{dt} = 0$,

$$\frac{d\delta(t)}{dt} = \frac{\omega_{\text{syn}} \, p_{mp.u.}}{2\text{H}} t + 0$$

$$\delta(t) = \frac{\omega_{\text{syn}} \, p_{mp.u.}}{4\text{H}} t^2 + \delta_0$$

At $t = 3$ cycles $= 0.05$ second,

$$\delta_1 = \delta(0.05 \text{ s}) = \frac{2\pi 60}{12}(0.05)^2 + 0.4179$$

$$= 0.4964 \text{ radian} = 28.44°$$

The accelerating area A_1, shaded in Figure 13.7, is

$$A_1 = \int_{\delta_0}^{\delta_1} p_m \, d\delta = \int_{\delta_0}^{\delta_1} 1.0 \, d\delta = (\delta_1 - \delta_0) = 0.4964 - 0.4179 = 0.0785$$

At $t = 0.05$ s the fault extinguishes and p_e instantaneously increases from zero to the sinusoidal curve in Figure 13.7. δ continues to increase until the decelerating area A_2 equals A_1. That is,

$$A_2 = \int_{\delta_1}^{\delta_2} (p_{\max} \sin \delta - p_m) \, d\delta$$

$$= \int_{0.4964}^{\delta_2} (2.4638 \sin \delta - 1.0) \, d\delta = A_1 = 0.0785$$

Integrating,

$$2.4638[\cos(0.4964) - \cos \delta_2] - (\delta_2 - 0.4964) = 0.0785$$

$$2.4638 \cos \delta_2 + \delta_2 = 2.5843$$

The above nonlinear algebraic equation can be solved iteratively to obtain

$$\delta_2 = 0.7003 \text{ radian} = 40.12°$$

Since the maximum angle δ_2 does not exceed $\delta_3 = (180° - \delta_0) = 156.05°$, stability is maintained. In steady-state, the generator returns to its initial operating point $p_{ess} = p_m = 1.0$ per unit and $\delta_{ss} = \delta_0 = 23.95°$.

Note that as the fault duration increases, the risk of instability also increases. The *critical clearing time*, denoted t_{cr}, is the longest fault duration allowable for stability.

To see this case modeled in PowerWorld Simulator, open case Example 13_4. Then select **Transient, Transient Stability Main Form** to view the Transient Stability Analysis Form. Notice that in the Transient Stability Events list a fault is applied to bus 1 at $t = 0$ s and cleared at $t = 0.05$ (three cycles later). To see the time variation in the generator 1 angle (modeled at bus 4 in PowerWorld Simulator), click the **Run Transient Stability** button. When the simulation is finished the Results page is displayed automatically. You can easily plot the generator angle δ by right clicking in the "Gen 4 #1 Rotor Angle" column and selecting **Plot Column**. The results are shown in Figure 13.7a. Notice that because this system is modeled without damping (i.e.,

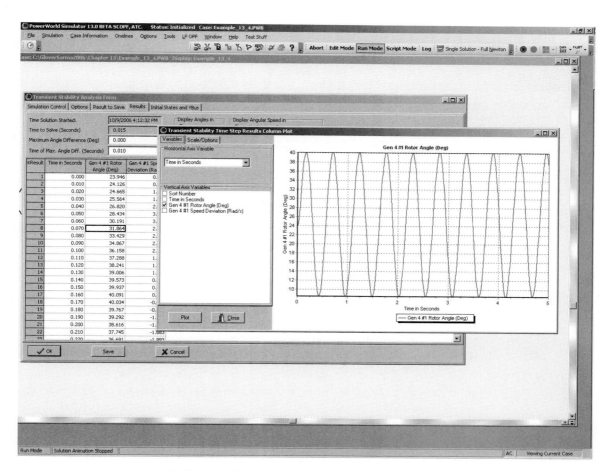

Variation in $\delta_1(t)$ for Example 13.4.

$D = 0$), the angle oscillations do not damp out with time. To rerun the example, select the Simulation Control tab on the top, left-hand portion of the Transient Stability Analysis Form. To change the duration of the fault, simply enter a different value for the time at which the fault is cleared, and again click the **Run Transient Stability** button.

To extend the example, right-click on the left generator on the Example 13_4 one-line diagram, select **Generator Information Dialog** to view the generator's dialog, click on the Stability tab, and set the "D" field to 1.0 to include a modest amount of generator damping. The results are as shown in Figure 13.7b. Notice that while the inclusion of damping did not significantly alter the maximum for $\delta_1(t)$, the magnitude of the angle oscillations is now decreasing with time. For convenience this modified example is contained in PowerWorld Simulator case Example 13_4b.

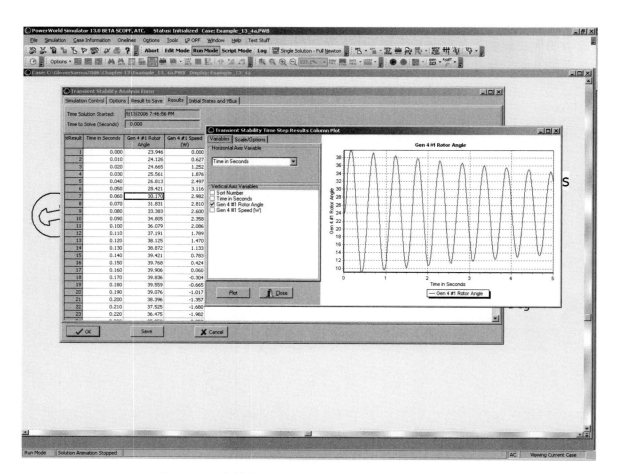

Variation in $\delta_1(t)$ for Example 13.4 with damping.

EXAMPLE 13.5   Equal-area criterion: critical clearing time for a temporary three-phase fault

Assuming the temporary short circuit in Example 13.4 lasts longer than 3 cycles, calculate the critical clearing time.

SOLUTION   The p–δ plot is shown in Figure 13.8. At the critical clearing angle, denoted δ_{cr}, the fault is extinguished. The power angle then increases to a maximum value $\delta_3 = 180° - \delta_0 = 156.05° = 2.7236$ radians, which gives the maximum decelerating area. Equating the accelerating and decelerating areas,

$$A_1 = \int_{\delta_0}^{\delta_{cr}} p_m \, d\delta = A_2 = \int_{\delta_{cr}}^{\delta_3} (P_{max} \sin \delta - p_m) \, d\delta$$

$$\int_{0.4179}^{\delta_{cr}} 1.0 \, d\delta = \int_{\delta_{cr}}^{2.7236} (2.4638 \sin \delta - 1.0) \, d\delta$$

Solving for δ_{cr},

$$(\delta_{cr} - 0.4179) = 2.4638[\cos \delta_{cr} - \cos(2.7236)] - (2.7236 - \delta_{cr})$$

$$2.4638 \cos \delta_{cr} = +0.05402$$

$$\delta_{cr} = 1.5489 \text{ radians} = 88.74°$$

From the solution to the swing equation given in Example 13.4,

$$\delta(t) = \frac{\omega_{syn} p_{mp.u.}}{4H} t^2 + \delta_0$$

Solving

$$t = \sqrt{\frac{4H}{\omega_{syn} p_{mp.u.}} (\delta(t) - \delta_0)}$$

FIGURE 13.8

p–δ plot for Example 13.5

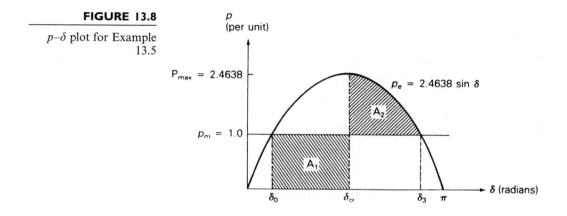

Using $\delta(t_{cr}) = \delta_{cr} = 1.5489$ and $\delta_0 = 0.4179$ radian,

$$t_{cr} = \sqrt{\frac{12}{(2\pi 60)(1.0)}(1.5489 - 0.4179)}$$

$$= 0.1897 \text{ s} = 11.38 \text{ cycles}$$

If the fault is cleared before $t = t_{cr} = 11.38$ cycles, stability is maintained. Otherwise, the generator goes out of synchronism with the infinite bus; that is, stability is lost.

To see a time-domain simulation of this case, open Example 13_5 in PowerWorld Simulator. Again select **Transient, Transient Stability Main Form** to view the Transient Stability Analysis Form, and click on the **Run Transient Stability** button to see the results for the fault cleared at 0.1897 s. However, this is actually not the critical clearing time for the PowerWorld simulation. The reason for the discrepancy is that PowerWorld Simulator, like practically all commercial transient stability analysis packages, does not make the approximation made after (13.1.17) that $\omega_1(t)$ is equal to ω_{msyn}. In reality at the time the fault is cleared the deviation in $\omega_1(t)$ is close to 12 rad/s, or a little more than 3%. Still, the impact on the critical clearly is slight with the actual critical clearing time equal to 0.1907 s. ∎

EXAMPLE 13.6 **Equal-area criterion: critical clearing angle for a cleared three-phase fault**

The synchronous generator in Figure 13.4 is initially operating in the steady-state condition given in Example 13.3 when a permanent three-phase-to-ground bolted short circuit occurs on line 1–3 at bus 3. The fault is cleared by opening the circuit breakers at the ends of line 1–3 and line 2–3. These circuit breakers then remain open. Calculate the critical clearing angle. As in previous examples, H = 3.0 p.u.-s, $p_m = 1.0$ per unit and $\omega_{p.u.} = 1.0$ in the swing equation.

SOLUTION From Example 13.3, the equation for the prefault electrical power, denoted p_{e1} here, is $p_{e1} = 2.4638 \sin \delta$ per unit. The faulted network is shown in Figure 13.9(a), and the Thévenin equivalent of the faulted network, as viewed from the generator internal voltage source, is shown in Figure 13.9(b). The Thévenin reactance is

$$X_{Th} = 0.40 + 0.20 \| 0.10 = 0.46666 \quad \text{per unit}$$

and the Thévenin voltage source is

$$V_{Th} = 1.0\underline{/0°}\left[\frac{X_{13}}{X_{13} + X_{12}}\right] = 1.0\underline{/0°}\frac{0.10}{0.30}$$

$$= 0.33333\underline{/0°} \quad \text{per unit}$$

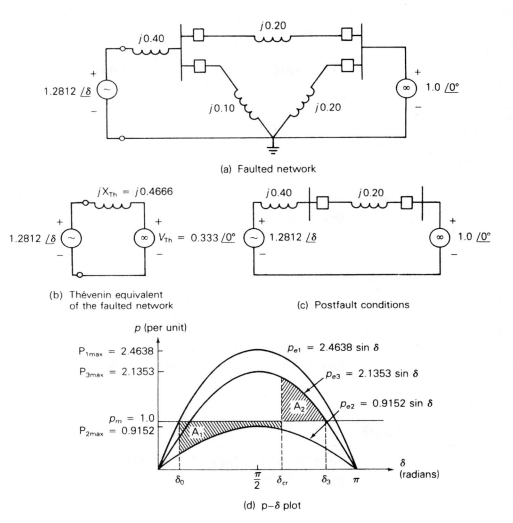

(a) Faulted network

(b) Thévenin equivalent
of the faulted network

(c) Postfault conditions

(d) p–δ plot

FIGURE 13.9   Example 13.6

From Figure 13.9(b), the equation for the electrical power delivered by the generator to the infinite bus during the fault, denoted p_{e2}, is

$$p_{e2} = \frac{E'V_{Th}}{X_{Th}} \sin \delta = \frac{(1.2812)(0.3333)}{0.46666} \sin \delta = 0.9152 \sin \delta \quad \text{per unit}$$

The postfault network is shown in Figure 13.9(c), where circuit breakers have opened and removed lines 1–3 and 2–3. From this figure, the postfault electrical power delivered, denoted p_{e3}, is

$$p_{e3} = \frac{(1.2812)(1.0)}{0.60} \sin \delta = 2.1353 \sin \delta \quad \text{per unit}$$

The p–δ curves as well as the accelerating area A_1 and decelerating area A_2 corresponding to critical clearing are shown in Figure 13.9(d). Equating A_1 and A_2,

$$A_1 = \int_{\delta_0}^{\delta_{cr}} (p_m - P_{2max} \sin \delta) \, d\delta = A_2 = \int_{\delta_{cr}}^{\delta_3} (P_{3max} \sin \delta - p_m) \, d\delta$$

$$\int_{0.4179}^{\delta_{cr}} (1.0 - 0.9152 \sin \delta) \, d\delta = \int_{\delta_{cr}}^{2.6542} (2.1353 \sin \delta - 1.0) \, d\delta$$

Solving for δ_{cr},

$$(\delta_{cr} - 0.4179) + 0.9152(\cos \delta_{cr} - \cos 0.4179)$$

$$= 2.1353(\cos \delta_{cr} - \cos 2.6542) - (2.6542 - \delta_{cr})$$

$$-1.2201 \cos \delta_{cr} = 0.4868$$

$$\delta_{cr} = 1.9812 \text{ radians} = 113.5°$$

If the fault is cleared before $\delta = \delta_{cr} = 113.5°$, stability is maintained. Otherwise, stability is lost. To see this case in PowerWorld Simulator open case Example 13_6. ∎

13.4

NUMERICAL INTEGRATION OF THE SWING EQUATION

The equal-area criterion is applicable to one machine and an infinite bus or to two machines. For multimachine stability problems, however, numerical integration techniques can be employed to solve the swing equation for each machine.

Given a first-order differential equation

$$\frac{dx}{dt} = f(x) \tag{13.4.1}$$

one relatively simple integration technique is Euler's method [1], illustrated in Figure 13.10. The integration step size is denoted Δt. Calculating the slope at the beginning of the integration interval, from (13.4.1),

$$\frac{dx_t}{dt} = f(x_t) \tag{13.4.2}$$

FIGURE 13.10

FIGURE 13.10

Euler's method

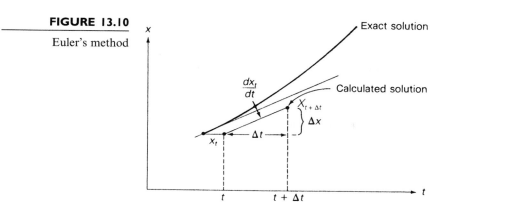

The new value $x_{t+\Delta t}$ is calculated from the old value x_t by adding the increment Δx,

$$x_{t+\Delta t} = x_t + \Delta x = x_t + \left(\frac{dx_t}{dt}\right)\Delta t \tag{13.4.3}$$

As shown in the figure, Euler's method assumes that the slope is constant over the entire interval Δt. An improvement can be obtained by calculating the slope at both the beginning and end of the interval, and then averaging these slopes. The modified Euler's method is illustrated in Figure 13.11. First, the slope at the beginning of the interval is calculated from (13.4.1) and used to calculate a preliminary value $\tilde{x}$ given by

$$\tilde{x} = x_t + \left(\frac{dx_t}{dt}\right)\Delta t \tag{13.4.4}$$

Next the slope at $\tilde{x}$ is calculated:

$$\frac{d\tilde{x}}{dt} = f(\tilde{x}) \tag{13.4.5}$$

FIGURE 13.11

Modified Euler's method

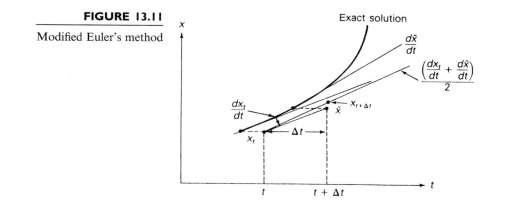

Then the new value is calculated using the average slope:

$$x_{t+\Delta t} = x_t + \frac{\left(\dfrac{dx_t}{dt} + \dfrac{d\tilde{x}}{dt}\right)}{2}\Delta t \tag{13.4.6}$$

We now apply the modified Euler's method to calculate machine frequency ω and power angle δ. Letting x be either δ or ω, the old values at the beginning of the interval are denoted δ_t and ω_t. From (13.1.17) and (13.1.18), the slopes at the beginning of the interval are

$$\frac{d\delta_t}{dt} = \omega_t - \omega_{\text{syn}} \tag{13.4.7}$$

$$\frac{d\omega_t}{dt} = \frac{p_{ap.u.t}\omega_{\text{syn}}}{2H\omega_{\text{p.u.}t}} \tag{13.4.8}$$

where $p_{ap.u.t}$ is the per-unit accelerating power calculated at $\delta = \delta_t$, and $\omega_{\text{p.u.}t} = \omega_t/\omega_{\text{syn}}$. Applying (13.4.4), preliminary values are

$$\tilde{\delta} = \delta_t + \left(\frac{d\delta_t}{dt}\right)\Delta t \tag{13.4.9}$$

$$\tilde{\omega} = \omega_t + \left(\frac{d\omega_t}{dt}\right)\Delta t \tag{13.4.10}$$

Next, the slopes at $\tilde{\delta}$ and $\tilde{\omega}$ are calculated, again using (13.1.17) and (13.1.18):

$$\frac{d\tilde{\delta}}{dt} = \tilde{\omega} - \omega_{\text{syn}} \tag{13.4.11}$$

$$\frac{d\tilde{\omega}}{dt} = \frac{\tilde{p}_{ap.u.}\omega_{\text{syn}}}{2H\tilde{\omega}_{\text{p.u.}}} \tag{13.4.12}$$

where $\tilde{p}_{ap.u.}$ is the per-unit accelerating power calculated at $\delta = \tilde{\delta}$, and $\tilde{\omega}_{\text{p.u.}} = \tilde{\omega}/\omega_{\text{syn}}$. Applying (13.4.6), the new values at the end of the interval are

$$\delta_{t+\Delta t} = \delta_t + \frac{\left(\dfrac{d\delta_t}{dt} + \dfrac{d\tilde{\delta}}{dt}\right)}{2}\Delta t \tag{13.4.13}$$

$$\omega_{t+\Delta t} = \omega_t + \frac{\left(\dfrac{d\omega_t}{dt} + \dfrac{d\tilde{\omega}}{dt}\right)}{2}\Delta t \tag{13.4.14}$$

This procedure, given by (13.4.7)–(13.4.13), begins at $t = 0$ with specified initial values δ_0 and ω_0, and continues iteratively until $t = T$, a specified final time. Calculations are best performed using a digital computer.

TABLE 13.1 Computer calculation of swing curves for Example 13.7

Case 1 Stable			Case 2 Unstable			
Time s	Delta rad	Omega rad/s	Time s	Delta rad	Omega rad/s	Program Listing
0.000	0.418	376.991	0.000	0.418	376.991	
0.020	0.426	377.778	0.020	0.426	377.778	10 REM EXAMPLE 13.7
0.040	0.449	378.547	0.040	0.449	378.547	20 REM SOLUTION TO SWING EQUATION
0.060	0.488	379.283	0.060	0.488	379.283	30 REM THE STEP SIZE IS DELTA
0.080	0.541	379.970	0.080	0.541	379.970	40 REM THE CLEARING ANGLE IS DLTCLR
0.100	0.607	380.599	0.100	0.607	380.599	50 DELTA + .01
0.120	0.685	381.159	0.120	0.685	381.159	60 DLTCLR = 1.95
0.140	0.773	381.646	0.140	0.773	381.646	70 J = 1
0.160	0.870	382.056	0.160	0.870	382.056	80 PMAX = .9152
0.180	0.975	382.392	0.180	0.975	382.392	90 PI = 3.1415927 #
0.200	1.086	382.660	0.200	1.086	382.660	100 T = 0
0.220	1.202	382.868	0.220	1.202	382.868	110 X1 = .4179
0.240	1.321	383.027	0.240	1.321	383.027	120 X2 = 2 * PI * 60
0.260	1.443	383.153	0.260	1.443	383.153	130 LPRINT "TIME DELTA OMEGA"
0.280	1.567	383.262	0.280	1.567	383.262	140 LPRINT "s rad rad/s"
0.300	1.694	383.370	0.300	1.694	383.370	150 LPRINT USING "#####.###"; T;X1;X2
0.320	1.823	383.495	0.320	1.823	383.495	160 FOR K = 1 TO 86
0.340	1.954	383.658	0.340	1.954	383.658	170 REM LINE 180 IS EQ(13.4.7)
Fault Cleared			0.360	2.090	383.876	180 X3 = X2 − (2 * PI * 60)
0.360	2.076	382.516				190 IF J = 2 THEN GOTO 240
0.380	2.176	381.510	Fault Cleared			200 IF X1 > DLTCLR OR X1 = DLTCLR THEN
0.400	2.257	380.638	0.380	2.217	382.915	PMAX = 2.1353
0.420	2.322	379.886	0.400	2.327	382.138	210 IF X1 > DLTCLR OR X1 = DLTCLR THEN
0.440	2.373	379.237	0.420	2.424	381.546	LPRINT "FAULT CLEARED"
0.460	2.413	378.674	0.440	2.511	381.135	220 IF X1 > DLTCLR OR X1 = DLTCLR THEN
0.480	2.441	378.176	0.460	2.591	380.902	J = 2
0.500	2.460	377.726	0.480	2.668	380.844	230 REM LINES 240 AND 250 ARE EQ(13.4.8)
0.520	2.471	377.307	0.500	2.746	380.969	240 X4 = 1 − PMAX * SIN(X1)
0.540	2.473	376.900	0.520	2.828	381.288	250 X5 = X4 * (2 * PI * 60) * (2 * PI * 60)/(6 * X2)
0.560	2.467	376.488	0.540	2.919	381.824	260 REM LINE 270 IS EQ(13.4.9)
0.580	2.453	376.056	0.560	3.022	382.609	270 X6 = X1 + X3 * DELTA
0.600	2.429	375.583	0.580	3.145	383.686	280 REM LINE 290 IS EQ(13.4.10)
0.620	2.396	375.053	0.600	3.292	385.111	290 X7 = X2 + X5 * DELTA
0.640	2.351	374.446	0.620	3.472	386.949	300 REM LINE 310 IS EQ(13.4.11)
0.660	2.294	373.740	0.640	3.693	389.265	310 X8 = X7 − 2 * PI * 60
0.680	2.221	372.917	0.660	3.965	392.099	320 REM LINES 330 AND 340 ARE EQ(13.4.12)
0.700	2.130	371.960	0.680	4.300	395.426	330 X9 = 1 − PMAX * SIN(X6)
0.720	2.019	370.855	0.700	4.704	399.079	340 X10 = X9 * (2 * PI * 60) * (2 * PI * 60)/(6 * X7)
0.740	1.884	369.604	0.720	5.183	402.689	350 REM LINE 360 IS EQ(13.4.13)
0.760	1.723	368.226	0.740	5.729	405.683	360 X1 = X1 + (X3 + X8) * (DELTA/2)
0.780	1.533	366.773	0.760	6.325	407.477	370 REM LINE 380 IS EQ(13.4.14)
0.800	1.314	365.341	0.780	6.941	407.812	380 X2 = X2 + (X5 + X10) * (DELTA/2)
0.820	1.068	364.070	0.800	7.551	406.981	390 T = K * DELTA
0.840	0.799	363.143	0.820	8.139	405.711	400 Z = K/2
0.860	0.516	362.750	0.840	8.702	404.819	410 M = INT(Z)
			0.860	9.257	404.934	420 IF M = Z THEN LPRINT USING
						"#####.###"; T;X1;X2
						430 NEXT K
						440 END

EXAMPLE 13.7   **Euler's method: computer solution to swing equation and critical clearing time**

Verify the critical clearing angle determined in Example 13.6, and calculate the critical clearing time by applying the modified Euler's method to solve the swing equation for the following two cases:

Case 1   The fault is cleared at $\delta = 1.95$ radians $= 112°$ (which is less than δ_{cr})

Case 2   The fault is cleared at $\delta = 2.09$ radians $= 120°$ (which is greater than δ_{cr})

For calculations, use a step size $\Delta t = 0.01$ s, and solve the swing equation from $t = 0$ to $t = T = 0.85$ s.

SOLUTION   Equations (13.4.7)–(13.4.14) are solved by a digital computer program written in BASIC. From Example 13.6, the initial conditions at $t = 0$ are

$$\delta_0 = 0.4179 \quad \text{rad}$$

$$\omega_0 = \omega_{syn} = 2\pi 60 \quad \text{rad/s}$$

Also, the H constant is 3.0 p.u.-s, and the faulted accelerating power is

$$p_{ap.u.} = 1.0 - 0.9152 \sin \delta$$

The postfault accelerating power is

$$p_{ap.u.} = 1.0 - 2.1353 \sin \delta \quad \text{per unit}$$

The computer program and results at 0.02 s printout intervals are listed in Table 13.1. As shown, these results agree with Example 13.6, since the system is stable for Case 1 and unstable for Case 2. Also from Table 13.1, the critical clearing time is between 0.34 and 0.36 s.   ∎

In addition to Euler's method, there are many other numerical integration techniques, such as Runge–Kutta, Picard's method, and Milne's predictor-corrector method [1]. Comparison of the methods shows a trade-off of accuracy versus computation complexity. The Euler method is a relatively simple method to compute, but requires a small step size Δt for accuracy. Some of the other methods can use a larger step size for comparable accuracy, but the computations are more complex.

To see this case in PowerWorld Simulator open case Example 13_7. In order to match the output shown in Table 13.1 the PowerWorld case has been modified to show the angles in radians as opposed to degrees, and the generator speed in actual radians per second as opposed to the deviation from synchronous speed. The results shown in PowerWorld differ slightly from those in the table because PowerWorld uses a more exact Runge-Kutta integration method.

13.5

MULTIMACHINE STABILITY

The numerical integration methods discussed in Section 13.4 can be used to solve the swing equations for a multimachine stability problem. However, a method is required for computing machine output powers for a general network. Figure 13.12 shows a general N-bus power system with M synchronous machines. Each machine is the same as that represented by the simplified model of Figure 13.2, and the internal machine voltages are denoted $E'_1, E'_2, \ldots, E'_M$. The M machine terminals are connected to system buses denoted $G1, G2, \ldots, GM$ in Figure 13.12. All loads are modeled here as constant admittances. Writing nodal equations for this network,

$$\begin{bmatrix} Y_{11} & Y_{12} \\ Y_{12}^T & Y_{22} \end{bmatrix} \begin{bmatrix} V \\ E \end{bmatrix} = \begin{bmatrix} 0 \\ I \end{bmatrix} \tag{13.5.1}$$

where

$$V = \begin{bmatrix} V_1 \\ V_2 \\ \vdots \\ V_N \end{bmatrix} \quad \text{is the } N \text{ vector of bus voltages} \tag{13.5.2}$$

$$E = \begin{bmatrix} E'_1 \\ E'_2 \\ \vdots \\ E'_M \end{bmatrix} \quad \text{is the } M \text{ vector of machine voltages} \tag{13.5.3}$$

FIGURE 13.12

N-bus power-system representation for transient stability studies

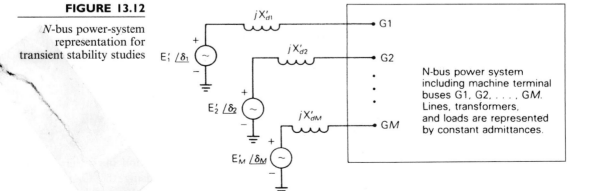

N-bus power system including machine terminal buses G1, G2, . . . , GM. Lines, transformers, and loads are represented by constant admittances.

$$I = \begin{bmatrix} I_1 \\ I_2 \\ \vdots \\ I_M \end{bmatrix} \quad \text{is the } M \text{ vector of machine currents} \atop \text{(these are current sources)} \tag{13.5.4}$$

$$\left[\begin{array}{c|c} Y_{11} & Y_{12} \\ \hline Y_{12}^{\mathrm{T}} & Y_{22} \end{array} \right] \quad \text{is an } (N + M) \times (N + M) \text{ admittance matrix} \tag{13.5.5}$$

The admittance matrix in (13.5.5) is partitioned in accordance with the N system buses and M internal machine buses, as follows:

$$Y_{11} \quad \text{is} \quad N \times N$$

$$Y_{12} \quad \text{is} \quad N \times M$$

$$Y_{22} \quad \text{is} \quad M \times M$$

Y_{11} is similar to the bus admittance matrix used for power flows in Chapter 7, except that load admittances and inverted generator impedances are included. That is, if a load is connected to bus n, then that load admittance is added to the diagonal element Y_{11nn}. Also, $(1/jX'_{dn})$ is added to the diagonal element Y_{11GnGn}.

Y_{22} is a diagonal matrix of inverted generator impedances; that is,

$$Y_{22} = \begin{bmatrix} \dfrac{1}{jX'_{d1}} & & & 0 \\ & \dfrac{1}{jX'_{d2}} & & \\ & & \ddots & \\ 0 & & & \dfrac{1}{jX'_{dM}} \end{bmatrix} \tag{13.5.6}$$

Also, the kmth element of Y_{12} is

$$Y_{12km} = \begin{cases} \dfrac{-1}{jX'_{dn}} & \text{if } k = Gn \text{ and } m = n \\ 0 & \text{otherwise} \end{cases} \tag{13.5.7}$$

Writing (13.5.1) as two separate equations,

$$Y_{11}V + Y_{12}E = 0 \tag{13.5.8}$$

$$Y_{12}^{\mathrm{T}}V + Y_{22}E = I \tag{13.5.9}$$

Assuming E is known, (13.5.8) is a linear equation in V that can be solved either iteratively or by Gauss elimination. Using the Gauss–Seidel iterative method given by (7.2.9), the kth component of V is

$$V_k(i+1) = \frac{1}{Y_{11kk}} \left[-\sum_{n=1}^{M} Y_{12kn}E_n - \sum_{n=1}^{k-1} Y_{11kn}V_n(i+1) - \sum_{n=k+1}^{N} Y_{11kn}V_n(i) \right]$$

$$(13.5.10)$$

After V is computed, the machine currents can be obtained from (13.5.9). That is,

$$I = \begin{bmatrix} I_1 \\ I_2 \\ \vdots \\ I_M \end{bmatrix} = Y_{12}^T V + Y_{22}E$$

$$(13.5.11)$$

The (real) electrical power output of machine n is then

$$p_{en} = \text{Re}[E_n I_n^*] \qquad n = 1, 2, \ldots, M$$

$$(13.5.12)$$

We are now ready to outline a computation procedure for solving a transient stability problem. The procedure alternately solves the swing equations representing the machines and the above algebraic power-flow equations representing the network. We use the modified Euler method of Section 13.4 to solve the swing equations and the Gauss–Seidel iterative method to solve the power-flow equations. We now give the procedure in the following 11 steps.

TRANSIENT STABILITY COMPUTATION PROCEDURE

STEP 1 Run a prefault power-flow program to compute initial bus voltages V_k, $k = 1, 2, \ldots, N$, initial machine currents I_n, and initial machine electrical power outputs p_{en}, $n = 1, 2, \ldots, M$. Set machine mechanical power outputs, $p_{mn} = p_{en}$. Set initial machine frequencies, $\omega_n = \omega_{\text{syn}}$. Compute the load admittances.

STEP 2 Compute the internal machine voltages:

$$E_n = E_n \underline{/\delta_n} = V_{Gn} + (jX'_{dn})I_n \quad n = 1, 2, \ldots, M$$

where V_{Gn} and I_n are computed in Step 1. The magnitudes E_n will remain constant throughout the study. The angles δ_n are the initial power angles.

STEP 3 Compute Y_{11}. Modify the $(N \times N)$ power-flow bus admittance matrix by including the load admittances and inverted generator impedances.

STEP 4 Compute Y_{22} from (13.5.6) and Y_{12} from (13.5.7).

STEP 5 Set time $t = 0$.

STEP 6   Is there a switching operation, change in load, short circuit, or change in data? For a switching operation or change in load, modify the bus admittance matrix. For a short circuit, set the faulted bus voltage [in (13.5.10)] to zero.

STEP 7   Using the internal machine voltages $E_n = \mathrm{E}_n\underline{/\delta_n}$, $n = 1, 2, \ldots$, M, with the values of δ_n at time t, compute the machine electrical powers p_{en} at time t from (13.5.10) to (13.5.12).

STEP 8   Using p_{en} computed in Step 7 and the values of δ_n and ω_n at time t, compute the preliminary estimates of power angles $\tilde{\delta}_n$ and machine speeds $\tilde{\omega}_n$ at time $(t + \Delta t)$ from (13.4.7) to (13.4.10).

STEP 9   Using $E_n = \mathrm{E}_n\underline{/\tilde{\delta}_n}$, $n = 1, 2, \ldots, M$, compute the preliminary estimates of the machine electrical powers $\tilde{p}_{en}$ at time $(t + \Delta t)$ from (13.5.10) to (13.5.12).

STEP 10   Using $\tilde{p}_{en}$ computed in Step 9, as well as $\tilde{\delta}_n$ and $\tilde{\omega}_n$ computed in Step 8, compute the final estimates of power angles δ_n and machine speeds ω_n at time $(t + \Delta t)$ from (13.4.11) to (13.4.14).

STEP 11   Set time $t = t + \Delta t$. Stop if $t \geq T$. Otherwise, return to Step 6.

EXAMPLE 13.8   **Modifying power-flow Y_{bus} for application to multimachine stability**

Consider a transient stability study for the power system given in Example 6.9, with the 184-Mvar shunt capacitor of Example 6.14 installed at bus 2. Machine transient reactances are $X'_{d1} = 0.05$ and $X'_{d2} = 0.025$ per unit on the system base. Determine the admittance matrices Y_{11}, Y_{22}, and Y_{12}.

SOLUTION   From Example 6.9, the power system has $N = 5$ buses and $M = 2$ machines. The second row of the 5×5 bus admittance matrix used for power flows is calculated in Example 6.9. Calculating the other rows in the same manner, we obtain

$$Y_{\mathrm{bus}} = \begin{bmatrix} (3.728 - j49.72) & 0 & 0 & 0 & (-3.728 + j49.72) \\ 0 & (2.68 - j26.46) & 0 & (-0.892 + j9.92) & (-1.784 + j19.84) \\ 0 & 0 & (7.46 - j99.44) & (-7.46 + j99.44) & 0 \\ 0 & (-0.892 + j9.92) & (-7.46 + j99.44) & (11.92 - j148.) & (-3.572 + j39.68) \\ (-3.728 + j49.72) & (-1.784 + j19.84) & 0 & (-3.572 + j39.68) & (9.084 - j108.6) \end{bmatrix} \text{ per unit}$$

To obtain Y_{11}, Y_{bus} is modified by including load admittances and inverted generator impedances. From Table 6.1, the load at bus 3 is $\mathrm{P}_{L3} + j\mathrm{Q}_{L3} = 0.8 + j0.4$ per unit and the voltage at bus 3 is $V_3 = 1.05$ per unit. Representing this load as a constant admittance,

$$Y_{\text{load }3} = \frac{P_{L3} - jQ_{L3}}{V_3^2} = \frac{0.8 - j0.4}{(1.05)^2} = 0.7256 - j0.3628 \quad \text{per unit}$$

Similarly, the load admittance at bus 2 is

$$Y_{\text{load }2} = \frac{P_{L2} - jQ_{L2}}{V_2^2} = \frac{8 - j2.8 + j1.84}{(0.959)^2} = 8.699 - j1.044$$

where V_2 is obtained from Example 6.14 and the 184-Mvar (1.84 per unit) shunt capacitor bank is included in the bus 2 load.

The inverted generator impedances are: for machine 1 connected to bus 1,

$$\frac{1}{jX'_{d1}} = \frac{1}{j0.05} = -j20.0 \text{ per unit}$$

and for machine 2 connected to bus 3,

$$\frac{1}{jX'_{d2}} = \frac{1}{j0.025} = -j40.0 \text{ per unit}$$

To obtain Y_{11}, add $(1/jX'_{d1})$ to the first diagonal element of Y_{bus}, add $Y_{\text{load }2}$ to the second diagonal element, and add $Y_{\text{load }3} + (1/jX'_{d2})$ to the third diagonal element. The 5×5 matrix Y_{11} is then

$$Y_{11} = \begin{bmatrix} (3.728 - j69.72) & 0 & 0 & 0 & (-3.728 + j49.72) \\ 0 & (11.38 - j29.50) & 0 & (-0.892 + j9.92) & (-1.784 + j19.84) \\ 0 & 0 & (8.186 - j139.80) & (-7.46 + j99.44) & 0 \\ 0 & (-0.892 + j9.92) & (-7.46 + j99.44) & (11.92 - j148.) & (-3.572 - j39.68) \\ (-3.728 + j49.72) & (-1.784 + j19.84) & 0 & (-3.572 + j39.68) & (9.084 - j108.6) \end{bmatrix} \text{ per unit}$$

From (13.5.6), the 2×2 matrix Y_{22} is

$$Y_{22} = \begin{bmatrix} \dfrac{1}{jX'_{d1}} & 0 \\ 0 & \dfrac{1}{jX'_{d2}} \end{bmatrix} = \begin{bmatrix} -j20.0 & 0 \\ 0 & -j40.0 \end{bmatrix} \text{ per unit}$$

From Figure 6.2, generator 1 is connected to bus 1 (therefore, bus $G1 = 1$ and generator 2 is connected to bus 3 (therefore $G2 = 3$). From (13.5.7), the 5×2 matrix Y_{12} is

$$Y_{12} = \begin{bmatrix} j20.0 & 0 \\ 0 & 0 \\ 0 & j40.0 \\ 0 & 0 \\ 0 & 0 \end{bmatrix} \text{ per unit} \qquad \blacksquare$$

To see this case in PowerWorld Simulator open case Example 13_8. To see the Y_{11} matrix go to the Transient Stability Analysis Form, and then select the **Initial States and YBus** tab. By default this case is set to solve a self-clearing fault at bus 4 that extinguishes itself after three cycles (0.05 s). Both generators are modeled with $H = 5.0$ p.u.-s and $D = 1.0$ p.u.

For the bus 4 fault Figure 13.13 shows the variation in the rotor angles for the two generators with respect to a 60 Hz synchronous reference frame. The angles are increasing with time because neither of the generators is modeled with a governor, and there is no infinite bus. While it is clear that the generator angles remain together, it is very difficult to tell from Figure 13.13 the exact variation in the angle differences. Therefore transient stability programs usually report angle differences, either with respect to the angle at a specified bus or with respect to the average of all the generator angles. The latter is shown in Figure 13.14 which displays the results from the PowerWorld Simulator Example 13_8 case.

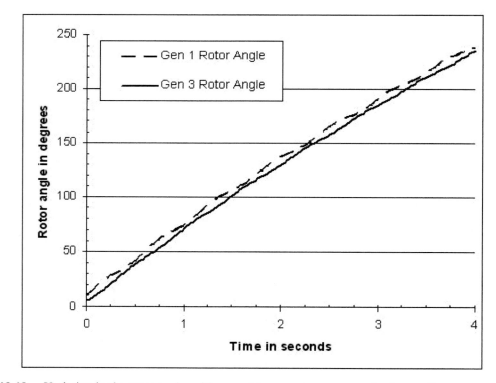

FIGURE 13.13 Variation in the rotor angles with respect to a synchronous speed reference frame.

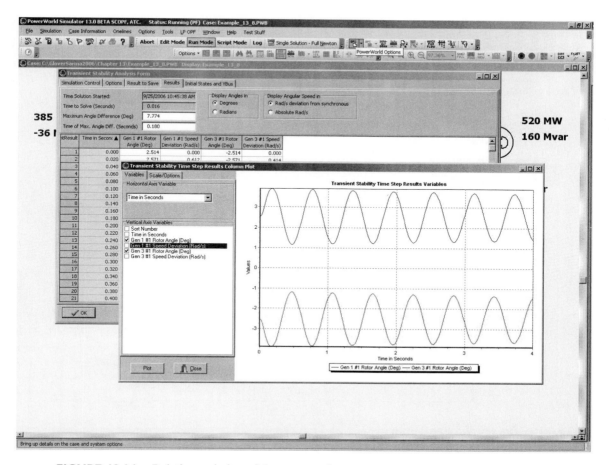

FIGURE 13.14   Relative variation of the rotor angles.

EXAMPLE 13.9   Stability results for 37 bus, 9 generator system

PowerWorld Simulator case Example 13_9 demonstrates a transient stability solution using the 37 bus system introduced in Chapter 6 with the system augmented to include classical models for each of the generators. By default the case models a transmission line fault on the 69 kV line from bus 44 (Lauf69) to bus 14 (Weber69) with the fault at the Lauf69 end of the line. The fault is cleared after 0.1 seconds by opening this transmission line. The results from this simulation are shown in Figure 13.15, with the largest generator angle variation occurring (not surprisingly) at the bus 44 generator. Notice that during and initially after the fault the bus 44 generator's angle increases relative to all the other angles in the system. The critical clearing time for this fault is about 0.262 seconds.

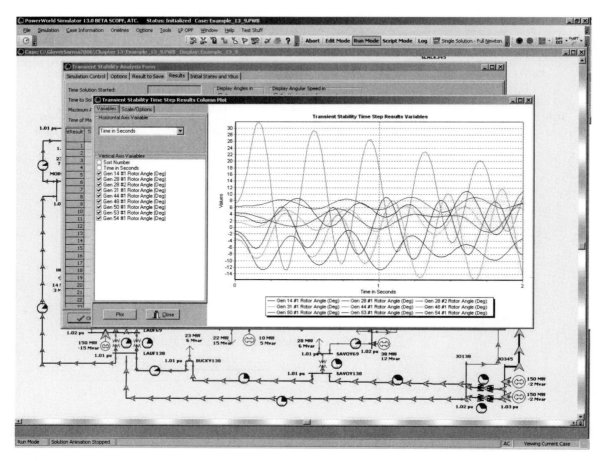

FIGURE 13.15   Rotor Angles for Example 13.9 case.

13.6

DESIGN METHODS FOR IMPROVING TRANSIENT STABILITY

Design methods for improving power system transient stability include the following:

1. Improved steady-state stability

    a. Higher system voltage levels

    b. Additional transmission lines

    c. Smaller transmission-line series reactances

  d. Smaller transformer leakage reactances

  e. Series capacitive transmission-line compensation

  f. Static var compensators and flexible ac transmission systems (FACTS)

2. High-speed fault clearing

3. High-speed reclosure of circuit breakers

4. Single-pole switching

5. Larger machine inertia, lower transient rectance

6. Fast responding, high-gain exciters

7. Fast valving

8. Braking resistors

We discuss these design methods in the following paragraphs.

1. Increasing the maximum power transfer in steady-state can also improve transient stability, allowing for increased power transfer through the unfaulted portion of a network during disturbances. Upgrading voltage on existing transmission or opting for higher voltages on new transmission increases line loadability (5.5.6). Additional parallel lines increase power-transfer capability. Reducing system reactances also increases power-transfer capability. Lines with bundled phase conductors have lower series reactances than lines that are not bundled. Oversized transformers with lower leakage reactances also help. Series capacitors reduce the total series reactances of a line by compensating for the series line inductance. The case study for Chapter 5 discusses FACTS technologies to improve line loadability and maintain stability.

2. High-speed fault clearing is fundamental to transient stability. Standard practice for EHV systems is 1-cycle relaying and 2-cycle circuit breakers, allowing for fault clearing within 3 cycles (0.05 s). Ongoing research is presently aimed at reducing these to one-half cycle relaying and 1-cycle circuit breakers.

3. The majority of transmission-line short circuits are temporary, with the fault arc self-extinguishing within 5–40 cycles (depending on system voltage) after the line is deenergized. High-speed reclosure of circuit breakers can increase postfault transfer power, thereby improving transient stability. Conservative practice for EHV systems is to employ high-speed reclosure only if stability is maintained when reclosing into a permanent fault with subsequent reopening and lockout of breakers.

4. Since the majority of short circuits are single line-to-ground, relaying schemes and independent-pole circuit breakers can be used to clear a

faulted phase while keeping the unfaulted phases of a line operating, thereby maintaining some power transfer across the faulted line. Studies have shown that single line-to-ground faults are self-clearing even when only the faulted phase is deenergized. Capacitive coupling between the energized unfaulted phases and the deenergized faulted phase is, in most cases, not strong enough to maintain an arcing short circuit [5].

5. Inspection of the swing equation, (13.1.16), shows that increasing the per-unit inertia constant H of a synchronous machine reduces angular acceleration, thereby slowing down angular swings and increasing critical clearing times. Stability is also improved by reducing machine transient reactances, which increases power-transfer capability during fault or postfault periods [see (13.2.1)]. Unfortunately, present-day generator manufacturing trends are toward lower H constants and higher machine reactances, which are a detriment to stability.

6. Modern machine excitation systems with fast thyristor controls and high amplifier gains (to overcome generator saturation) can rapidly increase generator field excitation after sensing low terminal voltage during faults. The effect is to rapidly increase internal machine voltages during faults, thereby increasing generator output power during fault and postfault periods. Critical clearing times are also increased [6].

7. Some steam turbines are equipped with fast valving to divert steam flows and rapidly reduce turbine mechanical power outputs. During faults near the generator, when electrical power output is reduced, fast valving action acts to balance mechanical and electrical power, providing reduced acceleration and longer critical clearing times. The turbines are designed to withstand thermal stresses due to fast valving [7].

8. In power systems with generation areas that can be temporarily separated from load areas, braking resistors can improve stability. When separation occurs, the braking resistor is inserted into the generation area for a second or two, preventing or slowing acceleration in the generation area. Shelton et al. [8] describe a 3-GW-s braking resistor.

PROBLEMS

SECTION 13.1

13.1 A three-phase, 60-Hz, 500-MVA, 13.8-kV, 4-pole steam turbine-generating unit has an H constant of 5.0 p.u.-s. Determine: (a) ω_{syn} and ω_{msyn}; (b) the kinetic energy in joules stored in the rotating masses at synchronous speed; (c) the mechanical angular acceleration α_m and electrical angular acceleration α if the unit is operating at synchronous speed with an accelerating power of 500 MW.

13.2   Calculate J in kg m^2 for the generating unit given in Problem 13.1.

13.3   Generator manufacturers often use the term WR^2, which is the weight in pounds of all the rotating parts of a generating unit (including the prime mover) multiplied by the square of the radius of gyration in feet. WR^2/32.2 is then the total moment of inertia of the rotating parts in slug-ft^2. (a) Determine a formula for the stored kinetic energy in ft-lb of a generating unit in terms of WR^2 and rotor angular velocity ω_m. (b) Show that

$$H = \frac{2.31 \times 10^{-4} WR^2 (rpm)^2}{S_{rated}} \quad \text{per unit-seconds}$$

where S_{rated} is the voltampere rating of the generator and rpm is the synchronous speed in r/min. Note that 1 ft-lb = 746/550 = 1.356 joules. (c) Evaluate H for a three-phase generating unit rated 800 MVA, 3600 r/min, with WR^2 = 4,000,000 lb-ft^2.

13.4   The generating unit in Problem 13.1 is initially operating at $p_{mp.u.} = p_{ep.u.} = 0.7$ per unit, $\omega = \omega_{syn}$, and $\delta = 12°$ when a fault reduces the generator electrical power output by 70%. Determine the power angle δ five cycles after the fault commences. Assume that the accelerating power remains constant during the fault. Also assume that $\omega_{p.u.}(t) = 1.0$ in the swing equation.

13.5   Repeat Problem 13.4 for a bolted three-phase fault at the generator terminals that reduces the electrical power output to zero. Compare the power angle with that determined in Problem 13.4.

13.6   A third generating unit rated 500 MVA, 15 kV, 0.90 power factor, 16 poles, with $H_3 = 3.5$ p.u.-s is added to the power plant in Example 13.2. Assuming all three units swing together, determine an equivalent swing equation for the three units.

SECTION 13.2

13.7   Given that for a moving mass $W_{kinetic} = 1/2\ Mv^2$, how fast would a 100,000 kg diesel locomotive need to go to equal the energy stored in a 60-Hz, 100-MVA, 60 Hz, 2-pole generator spinning at synchronous speed with an H of 3.0 p.u.-s?

13.8   The synchronous generator in Figure 13.4 delivers 0.75 per-unit real power at 1.05 per-unit terminal voltage. Determine: (a) the reactive power output of the generator; (b) the generator internal voltage; and (c) an equation for the electrical power delivered by the generator versus power angle δ.

13.9   The generator in Figure 13.4 is initially operating in the steady-state condition given in Problem 13.8 when a three-phase-to-ground bolted short circuit occurs at bus 3. Determine an equation for the electrical power delivered by the generator versus power angle δ during the fault.

SECTION 13.3

13.10   The generator in Figure 13.4 is initially operating in the steady-state condition given in Example 13.3 when circuit breaker B12 inadvertently opens. Use the equal-area criterion to calculate the maximum value of the generator power angle δ. Assume $\omega_{p.u.}(t) = 1.0$ in the swing equation.

13.11   The generator in Figure 13.4 is initially operating in the steady-state condition given in Example 13.3 when a temporary three-phase-to-ground short circuit occurs at point F. Three cycles later, circuit breakers B13 and B22 permanently open to clear the fault. Use the equal-area criterion to determine the maximum value of the power angle δ.

13.12   If breakers B13 and B22 in Problem 13.11 open later than 3 cycles after the fault commences, determine the critical clearing time.

13.13   Rework Problem 13.11 if circuit breakers B13 and B22 open after 3 cycles and then reclose when the power angle reaches 35°. Assume that the temporary fault has already self-extinguished when the breakers reclose.

13.14   The generator in Figure 13.4 is initially operating in the steady-state condition given in Problem 13.8 when circuit breaker B12 inadvertently opens. Use the equal-area criterion to calculate the maximum value of the generator power angle δ. Assume $\omega_{p.u.}(t) = 1.0$ in the swing equation.

13.15   The generator in Figure 13.4 is initially operating in the steady-state condition given in Problem 13.8 when a temporary three-phase-to-ground short circuit occurs at point F. Three cycles later, circuit breakers B13 and B22 permanently open to clear the fault. Use the equal-area criterion to calculate the maximum value of the generator power angle δ. Assume $\omega_{p.u.}(t) = 1.0$ in the swing equation.

13.16   If breakers B13 and B22 in Problem 13.15 open later than three cycles after the fault commences, determine the critical clearing time.

SECTION 13.4

13.17   Verify the maximum power angle determined in Problem 13.10 by applying the modified Euler's method to numerically integrate the swing equation. Write and run a computer program.

13.18   Open PowerWorld Simulator Case Problem 13_18. This case models the Example 13.4 system with damping at the bus 1 generator, and with a line fault midway between buses 1 and 2. The fault is cleared by opening the line. Determine the critical clearing time for this fault.

13.19   Open PowerWorld Simulator Case Example 13_8, and change the fault location from bus 4 to bus 2. This fault does not have a critical clearing time because the generators accelerate together. However, assuming the generators have over frequency protection with the relays set to trip a generator instantaneously any time its frequency is above 61.7 Hz (i.e., a speed deviation of 10.68 rad/s), determine the critical clearing time.

13.20   Verify the critical clearing time determined in Problem 13.12 by applying the modified Euler's method. Write and run a computer program.

13.21   In Problem 13.13, assume that the circuit breakers open at $t = 3$ cycles and then reclose at $t = 24$ cycles (instead of when δ reaches 35°). Determine the maximum power angle by applying the modified Euler method. Write and run a computer program.

SECTION 13.5

13.22   Consider the six-bus power system shown in Figure 13.16, where all data are given in per-unit on a common system base. All resistances as well as transmission-line capacitances are neglected. (a) Determine the 6×6 per-unit bus admittance matrix Y_{bus} suitable for a power-flow computer program. (b) Determine the per-unit admittance

matrices Y_{11}, Y_{12}, and Y_{22} given in (13.5.5), which are suitable for a transient stability study.

FIGURE 13.16

Single-line diagram of a six-bus power system (per-unit values are shown)

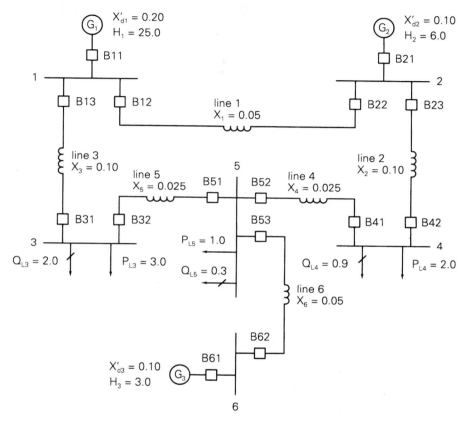

13.23 Modify the matrices Y_{11}, Y_{12}, and Y_{22} determined in Problem 13.22 for (a) the case when circuit breakers B12 and B22 open to remove line 1–2; and (b) the case when the load $P_{L4} + jQ_{L4}$ is removed.

13.24 With PowerWorld Simulator using the Example 13.9 case determine the critical clearing time (to the closest 0.01 s) for a transmission line fault on the transmission line between bus 47 (BOB138) and 53 (BLT138), with the fault occurring near bus 47. The fault is cleared by opening the line.

13.25 With PowerWorld Simulator using the Example 13.9 case determine the critical clearing time (to the closest 0.01 s) for a transmission line fault on the transmission line between bus 28 (JO345) and 31 (SLACK345), with the fault occurring near bus 28. The fault is cleared by opening the line.

13.26 With PowerWorld Simulator using the Example 13.9 case with PowerWorld Simulator determine the variation in the critical clearing time for the Example 13.9 fault as the output of the Lauf69 generator is varied between 0 and 150 MW in 50 MW increments (with changes in generator picked up at the swing bus).

CASE STUDY QUESTIONS

A. How is dynamic security assessment (DSA) software used in actual power system operations?

B. What techniques are used to decrease the time required to solve the DSA problem?

C. DSA software was not being used by the utilities involved in the August 14th, 2003 blackout. Do you think such software could have been helpful in preventing the blackout? Why or why not?

D. What in your opinion was the single most important cause of the August 14, 2003 blackout in the northeastern United States and Canada?

E. Is the United States presently becoming more prone to blackouts?

F. Will there be another blackout as large in scale as the August 14, 2003 blackout?

REFERENCES

1. G. W. Stagg and A. H. El-Abiad, *Computer Methods in Power Systems* (New York: McGraw-Hill, 1968).

2. O. I. Elgerd, *Electric Energy Systems Theory*, 2d ed. (New York: McGraw-Hill, 1982).

3. C. A. Gross, *Power System Analysis* (New York: Wiley, 1979).

4. W. D. Stevenson, Jr., *Elements of Power System Analysis*, 4th ed. (New York: McGraw-Hill, 1982).

5. E. W. Kimbark, "Suppression of Ground-Fault Arcs on Single-Pole Switched EHV Lines by Shunt Reactors," *IEEE Trans PAS*, *83* (March 1964), pp. 285–290.

6. K. R. McClymont et al., "Experience with High-Speed Rectifier Excitation Systems," *IEEE Trans PAS*, vol. PAS-87 (June 1986), pp. 1464–1470.

7. E. W. Cushing et al., "Fast Valving as an Aid to Power System Transient Stability and Prompt Resynchronization and Rapid Reload after Full Load Rejection," *IEEE Trans PAS*, vol. PAS-90 (November/December 1971), pp. 2517–2527.

8. M. L. Shelton et al., "Bonneville Power Administration 1400 MW Braking Resistor," *IEEE Trans PAS*, vol. PAS-94 (March/April 1975), pp. 602–611.

9. "Causes of the 14 August Blackout," *Power & Energy Continuity* (New York, NY: GDS Publishing Limited, www.gdsinternational.com, June 2004).

10. U.S.–Canada Power System Outage Task Force, *Final Report on The August 14, 2003 Blackout in the United States and Canada: Causes and Recommendations* (https://reports.energy.gov, April 2004).

11. P. W. Saver and M. A. Pai, *Power System Dynamics and Stability* (Prentice Hall, 1997).

12. P. Kundar, *Power System Stability and Control* (McGraw-Hill, 1994).

13. R. Schainker et al., "Real-Time Dynamic Security Assessment", *IEEE Power and Energy Magazine*, 4, 2 (March/April, 2006), pp. 51–58.

APPENDIX

Typical average values
of synchronous-machine
constants

Constant (units)	Type	Symbol	Turbo-Generator (solid rotor)	Water-Wheel Generator (with dampers)	Synchro-nous Condenser	Synchro-nous Motor
Reactances (per unit)	Synchronous	X_d	1.1	1.15	1.80	1.20
		X_q	1.08	0.75	1.15	0.90
	Transient	X_d'	0.23	0.37	0.40	0.35
		X_q'	0.23	0.75	1.15	0.90
	Subtransient	X_d''	0.12	0.24	0.25	0.30
		X_q''	0.15	0.34	0.30	0.40
	Negative-sequence	X_2	0.13	0.29	0.27	0.35
	Zero-sequence	X_0	0.05	0.11	0.09	0.16
Resistances (per unit)	Positive-sequence	R (dc)	0.003	0.012	0.008	0.01
		R (ac)	0.005	0.012	0.008	0.01
	Negative-sequence	R_2	0.035	0.10	0.05	0.06
Time constants (seconds)	Transient	T_{d0}'	5.6	5.6	9.0	6.0
		T_d'	1.1	1.8	2.0	1.4
	Subtransient	$T_d'' = T_q''$	0.035	0.035	0.035	0.036
	Armature	T_a	0.16	0.15	0.17	0.15

Adapted from E. W. Kimbark, *Power-System Stability: Synchronous Machines* (New York: Dover Publications, 1956/1968), Chap. 12.

Rating of Highest Voltage Winding kV	BIL of Highest Voltage Winding kV	Leakage Reactance per unit*	
Distribution Transformers			
2.4	30	0.023–0.049	
4.8	60	0.023–0.049	
7.2	75	0.026–0.051	
12	95	0.026–0.051	
23	150	0.052–0.055	
34.5	200	0.052–0.055	
46	250	0.057–0.063	
69	350	0.065–0.067	
Power Transformers 10 MVA and Below			
8.7	110	0.050–0.058	
25	150	0.055–0.058	
34.5	200	0.060–0.065	
46	250	0.060–0.070	
69	350	0.070–0.075	
92	450	0.070–0.085	
115	550	0.075–0.100	
138	650	0.080–0.105	
161	750	0.085–0.011	
Power Transformers Above 10 MVA		Self-Cooled or Forced-Air-Cooled	Forced-Oil-Cooled
8.7	110	0.050–0.063	0.082–0.105
34.5	200	0.055–0.075	0.090–0.128
46	250	0.057–0.085	0.095–0.143
69	350	0.063–0.095	0.103–0.158
92	450	0.060–0.118	0.105–0.180
115	550	0.065–0.135	0.107–0.195
138	650	0.070–0.140	0.117–0.245
161	750	0.075–0.150	0.125–0.250
230	900	0.070–0.160	0.120–0.270
345	1300	0.080–0.170	0.130–0.280
500	1550	0.100–0.200	0.160–0.340
765		0.110–0.210	0.190–0.350

*Per-unit reactances are based on the transformer rating

TABLE A.3 Characteristics of copper conductors, hard drawn, 97.3% conductivity

Circular Mils	AWG or B. & S.	No. of Strands	Dia. of Individual Strands (in)	Outside Diameter (in)	Breaking Strength (lb)	Weight (lb/mile)	Approx. Current Carrying Capacity* (amps)	GMR at 60 Hz (ft)	r_a 25°C dc	r_a 25°C 25 Hz	r_a 25°C 50 Hz	r_a 25°C 60 Hz	r_a 50°C dc	r_a 50°C 25 Hz	r_a 50°C 50 Hz	r_a 50°C 60 Hz	x_a 25 Hz	x_a 50 Hz	x_a 60 Hz	x'_a 25 Hz	x'_a 50 Hz	x'_a 60 Hz
1 000 000		37	0.1644	1.151	43 830	16 300	1300	0.0368	0.0585	0.0594	0.0620	0.0634	0.0640	0.0648	0.0672	0.0685	0.1666	0.333	0.400	0.216	0.1081	0.0901
900 000		37	0.1560	1.092	39 510	14 670	1220	0.0349	0.0650	0.0658	0.0682	0.0695	0.0711	0.0718	0.0740	0.0752	0.1693	0.339	0.406	0.220	0.1100	0.0916
800 000		37	0.1470	1.029	35 120	13 040	1130	0.0329	0.0731	0.0739	0.0760	0.0772	0.0800	0.0806	0.0826	0.0837	0.1722	0.344	0.413	0.224	0.1121	0.0934
750 000		37	0.1424	0.997	33 400	12 230	1090	0.0319	0.0780	0.0787	0.0807	0.0818	0.0853	0.0859	0.0878	0.0888	0.1739	0.348	0.417	0.226	0.1132	0.0943
700 000		37	0.1375	0.963	31 170	11 410	1040	0.0308	0.0836	0.0842	0.0861	0.0871	0.0914	0.0920	0.0937	0.0947	0.1759	0.352	0.422	0.229	0.1145	0.0954
600 000		37	0.1273	0.891	27 020	9781	940	0.0285	0.0975	0.0981	0.0997	0.1006	0.1066	0.1071	0.1086	0.1095	0.1799	0.360	0.432	0.235	0.1173	0.0977
500 000		37	0.1162	0.814	22 510	8151	840	0.0260	0.1170	0.1175	0.1188	0.1196	0.1280	0.1283	0.1296	0.1303	0.1845	0.369	0.443	0.241	0.1205	0.1004
500 000		19	0.1622	0.811	21 590	8151	840	0.0256	0.1170	0.1175	0.1188	0.1196	0.1280	0.1283	0.1296	0.1303	0.1853	0.371	0.445	0.241	0.1206	0.1005
450 000		19	0.1539	0.770	19 750	7336	780	0.0243	0.1300	0.1304	0.1316	0.1323	0.1422	0.1426	0.1437	0.1443	0.1879	0.376	0.451	0.245	0.1224	0.1020
400 000		19	0.1451	0.726	17 560	6521	730	0.0229	0.1462	0.1466	0.1477	0.1484	0.1600	0.1603	0.1613	0.1619	0.1909	0.382	0.458	0.249	0.1245	0.1038
350 000		19	0.1357	0.679	15 590	5706	670	0.0214	0.1671	0.1675	0.1684	0.1690	0.1828	0.1831	0.1840	0.1845	0.1943	0.389	0.466	0.254	0.1269	0.1058
350 000		12	0.1708	0.710	15 140	5706	670	0.0225	0.1671	0.1675	0.1684	0.1690	0.1828	0.1831	0.1840	0.1845	0.1918	0.384	0.460	0.251	0.1253	0.1044
300 000		19	0.1257	0.629	13 510	4891	610	0.01987	0.1950	0.1953	0.1961	0.1966	0.213	0.214	0.214	0.215	0.1982	0.396	0.476	0.259	0.1296	0.1080
300 000		12	0.1581	0.657	13 170	4891	610	0.0208	0.1950	0.1953	0.1961	0.1966	0.213	0.214	0.214	0.215	0.1957	0.392	0.470	0.256	0.1281	0.1068
250 000		19	0.1147	0.574	11 360	4076	540	0.01813	0.234	0.234	0.235	0.235	0.256	0.256	0.257	0.257	0.203	0.406	0.487	0.266	0.1329	0.1108
250 000		12	0.1443	0.600	11 130	4076	540	0.01902	0.234	0.234	0.235	0.235	0.256	0.256	0.257	0.257	0.200	0.401	0.481	0.263	0.1313	0.1094
211 600	4/0	19	0.1055	0.528	9617	3450	480	0.01668	0.276	0.277	0.277	0.278	0.302	0.303	0.303	0.303	0.207	0.414	0.497	0.272	0.1359	0.1132
211 600	4/0	12	0.1328	0.552	9483	3450	490	0.01750	0.276	0.277	0.277	0.278	0.302	0.303	0.303	0.303	0.205	0.409	0.491	0.269	0.1343	0.1119
211 600	4/0	7	0.1739	0.522	9154	3450	480	0.01579	0.276	0.277	0.277	0.278	0.302	0.303	0.303	0.303	0.210	0.420	0.503	0.273	0.1363	0.1136
167 800	3/0	12	0.1183	0.492	7556	2736	420	0.01559	0.349	0.349	0.349	0.350	0.381	0.381	0.382	0.382	0.210	0.421	0.505	0.277	0.1384	0.1153
167 800	3/0	7	0.1548	0.464	7366	2736	420	0.01404	0.349	0.349	0.349	0.350	0.381	0.381	0.382	0.382	0.216	0.431	0.518	0.281	0.1405	0.1171
133 100	2/0	7	0.1379	0.414	5926	2170	360	0.01252	0.440	0.440	0.440	0.440	0.481	0.481	0.481	0.481	0.222	0.443	0.532	0.289	0.1445	0.1205
105 500	1/0	7	0.1228	0.368	4752	1720	310	0.01113	0.555	0.555	0.555	0.555	0.606	0.607	0.607	0.607	0.227	0.455	0.546	0.298	0.1488	0.1240
83 690	1	7	0.1093	0.328	3804	1364	270	0.00992	0.699	0.699	0.699	0.699	0.765	Same as dc	Same as dc	Same as dc	0.233	0.467	0.560	0.306	0.1528	0.1274
83 690	1	3	0.1670	0.360	3620	1351	270	0.01016	0.692	0.692	0.692	0.692	0.757	Same as dc	Same as dc	Same as dc	0.232	0.464	0.557	0.299	0.1495	0.1246
66 370	2	7	0.0974	0.292	3045	1082	230	0.00883	0.881	0.882	0.882	0.882	0.964	Same as dc	Same as dc	Same as dc	0.239	0.478	0.574	0.314	0.1570	0.1308
66 370	2	3	0.1487	0.320	2913	1071	240	0.00903	0.873	Same as dc	Same as dc	Same as dc	0.955	Same as dc	Same as dc	Same as dc	0.238	0.476	0.571	0.307	0.1537	0.1281
66 370	2	1		0.258	3003	1061	220	0.00836	0.864	Same as dc	Same as dc	Same as dc	0.945	Same as dc	Same as dc	Same as dc	0.242	0.484	0.581	0.323	0.1614	0.1345
52 630	3	7	0.0867	0.260	2433	858	200	0.00787	1.112	Same as dc	Same as dc	Same as dc	1.216	Same as dc	Same as dc	Same as dc	0.245	0.490	0.588	0.322	0.1611	0.1343
52 630	3	3	0.1325	0.285	2359	850	200	0.00805	1.101	Same as dc	Same as dc	Same as dc	1.204	Same as dc	Same as dc	Same as dc	0.244	0.488	0.585	0.316	0.1578	0.1315
52 630	3	1		0.229	2439	841	190	0.00745	1.090	Same as dc	Same as dc	Same as dc	1.192	Same as dc	Same as dc	Same as dc	0.248	0.496	0.595	0.331	0.1656	0.1380
41 740	4	3	0.1180	0.254	1879	674	180	0.00717	1.388	Same as dc	Same as dc	Same as dc	1.518	Same as dc	Same as dc	Same as dc	0.250	0.499	0.599	0.324	0.1619	0.1349
41 740	4	1		0.204	1970	667	170	0.00663	1.374	Same as dc	Same as dc	Same as dc	1.503	Same as dc	Same as dc	Same as dc	0.254	0.507	0.609	0.339	0.1697	0.1415
33 100	5	3	0.1050	0.226	1505	534	150	0.00638	1.750	Same as dc	Same as dc	Same as dc	1.914	Same as dc	Same as dc	Same as dc	0.256	0.511	0.613	0.332	0.1661	0.1384
33 100	5	1		0.1819	1591	529	140	0.00590	1.733	Same as dc	Same as dc	Same as dc	1.895	Same as dc	Same as dc	Same as dc	0.260	0.519	0.623	0.348	0.1738	0.1449
26 250	6	3	0.0935	0.201	1205	424	130	0.00568	2.21	Same as dc	Same as dc	Same as dc	2.41	Same as dc	Same as dc	Same as dc	0.262	0.523	0.628	0.341	0.1703	0.1419
26 250	6	1		0.1620	1280	420	120	0.00526	2.18	Same as dc	Same as dc	Same as dc	2.39	Same as dc	Same as dc	Same as dc	0.265	0.531	0.637	0.356	0.1779	0.1483
20 820	7	1		0.1443	1030	333	110	0.00468	2.75	Same as dc	Same as dc	Same as dc	3.01	Same as dc	Same as dc	Same as dc	0.271	0.542	0.651	0.364	0.1821	0.1517
16 510	8	1		0.1285	826	264	90	0.00417	3.47	Same as dc	Same as dc	Same as dc	3.80	Same as dc	Same as dc	Same as dc	0.277	0.554	0.665	0.372	0.1862	0.1552

*For conductor at 75°C, air at 25°C, wind 1.4 miles per hour (2 ft/sec). frequency = 60 Hz

TABLE A.4 Characteristics of aluminum cable, steel, reinforced (Aluminum Company of America)—ACSR

Code Word	Circular Mils Aluminum	Aluminum Strands	Al Layers	Aluminum Strand Diameter (inches)	Steel Strands	Steel Strand Diameter (inches)	Outside Diameter (inches)	Copper Equivalent* Circular Mils or AWG	Ultimate Strength (pounds)	Weight (pounds per mile)	Geometric Mean Radius at 60 Hz (feet)	Approx. Current Carrying Capacity† (amps)	r_a 25°C dc	r_a 25°C 25 Hz	r_a 25°C 50 Hz	r_a 25°C 60 Hz	r_a 50°C dc	r_a 50°C 25 Hz	r_a 50°C 50 Hz	r_a 50°C 60 Hz	x_a 60 Hz	x_a' 60 Hz
Joree	2 515 000	76	4	0.1819	19	0.0849	1.880		61 700		0.0621									0.0450	0.337	0.0755
Thrasher	2 312 000	76	4	0.1744	19	0.0814	1.802		57 300		0.0595									0.0482	0.342	0.0767
Kiwi	2 167 000	72	4	0.1735	7	0.1157	1.735		49 800		0.0570									0.0511	0.348	0.0778
Bluebird	2 156 000	84	4	0.1602	19	0.0961	1.762		60 300		0.0588									0.0505	0.344	0.0774
Chukar	1 781 000	84	4	0.1456	19	0.0874	1.602		51 000		0.0534									0.0598	0.355	0.0802
Falcon	1 590 000	54	3	0.1716	19	0.1030	1.545	1 000 000	56 000	10 777	0.0520	1380	0.0587	0.0588	0.0590	0.0591	0.0646	0.0656	0.0675	0.0684	0.359	0.0814
Parrot	1 510 500	54	3	0.1673	19	0.1004	1.506	950 000	53 200	10 237	0.0507	1340	0.0618	0.0619	0.0621	0.0622	0.0680	0.0690	0.0710	0.0720	0.362	0.0821
Plover	1 431 000	54	3	0.1628	19	0.0977	1.465	900 000	50 400	9 699	0.0493	1300	0.0652	0.0653	0.0655	0.0656	0.0718	0.0729	0.0749	0.0760	0.365	0.0830
Martin	1 351 000	54	3	0.1582	19	0.0949	1.424	850 000	47 600	9 160	0.0479	1250	0.0691	0.0692	0.0694	0.0695	0.0761	0.0771	0.0792	0.0803	0.369	0.0838
Pheasant	1 272 000	54	3	0.1535	19	0.0921	1.382	800 000	44 800	8 621	0.0465	1200	0.0734	0.0735	0.0737	0.0738	0.0808	0.0819	0.0840	0.0851	0.372	0.0847
Grackle	1 192 500	54	3	0.1486	19	0.0892	1.338	750 000	43 100	8 082	0.0450	1160	0.0783	0.0784	0.0786	0.0788	0.0862	0.0872	0.0894	0.0906	0.376	0.0857
Finch	1 113 000	54	3	0.1436	19	0.0862	1.293	700 000	40 200	7 544	0.0435	1110	0.0839	0.0840	0.0842	0.0844	0.0924	0.0935	0.0957	0.0969	0.380	0.0867
Curlew	1 033 500	54	3	0.1384	7	0.1384	1.246	650 000	37 100	7 019	0.0420	1060	0.0903	0.0905	0.0907	0.0909	0.0994	0.1005	0.1025	0.1035	0.385	0.0878
Cardinal	954 000	54	3	0.1329	7	0.1329	1.196	600 000	34 200	6 479	0.0403	1010	0.0979	0.0981	0.0982	0.0982	0.1078	0.1088	0.1118	0.1128	0.390	0.0890
Canary	900 000	54	3	0.1291	7	0.1291	1.162	566 000	32 300	6 112	0.0391	970	0.104	0.104	0.104	0.104	0.1145	0.1155	0.1175	0.1185	0.393	0.0898
Crane	874 500	54	3	0.1273	7	0.1273	1.146	550 000	31 400	5 940	0.0386	950	0.107	0.107	0.107	0.108	0.1178	0.1188	0.1218	0.1228	0.395	0.0903
Condor	795 000	54	3	0.1214	7	0.1214	1.093	500 000	28 500	5 399	0.0368	900	0.117	0.118	0.118	0.119	0.1288	0.1308	0.1358	0.1378	0.401	0.0917
Drake	795 000	26	2	0.1749	7	0.1360	1.108	500 000	31 200	5 770	0.0375	900	0.117	0.117	0.117	0.117	0.1288	0.1288	0.1288	0.1288	0.399	0.0912
Mallard	795 000	30	2	0.1628	19	0.0977	1.140	500 000	38 400	6 517	0.0393	910	0.117	0.117	0.117	0.117	0.1288	0.1288	0.1288	0.1288	0.393	0.0904
Crow	715 500	54	3	0.1151	7	0.1151	1.036	450 000	26 300	4 859	0.0349	830	0.131	0.131	0.131	0.132	0.1442	0.1452	0.1472	0.1482	0.407	0.0932
Starling	715 500	26	2	0.1659	7	0.1290	1.051	450 000	28 100	5 193	0.0355	840	0.131	0.131	0.131	0.131	0.1442	0.1442	0.1442	0.1442	0.405	0.0928
Redwing	715 500	30	2	0.1544	19	0.0926	1.081	450 000	34 600	5 865	0.0372	840	0.131	0.131	0.131	0.131	0.1442	0.1442	0.1442	0.1442	0.399	0.0920
Flamingo	666 600	54	3	0.1111	7	0.1111	1.000	419 000	24 500	4 527	0.0337	800	0.140	0.140	0.141	0.141	0.1541	0.1571	0.1591	0.1601	0.412	0.0943
Rook	636 000	54	3	0.1085	7	0.1085	0.977	400 000	23 600	4 319	0.0329	770	0.147	0.147	0.148	0.148	0.1618	0.1638	0.1678	0.1688	0.414	0.0950
Grosbeak	636 000	26	2	0.1564	7	0.1216	0.990	400 000	25 000	4 616	0.0335	780	0.147	0.147	0.147	0.147	0.1618	0.1618	0.1618	0.1618	0.412	0.0946
Egret	636 000	30	2	0.1456	19	0.0874	1.019	400 000	31 500	5 213	0.0351	780	0.147	0.147	0.147	0.147	0.1618	0.1618	0.1618	0.1618	0.406	0.0937
Peacock	605 000	54	3	0.1059	7	0.1059	0.953	380 500	22 500	4 109	0.0321	750	0.154	0.155	0.155	0.155	0.1695	0.1715	0.1755	0.1775	0.417	0.0957
Squab	605 000	26	2	0.1525	7	0.1186	0.966	380 500	24 100	4 391	0.0327	760	0.154	0.154	0.154	0.154	0.1700	0.1720	0.1720	0.1720	0.415	0.0953
Dove	556 500	26	2	0.1463	7	0.1138	0.927	350 000	22 400	4 039	0.0313	730	0.168	0.168	0.168	0.168	0.1849	0.1859	0.1859	0.1859	0.420	0.0965
Eagle	556 500	30	2	0.1362	7	0.1362	0.953	350 000	27 200	4 588	0.0328	730	0.168	0.168	0.168	0.168	0.1849	0.1859	0.1859	0.1859	0.415	0.0957
Hawk	477 000	26	2	0.1355	7	0.1054	0.858	300 000	19 430	3 462	0.0290	670	0.196	0.196	0.196	0.196	0.216	Same as dc	Same as dc	Same as dc	0.430	0.0988
Hen	477 000	30	2	0.1261	7	0.1261	0.883	300 000	23 300	3 933	0.0304	670	0.196	0.196	0.196	0.196	0.216	Same as dc	Same as dc	Same as dc	0.424	0.0980
Ibis	397 500	26	2	0.1236	7	0.0961	0.783	250 000	16 190	2 885	0.0265	590	0.235	0.235	0.235	0.235	0.259	Same as dc	Same as dc	Same as dc	0.441	0.1015
Lark	397 500	30	2	0.1151	7	0.1151	0.806	250 000	19 980	3 277	0.0278	600	0.235	0.235	0.235	0.235	0.259	Same as dc	Same as dc	Same as dc	0.435	0.1006
Linnet	336 400	26	2	0.1138	7	0.0855	0.721	4/0	14 050	2 442	0.0244	530	0.278	0.278	0.278	0.278	0.306	0.306	0.306	0.306	0.451	0.1039
Oriole	336 400	30	2	0.1059	7	0.1059	0.741	4/0	17 040	2 774	0.0255	530	0.278	0.278	0.278	0.278	0.306	0.306	0.306	0.306	0.445	0.1032
Ostrich	300 000	26	2	0.1074	7	0.0835	0.680	188 700	12 650	2 178	0.0230	490	0.311	0.311	0.311	0.311	0.342	0.342	0.342	0.342	0.458	0.1057
Piper	300 000	30	2	0.1000	7	0.1000	0.700	188 700	15 430	2 473	0.0241	500	0.311	0.311	0.311	0.311	0.342	0.342	0.342	0.342	0.462	0.1049
Partridge	266 800	26	2	0.1013	7	0.0788	0.642	3/0	11 250	1 936	0.0217	460	0.350	0.350	0.350	0.350	0.385	0.385	0.385	0.385	0.465	0.1074

x_a Inductive Reactance (ohms per conductor per mile at 1 ft spacing, all currents)
x_a' Shunt Capacitive Reactance (megohms per conductor per mile at 1 ft spacing)

*Based on copper 97%, aluminum 61% conductivity.

†For conductor at 75°C, air at 25°C, wind 1.4 miles per hour (2 ft/sec), frequency = 60 Hz.

‡ "Current Approx 75% Capacity" is 75% of the "Approx. Current Carrying Capacity in Amps" and is approximately the current which will produce 50°C conductor temp. (25°C rise) with 25°C air temp. wind 1.4 miles per hour.

INDEX

A

ABCD parameters, 235–242, 245–248, 251–252
  approximations of medium and short transmission lines, 235–242
  lossless lines and, 251–252
  nominal π circuit, 237–238, 240–242
  transmission-line differential equations using, 245–248
  voltage regulation, 238, 240
ac circuits, 44–49
ac fault current, 359–360
Actual quantity, per-unit system, 109
Adaptive relaying, 490
Algebraic equations, 291–295, 296–301, 301–305
  direct solutions to for power flows, 291–295
  Gauss elimination, 291–295
  Gauss–Seidel method, 291, 296–301
  iterative solutions to for power flows, 296–301, 301–305
  Jacobi method, 296–301
  linear, 291–295, 296–301
  Newton–Raphson method, 301–305
  nonlinear, 301–305
Alkaline fuel cell (AFC), 37
All-aluminum alloy conductor (AAAC), 157, 167
All-aluminum conductor (AAC), 167
Aluminum alloy conductor steel reinforced (AACSR), 157
Aluminum conductor alloy reinforced (ACAR), 157, 167
Aluminum conductor carbon-fiber reinforced (ACFR), 159–160, 167
Aluminum conductor composite reinforced (ACCR), 160, 167–168
Aluminum conductor steel reinforced (ACSR), 157, 167, 736
Aluminum conductor steel supported (ACSS), 167
Aluminum-clad steel conductor (Alumoweld), 167
American Electric Power (AEP), 160–166

American Society for Testing and Materials (ASTM), 93
Ampere's law, 97, 176
Animation of power systems in PowerWorld Simulator, 25
Arcing faults, 356–358
Area control error (ACE), 581–582
Area frequency response characteristic, 578–579
Armature time constant, 363
Asymmetrical fault current, 359–360
Attenuation and lossy transmission lines, 654–655
Automated mapping/facilities management (AM/FM), 15
Automatic generation control (AGC), 15, 596–597
Autotransformers, 130–131
Available transfer capability (ATC), 288
Average power, 46–47

B

Back-to-back (BTB) dc links, 230, 232–233
Backup relays, 483
Balance beam relay, 526–527
Balanced three-phase circuits, 57–65, 65–69, 69–71
  Δ connections, 61–63
  Δ-Y conversion, 63–65
  equivalent line-to-neutral diagrams, 65
  generators, 65–67, 67–68
  impedance loads, 67, 68–69
  instantaneous power, 65–67
  line currents, 60–61
  line-to-line voltages, 59–60
  line-to-neutral voltages, 58
  motors, 67, 68
  power in, 65–69
  single-phase systems, versus, 69–71
  Y connections, 58, 68–69
Base quantity, per-unit system, 109
Basic insulation level (BIL), 668–671
Batteries, system protection and, 488
Battery energy storage (BESS), 395–396